Physical Constants and Data

Speed of light	c
Gravitational constant	G
Avogadro's number	$N_A = 6.022 \times 10^{26}$ particles/kmol
Boltzmann's constant	$k = 1.38066 \times 10^{-23}$ J/K
Gas constant	$R = 8314$ J/kmol$\cdot$K
	$= 1.9872$ kcal/kmol$\cdot$K
Planck's constant	$h = 6.6262 \times 10^{-34}$ J$\cdot$s
Electron charge	$e = 1.60219 \times 10^{-19}$ C
Electron rest mass	$m_e = 9.1095 \times 10^{-31}$ kg
	$= 5.486 \times 10^{-4}$ u
Proton rest mass	$m_p = 1.6726 \times 10^{-27}$ kg
	$= 1.007276$ u
Neutron rest mass	$m_n = 1.6749 \times 10^{-27}$ kg
	$= 1.008665$ u
Permittivity constant	$\varepsilon_0 = 8.85419 \times 10^{-12}$ C^2/N$\cdot$m^2
Permeability constant	$\mu_0 = 4\pi \times 10^{-7}$ N/A^2
Standard gravitational acceleration	$g = 9.80665$ m/s^2
	$= 32.17$ ft/s^2
Mass of earth	5.98×10^{24} kg
Average radius of earth	6.37×10^6 m
Average density of earth	5.57 g/cm^3
Average earth-moon distance	3.84×10^8 m
Average earth-sun distance	1.496×10^{11} m
Mass of sun	1.99×10^{30} kg
Radius of sun	7×10^8 m
Sun's radiation intensity at the earth	0.032 cal/cm$^2\cdot$s $= 0.134$ J/cm$^2\cdot$s

PRINCIPLES OF PHYSICS

Fourth Edition

Principles
of
Physics

F. BUECHE

Professor of Physics
University of Dayton

McGRAW-HILL BOOK COMPANY

New York St. Louis San Francisco Auckland
Bogotá Hamburg Johannesburg London
Madrid Mexico Montreal New Delhi Panama Paris
São Paulo Singapore Sydney Tokyo Toronto

PRINCIPLES OF PHYSICS

4 5 6 7 8 9 0   DODO   8 9 8 7 6 5 4

ISBN 0-07-008867-5

This book was set in Times Roman by York Graphic Services, Inc. The editors were John J. Corrigan, Jo Satloff, and James W. Bradley; the designer was Joseph Gillians; the production supervisor was Charles Hess. New drawings were done by J & R Services, Inc. The cover photograph was taken by Fern Logan.
R. R. Donnelley & Sons Company was printer and binder.

Library of Congress Cataloging in Publication Data

Bueche, F. (Frederick), date
    Principles of physics.

    Includes index.
    1. Physics. I. Title.
QC23.B8496   1982   530   81-12353
ISBN 0-07-008867-5   AACR2

Contents

Preface

It is usual for the author of a book to state in a preface his or her purpose for writing the book. I hope you, the reader, will not object too much if I do not conform exactly with custom. I would have liked, instead, to have talks with each instructor and student who will use the book. Although this is impossible, it is my hope that the following pages may serve the purpose of such conversation and aid us in understanding each other so that the text itself may prove of most value.

To the Instructor

The teacher of the precalculus course in general physics is presented with a formidable task. His or her students usually differ widely in mathematical maturity. Many of them have a fear of mathematics. Algebraic manipulations must therefore be used with extreme care. In addition, the scientific background and interests of these students are far from homogeneous.

Unlike the calculus course in general physics, the precalculus course has not been extended to three and four semesters as the emphasis on science has increased. Although the average size of the textbooks for this course has been constantly increasing, the number of class periods devoted to the subject has remained essentially constant. It is obvious that this trend cannot continue indefinitely. Most of us find that there is not enough time available to cover well the material in most books designed for this course. We should like to present more modern physics, but this usually means that in most texts we must sacrifice other material. Although it is common practice to omit certain sections in a text in order to obtain reasonable coverage, this is a difficult procedure in many instances. In particular, the better integrated the text, the more difficult the problem.

These two problems, the heterogeneous background of the students and the limited time available, have inspired the writing of this text, and it is believed that the presentation will materially aid the teacher in coping with these difficulties. I have assumed at the outset that students deserve better than a watered-down version of general physics. It has been my experience that these young people are fully as capable as those who take the calculus course in general physics and do excellent work in physics at this level. The majority of the basic laws of physics do not require calculus for understanding and for meaningful application to well-chosen problems.

You will notice as you read this text that mathematics is introduced "gently." Even the students who lack trigonometry are provided with the tools they will need. Although trigonometry is used extensively in the text, the trigonometric concepts involved are actually rather few and are easily taught in a course in general physics. In spite of this obvious attempt to dispel the usual fear of mathematics, the eventual level of mathematical facility in physics is not sacrificed. For the student to learn to swim well, it is not necessary to start with a sink-or-swim attitude.

Although modern physics concepts are introduced where possible throughout the whole text, separate chapters on these topics are required to unify and extend this material. I believe that the length of the text will make it possible for the teacher to cover these chapters. In order to make room for this material and to reduce the text to manageable size, considerable soul-searching was required. My guiding principles in this respect were the following:

1   The students are of normal intelligence and background.
2   Students taking such a course as this require a firm working grasp of the basic principles of physics.
3   High-school general science and "gee whiz" applications of physics have no place in a serious college-level course.

With the above principles as guidelines, much of the purely descriptive general science material has been removed. As you look through the text, you will see that this leaves the "physics" of the course essentially intact. We cannot reteach all of junior-high and high-school general science in a course at this level. We must restrict the course, in essential body at least, to the major principles of physics and their more important applications. You will see that enough material of practical importance still remains to maintain student interest. In fact, since this is the terminal course in physics for many students, a special effort has been made to prepare the student for the practical application of the principles of physics. It is believed that the resulting text can be well covered in the usual one-year course.

In this new edition we have made many changes. Foremost among these is the increased number of problems, nearly double that of the third edition. There are now nearly 1200 problems, carefully designed and of wide variety. They are categorized into three levels of difficulty, with the more difficult problems being indicated by a single or a double asterisk. Over 400 Questions and Guesstimates challenge the student to think through the meaning of the principles. They also serve as an excellent basis for class discussion. As in the previous edition, Minimum Learning Goals are stated for each chapter.

Many sections of the text have been rewritten to update them and to improve their teachability. For example, Chapter 1 has been modified so as to more gently ease the student into the study of physics. The chapters on Newton's laws, energy, electric field and potential, electromagnetic waves, and nuclear physics have been revised considerably. Moreover, many other changes of organization and approach have made the text an even better teaching tool.

In conformity with the shift to SI units, this edition employs primarily the SI system. British units are used only in the earliest sections of the text while the student is becoming accustomed to work in physics. Because they are relegated to a decidedly minor role, British units may be omitted entirely if the instructor so desires.

This new edition has been materially influenced by the many people who have made suggestions to me. Although it is impossible to list here the names of everyone to whom I am indebted in this regard, I want them to know that I appreciate their help. Particular thanks must be given to the following: George Barnes, University of Nevada, Reno; Peter S. Bartel, Wichita State University; Frank J. Blatt, Michigan State University; Jack Brennan, University of Central Florida; George L. Carr, University of Lowell; Robert Cole, University of Southern California; Peter P. Crooker, University of Hawaii at Manoa; Norman Gailar, University of Tennessee; Christopher E. Laird, Eastern Kentucky University; Robert N. Likes, Fort Lewis College; Laurence J. Logue, Southern Technical Institute; Edward Mooney, Jr., Youngstown State University; Kenneth Mucker, Bowling Green State University; D. L. Pursey, Iowa State University; Delbert L. Rutledge, Oklahoma State University; Barney Sandler, New York City Community College; Paul A. Smith, Coe College; Thorston F. Stromberg, New Mexico State University; R. H. Thorland, Litton Industries; William E. Vehse, West Virginia University; and John F. Ward, University of Michigan. Suggestions for the improvement of this new edition are welcomed.

To the Student

Before reading this, you might do well to "eavesdrop" a little by reading the To the Instructor section on the previous pages. You will find discussed there what I think you are like, what this course should be like, and some pitfalls you may encounter. I should like here to reemphasize a few points in regard to the difficulties for which you should be prepared.

First, a course in physics requires you to think logically about the way things in nature behave. You will be asked to memorize only a few facts and conclusions concerning nature. Your main chore is to reason out the behavior to be expected in new situations. That is to say, you must become adept at problem solving. Most often these problems are nonmathematical or involve only the simplest algebra. But to solve them, you will need to understand well the physical principles involved. Problem solving in physics relies mostly on your ability to reason a situation through; memorization of a formula or a fact is of only secondary importance.

The second point you should recognize has to do with the fact that physics and subjects such as history require entirely different methods of study. Your mind must be taught the methods of problem solving. These are methods that are learned well only through diligent, steady practice. You would not expect a violinist to give a good performance without consistent practice over a period of many weeks before the concert. Physical and mental skills take time to develop. Methods of problem solving and the requisite understanding cannot be crammed into your head the day before a test. Consistent, steady work is needed if you are to achieve a mastery of physics.

Despite these cautions, let me hasten to assure you that general physics is really not a course that only the most capable can safely expect to pass. It requires hard work. Memorizing will not get you through a good physics course with respectable grades. However, it has been my experience that the average college student in any branch of science can do average work in general physics. I have taught the material in this book to engineers, premeds, pre-dents, geologists, med-techs, and agriculture majors among others. Apart from their different mathematical background, there was little to choose between them. Good students who apply themselves do good work in general physics. If you are an average or below-average student, you will have to work steadily and hard. But as you gain facility in the solution of physics problems and in thinking in the way that physicists think, I believe you will find the work challenging and rewarding.

F. Bueche

1
Describing Measurements

At the heart of any science is measurement. Through measurements we discover the laws of nature. For the measurement results to be useful, we must state them in a way that is comprehensible to others. In this chapter we begin our study of physics by discussing some of the terms and methods we use to describe experimental results.

1.1 Discovery of Nature's Laws

We humans seem to be born with an innate sense of curiosity about the world around us. The parents of a small child will readily attest that they are driven to near distraction by the inquisitiveness of their child. As we grow older and more sophisticated, our curiosity manifests itself in less obvious behavior. But still, if we are to continue to grow in knowledge, our desire to learn about the world in which we live must not die. Discovery, knowledge, and progress are intimately linked with our inborn desire to explore.

Scientists are particularly fortunate individuals. Their life's work is to explore the avenues along which their curiosity about nature's ways leads them. Without curiosity, a scientist would be like an artist without talent. A scientist's quest is to learn nature's ways; in scientific jargon, we say that a scientist has as a goal the discovery of the laws of nature. These laws are intrinsic to the universe itself. Natural processes, whether they be the motion of an atom or the dynamics of the stars in the heavens, obey these laws. Humans can neither legislate nor alter them. *The laws of nature are concise descriptions of the behavior of the world around us.*   **Definition**

We discover the laws of nature through experiment. Many of these laws were known in qualitative form to our earliest ancestors. Even the earliest humans must have discovered that an object which is free to do so will fall toward the earth. They could therefore state this as a law of nature: When free to fall, all objects fall to the earth. But this law is woefully incomplete because it tells us so little about how objects fall.

Aristotle tried to extend the law by stating that heavy objects fall faster than lighter ones. He based his statement on observations (experimental results) typified by "a stone falls faster than a feather." Many centuries passed before Galileo showed Aristotle's proposed law to be misleading, if not completely wrong. Galileo, one of the pioneers in using well-designed experiments to test nature's behavior, showed the following to be consistent with his experiments: When not influenced by the air through which they fall, all objects fall to the earth with the same acceleration.

Galileo's proposed law was studied further by Newton, who showed that Galileo's "law of freely falling bodies" was deducible from a more comprehensive law, the law of universal gravitation. Newton pointed to the motions of the planets about the sun as large-scale experiments which indicate that each object in the universe exerts an attractive force on every other object. The equation that states this law of nature is called *Newton's law of universal gravitation*. In the first quarter of this century, Einstein showed in his *theory of relativity* that the gravitation law discovered by Newton is a portion of an even more comprehensive law. And still today scientists are searching for discoveries that will expand our understanding of gravity.

As we see, nature does not easily reveal these laws to us. *Our knowledge of nature's laws is restricted by our capacity to conceive and carry out revealing experiments as well as by our ability to decipher the clues that nature provides to us.* Moreover, the laws of nature as we presently understand them are open to revision. As we perform more experiments and see further facets of nature's behavior, we may find that our present understanding is incomplete. In rare cases, it may even be wrong. The ideal scientist must therefore be open-minded, not dogmatic. *Nature's laws are precise and immutable, but our present understanding of them may be incomplete or distorted.*

1.2 Measurement of Length

Experimental measurements are fundamental in science. Only by measuring precisely the way that nature behaves can we state her laws accurately. But to take measurements, we must have measuring sticks, or other devices, for describing our results. Let us begin our study of measurement by describing the basic unit we use for length.

The entire world uses a single basic unit in terms of which we measure lengths. It is the *meter*. You have certainly used a meterstick at one time or another; it is 1 meter (m) in length. For many years a bar of metal was kept securely near Paris, France, and the length of this bar was defined to be exactly 1 m long. Such a standard length has many drawbacks: It is not readily available to others in the world, its length changes slightly with temperature, and it might be harmed in war or an accident.

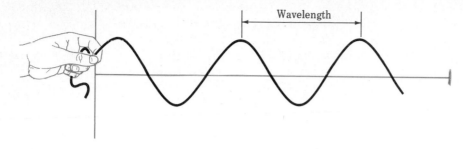

FIGURE 1.1
A wave on a string. Notice that a wavelength is the distance between adjacent crests.

To circumvent these and other difficulties, another nearly equivalent standard is now accepted. This modern standard is defined in terms of light. You learned in your precollege science courses that light is a wave. All simple waves can be likened to the wave on a string shown in Fig. 1.1. By moving the end of the string alternately up and down, the wave shown in the figure can be sent down the string. Of particular interest to us now is the distance between adjacent tops (or crests) of the wave. This distance is labeled *wavelength* in the figure. The meter length is defined in terms of a particular wavelength, the wavelength of a certain kind of light.

As a source of light, the definition uses a glowing tube of krypton gas. The gas in the tube is made to glow in the same way that a neon sign is made to glow: A high voltage is applied to the gas in the tube. See Fig. 1.2. Glowing krypton gas gives off several colors of light, each of a very precise wavelength. One of these colors, an orange-red, is used to define the meter. We shall see later that we can measure wavelengths extremely precisely and

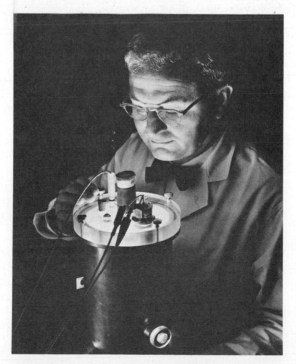

FIGURE 1.2
A U.S. National Bureau of Standards scientist adjusts a krypton 86 lamp. The wavelength of its orange-red light is used to define the meter length. (*Courtesy of National Bureau of Standards.*)

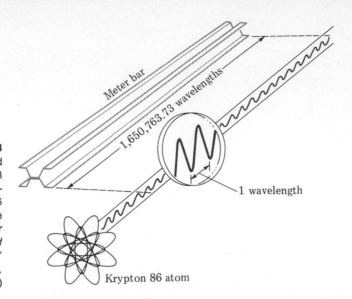

FIGURE 1.3
The meter length is defined to be 1,650,763.73 wavelengths of the orange-red light from a krypton 86 lamp such as the one shown in Fig. 1.2. (*After Weber, White, Manning, and Weygand, "College Physics," McGraw-Hill, New York, 1974.*)

that the wavelength for such a source is easy to duplicate. Therefore the following definition for the meter is convenient, precise, and immutable:

Definition *A* **meter** *is equal to the length of 1,650,763.73 wavelengths (in vacuum) of the orange-red light emitted by krypton 86.*

This definition is illustrated in Fig. 1.3. We shall learn that there are several different types (or isotopes) of krypton atoms and that the symbolism *krypton 86* tells us which isotope to use in the krypton lamp.

All other lengths are defined in relation to the meter. For example, 1 kilometer (km) is exactly 1000 m long, and 1 centimeter (cm) is exactly 0.01 m long. For those who still use the British system of units, the foot is also defined in relation to the meter: 1 foot (ft) = 0.3048 m. Other units in the British system are found by use of the facts that 1 yard (yd) = 3 ft, 1 inch (in) = $\frac{1}{12}$ ft, and 1 mile (mi) = 5280 ft.

1.3 Metric Prefixes and Scientific Notation

As you may have guessed from our discussion of the definition of the length standard, we emphasize the metric system in this text. We do so because all the scientific world does so and nearly all the world as a whole does so. Scientists worldwide make use of a precisely defined outgrowth of the metric system that is called (in French) the *Système International,* or SI for short. The meter as we have defined it is the SI unit for length.

One of the advantages of the metric system is its use of simple prefixes to denote multiples of a basic unit. For example, the prefix *kilo* always means 1000 whether it is in the word *kilometer* (1000 m), *kilogram* [1000 grams (g)], *kilowatt* [1000 watts (W)], or whatever. Similarly, the prefix *centi*

Prefix	Symbol	Value	Notation
tera	T	1,000,000,000,000	10^{12}
giga	G	1,000,000,000	10^{9}
mega	M	1,000,000	10^{6}
kilo	k	1,000	10^{3}
hecto	h	100	10^{2}
deka	da	10	10^{1}
		1.0	10^{0}
deci	d	0.1	10^{-1}
centi	c	0.01	10^{-2}
milli	m	0.001	10^{-3}
micro	μ	0.000001	10^{-6}
nano	n	0.000000001	10^{-9}
pico	p	0.000000000001	10^{-12}
femto	f		10^{-15}
atto	a		10^{-18}

always means $\frac{1}{100}$ while the prefix *milli* always denotes $\frac{1}{1000}$. A list of these and other prefixes is given in Table 1.1.

As you can see from the table, numbers such as 1,000,000,000 and 0.0000001 are very cumbersome to write. Because very large and very small numbers are commonplace in science, we often write numbers in *scientific notation*. In this notation, powers of 10 are used to simplify writing and computation. The notation makes use of the fact that the symbolism 6×10^{7} means to multiply 6 by 10 seven times. For example,

$$3.1 \times 10^{5} \text{ means } 310,000$$

and
$$0.0563 \times 10^{4} \text{ means } 563$$

Be sure you understand how these equivalent forms are obtained.

Very small numbers can be written conveniently if we note that 6×10^{-4} means to divide 6 by 10 four times. Hence

$$6 \times 10^{-4} \text{ means } 0.0006$$

$$373 \times 10^{-5} \text{ means } 0.00373$$

and
$$5.6 \times 10^{-3} \text{ means } 0.0056$$

Let us test your understanding of these notations by filling in the missing quantities in the following examples. In doing so, recall that $10^{0} = 1$.

Illustration 1.1 Fill in the blanks:

(a) $56,321,000 = 5.6321 \times 10^{7} = 56.321 \times 10^{6}$

(b) $43.7 \times 10^{5} = 4370000 \times 10^{0}$

(c) $0.0000516 = 5.16 \times 10^{-5}$

(d) $6.15 \times 10^{-2} = \underline{\quad .0615 \quad} \times 10^0$

Answers (a) 10^7, 10^6; (b) 4,370,000; (c) 10^{-5}; (d) 0.0615. Be sure you understand how these answers are obtained.

Illustration 1.2 A uranium nucleus is roughly a sphere of radius 8×10^{-15} m. Find the volume of the atom.

Reasoning Because the volume of a sphere is $\frac{4}{3}\pi r^3$, we have

$$\text{Volume} = \frac{4}{3}\pi(8 \times 10^{-15}\,\text{m})(8 \times 10^{-15}\,\text{m})(8 \times 10^{-15}\,\text{m})$$
$$= 2145 \times 10^{-15} \times 10^{-15} \times 10^{-15}\,\text{m}^3$$

But the law of exponents tells us that

$$10^a \times 10^b \times 10^c = 10^{a+b+c}$$

and so

$$\text{Volume} = 2145 \times 10^{-45}\,\text{m}^3 = 2.1 \times 10^{-42}\,\text{m}^3$$

1.4 Measurement of Time and Mass

Another quantity that we frequently wish to measure is time. *The SI unit for time is the second;* the minute, hour, and day are defined in the usual way in terms of the second. An atomic process has been chosen to define the time standard. A comparatively easily measured frequency of vibration associated with cesium atoms is utilized. *A **second** is defined to be the time taken for exactly 9,192,631,770 of these vibrations.* A very accurate cesium clock maintained by the National Bureau of Standards is shown in Fig. 1.4. It maintains time to an accuracy of better than 1 part in 1×10^{13}. Two such clocks synchronized at a certain instant are expected to differ by less than 1 second (s) after the passage of 300,000 years.

Definition

In addition to measuring length and time, we often need to measure a quantity called *mass*. This quantity measures the inertia of an object. If you kick a can filled with air, you can send it flying much more easily than if the can is filled with sand. The air-filled can has much less inertia than the sand-filled one. Because mass is a measure of inertia, we conclude that the mass of the air in the can is less than the mass of sand in a similar can filled with sand.

People often confuse mass with weight because the weight of an object is proportional to its mass. Massive objects weigh more than less massive ones provided both are weighed at the same place. We investigate the meaning of mass in detail in Chap. 3. For now we simply state that *the SI*

FIGURE 1.4
The heart of the U.S. National Bureau of Standards atomic clock system, a clock that has an accuracy of 0.000003 s in one year. (*Courtesy of National Bureau of Standards.*)

unit for mass is the **kilogram.** *A metal cylinder similar to the one shown in Fig. 1.5 is kept near Paris, and it is defined to have a mass of 1 kilogram (kg). The* mass of any other object can be determined by a comparison method, as we point out in Chap. 3.

Definition

1.5 Other SI Units

Altogether there are seven basic SI units. We have thus far listed three: the meter, second, and kilogram. These together with the other four are listed in Table 1.2. We discuss all but the candela in detail later. The candela is a unit used widely in photography and in engineering applications of light.

The magnitudes of all measurable physical quantities can be expressed in terms of these seven basic SI units. This fact will become apparent as we proceed with our studies. Although many different units of measure will be used, they can all be reduced to combinations of the units listed in the table.

TABLE 1.2
BASIC SI UNITS

Quantity	Name	Symbol
Length	Meter	m
Mass	Kilogram	kg
Time	Second	s
Temperature	Kelvin	K
Electric current	Ampere	A
Number of particles	Mole	mol
Luminous intensity	Candela	cd

FIGURE 1.5
The platinum-iridium cylinder shown here (prototype kilogram number 20) is a copy of the standard kilogram mass. This photograph was furnished by the National Bureau of Standards, whose responsibility it is to preserve this secondary standard of mass.

This is obvious for units such as the centimeter or milligram, and even for the units of speed such as kilometers per hour. But it is also true for units such as the kilowatt for power and the force unit we call the pound. All units of measure can be replaced by the units given in Table 1.2 or by combinations of them. In that sense, they are truly basic units.

1.6 Dimensional Analysis

Each unit given in Table 1.2 is associated with a physical **dimension.** For example, the meter is a measure of the dimension of length $[L]$. The kilogram is a mass dimension $[M]$, while the second is a time dimension $[T]$. Other quantities that we measure have dimensions which are combinations of these dimensions. For example, speed is measured in meters per second or miles per hour. These units obviously have the dimensions of length divided by time. Hence we can write

$$\text{Dimensions of speed} = \frac{\text{length dimension}}{\text{time dimension}}$$

or, in symbols,

$$[\text{Speed}] = \frac{[L]}{[T]} = [L][T]^{-1}$$

Square brackets are used to indicate dimensions.

Similarly, the dimensions of force units such as the pound, newton, and dyne are given by

$$[\text{Force}] = \frac{[M][L]}{[T]^2} = [M][L][T]^{-2}$$

Indeed, even the most complex physical quantities can be expressed in terms of the dimensions $[L]$, $[T]$, and $[M]$ together with the dimensions of the other basic SI quantities. For example, the dimensions of all forms of energy are $[M][L]^2[T]^{-2}$ while the dimensions of power are $[M][L]^2[T]^{-3}$.

Often an examination of dimensions in an equation can lead to useful information. Because only like quantities can be equated,[1] the dimensions in every term of an equation must be the same. For example, in the equation $s = vt$ (or, distance traveled = speed $\times$ time), the dimensions must be such that

$$[s] = [v][t]$$

Because

---

[1] Three cats can never be exactly equal to three apples, for example.

On the blackboard:

$$[Energy] = \frac{ML^2}{T^2} \qquad [c] = \frac{L}{T} \qquad [m] = M$$

$$Energy \stackrel{?}{=} m \longrightarrow \frac{ML^2}{T^2} \neq M \qquad Wrong!$$

$$Energy \stackrel{?}{=} mc \longrightarrow \frac{ML^2}{T^2} \neq M\left(\frac{L}{T}\right) \qquad Wrong!$$

$$Energy \stackrel{?}{=} mc^2 \longrightarrow \frac{ML^2}{T^2} = M\left(\frac{L}{T}\right)^2 \qquad Eureka!$$

FIGURE 1.6
Dimensional analysis can tell if an equation is possibly correct. (See S. Harris's cartoon book *What's so Funny about Science?* William Kaufmann, Inc., Los Altos, Calif., 1979, for the inspiration for this figure.)

$$[s] = [L] \qquad [v] = [L][T]^{-1} \qquad \text{and} \qquad [t] = [T]$$

we know that

$$[s] = [v][t] \qquad \text{becomes} \qquad [L] = \left[\frac{L}{T}\right] \times [T] = [L]$$

showing that the equation $s = vt$ is correct dimensionally.

However, suppose you thought that the correct equation were $v = st$. Then *dimensional analysis* would show that

$$[v] \stackrel{?}{=} [s][t] \qquad \text{becomes} \qquad \left[\frac{L}{T}\right] \stackrel{?}{=} [L] \times [T]$$

which is definitely incorrect. Hence dimensional analysis would point out at once that your starting equation was wrong. An amusing (but wrong) supposition concerning how Einstein might have used dimensional analysis to infer his famous equation $E = mc^2$ is shown in Fig. 1.6.

Illustration 1.3 The period of a pendulum (the time it takes for one swing) depends only on its length l and the acceleration due to gravity g. Knowing that $[g] = [L][T]^{-2}$, use dimensional analysis to find the relation among period, l, and g.

Reasoning The relation will be of the form

$$\text{Period} = (\text{constant})g^a l^b$$

where we are to find the exponents a and b. Because [period] = $[T]$, the dimensions of this equation are

$$[T] = [LT^{-2}]^a[L]^b = [L]^a[T]^{-2a}[L]^b$$

Because there is no $[L]$ on the left of the equation, we see that $[L]^a[L]^b = 1$. This means $[L]^{a+b} = 1$ and, because $x^0 = 1$, we see that

$$a + b = 0 \quad \text{or} \quad b = -a$$

Moreover, the factor $[T]$ on the left side must equal $[T]^{-2a}$ on the right. Hence

$$-2a = 1 \quad \text{or} \quad a = -\tfrac{1}{2}$$

We found above that $b = -a$, and so $b = \tfrac{1}{2}$

Placing these values in our original equation gives

$$\text{Period} = (\text{const})g^{-1/2}l^{1/2} = (\text{const})(l/g)^{1/2}$$

or $$\text{Period} = (\text{const})\sqrt{l/g}$$

1.7 Vector Quantities

Until now we have been discussing ways in which we express the magnitudes of quantities that we measure. We know how to describe distances, times, masses, and combinations of these quantities. But some quantities cannot be described by their magnitude alone because they have both direction and magnitude. For example, an airplane may be known to have a speed of 400 miles per hour (mi/h), but it makes a great deal of difference whether the plane is going east or west. The motion of the plane has direction as well as magnitude.

Definition   *Quantities that have no direction associated with them are called* **scalar quantities.** The number of pages in a book, the number of people in a city, and the amount of coffee in a bag are all scalar quantities. They can be specified completely without reference to a direction.

Other quantities require a specification of their direction if we are to know all about them. They have both direction and magnitude. Typical examples are the motion of an arrow, the straight-line path one must follow to go from Chicago to Miami, and the path a doctor's needle must take to reach a definite point within a person's body. These types of quantities are

Definition   called **vector quantities.** *Vector quantities have direction as well as magnitude.*

One of the most important features of vector quantities is that it is possible to represent them by pictures. For example, suppose a car goes 30 km east.[2] This can be pictured as shown in Fig. 1.7. The 30-km eastward displacement of the car is represented by an arrow. We call this arrow a

---

[2] 1 km = 0.6214 mi.

vector arrow (or simply *vector* for short). The direction of the arrow (or vector) is eastward and shows that the displacement is eastward. The length of the arrow is given a magnitude proportional to the magnitude of the 30-km displacement. For example, we might represent a distance of 1 km by an arrow length of 1 millimeter (mm). Then the arrow representing the 30-km displacement would be taken as (30 km) × (1 mm/km), or 30 mm long. Let us apply this concept of vector arrows to an example.

Illustration 1.4 Suppose you want to illustrate this statement: I went 30 km east and then 10 km north. How would you use vector arrows to do this?

Reasoning It is apparent that two vector arrows are involved here. They are drawn in Fig. 1.8. You obviously started at point *A*, went first to point *B*, and ended up at point *C*. A vector diagram such as this is a convenient way of picturing successive movements.

In addition, Fig. 1.8 shows clearly that the end point of the trip is not 30 km + 10 km = 40 km from the starting point. The straight-line distance from *A* to *C* is the length of the outlined arrow **R** shown in Fig. 1.9. *We call this straight-line distance from point A to point C the* **displacement** *from A to C. The displacement is defined to be a vector quantity. In this case its magnitude is the straight-line distance from A to C, and its direction is along the straight-line path from A to C.* Its direction is that of the vector arrow labeled **R** in Fig. 1.9.

The magnitude of the displacement from *A* to *C* can be found by applying the pythagorean theorem for a right triangle to the triangle in Fig. 1.9. We have that

$$\text{Magnitude of } \mathbf{R} \equiv R = \sqrt{(10)^2 + (30)^2} \text{ km}$$
$$= \sqrt{1000} \text{ km} = 10\sqrt{10} \text{ km} \qquad (1.1)$$

Before proceeding with this example, let us pause to define a few terms and symbols we shall be using.

1   The **displacement** *from one point to another is a vector. Its length is equal to the straight-line distance from the first point to the second point. Its direction is that of the straight-line path from the first to the second point.*
2   *When an object undergoes several different displacements in going from a starting point S to an end point E, the displacement from S to E is called the* **resultant displacement,** *and it is the sum of these several individual displacements.*
3   *A vector quantity such as a displacement can be represented by a symbol; for example, the symbol* **R** *was used in Fig. 1.9. This is boldface R. In writing the symbol for a vector by hand on paper or on a chalkboard, two different methods are frequently used to distinguish vectors. A wavy line is placed below the symbol:* R̰; *or an arrow is placed above it:* R⃗ . *Therefore, if we represent a displacement by* **D,** *you are being told by the symbol-*

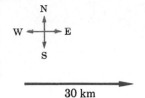

FIGURE 1.7
The vector arrow indicates a displacement of 30 km east.

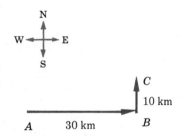

FIGURE 1.8
A vector diagram of a trip in which the traveler went 30 km east and then 10 km north.

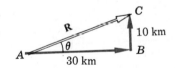

FIGURE 1.9
A displacement of 30 km east and 10 km north is equivalent to a resultant displacement of $10\sqrt{10}$ km in the direction shown.

Definition

Definition

Boldface Type Indicates Vector

ism that this is a vector and we shall be concerned with its direction. But sometimes we are interested in only the magnitude of a displacement. We then drop the boldface vector notation and represent it by R or D, for example. This same notation is applied to all vector symbols.

Let us now return to the situation shown in Fig. 1.9.

The vector diagram in Fig. 1.9 indicates the magnitude of the resultant displacement by the vector arrow from A to C. But, in addition to the magnitude of the displacement, the arrow gives the displacement's direction. We can describe the direction by saying, "The resultant vector is at an angle θ (theta) north of east." If the diagram had been drawn accurately to scale, with care being taken to make the angle between displacements AB and BC exactly $90°$, the length of the resultant could be measured directly on the diagram. The value so obtained should be the same as one would compute from Eq. (1.1) within the error of reading the ruler. A protractor could be used to find the value of θ.

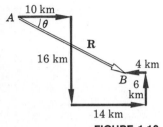

FIGURE 1.10
The path traveled by the car in Illustration 1.5. It started at point A and ended at point B.

Finding the resultant displacement (magnitude and direction) by use of ruler and protractor is referred to as the **graphical method** of solving the problem. It has general validity and can be used conveniently in many cases. The graphical method can be summarized as follows: To add several vectors graphically, place them end to end, with the tail of the second on the tip of the first. Place the tail of the third on the tip of the second, and so on. The resultant vector is an arrow with its tail at the tail of the first and its tip at the tip of the last vector. This method is applied in Illustration 1.5.

Illustration 1.5 If a car travels 10.0 km east, 16.0 km south, 14.0 km east, 6.0 km north, and 4.0 km west, what is the resultant displacement from the starting point?

Reasoning The appropriate vector diagram is shown in Fig. 1.10. Once again, the resultant displacement is shown as the outlined arrow. From its length you can measure the resultant displacement to be 22.4 km. The angle θ can be measured with a protractor and is $26.5°$ south of east.

It is not necessary that all the vectors be at right angles for our method to work. Suppose you want to add a 5-km south displacement to a 10-km northeast displacement. This is done in Fig. 1.11 and is obviously no more difficult than for displacements at right angles. Extension to several vectors is done exactly as in Fig. 1.10.

FIGURE 1.11
A vector diagram of a trip of 10 km northeast followed by one of 5 km south.

1.8 Vectors Other than Displacements

Vectors can be used to represent any quantity that has both magnitude and direction. For practical reasons, *we also require that the quantity obey the same mathematical laws which govern displacements.* All common quantities that have direction satisfy this requirement. For example, *forces are impor-*

tant vector quantities. A typical situation that illustrates the vector nature of forces is shown in Fig. 1.12.

In Fig. 1.12 we see a string pulling on a ring fastened to a post. *The force with which a string pulls on the object to which it is attached is called the* **tension** *in the string.* As indicated in the figure, we represent the tension in the string by **F**. This force exerted by the string on the ring has both magnitude (F) and direction (to the right). It can therefore be represented by a vector (or arrow) as shown. We choose a scale and draw the length of the arrow proportional to the magnitude of the force **F**. Since the force pulls to the right on the ring, the arrow drawn to represent it also points toward the right.

A convenient schematic diagram applicable to the situation of Fig. 1.12*a* is shown in Fig. 1.12*b*. Here the essential feature of part *a*, the force on the circular ring, is at once evident without extraneous detail. We shall make frequent use of schematic diagrams like this. They are called **free-body diagrams.**

Another example of a vector quantity is the velocity of a moving object, such as a car. We say that the car is going 20 meters per second (m/s) east. Hence this quantity can be represented by an arrow pointing eastward.

Although other quantities can also be represented as vectors, the three mentioned—displacements, forces, and velocities—are perhaps the most common. We see in the next section that all vectors can be treated in the same way. Consequently, when you have learned how to deal with displacement vectors, you will be able to handle force and velocity vectors with no difficulty.

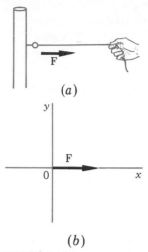

FIGURE 1.12
The pull of the string on the ring (the tension in the string) can be represented by the force vector **F**.

1.9 Rectangular Components of Vectors

Suppose that a person travels 20 km northeast. The appropriate vector is shown in Fig. 1.13. But it is clear that the same final point would have been reached if the person had gone east to point *B* and then directly north to point *C*. From this we see that the two colored vectors, when added, are equivalent to the single 20-km vector. *Several vectors which when added end to end give a single vector* (as *AB* and *BC* do to give *AC*) *are said to be* **components** *of the vector.*

We have already been dealing with vector components in previous sections when we added several vectors to obtain their resultant. The resultant is composed of the individual component vectors. Indeed, any given vector can be divided into arbitrary components subject to only one condition: The component vectors must add to give the original vector. Although such arbitrary selection of components is possible, it proves useful in most cases to take the component vectors along perpendicular directions. Let us now see how the concept of vector components is used to simplify the addition of vectors.

The component method for adding vectors is based on the fact that any vector can be thought of as being made up of components at right angles to

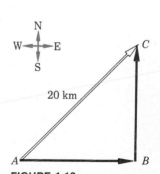

FIGURE 1.13
A displacement of 20 km northeast is resolved into component displacements *AB* east and *BC* north. Both *AB* and *BC* are components of vector *AC*.

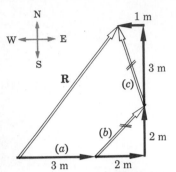

FIGURE 1.14
The sum of the several vectors is equivalent to the sum of the component vectors, as shown.

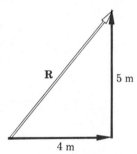

FIGURE 1.15
Adding the components of the vectors in Fig. 1.14 gives the same resultant as when the vectors themselves are added.

one another. In order to see how the addition is performed, let us refer to the displacement vectors shown in Fig. 1.14. The first vector (*a*) is all in the eastward direction. But the second vector (*b*) can be thought of as having two components as shown, 2 m toward the east and 2 m toward the north. In the same manner, the third vector (*c*) has a northward component of 3 m and a westward component of 1 m. But a 1-m westward displacement is simply a negative 1-m eastward displacement. That is, when you go 1 m westward, you are undergoing a -1-m eastward displacement. With this notation, we can use plus and minus signs to represent the direction of a vector component along a given line.

To find the resultant of the three displacement vectors (*a*), (*b*), and (*c*) in Fig. 1.14, we approach the problem in terms of component vectors. We see from the figure that the total displacement north is $0 + 2 + 3 = 5$ m, where the zero represents the fact that vector (*a*) has no northward component. We also see that the total eastward movement has been $3 + 2 - 1 = 4$ m. Since we have moved 4 m east and 5 m north, the resultant vector must have these components. It is shown in Fig. 1.15 and is seen to be the same as **R** in Fig. 1.14. Of course, the magnitude of **R** is $R = \sqrt{4^2 + 5^2}$ m.

Since the only thing that matters in finding **R** is how far east and how far north one has moved altogether, it does not matter in what order the movements were made. Hence, *the order in which we add vectors is of no importance.* We simply compute the total eastward movement and the total northward movement, and these totals are the east and north components of the resultant displacement. Therefore it is often convenient to replace a vector by its component vectors. To show in a diagram that this is what we do, the original vector is crossed off, leaving only its components. We illustrate this by placing slash marks on the original vector to be replaced by its components. That is why, in Fig. 1.14, vectors (*b*) and (*c*) are slashed out. During the computation of the resultant movement, we replaced these two vectors by their components.

Illustration 1.6 Add the displacement vectors shown in Fig. 1.16*a*.

Reasoning We shall use an *x* and *y* system of coordinates for our directions rather than east and north, but the method remains unchanged. For the present case we shall do the addition graphically even though the next section shows that using components is perhaps a better way of approaching the problem.

The actual graphical addition is carried out in Fig. 1.16*b*. Notice that the vectors are laid out tip to tail and that the resultant vector heads in the direction from the starting point to the end point, not vice versa. One might legitimately question whether the same resultant would be obtained if the vectors had been added in a different order. When this is tested in part *c* of the figure, we see that the resultant does not depend on the order in which the vectors are laid down. We should not be surprised that this is true since **R** simply gives the net result of all the *x* and *y* component displacements. It will not depend on the order in which the components are added.

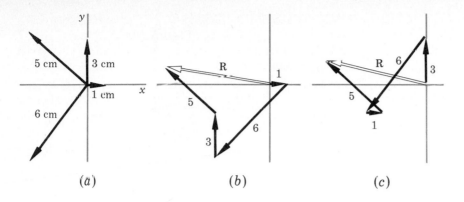

(a)   (b)   (c)

FIGURE 1.16
The four vectors shown in part *a* can be added as in parts *b* and *c* to obtain their resultant. Notice that the order in which the vectors are taken is of no importance.

1.10 Trigonometric Methods

The graphical method for adding vectors is satisfactory if you are good at drafting, are patient, and have a ruler and protractor. For most of us, though, it is more convenient to have a quicker, less cumbersome method for adding vectors. A more convenient method is available to those who spend a few minutes learning a few basic rules of trigonometry, which we give here.

If you know the east and north (or x and y) components of a group of vectors, the resultant is easily found. This is apparent from the fact that the x component of the resultant is the sum of all the x components of the vectors and the y component of the resultant is the sum of all the various individual y components. The virtue of trigonometry is that it provides an easy method for finding the components of vectors. Three quantities having to do with right triangles have been evaluated and are given in trigonometric tables, such as the one in Appendix 5. If we refer to Fig. 1.17, we can see what these quantities are. In relation to the right triangle shown in Fig. 1.17, we define three trigonometric quantities for the angle θ (theta):

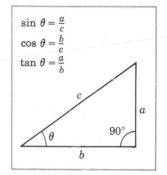

$$\sin \theta = \frac{a}{c} \qquad \cos \theta = \frac{b}{c} \qquad \tan \theta = \frac{a}{b}$$

FIGURE 1.17
In terms of this right triangle, $\sin \theta = a/c$, $\cos \theta = b/c$, and $\tan \theta = a/b$.

$$\sin \theta = \frac{a}{c} \qquad \cos \theta = \frac{b}{c} \qquad \tan \theta = \frac{a}{b}$$

Let us now refer to Fig. 1.18. From Fig. 1.18 and the definitions, it is apparent that the x component c_x of vector c is just $c \cos \theta$. Similarly, the y component c_y of vector c is $c \sin \theta$. If someone would tell us the value of $\cos \theta$ and $\sin \theta$, we could easily multiply the values by c and thereby obtain the x and y components of c. This is the purpose of the trigonometric tables. Let us now clarify this with some examples.

Illustration 1.7 Find the components of the vector shown in Fig. 1.19.

Reasoning The components are shown as c_x and c_y. From the definitions of the sine and cosine, we see from the figure that

FIGURE 1.18
The y component of vector c is $c \sin \theta$, and its x component is $c \cos \theta$.

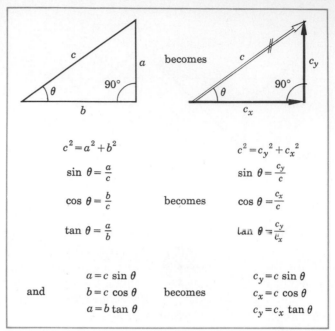

$$c^2 = a^2 + b^2 \qquad\qquad c^2 = c_y^2 + c_x^2$$

$$\sin \theta = \frac{a}{c} \qquad\qquad \sin \theta = \frac{c_y}{c}$$

$$\cos \theta = \frac{b}{c} \qquad \text{becomes} \qquad \cos \theta = \frac{c_x}{c}$$

$$\tan \theta = \frac{a}{b} \qquad\qquad \tan \theta = \frac{c_y}{c_x}$$

and
$$a = c \sin \theta \qquad\qquad c_y = c \sin \theta$$
$$b = c \cos \theta \quad \text{becomes} \quad c_x = c \cos \theta$$
$$a = b \tan \theta \qquad\qquad c_y = c_x \tan \theta$$

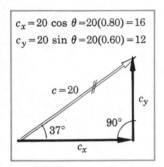

$$c_x = 20 \cos \theta = 20(0.80) = 16$$
$$c_y = 20 \sin \theta = 20(0.60) = 12$$

FIGURE 1.19
The y component of the 20-unit vector is $20 \sin 37°$, or 12, while the x component is $20 \cos 37°$, or 16.

$$\sin \theta = \frac{c_y}{20} \qquad \text{and} \qquad \cos \theta = \frac{c_x}{20}$$

where $\theta = 37°$ in this case. The trigonometric tables (or a calculator) tell us that $\sin 37° = 0.60$ and $\cos 37° = 0.80$. Substituting these values, we solve and find $c_y = 12$ and $c_x = 16$.

Illustration 1.8 Add the vectors shown in Fig. 1.20a by the component method.

Reasoning We have labeled the vectors a, b, c, and d, as shown. Vectors a and b have only one nonzero component each. In part b of the figure we find the components of vector c. Since $\sin \theta = $ (opposite side)/(hypotenuse), we have for the y component

FIGURE 1.20
With the component method, the vectors in part a can be added to give the resultant shown in part d.

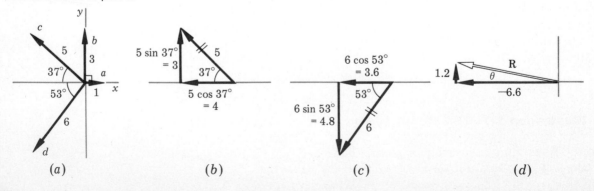

(a) $\qquad\qquad$ (b) $\qquad\qquad$ (c) $\qquad\qquad$ (d)

$$y \text{ component} = (\text{hypotenuse})(\sin 37°) = (5)(0.60) = 3$$

Similarly, $\cos \theta = (\text{adjacent side})/(\text{hypotenuse})$, which gives

$$\text{Adjacent side} = 5 \cos 37° = 4$$

But this is the magnitude of the x component. However, the x component is negative since it points in the $-x$ direction. The x component of the 5 unit vector is therefore -4.

A similar procedure, shown in Fig. 1.20c, can be used to find the components of the 6-unit vector. Its x and y components are -3.6 and -4.8. Now the components of all the vectors are known. Until you become familiar with the method, it is helpful to arrange the components in a table as follows:

Vector	x component	y component
a	$a_x = +1.0$	$a_y = 0$
b	$b_x = 0$	$b_y = +3.0$
c	$c_x = -4.0$	$c_y = +3.0$
d	$d_x = -3.6$	$d_y = -4.8$
Resultant	$R_x = -6.6$	$R_y = +1.2$

In preparing this table, notice how components pointing in the negative x and y directions are taken as negative. Now that we know the components of the resultant, we can draw it, as shown in Fig. 1.20d. By use of the pythagorean theorem,

$$R = \sqrt{R_x^2 + R_y^2} = \sqrt{(-6.6)^2 + (1.2)^2} = 6.7$$

We are not yet finished with this problem, since we still wish to find the value of the angle θ shown in Fig. 1.20d. This is easily done by referring to the definition of the tangent. Thus,

$$\tan \theta = \frac{R_y}{R_x} = \frac{1.2}{6.6} \approx 0.18$$

where $\approx$ is read "is approximately equal to." By examining the trigonometric table, we see that the angle whose tangent is 0.18 is approximately 10.5°. Therefore $\theta = 10.5°$ above the negative x axis. To obtain a more accurate value, we can use a process called **interpolation,** or we could consult a more complete table, such as those given in handbooks.[3] Many hand calculators also can be used to find trigonometric functions.

---

[3] *The Handbook of Chemistry and Physics* (Chemical Rubber Publishing Co.) contains a good set of mathematical tables. It also has a wealth of other information and is frequently consulted by persons working in the various sciences.

Galileo Galilei
(1564–1642)

Galileo, the son of a musician, initiated his career as a medical student, but soon abandoned it for mathematics and the physical sciences. When 25 years old, he was appointed professor at the University of Pisa, Italy, and 3 years later he accepted a professorship at the University of Padua. His investigations into the behavior of falling objects were carried out early in his career. When, in 1609, Galileo learned of the newly invented telescope, he constructed one for his own use. Pointing it at the heavens, he discovered many previously unknown luminous objects in the sky. These newly found stars as well as his closer glimpses of the moon and the planets cast doubt in his mind concerning the traditional view of the earth as the center of a perfect universe. His widely publicized views were ridiculed by many and confirmed and praised by others. In 1611 he was invited to Rome to receive many honors, including an audience with the Pope. But by 1616 his attacks on the church-sanctioned concept of an earth-centered universe drew the wrath of the clergy, and he found it wise to mute his discussion of the subject. After nearly 15 years of silence and work, he defiantly published his observations and conclusions, which still contradicted the church's doctrines. The result was trial and house arrest, until Galileo, by now an old man broken in health and spirit, denied his discoveries. Even so, he used the remaining decade of his life to complete his studies on motion. These are described in his greatest work, the *Dialogues Concerning Two New Sciences,* which was published in Holland after being smuggled out of Italy.

1.11 Addition of Forces

It is now a simple matter to add vectors of all kinds. We know that the order in which they are added is of no importance. We also now understand a powerful method for adding vectors, the rectangular-component trigonometric method.

Illustration 1.9 Consider the problem illustrated in Fig. 1.21*a*. Several people are pulling on ropes attached to a post, and the figure shows a top view of the post and ropes. The force in pounds exerted on the post by each rope is indicated in part *b* of the figure. Our problem is to find the net result of the various forces exerted on the post by the ropes.

   Reasoning The appropriate component table is shown below, where the 12-lb vector is considered first and the others are taken in turn, proceeding counterclockwise around the diagram (no particular reason for this, of course, since order doesn't matter):

Vector, lb	x component, lb	y component, lb
12	12.0	0.0
10	8.7	5.0
5	−3.0	4.0
8	−4.8	−6.4
Resultant	12.9	2.6

Notice that the 8-lb force pulls downward and to the left. Therefore its *x* and *y* components are negative. The pythagorean theorem now tells us that

$$R = \sqrt{(2.6)^2 + (12.9)^2}\ \text{lb} = 13.2\ \text{lb}$$

The resultant is shown in part *c* of Fig. 1.21. In addition, $\tan\theta = 2.6/12.9 \approx 0.20$, from which $\theta = 11.5°$.

   One should always examine the diagram for the resultant, to make sure that no error has been made in magnitude or direction. In this case, since the 10- and 12-lb forces are pulling mostly in the +*x* direction, and since the 5- and 8-lb forces tend to balance each other, it seems reasonable that the resultant pull on the post should be in the direction found for **R**.

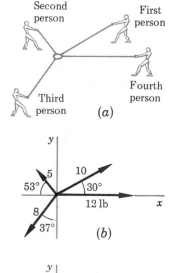

FIGURE 1.21
By adding the vectors shown in part *b* the resultant illustrated in part *c* is obtained.

1.12 Subtraction of Vector Quantities

Many physical situations lend themselves to analysis by vector subtraction. For example, if you walk 10 blocks east and then retrace your path by going 4 blocks west, you subtracted a 4-block displacement from a 10-block dis-

$$\overset{10}{\longrightarrow} \quad + \quad \overset{4}{\longleftarrow} \quad = \quad \overset{6}{\longrightarrow}$$

$$\overset{10}{\longrightarrow} \quad - \quad (\overset{4}{\longrightarrow}) \quad = \quad \overset{10}{\longrightarrow} \quad + \quad \overset{4}{\longleftarrow} \quad = \quad \overset{6}{\longrightarrow}$$

placement. Or if you wish, you could say you *added* the following displacements: a 10-block eastward displacement and a −4-block eastward displacement. The resultant displacement is 6 blocks eastward in either case.

With this equivalency of the two descriptions in mind, we see that *the subtraction of a vector is equivalent to the addition of the same vector, but with its direction reversed.* This is illustrated in Fig. 1.22. Notice that the following rule applies to vector subtraction:

To subtract a vector, reverse its direction and add it.

We can put this in mathematical symbolism by writing

$$\mathbf{A} - \mathbf{B} = \mathbf{A} + (-\mathbf{B})$$

where −**B** is simply vector **B** with its direction reversed.

Illustration 1.10 Subtract **B** from **A** in Fig. 1.23.

Reasoning To subtract **B**, we must first reverse its direction and add it to **A**. This is done in part *b* of the figure. The components of **A** are $A_x = 10 \cos 30° = 8.7$ m and $A_y = 10 \sin 30° = 5.0$ m. Then, from part *b* of the figure, noticing that $B_y = 0$, we have

and

$$R_x = -B_x + A_x = 6.0 \text{ m} + 8.7 \text{ m} = 14.7 \text{ m}$$

so

$$R_y = -B_y + A_y = 0 + 5.0 \text{ m} = 5.0 \text{ m}$$

$$R = \sqrt{R_x^2 + R_y^2} = 15.5 \text{ m}$$

The angle θ that **R** makes with the x axis is given by

from which

$$\tan \theta = \frac{R_y}{R_x} = \frac{5.0}{14.7} = 0.34$$
$$\theta = 19°$$

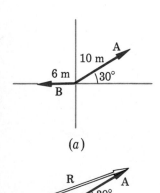

(a)

(b)

FIGURE 1.23
To find **A** − **B**, reverse the direction of **B** and add it.

1.13 Significant Figures

As we pointed out earlier, science is based on measurement. Nearly all the numbers we use in science are limited in their accuracy because of the imperfect nature of all measurements. With that fact in mind, it is important that we understand the limitations which experimental accuracy places on numbers that we compute. Let us now investigate how we handle numbers that are not exact.

The numerical value of every observed measurement is an approximation. Suppose that the length of an object is recorded as 15.7 cm. By convention, this means that the length was measured to the nearest tenth of a centimeter and that its exact value lies between 15.65 and 15.75 cm. If this measurement were exact to the nearest hundredth of a centimeter, it would have been recorded as 15.70 cm. The value 15.7 cm represents three significant figures (1, 5, 7), while the value 15.70 represents four significant figures (1, 5, 7, 0). A **significant figure** *is one which is known to be reasonably reliable.* Similarly, a recorded mass of 3.4062 kg means that the mass was determined to the nearest tenth of a gram and represents five significant figures (3, 4, 0, 6, 2); the last figure (2), being reasonably correct, guarantees the certainty of the preceding four figures.

Definition

Zeros may be significant, or they may merely serve to locate the decimal point. For instance, the statement that a car weighs 1800 lb does not indicate definitely the accuracy of the weighing. If it was weighed to the nearest 100 lb, the weight contains only two significant figures (1, 8) and may be written exponentially as 1.8×10^3 lb. If the car was weighed to the nearest 10 lb, the first zero is significant, but the second is not; the weight could be written as 1.80×10^3 lb, displaying the three significant figures. If the object was weighed to the nearest 1 lb, the weight could be written as 1.800×10^3 lb (four significant figures). Of course, if a zero stands between two significant figures, it is itself significant.

If a computation gives an answer 6.2795, only three of whose digits are significant, we round it off to 6.28. A number is rounded off to the desired number of significant figures by dropping one or more digits to the right. When the first digit dropped is less than 5, the last digit retained should remain unchanged. When the first digit dropped is more than 5, or when it is 5 followed by digits not all zeros, then 1 is added to the last digit retained. When the first digit dropped is 5 followed by zeros, 1 is added to the last digit retained if that digit is odd.

When we add or subtract numbers, the answer should be rounded off so as to retain digits only as far as the first column containing estimated figures. (Remember that the last significant figure is estimated.) This point is made clear in the following examples.

Add the following quantities expressed in meters:

25.340	58.0	4.20	415.5
5.465	0.0038	1.6523	3.64
0.322	0.00001	0.015	0.238
31.127 m (*Ans.*)	58.00381	5.8673	419.378
	= 58.0 m (*Ans.*)	= 5.87 m (*Ans.*)	= 419.4 m (*Ans.*)

When we multiply or divide numbers, the answer should be rounded off to contain only as many significant figures as are contained in the least exact factor. There are some exceptional cases, however. Consider the division 9.84 ÷ 9.3 = 1.06, to three places. By the rule given above, the answer should be 1.1 (two significant figures). However, a difference of 1 in the last place of 9.3 (9.3 ± 0.1) results in an error of about 1 percent, while a difference of 1 in the last place of 1.1 (1.1 ± 0.1) yields an error of roughly 10 percent. Thus the answer 1.1 is of much lower percentage accuracy than 9.3. Hence in this case the answer should be 1.06, since a difference of 1 in the last place of the least exact factor used in the calculation (9.3) yields a percentage of error about the same (about 1 percent as a difference of 1 in the last place of 1.06 (1.06 ± 0.01). Similarly, 0.92 × 1.13 = 1.04.

Summary

Measurements are fundamental in science. Through the data they provide, we learn the laws of nature. To describe our experimental data, we make use of standard units of measure. Three basic units in SI are the meter, second, and kilogram. The kilogram is the unit for mass.

In the metric system, units of the same quantity are related through powers of 10. Word prefixes are used to indicate these powers in such words as kilometer and milligram.

Vector quantities have both magnitude and direction. Scalar quantities have no direction. We can represent a vector quantity by an arrow called a vector. The length of the vector is proportional to the magnitude of the vector quantity. The direction of the vector shows the direction of the vector quantity.

To add several vectors graphically, place them end to end. Place the tail of the second on the tip of the first. Then place the tail of the third on the tip of the

second, and so on. The resultant vector is an arrow with its tail at the tail of the first and its tip at the tip of the last vector.

To add several vectors by the trigonometric component method, first find the rectangular components of each vector. This is done by the trigonometric method described in Sec. 1.10. The x component of the resultant is found by taking the algebraic sum of the x components of the vectors. Similarly, the y component of the resultant is found. Then

$$R = \sqrt{R_x^2 + R_y^2} \qquad \text{and} \qquad \tan \theta = \frac{R_y}{R_x}$$

where θ is the angle the resultant vector makes with the x axis.

To subtract a vector **B** from a vector **A,** first the direction of **B** is reversed, and then **B** is added to **A.** In symbols, **A** − **B** = **A** + (−**B**).

Minimum Learning Goals[4]

Upon completion of this chapter, you should be able to do the following:

1 State what is meant by a law of nature.
2 Give the definitions for the SI units of length, time, and mass.
3 Change a number from scientific notation to written-out form and vice versa.
4 Identify the power-of-10 value associated with each metric prefix.

5 Use dimensional analysis to test the validity of an equation.
6 Explain the difference between scalar and vector quantities and give examples of each.
7 Draw a vector diagram showing a series of displacements described to you.
8 Find the resultant of several vectors by means of a scale drawing.
9 Define sin θ, cos θ, and tan θ. Given an angle, use the trigonometric tables to find its sine, cosine, and tangent.
10 Use trigonometry to find the unknown sides or angles in a right triangle.

---

[4]Notice the word *minimum*. *All* students should achieve these goals.

11 Find the rectangular components of a given vector; or, given the components of a vector, find the vector and its angle.

12 Add several given vectors by use of the trigonometric component method.

13 Subtract one given vector from another given vector.

14 When provided with several numbers of known accuracy, round off the result after they are added, subtracted, multiplied, or divided.

Important Terms and Phrases

You should be able to define or explain each of the following:

Law of nature
Système International (SI)
Scientific notation
Metric prefixes
Dimensional analysis
Scalar quantity
Vector quantity
Displacement vector

Pythagorean theorem
Resultant vector
Graphical method for vector addition
Sine, cosine, and tangent
Rectangular components of a vector
Trigonometric method for vector addition
Force vector
Vector subtraction
Significant figures
Rounding off a number

Questions and Guesstimates

Questions marked with (E) at the end require you to make an estimate based on your experience.

1 Suppose a person proposes a law of nature that contradicts a law proposed by another person. How do you decide which, if either, person is correct?

2 In what ways are the laws of nature similar (or dissimilar) to the laws of a nation? Consider such factors as who obeys them, how frequently they are broken, who decides their interpretation, how easily they can be changed, etc.

3 The United States is slowly drifting into the use of the metric system. Most other countries have adopted it by law as the accepted measuring system. Why is the United States so backward in this respect? Should we adopt it?

4 Which of the following are vector quantities: (a) the population of Philadelphia; (b) the force a toe exerts on a football; (c) the airplane flight path from Denver to Portland; (d) the quantity of water in a glass; (e) the motion imparted to a ball by a bat; (f) the number of pages in this book?

5 What eastward displacement must be added to a 20-m eastward displacement if the resultant is to be zero? Repeat if the unknown eastward displacement is to be subtracted from the 20-m displacement.

6 Can two unequal-magnitude displacements be added to give a zero resultant? Can three? Show how three equal-magnitude vectors would have to be oriented relative to one another so as to give a zero resultant.

7 Represent each person in a city of 200,000 by a vector extending from toe to nose. Estimate the resultant of these vectors at (a) 12 noon and (b) 12 midnight. Repeat if the vector extends from the left ear to the right ear. (E) (You must estimate the average toe-to-nose distance of each person as well as how many persons are standing up.)

8 Suppose you have a meterstick and are asked to compute the volume of a cubical box. Which volume could you find more exactly, a box that is about 10 cm × 10 cm × 10 cm or one that is 0.5 cm × 20 cm × 10 cm?

9 An old saying has it that "A chain is only as strong as its weakest link." What analogous statement(s) can you make regarding experimental data used in a computation? Can you think of a pertinent situation where your statement does not apply?

Problems

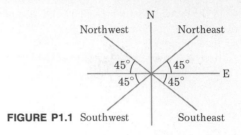

FIGURE P1.1 Southwest

The level of difficulty of the problem is indicated by asterisks. Vector problems labeled (G) at the end are to be done graphically. All others are to be done by using trigonometry.

1 Using scientific notation, write each of the following lengths in meters with one digit to the left of the decimal, as in 3.65×10^4: (a) 604 km; (b) 0.163 mm; (c) 47 nanometers (nm); (d) 0.00295 micrometer (μm); (e) 73 megameters (Mm).

2 Using scientific notation, write each of the following masses in grams with one digit to the left of the decimal, as in 6.21×10^{-3}: (a) 241 kg; (b) 20.6 nanograms (ng); (c) 0.0051 microgram (μg); (d) 17.3 gigagrams (Gg); (e) 0.135 picogram (pg); (f) 891,000 milligrams (mg).

3 Carry out the following computation: $(6.21 \times 10^{-3}) \times (3.74 \times 10^4) \div (5.40 \times 10^6)$. Write your answer in scientific notation with one digit left of the decimal.

4 Carry out the following computation: $(8.60 \times 10^5) \times (6.17 \times 10^{-2}) \div (1.79 \times 10^{-4})$. Write your answer in scientific notation with one digit to the left of the decimal.

5 Carry out the following computation: $(2.1 \times 10^4) \div 0.006543$. Give your answer in scientific notation.

6 Carry out the following computation: $0.026 \div 4724$. Give your answer in scientific notation.

7 The wavelength λ of a wave depends on the speed v of the wave and its frequency f. Knowing that $[\lambda] = [L]$, $[v] = [L][T]^{-1}$, and $[f] = [T]^{-1}$, decide which of the following equations is correct, $\lambda = vf$ or $\lambda = v/f$.

8 When we study about work, you will learn that work depends on the two quantities force F and distance d. Knowing that $[\text{work}] = [M][L]^2[T]^{-2}$, $[F] = [ML][T]^{-2}$, and $[d] = [L]$, decide which of the following equations is correct: work $= F/d$, work $= Fd$, or work $= d/F$.

9** The speed v of a wave on a string depends on the tension F in the string and the mass per unit length m/l of the string. Knowing that $[v] = [L][T]^{-1}$, $[m/l] = [M][L]^{-1}$, and $[F] = [ML][T]^{-2}$, determine the constants a and b in the following equation: $v = (\text{constant})F^a\,(m/l)^b$.

10** The frequency of vibration f of a mass m at the end of a spring that has a stiffness constant k is related to m and k by a relation of the form $f = (\text{constant})m^ak^b$. Use dimensional analysis to find a and b from the facts that $[f] = [T]^{-1}$, $[m] = [M]$, and $[k] = [M][T]^{-2}$.

11 My home is 6 blocks east and 8 blocks north from the building in which I work. What is the straight-line distance between these points? At what angle with east is this straight line? (G)

12 What is the actual straight-line distance between two points if one lies 20 km west and 16 km north of the other? What angle does this line make with east? (G)

13 If an airplane flies southwest for 100 km, how far west and how far south has it gone (see Fig. P1.1)? (G)

14 Detroit is approximately 650 km northwest of Washington, D.C. How much farther south is Washington than Detroit (see Fig. P1.1)? How much farther east? (G)

15 In being positioned on a milling machine table, a device is given the following displacements: 5.0 cm at 0°, 12.0 cm at 80°, 7.0 cm at 110°, 9.0 cm at 210°. All angles are measured counterclockwise from the x axis. (See Fig. P1.2 for the 7.0-cm displacement at 110°.) Find the magnitude and the angle of the resultant displacement. (G)

16 A treasure map gives the following directions: "Start at the large tree. Go 80 paces straight east, then 50 paces at 70° north of east, then 60 paces at 30° west of north, then 18 paces straight south to the treasure." How far from the tree and at what direction is the treasure? (G)

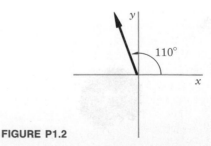

FIGURE P1.2

17 Point B is 300 km from point A and at an angle of 50° north of east. A road goes 40 km east from point A and then ends. Starting at the end of the road, how far and at what angle to the east must one travel in order to reach point B? (G)

18* To fly from St. Louis to Miami, a plane must fly 1670 km at an angle of 47° south of east. To fly from Ottawa to Miami, a plane must fly straight south for 2060 km. How far and in what direction must a plane fly to go from St. Louis to Ottawa? (G)

19 To go from Miami to Chicago, a plane must fly about 1780 km at an angle of 30° west of north. How far north of Miami is Chicago? How far west?

20 A force of 50 lb pulls at an angle of 110° measured counterclockwise to the $+x$ axis (see Fig. P1.2). What is the x component of this force? Its y component?

21* A certain force has a magnitude of 200 lb, and its x component is -80 lb. Find its y component and its direction. Two possible answers exist. Find both.

22* In an xy coordinate system, point P is 30 cm from the coordinate origin, and its y coordinate is -18 cm. Find P's x coordinate and its direction from the origin. Two answers are possible. Find both.

23 An object undergoes the following successive displacements in the xy plane: 25 cm at $\theta = 0°$ and 15 cm at $\theta = 110°$ with all angles measured counterclockwise from the $+x$ axis (as in Fig. P1.2). Find its resultant displacement.

24 Suppose you start at point A and walk first 230 m south and then 350 m at an angle of 30° west of north to a point we call B. Find the displacement from point A to point B.

25 Find the resultant of the forces shown in Fig. P1.3. Give the counterclockwise angle relative to the $+x$ axis. The SI force unit used in the figure is the newton (N), where 1 N = 0.225 lb.

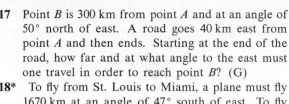

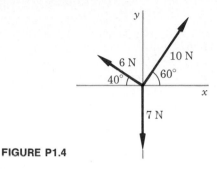

FIGURE P1.4

26 Find the resultant of the forces shown in Fig. P1.4. Give the counterclockwise angle relative to the $+x$ axis. The SI force unit used in the figure is the newton, where 1 N = 0.225 lb.

27 Solve Prob. 16 using the trigonometric method.

28 Solve Prob. 15 using the trigonometric method.

29* Solve Prob. 18 using the trigonometric method.

30* Solve Prob. 17 using the trigonometric method.

31* What third force must be added to the following two forces to give a zero resultant: 25 lb at 70° and 40 lb at 135°? Angles are measured as in Fig. P1.2.

32* What third force must be added to the following two forces to give a zero resultant: 60 lb at 120° and 40 lb at 200°? Angles are measured as in Fig. P1.2.

33** A shaft in a coal mine goes straight down for 85 m. From its lower end a horizontal tunnel goes 50 m west and then 30 m south to its end. How far is the end of the tunnel from the top of the shaft? What angle does a straight line from the top of the shaft to the end of the tunnel make with the vertical?

34** A boxlike room has a 2.5-m-high ceiling, and the floor is 6 m × 8 m. Find the length of the diagonal line from a ceiling corner to the opposite floor corner. What angle does the line make with the floor?

35* Two displacements $\mathbf{A}$ and $\mathbf{B}$ lie in the xy plane. Displacement $\mathbf{A}$ is 25 cm at an angle of 110° to the $+x$ axis (as in Fig. P1.2.), while displacement $\mathbf{B}$ is 15 cm in the $-x$ direction. Find the displacement that results by subtracting displacement $\mathbf{A}$ from displacement $\mathbf{B}$.

36* Two forces $\mathbf{A}$ and $\mathbf{B}$ act in the xy plane. Force $\mathbf{A}$ is 60 N and is directed along the $+y$ axis. Force $\mathbf{B}$ is 40 N and is directed at an angle of 30° below the $+x$ axis. If force $\mathbf{B}$ is subtracted from force $\mathbf{A}$, what is the resultant force?

37 Two steel rods have measured lengths of 27.03 and 26.9 cm. Find (a) the sum of their lengths and (b) the difference between their lengths.

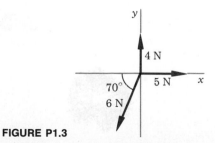

FIGURE P1.3

38 Two students measure the following volumes of water in two graduated cylinders: 258 and 3.15 cm³. Find (a) the sum of these two volumes and (b) their difference.

39 The third edition of this book has pages that are 18.9 cm × 23.32 cm. When tightly compressed, the 784-page book has a thickness of 2.9 cm for its pages. Find (a) the area of a page, (b) the volume occupied by the pages of the book, and (c) the thickness of a page.

40 A cylindrical can is completely filled with 43 identical spherical marbles. The diameter of each marble is 1.6 cm. The inside dimensions of the can are diameter = 7.65 cm and height = 3.90 cm. Find (a) the volume filled by the marbles in the can and (b) the empty space in the can.

41* After breaking his leg in a freak accident, a student finds himself in a hospital with traction applied to his leg. He notes that his condition is as shown in Fig. P1.5. Assume that the pulleys are frictionless; then the tension in the cord is everywhere the same, namely, 7.0 lb. How large is the force which stretches his leg? How large an upward force does the device exert on his foot and leg together?

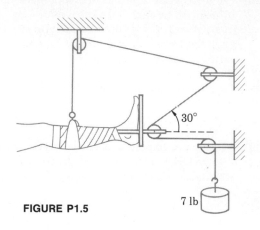

FIGURE P1.5

30°

7 lb

2
Uniformly Accelerated Motion

To describe the behavior of an object, often it is necessary to state how it is moving and how it will continue to move. In this chapter we discuss the methods we use to describe the motion of an object in terms of its velocity and acceleration. We shall find that these methods enable us to predict the motion of bodies in free fall. The role of forces in causing motion is the subject of Chap. 3.

2.1 Velocity and Speed

We all understand what is meant by the statement "the car was traveling at 50 miles per hour." This means that the car would travel 50 mi in 1 h if it continued at this speed. However, it is very unlikely that the driver would travel exactly at this speed for a full hour. If she actually did complete a 50-mi trip in 1 h, she would surmise that her *average* speed had been 50 mi/h. But we have no way of knowing from the data given that she didn't actually break the 55-mi/h speed limit and take time off for coffee along the way. Before delving further into this complication, let us define precisely what we mean by average speed.

Speed *is the ratio of the distance traveled to the time taken.* Its units are miles per hour, feet per second, meters per second, and so forth. If we travel 10 m in 2 s, our average speed is 5 m/s. We obtained this, perhaps unconsciously, by the equation

Definition

Average speed $= \dfrac{\text{distance traveled}}{\text{time taken}}$

or, in symbols,

$$\bar{u} = \frac{d}{t} \tag{2.1}$$

where the bar above the u indicates that it is an average value.

Let us see how the units of speed are arrived at. Carrying out the computation for a woman moving 10 m in 2 s, we find that her average speed is

$$\bar{u} = \frac{10 \text{ m}}{2 \text{ s}} = 5 \text{ m/s}$$

which we read as "5 meters per second." Similarly, if a snail travels 2 cm in 0.4 day, its speed is

$$\bar{u} = \frac{2 \text{ cm}}{0.4 \text{ day}} = 5 \text{ cm/day}$$

The units of speed are always a length unit divided by a time unit.

When a car moves along a straight road, its motion is in a certain direction. If we wish to combine the two concepts, speed and direction, we are speaking of a vector. It is customary to *define the* **average velocity vector** $\bar{\mathbf{v}}$ *in the following way:*

$$\bar{\mathbf{v}} = \frac{\text{displacement vector}}{\text{time taken}}$$

For example, suppose a car starts from a point A and at a time t later arrives at a point B. If the vector displacement from A to B is $\mathbf{s}$, we have for the definition of the average velocity

$$\bar{\mathbf{v}} = \frac{\mathbf{s}}{t} \tag{2.2}$$

Note that the velocity $\bar{\mathbf{v}}$ is in the same direction as the displacement $\mathbf{s}$.

For motion in a constant direction along a straight line, the length of the vector $\mathbf{s}$ would be simply the distance traveled d. As a result, the magnitude of the velocity *in this particular case* is equal to the speed since $\bar{u} = d/t = s/t = \bar{v}$. However, if the motion is not along a straight line, the total distance covered d may not be the same as the magnitude of the displacement $\mathbf{s}$.

A simple example of this is shown in Fig. 2.1*a*, which shows the path of a car traveling from A to B and then to C. We see that the total distance

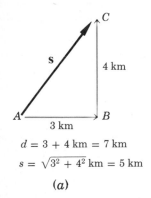

$d = 3 + 4 \text{ km} = 7 \text{ km}$

$s = \sqrt{3^2 + 4^2} \text{ km} = 5 \text{ km}$

(*a*)

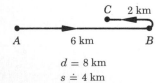

$d = 8 \text{ km}$

$s \doteq 4 \text{ km}$

(*b*)

FIGURE 2.1
Since the magnitude of the average speed is $\bar{u} = d/t$ and that of the average velocity is $\bar{v} = s/t$, the two quantities are not always equal.

covered d is 7 km. However, the displacement **s** has a magnitude of only 5 km. We therefore find

$$\bar{u} = \frac{d}{t} = \frac{7\ \text{km}}{t} \qquad \text{while} \qquad \bar{v} = \frac{s}{t} = \frac{5\ \text{km}}{t}$$

Clearly, the magnitudes of the speed and velocity are not the same in this case.

Even for motion along a straight line, these two quantities can differ. This is shown in Fig. 2.1b where the car goes from A to B and then back to C. In this case

$$\bar{u} = \frac{8\ \text{km}}{t} \qquad \text{while} \qquad \bar{v} = \frac{4\ \text{km}}{t}$$

Even so, much of our attention in this chapter is turned on situations in which motion occurs in a single direction. In such cases $\bar{u}$ and $\bar{v}$ will be equal.

2.2 Instantaneous Velocity and Speed

Let us consider an object that is moving along a straight line—the car in the upper part of Fig. 2.2, for example. We take the x coordinate along the line on which the car is moving, as shown. The motion of the car is therefore in either the $+x$ or $-x$ direction. (If we consider motion to the right as positive, then motion to the left is negative.) Suppose the car is traveling with constant speed (and velocity) in the $+x$ direction as shown. Let us say that the magnitude of the speed (and velocity) is 20 m/s.

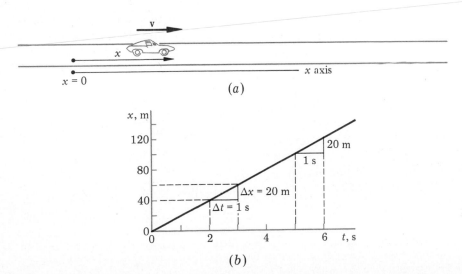

(a)

(b)

FIGURE 2.2
Motion along a straight line can be shown by a graph. In this case, the car's speed was constant at 20 m/s.

We can show the displacement of the car as a function of time by the graph shown in Fig. 2.2*b*. Because the car is moving in the *x* direction, we designate the magnitude of the displacement by *x* rather than by *s*. The direction of the displacement can be shown easily by appending plus or minus signs to *x*. If the car is to the left of $x = 0$, then its displacement might be -20 m, for example, where the negative sign tells us that the car has been displaced 20 m to the left of the $x = 0$ position. In the graph of Fig. 2.2, all the *x* values are positive, and so the car is displaced to the right of the $x = 0$ point.

The graph of Fig. 2.2 can be interpreted as follows: At time $t = 0$ the car was at the point $x = 0$. With the passage of time, the car's position changes. The colored straight line shown on the graph tells us where the car is at any particular instant. For example, at $t = 3$ s, the car was at $x = 60$ m, according to the graph, the car was 60 m from the $x = 0$ point.

We can attach meaning to the little triangles shown in the figure. Notice that the vertical side length of each is 20 m, while the horizontal side length is 1 s. The triangles therefore tell us that the car goes 20 m in the $+x$ direction for every second that passes. Using the symbolism Δx (read "delta *x*") to represent the small displacement that occurs in a short time interval Δt (read "delta *t*"), we can describe the motion of the car during a short time $\Delta t = 1.0$ s. From the tiny triangle, we see that during this 1-s interval the car underwent a displacement $\Delta x = +20$ m. Using these values, we can describe the speed and velocity of the car during this small time interval.

From the definition of average speed we have that, for this tiny 1-s-long interval,

$$\bar{u} = \frac{\text{distance traveled}}{\text{time taken}} = \frac{\Delta x}{\Delta t} = \frac{20 \text{ m}}{1.0 \text{ s}} = 20 \text{ m/s}$$

Similarly, for the magnitude of the average velocity during this time interval,

$$\bar{v} = \frac{\text{magnitude of displacement vector}}{\text{time taken}} = \frac{\Delta x}{\Delta t} = \frac{20 \text{ m}}{1.0 \text{ s}} = 20 \text{ m/s}$$

As we see, the magnitude of the average velocity equals the average speed in this case. Moreover, the graph tells us that the car is moving with constant speed. Each little Δx, Δt triangle we draw similar to the ones shown will always give a speed of 20 m/s. *An object traveling with constant speed and velocity along a straight line will always give a straight-line x-versus-t graph.*

Before leaving the example of Fig. 2.2, we should notice that $\Delta x/\Delta t$ is the slope of the line shown.[1] In this case at least, the magnitudes of the speed and velocity are equal to the slope of the *x*-versus-*t* graph. We shall soon see that this is always true.

---

[1] To find the slope of a curve at a point on it, construct a tangent line at the point. The slope equals the vertical rise of the tangent line divided by the concurrent horizontal run of the line. For example, the slope is $\Delta x/\Delta t$ in Fig. 2.2.

Illustration 2.1 Let us now refer to Fig. 2.3, which shows the motion of the car on the same straight road, but with nonconstant speed. Describe the car's motion.

Reasoning The graph tells us that the car is standing still at points A and C. Near these points on the graph, x is not changing with t. This must mean the car is not moving. However, near B the car is moving toward larger x's since x is increasing with time. Near B, the situation is much like that shown in Fig. 2.2. In fact, the average speed near B is

$$\bar{u} = \frac{\Delta x}{\Delta t} = \frac{50 \text{ m}}{20 \text{ s}} = 2.5 \text{ m/s}$$

The car's average *velocity* near B is 2.5 m/s in the $+x$ *direction*. (Remember, velocity is a vector.)

At point D, the situation is a little different. Near there, the car is moving toward smaller x values; the value of x is decreasing with time. In other words, the car is going in the reverse direction to that shown in Fig. 2.2a. Since x is becoming smaller in the region near D, the quantity Δx is negative since it is a *decrease* in x. The average speed of the car near D is

$$\bar{u} = \frac{\text{distance traveled}}{\text{time taken}} = \frac{|\Delta x|}{\Delta t} = \frac{100 \text{ m}}{20 \text{ s}} = 5.0 \text{ m/s}$$

Notice that speed is not concerned with direction. Therefore we use only the magnitude of the distance Δx, which we represent by $|\Delta x|$. The negative sign shows direction and is dropped when speed is computed.

But to compute the average velocity near D, we make use of the negative sign on Δx. We interpret the minus sign as a symbol which gives direction. Then

$$\text{Average velocity at } D = \frac{\Delta x}{\Delta t} = \frac{-100 \text{ m}}{20 \text{ s}} = -5.0 \text{ m/s}$$

The negative sign tells us the velocity is in the $-x$ direction.

A still different situation prevails at point E. There, the car is slowing down and is preparing to stop at F. If we now compute the average speed near E from the triangle shown, our result may not be the exact speed the car had at E. The computed value is only an average of the speeds during the interval for which this triangle is taken. However, if we take a very small time interval Δt centered on E, our result will be very close to the speed the car had at E.

The speed the car had as it passed point E in Fig. 2.3 was the car's speed for only an instant. Before the car reached E, it was moving faster than at E. Later it was moving slower than at E. We define the speed an object has at a certain instant to be the object's **instantaneous speed**. To find the car's instantaneous speed at point E, we must measure how long Δt it takes the

Definition

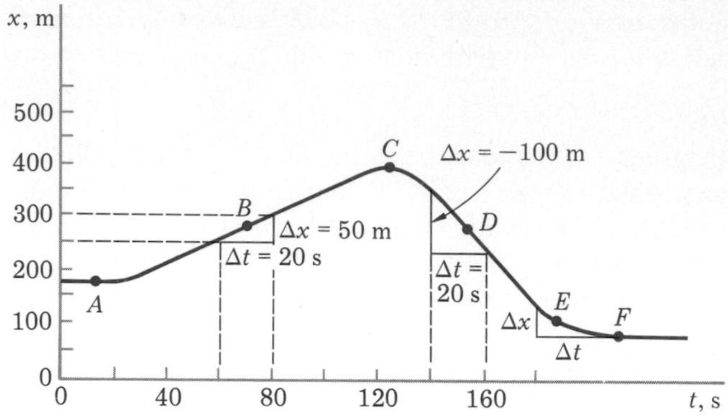

FIGURE 2.3

Can you show from the graph that the object is at rest at points *A* and *C*? That it is moving in opposite directions at points *B* and *D*?

car to pass through a very small distance Δx centered on point E. We can represent this statement in mathematical form as

$$\text{Instantaneous speed} \equiv u = \lim_{\Delta t \to 0} \frac{\Delta x}{\Delta t}$$

The notation $\lim_{\Delta t \to 0}$ (read "the limit as Δt approaches zero") tells us to evaluate $\Delta x/\Delta t$ when Δt is taken small enough for its value to be close to zero. Clearly, the car cannot possibly change its speed much in this tiny time. Therefore, the speed we measure will be very close to the speed at point E. It is the instantaneous speed at point E.

Since an object cannot change its direction much during an extremely short time interval, it follows that the measured displacement s of the object during this interval will be very nearly the same as the distance moved d during the interval. Hence the magnitude of the velocity computed for this tiny interval will be very close to the instantaneous speed. We therefore define an instantaneous velocity for an object in the following way. *Suppose in a very short time interval Δt the object undergoes a vector displacement $\Delta \mathbf{s}$.*

Definition *Then the **instantaneous velocity** of the object is defined to be*

$$\mathbf{v} = \lim_{\Delta t \to 0} \frac{\Delta \mathbf{s}}{\Delta t} \tag{2.3}$$

where the notation $\lim_{\Delta t \to 0}$ means to take the limiting value of $\Delta \mathbf{s}/\Delta t$ as Δt approaches zero magnitude. As pointed out above, *the magnitude of the instantaneous velocity $\mathbf{v}$ is equal to the instantaneous speed u.*

In our particular example, $\Delta s = \Delta x$, and so $v = \lim_{\Delta t \to 0} \Delta x/\Delta t$. As we see, instantaneous velocity is given by the ratio of Δx to Δt in a tiny triangle like those shown in Fig. 2.3. But if the triangle is taken small enough, the value of $\Delta x/\Delta t$ is the slope of the graph's curve, even near a point such as E. We conclude from this that *the instantaneous velocity in the x direction is equal to the slope of the x-versus-t curve at the instant in question.* Moreover, *the sign* *of the slope (+ or −) tells us whether the velocity is in the plus or in the minus x direction.*

2.3 Conversion of Units

Frequently we are given quantities in one set of units, and we need to express them in other units. We might, for example, know a speed limit in kilometers per hour and wish to find its equivalent in miles per hour. Or perhaps we need to know how many gallons of gasoline are equivalent to a certain number of liters of gasoline. To do this type of conversion, we make use of conversion factors. For example, it is known that

$$1 \text{ gallon (gal) (U.S.)} = 3.785 \text{ liters} \quad \text{or} \quad 1 \text{ liter} = 0.2642 \text{ gal}$$

To make use of this fact, we divide each side of the equation by 1 gal (or 1 liter), to obtain

$$1 = \frac{3.785 \text{ liters}}{1 \text{ gal}} \quad \text{or} \quad 1 = \frac{0.2642 \text{ gal}}{1 \text{ liter}}$$

These quantities are called *conversion factors.*

Notice that the conversion factor is equal to unity. When we multiply or divide any quantity by such a unit factor, the quantity remains unchanged; only the *unit* of the quantity changes. We see the utility of this in the following illustration.

Illustration 2.2   Change 30.0 liters to its equivalent in U.S. gallons.

Reasoning   We have

$$30.0 \text{ liters} = (30.0 \text{ liters}) \frac{0.2642 \text{ gal}}{1 \text{ liter}}$$

This is true because the multiplying factor, 0.2642 gal/liter, is equal to unity. But the unit *liters* can be canceled to give[2]

$$30.0 \text{ liters} = (30.0 \text{ liters}) \frac{0.2642 \text{ gal}}{1 \text{ liter}} = 7.93 \text{ gal}$$

By use of the conversion factor we have succeeded in finding that 30 liters is equivalent to 7.93 gal.

---

[2] Why didn't we use the factor 3.785 liters/gal?

We give a table of useful conversion factors on the inside front cover of this book. Let us now see a few typical uses of them.

Illustration 2.3

1 Change 36 in (inches) to centimeters.
Conversion factor: Because 1 in = 2.54 cm, the factor is 2.54 cm/in.
Method:

$$36 \text{ in} = (36 \text{ in}) \, \frac{2.54 \text{ cm}}{\text{in}} = 91 \text{ cm}$$

2 Change 270 cm to inches.
Method: Because 1 in = 2.54 cm, we have that

$$270 \text{ cm} = (270 \text{ cm}) \, \frac{1 \text{ in}}{2.54 \text{ cm}} = 106 \text{ in}$$

3 Change 60 acres to hectares (ha).
Conversion factors:

$$10,000 \text{ m}^2/\text{ha} \qquad 0.305 \text{ m/ft} \qquad 43,560 \text{ ft}^2/\text{acre}$$

Method:

$$60 \text{ acres} = (60 \text{ acres}) \left(\frac{43,560 \text{ ft}^2}{\text{acre}} \right) = 2,613,600 \text{ ft}^2$$

$$= (2,613,600 \text{ ft}^2) \left(\frac{0.305 \text{ m}}{\text{ft}} \right) \left(\frac{0.305 \text{ m}}{\text{ft}} \right)$$

$$= 243,100 \text{ m}^2$$

$$= (243,100 \text{ m}^2) \left(\frac{1 \text{ ha}}{10,000 \text{ m}^2} \right) = 24.3 \text{ ha}$$

We shall encounter many other uses of conversion factors as we proceed with our studies.

2.4 Acceleration

Definition *An object whose velocity is changing is said to be* **accelerating.** *Acceleration is defined as the change in velocity* (not speed!) *per unit time.* As an equation,

$$\text{Average acceleration} = \frac{\text{change in velocity}}{\text{time taken}}$$

or

$$\overline{\mathbf{a}} = \frac{\mathbf{v}_f - \mathbf{v}_0}{t} \tag{2.4}$$

where $\mathbf{v}_0$ is the starting velocity and $\mathbf{v}_f$ is the final velocity a time t later.

Since $\bar{a}$ is the difference of two vectors divided by a scalar, t, $\bar{a}$ must also be a vector. In this chapter we discuss motion in a straight line only, and so only the magnitude of $\bar{a}$, represented as $\bar{a}$, is usually considered. However, a plus or minus sign is appended to indicate whether it is in the positive or negative direction along the line.

Illustration 2.4 Suppose that a car starts from rest and accelerates to a velocity of 8.0 m/s along a straight-line path in 4.0 s, as shown by the black line in Fig. 2.4. Find its average acceleration.

Reasoning From the graph, $v_0 = 0$ and at $t = 4.0$ s, $v_f = 8.0$ m/s. Substituting in Eq. (2.4) (with vector notation omitted, since all motion is in the same direction) gives

$$\bar{a} = \frac{8.0 \text{ m/s} - 0}{4.0 \text{ s}} = \frac{8.0 \text{ m/s}}{4.0 \text{ s}} = \frac{2.0 \text{ m/s}}{1 \text{ s}}$$

which is read "2 meters per second each second" or "2 meters per second per second." In other words, the velocity of the car is increasing by 2 m/s each second, and this rate of increase in velocity is the quantity that we have called the acceleration. The above result for $\bar{a}$ is often written as 2.0 m/s^2 and read "2.0 meters per second squared."

In general, the units of $\bar{a}$ will be a distance unit divided by the product of two time units. The units could be miles per second squared, feet per second per hour, centimeters per second per minute, miles per hour per second, and so forth.[3] In any event, *the physical meaning of acceleration should be clear: It is the change in velocity per unit of time,* as shown in Fig. 2.4.

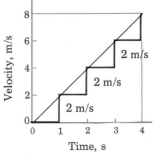

FIGURE 2.4
The black line shows the time variation of the velocity of a car undergoing a uniform acceleration of 2 m/s each second.

2.5 Uniformly Accelerated Motion

Let us refer once again to Fig. 2.4. Notice that the velocity of the car represented by the graph increases 2 m/s each second during the time covered by the graph. This means that the rate of increase of velocity, i.e., the acceleration, is uniform and is 2 m/s^2. Although uniform or constant accelerations are achieved only in certain special cases, we shall see that these cases are of considerable importance. Since motion involving nonuniform acceleration usually becomes quite difficult to handle, we restrict the discussion in this chapter to instances of constant acceleration. Under these conditions, the average and instantaneous accelerations are the same, since the acceleration is constant. Therefore we do not distinguish between them in the remainder of this chapter.

When the acceleration is constant, the speed and velocity of the moving object change uniformly with time. Because of this, *for constant accelera-*

---

[3]When you are solving problems, it is not a good idea to mix two time units in a problem.

tion, the *average velocity* is simply the average value of the initial and final velocities. Thus

$$\bar{v} = \frac{v_f + v_0}{2} = \tfrac{1}{2}(v_0 + v_f) \tag{2.5}$$

where v_f is the final velocity and v_0 is the initial velocity. In the special case when the acceleration is constant and equal to zero, the velocity of the object is not changing, and so $v_f = v_0$. In this case Eq. (2.5) leads to the result that $\bar{v} = \tfrac{1}{2}(v_0 + v_0) = v_0$, as we would expect.

We now have three equations that apply to uniformly accelerated motion, Eqs (2.2), (2.4), and (2.5). They are sufficient to deal with any ordinary situation where there is uniformly accelerated motion. Let us examine a typical case.

Illustration 2.5 Suppose that a car starts from rest and accelerates uniformly to a speed of 5.0 m/s in 10 s. Find its acceleration and the distance it traveled in this time.

Reasoning To begin the solution of this problem, let us first write down what is known and what is unknown:

$$v_0 = 0 \qquad v_f = 5.0 \text{ m/s} \qquad t = 10 \text{ s}$$
$$a = ? \qquad s = ?$$

From Eq. (2.4) we have at once

$$a = \frac{v_f - v_0}{t} = \frac{5.0 - 0}{10} \frac{\text{m/s}}{\text{s}} = 0.50 \text{ m/s}^2$$

From Eq. (2.5) we see that

$$\bar{v} = \frac{5.0 + 0}{2} = 2.5 \text{ m/s}$$

and finally, from the definition of average velocity, Eq. (2.2),

$$s = \bar{v}t = (2.5 \text{ m/s})(10 \text{ s}) = 25 \text{ m}$$

Illustration 2.6 As another example, suppose that a car traveling at 5.0 m/s is brought to rest in a distance of 20 m. Find its deceleration and the time taken to stop.

Reasoning Following the same procedure as before, we write down the knowns and unknowns:

$$v_0 = 5.0 \text{ m/s} \qquad v_f = 0 \qquad s = 20 \text{ m}$$
$$a = ? \qquad t = ?$$

We can first find the average velocity:

$$\bar{v} = \tfrac{1}{2}(v_0 + v_f) = 2.5 \text{ m/s}$$

Knowing this and the distance,

$$s = \bar{v}t \qquad \text{or} \qquad t = \frac{s}{\bar{v}}$$

we find

$$t = \frac{20 \text{ m}}{2.5 \text{ m/s}} = 8.0 \text{ s}$$

The acceleration is now obtained from Eq. (2.4):

$$a = \frac{0 - 5.0 \text{ m/s}}{8.0 \text{ s}} = -0.625 \text{ m/s}^2$$

The negative sign on the answer signifies that the car's velocity in the $+x$ direction is decreasing. The car is slowing down in this case.

These two illustrations show clearly that it is advantageous to follow a systematic approach to these problems. First, write down what is known and what is unknown. Then use the appropriate formula, Eq. (2.2), (2.4), or (2.5), to find the unknowns. There are some cases in which this is not as easily done as in the previous two illustrations. A case in point would be the following.

Illustration 2.7 A car starts from rest and accelerates at 4.0 m/s² through a distance of 20 m. How fast is it then going? How long did it take?

Reasoning We write down the knowns and unknowns:

$$v_0 = 0 \qquad a = 4.0 \text{ m/s}^2 \qquad s = 20 \text{ m}$$
$$v_f = ? \qquad t = ?$$

Our three equations are

$$s = \bar{v}t \qquad\qquad (2.2)$$
$$\bar{v} = \tfrac{1}{2}(v_0 + v_f) \qquad\qquad (2.5)$$
$$a = \frac{v_f - v_0}{t} \qquad\qquad (2.4)$$

Isaac Newton
(1642–1727)

AIP Niels Bohr Library

Before the age of 30, Newton had invented the mathematical methods of calculus, demonstrated that white light contained all the colors of the rainbow, and discovered the law of gravitation. His was a lonely and solitary life. His father died before his birth, and after the remarriage of his mother, he was raised by an aged grandmother. In 1661 he was admitted to Cambridge University, where he worked for the next 8 years, except for a year at home to escape the plague. During those years, he made his major discoveries, although none of them were published at that time. Nonetheless, his genius was recognized, and in 1669 he was appointed Lucasion Professor of Mathematics at Cambridge University, a position he retained until 1695. His major scientific work was completed prior to 1692, when he suffered a nervous breakdown. After his recovery, he determined to lead a more public life and soon became Master of the Mint in London. He was elected president of the Royal Society in 1703 and held that position until his death.

None of these equations can be used directly to obtain one of the unknowns. Each equation contains at least two unknowns, and we would have to solve simultaneous equations to obtain any unknowns. In the next section, we find two additional equations that will be useful for us in such situations.

2.6 Two Derived Equations for Uniformly Accelerated Motion

Illustration 2.7 can be solved easily if we first obtain two companion equations to the three given in (2.2), (2.4), and (2.5). These extra two equations are obtained by solving the three known equations simultaneously. Having done this once, we shall not have to repeat the process, but shall simply make use of the results.

If we substitute the value for $\bar{v}$ given by Eq. (2.5) for $\bar{v}$ in Eq. (2.2), we find

$$s = \tfrac{1}{2}(v_0 + v_f)t \tag{2.6}$$

Now we replace t by its value from Eq. (2.4) to give

$$s = \tfrac{1}{2}(v_f + v_0)\left(\frac{v_f - v_0}{a}\right) \tag{2.7}$$

After clearing fractions and rearranging, we have

$$v_f^2 - v_0^2 = 2as \qquad \text{or} \qquad v_f^2 = v_0^2 + 2as \tag{2.8}$$

(Notice, in passing, that this is just the equation we need to solve the previous example.)

The other equation we shall find useful is found by making a different substitution in Eq. (2.6). Replace v_f in that equation by its value obtained from Eq. (2.4). Thus

$$s = \tfrac{1}{2}v_0t + \tfrac{1}{2}(v_0 + at)t$$

which simplifies to

$$s = v_0t + \tfrac{1}{2}at^2 \tag{2.9}$$

We now have five equations available to use in the solution of problems involving **uniformly accelerated motion:**

$$s = \bar{v}t \tag{2.10a}$$

$$\bar{v} = \tfrac{1}{2}(v_f + v_0) \tag{2.10b}$$

$$v_f = v_0 + at \tag{2.10c}$$

$$v_f^2 = v_0^2 + 2as \tag{2.10d}$$

$$s = v_0t + \tfrac{1}{2}at^2 \tag{2.10e}$$

Equations of Uniform Acceleration

Returning now to the problem posed at the end of the last section, we had there that

$$v_0 = 0 \qquad a = 4.0 \text{ m/s}^2 \qquad s = 20 \text{ m}$$

$$v_f = ? \qquad t = ?$$

Using Eq. (2.10d), we have at once

$$v_f^2 = 0 + (2)(4.0 \text{ m/s}^2)(20 \text{ m}) = 160 \text{ m}^2/\text{s}^2$$
$$v_f = \pm\sqrt{160 \text{ m}^2/\text{s}^2} = \pm\sqrt{160} \text{ m/s} \approx \pm 12.6 \text{ m/s}$$

Only the positive sign is of concern to us here because we consider the motion to be in the positive direction. Therefore $v_f = 12.6$ m/s. (What is the meaning of the other answer, -12.6 m/s?) Now we can use Eq. (2.10c) to find t:

$$at = v_f - v_0$$
$$(4.0 \text{ m/s}^2)t = 12.6 \text{ m/s} - 0$$
$$t = 3.15 \text{ s}$$

Illustration 2.8 Find the time taken for a car to travel 98 m if it starts from rest and accelerates at 4.0 m/s².

Reasoning The knowns and unknown are

$$v_0 = 0 \qquad a = 4.0 \text{ m/s}^2 \qquad s = 98 \text{ m}$$
$$t = ?$$

The appropriate equation is

$$s = v_0 t + \tfrac{1}{2}at^2$$
$$98 \text{ m} = (0 \text{ m/s})t + \tfrac{1}{2}(4.0 \text{ m/s}^2)t^2$$
$$t = \sqrt{49} = 7.0 \text{ s}$$

You should check to see that the units of t are as stated.

Illustration 2.9 A car is moving at 60 km/h when it begins to slow down with a deceleration of 1.50 m/s². How long does it take to travel 70 m as it slows down?

Reasoning The knowns and unknown are

$$v_0 = (60 \text{ km/h})\left(\frac{1 \text{ h}}{3600 \text{ s}}\right)\left(\frac{1000 \text{ m}}{1 \text{ km}}\right) = 16.7 \text{ m/s}$$
$$s = 70 \text{ m} \qquad a = -1.50 \text{ m/s}^2 \qquad t = ?$$

We have taken the acceleration as negative because the velocity is decreasing in the direction of positive motion. Notice that we have changed the units of velocity to meters per second so that we won't be using two sets of distance and time units in the same problem. We must always make this kind of change.

If we now choose to solve this problem by using Eq. (2.10e), we run into trouble with algebra:

$$s = v_0 t + \tfrac{1}{2}at^2$$
$$70 = 16.7t + \tfrac{1}{2}(-1.50)t^2 = 16.7t - 0.75t^2$$

where you should supply the units. You can solve this quadratic equation by using the quadratic formula. However, it is usually simpler to find one of the other unknowns first so as to avoid the complicated algebra. For example, using Eq. (2.10d), we have

$$v_f^2 = v_0^2 + 2as = (16.7 \text{ m/s})^2 + (2)(-1.50 \text{ m/s}^2)(70 \text{ m})$$
$$= 279 \text{ m}^2/\text{s}^2 - 210 \text{ m}^2/\text{s}^2$$

from which

$$v_f = 8.3 \text{ m/s}$$

Now, using Eq. (2.10c),

$$v_f = v_0 + at$$
$$8.3 \text{ m/s} = 16.7 \text{ m/s} - (1.50 \text{ m/s}^2)t$$

from which $t = 5.6$ s.

2.7 A Note about Equations

Usually in physics we do not encourage "plugging" values into a multitude of memorized equations. The reason is simple. It is much easier for you to memorize a very few basic equations and, knowing what they mean, use them to solve all sorts of problems. There is no need to memorize a different equation for each situation. So far in this book we have asked that you memorize only the defining equations for the sine, cosine, and tangent.

In this chapter on uniformly accelerated motion, we might well ask you to memorize Eqs. (2.10b) and (2.10c) since they define average velocity and acceleration. Equation (2.10a) is a fundamental equation which we already knew in a qualitative way. Whether or not we should memorize Eqs. (2.10d) and (2.10e) will depend on several factors.

They are, of course, nothing new, since they are merely Eqs. (2.10a), (2.10b), and (2.10c) combined in various ways. However, the algebra involved in formulating them is not negligible. We saw in some of the examples that we must obtain these equations to solve certain motion problems. The question therefore reduces to the following: Shall we be using these two equations enough for it to be advisable to learn them? I think you will use all five of Eqs. (2.10) frequently enough to find it to your advantage to memorize them.

It should be remembered, though, that any equation worth memorizing is certainly worth understanding. It is important that you know the exact meaning of each of the symbols in these equations. In addition, Eqs. (2.10a), (2.10b), and (2.10c) summarize easily understood physical meanings and should not be looked on as a mere collection of algebraic expressions. We also have pointed out that Eqs. (2.10b) to (2.10e) apply only to situations where a is constant.

Sometimes it is worthwhile to examine the limiting forms of equations. Let us see what Eqs. (2.10) say when the acceleration is zero. First, look at Eq. (2.10c). It says that when $a = 0$, $v_f = v_0$. This is certainly true. If the acceleration is zero, the velocity is constant. The final velocity will be the same as the initial velocity.

Under the same condition of zero acceleration, Eq. (2.10b) says that

$$\bar{v} = \tfrac{1}{2}(v_f + v_0) = v_f = v_0$$

since v_f and v_0 are the same under these conditions. Of course, this relation is true, since if the acceleration is zero, the velocity is constant and the average velocity is the same as the initial and final velocities. Similarly, Eqs. (2.10d) and (2.10e) reduce to obviously true statements when $a = 0$.

2.8 Galileo's Discovery

Those who know the history of science are not surprised if a child gives the wrong answer to the question: Which will fall faster, a heavy ball or a light one? It does seem reasonable, doesn't it, that the heavier ball will fall faster than the light ball? Most children will give that answer.

In fact, even the great genius Aristotle (384–322 B.C.) gave this same answer. And he was not alone. It was commonly thought, until about A.D. 1600, that heavier bodies *do* fall more swiftly than light ones. After all, it is obviously true that a feather does not fall through the air nearly as fast as a stone. Aristotle himself showed by involved philosophical arguments that this should be true for all light bodies as compared with heavier ones. (In Aristotle's defense, however, we should state that he was considering the way the body falls in air, not vacuum.)

Natural scientists usually are not completely satisfied with philosophic proofs of statements which can be verified experimentally. It would be wonderful if our philosophic methods were so good that we need never doubt the conclusions to which they lead us. Then nearly all the time, effort, and money now devoted to expensive research could be diverted to other uses, and the problems of science could be solved by a few philosophers maintained in sumptuous solitude in velvet-carpeted ivory towers.

Galileo (1564–1642) had the temerity to question whether Aristotle had made a mistake in his philosophic arguments. It seems incongruous to us today that for nineteen centuries no one had troubled to carry out the simple experimental tests needed to answer this question. However, it remained for

Galileo to measure the speed with which objects fall.[4] He recognized that situations such as the feather-stone problem were not a real test, because of the pronounced effect of the rush of air past the feather as it fell. Hence, he experimented with relatively heavy objects, all the same size but differing in weight. His results are, of course, well known. He concluded that if one neglects air-friction effects, all bodies fall to earth with the same acceleration, independent of their weight. A simple experiment that shows this fact is illustrated in Fig. 2.5.

Galileo is often called the father of modern science, since he was instrumental in awakening the world to the fact that experimentation, when possible, is the most valid way of learning the facts of nature. Following in his footsteps, the great scientists Newton and Faraday, and many others, opened up a whole new area of knowledge with their experimental observations of nature's ways.

Even today, we sometimes find scientists following the path of Aristotle rather than that of Galileo. We still find, on occasion, that experiments disprove ideas which we supposed to be well founded. Sometimes experimental proofs are not undertaken because they involve methods beyond our capabilities. But in any event scientists must never forget that even the best minds make philosophic errors.

There is another side to the coin, though. In many fields, experimental proofs are not possible for one reason or another. We have no alternative in such cases but to rely on the conclusions of our most able philosophers. Even where experimental test is possible, it is sometimes necessary for scientists to postpone the test because of the pressure of other pursuits or prohibitive costs. We shall see in later chapters that some of the greatest scientific discoveries and theories were based not only on unverified but on wrong premises. The mere fact that the discoveries were made justifies the approach used. The scientist would certainly never have been able to prove the starting premise; so if he or she had wasted many years on that proof, the discovery might well have been delayed a century. Of course, we tend to remember those who were lucky in such an approach. Those who failed because of wrong and unproved premises are soon forgotten, even though their number is certainly very large.

2.9 Free Fall and the Acceleration Due to Gravity

Many experimenters, using highly refined techniques, have duplicated Galileo's experiments. It is not unlikely that in your laboratory work you will perform an experiment to check his results. We now recognize as undisputed fact that *a freely falling object will accelerate downward with essentially constant acceleration. We call this acceleration the acceleration due to grav-*

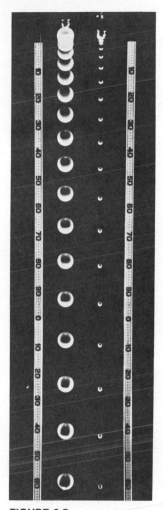

FIGURE 2.5
The falling objects are made visible at equal time intervals by means of a strobe light. Although the objects are of different size and mass, they fall in unison. (*Courtesy of Education Development Center.*)

---

[4]Galileo actually increased the time of fall by rolling spheres down an incline. By making the incline steeper and observing the effect of this action, he was able to draw quantitative conclusions about the behavior of freely falling bodies.

TABLE 2.1

ACCELERATION DUE TO GRAVITY g

Place	Elevation, m	g m/s^2	ft/s^2
Beaufort, N.C.	1	9.7973	32.143
New Orleans	2	9.7932	32.130
Galveston	3	9.7927	32.128
Seattle	58	9.8073	32.176
San Francisco	114	9.7997	32.151
St. Louis	154	9.8000	32.152
Cleveland	210	9.8024	32.160
Denver	1638	9.7961	32.139
Pikes Peak	4293	9.7895	32.118

ity, and we represent it by the symbol g. Although the exact value of g varies from place to place on earth, it is close to 9.8 m/s^2, which is the same as 980 cm/s^2 or 32 ft/s^2. Table 2.1 shows some typical values. These results, of course, are obtained only if proper care is taken to minimize the effects of the flow of air past the falling body.

Motion of a free body in an up-and-down direction will be uniformly accelerated motion; on earth this acceleration has the value $a = g = 9.8$ m/s$^2 = 32$ ft/s^2. Hence, in any such motion, Eqs. (2.10) apply with a replaced by g. Only one extra precaution must be taken in the use of these equations. It has to do with the vector nature of acceleration. *The acceleration due to gravity is always down. If we choose to call down the positive direction, a is a positive number,* 9.8 m/s^2. In any case where we choose the upward direction as positive, however, the pull of gravity will actually decrease the upward velocity of the body, and so $a = -9.8$ m/s^2. *For problems which involve both upward and downward motion,* such as a ball rising and then coming down again, *it is absolutely necessary at the outset to choose either up or down as the positive direction. This choice is arbitrary,* but once we have made it in a particular problem, we must retain that choice through the whole problem.

Illustration 2.10 A boy drops a stone from a bridge. If it takes 3.0 s for the stone to hit the water beneath the bridge, how high above the water is the bridge? Ignore air friction. (Notice here that our problem ends the instant *before* the stone hits the water. It is only during this interval that the stone is a freely falling body.)

Reasoning Taking down as positive, we have these knowns and unknown:

$$a = 9.8 \text{ m/s}^2 \quad t = 3.0 \text{ s} \quad v_0 = 0 \quad s = ?$$

Use

$$s = v_0 t + \tfrac{1}{2} a t^2$$
$$= 0 + \tfrac{1}{2}(9.8 \text{ m/s}^2)(9.0 \text{ s}^2) = 44 \text{ m}$$

Illustration 2.11 A boy throws a ball upward with a speed of 15.0 m/s. How high does it go? What is its speed just before the boy catches it again? How long was it in the air? Neglect the effects of air friction.

Reasoning Method 1: It is important that we define the "trip" we are interested in. Consider the motion from A to B in Fig. 2.6. Taking up as positive, we have the following knowns and unknown:

$$a = -9.8 \text{ m/s}^2 \qquad v_0 = 15.0 \text{ m/s} \qquad v_f = 0 \qquad s = ?$$

Use

$$2as = v_f^2 - v_0^2$$
$$-19.6s = 0 - 225$$

Supply the units in this equation, and show that

$$s = 11.5 \text{ m} \qquad \text{from } A \text{ to } B$$

Notice that at the top of the path, at B, the upward speed of the ball has decreased to zero. Since it has not yet started to fall downward again, its speed is zero at point B.

For the second part of the problem, we are concerned with the trip from B to C. This is a new trip, and so we have new knowns. Taking down as positive, we have

$$a = +9.8 \text{ m/s}^2 \qquad s = 11.5 \text{ m} \qquad v_0 = 0$$
$$v_f = ?$$

Use

$$2as = v_f^2 - v_0^2$$
$$(19.6 \text{ m/s}^2)(11.5 \text{ m}) = v_f^2 - 0$$
$$v_f = \sqrt{225 \text{ m}^2/\text{s}^2} = 15 \text{ m/s} \qquad \text{downward}$$

Notice that the final speed is the same as the speed with which the ball was thrown. This is an example of a famous basic law of physics, called the **law of conservation of energy,** which we examine in considerable detail in a later chapter.

Method 2: There is yet another way in which we could have solved this latter problem. Consider the full trip from A to B to C. The starting point is A, and the end point is C. Taking up as positive, we know

$$a = -9.8 \text{ m/s}^2 \qquad v_0 = 15 \text{ m/s} \qquad s = 0$$

FIGURE 2.6
The ball is thrown upward at point A with a speed of 15 m/s. Since the ball stops at point B, its upward speed there is zero.

We write that s, the vector displacement from the starting point A to the end point C, is zero, since the ball returned to the thrower. Use

$$2as = v_f^2 - v_0^2$$
$$0 = v_f^2 - (15 \text{ m/s})^2$$
$$v_f = \pm 15 \text{ m/s} \qquad \text{as before}$$

Notice that when the square root of a number is taken, the answer can be either plus or minus. For our trip, the minus sign is correct. Before investigating the reason for the choice of two answers, plus or minus, let us compute the time for the trip. Use

$$s = v_0 t + \tfrac{1}{2}at^2$$
$$0 = 15t - 4.9t^2$$

Supply the units to this equation, and show that

$$t = 0 \qquad t = 3.1 \text{ s}$$

Here again we have a choice of answers.

The reason for the choice of two answers in each case is not hard to determine. Our knowns for this method of solution are actually true for two trips, the one from A to B to C and the trip from A which lasts zero time. For this latter trip, all the known conditions are true, and the time taken for it is zero. In addition, the end velocity is still the same as the starting velocity.

Illustration 2.12 A girl throws a ball straight upward with a speed of 40 ft/s, as shown in Fig. 2.7. How long will it take to reach a point 20 ft above the ground on its way down again? Neglect the effects of the surrounding air.

Reasoning Taking up as positive, we have

$$a = -32 \text{ ft/s}^2 \qquad s = 20 \text{ ft} \qquad v_0 = 40 \text{ ft/s} \qquad t = ?$$

If we used $s = v_0 t + \tfrac{1}{2}at^2$, we would need the quadratic formula to find t. Instead, first find v_f, using

$$2as = v_f^2 - v_0^2$$
$$(-64 \text{ ft/s}^2)(20 \text{ ft}) = v_f^2 - 1600 \text{ ft}^2/\text{s}^2$$
$$v_f^2 = 320 \text{ ft}^2/\text{s}^2$$
$$v_f = \pm 17.9 \text{ ft/s}$$

We choose the negative sign, since we are interested in the ball on its way down at C. The positive sign represents the ball on its way up at point B. Now use

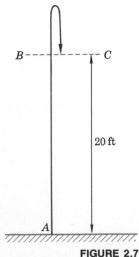

FIGURE 2.7
The ball is thrown upward from point A with a speed of 40 ft/s. How long does it take for the ball to reach point C?

$$v_f = v_0 + at$$
$$-17.9 \text{ ft/s} = 40 \text{ ft/s} - (32 \text{ ft/s}^2)t$$
$$32t = 57.9$$

Supply the units to this equation, and show that

$$t = 1.81 \text{ s}$$

Illustration 2.13   How fast must a ball be thrown straight upward if it is to return to the thrower in 3.0 s? Neglect the effects of the air on the ball.

Reasoning   Let us use the metric system in this problem. We take up as positive, and we notice that the displacement vector from beginning to end point has zero length. Then

$$a = -9.8 \text{ m/s}^2 \qquad t = 3.0 \text{ s} \qquad s = 0 \qquad v_0 = ?$$

We can use

$$s = v_0 t + \tfrac{1}{2}at^2$$

to give

$$0 = (v_0)(3.0 \text{ s}) - (4.9 \text{ m/s}^2)(9.0 \text{ s}^2)$$

We then find

$$v_0 = 14.7 \text{ m/s}$$

Summary

The average speed $\bar{u}$ of an object is the total distance traveled divided by the time taken for the trip. Average velocity $\bar{v}$ is the displacement vector from beginning to end of a trip divided by the time taken. Speed is a scalar, and velocity is a vector. They both have the units of distance divided by time.

When an object moves along a straight line, its motion can be shown by an x-versus-t graph. The slope of the graph line at any point gives the velocity and speed at the point. Instantaneous velocity is given by

$$v = \lim_{\Delta t \to 0} \frac{\Delta x}{\Delta t}$$

To change the units of a quantity, use is made of conversion factors. Each conversion factor is unity, and so multiplication by it does not change the value of a quantity. Units are carried through computations just as algebraic symbols.

If an object changes its velocity from $\mathbf{v}_0$ to $\mathbf{v}_f$ in a time t, its average acceleration is given by

$$\bar{\mathbf{a}} = \frac{\text{change in velocity}}{\text{time taken}} = \frac{\mathbf{v}_f - \mathbf{v}_0}{t}$$

Acceleration is a vector.

If an object moves along a straight line and its acceleration is constant, the average velocity of the object is given by

$$\bar{v} = \tfrac{1}{2}(v_f + v_0)$$

Using this equation and the definitions of average velocity

$$\bar{v} = \frac{s}{t}$$

and average acceleration

$$\bar{a} = a = \frac{v_f - v_0}{t}$$

we can derive two more useful equations:

$$v_f^2 = v_0^2 + 2as \qquad \text{and} \qquad s = v_0 t + \tfrac{1}{2}at^2$$

These are the five motion equations.

The earth pulls downward on each object with a force we call the object's weight. An object which is subject to only this one force is said to be freely falling. Freely falling objects experience an acceleration g toward the earth's center. The value of g, the acceleration due to gravity, is 9.8 m/s², which is the same as 32 ft/s².

Minimum Learning Goals

Upon completion of this chapter you should be able to do the following:

1 Define average speed and average velocity. Give examples which show that sometimes the magnitudes of speed and velocity are not the same.
2 Use a graph of x versus t to describe the behavior of an object. Interpret the slope of the graph in terms of instantaneous velocity and speed. Distinguish between average and instantaneous velocity.
3 Change a quantity from one set of units to another when conversion relations between the units are given.
4 Define average acceleration in words and by equation. Give several possible sets of units for it.

5 Draw a graph relating v to t for an object undergoing a given constant (uniform) acceleration.
6 Write the five uniform-motion equations. Define each symbol in them. Give the restrictions on the use of each equation.
7 Find one or more unknowns for an object undergoing uniform acceleration when sufficient data are given.
8 Define g, the acceleration due to gravity, and give its approximate value on earth.
9 Find one or more unknowns for an object undergoing free-fall vertical motion on the earth when sufficient data are given.

Important Terms and Phrases

You should be able to define or explain each of the following:

Average speed
Average velocity
Distance and displacement are different
Average acceleration

Instantaneous speed and velocity
Velocity in terms of slope
Free-fall acceleration g
Five uniform-motion equations
A positive direction must be chosen
Conversion factor

Questions and Guesstimates

1 Give an example of a case where the velocity of an object is zero but its acceleration is not zero.
2 Can the velocity of an object ever be in a direction other than the direction of acceleration of the object? Explain.
3 Sketch graphs of the velocity and the acceleration

as a function of time for a car as it strikes a telephone pole. Repeat for a billiard ball in a head-on collision with the edge of the billiard table.
4 Are any of the following true? (a) An object can have a constant velocity even though its speed is changing. (b) An object can have a constant speed

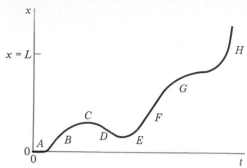

FIGURE P2.1

even though its velocity is changing. (c) An object can have zero velocity even though its acceleration is not zero. (d) An object subjected to a constant acceleration can reverse its velocity.

5 A rabbit enters the end of a drainpipe of length L. Its motion from that instant on is shown in the graph of Fig. P2.1. Describe the rabbit's motion in words.

6 A high school girl who is an average runner completes a 100-m dash by running twice around an indoor circular track which is 50 m in circumference. Estimate her average speed. Repeat for her average velocity. (E)

7 A stone is thrown straight up in the air. It rises to a height h and then returns to the thrower. For the time that the stone is in the air, sketch the following graphs: y versus t; v versus t; a versus t.

8 Under what condition is it wrong to say that the acceleration of an object is negative when the object is thrown upward? Does the sign of the acceleration depend at all on the direction of motion? Can the acceleration of an object be positive even though the object is slowing down?

9 The acceleration due to gravity on the moon is only about one-sixth that on earth. Estimate the ratio of the height to which you could throw a baseball on the moon to the corresponding height on earth. (E)

10 How could you best analyze the data given in Fig. 2.5 to obtain the value of g?

Problems

Unless otherwise stated, assume uniform acceleration and ignore air friction. Use $g = 9.8 \text{ m/s}^2 = 32 \text{ ft/s}^2$.

1 In 1904, Henry Ford set the world speed record for the mile. He drove the mile in a Ford "999" and took a time of 39.4 s. What was his average speed in miles per hour and in kilometers per hour?

2 In the 1932 Olympic Games, the 3000-m steeplechase was won by Volmari Iso-Hollo of Finland. He ran 3450 m in 10 min 33.4 s. (They ran an extra lap since someone made a counting error.) What was his average speed in meters per second and in miles per hour?

3 Sound travels through air with a speed of about 340 m/s. A professor at the front of a lecture hall shrieks in anguish at the answer given by a student 25 m away. How long after the start of the shriek does the student hear it?

4 Sonar-type depth gauges measure the length of time it takes a sound pulse to go from the water surface to the lake bottom and back to the surface. If a lake is 12.0 m deep and the speed of sound in water is 1450 m/s, how long does it take for a pulse sent down from the lake surface to return to the surface?

5* The earth's crust is still adjusting because of the loss of ice since the last Ice Age. The Lake Superior region is still rising at about 16 ft each 1000 yr. Find its speed due to this cause in centimeters per year. Find its average velocity in centimeters per century.

6* According to the theory of continental drift, South America was once attached to Africa. They have been drifting apart for about 200 million years. Their present separation is about 4000 mi. Assuming these numbers to be exact, what was the average rate of separation of the continents in centimeters per year?

7 A speck of dust is 10 cm from the center of a 33.3-revolution-per-minute (rev/min) record on a record turntable. The record starts to rotate, continues for 10.0 min (rotates 333 complete times), and stops. (a) What was the average speed of the dust speck? (b) What was its average velocity?

8 A hunter throws a boomerang with a speed of 16.0 m/s at a bird 30 m away. The boomerang misses the bird and returns to the hunter after 3.6 s. Assume its speed to remain constant. (a) How far did the boomerang travel? (b) What was its average velocity for the flight?

9* Refer to Fig. P2.2. It shows the motion of an object along a straight line. Find the average velocity

49

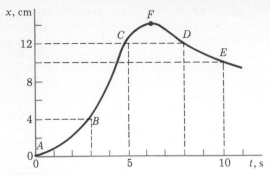

FIGURE P2.2

of the object during the following intervals: (a) A to E; (b) B to E; (c) C to E; (d) D to E; (e) C to D.

10* The motion of a bug along a wire along the x axis is shown in Fig. P2.2. Find the average velocity of the bug during the following intervals: (a) B to D; (b) D to E; (c) A to D; (d) A to B.

11* Person A can run at a top speed of 5.0 m/s while person B can run at a speed of only 3.0 m/s. They are to race 200 m. To make the race more even, A is required to start t seconds later than B. How large should t be if they are to end in a tie?

12* For the situation in Prob. 11, the handicap is to be made this way. Person B is to be given a head-start distance s, while A must run the full 200 m. Both start at the same time. How large should s be if the race is to end in a tie?

13 Change the following to centimeters: (a) 6.0 in; (b) 3.0 ft. (Remember, 1 in = 2.54 cm, 1 ft = 12 in, and 1 yd = 3 ft.)

14 Change the following to their equivalents in gallons: (a) 12.0 liters; (b) 4800 cm³; (c) 360 in³. You may need some of the following: 1 liter = 1000 cm³, 1 gal = 3.785 liters, 1 m³ = 35.3 ft³, and 1 liter = 61 in³.

15 Change the following speeds to meters per second: (a) 5.0 km/h; (b) 37 in/h; (c) 25 mi/h.

16 Find the following speeds in units of miles per hour: (a) 60 km/h; (b) 260 m/s; (c) 72 ft/ms.

17* Refer to Fig. P2.2 for the straight-line motion of an object. Find the instantaneous velocity of the object (a) at point F, (b) at point B, (c) at point E.

18* Figure P2.2 shows the motion of an object along the x axis. Find the instantaneous velocity of the object at (a) point C and (b) point D.

19 A car moving on a straight road accelerates from a speed of 4.1 m/s to a speed of 6.9 m/s in 5.0 s. What was its average acceleration?

20 An airplane in straight-line flight slows from a speed of 345 km/h to a speed of 217 km/h in 40 s. What was its average acceleration in kilometers per hour squared and in meters per second squared?

21 A car moving at 25 m/s skids to a stop in 14.0 s. Find the magnitude of the average acceleration of the car and the distance the car travels while stopping.

22 A car company claims that its car can accelerate from rest to a speed of 28 m/s in 20 s. Find the average acceleration of the car and the distance it goes in this time.

23* A bullet moving with speed 150 m/s strikes a tree and penetrates 3.5 cm before stopping. Find the magnitude of its acceleration and the time taken to stop.

24* In a TV tube, electrons shoot down the tube and strike the fluorescent material on the screen at the end of the tube. Their impact with the material causes light to be given off, thereby producing the picture we see. The electrons in the tube are accelerated from rest to speeds of about 200 million meters per second (2.0×10^8 m/s) in a distance of about 1.50 cm. What is the acceleration of an electron during the acceleration process? How long does the acceleration take?

25* A rough value for the deceleration of a skidding automobile is about 7.0 m/s². Using this value, how long does it take for a car going 30 m/s to stop after the skid starts? How far does the car go in this time?

26* A truck, initially traveling at 20 m/s, decelerates at 1.50 m/s². Find (a) how long it takes to stop, (b) how far it moves in that time, and (c) how far the truck moves in the third second after the brakes are applied.

27* A proton moving with speed 1.00×10^7 m/s passes through a 0.020-cm-thick sheet of paper and emerges with a speed of 2.00×10^6 m/s. Assuming uniform deceleration, find the deceleration and the time taken to pass through the paper.

28* The driver of a car that is going 25 m/s suddenly notices a train blocking the road. At the instant the brakes are applied, the train is 60 m away. The car decelerates uniformly and strikes the train 3 s later. (a) How fast was the car moving on impact? (b) What was the magnitude of its acceleration during the 3 s?

29* A train engine accelerates uniformly from rest and reaches a crossing 2.2 km from its starting point after 60 s. (a) How long after the engine reaches the crossing does the 137-m-long train clear the crossing? Assume the acceleration to remain constant and ignore the width of the crossing. (b) How fast would the train be moving as it clears the crossing?

30* A heavily loaded train blocks a crossing. Having nothing better to do, a waiting motorist notices that it takes 20 s for a railway car to pass through a distance equal to the length of one railway car just as the train starts from rest and begins to move. Find the acceleration of the train in terms of the length L of the railway car. Assuming the acceleration to remain the same, how long after the train starts will it be before the following 60 railway cars have passed?

31* Two boys start running straight toward each other from two points 100 m apart. One runs with a speed of 5.0 m/s, while the other moves at 7.0 m/s. How close are they to the slower one's starting point when they reach each other?

32* A truck traveling 15 m/s due west is on a collision course with a car going due east at 30 m/s. They are 400 m apart. How long does it take them to hit? How far are they from the original position of the truck when they finally collide? Assume their speeds to remain constant.

33** A car is traveling at 60 mi/h along a road parallel to a railroad track. How long does it take the car to pass a $\frac{1}{2}$-mi-long train traveling at 40 mi/h in the same direction? In the opposite direction?

34* Just as a car starts to accelerate from rest with acceleration 1.40 m/s^2, a bus moving with constant speed of 12.0 m/s passes it in a parallel lane. (a) How long will it be before the car overtakes the bus? (b) How fast will the car be going then? (c) How far will the car have gone at that point?

35** Two trains are headed toward each other on the same track with equal speeds of 20 m/s. When they are 2.0 km apart, they see each other and begin to decelerate. (a) If their accelerations are uniform and equal, what must be their magnitudes if the trains are to barely avoid collision? (b) If only one train slows with this accleration, how far will it go before collision occurs?

36** A police car is at rest alongside a road, monitoring passing cars, when one passes at a constant speed of 32 m/s. At 5 s later, the police car accelerates from rest with an acceleration of 1.60 m/s^2. (a) If it could maintain this acceleration, how far would the police car move before catching the car? (b) What would the police car's speed be then?

37** The driver of a car notes the car's speed to be 80 km/h in a 50-km/h zone. Just then, the driver notices a police car sitting behind a tree. Hoping

that the police are not using radar, the driver at once releases the accelerator. The police are actually timing cars between two check points 150 m apart. The first check point coincides with the point where the accelerator was released. What must be the deceleration of the car if the police are to time its speed as 50 km/h?

38 A flower pot slips off a window ledge that is 3.5 m above a woman sitting below. (a) How fast is the pot moving when it hits the woman? (b) How much time did the woman have to move after the man yelled just as he launched the pot on its journey?

39 A frightened diver hangs by his fingers from a diving board with his feet 5.0 m above the water. (a) How long after his fingers give out will he strike the water? (b) How fast will he be going then?

40 A stone is thrown straight up with an initial speed of 80 ft/s. How high does the stone go, and how long does it take to reach its highest point?

41 A girl throws a ball straight down from the top of a 50-ft building with a speed of 20 ft/s. How long does it take for the ball to reach the ground? How fast is it going just before it hits?

42 A bullet shot straight upward from a gun is found to rise to a height of 2.0 km. What is the least possible speed with which it could have left the gun? (In practice, air friction is important.)

43* A stone is thrown straight upward from the ground and goes as high as a nearby building. The stone returns to the ground 3.0 s after it was thrown. How high (in meters) is the building?

44* A girl is standing on the top edge of an 18-m-high building. She tosses a coin upward with a speed of 7.0 m/s. How long does it take for the coin to hit the ground 18 m below? How fast is the coin going just before it strikes the ground?

45* A physics student who always uses a calculator to evaluate 2×2 comes up with the following scheme to measure the height of a building. A timing mechanism is set up. It measures the time taken for an object dropped from the top of the building to fall the last 2.0 m before it hits the ground. By experiment, the object dropped from the building top is found to take 0.150 s to move this last 2.0 m. How high is the building?

46** A ball is thrown straight upward with a speed v from a point h meters above the ground. Show that the time taken for the ball to strike the ground is

$$\frac{v}{g}\left(1 + \sqrt{1 + \frac{2hg}{v^2}}\right)$$

47*   A monkey in a perch 20 m high in a tree drops a coconut directly above your head as you run with speed 1.5 m/s beneath the tree. (*a*) How far behind you does the coconut hit the ground? (*b*) If the monkey had really wanted to hit you, how much earlier should the coconut have been dropped?

48*   Two balls are dropped to the ground from different heights. One ball is dropped 2.0 s after the other, but they both strike the ground at the same time, 5.0 s after the first is dropped. (*a*) What is the difference in the heights from which they were dropped? (*b*) From what height was the first ball dropped?

49**   A boy wants to throw a can straight up in the air and then hit it with a second can. He wants the collision to occur 5.0 m above the throwing point. In addition, he knows that the time he needs between throws is 3.0 s. Assuming he throws both cans with the same speed, what must be the initial speeds of the cans?

50**   An elevator in which a woman is standing is moving upward at a speed of 4.0 m/s. If the woman drops a coin from a height 1.2 m above the elevator floor, how long does it take for the coin to strike the floor? What is the speed of the coin relative to the floor just before impact?

3
Newton's Laws

In Chap. 2 we discussed the concepts of velocity and acceleration without specifically considering what caused the object to move. The role of forces in causing motion was deferred until this chapter. We now investigate how forces cause accelerations. In carrying through this study, we state and discuss Newton's three laws of motion. These laws are of primary importance in physics. We also discuss a fourth very important law, first discovered by Newton, the law of universal gravitation.

3.1 Discovery of Physical Laws

A physical law is a statement of the way in which matter behaves. These are laws over which we have no control; they have existed and will exist forever. The purpose of all basic research in the physical sciences is to discover these laws. Understanding in science is equivalent to knowing the laws of nature and their consequences.

Incorrect or incomplete observations can lead to statements that are believed to be physical laws; but these incorrect statements are not laws of nature at all. For example, Aristotle believed that he had discovered a law of nature when he stated that heavier bodies accelerate toward the earth faster than lighter bodies. In fact, he had not discovered a law of physics. There is no such physical law. The actual law of nature which applies to this situation was discovered many centuries later by Galileo, as we have seen.

Even the law which Galileo discovered for the acceleration of falling bodies is now known to be far from complete and general. We all know that objects in a spaceship behave much differently from Galileo's falling weights. In an earth-satellite ship, to the occupant of the ship objects do not seem to fall at all. They appear to the occupant to be completely weightless. Of course, Galileo had no way of knowing this, and so it is only natural that his proposed law of nature was incomplete. Nor were his measurements accurate enough to show that the same body accelerates differently under gravity at various places on earth.

It is a general principle in science that no law of nature is ever fully and completely known. Scientists do their best to state nature's laws as they know them. However, it would be presumptuous to say that this or that law of nature, as now understood, will not have to be modified as we learn more about our universe. There is an excellent chance that nearly all the laws of nature which we propose and use in this text are correct in all essentials or that their limitations are known. We know this because the laws have been tested against experience in all manner of ways. However, there is always the chance that someday one of these laws as presently stated and understood will be found to disagree with some new and ingenious experiment. Our statement of the law will then be changed to conform with that new experience, as well as with the results we already know today. So when we state a physical law, it can be considered correct only in the light of present knowledge.

As you progress in your study of physics, you will notice that only rarely is an accepted idea concerning a physical law found to be wrong. It is, however, quite common to find that the accepted idea must be extended, amplified, and modified as our knowledge of the universe becomes wider. We shall see that the great reputation of the scientist who proposes a law sometimes does not ensure the correctness of the law. Even such a great scientist as Newton had some badly mistaken ideas about the way in which natural objects behave, as we shall see in our study of light. For this reason, the wise scientist does not accept a physical law because of the greatness or reputation of its promulgator alone. A law discovered by a young "unknown" through careful experiments is likely to be more reliable than the philosophic opinion of a "great" scientist.

Finally, we should state that *there is a hierarchy of physical laws. It is the ultimate aim of scientists to reduce to a minimum the number of physical laws needed to describe the universe.* We could have a different physical law for each different physical situation. This would be intolerable, since no one could remember such a large number of laws. However, many of these individual laws would be found to conform to a more general law which would encompass them all. For example, the individual laws

1   A 10-kg object accelerates in free fall on the earth at 9.8 m/s^2.
2   A 12-kg object accelerates in free fall on the earth at 9.8 m/s^2.

were encompassed in a single law by Galileo:

All bodies accelerate in free fall on the earth at approximately 9.8 m/s^2.

We shall see that Galileo's statement of a physical law is only a portion of a much more general law first stated by Newton some years later. *A statement of a law of nature which contains in it the substance of many lesser laws is preferable to any of the individual lesser laws.*

3.2 Newton's First Law of Motion

Isaac Newton (1642–1727) was one of the foremost physicists of all time. While still in his twenties, he invented the branch of mathematics known as calculus, discovered the law of gravitation, and found that sunlight is a combination of colors. We begin a study of his work by examining his three laws of motion. He first published these laws in 1687 in a classic compendium entitled *Principia Mathematica Philosophiae Naturalis*.

Newton's first law of motion, translated from the Latin, can be stated as:

Every body perseveres in its state of rest or of uniform motion in a straight line unless it is compelled to change that state by forces impressed thereon.

Newton's First Law

The first part of this law seems reasonable and in agreement with our everyday experience. We know full well that bodies at rest will remain at rest until some external force causes them to move. Everyone recognizes this fact. Great consternation arises when, as sometimes happens, a person claims to have seen a body move with no apparent force causing the motion. Physicists, together with most other rational persons, accept this part of the law as being self-evident. We state it as follows:

An object at rest will remain at rest unless a nonzero resultant force acts on it.

The second part of the law is somewhat more subtle:

A moving object will continue its motion along a straight-line path at constant velocity unless a nonzero resultant force acts on it.

Common experience seems to contradict this. We know that nothing continues to move forever without change. A ball rolled across the ground soon stops. A metal object sliding across a smooth table eventually slows and stops. There are many other similar cases.

However, each of the examples cited is not a valid test of Newton's law. A force is acting on each of these bodies that is trying to stop its horizontal motion. It is the force of friction. We all know that the more precautions we take to eliminate this force, the less rapidly the moving object is brought to rest. Newton mentally generalized this observation to the case where no friction forces exist and concluded that if this case were possible, the moving object would never stop. Although an example of perfect constant-velocity motion has never been attained, all our experience leads us to believe that Newton's conjecture, stated as a law of nature, is valid.[1]

3.3 Newton's Third Law of Motion

Because of its greater complexity, Newton's second law is discussed last. The **third law,** as Newton stated it, is as follows:[2]

Newton's Third Law

> To every action there is always opposed an equal reaction; or the mutual actions of two bodies upon each other are always equal and are directed oppositely.

We can restate the law in terms of two objects and the forces they exert on each other. In those terms we can say

> Whenever object *A* exerts a force on (another) object *B*, object *B* will exert a return force back on object *A*; the two forces are equal in magnitude, but opposite in direction.

This law is often called the *action-reaction* law, because it states that the reaction force is equal and opposite to the action force.

Several examples of the third law are shown in Fig. 3.1. We list below the action and reaction forces:

Example	Force 1	Force 2
Part *a*	Block pushes down on table	Table pushes up on block
Part *b*	Woman's head pushes down on man	Man's head pushes up on woman
Part *c*	Hand pulls rope to right	Rope pulls hand to left
Part *d*	Fist exerts force on jaw	Jaw exerts force on fist

You should notice that *Newton's third law deals with two distinct bodies. An action force is exerted on one (body A) by the other (body B), while a reverse reaction force is exerted by A on B.* From time to time we make use

[1] Galileo had stated a similar law, as Newton himself pointed out.
[2] This passage, together with the writings of many other great scientists, can be found in *Source Book in Physics,* by W. F. Magie, Harvard University Press, Cambridge, Mass., 1963. I think you might enjoy perusing this excellent book.

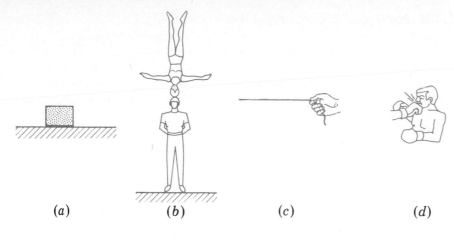

(a)   (b)   (c)   (d)

FIGURE 3.1
Notice that the action and reaction forces act on two different objects.

of this law to tell us the magnitude and direction of a force on a body when the force on some other body is known.

3.4 Definition of Force

We now wish to state and use Newton's second law of motion. The law relates three quantities in a precise way; they are force, mass, and acceleration. But before we can state that relation, we must define quantitatively the concept of force. The definition is made in terms of the acceleration given to an object by an unbalanced force acting on it.

Consider the idealized experiment shown in Fig. 3.2. A 1-kg mass is subjected to a resultant (or net) force $\mathbf{F}_{net}$. The mass might be the standard 1-kg mass defined in Chap. 1. We know, of course, that the unbalanced force $\mathbf{F}_{net}$ will cause the mass to accelerate in the direction of the force. The acceleration $\mathbf{a}$ of the mass can be measured in terms of the units of length and time defined in Chap. 1. We now use this experiment to define the SI unit of force. The definition is as follows:

Definition

The unbalanced **force** *($\mathbf{F}_{net}$) required to give a 1-kilogram mass an acceleration of* **a** *meters per second squared has magnitude a newtons* (N), *and its direction is that of* **a**.

In other words, a net force of 1 N gives a 1-kg mass an acceleration of 1 m/s².

Although the newton is the preferred force unit, two other force units are sometimes used. They are the dyne, where 1 dyne (dyn) = 10^{-5} N, and the pound, where

$$1 \text{ lb} = 4.4482 \text{ N}$$

These latter units are not members of the SI, and we avoid their use.

FIGURE 3.2
A net force of 1 N will give a 1-kg mass an acceleration of 1 m/s².

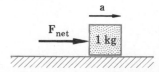

3.5 Newton's Second Law

Now that we have precise definitions for force, mass, and acceleration, we can investigate how these quantities are related. Let us use the idealized experiment shown in Fig. 3.3 for this purpose. In each case a net force of magnitude F_{net} is applied to an object of mass m to give it an acceleration a. Two variations of the experiment are shown.

At the left in Fig. 3.3 we see what happens when we give various masses the same acceleration, a_0. As indicated, the net force must be increased in proportion to the increased mass if the acceleration a_0 is to be preserved. Experiment therefore tells us that in a situation such as this

$$F_{net} \sim m \qquad \text{for } a \text{ constant}$$

where the symbol $\sim$ is read "is proportional to."

The other variation of this experiment, shown at the right in Fig. 3.3, is simply an extension of the experiment we used to define F_{net}. From it we see that the net force required to give a mass m_0 an acceleration changes in proportion to the acceleration. Hence we see that

$$F_{net} \sim a \qquad \text{for } m \text{ constant}$$

We can combine these two proportions into an equation by use of a proportionality constant. The result is

$$F_{net} = (\text{constant})ma$$

As a check, we notice that if a is held constant, then $F_{net} \sim m$. But if m is held constant, then $F_{net} \sim a$.

It remains to evaluate the proportionality constant. To do that, we recall that a force of 1 N gives a 1-kg mass an acceleration of 1 m/s². Thus the experimental equation becomes

$$1 \text{ N} = (\text{constant})(1 \text{ kg})(1 \text{ m/s}^2)$$

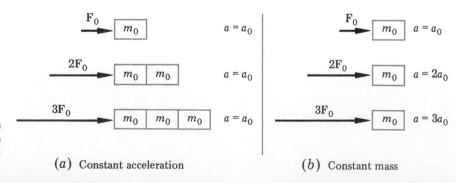

FIGURE 3.3
Newton's second law can be deduced from experiment.

(*a*) Constant acceleration

(*b*) Constant mass

from which we see that the constant is unity and the units are related through

$$1 \text{ N} = 1 \text{ kg} \cdot \text{m/s}^2$$

We arrive, then, at the following experimental result:

To give a mass m an acceleration $\mathbf{a}$, a net force $\mathbf{F}_{net}$ given by

$$\mathbf{F}_{net} = m\mathbf{a} \tag{3.1}$$

is required.

This relation, first proposed by Newton, is called *Newton's second law.* We should notice that $\mathbf{F}_{net}$ is the resultant force on the object whose mass is m. Moreover, the acceleration $\mathbf{a}$ is in the same direction as $\mathbf{F}_{net}$.

Because a force component in the x direction will cause an acceleration in the x direction, we can divide the equation $\mathbf{F}_{net} = m\mathbf{a}$ into three equivalent equations, one for each coordinate direction. We then have that

$$(F_{net})_x = ma_x$$
$$(F_{net})_y = ma_y$$
$$(F_{net})_z = ma_z$$

Often we are interested in the forces in a certain direction. Recalling the fact just stated, we then simply use $F_{net} = ma$ for the particular coordinate direction of interest.

Illustration 3.1 A 900-kg car is to be accelerated from rest to a speed of 12.0 m/s in 8.0 s. How large a force is required to accelerate it in this way?

Reasoning First we must find the acceleration of the car, and then we can use the acceleration in $F_{net} = ma$ to find the required force. We note that, for the car, $v_0 = 0$, $v_f = 12.0$ m/s, and $t = 8.0$ s. To find a, we make use of the motion equation $v_f = v_0 + at$:

$$a = \frac{v_f - v_0}{t} = \frac{(12.0 - 0) \text{ m/s}}{8.0 \text{ s}} = 1.50 \text{ m/s}^2$$

We now use Newton's second law to find the net force that caused the acceleration:

$$F_{net} = ma = (900 \text{ kg})(1.50 \text{ m/s}^2) = 13,500 \text{ N}$$

A force this large was required to accelerate the car. How was this force supplied to the car?

3.6 Inertia

Before proceeding further with application of Newton's second law, let us discuss three terms that will be important: *inertia, weight,* and *friction.* All these terms are familiar to you, but we wish to clarify their meanings.

Definition   *The* **inertia** *of an object is a measure of the difficulty you have in changing the state of motion of the object.* We recognize that a moving car is much more difficult to stop than is a bicycle moving with the same speed. Moreover, the car is much more difficult to deflect or start moving than the bicycle is. These facts are true because the car has a much larger inertia than the bicycle.

We could cite many other familiar examples of inertia. A football

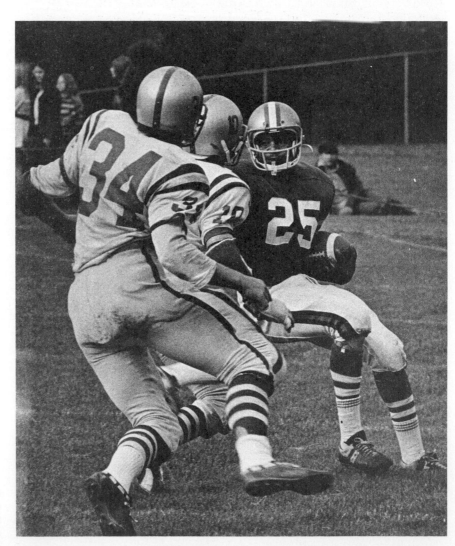

FIGURE 3.4
Inertia is important to these football players. (*Kirk Edwards, Monkmeyer.*)

team, for example, strives to have linemen that are very massive. The linemen have large inertia and are very difficult to set into motion; when an opposing player tries to push them aside, they are not easily moved. Further, a massive ball carrier has an advantage. When he is running, it is difficult for an opposing player to change his motion. Massive players have large inertia. If they are at rest, they are difficult to set into motion; if they are in motion, it is difficult to change their motion. See Fig. 3.4.

These and many other examples of inertia familiar to us are clarified by Newton's first and second laws. The first law is often called the *inertia law* because it tells us what properties an object has because of its inertia: an object at rest tends to remain at rest, while an object in motion tends to remain in motion. In other words, objects possess inertia.

The second law tells us how to measure inertia. If we are to change the state of motion of an object (start it moving, stop it, or deflect it), we must accelerate the object. But acceleration requires a force, according to Newton's second law. The magnitude of the force required for a given change in motion of an object is determined by the inertia of the object. For example, to reverse the motion of a cannonball moving at 5 m/s requires more force than is needed to reverse the motion of a pebble moving with the same speed. The cannonball has more inertia than the pebble. But the force required is larger in the case of the cannonball because the cannonball has more mass than the pebble. Newton's second law therefore tells us that the mass of an object is a measure of its inertia.

We have thus arrived at the following conclusion:

Mass is a measure of inertia.

Objects that have large mass also have large inertia. When we speak of the mass of an object, we are also speaking of its inertia. Because we have a quantitative way for specifying mass in terms of the standard 1-kg mass, we use the masses of objects in quantitative discussions. However, you should always keep in mind that mass is a measure of inertia.

3.7 Newton's Law of Gravitation

Mass is often confused with a totally different concept, the concept of weight. This is natural, perhaps, because massive objects also have large weight. The two concepts are obviously related. Let us now define exactly what we mean by the weight of an object on the earth.

The basic meaning of weight became clear with the discovery by Newton that each object exerts an attractive force on every other object. Newton made his discovery during his attempts to understand the motions of the planets orbiting the sun. His law of inertia seemed to be violated by their motion because the planets follow nearly circular, rather than

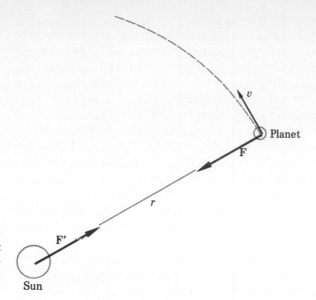

FIGURE 3.5
The sun and planet attract
each other with equal-
magnitude forces.

straight-line, paths. He could explain their motions if he assumed that the sun exerted an attractive force on each of them, as shown in Fig. 3.5.

The force **F** in Fig. 3.5 pulls the planet out of a straight-line path, as shown. We analyze this motion in more detail in Chap. 6. Newton was able to show that the orbital motion of the planets around the sun is due to a force F exerted on the planet by the sun, with F given by

$$F = G\frac{m_s m_p}{r^2} \tag{3.2}$$

where m_s and m_p are the masses of the sun and of the planet, respectively, and r is the distance between their centers. The quantity G is a constant whose value was not determined until much later.

However, if the sun attracts the planet, then the law of action and reaction tells us that the planet must attract the sun. This reaction force is shown as **F'** in Fig. 3.5. Because the action and reaction forces are equal and opposite, **F' = −F.**

Looking at still other astronomical bodies, Newton saw that the moon orbiting the earth also obeys an equation similar to Eq. (3.2). Other planets also have moons, and these, too, obey Eq. (3.2). Generalizing from these observations, Newton stated his *law of universal gravitation:*

Law of Universal Gravitation

Two spheres with masses m_1 and m_2 that have a distance r between centers attract each other with a radial force of magnitude

$$F = G\frac{m_1 m_2}{r^2} \tag{3.3}$$

If the objects are not spherical, the attraction still exists, but the value to be used for r in Eq. (3.3) is more complicated. However, if the objects are small compared to the distance of separation, then r can be taken equal to the distance between any two points on the objects, to a fair approximation. In any case, Newton concluded that each object attracts every other object with a gravitational force.

The value of the gravitational constant G in Eq. (3.3) is not predicted by theory and can be determined only by experiment. Its value was first found by Henry Cavendish in 1798 with a device called the Cavendish balance. One form of this device is shown schematically in Fig. 3.6. Two identical masses m are suspended from an extremely thin and delicate quartz fiber. The two large masses M can be moved close to the small masses m, and the attraction between M and m will cause the fiber to twist. After the system is calibrated so that the amount of force needed to produce a given twist is known, the attraction force between m and M can be computed directly from the observed twist of the fiber. Then, since m, M, r, and F are all known, these values can be substituted in Eq. (3.3) from which it is possible to solve for the single unknown, G.

In practice, this is an extremely delicate experiment since the attraction force is so very small. To measure the very small twist of the fiber, a beam of light is reflected from a mirror attached to the fiber system. By this means the slight twist of the fiber results in a more easily measured deflection of the light beam. (The use of a light beam in this way is referred to as an *optical lever*.) Because of the delicacy of the fiber, even the slightest movement of the air close to it will disrupt the measurements. Therefore great care is needed to eliminate air movement and vibration if reliable results are to be obtained. The value presently accepted for G is

$$G = 6.672 \times 10^{-11} \, \text{N} \cdot \text{m}^2/\text{kg}^2$$

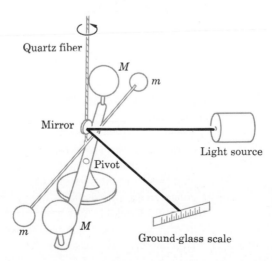

FIGURE 3.6
Schematic diagram of the Cavendish balance. Observe how the light beam is used to detect the twist of the fiber.

Armed with this understanding of the gravitational force, we are now able to understand the following definition of weight:

Definition   *The* **weight** *W of an object on earth is the gravitational force that the earth exerts on the object.*

The earth has an extremely large mass, and so the gravitational forces which the earth exerts on objects at its surface are sizable. For example, a person who weighs 100 lb (445 N) is pulled toward the center of the earth by a gravitational force of 100 lb.

Now that people have journeyed to the moon and soon, perhaps, will journey to distant planets, we extend the definition of weight to include these cases. We define the weight of an object on the moon (or other large body) to be the gravitational attraction force exerted on the object by that large body. Since the mass and radius of the moon were known, long ago it was possible to calculate that objects weigh only about one-sixth as much on the moon as on earth.

Illustration 3.2   Two coins, each having a mass of 8.0 g, are 200 cm apart. Find the gravitational force exerted by one on the other.

Reasoning   The coins are far enough apart and small enough that the gravitation law can be applied to them with $r = 200$ cm. The attractive force is

$$F = G\frac{m_1 m_2}{r^2} = \left(6.67 \times 10^{-11}\,\frac{\text{N} \cdot \text{m}^2}{\text{kg}^2}\right)\frac{(0.0080\,\text{kg})^2}{(2.0\,\text{m})^2}$$
$$= 1.07 \times 10^{-15}\,\text{N}$$

This is a force of only about 2×10^{-16} lb, and so the gravitational force between ordinary-size objects is usually negligible. Notice that in using the law of gravitation we change all quantities to the same SI units in which G is expressed, namely, kilograms and meters.

3.8 Relation between Mass and Weight

Weight is a force, the gravitational force, while mass is a measure of inertia. Even though they are completely different quantities, the fact that massive objects are heavy suggests that the two quantities are related. We now find the relation between weight and mass.

Let us consider the meaning of the experiment shown in Fig. 3.7. We see there an object of mass m falling freely under the pull of gravity. The unbalanced force F_{net} pulling the object downward is the pull of gravity, the weight W of the object. Moreover, in free fall, the object accelerates with

FIGURE 3.7
The unbalanced force on the object is *W*. It causes an acceleration *g*.

the free-fall gravitational acceleration g. Hence, for the situation shown in Fig. 3.7, $F_{net} = ma$ becomes

$$W = mg \qquad (3.4)$$

This is a very important equation. It tells us the relation between the mass and the weight of an object.

We see from Eq. (3.4) that because $g = 9.8 \text{ m/s}^2$ on earth,

Weight of 1 kg $= (1 \text{ kg})(9.8 \text{ m/s}^2) = 9.8 \text{ N}$

But on the moon, where $g = 9.8/6 \text{ m/s}^2$, we have

Weight of 1 kg $= (1 \text{ kg}) \left(\dfrac{9.8}{6} \dfrac{\text{m}}{\text{s}^2} \right) = 1.6 \text{ N}$

As we see, the weight of the standard 1-kg mass (as well as of any other mass) varies from place to place because the gravitational acceleration g varies from place to place. The mass of an object, however, is determined by the object's inertia, and this does not undergo change as the object is moved about.[3]

Now that we know the relation between mass and weight, we can also understand why it is possible to determine masses by weighing. If, *at a given point on earth,* an object weighs as much as the standard kilogram, then the object's mass is also 1 kg. Similarly, an object that weighs n times as much as the standard kilogram has a mass of n kilograms. But, if you understand what we have been saying, it should be clear that this method is valid only if the weighings are made at the same place. A 1-kg object does not weigh the same on the moon as the 1-kg standard mass weighs on earth.

Illustration 3.3   Use the law of gravitation to find the mass of the earth.

Reasoning   We know that the gravitational force which the earth exerts on an object at the earth's surface is simply the weight of the object. For example, a mass m weighs an amount $W = mg$. But this is the force with which the earth of mass M_{earth} attracts the object. Since the distance between centers of the two masses, M_{earth} and m, is simply the radius of the earth R_{earth}, Newton's law of gravitation can be written as

$$W = G \frac{M_{earth} m}{R_{earth}^2} \qquad (3.5)$$

---

[3]That this statement is not precisely correct was pointed out by Einstein in 1905. We discuss this fact in a later chapter.

Since $W = mg$, this becomes

$$mg = G\frac{M_{earth}m}{R_{earth}^2}$$

or, after we cancel m from each side.

$$g = G\frac{M_{earth}}{R_{earth}^2} \tag{3.6}$$

The acceleration due to gravity is $g = 9.8$ m/s^2 while $R_{earth} = 6.4 \times 10^6$ m. Using these together with the value of G, we find $M_{earth} = 6 \times 10^{24}$ kg. Can you devise a method that could be used to determine the masses of the moon, sun, and planets?

It is interesting to note that Eq. (3.6) tells us how the acceleration of gravity should depend on the distance R from the earth's center, provided we replace R_{earth} by R. Clearly, the larger the R, the smaller g will be. The fact that g depends on distance from the center of the earth has been checked by measuring the difference between g at sea level and on mountaintops. However, this change in g is very small because the distance involved is much smaller than the earth's radius. To good approximation, g is constant near the earth's surface. Now that space travel is an accomplished fact, it has been verified to high precision that *g varies inversely as the square of the distance from the center of the earth*. Notice, however, that *this is true only outside the surface of the earth*. Beneath the surface of the earth, Eq. (3.6) is no longer correct.

Before leaving this example, we should point out that Newton's law of gravitation encompasses Galileo's law of freely falling bodies. As we see from Eq. (3.6), the mass of the falling object has canceled out, and so the acceleration due to gravity will be the same for all bodies in free fall. Of course, Newton's law is much more comprehensive than Galileo's since it includes not only Galileo's results but also those of many others.

3.9 Friction Forces

One of the most commonly encountered types of force is the force of friction. It proves to be a force that must be reckoned with in nearly every use of Newton's second law. Therefore we digress for a moment from our discussion of Newton's laws to consider this important type of force.

Definition   **Friction forces** *are those forces that two surfaces in contact exert on each other to oppose the sliding of the one surface over the other.* You can learn quite a bit concerning friction forces by pushing your textbook across a table, as shown in Fig. 3.8. If you push gently at first, the book does not move. However, suppose you slowly increase your pushing force, as shown

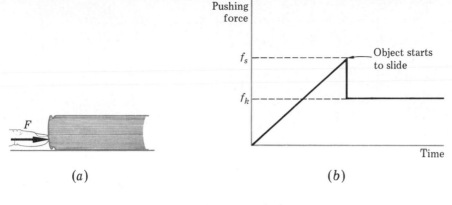

FIGURE 3.8
The friction force that opposes the motion of the book drops suddenly as the book begins to slide.

in the graph in Fig. 3.8*b*. When the pushing force reaches a certain critical value, the book suddenly begins to move. We see from this that there is a maximum friction force which resists the onset of sliding. It is equal in magnitude to the critical pushing force which causes the object to slide. We represent this friction force which resists the start of sliding by f_s. (The subscript s stands for "static," or stationary, since the object is at rest at the instant sliding begins.)

Once an object has begun to slide, a force much less than f_s will keep it moving. This tells us the friction force resisting the motion of the moving object is less than f_s. We represent it by f_k. (The subscript k stands for "kinetic," or moving.) You can easily verify what we are saying by pushing a book along a table. Try it and see.

The major reasons for this behavior can be seen from Fig. 3.9. As you see, the surfaces in contact are far from smooth. Their jagged points penetrate one another and cause the surfaces to resist sliding. Once sliding has begun, however, the surfaces do not have time to settle down onto each other completely. As a result, less force is required to keep them moving than to start their motion.

It is helpful in various circumstances to be able to estimate the magnitude of the friction force that impedes motion. For this purpose it is convenient to define a coefficient of friction μ (Greek mu) which, experiment shows, relates the friction force f to what is called the *normal force* F_N. (*Normal*

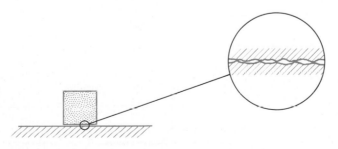

FIGURE 3.9
When highly magnified, the surfaces show considerable roughness; this makes it difficult for them to slide over each other.

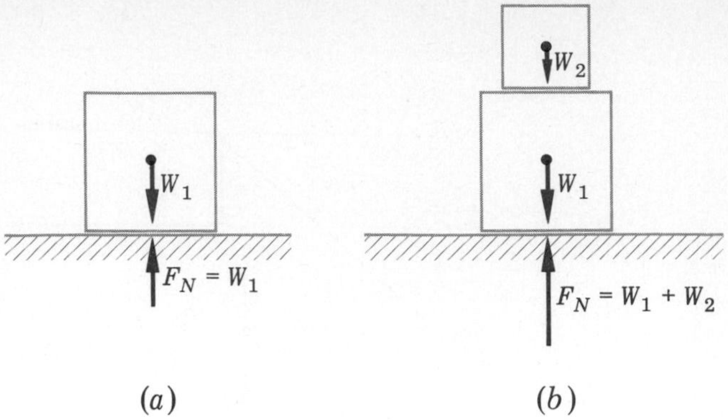

FIGURE 3.10
The normal force F_N is the
perpendicular force exerted
by the supporting surface
on the surface which it
supports.

(a) (b)

means *perpendicular* in this context.) We illustrate F_N in Fig. 3.10. The normal force is the perpendicular force that the supporting surface exerts on the surface which rests on it. In part a of the figure, the block pushes down on the supporting surface with a force equal to its weight. The supporting surface holds the object in place by pushing back with an equal, oppositely directed force, and so $F_N = W$ in this case. In part b, the force pushing down on the surface is the sum of the weights of the two blocks, and so the supporting force is $F_N = W_1 + W_2$ in this case.

Experiment shows that the friction force f impeding the motion of an object that is sliding along a surface obeys the following approximate relation:

$$f = \mu F_N \tag{3.7}$$

This equation is used to define the *coefficient of friction* μ. In words, the friction force increases in proportion to the normal force. It has been found that the coefficient of friction μ varies widely depending on the two surfaces in contact. A few typical values are given in Table 3.1.

The values given in Table 3.1 are designated kinetic (or dynamic) coefficients of friction because they apply only to an object that is already sliding across the surface. As we discussed in relation to Fig. 3.8, a somewhat larger

TABLE 3.1
KINETIC COEFFICIENTS OF FRICTION

Surface 1	Surface 2	μ
Wood	Snow	~0.06*
Brass	Ice	0.02–0.1
Metal	Metal (lubricated)	~0.07
Oak	Oak	0.25
Rubber	Concrete (wet)	0.5–0.9
Rubber	Concrete (dry)	0.7–1.0
Racing tires	Concrete	~1.5

*The sign ~ is read "approximately."

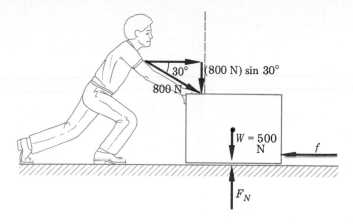

FIGURE 3.11
Notice that the normal
force equals
500 N + (800 N) sin 30°

friction force exists just before the object begins to slide, and the coefficient which applies to that case is usually called the *static* coefficient of friction. All these coefficients depend on the exact conditions of the surfaces, and so they can be used only to obtain an approximate friction force.

Illustration 3.4 Find the friction force f that opposes the motion of the 500-N box in Fig. 3.11 if the appropriate coefficient of friction is 0.60. The man pushes on the box as shown with a force of 800 N.

Reasoning We must first find the normal force F_N. The floor supports not only the weight of the box (500 N) but also the vertical component of the man's push. To provide this support, the floor must push upward with an equal force. Therefore

$$F_N = 500 \text{ N} + (800 \text{ N}) \sin 30° = 900 \text{ N}$$

Knowing that $f = \mu F_N$, we can now write

$$f = (0.60)(900 \text{ N}) = 540 \text{ N}$$

Hence the friction force that impedes the horizontal sliding motion of the box is 540 N. Notice the direction for f shown in the diagram. The friction force is always parallel to the contact surface and is in such a direction as to tend to stop the sliding of one surface over the other.

3.10 Application of Newton's Second Law

We now have the necessary background for the application of Newton's second law to a wide variety of situations. Before exhibiting its use through examples, let us point out the general procedure we shall follow. At the

P

W

(a)   *(b)*

outset we must recognize that $\mathbf{F}_{net} = m\mathbf{a}$ applies to an object of mass *m* which is subjected to a resultant force we represent by $\mathbf{F}_{net}$. Although many other forces may exist in a given situation, *only those forces that are experienced directly by the object are considered when we apply the second law.*

For example, consider the acrobat shown in Fig. 3.12*a*. Her mass is *m*, and we isolate her as the object to which we wish to apply $\mathbf{F}_{net} = m\mathbf{a}$. Only two forces act directly on her: the downward pull of gravity, equal to her weight *W*, and the upward pull *P* that the trapeze exerts on her leg. To show these forces clearly, we draw a *free-body* diagram, the diagram shown in part *b* of the figure. It eliminates all extraneous detail and shows only the forces acting on the object we are considering. By use of the diagram, we can determine the net force on the object and then write $\mathbf{F}_{net} = m\mathbf{a}$ for it.

Our procedure, then, for use of Newton's second law is as follows:

1   Isolate an object for discussion.
2   Draw the free-body force diagram for it.
3   Write $\mathbf{F}_{net} = m\mathbf{a}$. Because the newton was defined so as to have the unit $kg \cdot m/s^2$, if **F** is to be in newtons, then *m* must be in kilograms and **a** must be in meters per second squared.

Let us now illustrate this procedure with a few examples.

Illustration 3.5   How large must be the tension in the rope of Fig. 3.13*a* if the wagon is to be given an acceleration of 0.50 m/s^2? The wagon weighs 80 N. Assume friction forces to be negligible.

Reasoning   We isolate the wagon as the object for our discussion. The free-body diagram showing the forces that act on it is given in Fig. 3.13*b*. Because the floor supports the wagon, the 80-N upward push of the floor balances the downward pull of gravity on the wagon. The wagon will accelerate to the right because of the pull *T* due to the tension in the rope.

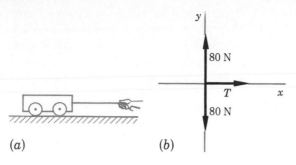

(a)   (b)

FIGURE 3.13
The free-body diagram for
the forces acting on the
wagon is shown in part b.

The net force acting on the wagon is T newtons to the right. Hence $F_{net} = ma$ becomes

$$T = m(0.50 \text{ m/s}^2)$$

The mass of the wagon can be found from the facts that $W = mg$ and $W = 80$ N. Therefore

$$m = \frac{W}{g} = \frac{80 \text{ N}}{9.8 \text{ m/s}^2} = 8.2 \text{ kg}$$

where use has been made of the fact that $1 \text{ N} = 1 \text{ kg} \cdot \text{m/s}^2$.

Substitution of this value for m in the equation for T yields

$$T = (8.2 \text{ kg})(0.50 \text{ m/s}^2) = 4.1 \text{ N}$$

Illustration 3.6 A child pulls with a force of 100 N on a wagon weighing 90 N, as shown in Fig. 3.14a. (a) At what rate will the wagon accelerate? (b) With how large a force is the ground pushing up on the wagon? Assume friction forces to be negligible.

Reasoning Isolating the wagon as the object for discussion, we draw its free-body diagram in Fig. 3.14b. The force P is the push of the ground upward on the wagon, the normal force. Notice that the pull of the rope has

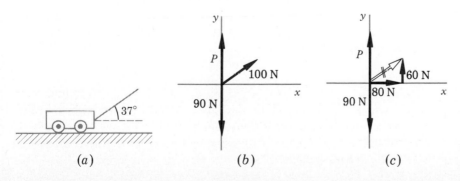

(a)   (b)   (c)

FIGURE 3.14
The rope pulling on the
wagon helps P balance the
weight of the wagon as well
as causing the wagon to
accelerate to the right.

both a vertical and a horizontal component. After this force is split into its x and y components, the force diagram is as shown in part c.

We expect the wagon to move neither upward nor downward. Hence both its vertical acceleration and the net force in the vertical direction must be zero. Adding the vertical forces in Fig. 3.14c and equating them to zero give

$$(F_{net})_{vertical} = P - 90 \text{ N} + 60 \text{ N} = 0$$

from which $P = 30$ N. This is the answer requested in part b.

The wagon will accelerate horizontally because of the net force in that direction. From the diagram, the net horizontal force is 80 N. Therefore $F_{net} = ma$ becomes

$$80 \text{ N} = \frac{90 \text{ N}}{9.8 \text{ m/s}^2} a$$

where m has been replaced by W/g. Solving for a gives $a = 8.7 \text{ m/s}^2$.

Illustration 3.7 A 20.0-kg object hangs by a rope from the ceiling of an elevator, as shown in Fig. 3.15a. What will be the tension in the rope if the elevator is (a) moving upward with constant speed, (b) accelerating upward at 3.0 m/s², (c) accelerating downward at 3.0 m/s², (d) accelerating downward at 9.8 m/s²?

Reasoning The tension T in the rope is defined to be equal to the force with which the rope pulls on the object at its end. We show in part b the free-body diagram for the object. The mass of the object is given as 20.0 kg, and so its weight is given by

$$W = mg = (20.0 \text{ kg})(9.8 \text{ m/s}^2) = 196 \text{ N}$$

In the previous examples, very little ambiguity in direction was involved. This problem is a little more complicated, and we should now be careful about directions. It is wise to choose a given direction as positive. Forces and accelerations will then be positive or negative depending on whether they are parallel or antiparallel to this direction.

(a) Applying $F_{net} = ma$ in this case of $a = 0$, taking up as the positive direction, we have that

$$T - 196 \text{ N} = (20.0 \text{ kg})(0 \text{ m/s}^2)$$

from which $T = 196$ N. The tension in the rope is the same as if the elevator were standing still.

(b) In this case $a = +3.0$ m/s² if we take up as the positive direction. Then $F_{net} = ma$ becomes

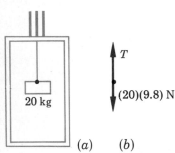

T

$(20)(9.8)$ N

20 kg

(a)   (b)

FIGURE 3.15
When the elevator is accelerating upward, T is larger than (20)(9.8) N. Under what condition will the object appear to be weightless?

$$T - 196 \text{ N} = (20.0 \text{ kg})(3.0 \text{ m/s}^2)$$

from which

$$T = 196 \text{ N} + 60.0 \text{ N} = 256 \text{ N}$$

This is a reasonable result because for the object to accelerate upward, T must be larger than the weight of the object.

(c) Taking up as positive, we have $a = -3.0 \text{ m/s}^2$, and so $\mathbf{F}_{net} = m\mathbf{a}$ becomes

$$T - 196 \text{ N} = (20.0 \text{ kg})(-3.0 \text{ m/s}^2)$$

giving

$$T = 196 \text{ N} - 60 \text{ N} = 136 \text{ N}$$

Is this a reasonable result?

(d) With up positive, $a = -9.8 \text{ m/s}^2$, and $\mathbf{F}_{net} = m\mathbf{a}$ becomes

$$T - 196 \text{ N} = (20.0 \text{ kg})(-9.8 \text{ m/s}^2)$$

from which

$$T = 196 \text{ N} - 196 \text{ N} = 0$$

This may seem to be a strange answer unless you realize that the elevator (and the object in it) is falling with the free-fall acceleration, 9.8 m/s^2. For this to be the case, the object must be unsupported, and so the tension in the cord holding it must be zero.

Illustration 3.8 A 3300-lb car is traveling at 38 mi/h when its brakes are applied and it skids to rest. The skidding tires experience a friction force about 0.7 times the weight of the car. How far does the car go before stopping?

Reasoning First we change all data to SI units:

$$\text{Weight} = (3300 \text{ lb})\left(\frac{4.45 \text{ N}}{1 \text{ lb}}\right) = 14{,}700 \text{ N}$$

$$\text{Mass} = \frac{W}{g} = \frac{14{,}700 \text{ N}}{9.8 \text{ m/s}^2} = 1500 \text{ kg}$$

$$\text{Speed} = \left(38\frac{\text{mi}}{\text{h}}\right)\left(1.61\frac{\text{km}}{\text{mi}}\right)\left(\frac{1 \text{ h}}{3600 \text{ s}}\right) = 0.0170 \text{ km/s}$$

This problem involves a typical combination of an $F = ma$ problem and a motion problem. We can use $\mathbf{F}_{net} = m\mathbf{a}$ to find $\mathbf{a}$. Then the five motion equations can be used to find the stopping distance.

We know that the car has a weight of 14,700 N. We are further told that the stopping force is

$$(0.70)(\text{weight}) = 10,300 \text{ N}$$

This is the unbalanced force acting on the car. Placing this value in $F = ma$ and taking it to be negative since we will take the direction of motion to be positive gives

$$-10,300 \text{ N} = (1500 \text{ kg})(a)$$

From this

$$a = -6.9 \text{ m/s}^2$$

Why is the acceleration negative?

Now we are able to solve the motion problem for the skidding auto. Our knowns and unknowns are

$$v_0 = 17.0 \text{ m/s} \qquad v_f = 0$$
$$a = -6.9 \text{ m/s}^2 \qquad s = ?$$

We can use $v_f^2 - v_0^2 = 2as$ to find $s = 20.9$ m.

It is instructive to notice that the stopping force we have used is a reasonable one. We therefore conclude that a car moving at 17 m/s (about 60 km/h) requires a distance of about 20 m to stop. Can you show that the stopping distance increases as the *square* of the initial speed of the car?

Illustration 3.9   The masses shown in Fig. 3.16 are tied to opposite ends of a massless rope, and the rope is hung over a massless and frictionless pulley. Find the acceleration of the masses. (This device is called **Atwood's machine.**)

Reasoning   Obviously the 10.0-kg mass will fall, and the 5.0-kg mass will rise. The frictionless pulley does nothing but provide a support for the rope. The tension in the rope is the same throughout its whole length. Call it T.

FIGURE 3.16
Since the 10-kg mass will fall as the pulley turns clockwise, the tension in the rope must be less than the weight of the 10-kg mass but more than the weight of the 5-kg mass.

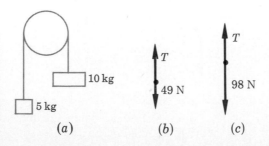

To solve a problem such as this, involving two or more bodies, we write $F = ma$ for each of the bodies separately. First, we must choose a positive direction for the motion. We know in this case that the pulley will turn clockwise, and so it is reasonable to take motions of the objects as positive when moving in that sense.

The free-body diagrams for the 10.0- and 5.0-kg masses are shown in parts b and c of the figure. Notice that the 10-kg body weighs $(10)(9.8)$ N and that its mass is 10 kg. Applying $\mathbf{F}_{net} = m\mathbf{a}$ to each of these bodies in turn and taking the clockwise direction as positive gives

$$98\text{ N} - T = (10.0\text{ kg})(a)$$
$$T - 49\text{ N} = (5.0\text{ kg})(a)$$

Adding the lower equation to the upper yields

$$49\text{ N} = (15\text{ kg})(a)$$
or
$$a = 3.3\text{ m/s}^2$$

where use has been made of the fact that $1\text{ N} = 1\text{ kg} \cdot \text{m/s}^2$. By substituting this value for a in either of the two force equations we find

$$T = 65\text{ N}$$

To check the answer, we see that the tension in the rope is larger than the weight of the lighter object, which thus will rise. In addition, the tension is smaller than the weight of the larger object, which thus will fall.

Illustration 3.10   The two objects shown in Fig. 3.17 are connected by a string, with one object hanging from a frictionless pulley and the other sitting on a table. We are told that the friction force retarding the motion of the object on the tabletop is 0.098 N. Find the acceleration of the bodies.

Reasoning   Sometimes when students are presented with this problem, they state that the objects will not move since the 200-g object is much lighter than the 400-g object. This is a fallacy, however. The weight of the

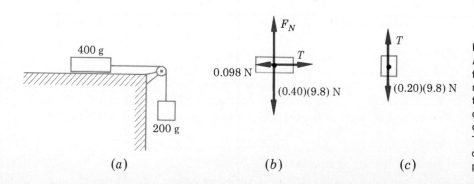

(a)   *(b)*   *(c)*

FIGURE 3.17
Although the 0.098-N friction force retards the motion of the 400-g object, the weight of the 200-g object is large enough to cause the objects to move. The weight of the 400-g object is balanced by the push of the table.

400-g object is downward. This force is balanced by the upward push of the table. The friction force is the only force pulling backward on the object, as is easily seen from the free-body diagram in part *b* of the figure.

Isolating each body in turn and writing $F = ma$ for each, after changing all quantities to SI units, we have (see Fig. 3.17*b* and *c*)

$$T - 0.098 = 0.40a$$

and $$(0.20)(9.8) - T = 0.20a$$

where the forces are in newtons and the masses in kilograms. In writing these we have taken the direction of motion which occurs as the 200 g falls downward to be positive. Notice also that since $W = mg$, the weight of a 0.20-kg object is $(0.20)(9.8)$ N, while its mass is 0.20 kg.

Adding the two equations and solving for *a* gives

$$a = 3.1 \text{ m/s}^2$$

Using the fact that $1 \text{ N} = 1 \text{ kg} \cdot \text{m/s}^2$, you should show that *a* really is in the units stated.

Substitution in either of the equations yields

$$T = 1.34 \text{ N}$$

Illustration 3.11 In an investigation of an accident, a police officer notices that one car shows skid marks on the dry, level pavement. The skid marks are measured to be 7.0 m long. Estimate a lower limit for the speed of the car before it started the skid.

Reasoning Let us assume that the car decelerated uniformly to rest. To find its near largest possible deceleration, let us assume (from Table 3.1) that the coefficient of friction between car and road is 0.90.

If the car has a mass *m*, its weight is *mg*. The normal force F_N must balance this, so $F_N = mg$. From $f = \mu F_N$ we then know that the friction force stopping the car is

$$f = 0.90mg$$

This is the unbalanced force on the car, and it causes the car to decelerate. To find the acceleration, we write $F = ma$ for the car with $F = \mu F_N = f$. Then $F_{\text{net}} = ma$ becomes

$$-0.90mg = ma$$

Notice that the mass of the car cancels, and we obtain $a = -0.90g$, where *g* is the acceleration due to gravity.

We can now solve the motion problem involving the skidding car. The knowns (or assumed knowns) are

$$v_f = 0 \qquad s = 7.0 \text{ m} \qquad a = -0.90g = -8.8 \text{ m/s}^2$$

We wish to find v_0, the initial velocity. To do so, we can use $v_f^2 = v_0^2 + 2as$. Then

$$0 = v_0^2 - (17.6)(7.0) \text{ m}^2/\text{s}^2$$

which gives

$$v_0 = 11.1 \text{ m/s}$$

In all probability, the car was traveling at an initial speed larger than this. Why can we conclude this?

Illustration 3.12 The box of mass m shown in Fig. 3.18 slides down the incline with acceleration a. A friction force f impedes its motion. Find the acceleration in terms of m, f, and θ, the angle of the incline.

Reasoning We show the free-body diagram for the box in Fig. 3.18b. As usual, W is the pull of gravity on the box, and F_N is the normal force due to the incline supporting the box. Because the box is sliding down the incline, the friction force f is directed upward along the incline so as to stop the sliding motion.

For convenience we take the x and y axes as shown in Fig. 3.18c, parallel and perpendicular to the incline. Because the direction of motion is along the incline, we are interested in the force components in the x direction. Therefore it is necessary to split the weight vector into its x and y components. Notice how this is done in the diagram. It is a technique you will need to master. Let us now proceed to write $F_{\text{net}} = ma$ for each coordinate.

Because we know that the box will not crash through the incline or float above it, no acceleration occurs in our y direction. Hence $F_{\text{net}} = ma$ gives $F_{\text{net}} = 0$ for the y direction, and so from Fig. 3.18c

$$F_N - W \cos \theta = 0 \qquad \text{or} \qquad F_N = W \cos \theta$$

FIGURE 3.18
In dealing with a body on an incline, it is convenient to take the x and y axes parallel and perpendicular to the incline as shown. The forces are then split into components along these axes.

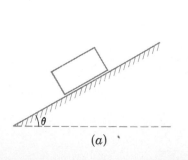

(a)

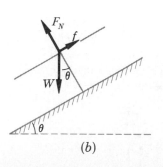

(b)

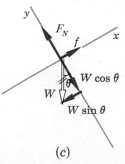

(c)

Writing $F_{net} = ma$ for the x direction, from Fig. 3.18c, we have

$$W \sin \theta - f = ma$$

Hence the acceleration of the box can be computed.

It is interesting to notice that since $W = mg$, the above equation becomes

$$mg \sin \theta - f = ma$$

and if there is no friction, so that $f = 0$,

$$a = g \sin \theta \qquad \text{no friction} \qquad (3.8)$$

We see that when there is no retarding friction present, the acceleration of a body on an incline does not depend on the nature of the body. This means that a child's wagon will move down a hill with the same acceleration as a coasting automobile if retarding friction forces are negligible.

In the limiting case, when $\theta = 0$ and the ground is flat, Eq. (3.8) says that the acceleration is zero, since $\sin \theta = 0$. Of course this is true. On the other hand, if $\theta = 90°$, that is, if the incline is straight up and down, the box will fall straight down. If there is no friction, the body should fall with its free-fall acceleration g. Equation (3.8) is also true in that limit, since $\sin 90° = 1$, and Eq. (3.8) becomes $a = g$.

Illustration 3.13 A particular 1200-kg automobile is capable of accelerating at 0.50 m/s² up an incline which rises 4.0 m in each 40 m. With retarding friction ignored, how large a force must be exerted on the car, pushing it up the incline?

Reasoning Refer to Fig. 3.19a and b. The force P pushing the car up the incline is the force desired. Obviously the unbalanced force in the x direction is $P - W \sin \theta$. From Newton's law we have, since $m = 1200$ kg and $W = mg$,

FIGURE 3.19
The force **P** is partly balanced by the component of the weight acting down the incline. Acceleration up the incline results from the unbalanced portion of **P**.

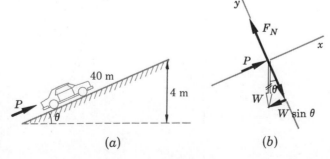

(a)   (b)

$$F_{net} = ma$$

$$P - W \sin \theta = ma$$

or $\quad P - (1200 \text{ kg} \times 9.8 \text{ m/s}^2)\left(\dfrac{4.0}{40}\right) = (1200 \text{ kg})(0.50 \text{ m/s}^2)$

from which

$$P = 1780 \ N$$

Can you describe how the wheels furnish this force to the car?

Illustration 3.14 A motor is to pull a 50-kg block up the incline shown in Fig. 3.20. The coefficient of friction between the block and incline is 0.70. What is the tension in the rope if the block is moving at constant speed?

Reasoning Notice here that the acceleration is zero. Hence there is no unbalanced force on the block. Since the sum of the x forces is zero, we see from Fig. 3.20b that

$$T - f - W \sin \theta = 0$$

or, since $W = mg$,

$$T = f + (50 \times 9.8 \text{ N})(\tfrac{6}{10})$$

But the friction force $f = \mu F_N$, and since the y forces must balance, $F_N = W \cos \theta$. Hence

$$f = (0.70)(50 \times 9.8 \text{ N})(\tfrac{8}{10})$$

Substitution of this value gives $T \approx 570$ N.

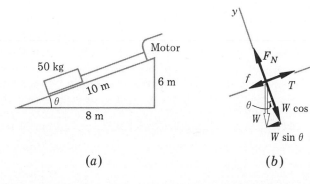

(a)　　　　　(b)

FIGURE 3.20
Since the block is moving up the incline with constant speed, the pull resulting from the motor must exactly balance the sum of the friction force and the component of the weight acting down the incline.

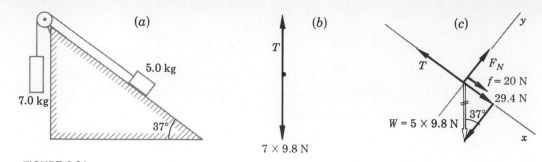

FIGURE 3.21
(a) The 7-kg weight falls downward, thereby pulling the other weight up the plane. (b) and (c) The free-body force diagrams for the two weights.

Illustration 3.15 Find the acceleration for the system shown in Fig. 3.21a. The friction force on the 5-kg block is 20 N.

Reasoning We isolate each body and write $F = ma$ twice, once for each body. The free-body diagrams are shown in parts b and c of Fig. 3.21. When friction forces are present, it is necessary to know the direction of motion, since f will oppose the motion. In this case, we surmise that the 7.0-kg mass will fall. Taking the direction of motion as the positive direction, we have the following $F = ma$ equations:

$$(7.0 \times 9.8 \text{ N}) - T = (7.0 \text{ kg})(a)$$
$$T - 29.4 \text{ N} - 20 \text{ N} = (5.0 \text{ kg})(a)$$

Adding the equations gives

$$19.2 \text{ N} = (12.0 \text{ kg})(a)$$

from which $a = 1.60$ m/s². The tension is found by substituting in either of the two starting equations. The result is $T = 57.4$ N.

* 3.11 British Units and $F_{net} = ma$

Although data given in the British system of units always can be converted to SI units for problem solution, some people prefer to work entirely with pounds and feet. This may be done if it is realized that $F_{net} = ma$ applies to any consistent set of units. It turns out that the pound-force unit is defined in such a way that $F_{net} = ma$ is correct provided a is measured in feet per second squared and $m = W/g$ is measured in pounds per ft/s². This latter unit, the unit for mass in the British system, is called the *slug*.

Definition
*The mass m in **slugs** of an object whose weight is W pounds, where the acceleration due to gravity is g feet per second squared, is given by W/g.*

---

*Optional section. This section may be omitted by those who use only SI units.

Frequently, those who work with the British system replace m by W/g and thus avoid use of the mass unit, the slug. Let us see how the British system is used in a typical problem.

Illustration 3.16 A 2000-lb car moving at 75 ft/s is to be slowed by a force of 620 lb. How far will it go before stopping?

Reasoning Taking the direction of motion as positive, we can write $F_{net} = ma$ as

$$F_{net} = \frac{W}{g}a$$

from which

$$-620 \text{ lb} = \frac{2000 \text{ lb}}{32 \text{ ft/s}^2}a$$

Solving yields $a = -9.9 \text{ ft/s}^2$.

Knowing $a = -9.9 \text{ ft/s}^2$ together with $v_0 = 75 \text{ ft/s}$ and $v_f = 0$, we can use $v_f^2 - v_0^2 = 2ax$ to find

$$x = \frac{v_f^2 - v_0^2}{2a} = \frac{-(75 \text{ ft/s})^2}{2(-9.9 \text{ ft/s}^2)} = 284 \text{ ft}$$

Summary

Newton's first law of motion has two parts. The first part tells us that an object at rest will remain at rest unless there is a nonzero resultant force acting on it. The second part states that the vector velocity of an object will not change unless there is a nonzero resultant force acting on the object.

Newton's third law of motion is the action-reaction law. It states that any force exerted by body A on body B is accompanied by an equal but opposite force exerted by B on A.

Newton's second law is summarized by $\mathbf{F}_{net} = m\mathbf{a}$. It applies to an object of mass m on which a resultant force $\mathbf{F}_{net}$ acts. The acceleration $\mathbf{a}$ of the object is in the same direction as $\mathbf{F}_{net}$.

The mass m of an object measures the inertia of the object. If m is large, the object has large inertia: it is difficult to set the object into motion; it is difficult to stop or deflect the object if it is in motion.

Every piece of mass in the universe attracts every other piece of mass. If two separated spheres have masses m_1 and m_2, the attractive force either one exerts on the other is given by Newton's law of gravitation:

$$F = G\frac{m_1 m_2}{r^2}$$

In this relation, G is a constant of nature, $6.67 \times 10^{-11} \text{ N} \cdot \text{m}^2/\text{kg}^2$. The distance between the centers of the spheres is r. This relation also applies to nonspherical masses which are tiny in comparison to their separation r.

The acceleration due to gravity g can be found from $F_{net} = ma$ and Newton's law of gravitation. If we assume the earth to be a sphere, a mass outside the earth at a distance r from the earth's center will fall with an acceleration $g = GM_{earth}/r^2$, where M_{earth} is the mass of the earth. A similar expression applies to free fall on the moon, with M_{earth} replaced by the mass of the moon. It is found that g_{moon} is about one-sixth as large as g_{earth}.

If the weight W of an object whose mass is m is measured where the free-fall acceleration is g, then $W = mg$. This is simply $F_{net} = ma$ rewritten for the case of a free-fall experiment.

When an object is sliding, it experiences a retarding force f, called the friction force. The supporting surface pushes on the sliding object with a force F_N, where F_N is directed perpendicular to the surface. It is then found that $f = \mu F_N$, where μ is called the coefficient of friction for the two surfaces involved.

Minimum Learning Goals

Upon completion of this chapter, you should be able to do the following:

1   State Newton's first law and give several examples to illustrate each part of it.
2   State Newton's third law. Point out the action-reaction pair of forces in any given simple situation.
3   List several given objects so that they are in sequence from smallest to largest inertia values. Select from the following list those quantities which, when changed, always cause comparable changes in inertia: size, shape, weight, speed, velocity, mass, motion.
4   State Newton's second law both in words and in equation form. Explain clearly what is meant by F_{net} in the law.
5   Give the equation which relates mass and weight. Explain how this equation is related to $F_{net} = ma$. Point out how W, g, and m vary as one goes from the earth to the moon.
7   Find the gravitational force exerted by one sphere on another provided m_1, m_2, r, and G are given. Repeat in the case of two tiny masses (point masses) of any shape provided r is much larger than the dimensions of the masses.
8   Compute F_{net}, m, or a for an object, providing sufficient data from the following list are given: forces acting on the object, weight, mass, enough motion data to compute a.
9   Determine the normal force acting on a sliding object, provided sufficient data are given. Use F_N to determine f or μ, provided one or the other is given.
10 Compute how far an object of known mass will slide in a given time when its initial speed and the forces acting on it are given.
11 Resolve the forces acting on an object supported by an incline into components along and perpendicular to the incline. Point out which component forces cause the object to slide along the incline.
12 Compute the acceleration of an object up or down an incline when the forces acting on the object are given.

Important Terms and Phrases

You should be able to define or explain each of the following:

Newton's three laws of motion
Newton's law of universal gravitation
Inertia
Free-body diagram
Weight = mg

Net force
Newton unit
Normal force
Friction force
Friction coefficient
Resolve the weight vector into components parallel and perpendicular to the incline.

Questions and Guesstimates

1   Why do you have a tendency to slide across a car seat as the car quickly turns a corner? Why will a carton of eggs fall off the seat if the car stops too quickly?

2   People sometimes say that "Newton's first law is a special case of the second law." What justification is there for this assertion?
3   Distinguish between mass, weight, and inertia.

4 A car at rest is struck from the rear by a second car. The injuries (if any) incurred by the drivers of the two cars will be of distinctly different character. Explain what will happen to each driver.

5 What happens to a spring in the seat of a car upon which a woman is sitting as the car goes over a large bump? Explain.

6 Clearly identify the action-reaction forces in each of the following: (a) a boy kicks a can; (b) a ball strikes a windowpane; (c) a parent spanks a child; (d) a Ping-Pong ball bounces on a table; (e) a boat tows a water skier.

7 Because objects weigh only one-sixth as much on the moon as on earth, you would almost certainly be able to lift Muhammad Ali if the two of you were on the moon. Could you easily stop him if he were running at a fair rate across the moon's surface?

8 It has been proposed that a space colony be established on one of the asteroids in our solar system. These tiny planetoids have masses only a small fraction of that of the moon. How would the lives of the inhabitants of the colony be changed because of gravitational effects?

9 Suppose the earth had its present geometrical size but had a mass 100 times larger. For reasons we shall see in later chapters, its atmosphere would differ from ours. But, for discussion, suppose that the atmosphere were the same as ours and that humanoid beings existed on it. How would their bodies have to differ from our own?

10 Why is it more dangerous for a diver to hit the concrete walk around the pool than to hit the water? Explain.

11 Suppose a brick is dropped from a height of several centimeters onto your open hand. Why may severe injury to your hand ensue if your hand is lying flat on a tabletop at the time, even though you can easily catch a falling brick without injury under other circumstances?

12 It is generally believed that, on the average, a drunk will be less injured after a fall from a window than a sober person would be. Can you explain why the belief might be valid?

13 When a Ping-Pong ball is dropped onto a tabletop, it bounces back up into the air. What knocked it upward?

14 People frequently experience a peculiar inner sensation (usually a disturbing feeling in the pit of the stomach) in a moving elevator. Discuss the cause of this effect.

15 Consider the large-type mops used to sweep the halls in a school. It is easy to slide the mop along if the mop handle makes only a small angle with the floor. But if the angle between handle and floor is too large, you can't push the mop along no matter how hard you push. Explain why. Can you find a relation between the critical angle for sliding and the friction coefficient between the floor and mop?

16 Even though the diameter of the planet Neptune is about 4 times as large as that of earth, the acceleration due to gravity is the same on Neptune as it is on the earth. What can you conclude about Neptune from these data?

17 Estimate the minimum distance in which a car can be accelerated from rest to 10 m/s if its motor is extremely powerful. (E)

18 When a high jumper leaves the ground, where does the force come from which accelerates the jumper upward? Estimate the force which must be applied to the jumper in a 2-m-high jump. (E)

19 Estimate the force that your ankles must exert as you strike the floor after jumping from the top of a 2.0-m-high ladder. Why should you let your legs flex in such a situation? (E)

20 Two 100-g balls hang side by side at the ends of two strings. If the strings are 2 m long and the balls are spaced 10 cm from center to center, give an order-of-magnitude estimate of the angle at which the strings will hang to the vertical if all except gravitational effects are ignored.

Problems

1 Show that on earth a 1-kg mass weighs 9.8 N; show that this is equivalent to 2.21 lb.

2 (a) What is the mass in kilograms of an object that weighs 1.00 N on earth? (b) What is the weight in newtons of a 1.00-lb object? (c) What is the mass in kilograms of a 1.00-lb object?

3 A 7.0-g bullet is given an acceleration of 40,000 m/s² as it is shot from a gun that has a 60-cm-long barrel. How large is the average force on the bullet during the acceleration process?

4 How large a horizontal force must be exerted on a 920-kg car to give it an acceleration of 3.0 m/s²

along a flat road? Assume friction forces to be negligible.

5 A water skier is being pulled at a constant speed of 12.0 m/s by a speedboat. The tension in the cable pulling the skier is 140 N. How large is the retarding force exerted on the skier by the water and air?

6 A 50-kg parachutist is gliding to the earth with a constant speed of 8.0 m/s. The parachute itself has a mass of 6.0 kg. (a) How much does the parachutist weigh? (b) How large a force upward does the air rushing past the parachutist and chute exert on them?

7 An advertisement claims that a certain 950-kg car can be accelerated from rest to a speed of 60 km/h in 8.0 s. How large a net force must act on the car to give it this acceleration?

8 A 1500-kg car is to be towed by another car. If the towed car is to be accelerated uniformly from rest to 2.0 m/s in 9.0 s, how large a force must the tow rope hold? Give your answer in both newtons and pounds.

9 Assume an 8.0-g bullet is accelerated uniformly from rest to a speed of 200 m/s in a 70-cm-long gun barrel. How large a force is required to propel it in this way?

10 A 1300-kg car moving at 20 m/s is to be stopped in a distance of 80 m. How large must the stopping force be on the car? Give your answer in both newtons and pounds. Assume uniform deceleration.

11 A neutron is an uncharged particle with a mass of 1.67×10^{-27} kg and a radius of the order of 10^{-15} m. Find the gravitational attraction between two neutrons whose centers are 1.00×10^{-12} m apart. Compare this with the weight of the neutron on earth.

12 Find the force of gravity that the moon exerts on a 45-kg student sitting next to you on earth. The moon's mass is 7.3×10^{22} kg, and its distance from the earth is 3.8×10^{5} km. Compare this to the weight of the student on earth.

13* Compare the pull of gravity on a spaceship at the surface of the earth with the gravitational pull when the ship is orbiting 1000 km above the surface of the earth. (The radius of the earth is 3960 mi, or 6370 km.)

14* The planet Jupiter has a mass 314 times larger than that of the earth. Its radius is 11.3 times larger than the earth's radius. Find the acceleration due to gravity on Jupiter.

15** A 50-kg astronaut is floating at rest in space 35 m from her stationary 150,000-kg spaceship. About how long will it take her to float to the ship under the action of the force of gravity?

16** The acceleration due to gravity on the moon is only one-sixth that on earth. Assuming the earth and moon to have the same average composition, what would you predict the moon's radius to be in terms of the earth's radius R_e? In fact, the moon's radius is $0.27R_e$.

17 For each of the situations shown in Fig. P3.1, the block weighs 40 N and $T = 20$ N. Find the normal force in each case.

18 For each of the situations shown in Fig. P3.2, the block weighs 35 N and $P = 25$ N. Find the normal force in each case.

19* Suppose in Fig. P3.2 the block weighs 20 N, $P = 30$ N, and the appropriate coefficient of friction is 0.40. (a) What is the value of the friction force in each case? (b) What is the value of the acceleration of the block?

20* In Fig. P3.1 the weight of the block is 70 N while $T = 60$ N and the appropriate coefficient of friction is 0.35. (a) What is the value of the friction force in each case? (b) What is the value of the acceleration of the block?

21** For the situation shown in Fig. P3.2b, the coefficient of friction appropriate when the block is just on the verge of slipping is 0.60. If the block weighs 200 N, at what value of P will the block begin to move?

22** In Fig. P3.1b the tension T is large enough to

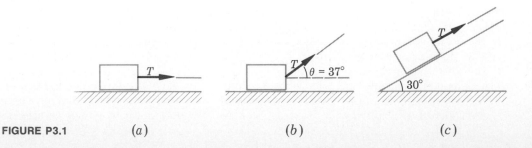

FIGURE P3.1   (a)   (b)   (c)

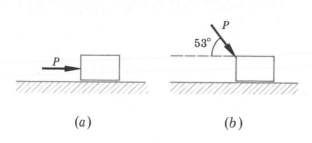

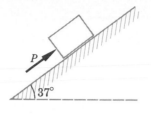

(a)   (b)   (c)   FIGURE P3.2

accelerate the block, but not large enough to lift it. The coefficient of friction between the block and surface is μ. Show that the block will undergo maximum acceleration if θ is chosen such that $\cos \theta + \mu \sin \theta$ is a maximum. Those who know calculus will be able to show further that this condition reduces to $\tan \theta = \mu$.

23* The friction force retarding the motion of a particular 150-kg box across the level floor is 400 N. (a) What is the coefficient of friction between the box and floor? (b) Assuming the friction coefficient to be unchanged, how large an acceleration can be given the box if it is pulled by a force of 750 N inclined at an angle of 53° above the horizontal?

24* (a) Find the acceleration of the 4.0-kg block shown in Fig. P3.3 if the coefficient of friction between the block and surface is 0.60. (b) Repeat if the 50-N force is pushing down on the block at an angle of 30° below the horizontal (i.e., if the force in the figure is reversed in direction).

25* A boy is running across a slippery floor at a speed of 4.0 m/s when he decides to begin sliding across the floor. If the coefficient of friction between his shoes and the floor is 0.150, how far will he slide before stopping?

26* What is the shortest distance in which a car going 30 m/s can stop on a level roadway if the maximum friction coefficient (i.e., the static coefficient) between its tires and the pavement is 0.75?

27* A 900-kg automobile traveling at 20 m/s collides with a tree and stops within 1.80 m. How large was the average retarding force exerted by the tree on the automobile?

28* An 8.0-g bullet enters a 2.0-cm-thick piece of wood with a speed of 140 m/s. It passes through the wood and emerges with a speed of 70 m/s. How large was the average force that resisted passage of the bullet through the wood?

29* If you pull straight upward on a 5.0-kg mass with a cord just capable of holding a 20.0-kg mass at rest, what is the maximum upward acceleration you can impart to the 5.0-kg mass?

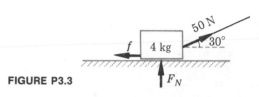

FIGURE P3.3

30* A 70-kg prisoner wishes to escape from a third-story window by going down a "rope" made of bedsheets tied together. Unfortunately, the "rope" can hold a load of only 600 N. How fast must the prisoner accelerate down the "rope" if it is not to break?

31* A 60-kg woman stands on an accurate spring scale inside an elevator. (The scale reads the force with which it pushes upward on the object that rests on the scale.) What does the scale read when the elevator is accelerating (a) upward at 3.0 m/s² and (b) downward at 3.0 m/s²?

32** A 100-g mass is hung from a thread; from the bottom of the 100-g mass a 200-g mass is hung by a second thread. Find the tensions in the two threads if the masses are (a) standing still; (b) accelerating upward at 20 cm/s²; (c) moving downward at a constant acceleration of 5.0 m/s²; (d) falling freely under the action of gravity.

33** A truck is moving at 8.0 m/s along a flat road when it decelerates to stop. In the rear of the truck is a large box. If the coefficient of friction between the box and truck bed is 0.55 when sliding just starts (i.e., its static coefficient of friction is 0.55), what is the maximum value the acceleration of the truck can have if the box is not to slide? What is the shortest distance in which the truck can stop without the box slipping?

34** An egg container rests on the horizontal surface of a car seat. The coefficient of friction between it and the car seat at the start of sliding (i.e., the coefficient of static friction) is 0.55. If the car is moving at 12.0 m/s, what is the shortest distance in which

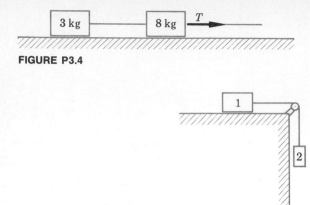

FIGURE P3.4

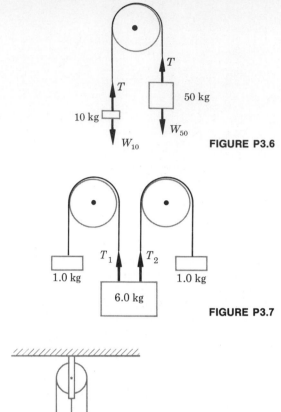

FIGURE P3.5

FIGURE P3.6

FIGURE P3.7

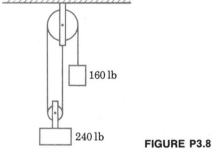

FIGURE P3.8

the car can stop without the eggs slipping from the seat?

35** In Fig. P3.4 the tension in the rope pulling the two blocks is 40 N. Find the acceleration of the blocks and the tension in the connecting cord if the friction forces on them are negligible. Repeat if the coefficient of friction between the blocks and surface is 0.30.

36* In Fig. P3.4, how large must be the tension T to give the blocks an acceleration of 0.40 m/s², assuming (a) friction forces are negligible and (b) the coefficient of friction between blocks and surface is 0.30? Also find the tension in the connecting cord in each case.

37* In Fig. P3.5 block 1 has a mass of 2.50 kg while block 2 has a mass of 1.60 kg. Ignoring friction forces, what would be the acceleration of the blocks and the tension in the connecting cord? Repeat if a friction force of 12.0 N retards the motion of block 1.

38* In Fig. P3.5, object 1 is a 3000-g mass while object 2 is a 2000-g mass. Upon release of the system, object 2 falls 50 cm in 1.50 s. How large a friction force opposes the motion of object 1? Assume no friction forces exist in the rest of the system.

39* Two 4-kg masses are held at the ends of a cord hung over a frictionless pulley in an Atwood's-machine arrangement. If 2000 g is now added to one of these masses, (a) what will be the tension in the cord and (b) how long will it take for the heavier mass to fall 2 m?

40* For the situation shown in Fig. P3.6, find the tension in the cord and the time needed for the masses to move 50 cm starting from rest. Assume the pulley to be frictionless and massless.

41* A 500-g mass is suspended from the end of a light spring. The spring stretches 10 cm more when 200 g is added to the 500-g mass. If the 200-g mass is suddenly removed, what will be the acceleration of the 500-g mass at the instant just after removal?

42** Refer to Fig. P3.7. Find T_1 and the acceleration of the 6-kg block.

43** A physicist is sealed in a closed boxcar on a train. She hangs an 8-lb object as a pendulum from the ceiling. As the train begins to accelerate, she notices that the pendulum hangs steadily at an angle of 10° with the vertical. How large is the acceleration of the train? Does the physicist really need to know the weight of the object to solve this problem?

44** In Fig. P3.8 the large weight is 240 lb, and the small weight is 160 lb. Ignoring the friction and

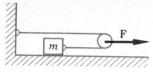

FIGURE P3.9

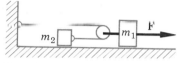

FIGURE P3.10

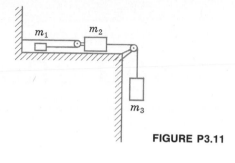

FIGURE P3.11

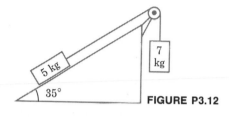

FIGURE P3.12

FIGURE P3.13

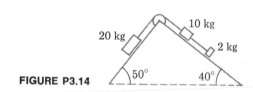

FIGURE P3.14

weight of the pulleys, find the tension in the string and the acceleration of each of the weights. Give answers in SI units. *Hint:* Notice that the large weight moves only half as fast as the small weight. Since the pulleys are frictionless and massless, the tension in the string is everywhere the same.

45** In Fig. P3.9 the pulley is assumed massless and frictionless. Find the acceleration of the mass m in terms of F if there is no friction between the surface and m. Repeat if the friction force on m is f.

46** In Fig. P3.10 assume there is negligible friction between the blocks and table. Compute the tension in the cord and the acceleration of m_2 if $m_1 = 300$ g, $m_2 = 200$ g, and $F = 0.40$ N. *Hint:* Note that $a_2 = 2a_1$.

47** Find the tensions in the two cords and the accelerations of the blocks shown in Fig. P3.11 if friction is negligible. Assume the pulleys to be massless and frictionless ($m_1 = 200$ g, $m_2 = 500$ g, and $m_3 = 300$ g).

48* A 2.0-kg block starts to slide from rest down a 37° incline. How far will it slide in the first 3.0 s if (*a*) friction is negligible and (*b*) μ between the block and surface is 0.40?

49* A 700-kg car is at rest on a hill that is inclined 15° to the horizontal. How far will the car move in the first 10.0 s after the brakes are released if (*a*) the car rolls freely down the hill and (*b*) a 1000-N friction force retards the car's motion?

50* Consider the two blocks shown in Fig. P3.12. Find the acceleration of the blocks and the tension in the cord if there is negligible friction.

51** Repeat Prob. 50 for $\mu = 0.20$.

52** Each of the two blocks in Fig. P3.13 experiences a friction force of 5 lb. How long will it take, start-

ing from rest, to pull the 32-lb block through a distance of 19 ft? Give your answer in SI units.

53** Find the tension in the various cords shown in Fig. P3.14 and the acceleration of the blocks if μ is 0.20 for each block.

The following problems are provided for those who wish to obtain facility in the use of British units.

54 Work Prob. 4 for a 2000-lb car and an acceleration of 9.0 ft/s².

55 Work Prob. 7 for a 2000-lb car and a speed of 60 ft/s.

56 Work Prob. 10 for a 2400-lb car with a speed of 60 ft/s stopping in 240 ft.

57 Work Prob. 19 with $P = 6.7$ lb and the weight of the block being 4.5 lb.

58 Work Prob. 27 for a 2000-lb car moving at 60 ft/s that stops in 5.4 ft.

59 Work Prob. 34 if the car is moving at 40 ft/s.

4
Work and Energy

One of the goals of scientists is to discover ways of unifying and simplifying the various facts and concepts in their field of study. In previous chapters we have discussed forces and their effect in causing motion. In principle, we can describe all motions in terms of the forces which cause them. However, a concept introduced in this chapter, the conservation of energy, greatly unifies and simplifies the description of motion in many instances. The principle of conservation of energy is a unifying concept important not only in mechanics but also in other branches of physics. In order to understand the principle, we first discuss the concept of work and how it leads to the concept of energy.

4.1 Definition of Work

When you sit at your desk studying this book, you are not doing work. This statement does not mean that you are lazy or that learning physics is an effortless process. It is simply stating a fact which arises from the definition of work as used by the scientist. There are so many colloquial ways in which the word *work* is used that it becomes particularly important to give it a precise meaning.

Does a baseball player work when he is playing baseball? Many people would say that since he is playing a game, he is not working. But what if he is a professional player paid to play baseball? Is the ground underneath a house doing work? It is holding the house. Is it, therefore, basically different in its function from a pillar holding the roof over the porch of the house? Yet some would insist that the pillar was doing work. Clearly, if we are to use the term *work* in physics, we need to define it in a precise way.

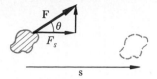

FIGURE 4.1
The work done by **F** during displacement **s** is $F_s s$.

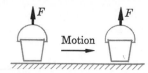

FIGURE 4.2
No work is done on the pail since there is no component of the force in the direction of the displacement.

Physicists and all other scientists define the work done by a force in the following way. Suppose a force **F** acts on an object and the object is given a displacement **s.** Suppose further that F_s is the component of **F** in the direction of **s** as shown in Fig. 4.1. *We then define the* **work done by the force F** *during the displacement* **s** *to be*

$$\text{Work done by } \mathbf{F} = F_s s \qquad (4.1)$$

Definition

Work

Once we accept this definition, it follows that the supporting force exerted by a post holding a load is not doing work; the load is not moved by the force, and so **s** is zero. Similarly, if you hold a pail of water stationary at your side for an hour, the supporting force of your hand has done no work on the pail. The pail has not been moved. Hence **s** is zero, and so is the work done by the supporting force.

Suppose that you carry the pail of water over level ground and pass two points $\frac{1}{2}$ mi apart. The work done on the pail by the supporting force is still zero according to our definition. As seen in Fig. 4.2, your hand pulls upward on the pail with the supporting force **F.** But the displacement of the pail is perpendicular to **F.** Hence F_s is zero, and so the work done during this displacement by the supporting force is zero.[1]

Illustration 4.1 As another illustration, suppose that you were to pull a sled along level ice, as illustrated in Fig. 4.3*a*. You exert a force F on the sled by means of the rope on which you are pulling. After you have pulled the sled through a distance s, how much work have you done?

Reasoning To answer this question, notice that the force F can be thought of as two forces, its components, as shown in Fig. 4.3*b*. Since the

---

[1] If the pail moved with constant velocity between the two points, it is easy to see that no work was done by horizontal forces on the pail. Recall that, for constant-velocity motion, the resultant force on the object is zero. Hence, the net horizontal force on the pail was zero, and so no work was done by horizontal forces. If the pail started and stopped along the way, some work was done in starting the motion. But an equal amount of work is done by the pail on the stopping agent when the motion stops, as we shall see in following sections. Therefore, the net work done by horizontal forces acting on the pail is also zero.

FIGURE 4.3
The vertical component of **F**
does no work, because the
sled does not move up or
down. All the work is done
by $F \cos \theta$, the component
parallel to the direction of
motion.

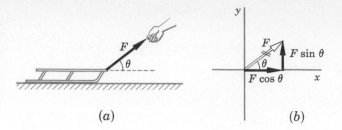

(a) (b)

vertical component, $F \sin \theta$, is perpendicular to the direction of motion, it does no work on the sled. The horizontal component, $F \cos \theta$, is parallel to the direction of motion, and so it does do work on the sled. The work done is

$$\text{Work} = (F \cos \theta)s = Fs \cos \theta \qquad (4.2)$$

Equation (4.2) can be used as a defining equation for work done, since it is Eq. (4.1) written in a different way.[2] It says merely that *the work done is equal to the following product: (the applied force)* $\times$ *(the displacement)* $\times$ *(the cosine of the angle between the force and the displacement).* Sometimes Eq. (4.2) is used instead of its equivalent, Eq. (4.1).

The SI unit for work is the SI force unit multiplied by the SI distance unit, namely the newton-meter (N · m). We give this unit a special name, the *joule* (J):

Definition   *A* **joule** *is the amount of work done by a force of one newton as it acts through a distance of one meter.*

In the British system, the unit for work is the foot-pound, as we see from the definition of work as $F_s s$. This unit has no other name. The foot-pound (ft · lb) is related to the joule (J) through the approximate relation

$$1 \, ft \cdot lb = 1.356 \, J \qquad or \qquad 1 \, J = 0.738 \, ft \cdot lb$$

Definition   *A* **foot-pound** *is the work done by a force of one pound as it acts through a distance of one foot.*

Another unit sometimes used for energy is the erg. *An* **erg** *is defined to be exactly* $10^{-7} \, J$.

---

[2] In physics the product $Fs \cos \theta$ is often written $\mathbf{F} \cdot \mathbf{s}$. This symbolism means, in words, "Take the product of the magnitude of the force vector with the magnitude of the displacement vector, and multiply this product by the cosine of the angle between the two vectors." With this symbolism the defining equation for work would be

$$\text{Work} = \mathbf{F} \cdot \mathbf{s}$$

Note that $\mathbf{F} \cdot \mathbf{s} = F(s \cos \theta) = (F \cos \theta)s = Fs \cos \theta$. The product of two vectors calculated in this manner is called the **dot** (or **scalar**) **product**.

Let us now summarize what we have learned about work.

The work done by a force **F** acting on an object during a displacement **s** is

$$F_s s \quad \text{or} \quad Fs \cos \theta$$

In these equivalent expressions, F_s is the component of **F** in the direction of the displacement **s.** The angle θ is the angle between **F** and **s.** The SI work unit is the joule.

Illustration 4.2 How much work do you do on an object of weight mg as you slowly lift it a distance h straight up? Repeat for the case where it is slowly lowered through this same distance.

Reasoning The situations are shown in Fig. 4.4. To slowly lift the object, we must pull up on it with a force equal to its weight mg. (A force slightly larger than this is needed to give it an initial acceleration upward. But once the object is moving, no unbalanced force is needed to continue the motion. As a result, a pulling force equal to the weight of the object is used except at the very first instant.) As we see in Fig. 4.4a, the lifting force is mg, equal to the object's weight. The displacement is h upward, and the lifting force is in the same direction. Therefore, from either Eq. (4.1) or (4.2) we have

$$\text{Work} = F_s s = Fs \cos 0° = mgh$$

This is the work we must do to lift the mass a distance h.

In Fig. 4.4b we show what happens when we lower the mass. Now **F** and **s** are in opposite directions. Therefore $F = mg$ and $\theta = 180°$. We then find from Eq. (4.2) that

$$\text{Work} = Fs \cos \theta = mgh \cos 180° = -mgh$$

Negative Work

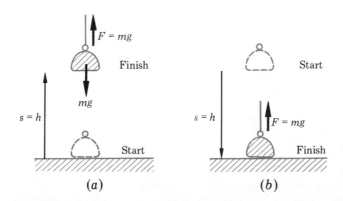

(a)   (b)

FIGURE 4.4
The work done by the lifting force in (a) is mgh, and in (b) it is $-mgh$.

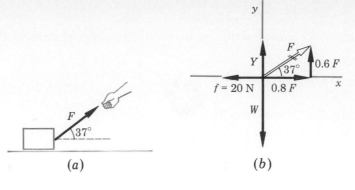

FIGURE 4.5
The horizontal component
of **F** does work on the
object, while the work done
by the vertical component is
zero.

Notice that negative work is done by a force when the displacement is opposite in direction to the force.

Illustration 4.3 As shown in Fig. 4.5, a box is pulled along the floor at constant speed by a force F. Let us say that the friction force opposing the motion is 20 N and the box has a mass of 30 kg. Find the work done by the pulling force as the object is moved 5.0 m.

Reasoning We know that the net force on the object is zero because it is moving with constant speed. Hence the x-directed forces must cancel, and so from Fig. 4.5b we can write

$$0.8F = 20 \text{ N}$$

To find the work done by F, we can use

$$\text{Work} = F_s s$$

But F_s is $0.8F$, and so we have

$$\text{Work} = (0.8F)(5.0 \text{ m}) = 100 \text{ N} \cdot \text{m} = 100 \text{ J}$$

Notice that the y component of **F** does no work. No displacement occurs in the y direction.

4.2 Power

The rate of doing work is called **power,** *and it is defined as the work done per unit time. As a formula, this would be*

$$\text{Power} = \frac{\text{work done}}{\text{time taken to do work}}$$

$$P = \frac{\text{work}}{t} \tag{4.3}$$

The units can be any unit of work divided by any unit of time. *The more common units are*

$$P \to \frac{J}{s} = \text{watts (W)}$$

$$P \to \frac{\text{ft} \cdot \text{lb}}{s}$$

Other units are also used at times. For example, 1 kilowatt (kW) is 1000 watts (W). *The unit* **horsepower** *(hp) is frequently used in the British system.* It is related to the other units for power through

$$1 \text{ hp} = 746 \text{ W} = 550 \text{ ft} \cdot \text{lb/s}$$

The power equation, Eq. (4.3), is used to define a frequently encountered unit used for work. Noting that

$$\text{Work} = (\text{power})(\text{time})$$

we measure power in kilowatts and time in hours. Then the work done by the source of the power has the units of kilowatts × hours, and this unit of work is called the *kilowatthour* (kWh). It can be related to the joule through[3]

$$1 \text{ kWh} = 3.60 \times 10^6 \text{ J}$$

Illustration 4.4 Suppose that the motor shown in Fig. 4.6 lifts the 200-kg object at a constant speed of 3.0 cm/s. What horsepower is being developed by the motor?

Reasoning It is convenient in situations such as this to notice that the power definition can be written as follows:

$$\text{Power} = \frac{\text{Work}}{t} = \frac{F_s s}{t} = F_s \frac{s}{t}$$

But s/t is the velocity component v_s with which the force is moving the load. Hence the power can be written as

$$\text{Power} = F_s v_s \tag{4.4}$$

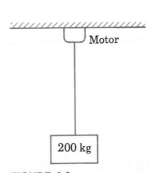

FIGURE 4.6
We wish to find the power output of the motor as it lifts the object with a speed of 3.0 cm/s.

---

[3] $1 \text{ kWh} = (1 \text{ kWh})\left(1000\frac{W}{kW}\right)\left(3600\frac{s}{h}\right) = 3.60 \times 10^6 \text{ W} \cdot \text{s} = 3.60 \times 10^6 \text{ J}.$

In this instance, $v_s = 0.030$ m/s, and the force is equal to the weight of the 200-kg object, namely, (200×9.8) N. Therefore,

$$\text{Power} = (1960 \text{ N})(0.030 \text{ m/s}) = 59 \text{ W}$$

Using the proper conversion factor gives

$$\text{Power} = (59 \text{ W}) \left(\frac{1 \text{ hp}}{746 \text{ W}} \right) = 0.079 \text{ hp}$$

4.3 Kinetic Energy

If an object can do work, we say that the object possesses energy. We define energy as follows:

Definition   **Energy** *is the ability to do work.*

There are many kinds of energy, and we begin our study of them by a consideration of kinetic energy.

An object which is moving can do work. For example, a hammer does work on a nail as it drives the nail into a piece of wood. Or a moving baseball does work as it breaks a window. There are many other examples we could cite where a moving object does work as it slows down. As we shall see, this is a very important fact, and we shall make much use of it. Let us now examine how work and motion are related.

To begin our discussion, let us see how an object's motion depends on the unbalanced force which acts on it. Suppose an object of mass m is subjected to a constant resultant force **F** in the direction of a displacement **s.** During this displacement, the force does work given by

$$\text{Work} = Fs$$

We wish now to relate this work to the change in motion of the object caused by the unbalanced force.

We know that the unbalanced force F is related to the acceleration of the object by $F = ma$. Therefore, the above equation for work becomes

$$\text{Work} = mas$$

Suppose the object had an original velocity v_0 and a final velocity v_f after the work was done. Then, since

$$v_f^2 - v_0^2 = 2as$$

we can substitute $\frac{1}{2}(v_f^2 - v_0^2)$ for as in the work equation to obtain

$$\text{Work} = \tfrac{1}{2}mv_f^2 - \tfrac{1}{2}mv_0^2$$

This is a very important relation. Let us summarize our result:

The work done by the resultant force on an object leads to a change in motion of the object:

$$\text{Work done by resultant force} = \tfrac{1}{2}mv_f^2 - \tfrac{1}{2}mv_0^2 \qquad (4.5)$$

Although this relation was derived for a constant resultant force, it proves to have general validity. Whenever an object is subjected to a resultant force, the work done by the force is related to the object's change in velocity by Eq. (4.5).

In obtaining Eq. (4.5) we tacitly assumed that the force speeded up the object. However, nothing in the derivation requires this. Equation (4.5) also applies if the object is being slowed down. We can easily see this by reference to Fig. 4.7. There we see a mass being slowed as it slides across a table. Notice that the stopping force, the force of friction f, is opposite in direction to the displacement. As a result, the work equation, namely

$$\text{Work} = Fs \cos \theta$$

where θ is the angle between $\mathbf{F}$ and $\mathbf{s}$, becomes in this case

$$\text{Work} = fs \cos 180° = fs(-1) = -fs$$

and so the work done by the stopping force is negative. Substituting in Eq. (4.5), we have

$$-fs = \tfrac{1}{2}mv_f^2 - \tfrac{1}{2}mv_0^2$$

and so v_f must be smaller than v_0; the object is slowed by the friction work done on it.

As we see from this discussion, the work done on an object is related to the quantity $\tfrac{1}{2}mv^2$. This quantity is given the name *kinetic energy* (KE). The name comes from the Greek words *kinetikos* (which means "for putting in motion") and *energeia* (which means "activity"). Let us now restate what we have found about kinetic energy:

Kinetic energy (*KE*) *is* $\tfrac{1}{2}mv^2$. *The work done on an object by the net (or resultant) force acting on it is equal to the change in kinetic energy caused by the force.*   **Definition**

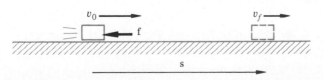

FIGURE 4.7
The friction force does negative work on the block. It slows it according to the relation $fs \cos \theta = \tfrac{1}{2}mv_f^2 - \tfrac{1}{2}mv_0^2$.

In interpreting this statement, we should recall that forces which speed up an object do positive work on the object (because the force is in the direction of motion). But forces that slow down an object are opposite in direction to the motion, and so they do negative work on the object. In particular, the work done on an object by friction forces is always negative because friction forces slow the motion.

We can summarize these considerations in equation form as follows:

$$\begin{pmatrix} \text{Change in KE} \\ \text{of an object} \end{pmatrix} = \begin{pmatrix} \text{work done on} \\ \text{object by } F_{\text{net}} \end{pmatrix}$$

Work-Energy Theorem

$$\tfrac{1}{2}mv_f^2 - \tfrac{1}{2}mv_0^2 = F_{\text{net}}s\cos\theta \tag{4.6}$$

This statement is called the **work-energy theorem** of mechanics. Let us now make use of this equation in a few examples.

Illustration 4.5 A 2000-kg car is traveling at 20 m/s and coasts to rest on level ground in a distance of 100 m. How large was the average friction force tending to stop it?

Reasoning The average friction force $\overline{f}$ acting on the object caused it to lose KE. Therefore

$$\text{Change in KE} = \text{work done by friction force}$$
$$\tfrac{1}{2}mv_f^2 - \tfrac{1}{2}mv_0^2 = \overline{f}s\cos\theta$$
$$-\tfrac{1}{2}(2000\text{ kg})(20\text{ m/s})^2 = (\overline{f})(100\text{ m})(-1)$$
$$\overline{f} = 4000\text{ kg} \cdot \text{m/s}^2 = 4000\text{ N}$$

where use has been made of the fact that $1\text{ N} = 1\text{ kg} \cdot \text{m/s}^2$.

Illustration 4.6 If the friction force on the car of Illustration 4.5 was constant at 4000 N, how fast was the car going after it had gone 50 m?

Reasoning The loss in KE of the car by the time it reached the 50-m mark was caused by friction. Therefore

$$\text{Change in KE} = \text{work done by friction force}$$
$$\tfrac{1}{2}mv_f^2 - \tfrac{1}{2}mv_0^2 = \overline{f}s\cos\theta$$
$$\tfrac{1}{2}(2000\text{ kg})v_f^2 - \tfrac{1}{2}(2000\text{ kg})(20\text{ m/s})^2 = (4000\text{ N})(50\text{ m})(-1)$$

Solving for v_f gives $v_f = 14.1$ m/s.

4.4 Gravitational Potential Energy

As we have seen, some objects are able to do work by virtue of their motion. They have kinetic energy. Other objects can do work because of their position or configuration. They are said to have *potential energy*. Let us begin our study of potential energy by discussing the energy an object has because of gravitational forces.

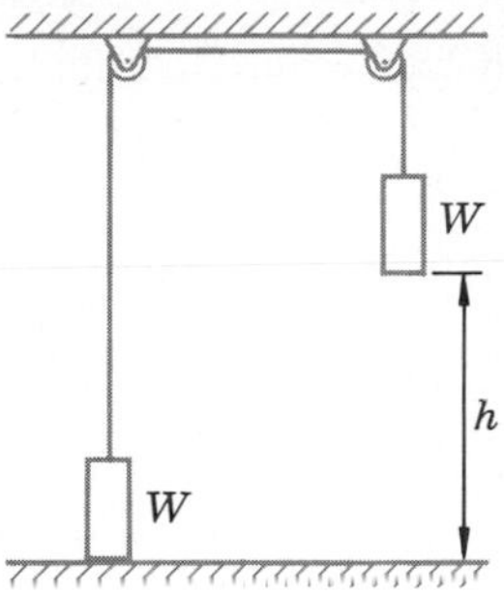

Consider the system shown in Fig. 4.8. The pulleys are assumed frictionless. Because each of the objects has the same weight, $W = mg$, if the top object is given a slight downward push to start it moving, it will fall slowly toward the floor with constant speed. At the same time, the lower weight will rise slowly. By the time the one object has fallen the distance h to the floor, the other object will have been lifted a distance h.

We now ask: How much work was done on the object at the left by the rope as the object was raised from the floor with constant speed? Because the tension in the rope was equal to the weight of the object, mg, the lifting work done by the rope is

$$\text{Lifting work done} = (\text{tension})(\text{distance}) = mgh$$

Who or what external agent did this work? The weight of the object on the right in Fig. 4.8 pulled the other object up, and so it did the work. We must therefore conclude that the right-hand object possessed the ability to do work when it hung in its original position above the floor. In summary, an object of mass m that can fall through a height h under the action of gravity possesses energy in the amount mgh. We define

$$\text{Gravitational potential energy (GPE)} = mgh \qquad (4.7)$$

Definition

where h is the height of the object, m is the mass of the body, and g is the acceleration due to gravity.[4] *The units of potential energy are the same as those for work,* i.e., joules or foot-pounds, as you may easily verify from Eq. (4.7). Notice in Eq. (4.7) that h is a difference between two heights. It measures how far the object has been lifted above a certain reference level.

In connection with this fact, we should mention one other important fact about potential energy. It may be stated this way:

The zero level for computing potential energy is arbitrary.

For example, it is a matter of personal choice from where you choose to measure the position of an object. One person might say an object was 50 cm above a table, and so its height is 50 cm. Another might say that since the tabletop is 90 cm from the floor, the object's height is 140 cm. Still a third might measure the object's position as being 60 cm below the ceiling of the room. For that person, its height is −60 cm. All these choices are possible and correct. As we shall see, *computations involve only the changes in potential energy of objects. This fact allows us to take the zero level of potential energy wherever it is convenient.*

FIGURE 4.8
As the object on the right falls, it does work by lifting the object on the left. The loss of GPE by the object on the right is compensated by a gain in GPE by the other object.

---

[4] For positions far from the earth, g is not equal to 9.8 m/s². If h is so large that g is not constant, we can no longer use mgh for PE.

4.5 Gravitational Force is a Conservative Force

We have already seen several situations in which work was done in lifting an object. To lift an object straight up at constant speed, a force equal to the weight of the object mg is required. As a result, the work done in lifting an object straight up through a distance h is mgh. We now show that even if the object is not lifted straight up, this same result is true.

Suppose we wish to lift the mass shown in Fig. 4.9a from the floor to the tabletop. How much work must be done? Let us lift it along the path shown by the line from A to B.

To compute the work done in lifting the mass from A to B, we approximate the actual path by the jagged path shown in b. By making the jag lengths very small, the two paths can be made identical for all practical purposes. We know that the lifting force is vertical. Therefore, it does no work in the tiny horizontal movements of the jagged path.[5] Work is done by the lifting force on only the vertical movements. When the object is raised, positive work is done. But when it is lowered (as near point C), negative work is done. The effect is that the downward movements cancel the work done on equivalent upward movements. We therefore conclude that the work done is dependent on only the net effect of all the vertical movements. In going from A to B, the object was lifted a net distance h. As a result, the work done is mgh.

But this is the same as the work done in lifting the object straight up from A through a distance h and then moving it sideways to point B. In fact, since the path shown from A to B is perfectly arbitrary, we conclude the following:

Definition

If point A is a distance h below point B, the work done against gravity in lifting a mass m from A to B is mgh. This result holds for any path taken between A and B.

Of course, if the object is lowered from point B to point A, the work done against the gravitational force is $-mgh$.

---

[5] Recall that only a negligibly small force is required to cause the horizontal motion.

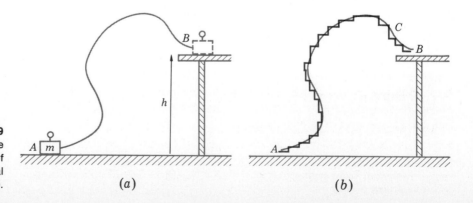

FIGURE 4.9
The path in (a) can be approximated by a series of horizontal and vertical steps, as shown in (b).

(a)   (b)

The gravitational force is an example of a **conservative force.** *A force is conservative if the work done in moving an object from point A to point B against the force is not dependent on the path taken for the movement.* We shall see later that the electrostatic and nuclear forces are also conservative. *Friction forces, on the other hand, are not conservative.* You can easily verify this by sliding your textbook from one point to another across a table. You obviously have to do more work when you slide it by way of a complicated, long path than when you follow a straight-line path. This would not be the case if the friction force were conservative.

Definition

Conservative Force

4.6 Interconversion of KE and GPE

Each time you toss an object into the air or drop an object you see an example of the interchange of KE and GPE. For example, when you toss a coin upward, as shown in Fig. 4.10, its original KE changes to GPE. This is quantitatively correct, as we now show.

At the instant the coin leaves your hand, it has an upward velocity v_0. It then rises to a height h, where its velocity is zero. To find h, we can use

$$v_f^2 - v_0^2 = 2as$$

to give

$$0 - v_0^2 = -2gh$$

from which $h = v_0^2/(2g)$. But its GPE at this height h is mgh. Substituting the above value for h gives

$$\text{GPE} = mgh = mg\frac{v_0^2}{2g} = \tfrac{1}{2}mv_0^2$$

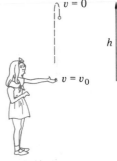

FIGURE 4.10
The KE of the coin is changed to PE as the coin rises. When it falls, the PE is changed back to KE.

As we see, the coin's original KE is changed to GPE as the coin rises.

It is of interest to note in this example that the change in KE (a loss in this case) is numerically equal to the coin's gain in GPE. A similar situation exists as a coin falls freely: as the coin falls, it loses GPE but gains KE. Again, the magnitudes of the energy changes are the same, but one is a loss while the other is a gain. This is an example of a much more sweeping generalization that we can make concerning energy. It is embodied in the law of conservation of energy which we shall discuss soon.

4.7 Other Forms of Energy

You will recall that energy is the ability to do work. With that definition in mind, it is obvious that there are many forms of energy. Coal, oil, gasoline, etc. possess energy because they can be burned to do work. The energy that can be liberated from materials by chemical reaction is called *chemical energy.* As another example, nuclei of atoms can be used to do work by use of nuclear reactors, and so they possess energy, which we call *nuclear energy*. Still another type of energy results from the fact that electric charges can do work; they possess *electrical energy*.

In mechanical systems, energy is frequently stored in elastic devices. The stretched rubber band used by a child to propel a paper wad has elastic potential energy stored in it. Similarly, the compressed spring in a pop-gun possesses potential energy because it can be used to shoot a pellet from the gun. As we see, there are many forms of energy.

Perhaps one of the most important forms of energy is *thermal energy* (TE), or, in colloquial parlance, *heat energy*. This energy allows hot steam in a steam engine to do work. It is also the energy resident in the hot gas that drives the pistons in an automobile engine. Thermal energy is present in all objects, and we study it in detail in later chapters.

Heat, or thermal energy, can be produced by rubbing, by friction. For example, primitive people sometimes started a fire by rubbing pieces of wood together. As another example, your skin suffers a floor burn when you accidentally slide across the gym floor. These are but a few of the many ways in which friction work leads to heating. We can trace this heating effect of friction forces to a very complicated mechanical action: The friction force acts on the molecules and atoms of the sliding surfaces in such a way as to set the atoms and molecules into violent motion. The energy associated with this random motion of molecules within a substance is what we call thermal energy. For now it is sufficient if we realize that *work done by friction forces results in the production of thermal energy.*

4.8 Law of Conservation of Energy

Despite the fact that there are numerous forms of energy, it has become evident through the results of many experiments[6] that the following law of nature exists:

Law of Energy Conservation   Energy can be neither created nor destroyed. When a loss occurs in one form of energy, an equal increase occurs in other forms of energy.

This statement is called the *law of conservation of energy.*[7] Notice that the law does *not* state that KE is conserved, that GPE is conserved, or that even the sum of KE and GPE is conserved. Losses of KE and GPE always result in gains in one or more other forms of energy.

The law of conservation of energy offers us a very powerful tool for understanding many physical situations. To use it, we note that it tells us the following:

---

[6] We shall learn more about these experiments as our study of physics progresses.
[7] The word *conservation,* when used in regard to a law of nature, should be interpreted to be synonymous with the phrase *exact constancy.*

For any process, the sum of all the energy changes (gains positive, losses negative) is zero.

Because the work done by friction forces is important in many applications, we will need to make use of the fact that

The magnitude of the work done by friction forces equals the gain in thermal energy. Friction forces always give a positive thermal-energy change.[8]

Let us now examine a few examples showing how we make use of this law.

Illustration 4.7 A 900-kg car moving at 20 m/s skids to a stop in a distance of 30 m. How large was the friction force between the car's wheels and the road?

Reasoning Only two energy changes took place: The car lost KE, while thermal energy (TE) was generated by the friction at the wheels. From the law of conservation of energy,

$$(\text{Change in KE}) + (\text{change in TE}) = 0$$
$$(\tfrac{1}{2}mv_f^2 - \tfrac{1}{2}mv_0^2) + |fs| = 0$$

where f is the friction force and s is the skidding distance. Notice that the TE term is always a positive change. We place fs between absolute value signs to emphasize this fact.

Because $v_f = 0$ and $v_0 = 20$ m/s, we have

$$0 - \tfrac{1}{2}(900 \text{ kg})(20 \text{ m/s})^2 + |f(30 \text{ m})| = 0$$

from which $f = 6000$ kg $\cdot$ m/s^2 = 6000 N.

Illustration 4.8 A 3.0-kg object falls 4.0 m. How fast is it going just before it hits the ground?

Reasoning If we assume negligible air friction, the only energy changes involve KE and GPE. From the law of energy conservation,

$$(\text{Change in KE}) + (\text{change in GPE}) = 0$$
$$(\tfrac{1}{2}mv_f^2 - \tfrac{1}{2}mv_0^2) + (mgh_f - mgh_0) = 0$$

or
$$\tfrac{1}{2}m(v_f^2 - v_0^2) + mg(h_f - h_0) = 0$$

---

[8]In some situations, as in abrasion, friction forces do work in breaking chemical bonds. We ignore these and other nonthermal effects of friction. They are almost always negligible.

But $v_0 = 0$ and because $h_f = 0$ and $h_0 = 4$ m, $h_f - h_0 = -4.0$ m. Notice that, as we said previously, only the difference in heights matters. In this case, the object lost GPE, so the GPE term will be negative (because $h_f - h_0$ is negative).

The above equation becomes

$$\tfrac{1}{2}(3.0 \text{ kg})(v_f^2 - 0) + (3.0 \text{ kg})(9.8 \text{ m/s}^2)(-4.0 \text{ m}) = 0$$

which yields $v_f = 8.85$ m/s. Notice that the mass of the object was not needed because it could be canceled from the equation.

Illustration 4.9 Suppose the object of the previous illustration was moving at only 6.0 m/s just before it hit. How large was the average friction force acting on it?

Reasoning In this case, the law of energy conservation tells us that

$$(\text{Change in KE}) + (\text{change in PE}) + (\text{change in TE}) = 0$$

But the TE term is simply $|fs|$, where $s = 4.0$ m in this case. We therefore have

$$\tfrac{1}{2}m(36 \text{ m}^2/\text{s}^2 - 0) + m(9.8 \text{ m/s}^2)(-4.0 \text{ m}) + |f(4.0 \text{ m})| = 0$$

Checking our signs, we see that the object gained KE and lost GPE and that the change in TE was positive, as it must always be when the change is caused by friction forces. Notice that the mass does not cancel in this situation. Replacing it by 3.0 kg, we find that $f = 15.9$ N.

Illustration 4.10 A 300-kg roller coaster starts from rest at point A in Fig. 4.11 and begins to coast down the track. If the retarding friction force is 20 N, how fast will the roller coaster be going at point B? At point C?

Reasoning Qualitatively, the roller coaster loses GPE while KE and TE are gained. From the law of energy conservation, for the trip from A to B,

FIGURE 4.11
The PE that the roller coaster has at A is at least partly changed to KE and work against friction forces as it moves to points B and C.

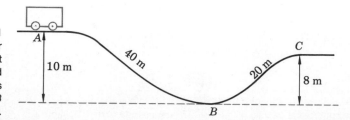

(Change in GPE) + (change in KE) + (change in TE) = 0

$$mg(h_B - h_A) + \tfrac{1}{2}m(v_f^2 - 0) + |fs| = 0$$

where s is the distance from A to B along the track. Using $m = 300$ kg, $h_B - h_A = -10.0$ m, $f = 20$ N, and $s = 40$ m, we can write this as

$$-29{,}400 \text{ J} + \tfrac{1}{2}(300 \text{ kg})v_f^2 + (20 \text{ N})(40 \text{ m}) = 0$$

Because GPE was lost, its term should be negative, as we find it to be. Solving gives $v_f = 13.8$ m/s.

For the trip from A to C we have

$$mg(h_C - h_A) + \tfrac{1}{2}m(v_f^2 - 0) + (20 \text{ N})(60 \text{ m}) = 0$$

with $h_C - h_A = (8.0 - 10.0)$ m $= -2.0$ m. Substituting and solving give $v_f = 5.6$ m/s.

Illustration 4.11 The 2000-kg car shown in Fig. 4.12 is at point A and moving at 20 m/s when it begins to coast. As it passes point B, its speed is 5.0 m/s. (*a*) How large is the average friction force which retards its motion? (*b*) Assuming the same friction force, how far beyond B will the car go before stopping?

Reasoning (*a*) In going from A to B, the car loses KE to GPE and in doing friction work. From the conservation law,

(Change in KE) + (change in GPE) + (change in TE) = 0

$$\tfrac{1}{2}m(v_B^2 - v_A^2) + mg(h_B - h_A) + |fs_{AB}| = 0$$

In our case $m = 2000$ kg, $v_B = 5.0$ m/s, $v_A = 20$ m/s, $h_B - h_A = 8.0$ m, and $s_{AB} = 100$ m. Solving gives $f = 2180$ N.

(*b*) In going from B to the point at which it stops, the car's KE at B is all lost doing friction work. From the law of conservation of energy, we have

(Change in KE) + (change in TE) = 0

$$\tfrac{1}{2}m(0 - v_B^2) + |fs_{BE}| = 0$$

where s_{BE} is the distance from B to the point at which it stops. Using $f = 2180$ N, we find that $s_{BE} = 11.5$ m.

FIGURE 4.12
The KE of the car when it is at A is partly lost to PE and work against friction forces as it moves to B.

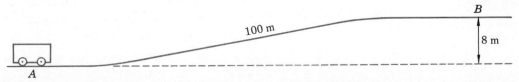

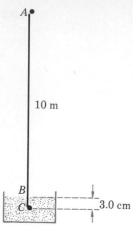

10 m

B

C

3.0 cm

FIGURE 4.13
All the PE of the ball is lost to friction work by the time the ball comes to rest at point C.

Illustration 4.12 A small, 2-kg object falls into a box of sand (see Fig. 4.13) from a height 10.0 m above the sand. It comes to rest 3.0 cm beneath the surface of the sand. How large was the average force exerted on it by the sand?

Reasoning We notice that all the GPE at A is changed to KE at B. This KE is then lost to friction work (thermal energy) as the object penetrates to C. Let us consider the trip from A to C. At both these points, KE = 0. The law of conservation of energy becomes

$$\text{(Change in GPE)} + \text{(change in TE)} = 0$$
$$mg(h_C - h_A) + |fs_{BC}| = 0$$

We know that $m = 2.0$ kg, $h_C - h_A = -10.03$ m, and $s_{BC} = 0.030$ m. Substituting and solving yields $f = 6600$ N.

Illustration 4.13 Consider the pendulum (a ball at the end of a string) shown in Fig. 4.14a. It starts from rest at point A and is released. How fast will the ball be moving (a) at point B, (b) at point C?

Reasoning We should notice at the outset that the pendulum cord does no work on the pendulum ball. The cord always pulls in a direction perpendicular to the direction of motion of the ball. Hence there is no component of the tension in the direction of the motion, and so the tension in the cord does no work.

As the ball swings back and forth, its GPE and KE keep varying. From the law of conservation of energy,

$$\text{(Change in GPE)} + \text{(change in KE)} = 0$$

When the ball loses GPE, it must gain a like amount of KE.
(a) For the trip from A to B, we have

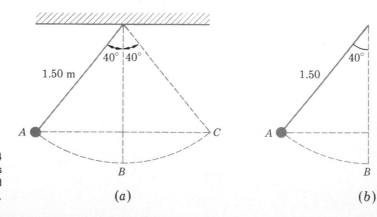

FIGURE 4.14
As the pendulum swings back and forth, its KE and PE keep interchanging.

(a)

(b)

104

$$mg(h_B - h_A) + \tfrac{1}{2}m(v_B^2 - v_A^2) = 0$$

Noticing that $h_B - h_A = -0.35$ m, $v_A = 0$, and m cancels from the equation, we have $v_B = 2.6$ m/s.

(b) For the trip from A to C, we have

$$mg(h_C - h_A) + \tfrac{1}{2}m(v_C^2 - v_A^2) = 0$$

But $h_C = h_A$, and so $v_C = v_A = 0$.

As we see, the pendulum is an interesting example of interconversion of GPE and KE. Over and over the GPE and KE interchange as the pendulum swings back and forth. At any point in its motion, the GPE lost in the fall from point A has been changed to KE. Of course, all this assumes friction forces to be negligible.

Illustration 4.14   How large a force is needed to accelerate a 2000-kg car from rest to a speed of 15.0 m/s in a distance of 80 m? Assume that an average friction force of 500 N opposes the motion.

Reasoning   We assume the car to move along level ground, so that its change in GPE is zero. Let us first solve this problem by use of the work-energy theorem. We have

$$\begin{pmatrix} \text{Change in KE} \\ \text{of object} \end{pmatrix} = \begin{pmatrix} \text{work done by} \\ F_{\text{net}} \text{ on object} \end{pmatrix}$$

$$\tfrac{1}{2}m(v_f^2 - v_0{}^2) = F_{\text{net}}s \cos \theta$$

In this case, two forces act on the car, the accelerating force F and the friction force of 500 N. Therefore the net force in the direction of motion is $F - 500$ N, and θ will be zero.

The above equation therefore becomes

$$\tfrac{1}{2}(2000 \text{ kg})(225 \text{ m}^2/\text{s}^2 - 0) = (F - 500 \text{ N})(80 \text{ m})$$

Solving gives $F = 3300$ N.

We could also solve this by use of energy conservation. However, we would need to recognize that the energy of the source of the accelerating force was decreased by an amount equal to the work which the source performed. The conservation law would then read

$$(\text{Change in KE}) + (\text{change in TE}) + (\text{change in source energy}) = 0$$

from which

$$\tfrac{1}{2}m(v_f^2 - v_0^2) + |fs| + (-Fs \cos \theta) = 0$$

Enrico Fermi
(1901–1954)

Nuclear reactors furnish an appreciable fraction of the world's energy. One of the pioneers in reactor development was Enrico Fermi. Born in Rome, Fermi took his doctorate in Pisa, and was appointed Professor of Theoretical Physics at the University of Rome in 1927. He had a remarkable talent for seeing and explaining things clearly, and for developing working approximations where it was not possible to be mathematically precise. He was also known for his extraordinary charm and his infectious zest for life. His work in the field of radioactivity included the discovery that the speed of nuclear reactions might possibly be controlled; this was a crucial step in the development of nuclear power. Fermi was awarded a Nobel Prize in 1938, and went directly from the awards ceremony in Stockholm to America, rather than return to Mussolini's Italy. After 3 years at Columbia University, Fermi went to the University of Chicago, where he directed the construction of an atomic pile in an abandoned squash court. When this pile produced a nuclear reaction that could be sustained and controlled, the atomic age was born.

Using $v_0 = 0$, $v_f = 15$ m/s, $f = 500$ N, $s = 80$ m, and $\theta = 0$, we get from this expression $F = 3300$ N.

4.9 Machines

Nearly any mechanical device which helps us do work can be called a machine. Most of these devices fall into one of three general categories. A very important class of machines makes it possible to lift a very heavy load by the application of a comparatively small force. Examples of this type of machine are the car jack, the claw hammer, and the compound pulley, to name only a few.

A second type of machine is used to move an object very swiftly even though the driving agent is moving comparatively slowly. An example of this would be a gear or pulley system which changes the rotation of an input shaft into the much faster rotation of an output shaft.

A third type of machine merely makes it more convenient for the work to be done. One such machine is the single pulley shown in Fig. 4.15. It is often more convenient for a person to pull down on a rope at A in order to lift weight W than it is to pull straight up on the weight. Another example of a machine of this general type would be a device which enables someone to move an object that is some distance away.

None of these machines is capable of doing more work than the driving agent does on the machine. We know this to be true without examining the machine in detail, since energy must be conserved in these systems. If more work could be done by the machine than was done on the machine to make it run, the machine would be creating the ability to do work within itself. This, of course, it cannot do. Moreover, since any practical machine will have some friction, energy will be lost doing friction work within the machine. Hence more work has to be done on the machine than is required to do the work without the machine. The purposes of machines are usually the three outlined above. Never does a machine do more work than is done on the machine to make it run.

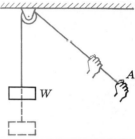

FIGURE 4.15
The type of machine shown here simply alters the direction of the applied force. By so doing, the machine makes it more convenient to lift the object.

4.10 Black-Box Machine

Certain characteristics are common to all machines. In order to describe the operation of machines in general, it is convenient to imagine that the machine is enclosed in a black box, so that its mechanisms will not distract us. This situation is shown in Fig. 4.16.

This machine has an input end, the rope at the left. When someone pulls on the rope with a driving force F, the machine runs. The output end of the machine is at the right; in this case, it does work by lifting the object having weight W.

To describe the characteristics of this machine completely, only two types of experiments need to be done with it. (1) We must measure how far the point of application of the input force moves, s_i, when the weight is lifted a distance s_o by the output end. That is, the rope on the left comes out of the machine a distance s_i when a length s_o of the rope on the right is taken up into the box. (2) We must measure how large a force we need to apply to the input end in order to lift a known load. The first of these experiments tells us how great a load the machine can lift if the applied force is F and if there is no energy loss within the machine. In particular, since the energy put into the machine Fs_i is all used by the machine to do useful work lifting the load, we have

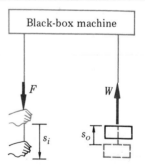

FIGURE 4.16
We can determine the IMA and AMA for this machine, as well as its efficiency, provided that F, W, s_i, and s_o are measured.

$$Fs_i = Ws_o \qquad \text{ideal, no friction}$$

From this it follows that

$$\frac{W}{F} = \frac{s_i}{s_o} \qquad \text{no friction}$$

Ideal Mechanical Advantage

Since the ratio of the load lifted W to the force required F is a measure of the weight-lifting advantage of the machine, we say that its **ideal mechanical advantage** (IMA) is given by the value this ratio would have if there were no friction, that is, s_i/s_o. Hence, for any machine

$$\text{IMA} = \frac{s_i}{s_o} \tag{4.8}$$

Equation (4.8) has the following meaning. If a given machine is examined and it is found that the input force moves through a distance of 10 cm when the load is lifted 2 cm, then

$$\text{IMA} = \frac{10 \text{ cm}}{2 \text{ cm}} = 5$$

This would mean that if the machine had no internal friction, it would be capable of lifting an object weighing 500 N if an input force of 100 N were applied. Notice that the machine cannot do more work than the driving agent furnishes, even in this ideal case. Although the output force is 5 times as large as the input force, the output force moves the object only one-fifth as far as the input force moves. Force is multiplied by the machine, but energy is not.

A real machine is not entirely friction-free, of course. We need yet another experiment to tell us about this aspect of the machine. It is the second experiment mentioned above, in which we measure the force F needed to lift the weight W. The ratio W/F in this case is the **true**, or **actual, mechanical advantage** (AMA) of the machine. We have, therefore,

Actual Mechanical Advantage

$$\text{AMA} = \frac{W}{F} \tag{4.9}$$

Of course, *the AMA will always be smaller than the IMA,* since, for the same applied force F, friction would reduce the load W which could be lifted.

Definition   *The **efficiency** of a machine or a process is defined to be the ratio of the output work to the input work.* Since these two quantities are Ws_o and Fs_i, respectively, we have

$$\text{Efficiency} = \frac{Ws_o}{Fs_i} = \frac{W/F}{s_i/s_o}$$

Hence $$\text{Efficiency} = \frac{\text{AMA}}{\text{IMA}}$$ (4.10) *Efficiency of a Machine*

where use has been made of Eqs. (4.8) and (4.9). If there is no friction, the AMA and IMA are identical, and the efficiency would be 1.00, or, as more often stated, 100 percent. On the other hand, if the AMA = 4 and the IMA = 5, the efficiency would be $\frac{4}{5}$, which is 0.80, or 80 percent.

Notice that the basic qualities of a machine—AMA, IMA, and efficiency—are defined in the same way for all machines. To investigate these quantities in particular situations, we must now examine the inner workings of various machines.

4.11 Simple Pulley

This is the machine shown in Fig. 4.17. It consists of a rope looped once over a wheel mounted on an axle. Since the rope is assumed not to stretch, if the force F pulls the rope out through a distance s_i, the object having weight W must rise a distance $s_o = s_i$. Therefore, for this machine, we have from Eq. (4.8)

$$\text{IMA} = \frac{s_i}{s_o} = 1.00$$

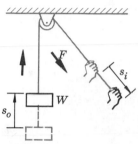

FIGURE 4.17
When used as shown, the single pulley has an IMA of unity.

The actual mechanical advantage of the machine would depend on the friction in the pulley. For a simple pulley such as this, the friction could be made quite small, and so the AMA would be nearly as large as the IMA. It could never be as large as 1.00, though. The efficiency of the machine would be somewhat less than 100 percent.

Illustration 4.15 For the system of Fig. 4.17, a force of 50 N is needed to lift a 40-N object. What is the efficiency of the machine?

Reasoning From the design of the machine, $s_i = s_o$, so its IMA = 1.00. The data given show AMA = 40 N/50 N = 0.80. Hence

$$\text{Efficiency} = \frac{0.80}{1.00} = 0.80, \text{ or } 80\%$$

4.12 Other Pulley Systems

In Fig. 4.18 we consider the case of a single movable pulley. Notice that for each meter that the load moves up, the force F must pull the rope up 2 m. This is the result of the fact that when the pulley moves 1 m closer to the

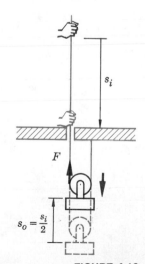

FIGURE 4.18
This pulley system has an IMA of 2, since the applied force moves twice as far as the load.

FIGURE 4.19
The IMAs for the systems
shown are 2, 3, and 4,
respectively. Note that
these are also the number
of ropes pulling up on the
load-bearing pulley.

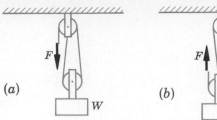

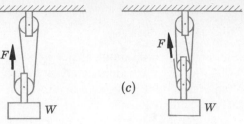

(a)   (b)   (c)

ceiling, the distances of both the right and left ropes to the ceiling will be 1 m shorter. This means that 2 m of rope must be pulled up through the hole to shorten the support by 1 m.

For this system, then, $s_i = 2s_o$, and so

$$\text{IMA} = \frac{s_i}{s_o} = 2.00$$

It is clear that if the system were not moving, both ropes would be pulling up on the pulley and weight. Hence, if the pulley were weightless, the tension in each rope would be $W/2$, so that their combined upward pull on the pulley would be W. In that case, F would be equal to $W/2$. Actually, of course, F must be large enough that the ropes also support the weight of the pulley and overcome the action of friction as well. Hence, it would be impossible for this machine to have an $\text{AMA} = 2.00$. If the pulley weighed nearly as much as W, the machine would have a very low efficiency, even though the friction in the pulley itself were very small. Nevertheless, it would be an improvement over a single rope in some respects.

More complicated pulley systems are shown in Fig. 4.19. Using the same reasoning as for the previous case, we find that the IMAs for systems (a), (b), and (c) are 2, 3, and 4, respectively. Notice that, in this case as well as in Fig. 4.18, the IMA is numerically equal to the number of ropes pulling up on the free pulley. This fact provides one with a simple rule of thumb for determining the IMA of block-and-tackle systems such as these.

4.13 Wheel and Axle

The device shown in Fig. 4.20 consists of an input rope wound around a big wheel of radius b. The load is held by the axle of the wheel, and the axle has a radius a, much smaller than b.

To find the IMA of this simple machine, consider what happens as the wheel and axle turn counterclockwise through one revolution. The input rope will unwind a length equal to the circumference of the wheel. Therefore

$$s_i = 2\pi b$$

The rope holding the load will wind up one turn on the axle, and this length is just the circumference of the axle. Thus

$$s_o = 2\pi a$$

We therefore have

$$\text{IMA} = \frac{s_i}{s_o} = \frac{b}{a}$$

This is a very useful machine, since it does not have too much inherent friction and yet b can be made much larger than a. It would not be uncommon for a to be 2 cm and b to be 50 cm. For this combination, the IMA is 25, a very large mechanical advantage for such a simple device.

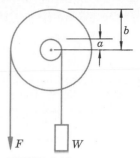

FIGURE 4.20
The wheel and axle have an IMA given by the ratio of the radius of the wheel to the radius of the axle.

Summary

The work done by the force **F** on an object as the object moves through a straight-line displacement **s** is given by work = $Fs \cos\theta$, where θ is the angle between **F** and **s**. Alternatively, in terms of the component of **F** along **s**, work = $F_s s$. The units of work are joules and foot-pounds. These units are related through 1 J = 0.738 ft · lb.

Power measures the rate of doing work. It is defined as the work done per unit time. Typical units are the watt (which is joules per second), the foot-pound per second, and the horsepower. Relations between these units are 746 W = 1 hp and 1.36 W = 1 ft · lb/s.

When an object has energy, it possesses the ability to do work. Kinetic energy (KE) is the ability to do work because of an object's motion. An object with mass m and velocity v has KE = $\frac{1}{2}mv^2$. It can do this much work against external forces as the object is brought to rest. The units of KE are the same as those of work. If an object is subjected to a resultant force which does work on the object, the work done on the object equals the change in KE of the object.

An object which is capable of doing work because it is pulled downward by a gravitational force is said to possess gravitational potential energy (GPE). Its potential energy is mgh; the object can do this much work while descending a distance h. The reference level for measuring the object's height is arbitrary.

If the work done in moving an object from point A to point B against a force is not dependent on the par-

ticular path taken for the movement, then the force is said to be conservative. The gravitational force is conservative. As a result, the work done against gravity in lifting an object from A to B is mgh independent of the path followed. Friction forces are not conservative.

The work-energy theorem can be stated as follows for any object:

$$\begin{pmatrix} \text{Change in KE} \\ \text{of object} \end{pmatrix} = \begin{pmatrix} \text{work done by } F_{\text{net}} \\ \text{on object} \end{pmatrix}$$

The law of conservation of energy states that energy can be neither created nor destroyed. When a loss occurs in one form of energy, an equal increase must occur in other forms of energy. Work done against friction forces produces heat energy equal in magnitude to the friction work done. Many forms of energy exist. Typical are chemical, electrical, and nuclear energy.

Machines are devices which help us do work. They never do more work than the energy given to them. To characterize a simple machine, one makes use of its ideal and actual mechanical advantages. If the input force F moves through a distance s_i when the output force W moves through a distance s_o, then IMA = s_i/s_o while AMA = W/F. The AMA is always less than the IMA. A machine's efficiency is defined by efficiency = (output work)/(input work). This is equivalent to eff = AMA/IMA.

Minimum Learning Goals

Upon completion of this chapter, you should be able to do the following:

1 Define work and its units. Give several examples of a force which does no work because (a) $s = 0$ or (b) $F_s = 0$. Give examples where a force does negative work.

2 Compute the work done on an object by a specified constant force when the object is moved through a given straight-line distance.

3 Define power and its units. Compute the power from data for the rate at which work is being done.

4 Convert between the units watts, foot-pounds per second, and horsepower.

5 Define KE. Compute its value provided m and v are given.

6 Compute the change in KE of an object which is subjected to a given resultant force acting through a known distance.

7 Define gravitational PE. Compute its value for an object provided the necessary data are given.

8 Explain what is meant by a conservative force. Point out the importance of the fact that the gravitational force is conservative.

9 Give several examples of situations in which KE and GPE are interchanged.

10 State the work-energy theorem in words and in equation form. Apply it to examples similar to those given in the text.

11 State the law of conservation of energy in your own words. Explain why the loss of KE in doing friction work does not contradict the law. Use the law to solve simple problems.

12 Compute the IMA, AMA, and efficiency of a black-box machine provided sufficient data are given.

13 Give the IMA of a simple machine such as a pulley system provided the construction of the machine is made clear.

Important Terms and Phrases

You should be able to define or explain each of the following:

Work $= F_s s = Fs \cos \theta$
Joule, foot-pound
Power $=$ work/time
Watt, horsepower
Kilowatthour
KE $= \frac{1}{2}mv^2$

Gravitational PE $= mgh$
Conservative forces
Work-energy theorem
Law of conservation of energy
IMA $= s_i/s_o$
AMA $= W/F$
Efficiency $=$ (output work)/(input work) $=$ AMA/IMA

Questions and Guesstimates

1 The moon and most manufactured earth satellites circle the earth in essentially circular paths with their center at the earth's center. How much work does the earth's gravitational force do on them?

2 Reasoning from the standpoint of KE, why is a loaded truck likely to be much more damaging than a Volkswagen in a collision with a massive stationary object? Assume equal initial speeds.

3 A ball hangs at the end of a thread, and the system swings as a pendulum. Describe what happens to the KE of the ball as the pendulum swings back and forth. Repeat for the GPE. How are the two

related? What happens to the energy of the system as the pendulum loses its energy? Does the pull of the thread on the ball do any work?

4 A person holds a bag of groceries while standing still talking to a friend. A car sits stationary with its motor running. From the standpoint of work and energy, how are these two situations similar?

5 A conscientious hobo in a boxcar traveling from Chicago to Peoria pushes on the front wall of the car all the way. Having at one time been a physics student, he thinks his pushing did a great deal of work since both F_s and s were large. Where was the flaw in his reasoning?

6 Since the earth is moving with respect to the sun, everything on it has KE, at least in the opinion of an observer who thinks the sun is at rest. Why did we not need to consider this when working the examples in this chapter?

7 Is KE a vector or a scalar quantity?

8 A ball of mass m is held at a height h_1 above a table. The tabletop is a height h_2 above the floor. One person says that the ball has a PE of mgh_1, but another person says that its PE is $mg(h_1 + h_2)$. Who is correct?

9 As a rocket reenters the atmosphere, its nose cone becomes very hot. Where did this heat energy come from?

10 If a piece of chalk falls off a table, its speed just before it hits the floor can be found by equating the original GPE to its total energy just before striking the floor. Show that the same result is obtained if the zero of GPE is taken as (a) the floor, (b) the tabletop, (c) the ceiling.

11 Automobiles, tractors, etc. have gear systems which can be changed by shifting. Considering these to be ideal machines, discuss why the shifting process is used.

12 It has been suggested that tides flowing in and out of harbors could be used as sources of energy. Another suggestion is to use the ocean waves for this purpose. Discuss the pros and cons of either proposal from a practical standpoint. Where does the energy come from in each case?

13 Reasoning from the interchange of kinetic and potential energy, explain why the speed of a satellite in a noncircular orbit about the earth keeps changing. Is its speed largest when it is at apogee (farthest point from the earth) or when it is at perigee (the closest point)?

14 About what horsepower is a human being capable of producing for a short period, as in climbing a flight of stairs? (E)

15 Estimate the amount of useful work (as defined in physics) which an average human being might perform in 1 day. For comparison purposes, a typical diet might furnish the person with 2000 kilocalories (kcal) ($\approx 8.4 \times 10^6$ J) of energy each day. Where does the rest of the energy go? (E)

16 Estimate the force a driver experiences when the car being driven hits another car head on. Assume both cars to be similar and traveling at 60 mi/h. Discuss the effect of seat belts, position of the person in the car, and similar factors. (E)

Problems

Many of these problems are solved most easily by energy methods. Use energy methods when appropriate.

1 A woman pushes a push-type lawnmower with a force of 200 N as the mower moves 25 m. The handle of the lawnmower makes an angle of 60° with the vertical. How much work does the woman do? Repeat if she pushes it 75 ft with a force of 40 lb.

2 In order to pull a certain child in a wagon, a force of 150 N at an angle of 20° above the horizontal sidewalk is required. How much work is done by the 150-N force in pulling the wagon 8.0 m? Repeat for a force of 30 lb and a distance of 50 ft.

3 A horizontal string is to be used to pull a 3.0-kg object along the floor at constant speed. If the coefficient of sliding friction between the floor and object is 0.60, how much work is done by the string in pulling the object 5.0 m?

4 A 20.0-kg crate is to be pushed 6.0 m along the floor at constant speed by a force inclined 30° below the horizontal. How much work is done by the pushing force if the friction force that retards the crate's motion is 140 N?

5 A 5.0-kg bucket is lowered slowly into the vertical shaft of a well by a rope-and-winch system. How much work is done on the bucket by the rope as the bucket is lowered 7.0 m?

6 A friction force of 2000 N acts on a car to stop its skidding wheels as the car moves 80 m along the

level ground. How much work does the friction force do on the car?

7* A 2,400-kg loaded coal cart is being pulled up a 10° incline at constant speed by a cable parallel to the incline. The friction force impeding the motion is 800 N. How much work is done on the cart by the cable as the cart is pulled 30 m?

8* By changing the angle of a variable-angle incline, it is found that a 500-g block slides down the incline with constant speed when the incline angle is 35°. How much work does the friction force do on the block as the block slides 80 cm?

9 The elevator in a certain building weighs 10,000 N when fully loaded. It is pulled upward at a constant speed of 0.50 m/s. What minimum power must the lifting motor produce in order to perform this task? Give your answer in horsepower.

10* A certain electric motor is rated at $\frac{1}{4}$ hp. By means of a pulley on its shaft, it pulls a belt at the rate of 300 cm/s. Assuming the tension in the low-tension side of the belt is zero, what is the maximum tension the motor can provide to the high-tension part of the belt?

11* What average horsepower does a 60-kg woman develop as she lifts herself 12.0 m in 20.0 s by climbing a long flight of stairs?

12* A certain 800-kg car is said to be able to accelerate from rest to a speed of 20 m/s in 9.0 s. Assuming this to be done on level road and with negligible friction work being done, what minimum horsepower must the car's engine be developing?

13* At high speeds the friction forces acting on a car increase in proportion to v^2, where v is the car's speed. Considering this to be the major factor involved, if a car is rated at 32 km (20 mi) per gallon of gas at 80 km/h (50 mi/h), what would its mileage rating be at 110 km/h (68 mi/h)?

14* At low speeds the friction force tending to stop a car is due mainly to energy loss in the rolling tires. For a typical compact car moving at 9.0 m/s, the total force resisting motion is about 200 N. (*a*) How much useful power must the motor provide to maintain this speed? Express your answer in horsepower. (*b*) At 36 m/s the combined rolling and air friction has risen to 1100 N for the same car. What horsepower is needed at this speed?

15 How large a force is required to accelerate an electron ($m = 9.1 \times 10^{-31}$ kg) from rest to a speed of 2×10^7 m/s in 0.50 cm?

16 How large a force is required to stop a 100-kg man in 1.5 m if he is moving at 2.0 m/s?

17* How large a friction force between its two rear tires and the pavement is needed to accelerate a 2000-kg car from rest to 20 m/s in 80 m? How long will it take? Assume constant acceleration. How large must μ be for this to be possible?

18* Assuming a coefficient of friction μ between the car and roadway, use energy methods to show that the stopping distance for a car traveling with speed v is given by $v^2/(2\,\mu g)$. The important quantities are speed squared and the friction coefficient.

19* Suppose that 10^{16} electrons strike the screen of a TV tube each second. If each electron is accelerated through a voltage large enough to give it a speed of 10^9 cm/s, starting from rest, how many watts of power are expended in maintaining this beam of electrons? (Note that $m_e = 9.1 \times 10^{-31}$ kg.)

20* An atom-smashing machine known as a Van de Graaff generator is capable of accelerating a beam of protons ($m_{proton} = 1.67 \times 10^{-27}$ kg) from rest to a speed of 10^9 cm/s. If the machine accelerates 10^{16} protons per second, how many watts of power is it producing? (A proton is a hydrogen atom from which an electron has been removed.)

21 An 80-kg hiker climbs a 600-m-high hill. How much work does the hiker do against gravity? Does this amount of work depend on the path the hiker takes? If the hike takes 90 min, what average horsepower was expended in this way by the hiker?

22 A 30,000-kg truck takes 40 min to climb from an elevation of 1800 m to 2900 m along a mountain roadway. (*a*) How much work did the truck do against gravity? (*b*) What average horsepower did the truck expend against gravity in the process?

23 A 5.0-kg mass is dropped from a window that is 12.0 m above the pavement. How fast is the mass moving as it strikes the top of a 170-cm-high car?

24* When an object is thrown upward, it rises to a height h. How high is the object, in terms of h, when it has lost two-thirds of its original KE?

25* A 2-kg object dropped from a height of 12 m is moving at 7 m/s just before it hits the ground. How large was the average friction force retarding its motion?

26* A 500-g ball is thrown straight upward with a speed of 18.0 m/s, and it rises to 14.0 m. How large was the average friction force that impeded its motion?

27* A motor is to lift a 900-kg elevator from rest at ground level in such a way that it has a speed of 3.0 m/s at a height of 20 m. (*a*) How much work will the motor have done? (*b*) What fraction of the total work went into kinetic energy?

28* Starting from rest, a locomotive pulls a series of boxcars up a 5° incline. After the train has moved 2000 m, its speed is 10.0 m/s. Assume the entire train has a mass of 7.0×10^5 kg. How much work must the locomotive do? What fraction of this work is done against gravity? Assuming the acceleration to be uniform, how long did the process take? What average horsepower did the locomotive expend?

29** An electric motor is to power a pump that lifts 500 g of water originally in a tank through a height of 2.0 m in 1 min. Assume the water is moving at 1.60 m/s when it exits at the top. What minimum output rating (in horsepower) should the motor have? The speed of the water in the tank is negligible.

30** Repeat Prob. 29 if the motor is operating on the moon where the acceleration due to gravity is only one-sixth that on earth.

31 A 1.8-g bullet with a velocity of 360 m/s strikes a block of wood and comes to rest at a depth of 6 cm. (*a*) How large is the average decelerating force? (*b*) How long does it take to stop the bullet?

32 A 100-kg ballplayer is running at 8.0 m/s when he is stopped in a distance of 1.50 m by an opposing player. (*a*) How large is the average force that the opposing player must exert to stop the runner? (*b*) How long does it take to stop the player?

33 The brakes on a 900-kg car are accidentally released when the car is on a hill. How fast will the car be moving when it reaches a point 12.0 m lower than its starting point, assuming an average friction force of 800 N retards the motion? Assume the road distance to be 30 m.

34 A 2000-kg car is traveling at 20 m/s up a hill when the motor stops. (*a*) If the car is a vertical distance of 8 m from the top of the hill at that instant, will the car be able to reach the top? (*b*) How far below the top of the hill could the car be and still reach the top? Ignore friction.

35 A 0.70-g bead slides along a wire, as shown in Fig. P4.1. If it starts from rest at point *A* and the friction forces are negligible, how fast will it be going at points *B*, *C*, and *D*?

36 Suppose the bead in Prob. 35 has an initial velocity of 2.0 m/s toward the right at point *A*. What will its speed be at point *D* if friction forces are negligible?

37* In Fig. P4.1 the distance along the wire from *A* to *D* is 400 cm. If the 0.70-g bead starts from rest at *A* and finally stops just as it reaches *D*, how large is the average friction force retarding its motion?

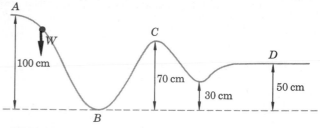

FIGURE P4.1

38* In Fig. P4.1, the 0.70-g bead starts from rest at point *C* and has a speed of 0.50 m/s when it reaches *D*, a distance of 2.0 m from *C* along the wire. How large was the average friction force retarding its motion?

39* A body of weight *W* is pulled up a frictionless incline of angle θ with speed *v*. Show that the work done on the body in time *t* is given by

$$\text{Work} = Wvt \sin \theta$$

40* A car of mass *m* rolls from rest down a hill of height *h* and length *L*. Show that when the car reaches the bottom of the hill, its speed is

$$v = \sqrt{2gh - \frac{2Lf}{m}}$$

where *f* is the average friction force retarding the motion.

41 A ball of mass *M* is suspended as a pendulum bob from a cord 4.0 m long. The cord is pulled to one side by a force on the mass until it makes an angle of 70° with the vertical, and the system is then released. With what speed is the mass moving as it passes directly underneath the point of suspension? (Ignore air friction.)

42 For the pendulum described in Prob. 41, how fast is the ball moving when the cord makes an angle of 20° with the vertical?

43* The pendulum of length *L* shown in Fig. P4.2 is released from point *A*. As it swings down, the string strikes the peg at *B*, and the ball swings through point *C*. (*a*) How fast is the ball moving as it passes through *C*? (*b*) If we neglect friction, the ball will approach a limiting speed as the string winds up on the pin. What is that speed? Assume the string does no work on the pendulum ball.

44* The masses in Fig. P4.3 are released from the position shown. (*a*) Find an expression for the speed

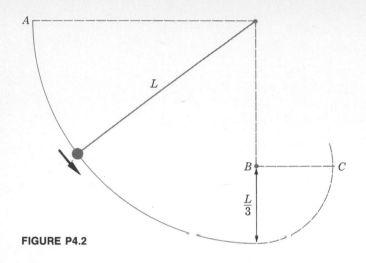

A

L

B ---- C

$\frac{L}{3}$

FIGURE P4.2

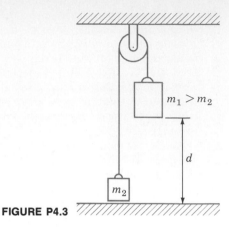

$m_1 > m_2$

d

m_2

FIGURE P4.3

of either mass just before m_1 strikes the floor. Ignore the mass and friction of the pulley. (b) Repeat if m_1 has a downward velocity v_0 at the instant shown in the figure.

45** A 3-kg block starts to slide up a 20° incline with an initial speed of 200 cm/s. It stops after sliding 37 cm up the incline, and then it slides back down. Assume the magnitude of the friction force impeding its motion to be constant. (a) How large is the friction force? (b) What is the block's speed as it reaches the bottom?

46** A man performs the following experiment. Using his 900-kg car on level ground, he finds that the car will accelerate from rest to 15.0 m/s in 10 s. The car will coast to rest from 15.0 m/s in 500 m. Compute the average horsepower delivered by the car. (Note that the friction work done in stopping is not the same as the friction work done in starting. Why? Assume instead that the average friction force is the same in the two cases.)

47 A 600-lb object is to be lifted by a pulley system using a force of 40 lb. The machine found suitable for this purpose lifts the load $\frac{1}{2}$ ft when the applied force moves 10 ft. Find the (a) IMA, (b) AMA, and (c) efficiency of the machine.

48 Estimating an efficiency of 75 percent, design the simplest pulley system capable of lifting 200 N by application of a 40-N force.

49 A force of 60 N is applied to a wheel-and-axle device in order to lift a 500-N load. If the radius of

the wheel is 25 cm and that of the axle is 2.0 cm, and if the diameter of the rope wound on the axle is 0.50 cm, find the (a) AMA, (b) IMA, and (c) efficiency of the system.

50 For a particular type of car jack, the operator moves her hand, i.e., the input force, through a distance of 60 cm for every 1 cm the load is lifted. (a) What is the IMA of the jack? (b) On the assumption of a 20 percent efficiency, how large an applied force is needed to lift 4000 N?

51 An electric motor is labeled as a 0.55-kW motor. On the assumption that it is 80 percent efficient, how many horsepower can it deliver?

52* A $\frac{1}{4}$-hp motor has an 8.0-cm-diameter pulley attached to its shaft. If the shaft rotates at 1800 revolutions per minute (rev/min), how large a load is the belt running on the pulley capable of pulling? Assume that the motor is 80 percent efficient and that the power input to the motor is $\frac{1}{4}$ hp.

53* A certain 50-W motor runs with a shaft speed of 1800 rev/min. By use of reducing gears the final, or output, shaft rotates at 18 rev/min. (a) If the machine is 30 percent efficient, with what force can the machine pull the belt on a 3.0-cm-radius pulley at the output shaft? (b) If the gear system on the motor were reversed so that the output shaft rotated at 180,000 rev/min, what force would be available to pull the belt on the same pulley? Assume the power input to the motor is 50 W.

5
Momentum and the Pressure of Gases

The law of conservation of energy, with which we dealt in the last chapter, is not the only conservation law known to physics. A second example, the law of conservation of momentum, is introduced in this chapter. It has far-reaching consequences. We use it in deriving a relation for the pressure of an ideal gas. Both the gas-pressure relation and the momentum-conservation law will be of importance to us in the following chapters.

5.1 Concept of Linear Momentum

An experience common to us all is that a moving object possesses a quality which causes it to exert a force on anyone who tries to stop it. The faster the object is traveling, the harder it is to stop. In addition, the more massive the object is, the more difficulty we have in stopping it. For example, an automobile moving at 2 m/s is much less easily stopped than a tricycle traveling at the same speed.

Newton called this quality of a moving body the *motion* of the body. Today, we term it the **linear momentum** of the body and define it by the relation

Definition

$$\text{Linear momentum} = m\mathbf{v} = \mathbf{p} \qquad (5.1)$$

Linear Momentum

In this expression, **v** is the vector velocity of the mass m. *Linear momentum is, therefore, a vector quantity* and has the direction of the motion, i.e., of the velocity. It is customary to use **p** to represent linear momentum.

Obviously, the momentum of a body is the result of forces which accelerated the body from rest to velocity v. Similarly, if we are to slow down the body, i.e., cause it to lose momentum, we must apply a retarding force to it. A large retarding force will cause the body to lose more momentum in a given length of time than a small retarding force will. We now wish to find the exact mathematical relation between force and change of momentum.

5.2 Newton's Second Law, Restated

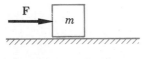

FIGURE 5.1
The net applied force **F** increases the momentum of the mass. Momentum has direction, and the increase will be in the direction of **F**.

Consider the acceleration of a body by a constant force **F** as shown in Fig. 5.1. The force is the net unbalanced force on the body and will cause a uniform acceleration in the direction of the force. Newton's second law gives

$$\mathbf{a} = \frac{\mathbf{F}}{m}$$

If the body had an original velocity $\mathbf{v}_0$ and the force acts for a time t, our motion equations yield

$$\mathbf{v} = \mathbf{v}_0 + \mathbf{a}t$$

Replacing **a** by $\mathbf{F}/m$ and rearranging leads to

$$\mathbf{F}t = m\mathbf{v} - m\mathbf{v}_0 \tag{5.2}$$

We see from this that the change in linear momentum of the object, $m\mathbf{v} - m\mathbf{v}_0$, is equal to the product of the applied force and the time for which the force was applied. This is actually closer to the way Newton wrote his second law than the form $F = ma$ which we previously gave. In some respects it is more general than the $F = ma$ form, since it can easily be extended to account for changes in mass as the body accelerates. For example, as a rocket accelerates, it loses mass because its fuel is being used up and ejected to provide thrust. We shall also see in Chap. 25 that the mass of an object increases at high speeds. If the mass changes with velocity, Eq. (5.2) should be written as

$$\mathbf{F}t = m\mathbf{v} - m_0\mathbf{v}_0 = \Delta\mathbf{p} \tag{5.3}$$

where $\Delta\mathbf{p}$ is the change in linear momentum. Notice that *the direction of the change in momentum is in the direction of the force. The units of momentum are the product of a force and a time unit* or, fully equivalent, a mass and a velocity unit. Examples are the newton-second or kilogram-meter per second.

Very often we wish to apply this concept of momentum to cases where the applied force is not constant. For example, when a bat hits a baseball, the force certainly varies from instant to instant during the collision. In such cases, we are usually not much interested in anything but an average force. We therefore define an average force by Eq. (5.2) or (5.3) as the steady force required to cause the observed change in momentum. Generally, the force obtained by application of Eq. (5.2) or (5.3) to collision processes will be an average force defined in this sense.

Since usually neither the time of collision nor the force is well known, it proves convenient to deal with their product Ft instead of either quantity alone. *The quantity Ft is called the* **impulse** *of the force.*

Definition

Illustration 5.1   A 1500-kg car traveling at a speed of 20 m/s reduces its speed to 15 m/s in 3.0 s. How large was the *average* retarding force?

Reasoning   From Eq. (5.2) or (5.3) we have, after dropping the vector notation because all the motion is along a straight line,

$$Ft = mv - mv_0$$
$$F(3.0 \text{ s}) = (1500 \text{ kg})(15 \text{ m/s}) - (1500 \text{ kg})(20 \text{ m/s})$$

From this we find $F = -2500$ N. The negative sign indicates that the force is opposite in direction to the direction of motion.

Illustration 5.2   An electron traveling at a speed of 2.6×10^8 m/s has a mass of 18.0×10^{-31} kg, which is about twice the rest mass of the electron, 9.1×10^{-31} kg. How large an average force F is needed to accelerate the electron to this speed from an initial speed of 1.0×10^8 m/s in 1.0×10^{-7} s? The mass of the electron at this slower speed is 9.5×10^{-31} kg. (We discuss the change in mass of particles moving with high speeds in Chap. 25.)

Reasoning   From Eq. (5.3) we have

$$Ft = mv - m_0 v_0$$
$$F(10^{-7} \text{ s}) = (18 \times 10^{-31} \text{ kg})(2.6 \times 10^8 \text{ m/s})$$
$$-(9.5 \times 10^{-31} \text{ kg})(1 \times 10^8 \text{ m/s})$$

or $F \approx (47 \times 10^{-16} - 9.5 \times 10^{-16}) \text{ N} \approx 37 \times 10^{-16} \text{ N}$

5.3 Conservation of Linear Momentum

Consider the collision of two particles shown in Fig. 5.2. The particles might be balls, spherical molecules, or something similar. Because we are assuming that no other particles exert forces on the two shown, Newton's third law

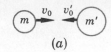

(a)

$F = -F'$

(b)

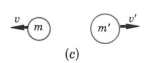

(c)

FIGURE 5.2

When two bodies collide, the forces on each are equal but oppositely directed. Momentum is conserved in a collision.

tells us that if there is a force on one of the particles shown, then there must be an equal but opposite reaction force on the other particle. This fact leads to an interesting and valuable conservation law.

Let us compute the change in momentum of the particle on the left in Fig. 5.2 as a result of the collision. From Eq. (5.2) or (5.3) we have for the *average* force

$$\mathbf{F}t = m\mathbf{v} - m\mathbf{v}_0$$

Similarly, for the particle on the right

$$\mathbf{F}'t = m'\mathbf{v}' - m'\mathbf{v}_0'$$

If we add these two expressions, we obtain

$$(\mathbf{F} + \mathbf{F}')(t) = (m\mathbf{v} - m\mathbf{v}_0) + (m'\mathbf{v}' - m'\mathbf{v}_0')$$

But since the vector force $\mathbf{F}$, the action force, is equal and opposite to the reaction force $\mathbf{F}'$, we have $\mathbf{F} = -\mathbf{F}'$ and the left side of this equation is zero. Hence

$$0 = (m\mathbf{v} - m\mathbf{v}_0) + (m'\mathbf{v}' - m'\mathbf{v}_0')$$

In words this says

0 = change in momentum of 1st ball + change in momentum of 2d ball

Consequently, the total change in momentum of the isolated two-particle system was zero.

We can extend this line of reasoning to much more complicated systems. To do so, we define what is called an isolated system. *An* **isolated system** *is a group of objects on which the net resultant force from outside is zero.* For such a group (or system) of objects, if one object within the group experiences a force, there must exist an equal but opposite reaction force on some other object in the group. As a result, the change in momentum of the group of objects as a whole is always zero.

These considerations are summarized in the **law of conservation of linear momentum,** stated as follows:

Law of Conservation of Linear Momentum

The total linear momentum of an isolated system of objects is a constant.

Even if the objects are not completely isolated, the law is often useful. For example, in a collision of two cars, the skidding of the wheels along the pavement causes external forces to act on the two-car system. Even so, the forces of one car on the other at the instant of collision are often extremely large, much larger than the skidding forces on the road. Therefore, the large changes in momentum that occur at the instant of collision are almost all the

Definition

result of the force of one car on the other. As a result, the law of conservation of momentum can still be applied to the two-car system at the instant of collision even though the system is not isolated.

In applying the conservation law, we must notice that the momentum of a body is a vector. To illustrate the importance of this, refer to Fig. 5.3. If we take the x direction to be positive, the total momentum of the system before collision (Fig. 5.3a) is

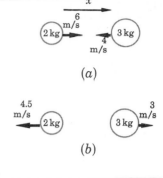

(a)

$$\text{Momentum before} = mv_0 + m'v_0'$$
$$= (2 \text{ kg})(6 \text{ m/s}) + (3 \text{ kg})(-4 \text{ m/s})$$
$$= 12 - 12 = 0$$

where the value for v_0' is taken negative because v_0' is in the negative x direction.

(b)

Notice that even though each of the individual bodies had momentum prior to the collision, the system as a whole had zero momentum. This is, of course, a very special case, chosen because it emphasizes so dramatically the fact that momentum is a vector. However, this particular case of zero total momentum is interesting in several other respects.

What must be true after the collision? The law of conservation of momentum tells us that the momentum of this isolated system will not be changed by the collision. Hence, in this case, the momentum after collision must still be zero. One possible way in which this could be achieved is shown in part b of Fig. 5.3. Notice that the momentum of each body has a magnitude of $9 \text{ kg} \cdot \text{m/s}$, one being positive, the other negative. This is definitely a possible solution for the problem, since momentum is conserved. We have the right, though, to ask whether this is the only possible solution to the problem.

It is a simple matter to show that the solution of Fig. 5.3b is not the correct solution in a certain case. Suppose that one of the bodies had a wad of chewing gum stuck on the side where the collision occurred. If the gum were sticky enough, the two bodies would remain stuck together after the collision. (If you don't like this way of fastening the bodies together, we could use magnets on the two bodies to hold them together after the collision. You can perhaps think of other ways this could be accomplished.) What can these bodies do if they stick together?

The law of conservation of momentum allows only one answer in this case. Since before collision the momentum of the system was zero, it must still be zero after collision. But now, since the bodies are stuck together, they must move as a unit, and their velocities will be in the same direction. Hence, there is no possibility at all that their momenta can cancel each other after the collision. We therefore conclude that the only way the total momentum can be zero after collision, if the bodies stick together, is for the bodies to be at rest. Therefore, in this case, the moving bodies will collide, stick together, and stop dead.

Of course, the above is a very special case. It is one of the simplest cases we can imagine, however. But even here, we are unable to give a completely definitive answer to the question: What happens when two objects with

FIGURE 5.3
The situations shown in (b) and (c) are physically possible results of the collision of the bodies shown in (a). In both instances the momentum is the same as before collision, namely, zero. Hence, momentum is conserved although KE is not.

equal and opposite momenta collide? The situation is even more intriguing if we notice that in neither solution given in Fig. 5.3 is the KE of the system the same after collision as before collision. It appears that *KE is usually not conserved in collisions.* We shall need to know how much of the KE was lost in the collision before we can find the correct answer to the question posed above.

5.4 Elastic and Inelastic Collisions

What happened to the KE of the two objects discussed in the last section when they collided and stuck together? Probably most of the energy was lost doing work on the gum between the bodies, causing it to flow under the applied force of the collision. This energy eventually appears as heat energy. In addition, probably some energy was given off in the form of sound waves. In any event, the KE was *entirely* transformed into other types of energy.

If the colliding objects had been tennis balls, only a portion of the KE would have been lost. Hence, the balls would have rebounded from each other. Even so, the energy lost to friction as the molecules in the ball move past one another as the balls distort, together with the energy lost in sound and by other means, will cause the KE to be less after impact than before impact. We term *a collision in which KE is lost an* **inelastic collision.**

Under certain special conditions, scarcely any energy is lost in the collision. *In the ideal case, when no KE is lost, the collision is said to be* **perfectly elastic.** An example is a hard ball dropped onto a hard, massive object, such as a marble floor, and rebounding to very nearly the same height as its starting point, with a negligible amount of energy lost in the collision with the floor.

These types of behavior in collision with a massive object are shown in Fig. 5.4. In fact, measuring the decrease in height on rebound is a common method for testing the resiliency of various materials. Obviously, if a ball dropped from height h rises to height h_2 on rebound, the difference in PE of the ball at these heights is the energy lost by the ball on collision. Even for a highly elastic object such as a solid rubber ball, the final height on rebound is only about 75 percent as large as the original height. (A Superball does better than this.)

Considerable heat is generated by friction among molecules when a ball distorts upon collision. Since this effect is essentially missing in a very hard material, it is not surprising that a small metal ball often loses less energy upon collision than a rubber ball. It therefore rebounds *better* than a rubber ball.

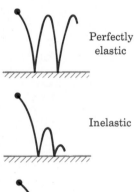

Perfectly
elastic

Inelastic

Inelastic

FIGURE 5.4
In a perfectly elastic
collision, KE is conserved.

Illustration 5.3 A 30,000 kg truck traveling at 10.0 m/s collides with a 1700-kg car traveling at 25 m/s in the opposite direction. If they stick together after the collision, how fast and in what direction will they be moving? (See Fig. 5.5.)

Reasoning Calling the x direction positive, we apply the law of conservation of momentum (KE is obviously not conserved in this collision). Let v be their combined speed after collision.

$$\text{Momentum before} = \text{momentum after}$$
$$(m_1 v_{01})_{\text{truck}} + (m_2 v_{02})_{\text{car}} = (m_1 + m_2)v$$
$$(30{,}000 \text{ kg})(10.0 \text{ m/s}) + (1700 \text{ kg})(-25 \text{ m/s}) = (31{,}700 \text{ kg})(v)$$

Notice that the car's velocity is negative. Solving for v gives $v = 8.1$ m/s. The positive sign for v indicates that the final motion is in the positive x direction, i.e., in the direction in which the truck was moving.

Notice that the truck was slowed down only slightly, while the direction of movement of the car was actually reversed. It is instructive to estimate the average force on the driver of the car during the collision, but this is left as an exercise. (If you wish to make this estimate, note that the impulse exerted on the driver equals the driver's momentum change. You will also need to estimate the time taken for the impact. One way to do this is to find the time it takes for the truck to move a distance about equal to the hood length of the car. Why?)

Illustration 5.4 Figure 5.6a shows an x-ray photo of a pistol just after the bullet has been fired. The hot combustion gases from the exploded gunpowder are accelerating the projectile part of the bullet down the barrel. If the masses of the gun and projectile are M and m, respectively, and the exit velocity of the projectile is v_b, find the recoil velocity of the gun v_g.

Reasoning If you look carefully at the figure, you can see a hand holding the gun. But the external force it exerts on the gun system is small in comparison to the internal forces exerted by the exploding powder.

(a) Before

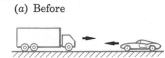

(b) After
$v = v'$

FIGURE 5.5
The collision of the truck and the car is inelastic in this case. Momentum must still be conserved, however.

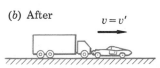

FIGURE 5.6
Before the gun was fired, its momentum was zero. Hence the sum of the momenta must still be zero after it is fired. (The photo was taken using an x-ray flash system.) (*Courtesy of Hewlett-Packard Co.*)

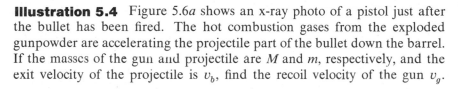

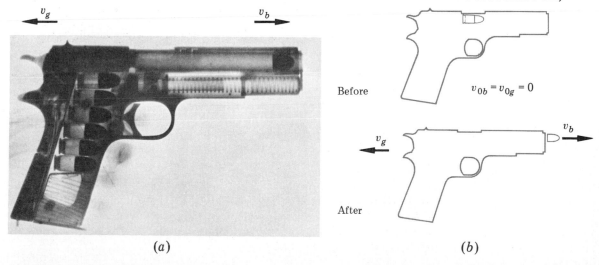

Before

$v_{0b} = v_{0g} = 0$

v_g

v_b

After

(a)

(b)

Therefore, for the instant of the explosion, we can assume that the gun is isolated and that momentum is conserved. The situation is shown in Fig. 5.6*b*. We have

$$\text{Momentum just before} = \text{momentum just after}$$
$$mv_{0b} + Mv_{0g} = mv_b + Mv_g$$

Since $v_{0b} = v_{0g} = 0$, this becomes

$$0 = mv_b + Mv_g$$

Therefore the recoil velocity of the gun is

$$v_q = -\frac{m}{M}v_b$$

Notice that the more massive the gun, the less the recoil will be.

Illustration 5.5 In the previous illustration suppose the ratio $m/M = 1/150$ and $v_b = 200$ m/s. Suppose, further, that the gun recoils 2.5 cm as the hand holding it brings it to rest. What average force does the gun exert on the hand? Take the mass of the gun to be 1.25 kg.

Reasoning After the discharge, the gun has KE given by

$$\text{Gun's KE} = \tfrac{1}{2}Mv_g^2 = \tfrac{1}{2}M\left(\frac{m}{M}\right)^2 v_b^2 = 0.889\,M \quad \text{joules}$$
$$= 1.11\,\text{J}$$

This energy is lost doing work against the hand which holds it. Taking the average restraining force exerted by the hand to be F and the recoil distance to be 0.025 m, we have

$$\text{Work done} = \text{gun's KE}$$
$$(F)(0.025\text{ m}) = 1.11\text{ J}$$

from which

$$F = 44\text{ N}$$

Can you list some of the factors on which the kick of a rifle or shotgun depends?

Illustration 5.6 A 40-g ball traveling to the right at 30 cm/s collides head on with an 80-g ball that is at rest. If the collision is perfectly elastic, find the velocity of each ball after collision. (By "head on" we mean that all motion takes place on a straight line.)

Reasoning   During the collision, momentum is conserved. Hence, letting the velocities of the 40- and 80-g balls after collision be v and v', respectively, gives

$$\text{Momentum before} = \text{momentum after}$$
$$(0.040 \text{ kg})(0.30 \text{ m/s}) + 0 = (0.040 \text{ kg})v + (0.080 \text{ kg})v'$$
$$2v' + v = 0.30 \text{ m/s}$$

We need yet another equation since we have two unknowns, v and v'. Since the collision was perfectly elastic, KE was also conserved. Therefore

$$\text{KE before} = \text{KE after}$$
$$\tfrac{1}{2}(0.040 \text{ kg})(0.30 \text{ m/s})^2 + 0 = \tfrac{1}{2}(0.040 \text{ kg})v^2 + \tfrac{1}{2}(0.080 \text{ kg})v'^2$$

or
$$2v'^2 + v^2 = 0.090 \text{ m}^2/\text{s}^2$$

Solving for v in the momentum equation and substituting in the KE equation yields

$$6v'^2 - (1.20 \text{ m/s})v' = 0$$

from which $v' = 0$ and 0.20 m/s. Hence there are two possible answers for v', namely, 0 and 0.20 m/s. Substitution of these values in the first equation gives 0.3 m/s and -0.10 m/s for v.

The first of these answers implies that the first ball went right through the second. Since this is a physical impossibility, we discard this answer. We therefore find that after the collision the first ball is going to the left at 10 cm/s while the other ball is going to the right at 20 cm/s.

Illustration 5.7   A 10-g pellet of unknown speed is shot into a 2.000-kg block of wood suspended from the ceiling by a cord. The pellet hits the block and becomes lodged in it. After the collision, the block and pellet swing to a height 30 cm above the original position (see Fig. 5.7a, b, and c). What was the speed of the pellet? (This device is called a **ballistic pendulum**.)

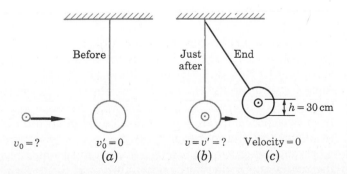

$v_0 = ?$ $\qquad$ $v_0' = 0$ $\qquad$ $v = v' = ?$ $\quad$ Velocity $= 0$

(a) $\qquad\qquad$ (b) $\qquad\qquad$ (c)

FIGURE 5.7
The momentum is the same in (a) and (b) but not in (c). KE is changed to PE in going from (b) to (c).

Reasoning *KE is not conserved on impact*. However, *after impact,* the KE of the pellet plus that of the block is transformed into PE. Hence, in going from Fig. 5.7*b* to *c* we can write (remember, this is all *after* the collision)

$$\text{KE at bottom} = \text{PE at top}$$
$$\tfrac{1}{2}(m_1 + m_2)v^2 = (m_1 + m_2)(gh)$$
$$\tfrac{1}{2}(2.000 + 0.010)(v^2) = (2.000 + 0.010)(9.8)(0.30)$$
$$v \approx 2.4 \text{ m/s}$$

This is the speed of the block and pellet just after collision. You should supply the units in the above equation and verify the units given for v.

Now in going from part *a* to part *b* in Fig. 5.7, a collision occurred. Momentum was conserved, but KE was not. We have, writing v_0 for the original velocity of the pellet,

$$\text{Momentum before} = \text{momentum after}$$
$$(0.010 \text{ kg})(v_0) + 0 = (2.000 + 0.010)(2.4) \text{ kg} \cdot \text{m/s}$$
$$v_0 = 480 \text{ m/s}$$

Hence, the original speed of the pellet was 480 m/s.

5.5 Rocket and Jet Propulsion

One of the most spectacular examples of the use of the law of momentum conservation occurs in the jet propulsion of rockets, spaceships, and jet airplanes. Operation of all these devices depends on the fact that recoil occurs when an object shoots part of its mass out from itself. All these devices resemble a gun or a cannon which recoils when it shoots a projectile.

In jet engines and rockets, fuel is burned, and in the process very hot gases are formed. The swiftly moving gas molecules are shot out the rear of the jet or rocket engine, much as a stream of bullets would be shot from a fantastically fast repeating gun. Like the gun recoiling, the rocket and aircraft are made to recoil in the direction opposite to the motion of the gas. Since the gas molecules were given momentum in the direction toward the rear, the rocket must acquire an equal momentum in the opposite direction (forward) because momentum is conserved.

A close examination of this kind of propulsion system shows that the interior of the engine pushes on the hot gas molecules in such a way that they are shot preferentially in a rearward direction. But in the process, according to Newton's law of action and reaction, the molecules exert a force forward on the engine, thereby throwing the rocket forward. Both these forces occur *within the engine itself.* No force is exerted on the craft from outside. The craft is not propelled by interaction of the expelled hot gases with the atmosphere, and, in fact, a rocket operates best in outer space, where there is no air. Air exerts a friction force which retards the motion of the rocket and is therefore undesirable.

Illustration 5.8 A Centaur rocket shoots hot gas from its engine at a rate of 1300 kg/s. The speed of the molecules is 50,000 m/s relative to the rocket. How large a forward push (or **thrust**) is given to the rocket by the exiting gases?

Reasoning The impulse exerted on the gas thrown out each second will be

$$Ft = mv - mv_0$$

If we take $t = 1$ s, m is the mass of gas expelled in 1 s (1300 kg), $v = 50,000$ m/s, and $v_0 = 0$. Putting in the values gives $F = 65,000,000$ N, or about 15 million pounds. This is the action force needed to expel the hot gases. An equal and opposite reaction force exerts a forward thrust on the rocket.

5.6 Momentum Components

Because momentum is a vector, we can speak of its components. For example, the object shown in Fig. 5.8 has a momentum **p,** as shown. But the momentum vector is composed of the three components labeled $p_x, p_y,$ and p_z, and the vector sum of these components is fully equivalent to the original vector **p**. Therefore we can use the components of **p** in place of **p** itself for discussion purposes.

Similarly, if a system of objects has a total momentum **p,** then the total momentum can be replaced by its components. If the system is isolated, the law of conservation of momentum tells us there will be no change in both the total momentum and its components. We can therefore state that *the components of the total momentum are conserved for an isolated system.* Let us see how we can make use of this fact.

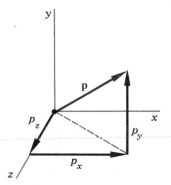

FIGURE 5.8
The momentum vector can be replaced by its components.

Illustration 5.9 In nuclear reactors, uranium nuclei undergo fission (explode) and thereby produce energy. Suppose a nucleus of mass m explodes into three equal-mass pieces, as shown in Fig. 5.9. The original nucleus was at rest, while two of the pieces have the velocities shown. Find the velocity components for the third piece.

Reasoning The speeds of two of the fragments are v_0, and we wish to find the velocity of the third piece. To do so, we call the unknown velocity **v,** and its components are v_x, v_y, and v_z. Applying the law of momentum conservation in component form and recognizing that the original nucleus had zero momentum, we have

(Momentum before)$_x$ = (momentum after)$_x$ →
$$0 = (\tfrac{1}{3}m)v_0 + (\tfrac{1}{3}m)(0.8v_0) + (\tfrac{1}{3}m)(v_x)$$

FIGURE 5.9
The original nucleus of
mass m was at rest. What
is the velocity of the third
fragment?

from which we find $v_x = -1.8v_0$,

$(\text{Momentum before})_y = (\text{momentum after})_y \rightarrow$
$$0 = 0 + (\tfrac{1}{3}m)(0.6v_0) + (\tfrac{1}{3}m)(v_y)$$

from which $v_y = -0.6v_0$, and

$(\text{Momentum before})_z = (\text{momentum after})_z \rightarrow 0 = 0 + 0 + (\tfrac{1}{3}m)(v_z)$

from which $v_z = 0$.

We therefore conclude that the third piece had a velocity with components $v_x = -1.8v_0$, $v_y = -0.6v_0$, and $v_z = 0$.

Illustration 5.10 A ball moving at 5.0 m/s collides with an equal-mass ball that is at rest as shown in Fig. 5.10a. After collision, one of the balls has the velocity shown in Fig. 5.10b. What are the velocity components for the other ball after collision?

Reasoning We know that the second ball will also be moving in the xy plane. Why?

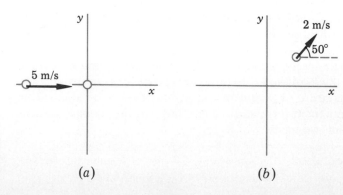

FIGURE 5.10
What was the second ball
doing after collision?

(a)   (b)

Writing the conservation of momentum for the x and y coordinates gives

x coordinate: $m(5.0 \text{ m/s}) + 0 = m(2.0 \cos 50° \text{ m/s}) + mv_x$

y coordinate: $0 + 0 = m(2.0 \sin 50° \text{ m/s}) + mv_y$

The first equation gives $v_x = 3.7$ m/s, while the second gives $v_y = -1.53$ m/s. From this we see that the second ball moves at an angle θ below the $+x$ axis where $\tan \theta = 1.53/3.7$, or $\theta = 22°$. Its velocity magnitude is $\sqrt{(1.53)^2 + (3.7)^2}$ m/s, or 4.0 m/s.

5.7 Pressure of an Ideal Gas

In this section we compute the pressure on the walls of a container resulting from the bombardment of the walls by the gas molecules. Qualitatively, the picture is very simple. In Fig. 5.11, the gas molecule shown will be bouncing around inside the container. Following the dotted path shown, it will bounce off the wall. Of course, it exerts a force on the wall during the time of collision. The equal and opposite reaction force of the wall on it is responsible for its momentum change.

Since there are such a large number of molecules in a gas at most reasonable pressures, there will be billions of collisions with the wall in a second or less. Hence, the force due to these collisions will appear about constant. Moreover, the average force will be perpendicular to the wall. *We define the perpendicular force per unit area of the wall to be the* **pressure** *of the gas.* Pressure is force per unit area. In symbols, the pressure P is

Definition

$$P = \frac{F}{A} \tag{5.4}$$

where F is the force on and perpendicular to the area A. The units of pressure are newtons per square meter, pounds per square inch, etc. The SI unit, the newton per square meter, is given the special name **pascal** (Pa), where 1 Pa $= 1$ N/m^2.

The exact computation of the pressure of a gas on the container wall is an extremely complicated procedure for any real gas. However, it turns out that even very simplified versions of the actual physical situation give nearly the correct answer for most real gases. It is common procedure in physics to carry through a grossly simplified computation for a physical situation first. As scientists grow in their understanding of the physical processes, they refine their computation, using more and more realistic pictures or models of the true physical situation. Many years may go by before the original, unrealistic model is successfully modified to conform more exactly to the actual process as evidenced by experiment. These delays are usually the result of the scientist's inability to devise a more realistic model which would be susceptible to mathematical analysis. In addition, the true physical situa-

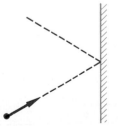

FIGURE 5.11
Huge numbers of gas molecules colliding with the wall in the manner shown give rise to the observed pressure of the gas.

Discovery of the Neutron and Neutrino

The conservation law for linear momentum has been used to learn much about the basic particles that exist in our universe. Two of the earliest uses of the law in this way resulted in the discovery of the neutron and the neutrino. Both these particles are uncharged. Although the neutron has a mass nearly equal to the mass of the proton (1.67×10^{-27} kg), the neutrino has such a small mass that we are still uncertain as to its exact magnitude. Its mass may even be zero. (The name *neutrino* means "tiny neutron" in Italian; it was named by the Italian physicist Enrico Fermi.)

Discovery of the free neutron first took place in 1930 when a very penetrating beam of particles was noticed to be emitted by radioactive beryllium atoms. The nature of these particles was first determined in 1932 by James Chadwick. Direct observation of the particles was not then possible because they are uncharged and are difficult to capture or detect. Instead, Chadwick let the particles collide with hydrogen atoms and with nitrogen atoms. The motions of these atoms can be measured, as we shall see in later chapters. Upon collision of a particle with an atom, the atom is given energy and momentum. Because such collisions are perfectly elastic, the law of conservation of KE equation as well as the law of conservation of momentum equation can be written for the collisions with hydrogen and nitrogen atoms. Because the energy and momentum of the hydrogen and nitrogen atoms could be measured, Chadwick had enough data to solve the energy and momentum equations for the mass of the incoming particle, the neutron. It was in this way that he found the neutron mass to be 1.67×10^{-27} kg.

The neutrino, on the other hand, is far more elusive than the neutron because of its lack of both charge and mass. A neutrino can pass through the earth with very little chance of being stopped. Its presence became known through examination of "beta decay," a process in which a radioactive atom emits an electron. Like a gun recoiling as it discharges a bullet, the atom should recoil as it throws out an electron. But careful measurements showed that when the atom and electron alone are considered, energy and momentum could not be conserved in the process. Hence it was proposed that another particle, then undetected, was emitted with the electron. By using the conservation laws, it was surmised that the particle must be uncharged, be nearly massless, and travel at a speed close to the speed of light. It was not until the 1950s that this elusive particle, the neutrino, was confirmed to exist. But scientists throughout the world accepted its existence much earlier based on their faith in the laws of momentum and energy conservation.

tion itself is often obscure, and the scientists must first conceive a model for the physical process which could give rise to the experimentally observed facts.

In the present case, the pressure of the common gases, even quite unrealistic models of the physical situation yield essentially correct results, i.e., results which agree with experiment. We shall find the result for the pressure of an ideal gas by using a very simple (and unrealistic) model. However, our result will be correct in all essentials.

Suppose, first, that we have a gas in a container, there being ν_0 (read as "nu subzero") molecules per unit volume. Suppose, further, that one-third of the molecules are traveling in each of the three directions x, y, and z. We take their speeds to be identical, v. Of the third of the molecules traveling in the x direction, half will be headed toward the $+x$ direction and half will be going in the $-x$ direction. Hence, one-sixth of the molecules will be traveling in the $+x$ direction, and each will have a speed v. We assume, further, that the molecules do not collide with one another.

Consider the situation shown in Fig. 5.12, where a particle is traveling in the $+x$ direction toward the wall of area A. The time taken for this particle to reach the wall, a distance s away, and hit it will be s/v. As shown in the figure, the width of the box is d. Therefore in time $t = d/v$, all the molecules in the box which were traveling in the $+x$ direction will hit the area A. Since the volume of the box, and therefore of the gas, is Ad, and since there are $\frac{1}{6}\nu_0$ molecules per unit volume traveling in the $+x$ direction, there will be a total of $(\frac{1}{6})(\nu_0)(Ad)$ molecules traveling in the $+x$ direction. All these will hit the right-hand wall of the box in time d/v.

If we assume the collisions with the walls to be perfectly elastic, when any single molecule hits the wall, its velocity will completely reverse. Hence, for each molecule hitting the wall

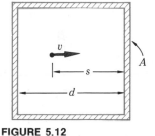

FIGURE 5.12
One-sixth of the molecules in the box are assumed to be traveling in the direction shown. They will all strike the wall in time d/v. We assume the number of molecules in unit volume to be ν_0.

$$\text{Change in momentum} = 2mv$$

where m is the mass of the molecule. That is, the momentum of the particle was changed by an amount mv when it was stopped by the wall, and then it was given an additional reverse momentum of mv, making a total change of $2mv$.

Since there will be $(\frac{1}{6})(\nu_0)(Ad)$ molecules hitting the wall in time $t = d/v$, the total momentum change caused by the wall in this time is

$$(2mv)(\tfrac{1}{6})(\nu_0)(Ad)$$

We know that the impulse Ft on the wall is equal to this change in momentum, where F is the average force exerted on the wall by the molecules.[1] Using the fact that $t = d/v$, we have

$$F\frac{d}{v} = (2mv)(\tfrac{1}{6})(\nu_0)(Ad)$$

---

[1] Notice that t, as used here, is *not* the time of impact for a single molecule. It is the time taken for a large number of molecules to strike the wall and, as used here, is given by d/v.

After simplification of this expression we have the following equation for the average pressure on the wall:

$$P = \frac{F}{A} = (\tfrac{1}{3})(\nu_0)(mv^2) \tag{5.5}$$

*Pressure Is Proportional
to KE of Molecule*

In terms of the kinetic energy of the molecule, this is

$$P = (\tfrac{2}{3})(\nu_0)(\tfrac{1}{2}mv^2) \tag{5.6}$$

Notice that the pressure of a gas is equal to two-thirds of the kinetic energy per unit volume, since ν_0 is the number of molecules in unit volume. We shall see in Chap. 9 that gas pressure can also be used as a measure of the temperature of a gas. Hence it appears that we shall find temperature and kinetic energy to be intimately related. In fact, we shall learn that a hot object is one whose constituent molecules possess a great deal of kinetic energy.

Although Eq. (5.5) was obtained by using a greatly oversimplified model, the result is general. *As long as a gas is far removed from conditions under which it will liquefy, Eq. (5.6) is found to be applicable, to good approximation. A gas which obeys this equation exactly would be an* **ideal gas,** *and Eq. (5.6) is one form of the* **ideal-gas law.** More is said about this in Chap. 9. It is fortuitous that the crude method of computation we have used (and which was used by the early workers in physics) leads to a result which agrees so well with experiment. In addition to the artificiality of the assumptions of equal speeds, perpendicular motions, and elastic collision, the assumption that the molecules do not collide with one another is far from true. *In air at ordinary pressures, a molecule collides with another after traveling a distance of about 10^{-5} cm.*

Illustration 5.11 *Standard atmospheric pressure is 1.01×10^5 Pa. At standard pressure and temperature (0°C), there are 2.7×10^{25} molecules in 1 m³ of volume. Find the average speed of the nitrogen molecules in air under standard conditions. One molecule of nitrogen has a mass of 4.7×10^{-26} kg.*

Reasoning We make use of Eq. (5.6). From the data we have $\nu_0 = 2.7 \times 10^{25}$ m⁻³, $m = 4.7 \times 10^{-26}$ kg, and $P = 1 \times 10^5$ N/m², all in the proper SI units. Hence, from Eq. (5.6)

$$1 \times 10^5 \text{ N/m}^2 = \tfrac{1}{3}(2.7 \times 10^{25} \text{ m}^{-3})(4.7 \times 10^{-26} \text{ kg})(v^2)$$

from which, remembering that $1 \text{ N} = 1 \text{ kg} \cdot \text{m/s}^2$, we have

$$v \approx 490 \text{ m/s}$$

Illustration 5.12 Commercial jet aircraft often fly at altitudes in excess of 7000 m. At that altitude, air pressure is only about half as great as it is at sea level. Nearly all aircraft are pressurized to compensate for this. What effect would it have on the occupants of an aircraft at this altitude if the cabin were at the same pressure as the atmosphere?

Reasoning From Eq. (5.6) we know that P is proportional to v_0, where v_0 is the number of molecules per unit volume. At an altitude of 7000 m, the pressure (and v_0) is only one half as great as at sea level. Consequently, each time a person takes in a breath, only about half as many oxygen molecules enter the lungs as compared to at sea level. The oxygen supplied to the body is cut in half. To compensate for this, a person needs to breathe much faster than ordinary. Often, this in itself is very tiring. But if it is not done, the person will be extremely limited in body activity. This effect is often noticed by visitors to mountainous regions. The natives of regions at high altitude have undergone bodily adaptations, such as increased red-blood-cell-count, which allow them to function better with this smaller supply of oxygen.

Summary

The linear momentum of an object of mass m and velocity $\mathbf{v}$ is $m\mathbf{v}$. It is a vector quantity and is often represented by $\mathbf{p}$.

When an object is subjected to a force $\mathbf{F}$ for time t, an impulse $\mathbf{F}t$ has been applied to it. The impulse causes a change in momentum of the object. We have $\mathbf{F}t = m\mathbf{v} - m_0\mathbf{v}_0$. This is an alternate way for stating Newton's second law of motion.

A group of objects on which no resultant external force acts can be considered isolated as far as its translational motion is concerned. For an isolated system, the linear momentum of its constituent objects is constant. This is the law of conservation of linear momentum.

It is of particular value in analyzing collisions.

In a perfectly elastic collision, KE is conserved. More often, much of the KE is lost to heat and other forms of energy during the collision. Such a collision is said to be inelastic.

The pressure P on a surface of area A due to a force F perpendicular to the surface has magnitude $P = F/A$. A gas in a container exerts a pressure on the walls by virtue of the collisions of the gas molecules with the walls. The pressure due to an ideal gas containing v_0 molecules per unit volume is $P = (\frac{2}{3})(v_0)(\frac{1}{2}mv^2)$. In this expression v is an average speed for a molecule of mass m.

Minimum Learning Goals

Upon completion of this chapter you should be able to do the following:

1 Define linear momentum and compute it for an object provided sufficient data are given. Give examples showing how momentum depends on velocity and/or mass.

2 Find the change in momentum for an object subjected to a given average force for a stated time. Define impulse and show how it is related to linear momentum. Compute the impulse required to cause a given change in momentum.

3 State the law of conservation of linear momentum, being careful to point out the importance of the words *isolated system*. Explain why, under certain conditions, the law is useful even though the system is not completely isolated.

4 Analyze the collision of two objects which stick together on impact, i.e., situations similar to Illustration 5.3.

5 Analyze situations in which an object originally at rest explodes into two pieces or recoils, i.e., similar to Illustration 5.4.

6 Analyze situations where two objects move along a straight line, undergo a perfectly elastic collision, and then continue to move along the same straight line.

7 Explain the difference between an elastic and an inelastic collision. In your explanation, describe how a bounding ball would behave in each case. Give plausible reasons for the fact that KE is not conserved in most collisions.

8 Analyze situations in which an object strikes a stationary object and becomes embedded in it and the remaining KE is changed to GPE or lost doing friction work. The ballistic pendulum would be a typical situation.

9 Explain the principle of operation of rockets, jet engines, and similar devices propelled by means of a jet.

10 Define pressure and explain why a gas in a container exerts pressure on the container walls. In your explanation, make it clear why the pressure should depend on v, m, and v_0.

11 Given the relation among P, v_0, v, and m, be able to use it to solve simple problems such as that typified by Illustration 5.11.

Important Terms and Phrases

You should be able to define or explain each of the following:

Linear momentum: $\mathbf{p} = m\mathbf{v}$
Impulse: $\mathbf{F}t = m\mathbf{v} - m_0\mathbf{v}_0$
Isolated system
Conservation of linear momentum
Elastic and inelastic collisions

Recoil
Ballistic pendulum
Jet propulsion
Pressure: $P = F/A$
Ideal gas
Ideal-gas law: $P = (\frac{2}{3})(v_0)(\frac{1}{2}mv^2)$

Questions and Guesstimates

1 When a large cannon is fired, it recoils for some distance against a cushioning device. Why is it necessary to make the support so that it "gives" in this way?

2 A wad of gum is shot at a block of wood. In which case will the gum exert the larger impulse on the block, if it sticks to it or rebounds from it?

3 When a balloon filled with air is released so that the air escapes from it, the balloon shoots off into the air. Explain why this happens. Would it happen also if it were released in a vacuum?

4 Explain why a rocket is capable of accelerating even in outer space where there is no air against which it can push.

5 An inventor constructs a sailboat with a large electric fan mounted on it. He directs the fan at the sail and blows air at the sail, expecting thereby to move in the direction of this artificial wind. To his surprise, the boat moves slowly in the opposite direction. Can you tell him why it does so?

6 A ball dropped onto a hard floor has a downward momentum, and after it rebounds, its momentum is upward. Clearly the momentum of the ball is not conserved in the collision, even though the ball may rebound to the height from which it was dropped. Does this contradict the law of momentum conservation?

7 Reasoning from the impulse equation, explain why it is unwise to hold your legs rigidly straight when you jump to the ground from a wall or table. How is this related to the commonly held belief that a drunken person has less chance of being injured in a fall than one who is sober?

8 Explain, in terms of the impulse equation, the principle of operation of impact-absorbing car bumpers and similar impact-absorbing devices.

9 A baseball player has the following nightmare. He is accidentally locked in a railroad boxcar. Fortunately, he has his ball and bat along. To start the car moving, he stands at one end and bats the ball toward the other. The impulse exerted by the ball as it hits the end gives the car a forward motion. Since the ball always rebounds and rolls along the floor to him, the player repeats this process over and over. Eventually the car attains a very high speed, and the player is killed as the boxcar collides with another car sitting at rest on the track. Analyze this dream from a standpoint of the physics involved.

10 Explain how a Mexican jumping bean can cause itself to jump.

11 Contrive a device which, momentarily at least, can have kinetic energy but no momentum. Is it possible to design a device having momentum but no kinetic energy?

12 Two blocks of unequal mass are connected by a spring, with the whole system lying on an essentially frictionless table. The blocks are pushed together and tied together with a string so that the spring is compressed. If now the string is cut or burned, describe the motion of the blocks.

13 A 70-kg woman jumps from a roof 10 m above the ground. (*a*) Approximately what is her speed just before she strikes the ground? (*b*) She lands on her feet, but allows her legs to "give." About how long does it take her to come to rest? (*c*) About how large an average force does the ground exert on

her? (*d*) Why should you always let your legs "give" when you land from a jump? (E)

14 A 10-kg child falls from a window and is caught by a woman 15 m below. Estimate the force experienced by the child as he is caught. (E)

15 A hose squirts a horizontal beam of water against a window at a speed of 5 m/s and a rate of 30 g/s. About how large a force is exerted by it on the window? (E)

16 The wind is blowing against a sign with a speed of 30 m/s. If the mass in 1 m³ of air (the density) is 1.29 kg/m³, estimate how large a force is exerted on a unit area of the sign by the wind. (E)

17 Estimate the force exerted by a woman's head on her neck if her stationary car is hit from the rear by a loaded truck going 20 mi/h (9 m/s). Why does this type of accident often lead to the so-called whiplash injury? You will need to make an estimate of the mass of the woman's head and the time taken for her head to be accelerated. (E)

18 Suppose you lay your hand flat on a tabletop and then drop a 1.0-kg laboratory mass squarely on it from a height of 0.50 m. Estimate the average force exerted on your hand by the mass. Why is injury very likely in this case even though you can catch the mass easily when it is dropped from this height? (E)

Problems

1 (*a*) What is the linear momentum of an 800-kg car that is moving at 90 km/h? (*b*) Repeat for a 0.75-g fly moving at 30 cm/s. Express your answers in kilogram-meters per second.

2 Find the linear momentum of an 800-g object undergoing free fall after it has fallen 60 cm.

3 A mass m undergoes free fall. What is its linear momentum after it has fallen a distance h?

4 Show that the linear momentum p and KE of a mass m are related through $KE = p^2/(2m)$.

5 How large a force is needed to accelerate a 1200-kg car from rest to 50 m/s in 7.0 s? How large was the accelerating impulse?

6 A 950-kg car moving at 80 m/s is to be stopped in 3.0 s. How large a stopping force will be required? How large an impulse would be imparted to the car by the stopping force?

7 By use of the impulse equation, (*a*) determine how large an average force is required to stop a 1400-kg car in 5.0 s if the car's initial speed is 25 m/s.

(*b*) Assuming uniform deceleration, how far would the car go in this time?

8 A 1500-kg car going 20 m/s strikes a wall and stops in 3.0 m. Assuming uniform deceleration, how long did it take the car to stop? Use the impulse equation to find the stopping force which acted on it.

9 A 120-g ball moving at 18 m/s strikes a wall perpendicularly and rebounds straight back with this same speed. The center of the ball moves 0.27 cm farther toward the wall after touching it. Assuming that the deceleration produced by the wall is uniform, show that the time of contact with the wall is 2×0.00030 s. How large is the average force which the ball exerts on the wall?

10 A car of mass M and speed v strikes a tree head on and stops. (*a*) What is the magnitude of the impulse that the tree exerts on the car? (*b*) If the time taken to stop is t_s, how large an average force is exerted on the car by the tree? (*c*) Evaluate the

average force for a 2000-kg car with speed 5.0 m/s which stops in 0.40 s.

11 While waiting in his car at a stoplight, an 80-kg man and his car are suddenly accelerated to a speed of 5.0 m/s as the result of a rear-end collision. Assuming the time taken to be 0.30 s, find (a) the impulse on the man and (b) the average force exerted on him by the back of the seat of his car.

12* A 75-g egg is dropped from a height of 80 cm onto the hard floor. It crushes and stops in 0.020 s. (a) How large an impulse did the floor exert on the egg to stop its motion? (b) How large an average force did the floor exert on the egg? Neglect the force supplied by the floor to support the egg's weight. Can you show that the stopping time given is plausible?

13* A 250-g ball is thrown at a wall with a speed of 20 m/s, and it rebounds straight back with a speed of 14.0 m/s. (a) How large an impulse was exerted on the ball by the wall? (b) If the time of contact with the wall was 4.0×10^{-4} s, what average force did the wall exert on the ball? How could you estimate the contact time in an actual situation?

14* A 180-g ball is dropped onto a hard floor from a height of 120 cm. The ball rebounds to a height of 85 cm. (a) How large an impulse was imparted to the ball by the floor? (b) Assuming the contact time between the floor and ball was 2.0×10^{-4} s, how large an average net force was exerted on the ball?

15* An unlucky bystander finds himself in the center of a shoot-out between the good guys and bad guys. A 5.0-g bullet moving at 100 m/s strikes him and lodges in his shoulder. Assuming the bullet undergoes uniform deceleration and stops in 6.0 cm, find (a) the time taken to stop, (b) the impulse on the shoulder, and (c) the average force experienced by the man.

16** A stream of water from a hose is hitting a window. The window is vertical, the stream is horizontal, and the water stops when it hits. About 10 cm³ (that is, 10 g) of water with speed 2.0 m/s strikes the window each second. Find (a) the impulse on the window exerted in time t and (b) the force exerted by the stream on the window.

17 During a switching operation, a train car of mass M_1 coasting along a straight track with speed v strikes and couples to a car of mass M_2 sitting at rest. Find their speed after coupling.

18 While engaging in target practice, a woman shoots a 3.0-g bullet with horizontal velocity 250 m/s into a 5.0-kg watermelon sitting on top of a post. The

bullet lodges in the watermelon. With what speed does the watermelon fly off the post?

19 Two identical balls collide. Ball 1 is traveling to the right at 10 m/s, and ball 2 is standing still. Find the direction and magnitude of their velocity if they stick together after collision.

20 As shown in Fig. P5.1, a 20-g bullet with speed 5000 cm/s strikes a 7000-g block resting on a table. The bullet embeds in the block after collision. Find (a) the speed of the block after collision and (b) the friction force between the table and block if the block moves 1.5 m before stopping.

21 While coasting along a street at a constant velocity of 0.50 m/s, a 20-kg girl in a 5-kg wagon sees a vicious dog in front of her. She has with her only a 3.0-kg bag of sugar which she is bringing from the grocery, and she throws it at the dog with a forward velocity of 4.0 m/s relative to her original motion. How fast is she moving after she throws the bag of sugar?

22* Near the Fourth of July, a very foolish boy places a firecracker (of negligible mass) in an empty soup can (mass, 40 g) and plugs the end with a wooden block (mass, 200 g). After igniting the firecracker, he throws the can straight up, and it explodes at the top of its path. If the block shoots out with a speed of 3.0 m/s, how fast will the can be going?

23** A 500-g pistol lies at rest on an essentially frictionless table. It accidentally discharges and shoots a 10.0-g bullet parallel to the table. How far has the pistol moved by the time the bullet hits a wall 5.0 m away?

24* A railroad flatcar of mass M is coasting along a track at speed v when a large machine of mass m topples off a platform and falls straight down onto the car. How fast is the car moving after the machine comes to rest on it?

25* In Fig. P5.1 a horizontal force of 0.70 N is required to pull the 5000-g block across the table at constant speed. Find the speed of the 20-g bullet shown if the bullet embeds in the block and causes the block to slide 1.50 m before coming to rest.

26* A 2.0-kg block rests over a small hole on a table. A woman beneath the table shoots a 15.0-g bullet through the hole into the block, where it lodges. How fast was the bullet going if the block rises 1.30 m above the table?

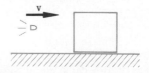

FIGURE P5.1

136

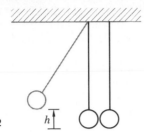

FIGURE P5.2

27* A 60-kg astronaut becomes separated in space from her spaceship. She is 15.0 m away from it and at rest relative to it. In an effort to get back, she throws a 500-g wrench with a speed of 8.0 m/s in a direction away from the ship. How long does it take her to get back to the ship?

28* A 2.0-kg melon is balanced on a bald man's head. His wife shoots a 50-g arrow at it with speed 30 m/s. The arrow passes through the melon and emerges at 18 m/s. Find the speed of the melon as it flies off the man's head.

29* Refer to Fig. P5.2. The pendulum on the left is pulled aside to the position shown. It is then released and allowed to collide with the other pendulum, which is at rest. (a) What is the speed of the ball on the left just before collision? After collision, the two stick together. (b) How high, in terms of h, does the combination swing? Assume the two balls have equal masses.

30* Refer to Fig. P5.2. The two pendulum balls have different masses; the ball on the left has a mass m_1. When the ball on the left is let swing from the height shown, it strikes and sticks to the ball on the right. The combination then swings to a height $h/3$. Find the mass m_2 of the ball on the right in terms of m_1.

31* In nuclear reactors, many fast neutrons are produced. To slow them, they are allowed to collide with other particles of comparable mass. Suppose a neutron with speed v collides head on with a proton which is at rest. The masses of these two particles are nearly identical. If the collision is perfectly elastic, what is the speed of the neutron after the collision?

32* In Fig. P5.2 the two masses are identical. The mass on the left is displaced as shown and released. It collides perfectly elastically with the other mass. (a) How high, in terms of h, does the mass on the right swing after collision? (b) The mass on the left?

33* A neutron ($m = 1.67 \times 10^{-27}$ kg) moving with speed v_0 strikes a stationary particle of unknown

mass and rebounds elastically straight back along its original path with speed $v_0/3$. What is the mass of the particle it struck?

34* A neutron ($m = 1.67 \times 10^{-27}$ kg) moving with speed v_0 strikes a gold nucleus ($m = 197 \times 1.67 \times 10^{-27}$ kg) and rebounds straight back in a perfectly elastic collision. Find the speed of the gold nucleus after the collision if it is free to move and was originally at rest.

35* According to a police report, a car was sitting at rest waiting for a stoplight when it was hit from the rear by an identical car. Both cars had their brakes on, and from their skid marks, it is surmised that they skidded together about 8.0 m in the original direction of travel before coming to rest. Assuming a stopping force of about 0.7 times the combined weights of the cars (that is, $\mu = 0.7$), what must have been the approximate speed of the oncoming car?

36 In Fig. P5.2, both masses are displaced to a height h, one to the left, as shown, and the other to the right. They are released simultaneously and undergo a perfectly elastic collision at the bottom. How high does each swing after collision? Both masses are identical.

37** As shown in Fig. P5.2, the mass on the left is pulled aside and released. Its speed at the bottom is v_0. It then collides with the mass on the right in a perfectly elastic collision. Find the velocities of the two masses just after collision if the mass on the left is 3 times as large as the mass on the right.

38** Repeat Prob. 37 if the mass on the left is one-third the mass on the right.

39** An electron ($m = 9 \times 10^{-31}$ kg) traveling at 2.0×10^7 m/s undergoes a head-on collision with a hydrogen atom ($m = 1.67 \times 10^{-27}$ kg) which is initially at rest. Assuming the collision to be perfectly elastic and the motion to be along a straight line, find the final velocity of the hydrogen atom.

40* The device shown in Fig. P5.3 is sold as a novelty. All the masses are identical. When one of the masses is pulled back, as shown, and released, the mass at the opposite end flies out while all the oth-

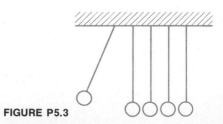

FIGURE P5.3

ers retain their position. (*a*) Show that this result is to be expected if the collisions are perfectly elastic. (*b*) What will happen if two masses are pulled back rather than one?

41* A particle of mass *m* is at rest at the origin of coordinates when it suddenly explodes into three equal pieces. One piece shoots out along the *x* axis with speed v_0; the other shoots out along the *y* axis with speed v_0. What does the third fragment do?

42* Two equal-mass cars are moving along perpendicular streets with equal speeds, v_0. At the corner made by the intersection of the two streets, the cars collide and stick together. What are the direction and speed of the motion of the wreckage just after the collision?

43** A ball of mass m_0 is moving with speed v_0 along the +*x* axis toward the origin of coordinates. It strikes a glancing blow on a ball of mass $m_0/3$ at rest at the origin. After collision, the incoming ball is moving toward the left with speed $v_0/2$ at an angle 37° above the −*x* axis. Find the speed and direction of motion of the other ball. There is something strange about this collision. What is it?

44** Repeat Prob. 43 if the incoming ball is reflected back to the right at an angle of 37° to the +*x* axis

with a speed of $v_0/5$. There is something strange about this collision. What is it?

45 Suppose that one-half the molecules of a gas are removed from a container without changing the average KE per molecule within the container. By what factor does the gas pressure change?

46 A number of molecules are contained at atmospheric pressure in a chamber closed by a piston device, as shown in Fig. P5.4. If the piston is pushed down in such a way that the KE of the molecules is not changed appreciably, what is the pressure in the container when the volume is only one-fourth its original size?

47* Air in a certain room has a pressure of 0.98×10^5 Pa, and the mass of 1 m³ of the air is 1.28 kg/m³. Find the average speed of the air molecules from these data.

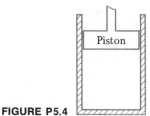

FIGURE P5.4

6
Motion In a Plane: Rotational Motion

Previously we were concerned primarily with motion along a straight line. We now wish to extend our discussion to two-dimensional motion, motion confined to a plane. Two types of planar motion are of great importance, motion in a circle and projectile motion. Both are described in this chapter.

6.1 Angular Distance θ

In order to describe the motion of an object along a line, we needed a coordinate along the line, and often we took it to be the x coordinate. Now we are discussing the motion of an object on a circular path, and we need a coordinate to describe its position on the circle. To see how we find one, let us refer to Fig. 6.1.

The angle θ shown in the figure is measured from a reference line or point, as indicated. In Fig. 6.1a we see a wheel with a ball attached to it. The wheel is then rotated through an angle θ to the new position, shown in Fig. 6.1b. There are three common ways in which we express the angle θ.

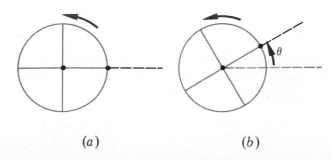

(a)   (b)

FIGURE 6.1
The angle θ describes the angular distance through which the wheel has turned.

One way is to measure θ in degrees. If the wheel turns through a full circle, θ is 360°. This way of measuring how far the wheel has rotated is well known to you. The second way for measuring θ is also common in everyday life. When a wheel turns through a full circle, we say that $\theta = 1$ rev (revolution). Clearly,

$$1 \text{ rev} = 360°$$

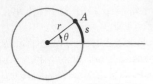

FIGURE 6.2

In radian measure, $\theta = s/r$.

The third method for measuring θ is of particular value in science. This way of measuring angles, called **radian measure,** is defined in terms of a ratio of two lengths. If we refer to Fig. 6.2, we see that the point A on the wheel has moved through an arc length s as the wheel turned through the angle θ. Calling the radius of the wheel r, we define

Angles in Radians

$$\theta \text{ in radian measure} = \frac{\text{arc length}}{\text{radius}}$$

$$\theta = \frac{s}{r} \quad \text{radians} \tag{6.1}$$

Notice that θ is simply a ratio. In taking the ratio, we divide a length by a length, and so the result, the angle, is dimensionless. Even so, we assign a unit to the angle despite the fact that we cannot expect the unit to follow through equations as a dimension-specifying unit would do. We say, for example, "the angle is π radians (rad) or 180° or $\frac{1}{2}$ rev" to make it clear how we are measuring angles. Since $s = 2\pi r$ for one full turn of the wheel, we see that

$$1 \text{ rev} = 360° = 2\pi \text{ rad}$$

Illustration 6.1 A certain angle is 70°. Find its equivalent in radians and in revolutions.

Reasoning The relations among the various angular measures tell us that

$$70° = (70 \text{ deg}) \left(\frac{2\pi \text{ rad}}{360 \text{ deg}} \right) = 1.22 \text{ rad}$$

and

$$70° = (70 \text{ deg}) \left(\frac{1 \text{ rev}}{360 \text{ deg}} \right) = 0.194 \text{ rev}$$

Illustration 6.2 As shown in Fig. 6.3, a string is wound around a wheel that has a radius of 20 cm. Through how large an angle does the wheel turn to unwind 30 cm of string?

Reasoning If you examine Fig. 6.3, you can convince yourself that the following is true: As a point on the rim of the wheel turns through an arc length s, the length of string unwound is also s. Further, Eq. (6.1) tells us that

$$\theta \text{ in radians} = \frac{s}{r}$$

In the present case, $s = 30$ cm and $r = 20$ cm. Therefore

$$\theta \text{ in radians} = \tfrac{30}{20} = 1.50 \text{ rad}$$

or
$$\theta = (1.50 \text{ rad}) \left(\frac{360 \text{ deg}}{2\pi \text{ rad}}\right) = 86°$$

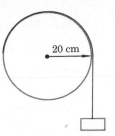

FIGURE 6.3
How far must the wheel turn in order to unwind 30 cm of string?

6.2 Angular Velocity

When we state that a wheel is rotating at 800 rev/min, we are giving its *angular speed*. We are telling how far it rotates in a given time. The **average angular velocity** of a rotating wheel is defined as

$$\bar{\omega} = \frac{\theta}{t} \qquad (6.2) \qquad \textit{Average Angular Velocity}$$

where θ is the angle through which the wheel rotates in time t (ω is Greek omega). As we see, *the units for ω are those of an angle divided by a time.* For example, the units might be degrees per second, revolutions per minute, radians per second, etc.

In advanced work, direction is assigned to rotational quantities such as angular velocity, and so they are vector quantities. For example, the direction of the angular velocity for a wheel rotating like those in Figs. 6.1 and 6.2 is taken perpendicular to the page, out of the page. If the wheels had been rotating in the opposite direction, then the angular velocity vector would point into the page. Because we are concerned with rotations only in a plane, the vector nature of rotational quantities can be expressed by plus and minus signs to indicate the vector to be directed out of or into the plane. For the present, the vector nature of rotational quantities is unimportant for us.

You will notice that average angular velocity is defined in such a way that it is analogous to average linear velocity. We knew that $\bar{v} = s/t$, where s is the linear distance traveled in time t. The analogous angular quantity, $\bar{\omega} = \theta/t$, has the linear distance s replaced by the angular distance θ. We will, in fact, see that each of our five linear-motion equations has an analog in circular motion.

Just as we did in linear motion, we make a distinction between average and instantaneous velocity. You will recall that instantaneous velocity is obtained by measuring the distance moved in such a small time that the velocity could not change appreciably during that time. This is written mathematically in the following way, where ω is the **instantaneous angular velocity**:

$$\omega = \lim_{\Delta t \to 0} \frac{\Delta \theta}{\Delta t} \qquad (6.3)$$

In this expression, $\Delta\theta$ is the small angular distance moved by the wheel in the small time Δt, and $\lim_{\Delta t \to 0}$ tells us to take the value of the ratio as the time interval Δt approaches zero, as discussed in Chap. 2.

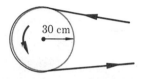

FIGURE 6.4
As the wheel rotates, it winds up one end of the belt and unwinds the other.

Illustration 6.3 The wheel in Fig. 6.4 is turning at 240 rev/min. (*a*) Through how large an angle will it turn in 10 s? (*b*) If the wheel has a radius of 30 cm, how far will a belt that runs on this wheel be pulled in this time?

Reasoning (*a*) Changing all time units to minutes, we have from Eq. (6.2), since $\bar{\omega} = 240$ rev/min and $t = 10$ s,

$$\theta = \bar{\omega}t = (240 \text{ rev/min})(\tfrac{10}{60} \text{ min}) = 40 \text{ rev} = 80\pi \text{ rad}$$

(*b*) We recall that Eq. (6.1) can be used to tell us how much string or belt a wheel winds. As in Illustration 6.2, the amount wound in time t is

$$s = \theta r$$

provided θ is measured in radians. Since we found $\theta = 80\pi$ rad and we were told that $r = 30$ cm, we have

$$s = (80\pi \text{ rad})(0.30 \text{ m}) = 75.4 \text{ m}$$

Notice that the term *radian* does not carry through the equation. As we saw from the definition, it is not a unit in the usual sense of the word.

6.3 Angular Acceleration α

The average linear acceleration was defined in Chap. 2 by the relation

$$a = \frac{v_f - v_0}{t}$$

It measured the rate at which the moving object was speeding up or slowing down. The quantity $v_f - v_0$ was the change in velocity during time t. You will recall that typical units for acceleration are meters per second squared and feet per second squared.

In the case of rotating objects, we are often interested in how they speed up or slow down. Hence we are concerned with angular acceleration, i.e., the rate of change of angular velocity. We *define the* **average angular acceleration** α (alpha) of a rotating wheel or any other object by the relation

$$\alpha = \frac{\omega_f - \omega_0}{t} \qquad (6.4)$$

Its units will be those of angular velocity divided by time. For example, if t is measured in seconds and ω is measured in radians per second, the angular acceleration is expressed in radians per second per second. Although it is not wrong to measure ω in radians per second while t is in minutes, α therefore having units of radians per second per minute, it is generally preferable to use the same unit for t.

If the angular acceleration is uniform, we know, as in the case of linear motion, that the average angular velocity is given by

$$\bar{\omega} = \tfrac{1}{2}(\omega_f + \omega_0)$$

Illustration 6.4 A wheel starts from rest and attains a rotational velocity of 240 rev/s in 2.0 min. What is its average acceleration?

Reasoning Known are

$$\omega_0 = 0 \qquad \omega_f = 240 \text{ rev/s} \qquad t = 2.0 \text{ min} = 120 \text{ s}$$

Using Eq. (6.4), we have

$$\alpha = \frac{\omega_f - \omega_0}{t} = \frac{(240 - 0) \text{ rev/s}}{120 \text{ s}} = 2 \text{ rev/s}^2$$

6.4 Angular-Motion Equations

The three angular-motion equations thus far written *are exactly the same as the analogous linear-motion equations, except that* θ, ω, *and* α *replace* s, v, *and* a, *respectively.* Since the remaining two linear-motion equations were obtained by algebraic manipulation of the first three, it follows that analogous equations will also apply to angular motion. We list all five equations here, together with their linear counterparts:

Linear	Angular	
$s = \bar{v}t$	$\theta = \bar{\omega}t$	(6.5)
$v_f = v_0 + at$	$\omega_f = \omega_0 + \alpha t$	(6.6)
$\bar{v} = \tfrac{1}{2}(v_f + v_0)$	$\bar{\omega} = \tfrac{1}{2}(\omega_f + \omega_0)$	(6.7)
$2as = v_f^2 - v_0^2$	$2\alpha\theta = \omega_f^2 - \omega_0^2$	(6.8)
$s = v_0 t + \tfrac{1}{2}at^2$	$\theta = \omega_0 t + \tfrac{1}{2}\alpha t^2$	(6.9)

The last four equations, (6.6) to (6.9), are not always applicable. It will be recalled that Eqs. (6.6) and (6.7) were true only for constant acceleration. Since Eqs. (6.8) and (6.9) make use of these equations in their derivation, Eqs. (6.7) to (6.9) are restricted to cases of uniform acceleration.

Illustration 6.5 An electric fan is turning at 3.0 rev/s when it is turned off. It coasts to rest in 18.0 s. Assuming its deceleration to be uniform, what was its deceleration? How many revolutions did it turn through while coming to rest? (In practice, the deceleration would not be constant.)

Reasoning This is a typical angular-motion problem. The knowns and unknowns are

$$\omega_0 = 3.0 \text{ rev/s} \qquad \omega_f = 0 \qquad t = 18.0 \text{ s}$$
$$\alpha = ? \qquad \theta = ?$$

From the defining equation for α [Eq. (6.6)]

$$\alpha = \frac{\omega_f - \omega_0}{t} = \frac{(0 - 3.0) \text{ rev}}{18.0 \text{ s}^2} = -0.167 \text{ rev/s}^2$$

To find θ, let us use Eq. (6.9):

$$\theta = \omega_0 t + \tfrac{1}{2}\alpha t^2$$
$$= (3.0 \text{ rev/s})(18.0 \text{ s}) + \tfrac{1}{2}(-0.167 \text{ rev/s}^2)(18.0 \text{ s})^2 = 27 \text{ rev}$$

6.5 Tangential Quantities

When a wheel unwinds a string or rolls along the ground, both rotational and linear motions occur. We wish now to find how these two types of motion are related. The relation between linear and angular distances, s and θ, was inherent in Eq. (6.1), the definition of angular measure. To see this, let us look at Fig. 6.5.

As we see in Fig. 6.5a, a point on the rim of the wheel traces an arc

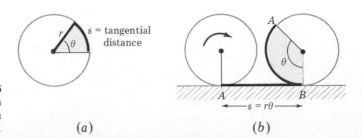

FIGURE 6.5
As the wheel turns through an angle θ, it lays out a tangential distance $s = r\theta$.

(a)

(b)

length s as the wheel turns through the angle θ. From Eq. (6.1), the definition of radian measure, we have that

$$s = r\theta$$

provided θ is measured in radians. We call the distance s a **tangential distance** *since it is measured tangential to the wheel's rim.*

If we look at Fig. 6.5b, we see that the wheel rolls a linear distance $s = r\theta$. This fact allows us to relate linear motion to angular motion for a rolling wheel. We have already seen in Figs. 6.3 and 6.4 that a tangential length $s = r\theta$ of belt or string is wound by a wheel as it rotates through an angle θ. In all cases like this

$$\text{Tangential distance} = s = r\theta \qquad (6.10)$$

Relation between Tangential and Angular Distances

Once again, *we emphasize that θ must be measured in radians.*

Suppose a wheel is rotating with a constant angular speed ω. As it does so, the angle θ keeps changing, as shown in Fig. 6.6a. From the definition of ω we have

$$\omega = \frac{\theta}{t}$$

where θ is the angle turned in time t. But θ is related to the tangential distance through $s = r\theta$, and so after substitution we obtain

$$\omega = \frac{s/r}{t} = \frac{s}{t}\frac{1}{r}$$

But s/t is simply the speed of point A on the rim of the wheel due to the turning motion of the wheel. Or, alternatively, it is the linear speed with which the wheel's center point is moving in Fig. 6.6b. (Can you see why? If not, study Fig. 6.5b.) We call this the tangential velocity v_T. Substituting $v_T = s/t$ and rearranging, we find

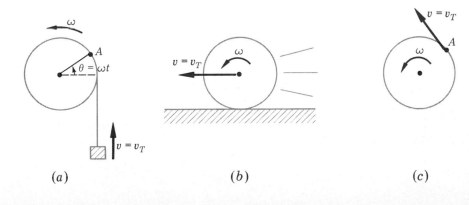

(a)   (b)   (c)

FIGURE 6.6
The angular velocity ω is related to the tangential velocity v_T by $v_T = \omega r$. In this relation, ω must be in radians.

$$\text{Tangential velocity} = v_T = \omega r$$

This is also the tangential velocity of point A in Fig. 6.6c.

If ω is increasing for a rotating wheel, then v_T must also be increasing. The angular acceleration α is given by

$$\alpha = \frac{\omega_f - \omega_0}{t}$$

where $\omega_f - \omega_0$ is the change in the angular velocity ω during time t. But since $\omega = v_T/r$, we can write this as

$$\alpha = \frac{v_{Tf} - v_{T0}}{rt} \qquad \text{or} \qquad \frac{v_{Tf} - v_{T0}}{t} = \alpha r$$

But this is simply the rate of change of tangential velocity, or the **tangential acceleration** a_T. Therefore

$$\text{Tangential acceleration} = a_T = \alpha r \qquad (6.11)$$

This is also the **linear acceleration** *of the center of a rolling wheel or of an unwinding string.* Can you show this from a consideration of the fact that acceleration is the rate of change of velocity—tangential velocity in this case?

Illustration 6.6 A car with 80-cm-diameter wheels starts from rest and accelerates uniformly to a velocity of 20 m/s in 9.0 s. Find the angular acceleration and final angular velocity of one of its wheels.

Reasoning We know that the linear velocity and acceleration of the center of the wheel are given by the tangential relations to be

$$v_T = \omega r \qquad \text{and} \qquad a_T = \alpha r$$

In our case, the final value for v_T was 20 m/s. Therefore

$$\omega_f = \frac{v_{Tf}}{r} = \frac{20 \text{ m/s}}{0.40 \text{ m}} = 50 \text{ rad/s}$$

Notice how we must insert the proper angular measure since angular units do not carry through the equations. Why is it radians per second and not revolutions per second?

To find the acceleration, we first solve the linear-motion problem. Our knowns and unknowns are

$$v_0 = 0 \qquad v_f = 20 \text{ m/s} \qquad t = 9.0 \text{ s} \qquad a = ?$$

We have

$$a = \frac{v_f - v_0}{t} = \frac{20 - 0}{9.0} \text{ m/s}^2 = 2.22 \text{ m/s}^2$$

Then, since this value of a is really a_T, we can write

$$\alpha = \frac{a_T}{r} = \frac{2.22 \text{ m/s}^2}{0.40 \text{ m}} = 5.55 \text{ rad/s}^2$$

Notice that, once again, we must furnish the proper angular units (radians) for our result.

6.6 Radial Quantities and Centripetal Force

Until now we have been concerned simply with motion on a circle. We have said nothing about the forces which cause an object to follow a circular path. We shall see in this section that a radial force, i.e., a force directed along a radius, is needed if an object is to move on a circle. This radial force causes the object to undergo a radial acceleration.

It is not natural for an isolated object to travel in a circle. Newton knew this when he formulated his laws of motion. He stated, in effect, that a body will travel *in a straight line* unless forced to do otherwise. We all know that a ball twirled in a horizontal circle at the end of a string would not continue in a circular path if the string broke. Careful observation will show at once that if the string breaks when the ball is at point *A* in Fig. 6.7b, the ball will follow the straight-line path *AB*, just as Newton said it would.

The fact is that unless a string or some other mechanism pulls the ball toward the center of the circle with a force *F*, as shown in Fig. 6.7, the ball will not continue along the circular path. *This force needed to bend the normally straight path of the particle into a circular path is called the* **centripetal force.** Notice that *it is a pull on the body and it is directed toward the center of the circle. It is a radial force.*

Newton's laws of motion tell us another fact which it is sometimes

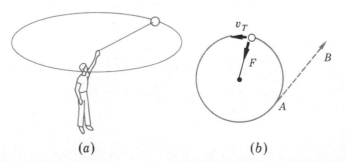

(a)   *(b)*

FIGURE 6.7
The tension *F* in the string holds the ball in a circular path. If the string breaks when the ball is at *A*, the ball will continue in a straight line toward *B*, provided gravity is neglected.

difficult to accept. Since the only thing pulling on the ball in Fig. 6.7 is the string (neglect gravity for now), the centripetal force F exerted by the string on the ball is an **unbalanced force.** By Newton's second law,

$$\mathbf{F} = m\mathbf{a}$$

and since the unbalanced force F is not zero, the ball must be accelerating. This is true even though the ball is going around the circle with constant angular speed! Since $\mathbf{F}$ and $\mathbf{a}$ are vectors and m is a scalar, the acceleration predicted above must be in the direction of the force and hence must be directed toward the center of the circle. Clearly, the acceleration predicted from Newton's law is not in the correct direction to either speed up or slow down the ball in its circular orbit. To learn the nature of the acceleration produced by the centripetal force, we must return to the basic concepts involved. Let us consider the definitions of velocity and acceleration.

First, we defined velocity to be a vector; second, we defined acceleration to be the rate of change of that vector velocity. Notice that the velocity of an object, defined to be a vector, can change even though the object's speed, a mere number, does not change. For an example of this, consider a car that travels at 50 km/h toward the west and then turns and continues with a speed of 50 km/h, but now traveling south. The appropriate velocity vectors are shown in Fig. 6.8a and b.

Clearly the vector velocity has been changed. To find the change in the velocity vector, we must find the velocity vector $\Delta\mathbf{v}$ which must be added to the original velocity vector $\mathbf{v}_0$ to obtain the final velocity vector $\mathbf{v}$. It is shown in Fig. 6.8c.

Notice that the speed of the moving car has not changed. Its velocity has changed because its direction of motion has changed. By the definition of average acceleration,

$$\mathbf{a} = \frac{\Delta\mathbf{v}}{t}$$

where t is the time taken for the change $\Delta\mathbf{v}$ in velocity.

When an object travels in a circle, its direction, and therefore its velocity, is constantly changing. Even though the speed is not changing, its direction of motion is changing under the action of the centripetal force. Let us

FIGURE 6.8
To change a velocity $\mathbf{v}_0$ of
50 km/h west, as shown in
(a), to a velocity $\mathbf{v}$ of
50 km/h south, as shown in
(b), one must add to $\mathbf{v}_0$ a
velocity change $\Delta\mathbf{v}$, as
shown in (c).

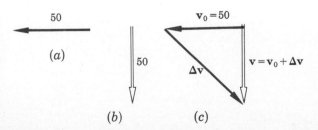

now examine this case more closely in order to obtain a mathematical expression for the force needed to hold an object in a circular path.

Consider what happens when the ball travels from A to B in Fig. 6.9a. Since the velocity of the ball has changed its direction but not its magnitude, the change in velocity is as shown in Fig. 6.9b. Hence the acceleration of the ball will be

$$a = \frac{\Delta v}{\Delta t}$$

where Δt is the time taken for the ball to travel from A to B. Eventually we take the limit as Δt approaches zero, in which case a will be the instantaneous acceleration.

Since the speed of the ball is v and the distance it travels is s, as shown in Fig. 6.9a, the time taken is just

$$\Delta t = \frac{s}{v}$$

from which we have, after substitution in the equation for a,

$$a = \frac{v \, \Delta v}{s}$$

Let us now compare the triangle OAB with the triangle in Fig. 6.9b to show that the two triangles are similar. Since the radius of a circle is perpendicular to its tangent, we know that OA is perpendicular to v_0 and OB is perpendicular to $\mathbf{v}$. Therefore $\angle AOB$ equals the angle between v_0 and $\mathbf{v}$. Further, because $v_0 = v$ and $OA = OB$, both triangles are isosceles. Invoking the geometry theorem that says "Two isosceles triangles are similar if the angles between their two equal legs are the same," we conclude that triangle AOB in Fig. 6.9a is similar to the triangle in Fig. 6.9b. We can therefore write

$$\frac{\Delta v}{v} = \frac{\overline{BA}}{r}$$

But if point A is close to point B on the circle, as it will be if $\Delta t \to 0$, the arc from B to A is nearly the same length as the line $\overline{BA}$. To that approximation

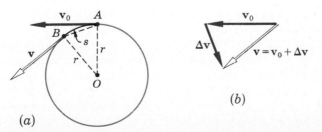

(a)

(b)

FIGURE 6.9
The vector triangle in (b) is similar to triangle OAB in (a), since the two triangles have sides that are mutually perpendicular.

we can write $\overline{BA} = s$, and after substituting and rearranging terms, we have

$$\Delta v = \frac{sv}{r}$$

in the limit as $\Delta t \rightarrow 0$.

Placing this value for Δv in the equation for the acceleration a yields

Centripetal Acceleration

$$a = \frac{v^2}{r} \qquad (6.12)$$

where a is now the instantaneous acceleration, since it is evaluated in the limit as $\Delta t \rightarrow 0$. *This acceleration, which is caused by the centripetal force, is called the* **centripetal acceleration.** Finally, *the centripetal force is given by* $F = ma$, and we have

Centripetal Force

$$F_{\text{centripetal}} = \frac{mv^2}{r} \qquad (6.13)$$

We can summarize our result as follows:

> The force required to hold an object of mass m and speed v in a circular path of radius r is called the **centripetal force** and is equal to mv^2/r.

The centripetal acceleration is directed radially inward, a fact we can show in the following way. Because the triangles in Fig. 6.9 are similar and have two mutually perpendicular sides, the third sides must also be perpendicular. Hence $\Delta \mathbf{v}$ is perpendicular to AB, and so $\Delta \mathbf{v}$ is parallel to the perpendicular bisector of AB. But the acceleration of the object moving on the circle is parallel to $\Delta \mathbf{v}$, and so, for $AB \rightarrow 0$, the centripetal acceleration is directed radially toward the center of the circle. We therefore conclude that *the instantaneous acceleration of an object traveling with uniform speed in a circle is directed inward toward the center of the circle. That is why it is sometimes called the* **radial acceleration.** *The centripetal force has this same direction.*

Equation (6.13) tells us that the force needed to hold an object in a circular path is proportional to both the square of the velocity of the object and its mass. Hence, the larger the mass of the object and the faster it is circling, the greater must be the tension in the rope which constrains it to move in a circle. This is a common experience for most of us. Even the fact that the centripetal force increases as the radius of the circle decreases is easily demonstrated by whirling a piece of chalk tied to a string around a circle. Remember while doing this experiment, though, that if v is held constant in magnitude, it will take the piece of chalk less time to travel around a small circle than around a large one. Hence the angular speed of the particle will be greater for a small circle even if the linear speed of the object is held constant.

Illustration 6.7 A 1200-kg car is turning a corner at 8.0 m/s, and it travels along an arc of a circle in the process. If the radius of the circle is 9.0 m, how large a horizontal force must be exerted by the pavement on the tires to hold the car in the circular path? (See Fig. 6.10.)

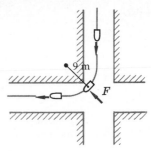

Reasoning The force required is the centripetal force:

$$F = m\frac{v^2}{r} = (1200 \text{ kg})\left(\frac{64 \text{ m}^2/\text{s}^2}{9.0 \text{ m}}\right) = 8530 \text{ N}$$

This force must be supplied by the friction force of the pavement on the wheels. Can you show that the coefficient of friction must be at least 0.73 in order to provide such a large force? If the pavement is wet so that there is little friction between the tires and the road, perhaps the friction force on the tires will not be this large. In that event, the car will skid out of a circular path (into more nearly a straight line) and may not be able to make the curve.

FIGURE 6.10
For the car to turn the corner as shown, the friction force F between the tires and pavement must furnish the centripetal force needed to hold the car in a circular path.

Illustration 6.8 A ball tied to the end of a string is swung in a vertical circle of radius r under the action of gravity, as shown in Fig. 6.11. What will be the tension in the string when the ball is at point A of the path, as shown in Fig. 6.11, if the ball's speed is v at that point?

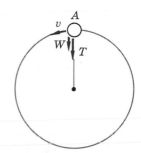

Reasoning For the ball to travel in a circle, the forces acting on the ball must furnish the required centripetal force. In this case, at point A two forces act on the ball—the pull T of the string and the pull of the earth (the weight W of the ball). Both forces are radial at A, and so their vector sums must provide the required centripetal force. We therefore have

$$T + W = \frac{mv^2}{r}$$

which gives the tension in the string as

$$T = \frac{mv^2}{r} - W = m\left(\frac{v^2}{r} - g\right)$$

FIGURE 6.11
When the ball is in the position shown, its weight provides part of the necessary centripetal force.

Notice that if $v^2/r = g$, the tension in the string will be zero. Then the centripetal force is just equal to the weight. This is the case where the ball just makes it around the circle. If the speed v of the ball is less than the value given by this relation, that is, $v < \sqrt{rg}$, the required centripetal force is less than the ball's weight. In that case, the ball will fall down out of the circular path. To test your understanding of this, describe what would happen to the people shown in Fig. 6.12 if the roller coaster were moving too slowly at the top of the loop.

FIGURE 6.12
Describe the sensations of
the riders at various points
on their path. (*Leeanne
Schmidt, Design
Photographers
International*.)

6.7 A Common Misconception

People are likely to jump to completely erroneous conclusions when interpreting their experiences. For example, a man seated in the center of a car seat sometimes thinks that he has been pushed to the side of the car as it rounds a corner. He might even assert that the force pushing him sideways was so great that it threw him against the side hard enough to injure him. This is nonsense, of course. There was no mysterious ghost pushing him toward the side of the car. Certainly no material object was pushing him in that direction. He must therefore be mistaken.

The same man would not claim that a mysterious force suddenly acted on him to throw him violently against the dashboard of the car as it suddenly stopped. He knows that his forward momentum could be lost only if some force retarded his motion Hence, when the car suddenly stopped, he continued going forward until the dashboard of the car began to exert a force on him to stop him from moving forward. This is merely an example of Newton's idea that things continue in motion until a force acts on them to stop them.

Similarly with the car turning the corner; the friction between the pavement and the tires pushed horizontally on the car and altered its straight-line motion. It is too bad about the man sitting in the middle of the nearly frictionless seat. The friction force between the seat of his pants and the seat was too small to alter his straight-line motion. Hence he slid along in a straight line until he hit the side of the car, which then exerted a force on him so that he could travel in the same curved path as that followed by the automobile.

6.8 Orbital Motion

Perhaps the most majestic examples of circular motion are found in the heavens. The earth and other planets travel around the sun in nearly circular paths; the earth's moon follows a nearly circular path around the earth, and the moons of other planets do much the same for their planets. In addition, planets created by people, artificial satellites, trace nearly circular paths around the earth. Let us now examine this type of motion, *orbital motion*.

In Fig. 6.13 we see the moon or some other satellite following a circular path around the earth. The satellite's mass is m_s, and its speed is v. We represent the orbit radius by r and the mass of the earth by m_e. A centripetal force of magnitude $m_s v^2/r$ is required to hold the satellite in orbit. This force is furnished by the gravitational attraction of the earth for the satellite, namely,

$$\text{Gravitational force} = G\frac{m_s m_e}{r^2}$$

as described in Sec. 3.7. Equating the gravitational force to the required centripetal force gives

$$G\frac{m_s m_e}{r^2} = \frac{m_s v^2}{r} \qquad (6.14)$$

It is of interest to notice that m_s, the satellite mass, cancels from this equation. We therefore conclude that *the mass of the satellite is unimportant in describing the satellite's orbit.* At a given orbital speed v and radius r, the moon will orbit in the same way as a baseball would orbit. Thus any satellite orbiting at radius r must have the following speed, given by Eq. (6.14):

$$v = \sqrt{\frac{Gm_e}{r}} \qquad (6.15) \quad \textit{Orbital Speed}$$

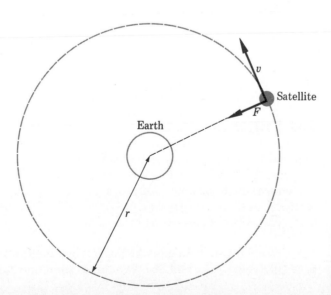

FIGURE 6.13
The centripetal force on the satellite is supplied by the gravitational attraction of the earth.

Illustration 6.9 Radio and TV signals are sent from continent to continent on earth by "bouncing" the signals from *synchronous satellites*. These satellites circle the earth once each 24 h, so if the satellite circles eastward above the equator, it stays over the same spot on the earth because the earth is rotating at this same rate. (*a*) What is the orbital radius for a synchronous satellite? (*b*) What is its speed?

Reasoning The satellite circles the earth once each 24 h, and so its speed is given by

$$v = \frac{\text{distance gone}}{\text{time taken}} = \frac{2\pi r}{t}$$

where r is the orbital radius and $t = 24$ h. Substituting for v in Eq. (6.14), we have

$$\frac{4\pi^2 r^2}{t^2} = \frac{Gm_e}{r}$$

from which

$$r^3 = \frac{Gm_e t^2}{4\pi^2}$$

But $G = 6.67 \times 10^{-11} \text{ N} \cdot \text{m}^2/\text{kg}^2$, $m_e = 5.98 \times 10^{24}$ kg, and

$$t = 24 \text{ h} = (24 \text{ h}) \left(\frac{60 \text{ min}}{\text{h}}\right)\left(\frac{60 \text{ s}}{1 \text{ min}}\right) = 86{,}400 \text{ s}$$

Substituting these values yields $r = 4.2 \times 10^7$ m.

We can now substitute this value for r in the expression for v to give

$$v = \frac{2\pi r}{t} = \frac{2\pi(4.2 \times 10^7 \text{ m})}{86{,}400 \text{ s}} = 3100 \text{ m/s}$$

6.9 Projectile Motion

As we saw in the preceding section, an object will circle the earth if it has the proper speed for its particular orbital radius. At the surface of the earth, this speed turns out to be about 7900 m/s. But if an object is shot parallel to the earth with a speed smaller than this, it will fall to the earth, as shown in Fig. 6.14. Let us now examine the motion of projectiles near the surface of the earth.

Consider what happens when a baseball is thrown horizontally by the baseball pitcher. The ball does two things simultaneously after it leaves the

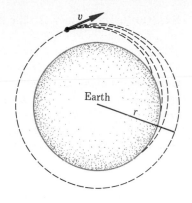

FIGURE 6.14
If an object is shot fast enough parallel to the earth, it will circle the earth. (Newton was probably the first to recognize this fact.)

pitcher's hand: It falls to the earth under the action of gravity, and it travels horizontally with constant speed until it strikes something (provided air-friction effects can be ignored). To see this, let us refer to Fig. 6.15.

Suppose that the ball leaves the pitcher's hand at point A and that its velocity at that instant is completely horizontal. Call its velocity at that instant v_H. According to Newton's law, there will be no acceleration in the horizontal direction unless a horizontally directed force acts on the ball. But if we ignore friction with the air, the only force acting on the ball once it is free from the hand of the pitcher is the force of gravity. There is no horizontal force acting on the ball. Hence, its horizontal velocity will remain unchanged and will be v_H until the ball hits something.

It should be clear that the horizontal motion of the ball is extremely simple. The ball moves with constant horizontal velocity component v_H until the ball strikes something. For the horizontal motion we therefore have (since the initial velocity v_0 is equal to the final velocity v_f and average velocity $\bar{v}$)

Horizontal:
$$v_0 = v_f = \bar{v} = v_H$$
$$a = 0 \qquad s = \bar{v}t = v_H t$$

The vertical motion of the ball is not much more complicated. It will accelerate downward under the force of gravity, and hence $a = g$. This vertical motion is exactly the same as we discussed for a freely falling body

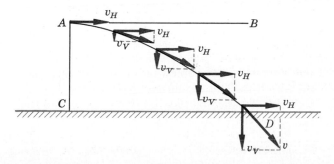

FIGURE 6.15
The initial horizontal speed of the ball v_H does not change. As the ball moves with this speed to the right, it also falls under the action of gravity, as shown by the vectors representing the vertical component of the velocity. Note that $v = \sqrt{v_H^2 + v_V^2}$ and is tangent to the trajectory.

in Chap. 2. The solution to the problem of the vertical motion of the ball is not new to us, therefore, and we should be able to carry it through with no difficulty.

Our procedure, then, is to recognize that the free motion of a ball, bullet, or any other projectile contains within it two separate problems. The horizontal problem is motion with constant velocity and is therefore quite simple. The vertical motion is identical to the motion of a free body in a vertical line. We compute each portion of the projectile-motion problem separately. The solutions to the individual problems are then combined to obtain the answer requested. In effect, we shall have found the horizontal v_H and vertical v_V components of the velocity of the projectile. The magnitude of the resultant velocity is found by using the usual method for vectors at right angles, $v = \sqrt{v_H^2 + v_V^2}$. This procedure is best seen by means of an example.

Illustration 6.10 Consider the situation in Fig. 6.15. Suppose that the ball leaves the hand of the pitcher, traveling horizontally with a velocity of 15 m/s, and suppose that it is 2.0 m above the ground at that instant. Where will it hit the ground? (That is, how far is point D from point C?)

Reasoning We start the solution by splitting the problem into two parts:

Horizontal	Vertical (down positive)
$v_0 = v = \bar{v} = 15$ m/s $s_H = \bar{v}t = 15t$	$v_0 = 0$ $a = 9.8$ m/s^2 $s_V = 2.0$ m
To find t, we solve the vertical problem	To find t, we use $s_V = v_0 t + \frac{1}{2}at^2$ $2.0 = 4.9t^2$ $t = 0.64$ s

Now that the time of flight, i.e., the time to drop to the ground, has been found from the vertical problem, the result can be used in the horizontal problem. One has

$$s_H = (15 \text{ m/s})(0.64 \text{ s}) = 9.6 \text{ m}$$

In other words, the ball travels only 9.6 m in a horizontal direction before the pull of gravity causes it to fall to the ground. Actually, one usually throws a ball somewhat upward[1] if one wants it to travel any great distance. If the ball has an initial upward component to its velocity, it will take longer for the ball to fall to the ground, and so the ball will have more time to travel in a horizontal direction.

---

[1] But not at an angle to the horizontal greater than 45°. Why not?

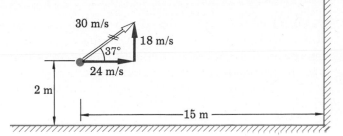

Illustration 6.11 A ball is thrown with a velocity of 30 m/s at an angle of 37° above the horizontal. It leaves the pitcher's hand 2.0 m above the ground and 15.0 m from a wall, as shown in Fig. 6.16. (*a*) At what height above the ground will it hit the wall? (*b*) Will it still be going up just before it hits, or will it already be on its way down?

Reasoning Consider the trip from the starting point to the wall. For part *a*, because the horizontal and vertical components of the velocity are 30 cos 37° and 30 sin 37° m/s, respectively,

Horizontal	Vertical (up positive)
$v_0 = v = \bar{v} = 24$ m/s	$v_0 = 18$ m/s
$s = 15$ m	$t = \frac{5}{8}$ s
$\quad = \bar{v}t$	$a = -9.8$ m/s^2
$15 = 24t$	$s = ?$
$t = \frac{5}{8}$ s	$s = v_0t + \frac{1}{2}at^2$
We now use this value	$\quad = (18)(\frac{5}{8}) - (4.9)(\frac{5}{8})^2$
in the vertical problem	$\quad = 11.2 - 1.9$
	$\quad = 9.3$ m above the starting point

Hence, the ball strikes the wall 9.3 m above the starting point, or 11.3 m above the ground.

For part *b*, let us continue the vertical part of the problem and find the vertical component of the velocity of the ball just before it hits the wall. We have

$$v_y = v_0 + at = 18 \text{ m/s} - (9.8 \text{ m/s}^2)(\tfrac{5}{8}\text{ s}) = +11.9 \text{ m/s}$$

Since the upward direction was taken as positive, a positive velocity means an upward velocity. Hence, the ball was on its way up when it hit.

The total velocity of the ball just before it strikes has an upward component of 11.9 m/s, and a horizontal component of 24 m/s. These components can be added in the usual way to find the magnitude of the velocity at that point. It is

$$v_y = \sqrt{(11.9)^2 + (24)^2} \text{ m/s} = 27 \text{ m/s}$$

6.10 Weightlessness

We often hear that objects appear to be weightless in a spaceship circling the earth or on its way to a distant point in space. Let us examine this effect in some detail. First, we should restate our definition of weight. It was defined to be the pull of gravity on the object in question. *On the earth, the weight of an object is the gravitational pull of the earth on the object.* Similarly, *an object's weight on the moon is taken to be the gravitational pull of the moon on the object.*

Ordinarily we measure the weight of an object by placing it on a scale. Or if only a rough measure is needed, *we simply notice the force the object exerts on our hand when we hold it fixed.* Usually the force read by the scale, i.e., the force exerted by the object on the scale, and the force on one's hand are equal to the pull of gravity on the object, i.e., to the weight of the object. This is not always true, however, as we shall now see, and so *we reserve the phrase* **apparent weight** *for the reading of the scale and the force on our hand together with other common nonbasic ways for judging the weight of an object.*

To illustrate this point, let us consider the apparent weight of an object of mass m in an elevator. This question is discussed in Chap. 3, but let us consider it again here. In Fig. 6.17a, if the elevator is at rest, Newton's second law tells us that since the acceleration is zero, the resultant force on the object is zero. Calling the gravitational force on the body (its weight) W and the tension in the string holding the object T, we have

$$T - W = 0 \qquad \text{or} \qquad T = W$$

when $a = 0$. In this instance, the tension in the string is W, and the apparent weight of the object, the reading of the scale, is equal to its actual weight W.

This same situation will prevail *as long as a = 0.* Under that condition $T = W$, and the apparent weight is equal to the actual weight. Even if the elevator is moving up or down at constant speed, the acceleration is still zero and the apparent weight will equal the actual weight.

Let us now examine the situation shown in Fig. 6.17b. The elevator is accelerating downward in this case. If we apply Newton's second law as before, we find

$$W - T = ma$$

which gives

$$T = W - ma$$

Notice that the tension in the string (and therefore the scale reading) is less than W by the amount ma. To the person in the accelerating elevator, the object appears to weigh less than W. Its apparent weight is $W - ma$.

The most spectacular observation occurs when the elevator is freely

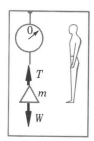

$a = 0$
$T = W$

(a)

a downward
$W - T = ma$
or $T = W - ma$

(b)

FIGURE 6.17
The weight of an object in an elevator seems to a person in the elevator to vary, depending on the motion of the elevator.

falling, so that $a = g$, the acceleration in free fall. Then, since $W = mg$ and since $a = g$ for a freely falling body, the tension in the string

$$T = W - ma$$

becomes

$$T = mg - mg = 0$$

The object appears weightless in a freely falling elevator! If we think about it a little, this is not strange at all. Since the elevator (and everything in it) is supposedly accelerating with the acceleration of free fall, by the very definition of free fall, there can be no force supporting the objects (elevator and everything in it) or in any way retarding their free fall. Hence all support forces on the elevator and everything in it must be zero. The tension in the cord supporting the object must be zero. All objects within the elevator appear to be weightless.

We see from these considerations that *apparent weights of objects are not necessarily equal to the true weights of the objects in accelerating systems. In particular, if the system is freely falling,*[2] *all support forces must be zero and all objects appear to be weightless.* This means that *whenever a spaceship is falling freely in space,* i.e., when its rocket engines are not being operated, *everything within this freely falling system will appear to be weightless.* It does not matter where the object is, whether it is falling under the force of attraction of the earth, the sun, or some distant star; as long as it is freely falling, everything in it will appear weightless.

An earth or moon satellite ship is simply an example of a freely falling ship. At first this statement may surprise you, but it is easily seen to be correct. Consider the behavior of a projectile shot parallel to the horizontal surface of the earth in the absence of air friction. (At satellite altitudes, the air is so thin as to be almost negligible.) The situation was shown in Fig. 6.14. As shown, the projectile is thrown at successively larger speeds, and we see that during its free fall to the earth, the curvature of the path decreases with increasing horizontal speed. If the object is thrown fast enough parallel to the earth, the curvature of its path will match the curvature of the earth, as shown. In this case, the object (a spaceship perhaps) will simply circle the earth. The satellite is accelerating toward the center of the earth at all times since it circles the earth. Its radial acceleration is simply g, the free-fall acceleration. In effect, the satellite is falling toward the center of the earth at all times, but the curvature of the earth prevents the satellite from hitting. Since the satellite is in free fall, all objects within it appear weightless.

---

[2] Recall that a freely falling object is one which is subject to only one type of unbalanced external force, the gravitational pull of the earth and similar bodies.

Summary

Angles are measured in degrees, revolutions, and radians; 1 rev = 360° = 2π rad. We represent angles by θ. The time rate of change of θ is called the angular velocity ω. The angular acceleration α is the time rate of change of ω.

As in linear motion, five angular-motion equations apply if α is constant:

$$\theta = \bar{\omega} t \qquad \alpha = \frac{\omega_f - \omega_0}{t} \qquad \bar{\omega} = \tfrac{1}{2}(\omega_0 + \omega_f)$$

$$\omega_f^2 - \omega_0^2 = 2\alpha\theta \qquad \theta = \omega_0 t + \tfrac{1}{2}\alpha t^2$$

The first equation defines average angular velocity, and the second defines average angular acceleration.

When a wheel turns on its axle, a point on its circumference can be described in terms of tangential quantities. The tangential distance moved by the point s when the wheel turns through an angle θ is given by $s = r\theta$, where r is the radius of the wheel. Similarly, the tangential velocity and acceleration of the point are given by

$$v_T = r\omega \qquad a_T = r\alpha$$

In all these equations relating angular to tangential quantities, radians must be used for the angular measure.

When a wheel on an axle winds a string on its rim, the displacement, speed, and acceleration of a point on the string are given by the tangential quantities s, v_T, and a_T. Similarly, when a wheel rolls without slipping along a level surface, the displacement, speed, and acceleration of the center of the wheel are given by s, v_T, and a_T, respectively.

A centripetal force is required to cause an object to move in a circular path. This force is needed to pull the object from its normal straight-line motion. When following a circle, the vector velocity of an object is constantly changing. It is undergoing a radial acceleration toward the center of the circle. This acceleration is given by v^2/r, and the centripetal force required to cause this acceleration is mv^2/r.

Projectile motion, without friction, is analyzed in terms of its component motions. For the horizontal motion along the earth, the acceleration is zero. As a result, $v_0 = v_f = v_H$ and the five motion equations reduce to $s = v_H t$. The vertical component of the motion is treated independently and is exactly the same as that which would occur in the absence of the horizontal motion.

The force needed to support an object is not always equal to the weight of the object. We call the required supporting force the apparent weight of the object. If the object and its support are accelerating, the weight and apparent weight usually differ. In the special case of free fall, the apparent weight is zero. As a result, objects appear weightless when coasting in space. Similarly, since an earth satellite is freely falling, objects within it appear weightless.

Minimum Learning Goals

Upon completion of this chapter, you should be able to do the following:

1 Convert an angle in degrees, radians, or revolutions into each of the other units.

2 Make use of the equation $\theta = \bar{\omega} t$ in simple situations in which two of the three quantities θ, $\bar{\omega}$, t are given.

3 Compute the angular acceleration of a wheel when ω_0, ω_f, and t are given.

4 Write the five angular-motion equations and define each quantity in them. State the major restriction on their use.

5 Use the five angular-motion equations to solve problems.

6 Explain how θ for a rotating wheel is related to the distance the wheel rolls or the amount of string which winds onto the wheel. When θ or s is given for a wheel of known radius, be able to find the other.

7 Explain the difference between tangential velocity and angular velocity. When one is given for a wheel of known radius, find the other. Explain why tangential velocity is important for a rolling wheel and for a wheel which is winding something on its rim.

8 Explain what is meant by a centripetal force and why it must be furnished to an object if the object is to follow a circular path. State the relation for centripetal force, and be able to use it in simple

situations where an object is moving in a horizontal or vertical circle.

9 Using a diagram, show why an object traveling on a circular path is accelerating even though the object's angular speed is constant. State the direction and value for the radial acceleration.

10 Find the distance traveled over level ground by a projectile shot at a known angle and speed from a given height above the ground.

11 Calculate where a projectile will strike a distant wall when the initial velocity and position of the projectile are given.

12 Calculate the supporting force needed on an object of known mass if the object is moving with constant speed; is accelerating upward; is accelerating downward. Explain what is meant by apparent weight in such circumstances, and show why it differs from the object's weight.

13 Explain why an object orbiting the earth (or in some similar situation) is said to be freely falling. Use your explanation to point out why objects appear weightless under certain circumstances.

Important Terms and Phrases

You should be able to define or explain each of the following:

Radian measure
Angular analogs to s, v, a
Five angular-motion equations
Tangential quantities: $s = r\theta$, $v_T = r\omega$, $a_T = r\alpha$
Centripetal acceleration
Centripetal force

Even though an object is going around a circle at constant speed, it is accelerating
Apparent weight and weightlessness
An earth satellite is continuously falling toward the earth
Projectile motion consists of two separate motions occurring simultaneously

Questions and Guesstimates

1 When mud flies off the tire of a bicycle, in what direction does it fly? Explain.

2 Figure P6.1 shows a simplified version of a cyclone-type dust remover. It is widely used to purify industrial waste gases before venting them to the atmosphere. The gas is whirled at high speed around a curved path, and the dust particles collect at the outer edge and are removed by a water spray or by other means. Explain the principle behind this method for removing particulate matter from dirty air.

3 Discuss the principle of the spin-dry cycle in an automatic washing machine.

4 An insect is sitting on a smooth, flat wheel which can be rotated about a vertical axis perpendicular to the plane of the wheel and through its center. Describe qualitatively the motion of the insect as the wheel begins to rotate. Assume that the insect is quite close to the axis and that there is some, but not much, friction between the bug and the wheel. (The wheel might be a phonograph turntable, for example.)

FIGURE P6.1

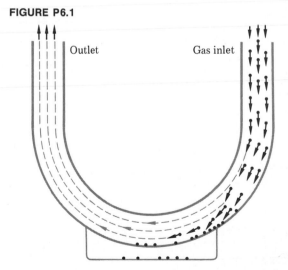

Outlet   Gas inlet

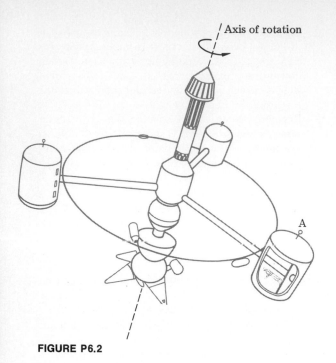

Axis of rotation

FIGURE P6.2

5 A person in an earth satellite is said to be weightless. Explain what is meant by this. Why do objects not fall closer to the earth in the ship since they experience the pull of gravity?

6 Figure P6.2 is an artist's conception of a rotating space station. Assuming the rotation to take place as shown, did the artist choose the proper direction for "up" in the space capsule at *A*? Explain.

7 A woman weighs herself daily on a spring-operated bathroom scale. Suppose that the earth stopped rotating about its axis. Would she weigh more, less, or the same? Would it matter where she lived on the earth?

8 A bug sits on the very top of a freshly waxed bowling ball. It loses its footing and slides freely down the surface of the ball. Explain why the bug will leave the surface *before* it falls halfway down the ball.

9 From the fact that the moon circles the earth at a radius of about 3.8×10^8 m, estimate the mass of the earth.

10 Airplane enthusiasts sometimes hold meets where they try to show their skills. One event is to drop a sack of sand exactly in the center of a circle on the ground while flying at a predetermined height and speed. What is so difficult about that? Don't they just drop the sack when they are directly above the circle?

11 Parents should be prepared for nearly anything. Suppose your child wants to find out how fast a slingshot can shoot a stone. Devise a method for finding out. Assume the only tool you have is a meterstick.

12 If you want to hit a distant, stationary object with a rifle, you do not aim the gun with the object in line with the hole in the barrel of the gun. How is the gun aimed?

13 About how fast can a car be going if it must negotiate a turn from one street onto a perpendicular street? Assume both streets to be made of concrete and to be of average size, carrying one lane of traffic each way. (E)

14 During the 1970 flight of *Apollo* 13 to the moon, serious trouble developed when the ship was about halfway, and it returned to earth without executing its moon mission. However, the ship continued toward the moon, passed on the other side of it, and only then returned to earth. Why didn't they simply turn around at the time the decision was made to return instead of continuing toward the moon?

15 Suppose that a huge mass, much larger than the mass of the solar system or our galaxy, is created at this instant. The large gravitational force it would exert on the solar system would cause us to accelerate toward this distant mass. After the first few seconds, what long-term effects would we notice on earth because of this acceleration? Assume the earth's acceleration due to this cause to be of the order of 10 m/s².

Problems

1 For each of the following angles, give the equivalents in degrees, revolutions, and radians: (*a*) 25°; (*b*) 8.1 rad; (*c*) 0.73 rev.

2 For each of the following angles, give the equivalents in degrees, revolutions, and radians: (*a*) 123°; (*b*) 1.50 rad; (*c*) 2.64 rev.

3 The moon has a diameter of 3480 km and is 3.8×10^8 m from the earth. (a) How large an angle in radians does the diameter of the moon subtend to a person on earth? (b) If the diameter of the earth is 1.28×10^4 km, what is the angle subtended by the earth to a person on the moon?

4 A tiny laser beam is directed from the earth to the moon. If the beam is to have a diameter of 2.50 m at the moon, how small must the divergence angle be for the beam?

5 A phonograph record rotates at $33\frac{1}{3}$ rev/min. What is its angular speed in (a) degrees per second and (b) radians per second?

6 A pulley on a motor is rotating at 1400 rev/min. What is its angular speed in (a) radians per second and (b) degrees per second?

7 The minute hand on a clock turns through 90° in 15 min. What is the angular velocity of the hand in radians per second?

8 The hour hand on a clock turns through $(360/12)°$ in 1 h. What is the angular velocity of the hand in radians per second?

9 A record turntable accelerates from rest to an angular speed of 45 rev/min in 1.60 s. What is its average angular acceleration in revolutions per second squared? In radians per second squared?

10 A record turntable turning at $33\frac{1}{3}$ rev/min coasts to rest in 15.0 s. What is the average angular acceleration of the record in revolutions per minute squared? In radians per second squared?

11* In a certain large motor, it takes 20 s for the motor shaft to accelerate uniformly from rest up to its operating speed of 16.0 rev/s. Find (a) the angular acceleration of the motor shaft (revolutions per second squared) and (b) the number of revolutions it turns in this time.

12* How large an angular acceleration in radians per second squared, must be given a wheel if it is to be accelerated from rest to a rotational speed of 520 rad/s after completing 7.5 rev?

13* The time taken for a certain roulette wheel to coast to rest is 15.0 s. If the wheel turns through 8.5 rev in that time, how fast was it originally turning? (Assume uniform deceleration.)

14* A motor whose shaft is turning at 400 rev/min is speeded up until its speed is 600 rev/min. The change takes 20 s. Assume the acceleration to be uniform. (a) What was its angular acceleration in radians per second squared? (b) Through what angle, in degrees, did it turn in this time?

15* There is a speck of dust 5.0 cm from the center of a $33\frac{1}{3}$-rev/min phonograph record. (a) When the

phonograph is running, what is the speed, in radians per second, of the record? (b) How fast, in centimeters per second, is the dust speck moving?

16 The minute hand on a clock is 6.0 cm long. When the clock is running normally, (a) what is the angular speed of the clock hand, in radians per second, and (b) how fast, in meters per second, is the tip of the hand moving?

17 A 22-cm-diameter bowling ball rolls 12.0 m along the floor without slipping. Through how many revolutions did it roll?

18 Through how many revolutions must the 60-cm-diameter wheel of a car turn as the car travels 2.5 km?

19* The radius of the earth is 6.37×10^6 m. (a) How fast, in meters per second, is a tree at the equator moving because of the earth's rotation? (b) A polar bear at the North Pole?

20* The earth orbits the sun in 365 days. What is the speed, in meters per second, of the earth in the orbit? The earth-sun distance is 1.50×10^{11} m.

21 A 10.0-cm-diameter wheel that is turning at a rate of 0.40 rev/s winds a string on its rim. How long a piece of string is wound on the wheel in 30 s?

22 A wheel turning at 1800 rev/min has a diameter of 5.0 cm. If a string is wound on the wheel, how much string does the wheel wind up in 4.0 s?

23 A vehicle is traveling along the road at 20 m/s. If the diameter of its wheels is 80 cm, how fast are the wheels rotating in revolutions per second, radians per second, and degrees per second?

24 A 50-cm-diameter wheel comes loose from a car that is going 30 m/s, and the wheel rolls alongside the car. Find the angular speed of the wheel in revolutions per second, radians per second, and degrees per second.

25* A bicycle with 60-cm-diameter wheels is coasting at a speed of 5.0 m/s. It decelerates uniformly and stops in 20 s. (a) How far did it go in this time? (b) Through how many revolutions did each wheel turn as the bicycle was coming to a stop?

26* A car with 80-cm-diameter wheels starts from rest and accelerates uniformly to 15 m/s in 30 s. Through how many revolutions did each wheel turn in this time?

27* A motor turning at 1800 rev/min coasts uniformly to rest in 15 s. (a) Find its angular deceleration and the number of revolutions it turned before stopping. (b) If the motor had a 5.0-cm-radius wheel attached to its shaft, what length of belt did the wheel wind in the time taken for it to stop?

28* Two gear wheels which are meshed together have

radii of 0.50 and 0.15 cm. Through how many revolutions does the smaller turn when the larger turns through 3 rev?

29 What is the angular acceleration of a 60-cm-diameter wheel on a vehicle as the vehicle undergoes an acceleration of 0.50 m/s²?

30 An object is being lifted by a cord wound on the rim of a 20-cm-diameter wheel. If the wheel is accelerating at 0.30 rad/s², what is the acceleration of the object in meters per second squared?

31 How large a force is needed to hold a 900-kg car in an arc of 6.0-m radius when the car is turning a corner at 12.0 m/s?

32 A 500-g mass at the end of a string is whirled around in a horizontal circle of 75-cm radius. If its speed in the circle is 7.0 m/s, what must be the tension in the string that holds the mass?

33* In a thrilling ride at an amusement park, carts on a horizontal surface are whipped about at the end of a long rod. Suppose the cart and occupants weigh 2000 N and the rod is 3.0 m long. How large a stretching force (tension) must the rod withstand as it whips the cart around a 3.0-m-radius circle at a rate of 1 rev in 2.0 s?

34* A 20-mg bug sits on the smooth edge of a 25-cm-radius phonograph record as the record is brought up to its normal rotational speed of 45 rev/min. How large must the coefficient of friction between the bug and record be if the bug is not to slip off? (It is a very compact bug, and so air friction can be ignored.)

35* In a certain research device, a person is subjected to an acceleration of 5g, that is, 5 times the acceleration due to gravity. This is done by rotating the person in a horizontal circle at very high speed. The seat in which the subject is strapped is 8.0 m from the rotational axis. How fast is the person rotating, in revolutions per second, if the centripetal force on the person is 5 times the person's weight?

36* An old trick is to hold a pail of water with your hand and swing it in a vertical circle. If the rotation rate is large enough, water will not fall out of the pail when the pail is upside down at the top of its path. What is the minimum speed your hand must have at the top of the circle if the trick is to succeed? Assume your arm to be 0.60 m long.

37* When constructing a roller coaster, the designer wishes the riders to experience weightlessness as they round the top of one hill. How fast must the car be going if the radius of curvature at the hilltop is 20 m?

38* A Ferris wheel is a large, vertical, circular device which carries riders in horizontal seats around the vertical circle. One such wheel has a radius of 20 m. (a) How fast, in revolutions per second, must the wheel be turning if the rider is to push down on the seat with a force 1½ times as large as the rider's weight when the rider is at the bottom of the circle? (b) How fast must the wheel be turning if the rider is to exert no force on the seat at the top of the circle?

39* In one model of the hydrogen atom (the Bohr model), an electron is pictured rotating in a circle (with a radius of 0.5×10^{-10} m) about the positive nucleus of the atom. The centripetal force is furnished by the electrical attraction of the positive nucleus for the negative electron. How large is this force if the electron is moving with a speed of 2.3×10^6 m/s? (The mass of an electron is 9×10^{-31} kg.)

40* A certain disk oriented horizontally starts from rest and begins to rotate about its axis (vertical) with an acceleration of 0.50 rev/s². After 20 s, a 3.0-g mass cemented to the rim of the wheel breaks loose from the wheel. How large was the force holding the mass in place? The wheel's radius is 40 cm.

41 The moon orbits the earth in an approximately circular path of radius 3.8×10^8 m. It takes about 27 days to complete one orbit. What is the mass of the earth as obtained from these data?

42 With what speed, in meters per second, must a satellite travel in a circular orbit about the earth if its orbital radius is to be 7.0×10^6 m? At this radius, the satellite would be about 400 mi above the earth.

43* In an ultracentrifuge a solution is rotated with an angular speed of 3000 rev/s at a radius of 10 cm. How large is the radial acceleration of each particle in the solution? Compare the centripetal force needed to hold a particle of mass m in the circular path with the weight of the particle mg.

44* The red blood cells and other particles suspended in blood are too light to settle out easily when the blood is left standing. How fast, in revolutions per second, must a sample of blood be rotating at a radius of 10 cm in a centrifuge if the centripetal force needed to hold one of the particles in a circular path is 10,000 times the weight of the particle, mg? Why do the particles separate from the solution in a centrifuge?

45* A certain car of mass m has a maximum friction force of 0.7mg between it and the pavement as it rounds a curve on a flat road ($\mu = 0.7$). How fast can the car be moving if it is to successfully negotiate a curve of 15-m radius?

46** The rotational speed of the earth is 1 rev/day, or 1.16×10^{-5} rev/s, and the earth's radius is 6.37×10^6 m. If a man at the equator were standing on a spring scale, by what percentage would his apparent weight increase if the earth were to stop rotating? A man at the north pole?

47** A 2.0-kg ball hangs as a pendulum from a cord 10.0 m long. If the pendulum is pulled to one side so that it makes an angle of 37° with the vertical and is then released, find the tension in the cord (*a*) just after its release and (*b*) when it passes through the bottom of its swing.

48** As shown in Fig. P6.3, a boy on a rotating platform holds a pendulum in his hand. The pendulum is at a radius of 6.0 m from the center of the platform. The rotational speed of the platform is 0.020 rev/s. It is found that the pendulum hangs at an angle θ to the vertical as shown. Find θ.

49** The bug shown in Fig. P6.4 has just lost its foot-

ing near the top of the bowling ball. It slides down the ball without appreciable friction. Show that it will leave the surface of the ball at the angle θ shown, where θ is given by $\cos \theta = \frac{2}{3}$.

50** Figure P6.5 shows a possible design for a space colony. It consists of a 6-km-diameter cylinder of length 30 km floating in space. Its interior is provided with an earthlike environment. To simulate gravity, the cylinder spins on its axis, as shown. What should be the rate of rotation of the cylinder, in revolutions per hour, so that a person standing on the land mass will press down on the ground with a force equal to his or her weight on earth? (For details, see G. K. O'Neill, *Physics Today*, September 1974, p. 32 and February 1977, p. 30.)

51** A particle is to slide along the horizontal circular path on the inside of the funnel shown in Fig. P6.6. The surface of the funnel is frictionless. How fast must the particle be moving, in terms of r and θ, if it is to execute this motion?

52** In Fig. P6.7 the mass m is held by two strings, and the system is rotating with angular velocity ω.

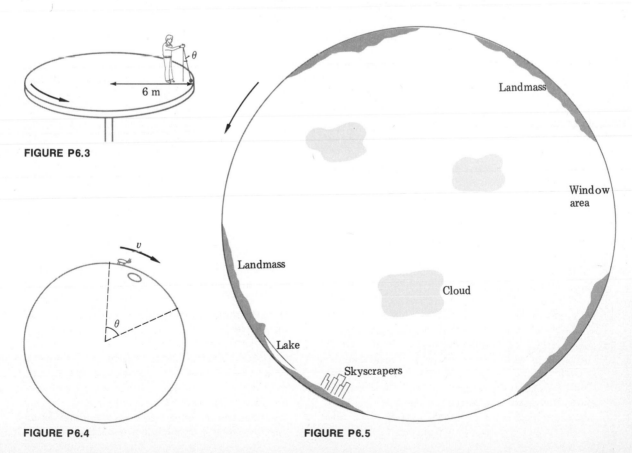

FIGURE P6.3

FIGURE P6.4

FIGURE P6.5

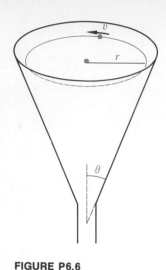

FIGURE P6.6

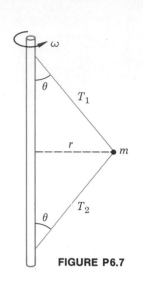

FIGURE P6.7

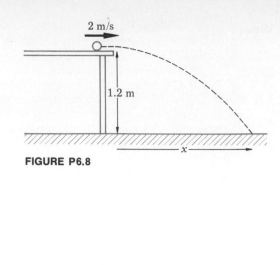

FIGURE P6.8

FIGURE P6.9

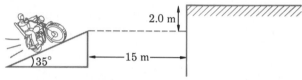

Find the tensions in the two strings in terms of m, ω, r, and θ.

53 A marble rolls off the edge of a horizontal table 1.20 m high, as shown in Fig. P6.8. Find the distance labeled x if the marble is rolling at 2.00 m/s. What are the vertical and horizontal components of its velocity just before it hits the floor?

54 A fire hose shoots water horizontally from the top of a tall building toward the wall of a building 20 m away. If water leaves the hose at 5.0 m/s, how far below the hose level does the water strike the wall? *Hint:* Consider the water to consist of a series of particles shooting along the stream.

55 An electron ($m = 9.1 \times 10^{-31}$ kg) traveling horizontally leaves an electron gun at the end of a TV tube with a speed of 10^8 cm/s. If the fluorescent screen on the opposite viewing end of the tube is 40 cm away, how far below its original level has the electron fallen by the time it hits the screen?

56* At a circus, the "human cannonball' is shot out of the cannon with a speed of 18 m/s. The cannon barrel is pointed 40° above the horizontal. How far from the end of the cannon should the net (used to catch the person) be placed? Assume the net is at the same level as the end of the barrel. If the person is accelerated uniformly from rest to 18 m/s in

3.0 m in the barrel, how large a force must push on the person? Give your answer in terms of W, the weight of the person.

57* A ball is thrown at 20 m/s at an angle of 30° below the horizontal from a bridge 30 m above the water. (*a*) Where, relative to a point on the water directly below the throwing point, does the ball hit the water? (*b*) For how long is the ball in the air?

58* Repeat Prob. 57 for a ball thrown at an angle of 30° above the horizontal.

59** As shown in Fig. P6.9, a stunt driver wishes to shoot off the incline and land on the platform. How fast must the motorcycle be moving if the stunt is to succeed?

60** A projectile is shot from the ground with a velocity v_0 at an angle θ above the horizontal level ground. It returns to the ground at a distance R from the shooting point. Show that the range R of the projectile is given by

$$R = \frac{2v_0^2 \sin \theta \cos \theta}{g}$$

provided friction forces are negligible. Using the trigonometric formula $2 \sin \theta \cos \theta = \sin 2\theta$, show that the range is maximum when $\theta = 45°$.

7
Motion of Rigid Bodies

The translational motion of an object depends on the net force acting on the object. We see in this chapter that the rotational motion of an object depends on the net torque which acts on the object. In place of $F_{net} = ma$, which we found so useful in describing translational motion, there is a relation between torque and angular acceleration that applies to rotation. Moreover, objects possess rotational inertia as well as translational inertia, and so a quantity analogous to mass is of importance for rotation. In this chapter we examine the combined translational and rotational behavior of rigid objects subjected to forces and torques.

7.1 Equilibrium of a Point Object

An object is said to be *in equilibrium* if it is undergoing neither linear nor angular acceleration. Of course, this means that an object which is at rest and remains at rest is in equilibrium; this is the colloquial meaning of the phrase. But in science we extend the meaning so that it applies to all nonaccelerating objects. Let us now determine what conditions must be satisfied if an object is to be in equilibrium.

At this time we restrict the discussion to *point objects*. These are objects whose dimensions are so small that rotation of the object can be ignored. Later we extend the discussion to include objects for which rotation is of importance. But for now we consider only the translational motion of an object, such as the one shown in Fig. 7.1.

If the object shown in Fig. 7.1 is to be in translational equilibrium, then its acceleration a must be zero. Newton's second law, $F_{net} = ma$, then tells us that the condition for such a situation to exist is that the net force acting on the object must be zero. We can state this *condition for translational equilibrium* as follows:

> For an object to be in **translational equilibrium,** $F_{net} = 0$ where F_{net} is the resultant external force acting on the object.

As we shall see in applications of this condition, frequently we write $F_{net} = 0$ as three equations involving the components of the vector forces. If $F_{not} = 0$, then the sum of all the x-component forces acting on the object must be zero, and similarly for the y- and z-component forces. Using the summation sign ΣF_x to represent the words *the sum of all the x-directed forces,* we have the following alternate way of stating the condition for translational equilibrium:

> For an object to be in translational equilibrium, $\Sigma F_x = 0, \Sigma F_y = 0$, and $\Sigma F_z = 0$ for the forces that act on the object. Or, $\mathbf{F}_{net} = \Sigma \mathbf{F} = 0$.

Let us now show how we make use of this condition.

Illustration 7.1 The object shown in Fig. 7.2 is subjected to three forces, as shown. As indicated, $F_1 = 200$ N, $F_2 = 300$ N, and F_3 is unknown in both magnitude and direction. Find F_3 if the object is to remain at rest.

Reasoning Because the object is to remain in translational equilib-

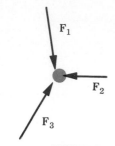

FIGURE 7.1
What condition must the force on the object satisfy if the object is to be in translational equilibrium?

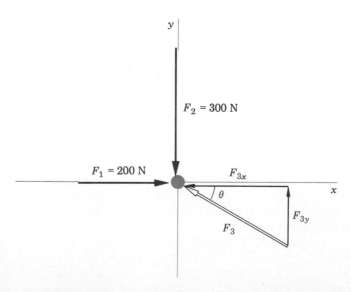

FIGURE 7.2
Find F_3 if the object is to remain in translational equilibrium.

rium, we know that the forces acting on it must satisfy the conditions $\Sigma F_x = \Sigma F_y = \Sigma F_z = 0$. There are no z-directed forces acting on the object, so we are concerned with only the following:

$$\Sigma F_x = 0 \quad \text{or} \quad F_1 - F_{3x} = 0$$

and

$$\Sigma F_y = 0 \quad \text{or} \quad -F_2 + F_{3y} = 0$$

where we have written the second equation in each case by referring to the figure.

We find from these two equations that

$$F_{3x} = F_1 = 200 \text{ N}$$

and

$$F_{3y} = F_2 = 300 \text{ N}$$

Therefore

$$F_3 = \sqrt{F_{3x}^2 + F_{3y}^2} = \sqrt{200^2 + 300^2} \text{ N} = 360 \text{ N}$$

To find the angle θ, we note that

$$\tan \theta = \frac{F_{3y}}{F_{3x}} = \frac{300 \text{ N}}{200 \text{ N}} = 1.50$$

from which $\theta = 56°$.

Illustration 7.2   For the 20.0-kg mass in Fig. 7.3 to hang in equilibrium, as shown, what must be the tension in the light horizontal cable that holds it?

Reasoning   Everything shown in part a is in equilibrium, and so we

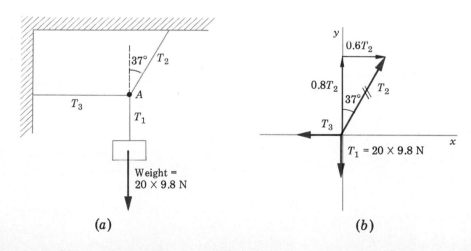

(a)

(b)

FIGURE 7.3
The free-body diagram for
the knot at A is shown in
part b.

could apply our equilibrium condition to many things. For example, if we applied it to the mass, we would learn that the tension T_1 is 20×9.8 N. This follows from the fact that only T_1 and the weight act on the mass, and so these two forces must be equal and opposite if the mass is to hang at rest. In addition, the cables pull on the wall and ceiling. But we are not interested in these forces.

It proves convenient in situations such as this to isolate the knot at the junction of the three cables at A as the object for discussion. The knot experiences three separate forces due to the tensions T_1, T_2, and T_3 in the cables that pull on it. (Recall that the tension in a rope is the force which the rope exerts on the object that holds the rope's end.) We show these three forces in the free-body diagram for the knot shown in part b of the figure.

Using the free-body diagram, we can write the equilibrium conditions for the knot:

$$\Sigma F_x = 0 \qquad \text{or} \qquad -T_3 + 0.60T_2 = 0$$

$$\Sigma F_y = 0 \qquad \text{or} \qquad 0.80T_2 - T_1 \quad = 0$$

But $T_1 = 196$ N, and so the second equation yields $T_2 = 245$ N. Placing this in the first equation yields the quantity requested, $T_3 = 147$ N.

7.2 Torques

To be in equilibrium, objects that can rotate require not only that $F_{\text{net}} = 0$, but also that a second condition be satisfied. We can see this if we refer to the simple experiment shown in Fig. 7.4a. We see there a meterstick pivoted at its center so that it is free to rotate. The stick remains at rest at any angle

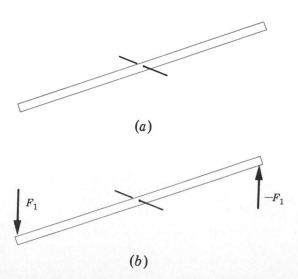

(a)

FIGURE 7.4
Even though $\mathbf{F}_{\text{net}} = 0$ in part
a and b, the meterstick will
not remain at rest in part b.

F_1

$-F_1$

(b)

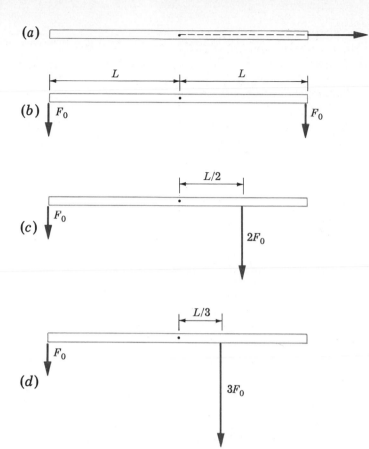

FIGURE 7.5
The meterstick is in equilibrium in each of the cases shown. We generalize to state that a force kF_0 placed at L/k balances a force F_0 placed a distance L from the pivot.

we place it. Obviously, the downward pull of gravity on the stick is balanced by the upward force which the pivot pin exerts on it. Our condition for translational equilibrium is therefore satisfied.

But if equal, oppositely directed forces are applied to the stick, as in part *b*, the equilibrium condition is still satisfied because the two new forces cancel each other. In spite of that fact, we know that these two balanced forces will cause the stick to rotate about the pivot, to undergo angular acceleration. Clearly, the stick is not in equilibrium in Fig. 7.4*b* even though the condition for translational equilibrium is satisfied. We need yet a second condition to ensure rotational equilibrium, i.e., to ensure that there will be no rotational acceleration.

To learn more about how forces cause rotation, let us refer to Fig. 7.5. We see there several simple experiments involving a center-pivoted meterstick that tell us a great deal about this subject. In each case shown, the meterstick is found to be in equilibrium. Referring to each part in turn, we can draw the following conclusions: (*a*) A force, the line of which passes through the pivot (or axis), causes no turning effect about that axis. (*b–d*) A perpendicular force F_0 placed at a distance L from the pivot balances a force kF_0 placed at a perpendicular distance L/k from the pivot, where k is a constant. These conclusions are confirmed by a multitude of experiments.

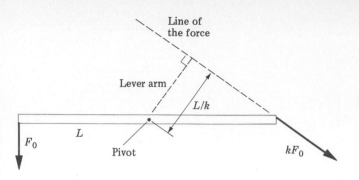

FIGURE 7.6
The stick is at equilibrium.
The turning effect (torque)
of a force equals the
product of the force and its
lever arm.

We see from these experiments that the turning effect of a force is dependent on both the magnitude of the force and the distance from the pivot of the line along which the force acts. This latter point can be clarified by experiments similar to the one shown in Fig. 7.6. Notice that the distance labeled *lever arm* in the figure is of importance. The results of these experiments can be summarized this way:

Definition   *The turning effect of a force, the* **torque,** *is equal to the product of the force and its lever arm,*

where

Definition   *The* **lever arm** *of a force is the length of the perpendicular dropped from the axis (or pivot) to the line of the force.*

It is customary to represent torque by τ, Greek tau.

If we refer to Fig. 7.5*a*, we see that the lever arm is zero, and so the torque (or turning effect) is

$$\text{Torque} = (\text{force}) \cdot (\text{lever arm}) = 0$$

Also, the torques in parts *b*, *c*, and *d* are all the same because

$$\text{Torque} = F_0 L = 2F_0 \frac{L}{2} = 3F_0 \frac{L}{3} = kF_0 \frac{L}{k}$$

Let us review what we have learned about torque. The torque, or turning action, of a force in respect to a pivot (or an axis) is equal to the product of the force and its lever arm. The lever arm is the length of a perpendicular dropped from the pivot to the line of the force.

Illustration 7.3   Find the lever arms for the forces shown in Fig. 7.7 with respect to the indicated axis.

Reasoning   To find the lever arm, drop a perpendicular to the line of

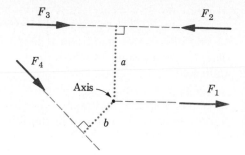

FIGURE 7.7
Find the lever arms for the forces with respect to the pivot shown.

the force from the pivot. In doing this, the lever arms are the lines indicated by dotted lines in the figure. For F_1 the lever arm is zero because the line of the force goes through the pivot.

The lever arms for F_2 and F_3 are the same because both forces act along the same line. This shared lever arm is labeled a in the figure. The lever arm for F_4 is obtained in a similar way and is labeled b.

7.3 Condition for Rotational Equilibrium

Now that we have mastered the concept of torque, we are prepared to state the condition that must prevail if an object is to remain in rotational equilibrium. For equilibrium to prevail, the net effect of all the torques acting on an object must be zero. To state this condition in compact form, we assign direction to torque, as shown in Fig. 7.8. We see there a meterstick pivoted

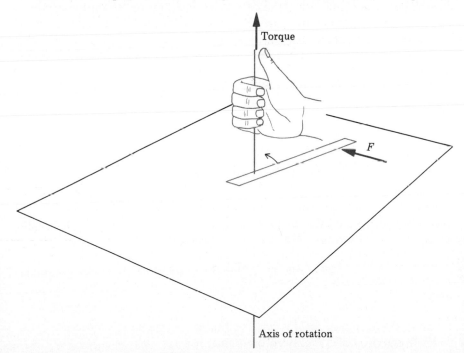

FIGURE 7.8
The force **F** causes a counterclockwise rotation about the axis shown. To find the direction of the torque, we use the right-hand rule. Notice that the torque direction is perpendicular to the plane of rotation.

at one end. It is being given a counterclockwise rotation by the force F. (Assume the stick to be the hand of a clock. It rotates in a direction opposite to that of a clock's hand and is therefore said to rotate counterclockwise.) The direction assigned to the torque caused by F is upward along the axis as indicated.

There is a simple *right-hand rule* that people remember for assigning direction to torque and many other quantities in physics. As shown in the figure, direction is assigned in the following way:

Right-Hand Rule

Grasp the axis of rotation in the right hand with fingers circling the axis in the same sense as the rotation. Then the thumb of the hand points along the axis in the direction of the quantity concerned, the torque in this instance.

Using this rule to provide direction to torque, we can assign to it both a magnitude and a direction. Therefore *torque is a vector quantity*.

The *condition for rotational equilibrium* can now be stated in terms of the net torque acting on an object:

If an object is to remain in **rotational equilibrium,** then the vector sum of the torques that act on it must be zero:

$$\Sigma \tau = 0$$

This general condition can be simplified in the applications we consider. For situations such as the one shown in Fig. 7.8, the object can only turn about a fixed axis. In these cases, the direction of the torque and the rotation can be specified by stating whether the torque is clockwise or counterclockwise. The object shown in the figure is undergoing a counterclockwise torque and rotation as we have said earlier. It is customary (although arbitrary) to take counterclockwise torques as positive and clockwise torques as negative. Then the equation $\Sigma \tau = 0$ can be written, with plus and minus signs to show the directions of the torques.

7.4 Rigid Objects in Equilibrium

Now that we know the two necessary conditions for equilibrium, and there are only two, we are able to treat all ordinary situations in which a rigid object is at equilibrium. The two conditions are

$$\Sigma F_x = 0 \qquad \Sigma F_y = 0 \qquad \Sigma F_z = 0 \tag{7.1}$$

and
$$\Sigma \tau = 0 \tag{7.2}$$

In using these equations, it is wise to follow a systematic procedure. The following approach is highly recommended.

1 Isolate a body for discussion.
2 Draw a free-body diagram showing the forces that act on the isolated body.
3 Split the forces into their components.
4 Write Eqs. (7.1) and (7.2).
5 Solve the equations for the unknowns.

Let us illustrate the procedure by the following example.

Illustration 7.4 In Fig. 7.9 the weight W is supported by a beam of length L and negligible weight. Find the tension T in the supporting cord in terms of W.

Reasoning First, we isolate the beam as the object for discussion. Second, its free-body force diagram is drawn in part b of the figure. Because we know very little about the forces on the beam at its hinged end, we represent the force there in a general way by giving its components, H and V. Third, we see that only T needs to be split into components, and this is done in the figure.
We now write the equilibrium equations:

$$\Sigma F_x = 0 \qquad \text{becomes} \qquad 0.50T - H = 0$$

$$\Sigma F_y = 0 \qquad \text{becomes} \qquad V + 0.87T - W = 0$$

To write the torque equation, we must choose a pivot or an axis. It seems natural to take the axis at point P, at the hinge. (We shall soon see, however, that we have a much wider choice than this.) For an axis at P, the lever arms for H and V are zero, so they cause no torque. The torque due to the weight W is clockwise (and therefore negative) and has a magnitude WL because L is its lever arm.
We choose to find the torque due to T by using its components. The

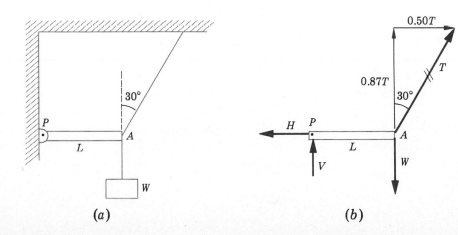

(a)   (b)

FIGURE 7.9
The beam is isolated as the object for discussion, and its free-body diagram is shown in part b. We assume the beam's weight to be negligible.

component $0.87T$ causes a counterclockwise (positive) torque of magnitude $(0.87T)(L)$. Notice that the other component, $0.50T$, really acts at the end of the beam, at A. Its line therefore goes through the axis, and thus its lever arm and torque are zero.

With these considerations in mind, taking an axis at P, we find that

$$\Sigma \tau = 0 \quad \text{becomes} \quad -WL + (0.87T)(L) = 0$$

This latter equation can be solved for T to give $T = 1.15W$. This value can be used in the previous equations to find V and H if one wishes to do so.

7.5 Contor of Gravity

Two possible complications were deftly avoided in Illustration 7.4. One involved the choice of axis, and we discuss this matter in the next section. The other was avoided by assuming the beam to have negligible weight. Because that will not be true in general, we must now consider what effect the weight of an object will have when we write the torque equation for the object. In particular, where can we consider the pull of gravity to be applied to an object so that we can state its lever arm?

Of course, gravity pulls on all parts of each object. But it turns out that, for torque purposes, the pull of gravity (the object's weight) appears to act at a point. We call this point the *center of gravity* of the object. Let us now see how this point can be located experimentally.

Suppose we wish to locate the center of gravity of the object shown in Fig. 7.10. We first support it from a string and pivot as in part a. The weight of the object causes the object to take the equilibrium position shown. At equilibrium, the torques about the pivot must add to zero. Only two forces act on the object—the upward pull of the string and the downward pull of

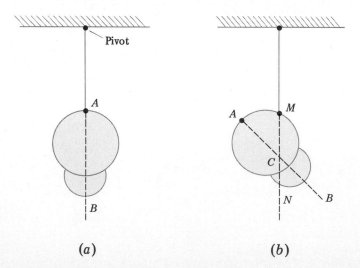

FIGURE 7.10
An experimental method for determining the center of gravity of an object.

(a) (b)

gravity. The torque due to the string is zero. (Why?) Therefore, the torque due to the pull of gravity must also be zero. This could be true only if the lever arm for the weight were zero, and this can be true only if the force of the weight acts along the line *AB* so its line passes through the pivot. Hence we learn from part *a* that the pull of gravity, the weight of the object, acts at a point somewhere along line *AB*.

But we can repeat this experiment for another suspension point, as in Fig. 7.10*b*. By using the same reasoning, the pull of gravity acts at a point along the line *MN*. We see, then, that the point at which the weight acts is on both lines *MN* and *AB*. It must therefore act at their point of intersection, point *C* in the figure. A check may be made by using a third suspension point. Its vertical line also will intersect the others at point *C*. We therefore conclude point *C* to be the center of gravity for this object.

> *The **center of gravity** of an object is the point at which the weight of an object can be considered to act for the purpose of computing torques.*   **Definition**

7.6 The Position of the Axis Is Arbitrary

Often an object at equilibrium has an obvious axis, and it is common to use that point as the axis for computation of torques. In other situations, no obvious axis exists. We see in this section that we are free to choose an axis anywhere that is convenient when we apply the torque equation to an object at equilibrium.

Consider the situation shown in Fig. 7.11 where a sign painter of weight W_p stands in equilibrium on a uniform board of weight W_b and length L. The center of gravity of the uniform board is at its center, and so W_b is shown to act at the board's center in Fig. 7.11*b*. We shall try to show that the final form of the torque equation we write for this equilibrium situation does not depend on the position we choose for the axis.

Let us choose the axis at the point *A* shown. You should verify that the torque equation $\Sigma\tau = 0$ becomes

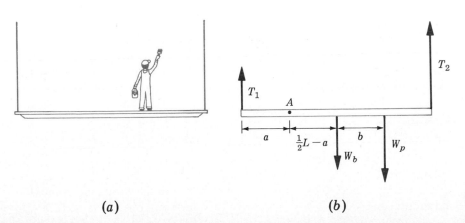

(a)   (b)

FIGURE 7.11
The position of the axis *A* is arbitrary.

$$-T_1(a) - W_b(\tfrac{1}{2}L - a) - W_p(\tfrac{1}{2}L - a + b) + T_2(L - a) = 0$$

Let us group the terms involving the arbitrary length a. Then

$$-a(T_1 - W_b - W_p + T_2) - \tfrac{1}{2}W_bL - W_p(\tfrac{1}{2}L + b) + T_2L = 0$$

But we can easily show that the factor multiplying a is zero *provided the system is at equilibrium.* At equilibrium,

$$\Sigma F_y = 0 \qquad \text{so} \qquad T_1 + T_2 - W_b - W_p = 0$$

This tells us the term in question is zero. Therefore, the torque equation becomes

$$-\tfrac{1}{2}W_bL - W_p(\tfrac{1}{2}L + b) + T_2L = 0$$

which is independent of a and the position chosen for the axis. In this case at least, the position chosen for the axis is arbitrary.

Although we have obtained this result for a specialized situation, it is possible to give a more general proof of the proposition. We therefore have the following general result:

> In writing the torque equation for a body in equilibrium, the position chosen for the axis is arbitrary.

In practice, we usually choose the axis so that the line of an unknown force will pass through the axis. Then the torque due to that force will be zero, and its unknown value will not appear in the torque equation. Let us now continue with examples of equilibrium problems.

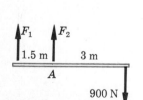

(a)

Illustration 7.5 As shown in Fig. 7.12, a 900-N man is about to dive from a diving board. Find the forces exerted by the pedestals on the board. Assume the board has negligible weight. (900 N is about 200 lb.)

Reasoning We isolate the board and draw the forces on it, as shown in part *b* of the figure. A little reflection will convince you that we have drawn the force F_1 in the wrong direction. This is a deliberate error, and we shall see how it affects the result. Choosing point A as pivot, we have

$\Sigma \tau = 0:$ $\qquad -(900 \text{ N})(3 \text{ m}) - (F_1)(1.5 \text{ m}) = 0$

$\Sigma F_y = 0:$ $\qquad\qquad\qquad F_1 + F_2 - 900 \text{ N} = 0$

Solving these gives $F_1 = -1800$ N and $F_2 = 2700$ N. Notice that the negative sign found for F_1 tells us our force was drawn in the wrong direction. Its magnitude is 1800 N.

F_1 F_2

1.5 m 3 m

A

900 N

(b)

FIGURE 7.12
A 900-N man stands at the end of a light diving board. We guess, incorrectly, that the forces exerted by the two pedestals on the board are as shown.

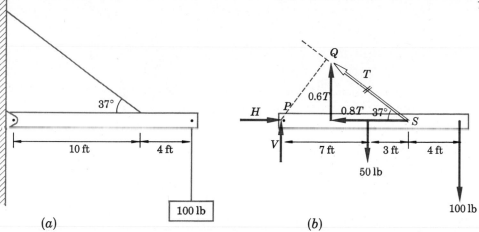

(a)

(b)

FIGURE 7.13
The forces acting on the
beam in part a are shown in
detail in part b. Notice that
the force component 0.6T
pulls on the board at point
S, and so its lever arm is
10 ft.

Illustration 7.6   For the uniform 50-lb beam shown in Fig. 7.13, how large is the tension in the supporting cable, and what are the components of the force exerted by the wall on the beam?

Reasoning   If we isolate the beam for consideration, the forces acting on it are shown in Fig. 7.13b. Notice that the weight of the beam, 50 lb, is taken to act at the board's center of gravity. Further, notice how the tension in the cable has been replaced by its components. We can eliminate the force components at the wall, H and V, by taking point P as the axis. Then

$\Sigma \tau = 0$: $(0.6T)(10) - (50)(7) - (100)(14) = 0$

$\Sigma F_x = 0$: $H - 0.8T = 0$

$\Sigma F_y = 0$: $V + 0.6T - 50 - 100 = 0$

First solving for T in the torque equation and then substituting in the other two equations give

$$T = 292 \text{ lb} \qquad H = 234 \text{ lb} \qquad V = -25 \text{ lb}$$

What does the minus sign on V tell us?

Illustration 7.7   A person holds a 20-N weight, as shown in Fig. 7.14a. Find the tension in the supporting muscle and the component forces at the elbow.

Reasoning   The system can be replaced by the simplified model shown in part b. We assume the lower arm to weigh 65 N. The dimensions given are typical. Notice that the situation is very similar to that of the previous illustration, the case of a beam supported by a cable. The appro-

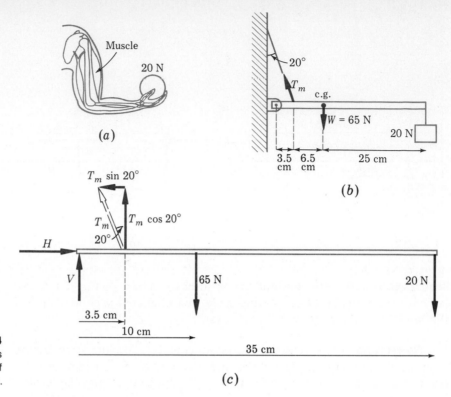

FIGURE 7.14
We can analyze the forces
in the human arm by use of
the models in parts *b* and *c*.

priate free-body diagram is shown in part *c*. We use it to write the equilibrium conditions:

$$\Sigma F_x = 0: \qquad\qquad\qquad H - T_m \sin 20° = 0$$

$$\Sigma F_y = 0: \qquad\qquad\qquad V + T_m \cos 20° - 65 - 20 = 0$$

$$\Sigma \tau = 0: \qquad (T_m \cos 20°)(0.035) - (65)(0.10) - (20)(0.35) = 0$$

where the left end of the arm (the elbow) is taken as the axis. Notice that the
force $T_m \sin 20°$ actually acts through the axis, and so its torque is zero.
Solving these equations, we find

$$T_m = 410 \text{ N} \qquad H = 140 \text{ N} \qquad V = -300 \text{ N}$$

All these forces are much larger than the weight being held. Can you show
that T_m becomes very large as the arm is outstretched? Why is it very tiring
to hold a weight in your outstretched hand?

Illustration 7.8 A uniform 50-lb ladder leans against a smooth wall, as
in Fig. 7.15. (By the term **smooth,** we mean the force at the wall is perpendicular to the wall surface. No friction exists.) If a 100-lb boy stands on the
ladder as shown, how large are the forces at the wall and at the ground?

Reasoning If we isolate the ladder, the forces acting on it are as shown in part b of the figure. We then have (taking point A as the axis)

$\Sigma F_x = 0$: $H - P = 0$

$\Sigma F_y = 0$: $V - 50 - 100 = 0$ or $V = 150$ lb

$\Sigma \tau = 0$: $(P)(0.8 \times 20) - (50)(0.6 \times 10) - (100)(0.6 \times 15) = 0$

In this problem, the lever arms should be noticed particularly. By definition, the lever arm is the perpendicular dropped from the axis A to the line of the force. The lever arms are AB, AC, and AD for the forces 50, 100, and P, respectively. Solving the equations simultaneously gives

$$V = 150 \text{ lb} \qquad P = H = 75 \text{ lb}$$

Illustration 7.9 Consider the woman lifting a 60-N bowling ball, as shown in Fig. 7.16a. Find the tension in her back muscle and the compressional force in her spine. Assume the upper part of her body to weigh 250 N.

Reasoning The horizontal upper portion of her body can be represented as a beam as shown in Fig. 7.16b. To a rough approximation, the dimensions can be taken as shown. When the pivot point is taken as shown, the equilibrium equations become

$\Sigma F_x = 0$: $H - T_m \cos 12° = 0$

$\Sigma F_y = 0$: $T_m \sin 12° + V - 60 - 250 = 0$

$\Sigma \tau = 0$: $(T_m \sin 12°)\left(\dfrac{2L}{3}\right) - (250)\left(\dfrac{L}{2}\right) - (60)(L) = 0$

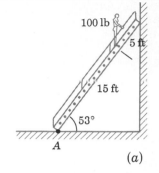

(a)

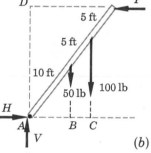

(b)

FIGURE 7.15
A 100-lb boy stands on the 50-lb ladder, as shown. On the assumption that the wall is smooth, the forces acting on the ladder are shown in part b.

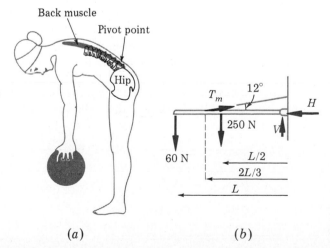

(a)   (b)

FIGURE 7.16
The forces in the woman's back can be found using the model shown in part b.

How do the factors containing T_m arise? Solving these equations yields

$$T_m = 1335 \text{ N} \qquad V = 32 \text{ N} \qquad H = 1305 \text{ N}$$

The compression in the lower portion of the spine will be H. Notice that $H = 1305 \text{ N} \approx 290 \text{ lb}$. The tension in the back muscle is even larger than this. Although she is lifting only 60 N (about 15 lb), the forces are already large. Why should you not lift a very heavy object in the way shown? How should you lift a heavy object?

7.7 Stable, Neutral, and Unstable Equilibrium

As we have seen, a rigid object is in equilibrium if $\Sigma \mathbf{F} = 0$ and $\Sigma \tau = 0$. However, there are three distinct types of static equilibrium. They are illustrated in Fig. 7.17. It is reasonably obvious how the stable and unstable forms obtain their names. If the stable cone is tilted slightly away from the position shown, it will return to position when released. But the unstable cone will topple over if even slightly tilted. The characteristic that distinguishes the third case, the case of neutral equilibrium, is that the object has no special preference for any one position.

An object is in **stable equilibrium** *with respect to a slight disturbance if after the cause of the disturbance is removed, the object returns to its nondisturbed condition.* For example, in Fig. 7.18a, the cone has been slightly disturbed. When released, the cone will fall back to its original position. This return is caused by the torque, due to the weight mg, about the point of contact A. In passing we should also note that the center of gravity of the object was raised during the disturbance. As a result, the disturbed position has a higher PE than the stable position.

An object is in **unstable equilibrium** *with respect to a slight disturbance if after the cause of the disturbance is removed, the object by itself increases the disturbance.* For example, in Fig. 7.18b, the cone will continue to fall over if released. It is pulled over by the torque, due to the weight mg, about the point of contact A. Moreover, we should notice that the disturbance *lowers* the center of gravity and PE for the object in part b.

An object is in **neutral equilibrium** *if it is indifferent to a disturbance.* For example, in part c, the cone will remain in any position to which it is rolled. In this case the torque due to mg is zero about the points of contact as axis. The line of the force goes through the contact line. In addition, this

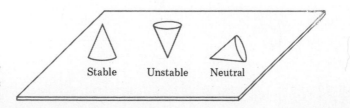

FIGURE 7.17
The three forms of static equilibrium.

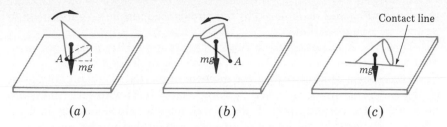

Contact line

(a)   *(b)*   *(c)*

type of disturbance neither raises nor lowers the object's center of gravity. The PE of the object is not changed by the disturbance.

This particular physical example is not unique. All rigid bodies behave in a similar way. We can therefore make the following statement in regard to static equilibrium:

> An object's equilibrium can be characterized as (1) stable, (2) unstable, or (3) neutral depending on whether a slight disturbance (1) raises, (2) lowers, or (3) leaves unchanged the PE of the object.

Static Equilibrium

Indeed, we shall see that this statement applies to other forms of PE in addition to gravitational PE.

7.8 More Analogies between Linear and Rotational Motion

Anyone who has ever spun a wheel on an axle knows that the wheel possesses a quality closely akin to inertia. The fact that a body possesses inertia makes it necessary for us to exert a force on it if it is to be set in motion. Similarly, if the body is already in motion, a force is required to bring it to rest. This is evident in rotational motion as it is in translational motion.

A spinning fan blade continues to spin for some time after the fan motor has been shut off. The person who tries to stop the blade with a finger is likely to find that the blade has enough rotational inertia to injure the finger. It is also apparent that the fan blade resists being put into rotation, since it takes several seconds for the motor to get the fan up to speed. These typical evidences of inertia are present to a greater or lesser extent in all rotating objects.

It was found in earlier chapters that the force needed to overcome inertia effects and to set a body in linear motion is expressed by Newton's law,

$$F_{\text{net}} = ma$$

In the last chapter, we saw that the equations of linear motion have analogous rotational counterparts. Clearly, for rotational motion, the linear acceleration *a* in Newton's law should be replaced by the angular acceleration α. We might even guess, since torques cause bodies to rotate, that the unbalanced force F should be replaced by the unbalanced torque τ. This is seen to be true in a later section.

However, it is not quite so obvious what should be done with m to obtain the rotational form of Newton's law. We shall soon see that it must be replaced by a quantity called the **moment of inertia.** This quantity must be dependent on mass, since we all know that objects with large mass are more difficult to set into rotation than small masses. However, we shall see that the geometry of the object is also of considerable importance.

Finally, the center of mass of a rotating wheel mounted on a stationary axle has zero velocity, and so the wheel has no kinetic energy of translation. Nor does it possess any translational momentum. Yet we know that a rotating flywheel can do work because of its rotation. For example, it could wind up a rope with a weight hanging from it. Therefore, it does possess kinetic energy, i.e., energy of motion. In the following portions of this chapter, we find an expression for this KE of rotation, as well as for the rotational momentum of the rotating object.

7.9 Relation between Torque and Angular Acceleration

Consider the situation shown in Fig. 7.19. A relatively large mass m is held to an axle at O by a very light rod. Assume that the bearing at the pivot point O is frictionless, and further assume that the mass of the rod is negligible. Let us take the system to be in a horizontal plane so that gravity will cause no torque about the pivot point.

The force F acting as shown on the mass m will not be opposed by any other force, since there is no friction in the pivot. It then is an unbalanced force acting on m, and it will accelerate the mass according to

$$F = ma$$

However, in so doing, it will cause the mass to rotate about the pivot point.

Since we saw in the last chapter that the linear acceleration a was related to the angular acceleration α by the equation

$$\alpha r = a$$

we have at once

$$F = mr\alpha$$

Since the relation between α and a was true only in radian measure, we must measure angles in radians when using this equation.

If the force F in Fig. 7.19 had not been perpendicular to the bar, only the component of F acting on the mass in a direction perpendicular to the

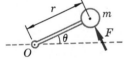

FIGURE 7.19
The force F causes a torque about the axis O and gives the mass m an angular acceleration about the pivot point.

bar would cause rotation of the object. This is merely another way of saying that rotations are caused by torques, since a torque is defined to be the product of a force and its lever arm. Hence, it would be better to write our equations for rotations in terms of torques. We may do this by multiplying both sides of the last equation by r, since the torque is just Fr. Thus

$$Fr = \tau = \text{torque} = mr^2\alpha \qquad \alpha \text{ in radians per second squared} \quad (7.3)$$

Equation (7.3) is one form of the rotational analog to Newton's equation $F = ma$. As we expected, F is replaced by τ, the torque on the system, and a is replaced by α, the angular acceleration. In place of mass m, we have what is called the moment of inertia for this system, mr^2. Notice that the rotational inertia of the system depends not only on m but also on r. More is said about this later.

In practice, most rigid rotating bodies have the mass distributed at many different distances r from the pivot point. This is shown in Fig. 7.20a. We can easily modify Eq. (7.3) to apply in this case as well. The only change is to split up the rigid object into many little pieces at various radii and write Eq. (7.3) for each mass individually.

In Fig. 7.20a, several of the small pieces of mass from which the object is constituted are illustrated. The torque about the pivot acting on the first mass, m_1, is the result of all forces acting on that mass, whether due to forces external to the object or to forces exerted by the masses adjacent to m_1. Let us call this torque τ_1. Applying Eq. (7.3) to mass m_1, we have

$$\tau_1 = m_1 r_1^2 \alpha_1$$

where r_1 is the distance from m_1 to the pivot point. A similar equation can be written for each mass in the object. Let us write the sum of these equations, to give

$$\tau_1 + \tau_2 + \cdots + \tau_N = m_1 r_1^2 \alpha_1 + m_2 r_2^2 \alpha_2 + \cdots + m_N r_N^2 \alpha_N \quad (7.4a)$$

where it is assumed there are N masses composing the object. This equation can be written more concisely by abbreviating Eq. (7.4a) as follows:

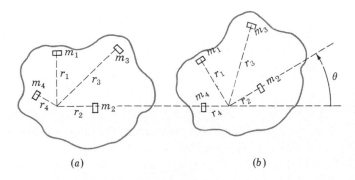

(a)   (b)

 FIGURE 7.20
Each small piece of mass within a large, rigid body undergoes the same angular acceleration about the pivot point.

$$\sum_{i=1}^{N} \tau_i = \sum_{i=1}^{N} m_i r_i^2 \alpha_i \qquad (7.4b)$$

where the symbol $\sum_{i=1}^{N}$ means we are taking a sum of terms. The first term has i replaced by 1. The second has i replaced by 2, etc. Finally, the last term has i replaced by N.

Since the body under discussion is rigid, each mass within it will rotate about the axis in the same time. As indicated in Fig. 7.20b, each mass rotates through the same angle as the body rotates. Consequently, the angular velocity and angular acceleration will be the same for each. Therefore each of the α values in Eq. (7.4b) is the same and can be factored out of the sum to give

$$\sum_{i=1}^{N} \tau_i = \alpha \sum_{i=1}^{N} m_i r_i^2 \qquad (7.4c)$$

Let us now investigate the meaning of the left-hand side of Eq. (7.4c). It is the sum of all the torques acting on the individual masses. In order to express it in terms of the torque exerted by external forces on the body as a whole, we consider the following fact: If we did not wish the object to rotate at all under the action of external torques, we could stop the body from rotating in two ways. We could apply an equal and opposite external torque to the body in order to cancel the original external torque. Or we could apply small torques to each of the masses so as exactly to cancel each of the individual torques acting on them. Therefore we see that an external torque on the body as a whole is equivalent to N small torques applied to the individual masses of which the body is made. As a result, we see that the left-hand side of Eq. (7.4c) is equivalent to the external torque τ applied to the object. We can therefore write Eq. (7.4c) as

$$\tau = \alpha \sum_{i=1}^{N} m_i r_i^2 \qquad (7.4d)$$

Definition *It is customary to define a quantity I, the **moment of inertia** of the body, as*

Moment of Inertia

$$I \equiv \sum_{i=1}^{N} m_i r_i^2 \qquad (7.5)$$

where r_i is the distance of mass m_i from the pivot point or axis. When this is placed in Eq. (7.4d), we find

$$\tau = I\alpha \qquad (7.6)$$

This equation is the rotational analog of the relation $F = ma$ in the case of extended rigid bodies. We see that torque replaces force and angular acceleration replaces linear acceleration. *The inertial effect corresponding to mass is contained in I, the moment of inertia.* To see the meaning of the moment of inertia more clearly, let us rewrite Eq. (7.5) for a special case. Suppose that the object were split into pieces such that each of the m_i was the same. Then Eq. (7.5) would become

$$I = mr_1^2 + mr_2^2 + mr_3^2 + \cdots + mr_N^2$$

or $$I = m(r_1^2 + r_2^2 + r_3^2 + \cdots + r_N^2)$$

This may be written

$$I = Nm\frac{r_1^2 + r_2^2 + r_3^2 + \cdots + r_N^2}{N}$$

But Nm is just the total mass of the object M, and the second term is just the average value of the various r^2 values for the masses composing the body. We represent this average squared radius by K^2, and *we call K the* **radius of gyration** *of the body; K^2 is an average measure of the square of the distance from the pivot point to the various pieces of matter composing the object.* In symbols we have

Radius of Gyration

$$I = MK^2$$

From the definition of I we see that it is a measure of the rotational inertia of the object. It is proportional to the mass of the object. The rotational inertia is also proportional to the average square of the distance from the pivot to the pieces of mass comprising the body. Notice that because I depends on the radii of its constituent masses from the axis, I depends on the position chosen for the axis.

7.10 Moments of Inertia for Various Bodies

As defined in the last section, *the moment of inertia of a body may be considered to be the product of the mass of the body and the average of the square of the distance of the mass from the pivot point, or axle.*

In certain cases, the square root of this distance term, the *radius of gyration* for a body, is very easily found. For example, in Fig. 7.21 the four masses shown mounted on an essentially massless rigid frame present a rather simple situation. Since $r_1 = r_2 = r_3 = r_4 = b$, the average distance from the pivot point at the center to the various pieces of mass is just b. Hence, for this device, I would be just Mb^2, where M would be the sum of m_1, m_2, m_3, and m_4.

Another interesting case is that of a hoop, or rim. This is shown in Fig.

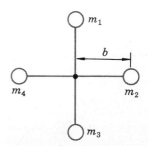

FIGURE 7.21
The radius of gyration of this body for a pivot at its center is just *b*. The size of the spheres is exaggerated.

TABLE 7.1
MOMENTS OF INERTIA

Object (axis as shown)		I	Radius of gyration K
Hoop	Axis 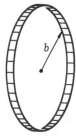 *b*	mb^2	b
Solid disk	*b*	$\frac{1}{2}mb^2$	$b/\sqrt{2}$
Solid sphere	*b*	$\frac{2}{5}mb^2$	$b\sqrt{\frac{2}{5}}$
Solid cylinder (radius *b*)		$\frac{1}{2}mb^2$	$b/\sqrt{2}$
Solid cylinder (thin) (length *L*)		$\frac{1}{12}mL^2$	$L/\sqrt{12}$

FIGURE 7.22
The radius of gyration of a hoop about its own axis is equal to the radius of the hoop.

7.22. Notice once again that all the mass is at a distance b from the pivot point. If m is the mass of the hoop, I would be just mb^2.

There are many other fairly simple cases in which I can be found without too much trouble. Several of these are shown in Table 7.1. Notice in this respect that the gyration radius K is indeed a measure of the average distance of the mass from the axle, or pivot point. For the thin hoop, $K = b$, since all the mass is at a distance b from the axis. However, for a disk, the value of K will be less than b, since only the outermost mass is a distance b from the axis. In this case $K = b/\sqrt{2}$.

The units of I are mass units times the square of distance units. Also, it should be pointed out again that the use of the relation $a = r\alpha$ in deriving Eq. (7.6) restricts the units used in that equation. The angular acceleration α should be measured in radians per second per second.

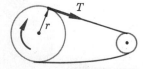

FIGURE 7.23
Angular acceleration is imparted to the large wheel by the torque resulting from the tension T in the upper belt. Notice that the lower belt is slack.

Illustration 7.10   The large 80-kg wheel shown in Fig. 7.23 has an actual radius of 25 cm and a radius of gyration of 20 cm. It is driven by the belt shown running to a small pulley on a motor. Find the tension in the belt necessary to uniformly accelerate the wheel to 2.0 rev/s in 20 s.

Reasoning   Before we can apply $\tau = I\alpha$, we must find α from the motion problem. Known are

$$\omega_0 = 0 \qquad t = 20\ \text{s} \qquad \omega_f = 2.0\ \text{rev/s} = 4\pi\ \text{rad/s}$$

We use

$$\omega_f = \omega_0 + \alpha t$$
$$4\pi \text{ rad/s} = (20 \text{ s})(\alpha) \qquad \alpha = 0.20\pi \text{ rad/s}^2$$

(Notice that we changed to radian measure so that α would be in proper units for the torque equation.)

Now we need I, and this is simply mK^2, where K is the radius of gyration. Hence, $\tau = I\alpha$ becomes

$$\tau = (80 \text{ kg})(0.20 \text{ m})^2(0.20\pi \text{ rad/s}^2) = 0.64\pi \text{ N} \cdot \text{m}$$

However, torque is force times lever arm. From the figure, the force is the tension T in the belt, and the lever arm is the radius of the wheel. Therefore

$$\tau = (T)(0.25 \text{ m})$$

giving
$$T = 2.56\pi \text{ N} \approx 8.0 \text{ N}$$

Illustration 7.11 The 3-kg object shown in Fig. 7.24 hangs from a rope wound on a 40-kg wheel. The wheel has an actual radius of 0.75 m and a radius of gyration of 0.60 m. Find (a) the angular acceleration of the wheel; (b) the distance the weight will fall in the first 10 s after it is released. Notice that the 3-kg object weighs 3×9.8 N.

Reasoning Problems involving two bodies must be solved by isolating each in turn

(a) For the body hanging from the rope we have that the unbalanced force acting on it is 29.4 N $-$ T, as shown in Fig. 7.24b. Using Newton's law, $F = ma$, we have

$$29.4 \text{ N} - T = (3 \text{ kg})(a) \qquad (7.7)$$

Isolating the wheel as in part c of the figure, we have

$$\tau = I\alpha$$
$$T(0.75 \text{ m}) = (40 \text{ kg})(0.60 \text{ m})^2(\alpha)$$
or
$$T = (19.2 \text{ kg} \cdot \text{m})(\alpha) \qquad (7.8)$$

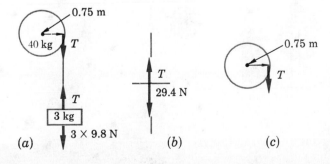

(a) (b) (c)

FIGURE 7.24
As the 3-kg object accelerates under the pull of gravity, the tension in the rope imparts an angular acceleration to the wheel.

The general idea is to solve these two equations, (7.8) and (7.7), simultaneously. To do this, we make use of the fact that $a = r\alpha$, which in this case gives $a = (0.75 \text{ m})(\alpha)$. Equation (7.7) then becomes

$$29.4 \text{ N} - T = (2.25 \text{ kg} \cdot \text{m})(\alpha)$$

Substitution from Eq. (7.8) gives

$$29.4 = 21.45\alpha$$

from which

$$\alpha = 1.37 \text{ rad/s}^2$$

You should carry through the units of the problem and verify the units of the answer.

(**b**)   We make use of the usual linear-motion equations together with the fact that

$$a = r\alpha \approx 1.03 \text{ m/s}^2$$

Known are

$$a = 1.03 \text{ m/s}^2 \qquad v_0 = 0 \qquad t = 10 \text{ s}$$

Using

$$s = v_0 t + \tfrac{1}{2}at^2$$

gives

$$s \approx 51.5 \text{ m}$$

If the distance had been measured and were known in a situation such as this, we could have reversed the procedure and computed the moment of inertia of the wheel. This type of experiment is sometimes used to determine moments of inertia and radii of gyration.

7.11 Kinetic Energy of Rotation

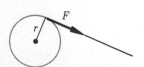

FIGURE 7.25
The person pulling on the rope with force F does work as the rope is unwound. This work results in increased rotational KE of the wheel.

Consider an experiment in which a rope wrapped around a wheel is used to accelerate the wheel from rest. This situation is shown in Fig. 7.25, where the tension in the rope is designated F. We now ask: How much work does the person pulling on the rope do on the wheel as the wheel is set into rotation?

If the rope is pulled out a distance s, the work done on the wheel is clearly

$$\text{Work done} = Fs$$

But $s = r\theta$, and so this becomes

where θ is the angle through which the wheel has turned.

Working an angular-motion problem, we can use

$$2\alpha\theta = \omega_f^2 - \omega_0^2 = \omega_f^2 - 0$$

to give $\theta = \omega_f^2/(2\alpha)$. Placing this in Eq. (7.9) gives

$$\text{Work done} = \frac{\tau\omega_f^2}{2\alpha}$$

But from $\tau = I\alpha$, we can replace τ/α to give

$$\text{Work done} = \tfrac{1}{2}I\omega_f^2$$

This is a very important relation. It states that the amount of work which was done on the wheel ended up in rotational energy of the wheel and is $\tfrac{1}{2}I\omega^2$. We *define the* **kinetic energy of rotation** *of the wheel to be* $\tfrac{1}{2}I\omega^2$.   **Definition**

The total KE of an object which is both translating and rotating is   *Total Kinetic Energy*
$\tfrac{1}{2}mv^2 + \tfrac{1}{2}I\omega^2$, where I must be the moment of inertia about the mass center[1] and v is the velocity of the mass center.

We therefore have the following two relations for KE, one for translational KE, the other for rotational KE:

$$\text{Translational KE} = \tfrac{1}{2}mv^2$$
$$\text{Rotational KE} = \tfrac{1}{2}I\omega^2 \qquad (7.10)$$

As we see, the linear velocity v in translational KE $= \tfrac{1}{2}mv^2$ has been replaced by the angular velocity in the expression for rotational KE. And, as we have stated before, the rotational inertia of the object is represented by I, the moment of inertia. Therefore m and I are analogous quantities for linear and rotational motion, respectively.

The law of conservation of energy applies to all forms of energy, rotational KE included. When the rotational KE of an object changes, compensating changes must occur in other forms of energy. We make use of that fact in the following illustration.

Illustration 7.12 The spherical ball of radius r and mass m shown in Fig. 7.26 starts from rest at the top of the incline of height h and rolls down

---

[1] The center of mass and center of gravity of an object coincide except in that rare instance where g, the free-fall acceleration, varies over the expanse of the object.

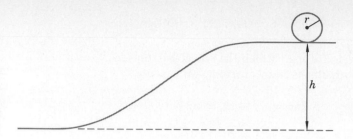

FIGURE 7.26
As the ball rolls to the
bottom of the hill, its PE is
changed to KE of rotation
and translation.

it. How fast is it moving when it reaches the bottom? (Assume that it rolls smoothly and that friction energy losses are negligible.)

Reasoning This problem is easily solved if we make use of the law of conservation of energy. In this instance, the ball originally had GPE, and this is changed to KE of translation and rotation. We have

Change in KE of translation + change in KE of rotation

+ change in GPE = 0

which becomes

$$\tfrac{1}{2}m(v_f^2 - v_0^2) + \tfrac{1}{2}I(\omega_f^2 - \omega_0^2) + mg(h_f - h_0) = 0$$

But v_0, ω_0, and h_f are all zero. Further, $I = \tfrac{2}{5}mr^2$ for a sphere, and so the above equation becomes

$$\tfrac{1}{2}mv_f^2 + \tfrac{1}{5}mr^2\omega_f^2 - mgh_0 = 0$$

We wish to find v_f, and so we eliminate ω_f through the relation $v = r\omega$. Further, we were told $h_0 = h$, and so we have

$$\tfrac{1}{2}v_f^2 + \tfrac{1}{5}v_f^2 - gh = 0$$

Solving for v_f, we find it to be

$$v_f = \sqrt{\frac{10gh}{7}}$$

It is interesting to notice that the radius of the sphere cancels. Moreover, the translational-KE term is $\tfrac{1}{2}v_f^2$ while the rotational-KE term is $\tfrac{1}{5}v_f^2$. Therefore the translational KE of the ball is 2.5 times larger than its rotational KE. But if the rolling object had been a hoop, for which $I = mr^2$, the kinetic energy would have been equally apportioned between translational and rotational KE.

7.12 Conservation of Angular Momentum

In view of the many analogies found thus far between linear and rotational phenomena, it should come as no surprise that linear momentum has a rotational counterpart. Rotational, or angular, momentum is associated with the fact that a rotating object persists in rotating. As you might expect from the fact that linear momentum is given by mv, the defining equation for **angular momentum** is

Definition

$$L = \text{angular momentum} = I\omega \qquad (7.11)$$

where L is used to represent angular momentum.

We also give direction to angular momentum so that it, like linear momentum, is a vector. Its direction is assigned much as we previously assigned direction to other rotational quantities in Fig. 7.8. The direction assigned to both angular velocity ω and angular momentum L is the same and is shown in Fig. 7.27. If you refer to Fig. 7.8 and the right-hand rule shown there, you can see that the direction of L and of ω satisfies the rule.

The angular momentum of an object or a system obeys a conservation law much like the one obeyed by linear momentum. *The* **law of conservation of angular momentum** *may be stated as follows:*

If no net torque acts on a body or system, its angular momentum will remain constant in both magnitude and direction:

$$I\omega = \text{const} \qquad \text{if} \qquad \Sigma\tau = 0$$

Notice that not only is the magnitude of the angular momentum of an object constant if no unbalanced torque acts on the object, but also the *direction* of its angular-momentum vector does not change. This is equivalent to saying that *the axis of rotation of a spinning object will not alter its orientation unless a torque acts on it to cause it to alter.* You may demonstrate this orientation effect using a simple gyroscope or a swiftly spinning wheel. For example, if a large wheel is set rotating about a north-south axis, the wheel will not change its orientation readily unless very large forces are applied to it. When a torque is applied to a rotating system such as this, the resulting motion of the system is interesting since it appears to contradict what we would expect to happen. Although the analysis of these effects is too complicated for us to pursue in this course, the effects themselves are easily demonstrated, and your instructor may show you some.

The axis of rotation of an object will not change its orientation unless an external torque causes it to do so. This fact is of great importance for the earth as it circles the sun. No sizable torque is experienced by the earth since the major force on it, the pull of the sun, is a radial force. The earth's axis of rotation therefore remains fixed in direction with reference to the

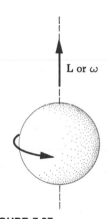

L or ω

FIGURE 7.27
The angular momentum and angular velocity have the direction shown for the rotating sphere. Can you use the right-hand rule to find this direction?

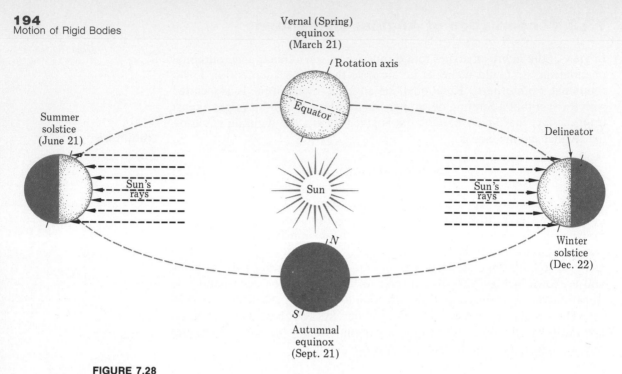

Vernal (Spring)
equinox
(March 21)

Rotation axis

Equator

Summer
solstice
(June 21)

Sun's
rays

Sun

Delineator

Sun's
rays

Winter
solstice
(Dec. 22)

N

S

Autumnal
equinox
(Sept. 21)

FIGURE 7.28
As an example of the
conservation of angular
momentum, the earth's
rotational axis retains its
same orientation relative to
the rest of the universe.
The dates shown are
approximate.

universe around us, i.e., with respect to the distant stars. We can see this behavior in Fig. 7.28.

Notice that the earth's path around the sun is nearly circular. But the rotational axis of the earth is not perpendicular to the plane defined by this orbit. Instead, the axis makes a fixed angle to the plane and, because of the conservation of angular momentum, maintains this orientation as the earth circles the sun. As you can see from the figure, the North Pole of the earth is in continuous daylight during summer (July). In midwinter (January), the North Pole experiences continual darkness. Of course, the reverse is true for the South Pole. The seasons we observe are a less striking example of this same effect.

Many other examples of the conservation of angular momentum can be seen in the universe about us. By use of it, the rotation of planets in their orbits and the motions of the stars in the heavens can be predicted. This same law is influential in determining the behavior of atoms in molecules and of electrons in atoms. Its scope is unlimited. It applies to both the smallest and the largest objects in the universe.

Illustration 7.13 Consider the earth satellite circling the earth as shown in Fig. 7.29. Find the ratio of its speed at perihelion to that at aphelion.

Reasoning The satellite circles the earth in an ellipse with the center

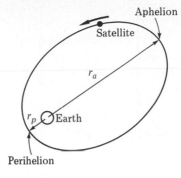

FIGURE 7.29
Find the ratio of the
satellite's speed at
perihelion to that at
aphelion.

of the earth at one focus of the ellipse. Since the earth's force on the satellite is radial, the angular momentum of the satellite about the earth's center as axis must be conserved. When we denote perihelion and aphelion by subscripts p and a, respectively, the conservation of angular momentum tells us that

$$I_p \omega_p = I_a \omega_a$$

But the moment of inertia of a point mass m (the satellite) at a distance r from the rotational axis is simply mr^2. Therefore this relation becomes

$$mr_p^2 \omega_p = mr_a^2 \omega_a$$

This gives

$$\frac{\omega_p}{\omega_a} = \left(\frac{r_a}{r_p}\right)^2$$

We know that $v_T = \omega r$. Since the velocity is tangential at both perihelion and aphelion, we therefore have

$$\frac{v_p/r_p}{v_a/r_a} = \left(\frac{r_a}{r_p}\right)^2$$

which simplifies to the result requested,

$$\frac{v_p}{v_a} = \frac{r_a}{r_p}$$

Illustration 7.14 The earth rotates on its axis once each day. Suppose that by some process the earth contracts so that its radius is only half as large as at present. How fast would it be rotating then?

Reasoning Let us assume as an approximation that the earth is a uniform-density sphere in each case. The moment of inertia of the earth will change from $I_0 = \frac{2}{5}MR_0^2$ to $I_f = \frac{2}{5}MR_f^2$, where M is the mass of the earth and $R_0 = 2R_f$.

Since no outside torque is applied to the earth during the shrinking process, its angular momentum must be conserved. Hence,

$$I_0\omega_0 = I_f\omega_f$$

where $\omega_0 = 1$ rev/day and ω_f is the final speed of rotation. Putting in the values of I_0 and I_f, we find

$$\tfrac{2}{5}MR_0^2\omega_0 = (\tfrac{2}{5}M)(\tfrac{1}{4}R_0^2)(\omega_f)$$

From this we find

$$\omega_f = 4\omega_0 = 4 \text{ rev/day}$$

In other words, the length of a day would be reduced to 6 h.

Summary

The turning action of a force is measured by the torque τ due to the force. Torque about a given axis or pivot point is the force multiplied by its lever arm. The lever arm is the length of a perpendicular dropped from the axis to the line of the force.

To be in equilibrium, the following conditions must apply to an object:

$$\Sigma F_x = \Sigma F_y = \Sigma F_z = 0 \qquad \text{and} \qquad \Sigma\tau = 0$$

In taking the sum of the torques, counterclockwise torques are positive and clockwise torques are negative.

When the torque equation is written for an equilibrium situation, the axis may be taken wherever convenient. Usually the axis is taken so that the line of an unknown force passes through it. In this way, the unknown force is made absent from the torque equation.

Static equilibrium can be characterized as (1) stable, (2) unstable, or (3) neutral depending on whether a slight disturbance (1) raises, (2) lowers, or (3) leaves unchanged the PE of the object.

Unbalanced external torques give rise to angular accelerations of objects in accordance with the equation $\tau = I\alpha$. The quantity I measures the rotational inertia of the object about the same axis as that taken for the torque. It is called the moment of inertia of the object and is given by $I = \Sigma m_i r_i^2$. Or, in terms of the total mass M of the object, $I = MK^2$, where K is the radius of gyration. Values of I for simple objects are given in Table 7.1.

A rotating object has rotational KE equal to $\frac{1}{2}I\omega^2$. The total KE of an object which is simultaneously rotating and translating is $\frac{1}{2}mv^2 + \frac{1}{2}I\omega^2$. In this expression, both v and I must refer to the mass center.

The angular momentum of a rotating object is $I\omega$. In the absence of external, unbalanced torques on the system, the law of conservation of angular momentum states that the angular momentum of the system is constant, both in magnitude and in direction.

Angular momentum, angular velocity, and torque are all defined as vectors. Their direction is given by the right-hand rule shown in Figs. 7.8 and 7.27.

Minimum Learning Goals

Upon completion of this chapter, you should be able to do the following:

1 Point out the center of gravity for any uniform, simple object such as a sphere, hoop, rod, cube, etc.

2 Perform an experiment to locate the center of gravity of any simple, rigid, irregular object.

3 Locate the lever arm for a given force with reference to a given axis.
4 Compute the torque about a given axis due to a given force.
5 State the necessary conditions for equilibrium both in words and by equation.
6 Solve simple problems involving equilibrium.
7 Point out whether a given object is in stable, unstable, or neutral equilibrium.
8 Give the moment of inertia of a point mass m at a given distance from an axis. State in words how this result is related to the definition of moment of inertia for a complex object.
9 Find one of the following quantities if both the others are given: M, I, K.

10 Write the rotational analog to $F = ma$ and define each quantity in it. Use the relation to solve simple problems in rotational acceleration.
11 Give the formula for rotational KE.
12 Find the total KE of a rolling object provided its radius, its speed, and I are given.
13 Solve simple situations involving the conservation of energy for rolling and rotating objects.
14 State the law of conservation of angular momentum. Give the formula for angular momentum. Use the law in simple problems.
15 State the direction of the following vectors in simple situations: torque, angular momentum, angular velocity.

Important Terms and Phrases

You should be able to define or explain each of the following:

Center of gravity
Lever arm
Torque
Condition for equilibrium: $\Sigma\tau = 0, \Sigma\mathbf{F} = 0$
The position of the axis is arbitrary
Three types of static equilibrium

Moment of inertia: $I = \Sigma m_i r_i^2$
$\tau = I\alpha$
Right-hand rule
Radius of gyration
Kinetic energy of rotation: $\frac{1}{2}I\omega^2$
Total KE = $(\text{KE})_{\text{trans}} + (\text{KE})_{\text{rot}}$
Angular momentum: $I\omega$
Law of conservation of angular momentum

Questions and Guesstimates

1 Suppose you are given some string, a meterstick, a 1-kg mass, and an unknown mass of the order of a few kilograms. How could you use the principles of this chapter to evaluate the unknown mass?
2 There is a right and a wrong way to lift a heavy weight. Refer to Fig. 7.16, and use it to explain why the method shown there is the wrong way. What is the right way? Explain.
3 Explain how a tack puller or a claw hammer generates large forces to accomplish its task. Repeat for a nutcracker.
4 Hold your body rigid with your feet together, and try to lean forward at an angle θ to the vertical. Notice how small θ must be if you are to retain your balance. Explain what limits the size of the angle. Why are people with big feet able to lean farthest?
5 Slender people are less apt to have back trouble than obese people. Explain how this fact influences posture, muscle strain, and similar factors.

FIGURE P7.1

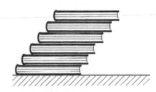

6 As shown in Fig. P7.1, a large number of books are piled on top of one another at uniform offsets. Discuss the conditions which influence whether the pile will topple.
7 You wish to mow a steep hill using a riding lawn mower or tractor. Sketch the situation for the mower going up the hill. Under what conditions will this be safe? Repeat for going down the hill. Repeat for riding sideways along the hill. In answering this, you will have to estimate the position of the center of gravity of mower and rider. Why?

8 A common trick is to prop open a door by placing a wedge of wood in the crack next to the hinge. Why does this usually ruin the hinge?

9 A large irregular rock is to be lifted by a crane. How must the cables supporting the rock be placed if the rock is not to rotate as it is lifted from the level ground on which the rock rests?

10 In order to keep a football or any other projectile from wobbling, the projectile is caused to spin about an axis in line with the direction of motion. Explain why the spin is of aid in this respect.

11 A "do-it-yourselfer" builds a helicopter with a single propeller on a vertical axis. In its maiden flight, the operator becomes sick because the whole helicopter tends to spin about a vertical axis. What went wrong? How is this difficulty overcome in more sophisticated machines?

12 Which bicycle wheel would be easier to stop rotating about its axle, one with a tire filled with air or one filled with water? Explain.

13 Which has the larger speed after rolling down the same incline, a sphere or a hoop (they have equal mass and radius)? Explain. Discuss the case of a hoop and a disk that have identical radius and mass.

14 Suppose that the sun's attraction for the earth suddenly doubled. What can you say about the rate of rotation and orbit of the earth about the sun?

15 Suppose that an internal explosion suddenly opened a huge cavity in the earth by pushing the earth's surface outward. How would this affect the rotation of the earth about its axis and about the sun?

16 We maintain that angular momentum is conserved. Isn't this contradicted by the fact that almost all rotating objects eventually slow and stop?

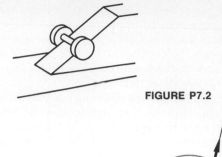

FIGURE P7.2

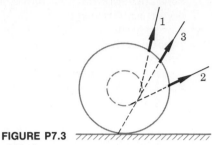

FIGURE P7.3

17 The spool shown rolling down the incline in Fig. P7.2 takes off with greatly increased translational motion as soon as the large-diameter disks on the sides of the spool touch the floor. Explain why, paying particular attention to the way the KE is apportioned.

18 The spool shown in Fig. P7.3 moves to the left when the cord is in position 1 and to the right when the cord is in position 2. Why? What happens when the cord is in position 3?

19 When a car rolls down a hill, about what fraction of its total KE is the rotational energy of the wheels? (E)

20 Estimate your moment of inertia (when standing) about a vertical axis through your center of gravity. Repeat for a horizontal axis directed into your midsection.

Problems

1 Three forces pull on a point object: 200 N at 37°, 300 N at 90°, and a force F. The angles are specified as shown by the angle θ in Fig. P7.4. Find F if the object is to remain at equilibrium.

2 A point object is subjected to the following forces: 150 N at 120°, 80 N at 270°, and a force F. The angles are specified as shown by the angle θ in Fig. P7.4. Find F if the object is to remain at equilibrium.

3 Two boys are fighting for a child's wagon in which another child sits. One boy pulls forward on the wagon handle with a force of 500 N with the han-

FIGURE P7.4

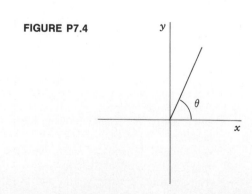

dle positioned 30° above the horizontal. With what force must the second boy pull straight back on the wagon if it is not to move?

4 A girl is trying to shove a chair from one room to another by pushing on it with a force of 300 N directed 20° below the horizontal. The chair is found then to move across the floor with constant speed. How large is the friction force between the chair and floor?

5* A 350-N child hangs from the center of a clothesline, as shown in Fig. P7.5. Find the tension in the clothesline.

6* The clothesline in Fig. P7.5 will break if the tension in it exceeds 500 N. How heavy must the child shown there be to break the line? Assume the angle is as shown.

7* In Fig. P7.6 the weight $W = 2000$ N. What is the tension in (*a*) the horizontal rope and (*b*) the rope running to the ceiling?

8* The tension in the horizontal cord in Fig. P7.6 is 400 N. How large is the weight W?

9* The tension in the left-hand rope in Fig. P7.7 is 60 N. Find W and the tension in the rope at right.

10* In Fig. P7.7 the weight $W = 240$ N. Find the tension in the rope at the right and the rope at the left.

11* For the equilibrium situation shown in Fig. P7.8, find W_2 and W_3, assuming that $W_1 = 400$ N and that the pulleys are frictionless.

12* Find W_1 and W_3 in Fig. P7.8 if the system is at equilibrium with $W_2 = 500$ N. The pulleys are frictionless.

13** Assuming the pulleys in Fig. P7.9 to be frictionless and the system to be at equilibrium, find W_1, W_2, T_1, and T_2 if $W_3 = 500$ N.

14** In Fig. P7.9, the pulleys are frictionless and $W_1 = 200$ N. Find the values of W_2 and W_3 for this equilibrium situation.

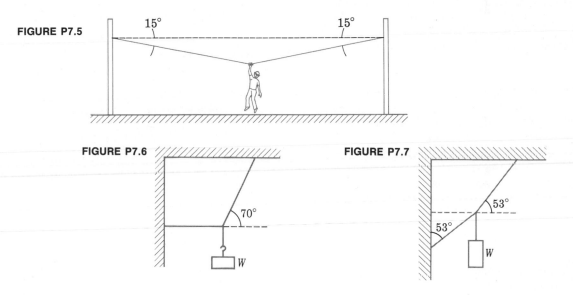

FIGURE P7.5

FIGURE P7.6

FIGURE P7.7

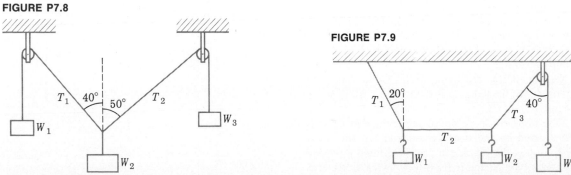

FIGURE P7.8

FIGURE P7.9

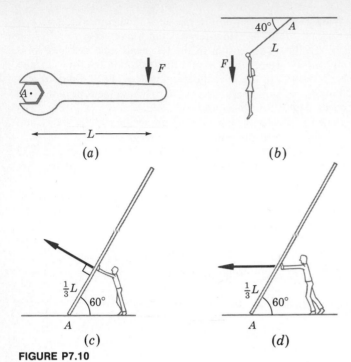

(a)

(b)

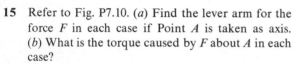

(c)

(d)

FIGURE P7.10

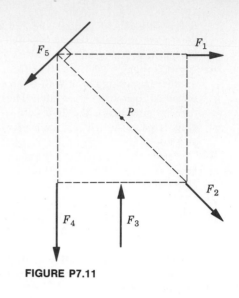

FIGURE P7.11

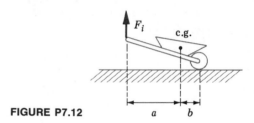

FIGURE P7.12

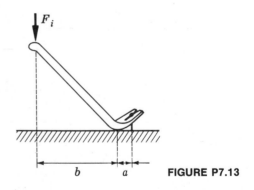

FIGURE P7.13

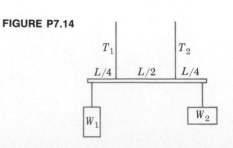

FIGURE P7.14

15 Refer to Fig. P7.10. (a) Find the lever arm for the force F in each case if Point A is taken as axis. (b) What is the torque caused by F about A in each case?

16 For each force shown in Fig. P7.11, what is the (a) lever arm and (b) torque about point P as axis? The dotted square has sides of length L.

17 How large a force F_i applied to the wheelbarrow handles in Fig. P7.12 will be able to lift a 700-N load at the center of gravity indicated? Give your answer in terms of a and b.

18 For the nail puller shown in Fig. P7.13, how large a force is applied to the nail when the force on the handle F_i is 250 N? Express your answer in terms of a and b.

19* The two vertical ropes shown in Fig. P7.14 support a uniform 8.0-kg plank and the weights as shown. If T_1 is 210 N and the mass of W_2 is 25 kg, find W_1 and T_2.

20* In Fig. P7.14, the 8.0-kg plank is uniform, the mass of W_1 is 40 kg, and the mass of W_2 is 30 kg. Find T_1 and T_2.

21* The uniform 25-kg plank shown in Fig. P7.14 is supported by the two ropes shown. If the ropes supporting the plank can each withstand a tension of only 1000 N, and if W_2 is to be twice as heavy as

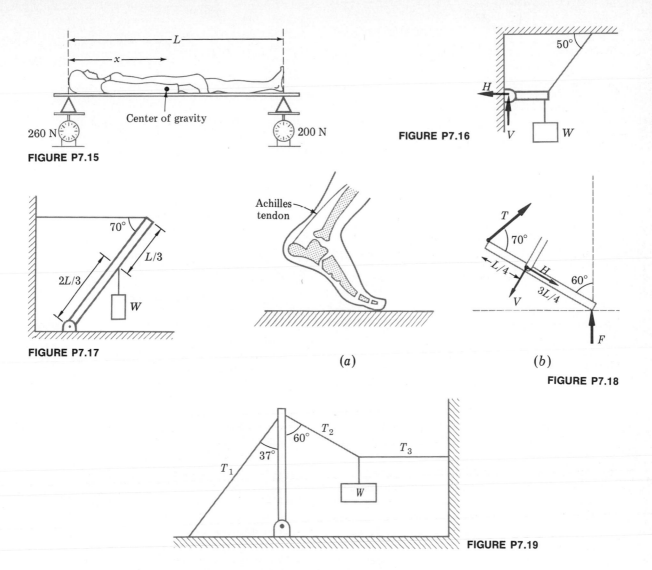

FIGURE P7.15

FIGURE P7.16

Achilles tendon

(a)

(b)

FIGURE P7.17

FIGURE P7.18

FIGURE P7.19

W_1, what is the greatest value W_1 can have? Assume the ropes holding W_1 and W_2 are very strong.

22* To locate a woman's center of gravity, she is placed on two scales as shown in Fig. P7.15. The scale on the left reads 260 N while that on the right reads 200 N. Find the distance x indicated to the center of gravity in terms of L. Assume scale readings have been corrected by subtracting the readings when the woman was not in place.

23* In Fig. P7.16, the beam is uniform and weighs 200 N. Find (a) the tension in the rope and (b) the H and V component forces exerted by the pin if $W = 800$ N.

24* The uniform 500-N beam shown in Fig. P7.17 supports a load, as shown. (a) How large can the

load be if the horizontal rope is able to hold 2000 N? (b) What are the components of the force at the base of the beam?

25** When you stand on tiptoe, the situation is much like that shown in Fig. P7.18. We can replace the actual situation by the model shown in part b. The force F, the push of the floor, will be equal to the person's weight if the person is standing on one foot. Find (a) the tension in the Achilles tendon and (b) the forces H and V at the ankle in terms of F for the situation shown.

26** In Fig. P7.19, the beam weighs 750 N, and the value of T_3 is 870 N. Find T_1, T_2, W, and the force with which the beam pushes down on the frictionless pin at its base.

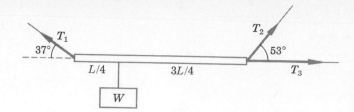

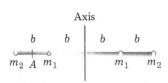

27* The uniform board in Fig. P7.20 weighs 120 N and supports a weight $W = 700$ N. Find the tensions in the three supporting ropes.

28** A uniform, 150-N ladder of length L leans against a smooth wall in such a way that it makes an angle of 53° with the floor. A 60-kg man wishes to climb to a point two-thirds of the way up the ladder. How large must the coefficient of friction be between the floor and ladder if the ladder is not to slip?

29 What is the moment of inertia of the earth with respect to the sun as axis? Note that $M_{\text{earth}} = 6 \times 10^{24}$ kg, $M_{\text{sun}} = 2 \times 10^{30}$ kg, and earth-sun distance $= 1.5 \times 10^{11}$ m.

30 The four point masses shown in Fig. P7.21 are attached rigidly to a bar of negligible mass. (*a*) Find the moment of inertia of the system with respect to the axis shown. (*b*) Repeat for a similar axis through point A, midway between m_1 and m_2. (*c*) What is the radius of gyration in each case? Use $m_1 = 0.50$ kg, $m_2 = 2.0$ kg, and $b = 0.50$ m.

31 A very light rod is placed along the x axis. It has a mass $m_1 = 2.0$ kg at $x = 0$, a mass $m_2 = 1.50$ kg at $x = 50$ cm, and a mass $m_3 = 3.0$ kg at $x = 100$ cm. Find the moment of inertia of the system about a pivot point at (*a*) $x = 0$ and (*b*) $x = 2.0$ m.

32* Two hoops are mounted on a frame of negligible mass, as indicated in Fig. P7.22. The inner hoop has a mass M_1 and a radius a, and the outer hoop has a mass M_2 and a radius b. Find the moment of inertia and radius of gyration for rotation about an

axis through the center perpendicular to the page if $M_2 = \frac{1}{2}M_1$ and $b = 2a$.

33 Find the moment of inertia of a solid disk having a radius of 10 cm and a mass of 200 g.

34 A hooplike wheel weighs 50 N and has a radius of 40 cm. Find its approximate moment of inertia.

35 How large a torque is required to give an angular acceleration of 4.0 rad/s² to a wheel having a moment of inertia of 0.20 kg · m²?

36* A tangential force of 30 N applied to the rim of a 20-cm-radius wheel gives the wheel an acceleration of 0.150 rev/s². What is the moment of inertia of the wheel about its axis?

37* A wheel having $I = 0.20$ kg · m² is spinning at $120/\pi$ rev/min when the power source is shut off. It coasts uniformly to rest in 100 s. How large was the average torque which stopped the wheel?

38* (*a*) How large a torque is required to accelerate a 50-kg wheel having a 25-cm radius of gyration from rest to $1/\pi$ rev/s in 20 s? (*b*) How far will the wheel turn in this time?

39* (*a*) How long will it take to accelerate a 50-kg, 30-cm-radius solid disk (rotating about its usual axis) from rest to 4.0 rad/s, provided a force of 3.0 N acts tangential to its rim? (*b*) Through how many revolutions will it rotate in this time?

40* A 6.0-cm-radius, 7.0-kg solid sphere is mounted on an axle through its center. A string wound around it supplies a tangential force of 5.0 N to it for 3.0 s. If the sphere starts from rest, how fast will it be rotating (in revolutions per second) at the end of the 3 s?

41* A 5.0-cm-radius wheel is mounted on a horizontal axis. A string is wound on the rim of the wheel, and a 50-g mass is suspended from it. After the mass is released, the system accelerates in such a way that the mass drops 2.0 m in 14.0 s. What is the moment of inertia of the wheel? What was the tension in the cord as the mass was falling?

42* A cylinder of radius 20 cm is mounted on a horizontal axis coincident with the cylinder axis. A cord is wound on the cylinder, and a 70-g mass is hung from it. After being released, the mass drops 150 cm in 12.0 s. Find the moment of inertia of the cylinder. What was the tension in the cord while the mass was falling?

43* A wheel mounted on an axle has a string wound on its rim. The moment of inertia of the wheel is $0.050 \text{ kg} \cdot \text{m}^2$. The wheel is accelerated from rest by a force of 30 N which pulls the end of the string a distance of 75 cm. What will be the final speed of rotation (in revolutions per second) of the wheel?

44* A wheel of radius 8.0 cm with $I = 0.070 \text{ kg} \cdot \text{m}^2$ is turning at 3.0 rev/s when a tangential friction force of 0.90 N is applied to the rim. Through how many revolutions will the wheel turn before stopping?

45* A 90-g mass is suspended from the rim of an 80-cm-radius wheel by a string wound around the wheel. The wheel has $I = 0.070 \text{ kg} \cdot \text{m}^2$ and is mounted on a friction-free axle. The wheel is accelerated from rest by letting the mass fall 120 cm. How fast (in revolutions per second) is the wheel then turning?

46* A wheel with $I = 600 \text{ kg} \cdot \text{cm}^2$ is spinning at 3.0 rev/s when a mechanism is engaged that causes a cord on the rim of the wheel to lift a 5.0-kg mass as the wheel coasts to rest. How high can the wheel lift the mass?

47* A hoop has a radius of 5.0 cm. It starts from rest and rolls down a slope. (*a*) What is its linear speed when it reaches a point which is 30 cm lower than its starting point? (*b*) How fast is it rotating (in revolutions per second) at that time?

48* Repeat Prob. 47 if the object is a wheel which has a radius of gyration of 4.0 cm and an actual radius of 5.0 cm.

49* A steel ball bearing of 0.50-cm radius is rolling along a table at 20 cm/s when it starts to roll up an incline. How high above the table level will the ball bearing rise before stopping? Ignore friction losses.

50* A uniform sphere and a uniform disk are rolled down an incline from the same point. Find the

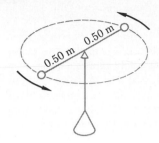

FIGURE P7.23

ratio of the disk's speed to that of the sphere at the bottom of the incline. Ignore friction losses.

51 As shown in Fig. P7.23, two identical small balls, each of mass 3.0 kg, are fastened to the ends of a light, metal rod 1.00 m long. The rod is pivoted at its center point and is rotating at 7.0 rev/s. An internal mechanism is capable of moving the balls in toward the pivot. (*a*) Find the moment of inertia of the original device. (*b*) If the balls are suddenly moved in until they are 25 cm from the pivot, what is the new speed of rotation?

52 It is surmised that the sun was formed in the gravitational collapse of a dust cloud which filled the space now occupied by the solar system and beyond. Assuming the original cloud to be a uniform sphere of radius R_0 with an average angular velocity of ω_0, how fast should the sun be rotating now? For present purposes, ignore the small mass resident in the planets and assume the sun to be a uniform sphere of radius R_s.

53 A woman stands over the center of a horizontal platform that is rotating freely at 2.0 rev/s about a vertical axis through the center of the table and straight up through the woman. She holds two 5-kg masses in her hands close to her body. The combined moment of inertia of table, woman, and masses is $1.2 \text{ kg} \cdot \text{m}^2$. The woman now extends her arms so as to hold the masses far from her body. In so doing, she increases the moment of inertia of the system by $2.0 \text{ kg} \cdot \text{m}^2$. (*a*) What is the final rotational speed of the table? (*b*) Was the kinetic energy of the system changed during the process? Explain.

54* An ice skater moving past a post with speed v_0 grabs onto the end of a rope tied to the post. The original length of the rope is L_0, but the rope shortens as the skater circles the very thin post, thereby winding up the rope. Assuming that the skater coasts and does not try to stop, how fast will he be

203

moving when the length of rope (the circle radius) is L? Assume the post radius is much smaller than L.

55* Refer to Fig. P7.21. Prove that the moment of inertia for rotation about an axis parallel to the one shown but a distance d away from it is given by $I = Md^2 + I_0$, where M is the total mass and I_0 is the moment of inertia about an axis through the center of mass. Neglect the mass of the bar. (This relation is a general one and is called the parallel-axis theorem.)

56* A solid wheel rotating about an axis through its center has a moment of inertia I_0. Two identical masses are placed at opposite points on the rim of the wheel at opposite ends of a diameter. Calling the wheel radius a and each mass M, show that the wheel now has a moment of inertia given by $I = I_0 + 2Ma^2$.

57* A children's merry-go-round in a park consists of an essentially uniform, 150-kg solid disk rotating about a vertical axis. The radius of the disk is 6.0 m, and a 90-kg teacher is standing on it at its outer edge when it is rotating at 0.20 rev/s. How fast will the disk be rotating if the teacher walks 4.0 m toward the center along a radius? *Hint:* Momentum must be conserved.

58* Suppose that the merry-go-round described in Prob. 57 has no one on it, but is moving at 0.20 rev/s. If a 90-kg teacher quickly sits down on the edge of it, what will be its new speed?

8
Mechanical Properties of Matter

The action of forces and torques on rigid bodies has been discussed in the past several chapters. We now wish to investigate some of the properties of bodies and materials which deform or flow when subjected to forces. This we do in the present chapter. In addition, an introduction to the static and dynamic properties of liquids is presented.

8.1 Three States of Matter

We are all familiar with the three general classifications of matter: solids, liquids, and gases. In **gases** the molecules are essentially free from one another and travel through space, colliding but not sticking together. Their individual collisions with the walls of a container in which they are placed give rise to a pressure on the walls. The pressure of a gas was computed in Chap. 5 and was found to depend directly on the KE of the molecules. The KE of the extremely light gas molecules is large enough for the difference in PE of a molecule at the top and bottom of any room-size container to be negligible in comparison with the KE. As a result, the *gas molecules fill the entire container in which they are placed.*

The situation is quite different in **liquids.** Although we know the molecules in a liquid are in continuous motion, they do not possess enough KE to overcome the attractive forces of the neighbor molecules. They are unable to break loose completely from one another, and they therefore exist as a fluid aggregate of molecules. From time to time, a molecule may acquire enough KE to tear itself loose from the surface of the liquid. This is what happens when molecules evaporate from the surface of a substance. More is said about this in a later chapter. *Liquids and gases both undergo flow quite easily. They are therefore grouped in a classification called fluids. If we use that classification system, there are two groups of substances, fluids and solids.*

Solids are much like liquids in many ways. The chief difference is largely one of degree rather than of kind. *While the molecules in a liquid are still able to slip by their neighbors, the molecules in a solid are essentially locked in place.* The forces between molecules in a solid are so strong that, to a good approximation, each molecule is held tightly in place by its neighbors. The substance is hard and rigid, because the molecules are unable to slide past one another. This may be seen diagrammatically in Fig. 8.1.

If the molecules such as *A* and *B* in the figure are tightly bound together, they will not easily be torn apart. The forces shown in the figure will be unable to separate the molecules, and the material will not flow under their action. However, if the attractive forces between molecules *A* and *B* are relatively small, the applied forces will cause the molecular layers to slide past each other. In this case the material will be fluid.

It is sometimes difficult to say whether a substance should be called a solid or a liquid. For example, if molasses is cooled, it becomes very "thick," or viscous (hence the phrase *slower than molasses in January*). At still lower temperatures it becomes quite hard. Yet is it a solid, or is it merely a very viscous liquid? Another example is window glass. Modern window glass is harder than that used many years ago. It is not uncommon to find a very old window pane that is thicker at the bottom than at the top. It has obviously flowed ever so slowly during the passage of the years. Many plastics behave in this way also.

It is clear from these examples that *the border line between solids and liquids is not a sharp one.* We see in the next section that there are actually two types of solids, one being merely a very viscous liquid.

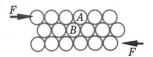

FIGURE 8.1
The ease with which the molecular layers will flow over one another under the action of the forces shown depends on the attractive force between such molecules as *A* and *B*.

8.2 Crystalline and Glassy Solids

Definition
A **crystal** *is a collection of atoms or molecules in which each atom is placed precisely in a definite pattern with respect to its neighbors. This pattern is repeated over and over throughout the crystal.* An example would be the sodium chloride crystal (table salt), a small portion of which is pictured in Fig. 8.2a.

Notice that the sodium and chloride atoms are placed in perfect geometric order on the corners of a cube. The whole crystal consists of a multitude of these cubes packed in essentially perfect order. A sodium atom anywhere in the crystal has its neighbor atoms arranged exactly the same as every other sodium atom does.

Many crystals do not have as simple a structure as the cubic-lattice form of the sodium chloride crystal. However, in any crystal there is a

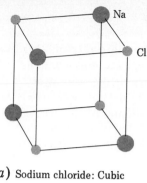

(a) Sodium chloride: Cubic

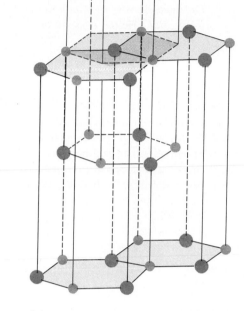

(b) Cesium chloride:
Interpenetrating cubes

(c) Boron nitride:
Interpenetrating structures

FIGURE 8.2
Typical crystal lattices.

definite, precise arrangement pattern of the atoms. This pattern is repeated over and over, much in the way that the pattern in many types of wallpaper repeats itself. In the crystal, though, the pattern is three-dimensional rather than existing only in a plane. Other typical examples are shown in Fig. 8.2.

In highly viscous liquids, so viscous that they are hard and brittle, the situation is much different. The molecules are no more ordered than they are in the molten liquid. Some slight amount of order may exist in the neighborhood of a given molecule, but this order does not persist throughout the liquid. In these liquids, the molecules are locked in place, but are not arranged in any precise pattern. *Supercooled liquids such as these are called* **glasses.**

Definition

Although most window-type glasses are noncrystalline, supercooled liquids, many other substances can exist as a glass. Many plastics are amorphous, i.e., noncrystalline, glasses. However, some plastics, such as polyethylene, are partly crystalline and partly amorphous. It is often difficult to determine whether a given solid material is crystalline or amorphous. Usually this is best decided by using x-rays in a manner which is briefly touched on later in this book. In many situations, especially involving plastics, amorphous and crystalline regions are so well intermixed that the state of the substance is very complex.

8.3 Density and Specific Gravity

There are many ways in which a substance can be characterized. One of the most useful tells what quantity of substance occupies a unit of volume. This method of characterization is termed the **density** of the substance.

Definition *The* **mass density** *of a material is defined to be the mass of a unit volume of the material.* Hence mass density = mass/volume, or, in symbols,

$$\text{Mass density} = \rho = \frac{m}{V} \tag{8.1}$$

where ρ is Greek rho. Clearly, the units of mass density are grams per cubic centimeter, kilograms per cubic meter, etc. The concept of mass density is seldom employed in the British system of units.

Those scientists and engineers who have not yet fully embraced the SI often make use of a quantity called the **weight density,** defined as the weight of a unit volume of substance, or, in equation form,

$$\text{Weight density} = D = \frac{W}{V}$$

Because $W = mg$, we see that weight density is related to mass density through

$$D = \rho g \tag{8.2}$$

In this text we do not use weight density.

Many materials expand as they are heated. This is the result of the fact that the molecules are vibrating through larger distances at the higher temperature, and hence their average separation distance is increased. Because the mass in a unit volume will change if the molecules move farther apart, the density of a substance will change with temperature. Although the densities of most substances decrease with increasing temperature, there are several common exceptions in which the density actually increases as the temperature is raised through a certain temperature range. Water in the range 0 to 4.0°C is such a substance.[1] A representative list of densities is given in Table 8.1. In using these data it is convenient to know that

$$1 \text{ kg/m}^3 = 10^{-3} \text{ g/cm}^3$$

Definition *The* **specific gravity,** *or* **relative density,** *of a substance is defined to be the ratio of the density of the substance to the density of water.* A temperature at which this ratio is to be taken must be specified. One usually takes the ratio at 3.98°C, where the density of water is 1000 kg/m³. In this case, the specific

---

[1] In this temperature range, the water molecules exhibit order over short distances. The order breaks up with increasing temperature, and this allows the molecules to pack together more densely.

TABLE 8.1
DENSITIES OF SUBSTANCES

Substance	Temperature, °C	ρ, kg/m³
Air (normal pressure)	0.0	1.29
Benzene	20.0	879
Water	20.0	998
Water	3.98	1,000
Bone	20.0	~1,800
Aluminum	20.0	2,700
Iron	20.0	7,860
Copper	20.0	8,920
Lead	20.0	11,340
Mercury	0.0	13,600

gravity is equal numerically to the mass density in grams per cubic centimeter.

8.4 Hooke's Law

Another way of characterizing a material is in terms of its deformability. Two major types of deformation occur. In one, the substance flows under the action of a force. This behavior is characteristic of fluids. The other type of deformation, which is only temporary, is elastic in nature, like the stretching of a spring. When the deforming force is removed, the deformation returns to zero. Let us now examine this latter type of deformation.

If a rigid bar such as that shown in Fig. 8.3 is subjected to a stretching force by hanging a weight on it, the bar will stretch a distance ΔL. For most solid materials, a graph relating applied load F to elongation ΔL will appear much like Fig. 8.4. In other words, if the applied load is doubled, the amount of stretch ΔL will double. Expressed as an equation, this becomes

$$F = (\text{const})(\Delta L) \tag{8.3}$$

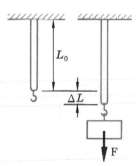

FIGURE 8.3
The amount ΔL that the bar stretches under load W is proportional to the load as well as to L_0, provided that Hooke's law is satisfied.

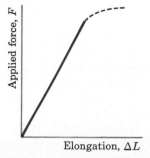

FIGURE 8.4
A typical stress-strain curve. Hooke's law applies in the linear region.

where F is the applied force. In addition, if the force is not too large, the bar will return to its original length when the load is removed. The bar is said to be **elastic** in this range of loads. Equation (8.3) is a statement of **Hooke's law:**

Hooke's Law   The distortion is proportional to the distorting force.

We further emphasize that *in the elastic range, the distortion returns to zero when the distorting force is removed.*

However, if too large a load is applied to the bar, the graph shown in Fig. 8.4 will begin to deviate from a straight line—perhaps as shown by the dashed curve. Moreover, the bar will not usually return completely to its original length if the load is now removed. We say that the bar has been stretched beyond its **elastic limit** in this case. The bar may even break under the tension if the load is increased much beyond the elastic limit.

In terms of the molecules or atoms in the bar, the applied stretching, or tensile, force is merely pulling them apart. In Fig. 8.5, the tensile force has separated the atoms slightly along the line of the force. These very small separations between the atoms, when added along the length of the bar, give rise to the observed elongation of the bar. As long as these separations between atoms are kept small, the separation distance is proportional to the applied stretching force and Hooke's law, Eq. (8.3), applies. However, the range of validity of Hooke's law varies widely from substance to substance and is usually difficult to predict without actually plotting a graph such as shown in Fig. 8.4.

FIGURE 8.5
A tensile force on a bar tends to separate the molecules in the way shown.

8.5 Stress and Strain

Hooke's law can be stated in a more useful form. The statement can be made applicable to many different situations if we make the statement in terms of stress and strain.

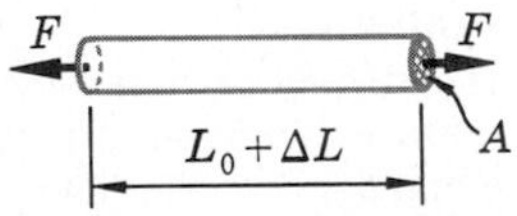

Definition

The terms *stress* and *strain* have precise meanings in physics. Although we often employ such phrases as the *stresses of everyday life* and the *strain of taking a test,* such uses are colloquial. They do, however, bear some relation to the exact definition of the words as used by scientists, as we shall now see.

Stress *is defined to be the applied force per unit of area to which the force is applied.* As an example, consider the rod shown in Fig. 8.6. Here a tensile (or stretching) force F is applied to the end surface of the bar. The cross-sectional area of the bar is A, as indicated. We define the stress on the bar to be

FIGURE 8.6
The stress is F/A, and the strain is $\Delta L/L_0$.

Stress

$$\text{Stress} = \frac{\text{force}}{\text{area}} = \frac{F}{A} \qquad (8.4)$$

Notice in particular that the area involved is the cross-sectional area, not the area of the outside of the bar.

The SI units of stress are newtons per square meter, and so they are the same as the units for pressure. As we did for pressure, we call this combined unit the pascal (Pa):

$$1 \text{ Pa} = 1 \text{ N/m}^2$$

Definition

*We define the **strain** of an object to be the distortion of the object divided by the original dimension before distortion.* For example, in Fig. 8.6, the strain in the bar is a measure of how much the bar has been stretched:

$$\text{Strain} = \frac{\text{elongation}}{\text{original length}} = \frac{\Delta L}{L_0} \qquad (8.5)$$

Strain

The elongation alone is not a good measure because it varies with the original length of the bar. This follows from the fact that a bar twice as long will have twice as many atomic separations that lengthen under a given stress. There are no units for the strain; since it is simply the ratio of two lengths, the units cancel. We shall see in later sections that there are many types of strain, depending on the geometry of the system. In this case we are speaking about tensile strain. If the bar were being compressed, the strain would be the ratio of the decrease in length of the bar to its original length.

We are now able to restate Hooke's law. Since the law states that $F \propto$ distortion, we can replace F by the stress, since the two are proportional. Similarly, we can replace the distortion by the strain. Then, Hooke's law becomes

$$\text{Stress} \propto \text{strain}$$

Hooke's Law

In this form, the law may be applied to many situations other than the stretching of a bar. For example, Hooke originally proved its applicability to the stretching, bending, and twisting of springs and other objects.

8.6 Concept of Modulus

If a large stress is needed to give a small strain to an object, the object is hard and rigid. *We measure this quality of hardness, or rigidity, by the **modulus** of elasticity of the material.* There are several different kinds of moduli, depending on the exact way the material is being stretched, bent, or otherwise distorted. Several of these moduli are discussed in this section. They are all defined by the following relation for special geometries:

$$\text{Modulus} = \frac{\text{stress}}{\text{strain}}$$

Modulus

Since strain has no units, the modulus has the units of stress.

TABLE 8.2
YOUNG'S MODULUS Y AND BULK COMPRESSIBILITY k

Material	Young's Modulus,* 10^{10} N/m^2	Compressibility, 10^{-11}m^2/N
Tungsten	35	0.5
Steel	19–20	0.6
Iron (wrought)	18–20	0.7
Femur bone	~1.4	—
Copper	10–13	0.8
Brass (cold-rolled)	9	1.6
Iron (cast)	8–10	1.0
Aluminum	5.6–7.7	1.5
Polystyrene	~0.14	20
Water	—	50
Benzene		100

*The label 10^{10} N/m^2 means that the numbers listed are in units 10^{10} times larger than newtons per square meter. Therefore, Y for tungsten is 35×10^{10} N/m^2.

Young's Modulus   This modulus is used to describe situations such as the stretching of the bar shown in Fig. 8.6. Young's modulus would be of interest if you wished to compute how much a wire or a rod would stretch under a tensile force. By definition,

$$\text{Young's modulus} = Y = \frac{F/A}{\Delta L/L_0} \tag{8.6}$$

Typical values of Y for various materials are given in Table 8.2.

Shear Modulus   Suppose that we try to distort a cube of material in the manner shown in Fig. 8.7. A force F is applied parallel to the top face of the cube, the face having an area A. In this case, the stress is still F/A, and the strain is $\Delta L/L_0$, but observe how these symbols are defined in the figure. We have

$$\text{Shear modulus} = \frac{F/A}{\Delta L/L_0} \tag{8.7}$$

Although this equation is identical to Eq. (8.6) for Young's modulus, notice that the symbols are defined differently in the two cases.

For many substances it turns out that the shear modulus is approximately one-third as large as Young's modulus. Exceptions to this rule occur chiefly in cases where the volume of the material changes upon deformation or the material shows different properties in different directions.

The reciprocals of the shear and Young's (tensile) moduli find frequent use. They measure how *easily* the material deforms rather than how hard

FIGURE 8.7
Here ΔL is exaggerated so it can be seen. The shear modulus is given by $(F/A)/(\Delta L/L_0)$.

the material is to deform. They are given the descriptive names **shear compliance** and **tensile compliance.**

Bulk Modulus Suppose that a block having volume V_0 is subjected to a pressure increase ΔP on all sides, as shown in Fig. 8.8. The cube will shrink in volume by an amount ΔV. In this case, the strain is defined to be $\Delta V/V_0$, and the stress is the applied pressure ΔP. As for the other moduli, the **bulk modulus** E is defined to be the ratio of stress to strain. Therefore

$$E = \text{bulk modulus} = \frac{\Delta P}{\Delta V/V_0} \qquad (8.8)$$

The reciprocal of this quantity, the compressibility, is discussed below.

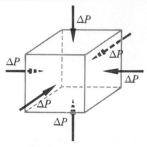

FIGURE 8.8
The cube of original volume V_0 will contract by an amount ΔV under the action of the increased external pressure ΔP.

Bulk Compressibility k *While a material's bulk modulus is a measure of how hard it is to compress the material, the compressibility k is a measure of how easily it is compressed. The* **compressibility** *k is just the inverse, or reciprocal, of the bulk modulus.* Usually, the equation by which it is defined is written as

$$\frac{\Delta V}{V_0} = k\,\Delta P$$

It has the units of reciprocal pressure. Some typical values are given in Table 8.2. Notice that liquids have a much higher compressibility than crystalline solids. This reflects the fact that the molecules in a liquid are fairly widely separated and compression simply pushes them closer together.

8.7 Pressure in a Fluid

You and I walk and live at the bottom of a vast sea of air. Our bodies are constantly under pressure from the great weight of the tremendous height of air above us. As we shall see in the next chapter, each square centimeter of our bodies experiences a force of about 10 N. And yet we are not even aware in most cases that the force exists. Why is this?

The body, in some respects, is like a paper bag filled with only air. There are cavities within us, the lungs, for example; and these body cavities are, in a rough sense, equivalent to the inside of the paper bag. When we close the top of the bag, it does not collapse under the pressure of the air on it. The air inside the bag exerts the same force outward on the bag as the outside air exerts trying to collapse the bag (see Fig. 8.9). These forces balance, and hence the bag appears as though there were no air pressure being exerted on it. This is also true of our bodies.

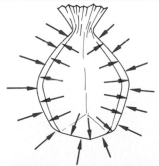

FIGURE 8.9
The sack does not collapse because the air pressure on the inside balances the pressure of the air on the outside.

FIGURE 8.10
As the pump removes the
air inside the metal can, the
can collapses under the
unbalanced forces due to
the atmospheric pressure
outside it.

However, if the air within the sack were removed, there would be no outward-directed forces against the inside of the bag, and it would collapse. The great pressure of the air around us is dramatically shown when the air is pumped out of a metal can. Unless the can is extremely rigid, it will collapse under the unbalanced pressure of the air in which we live. This is shown in Fig. 8.10.

In this age of space flight, and even with jet airplanes at high altitudes, the reverse of the phenomenon just discussed is of life-and-death importance. Suppose that the bag of Fig. 8.9 were in a spaceship which, by some accident, suffered a pressure failure. The air pressure within the ship would drop to zero, and the outside forces on the bag would no longer exist. As a result, the pressure of the air on the inside of the bag would no longer be balanced. The bag would explode. It is clear that human beings are also in dire peril in such a situation.

Now let us turn to the pressures resulting from columns of fluids. A simple case in point is shown in Fig. 8.11. The liquid of height h is maintained in a tube of cross sectional area A, as shown. Our aim is to compute the pressure on the bottom area A *due to the fluid above it*.

The definition of pressure is given in Chap. 5. It is the force per unit area, or, in symbols,

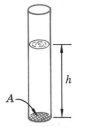

FIGURE 8.11
The force on the bottom
of the cylinder due to the
fluid is equal to the weight
of the fluid.

$$P = \frac{F}{A} \tag{8.9}$$

where F must be perpendicular to A. The force on the bottom of the container due to the fluid is just equal to the weight of the fluid. But the mass of the fluid being supported by the container bottom is given by

$$\text{Mass} = (\text{mass per unit volume})(\text{volume}) = (\rho)(hA)$$

where ρ is the density of the fluid and hA is its volume.

To find the weight of the fluid, we recall that weight $= mg$, and so we have

$$\text{Weight supported by } A = [\rho(hA)](g)$$

This is the force exerted on area A by the fluid above it. We substitute this value for F in the equation for pressure, Eq. (8.9), to obtain

$$P = \frac{\rho h A g}{A} = \rho g h$$

In summary, we find that *the pressure P due to a height h of fluid whose density is ρ is given by*

$$P = \rho g h \tag{8.10}$$

Illustration 8.1   Find the pressure due to a column of water 34 ft high.

Reasoning   We have

$$h = 34 \text{ ft} = (34 \text{ ft})\left(0.305 \frac{\text{m}}{\text{ft}}\right) = 10.4 \text{ m}$$

and we know that $\rho = 1000 \text{ kg/m}^3$ for water. Therefore, the pressure P due to this height of water is

$$\begin{aligned} P = \rho g h &= (1000 \text{ kg/m}^3)(9.8 \text{ m/s}^2)(10.4 \text{ m}) \\ &= 1.02 \times 10^5 \text{ N/m}^2 = 1.02 \times 10^5 \text{ Pa} \\ &= 10 \text{ N/cm}^2 \end{aligned}$$

Notice that this is just equal to the pressure at the surface of the earth resulting from the weight of the air above the earth, given earlier in this section. *The pressure of the air on the surface of the earth is equivalent to the pressure due to a 34-ft (10.4-m) column of water.*

Illustration 8.2   Find the pressure due to a 76-cm-high column of mercury.

Reasoning   From Table 8.1, the density of mercury is $13,600 \text{ kg/m}^3$. Because $P = \rho g h$,

$$P = (13,600 \text{ kg/m}^3)(9.8 \text{ m/s}^2)(0.76 \text{ m}) = 1.01 \times 10^5 \text{ Pa}$$

This pressure, 101 kPa, or 14.7 lb/in^2, is the familiar value for atmospheric pressure. We conclude that *a 76-cm-high column of mercury gives a pressure equivalent to atmospheric pressure.*

8.8 Properties of Pressure in Fluids

The pressure in a fluid acts in all directions. This is easily seen if we consider the experiment illustrated in Fig. 8.12. If we take a small piece of tissue paper and place it near the bottom of a tank of water, as shown in the figure, the tissue paper will not be bent or moved appreciably by the pressure of the still water. We must therefore conclude that the force resulting from the pressure of the water on one side of the paper is balanced by the equal and opposite force on the other side. Since this is true no matter what the orientation of the paper, the pressure in a liquid at a given point must be the same in one direction as in the reverse direction.

Since the pressure due to the liquid at a point beneath the surface of the liquid in Fig. 8.12 is $\rho g h$ from Eq. (8.10), all points at a given depth will have the same pressure, and this pressure will be exerted on any surface at this depth, no matter what its orientation. This is all quite obvious for the case of Fig. 8.12. But what about the case shown in Fig. 8.13? Is the pressure the same at point A as it is at B and C?

The pressure must be the same at each of these points, or the liquid would flow in one direction or the other in the lower pipe. Clearly, then, the shape of the vessel is of no importance. Equation (8.10) for the pressure due to a height h of fluid is correct no matter what type of container encloses the fluid. Try to answer for yourself why the small total force on the bottom of tube B is able to support a large amount of fluid while the same force at A supports less fluid. *Hint:* Does the funnel support any of the liquid?

Suppose, now, that we have a liquid in a closed container like that shown in Fig. 8.14. This device is actually one modification of a hydraulic press. The two pistons shown have cross-sectional areas A_1 and A_2. If there are no forces on the pistons, and if the weight of the pistons is negligible, the liquid will stand at the same height in each of the tubes.

When an external force F_1 is applied to piston 1, the other piston would be pushed up unless a force F_2 were applied to it. If F_2 were made just large enough for no movement to occur, the liquid in the container would remain at rest. Clearly, then, the extra pressure at piston 1 must be balanced by equal pressures everywhere within the fluid, for if this were not true, the unbalanced pressure would cause the liquid to flow. This is an example of **Pascal's principle**, which may be stated as follows:

Pascal's Principle

If a pressure is applied to a confined liquid, the pressure is transmitted to every point within the liquid.

Of course, the liquid must remain at rest for this to be true.

It is instructive to compute the force F_2 needed to balance the force F_1. The additional pressure in the liquid resulting from F_1 is

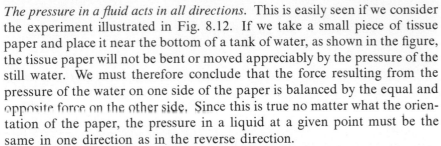

FIGURE 8.12
Since the sheet of paper can exist at equilibrium in the positions shown, we conclude that the pressure at a point within a fluid is the same in all directions.

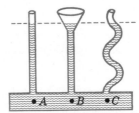

FIGURE 8.13
Why does the liquid stand at the same height in all three tubes?

$$\Delta P = \frac{F_1}{A_1}$$

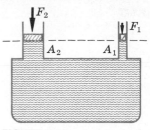

This increase in pressure, according to Pascal's principle, is also exerted on piston 2. This causes the liquid to exert a force F_2 on the piston, where

$$F_2 = \Delta P A_2$$

Substituting for ΔP,

$$F_2 = \frac{A_2}{A_1} F_1$$

If the area of the second piston is much larger than the area of the first, F_2 will be much greater than F_1. Hence, a device such as this is capable of lifting a large weight by the exertion of a small force. You should convince yourself, though, that the small force F_1 must still do an amount of work $F_2 s$ to lift the second piston a distance s. If this were not the case, the device could be made into a perpetual-motion machine.

8.9 Archimedes' Principle

As you know, objects often float on fluids. Even if they sink, they appear to weigh less than when they are not submerged. These effects reflect the fact that *an upward force helps to support a submerged object. We call this a* **buoyant force.**

The buoyancy principle, first discovered by Archimedes, is as follows:

A body partially or wholly immersed in a fluid is buoyed up by a force equal to the weight of the fluid which it displaces. This is known as **Archimedes' principle.**

Archimedes' Principle

For example, if an object with volume V is submerged in a fluid, the buoyant force (BF) on it will equal the weight of the displaced fluid, namely, the weight of the volume V of fluid. The mass of the displaced fluid is given from the definition of density to be

$$m = \rho_{\text{fluid}} V$$

and so its weight mg is simply $\rho_{\text{fluid}} V g$. This is the BF experienced by the submerged object.

In certain cases it is a fairly simple matter to show that Archimedes' principle is true. Consider a cylindrical piece of material immersed in a fluid, as shown in Fig. 8.15. The BF on the cylinder will be equal to the difference between the force on the bottom $P_2 A$ and the force on the top $P_1 A$:

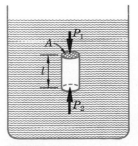

$$BF = P_2 A - P_1 A = A(P_2 - P_1)$$

However, $P_2 - P_1$ is simply the pressure difference due to the height $h = l$ of fluid shown. We know this to be $\rho_f g l$, where ρ_f is the fluid density, and so

$$BF = A\rho_f g l = \rho_f g V$$

where the volume of the displaced fluid V is substituted for the equal volume of the submerged cylinder, Al.

But from the definition of density, $\rho = m/V$, we see that $\rho_f V$ may be replaced by the mass of the displaced fluid m_f, to give

$$BF = m_f g = \text{weight of displaced fluid}$$

As we see, the buoyant force equals the weight of the displaced fluid, in conformity with Archimedes' principle.

Illustration 8.3 A cube of wood 5.0 cm on each edge floats in water with three-fourths of the wood submerged. What is the weight of the cube? The mass of the cube?

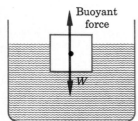

FIGURE 8.16
What must be true if the
cube is to be in equilibrium?

Reasoning The situation is shown in Fig. 8.16. Because the cube is floating, it is at equilibrium. Clearly, its weight is balanced by the buoyant force. Therefore,

$$W = BF$$

Now the volume of the displaced fluid is equal to the volume of the cube below the surface,

$$V_f = \tfrac{3}{4}(0.050 \text{ m})^3 = 9.4 \times 10^{-5} \text{ m}^3$$

From the expression for density $\rho = m/V$, the mass of the displaced fluid is

$$m_f = \rho_f V_f$$

Because the buoyant force is equal to the weight of the displaced fluid, it is therefore given by

$$BF = m_f g = \rho_f V_f g$$

Using the values $\rho_f = 1000 \text{ kg/m}^3$, $V_f = 9.4 \times 10^{-5} \text{ m}^3$, and $g = 9.8 \text{ m/s}^2$ gives the buoyant force to be 0.92 N. We found above that $W = BF$, and so this is also the weight of the wooden cube.

To find the mass of the cube, we recall that $W = mg$ and write

$$m = \frac{W}{g} = \frac{0.92 \text{ N}}{9.8 \text{ m/s}^2} = 0.094 \text{ kg}$$

Illustration 8.4 A 20.0-g piece of metal has a density of 4000 kg/m³. It is hung in a jar of oil ($\rho_f = 1500$ kg/m³) by a thread, as shown in Fig. 8.17. What is the tension in the thread (i.e., the apparent weight of the metal)?

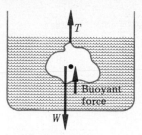

FIGURE 8.17
The weight of the object balances the sum of the buoyant force and the tension in the cord.

Reasoning Because the object in Fig. 8.17 is at equilibrium, we can write

$$T - W + BF = 0 \quad \text{or} \quad T = W - BF$$

To find the BF, we note that

$$BF = (\text{mass of displaced fluid})(g)$$

But from $\rho = m/V$, this becomes

$$BF = (\rho_f V_f)(g)$$

Using this value for the BF in the expression for T and recalling that $W = mg$, we have

$$T = mg - \rho_f V_f g$$

But the volume of the displaced fluid is equal to the volume of the metal in this instance, and so $V_m = V_f$. However, from the definition of density, $\rho = m/V$, we have $V_f = V_m = m/\rho_m$, which can be substituted in the expression for T to give

$$T = mg - (\rho_f)\left(\frac{m}{\rho_m}\right)(g)$$

This simplifies to

$$T = mg\left(1 - \frac{\rho_f}{\rho_m}\right)$$

We know that $m = 0.0200$ kg, $g = 9.8$ m/s², $\rho_f = 1500$ kg/m³, and $\rho_m = 4000$ kg/m³. Substituting these values gives T to be 0.1225 N.

It is instructive to notice the meaning of this result for T, the apparent weight. We see that the apparent weight of a fully submerged object is equal to a fraction $1 - \rho_f/\rho_m$ of the true weight. If the fluid has the same density as the object, the apparent weight is zero. Can you justify this fact?

8.10 Density Determinations

Let us now spend a short time discussing how densities of materials are determined. You will recall that $\rho = m/V$. We must therefore know m and V for an object in order to find its density. The mass of the object is usually easily found by weighing, and so the difficult part of density determination is finding its volume. In certain special cases, the volume can be measured by using calipers or calibrated volumetric flasks. When this is not possible, volumes often are found by use of Archimedes' principle.

It is possible to find the volume of an object if the BF on it is known when the object is submerged in a fluid of known density. This follows from the fact that

$$\text{BF} = (\text{mass of displaced fluid})(g) = m_f g$$

But because $\rho = m/V$, we have $m_f = \rho_f V_f$, and so

$$\text{BF} = \rho_f V_f g$$

Therefore, knowing the BF and the density of the fluid, we can find V_f and its equal, the volume of the object. Then, because both V and m for the object are known, its density can be found at once from $\rho = m/V$.

Illustration 8.5 A piece of rock is weighed, and its mass is found to be 9.173 g. It is then weighed again when submerged in a fluid of density 873 kg/m³ and is found to have an apparent mass of 7.261 g. Find the density of the material of the rock.

Reasoning We know the mass of the object to be $m_0 = 9.173 \times 10^{-3}$ kg. We need its volume.

The actual weight of the object is

$$W_0 = m_0 g$$

Its apparent weight when submerged, corresponding to its apparent mass m_a, is given by

$$W_a = m_a g$$

The difference between these two values is the buoyant force,

$$\text{BF} = W_0 - W_a = (m_0 - m_a)(g)$$

However, another expression for the BF can be found from the fact that it is equal to the weight of a volume V of the fluid, where V is the volume of the displaced fluid and of the object as well. We know that the mass of the displaced fluid m_f is just $\rho_f V$, and so the weight of the displaced fluid is $\rho_f V g$ and this is the BF.

Equating our two expressions for BF, namely,

$$\text{BF} = \rho_f V g \quad \text{and} \quad \text{BF} = (m_0 - m_a)(g)$$

yields

$$V = \frac{m_0 - m_a}{\rho_f}$$

Now that we know both the mass and the volume of the object, we find its density to be

$$\rho = \frac{m_0}{V} = \frac{m_0}{(m_0 - m_a)/\rho_f} = \rho_f \frac{m_0}{m_0 - m_a}$$

Placing in the given values, $\rho_f = 873 \text{ kg/m}^3$, $m_0 = 9.173 \times 10^{-3} \text{ kg}$, and $m_a = 7.261 \times 10^{-3} \text{ kg}$, gives $\rho = 4190 \text{ kg/m}^3$.

Illustration 8.6 An object that has a mass of 24.0 g has an apparent mass of 16.0 g when submerged in water and of 12.0 g when submerged in another fluid. What is the density of the fluid?

Reasoning We found in the previous illustration that the volume of the object can be found from its apparent mass in a fluid of known density, water. The pertinent relation was (the fluid is now water)

$$V = \frac{m_0 - m_{aw}}{\rho_w} = \frac{8.0 \times 10^{-3} \text{ kg}}{1000 \text{ kg/m}^3} = 8.0 \times 10^{-6} \text{ m}^3$$

This is also the volume of the displaced fluid when the object is submerged in the unknown fluid.

But now we can apply this same relation for the volume to the experiment in which the object was submerged in the fluid. We have

$$V = \frac{m_0 - m_{af}}{\rho_f}$$

Solving for ρ_f, we have

$$\rho_f = \frac{m_0 - m_{af}}{V} = \frac{(24.0 - 12.0) \times 10^{-3} \text{ kg}}{8.0 \times 10^{-6} \text{ m}^3} = 1500 \text{ kg/m}^3$$

8.11 Barometer

Let us now return to the discussion of fluid pressure and its effects. The air surrounding us, the atmosphere, is a very important fluid. The pressure within it is of considerable importance to us. As we shall see, atmospheric pressure is an influential variable in many processes. In this section we discuss ways in which atmospheric pressure is measured.

We are all familiar with the fact that increasing barometric pressures usually precede fair weather, while decreasing and low barometric pressures are characteristic of stormy weather. The barometric pressure is included in many weather reports and is one of the primary pieces of data in predicting weather. Not only is it important in this application, but also it is used for many other purposes. For example, the boiling points of liquids change depending on the pressure, and so the barometric pressure is often needed in the laboratory. Let us now examine the operation of the barometer.

Consider the situation shown in Fig. 8.18*a*. The beaker is filled with mercury, and an open glass tube is immersed in it. Since both the mercury surfaces, inside and outside the tube, are open to the air, the air pressures on them are equal. The air pressure on the surfaces is the result of the weight of the air above the earth.

If an ideal vacuum pump is connected to the end of the tube, as shown in Fig. 8.18*b*, the pump will remove the air from the tube, and so the pressure on the mercury surface within the tube will be reduced to zero. (Recall that pressure exerted by a gas on a surface is the result of the collisions of the gas molecules with the surface. If no molecules are present, the pressure is obviously zero. This is what we mean by a perfect vacuum.) There will now be an unbalanced force on the mercury tending to push it up the tube. The mercury will rise to such a height that the pressure at level *A* due to the mercury column is equal to the pressure *P* of the atmosphere. This follows from the fact that the pressure at a given level in a liquid must be everywhere the same, or else the liquid would flow to equalize the pressure.

Finally, the top of the evacuated tube is sealed off, as shown in Fig. 8.18*c*, so that the pressure on the surface of the mercury inside the tube will always be zero. Since the pressure at the bottom of the column of mercury just equals atmospheric pressure, we have

$$P = \rho g h$$

Since mercury has a known density, the atmospheric pressure *P* is easily found by measuring the height of the mercury column needed to balance the atmospheric pressure. In fact, more often that not, the pressure is quoted as so many inches or centimeters of mercury (Hg). *Standard atmospheric pressure is taken to be 760 mmHg.* Of course, *this pressure is* really

FIGURE 8.18
When the tube is evacuated, the mercury rises until $\rho g h = P$. Hence the device, a barometer, is capable of measuring pressure.

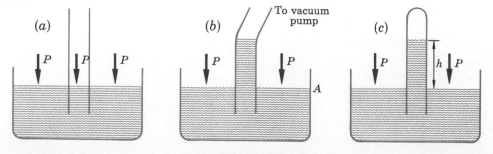

(a) (b) To vacuum pump (c)

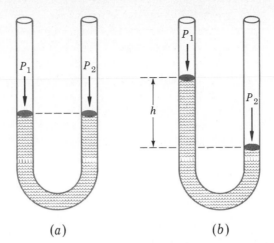

FIGURE 8.19
In a manometer, the
pressure difference $P_2 - P_1$
is measured by height h.

(a)   (b)

$\rho g h = (0.76)(13,600)(9.8)\ \mathrm{N/m^2} = 1.01 \times 10^5\ \mathrm{Pa}$. Why? Since 1 in is 2.54 cm, 760 mmHg is equivalent to about 30 inHg.

Commercial mercury barometers are more refined than the simple device shown in Fig. 8.18c. They have an accurate scale beside the mercury column and special devices to adjust the level of the mercury in the cup. Basically, though, they are as shown. There are other types of barometers based on different principles, but for accurate work the mercury barometer is preferred. However, it must be at least 76 cm long, and so there is often good reason to replace it by a smaller, but less accurate, device.

Another device often used to measure gas pressures precisely is called a **manometer** (Fig. 8.19). Although it has many variations, a manometer is basically a U-shaped tube partly filled with a liquid such as mercury. If the mercury stands at equal levels in the two tubes, as shown in part a, we know that the two gas pressures P_1 and P_2 above the columns must be equal. However, if P_2 is larger than P_1, the columns will adjust as in part b. The difference in heights h of the two columns, measured in centimeters, gives the pressure difference $P_2 - P_1$ in centimeters of mercury. Usually one column is open to the atmosphere; let us say P_1 is atmospheric pressure. Then P_2 is found by adding the barometric pressure to h. For small pressure differentials, it is often convenient to use a liquid which is less dense than mercury. If a liquid of density ρ is used to replace mercury, the difference in levels of the liquid will be increased by a factor $13,600/\rho$, where ρ is in kilograms per cubic meter. Can you explain why?

Illustration 8.7 A simple lung test is to have the patient blow with full force into one end of a manometer, as shown in Fig. 8.20. Suppose in a certain case a mercury manometer is used and the fluid level stands as shown. What is the pressure within the patient's lungs? (Ordinarily, a mercury manometer would not be used because mercury vapor is dangerous on repeated exposure.)

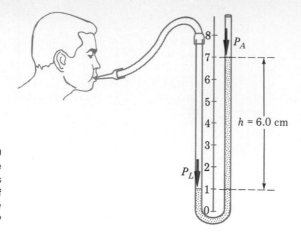

FIGURE 8.20
By blowing into the
manometer, the person is
able to support a column of
fluid 6.0 cm high. How
large is P_L?

Reasoning Let us call the air pressure in the lungs P_L. The pressure exerted on the left side of the manometer is very nearly P_L. This pressure balances the pressure of the atmosphere plus the pressure due to the height $h = 6.0$ cmHg. Therefore

$$P_L = 6.0 \text{ cmHg} + P_A$$

Normally, P_A is about 76 cmHg, and so $P_L = 82$ cmHg. (Of course, in accurate work P_A would be read by means of a barometer.) Since the pressure due to a height h of fluid is $\rho g h$, we have

$$\begin{aligned} P_L = 82 \text{ cmHg} &= (0.82 \text{ m})(13{,}600 \text{ kg/m}^3)(9.8 \text{ m/s}^2) \\ &= 1.093 \times 10^5 \text{ Pa} = 1.08 \text{ atm} \end{aligned}$$

The conversions between these and other units of pressure are given in the table of conversion factors inside the front cover.

8.12 Fluids in Motion

Until now we have been discussing fluids that were at rest. Moving fluids are also of great importance. We can learn a great deal by examining the flow of fluids through pipes. Typical behavior is shown in Fig. 8.21.

As we see in part *a* of the figure, fluid does not simply move down a pipe in a pluglike form. Instead, the fluid close to the walls of the pipe moves scarcely at all. As shown by the velocity profile, the fluid near the center of the pipe moves most swiftly. This variation in speed across the pipe's cross section causes the fluid near the center of the pipe to rub past the outer portion of the fluid. As a result, friction energy loss occurs in the flow process. More is said about this later.

Part b of the figure shows the streamlines of simple flow. These lines show

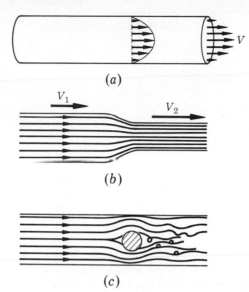

(a)

(b)

(c)

FIGURE 8.21
Examples of various features of flow in a tube: (a) velocity profile; (b) streamlines; (c) turbulent flow.

the path a tiny particle in the fluid follows as it moves along the pipe. *Flow such as this is termed* **laminar flow.** Notice also in this figure that the flow velocity changes as the cross-sectional area changes. The fluid velocity is lower in the large cross-sectional region.

Part *c* of the figure shows what happens if the flow becomes too swift past an obstruction. The smooth flow lines no longer exist. As the fluid rushes past the obstacle, it starts to swirl in erratic motion. No longer can one predict the exact path a particle will follow. *This region of constantly changing flow lines is said to consist of* **turbulent flow.** As you might expect, frequently more energy is lost to friction effects in turbulent flow than in laminar flow. Turbulent flow also occurs in a uniform tube even when an obstacle is not present, provided the flow rate is large enough.

8.13 Viscosity

We saw in the last section that *friction effects occur in a flowing fluid. This effect is described in terms of the* **viscosity** *of the fluid. Viscosity measures how much force is required to slide one layer of the fluid over another layer. Substances that do not flow easily,* such as thick tar or syrup, *have large viscosity. Substances* (like water) *which flow easily have small viscosity.*

To give quantitative meaning to viscosity, we refer to the hypothetical, shear-type experiment shown in Fig. 8.22. There we see two parallel plates, each of area A, separated by a distance l. The region between the plates is filled with a fluid whose viscosity we shall denote by η (Greek eta). In order to move the top plate with speed v relative to the bottom one, a force is required. The force will be large if the fluid has a large viscosity. *We define viscosity η by*

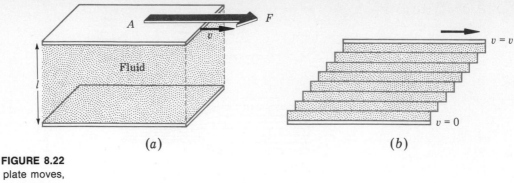

FIGURE 8.22
As the upper plate moves,
layers of the fluid slide over
one another.

Viscosity

$$\eta = \frac{F}{A}\frac{l}{v} \tag{8.11}$$

We see from its defining equation that the SI unit for viscosity is new-ton-seconds per square meter. This unit is given the special name *poiseuille* (Pl). Other common units for viscosity are the poise (P), where 1 P = 0.1 Pl, and the centipoise (cP). This latter unit is easily remembered because it is equal to the millipoiseuille: 1 cP = 1 mPl. Typical viscosities are given in Table 8.3.

We can gain further insight into the meaning of viscosity by examining Fig. 8.22b. Notice that the fluid layers next to the two plates remain attached to the plates. We can think of the fluid between the plates as consisting of many thin layers, many more than shown. As the upper plate moves, these layers must slide over one another. In a high-viscosity fluid, the layers do not slide easily. A large amount of friction work is done as the layers are made to slide past one another. It is for this reason that *work done against viscous forces is equivalent to friction work.*

TABLE 8.3
VISCOSITIES OF LIQUIDS
AND GASES AT 30°C

Material	Viscosity,* mPl or cP
Air	0.019
Acetone	0.295
Methanol	0.510
Benzene	0.564
Water	0.801
Ethanol	1.00
Blood plasma	~1.6
SAE No. 10 oil	200
Glycerin	629
Glucose	6.6×10^{13}

*1 mPl = 10^{-3} Pl = 1 cP.

8.14 Poiseuille's Law

It is often of value to know what volume of fluid will flow through a pipe in unit time. Let us call this quantity the **flow rate** and represent it by the symbol Q. From its definition,

$$Q = \text{volume flowing through pipe per second}$$

Let us refer to Fig. 8.23 and see if we can guess how Q should depend on the variables shown there.

It is reasonable to think that the fluid will flow faster if the pressure differential $P_1 - P_2$ is large. We might guess that Q would be proportional to the driving pressure $P_1 - P_2$. The longer the pipe, the greater the resistance to flow should be. Therefore, Q should vary inversely with L. But the larger the pipe's cross section, the greater the flow rate should be. As a result, Q should increase as R increases. Finally, the larger the viscosity η, the smaller the rate of flow should be.

An exact mathematical relation for Q in terms of $P_1 - P_2$, L, R, and η was first found by Poiseuille (after whom the viscosity unit was named). He showed that *for laminar flow through a pipe of length L and radius R*

$$Q = \frac{\pi R^4 (P_1 - P_2)}{8 \eta L} \tag{8.12}$$

This is called **Poiseuille's law.** Notice that Q increases as the fourth power of R, the pipe's radius.

Illustration 8.8 At $30°C$, water has a viscosity of 0.801 mPl. How much water will flow each second through a 20-cm-long capillary that has a radius of 0.15 cm, if the pressure differential across the tube is 3.0 cmHg?

Reasoning We have the following quantities:

$$L = 0.20 \text{ m} \qquad R = 0.15 \times 10^{-2} \text{ m}$$

$$\eta = 0.801 \text{ cP} = 0.801 \times 10^{-3} \text{ Pl}$$

$$P_1 - P_2 = 3.0 \text{ cmHg} = (\tfrac{3.0}{76})(1.01 \times 10^5 \text{ N/m}^2) = 0.40 \times 10^4 \text{ Pa}$$

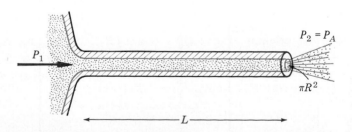

$P_2 = P_A$

P_1

πR^2

L

FIGURE 8.23
The flow rate is proportional to $(P_1 - P_2)R^4/\eta L$, according to Poiseuille's law.

The conversion in pressure units is made here by noting that 1 atm = 76 cmHg = 1.01×10^5 N/m². Substituting these values in Eq. (8.12) yields

$$Q = 5.0 \times 10^{-5} \, \text{m}^3/\text{s} = 50 \, \text{cm}^3/\text{s}$$

Illustration 8.9 A capillary viscometer consists of a vertical capillary tube through which a repeatable quantity of liquid flows in a measured time. The driving pressure is furnished by the weight of the liquid. In a certain viscometer at 30°C, the benzene flow time is 206.3 s, while a very dilute solution in benzene has a flow time of 309.7 s. Find the viscosity of the solution.

Reasoning The dilute solution will have very nearly the same density as benzene. Hence, the driving pressure $P_2 - P_1$ will be the same for both solution and solvent. According to Poiseuille's law, the only factor influencing Q which varied in the two experiments was η. Therefore, we have

$$\frac{\eta \text{ for solvent}}{\eta \text{ for solution}} = \frac{\text{flow time for solvent}}{\text{flow time for solution}}$$

From Table 8.3, the viscosity of benzene at 30°C is 0.564 mPl. Therefore, upon substitution of the known values,

$$\frac{0.564 \text{ mPl}}{\eta \text{ for solution}} = \frac{206.3}{309.7}$$

giving

$$\eta \text{ for solution} = 0.847 \text{ mPl}$$

Illustration 8.10 Older people often develop blood-circulation problems because of deposits building up in their arteries. By what factor is the blood flow reduced in an artery if the artery radius is cut in half? Assume the same pressure differential in the two cases.

Reasoning Poiseuille's law tells us that the volume of blood Q flowing through the artery each second is related to R by

$$Q \propto R^4$$

In the original artery, $Q_0 = (\text{const})(R_0^4)$, but in the constricted artery, $Q = (\text{const})(R_0/2)^4$. Taking the ratio Q/Q_0, we find $Q/Q_0 = \frac{1}{16}$. The flow rate is reduced by a factor of 16. It is clear from this strong dependence of Q on R why blood-circulation difficulties result from arterial deposits.

8.15 Viscous Drag; Stokes' Law

Because liquids and gases have nonzero viscosity, a force is required if an object is to be moved through them. Even the small viscosity of the air causes a large retarding force on an automobile as it travels at high speed. If you stick your hand out of the window of a fast-moving car, you easily recognize that considerable force must be exerted on your hand to move it through the air.[2]

These are typical examples of the following fact:

An object moving through a fluid experiences a retarding force called a *drag force*. The drag force increases with increasing speed of the object.

The exact value of the drag force is difficult to calculate, even in the simplest cases. Indeed, a great amount of money is being invested by automotive companies to evaluate the drag forces on various car designs. Invariably the force must be found experimentally by using models in a wind tunnel. However, the costly effort is fully justified by the increased efficiency of models that have the smallest drag forces.

Even for the simplest objects moving through a fluid the drag force is difficult to compute. However, the case of a sphere moving through a fluid is of great importance, and we examine it in some detail. Our approach is based on the use of dimensional analysis, and even though we can obtain only the form of the drag force, the calculation nevertheless proves to be instructive.

Suppose a sphere of radius R is pulled at constant speed v through a fluid of viscosity η. We seek the force F that is required to pull the sphere through the fluid. It is equal and opposite to the drag force on the sphere.

If we think about the situation, we can easily conclude that F depends in some way on η, v, and R. No other variables appear to be involved. We therefore propose that F is given by a relation of the following form:

$$F = (\text{const})(\eta^a v^b R^c)$$

where a, b, and c are constants. The constant in parentheses is a mere number, and so it is dimensionless.

We now wish to write the dimensional equation equivalent to the equation for F. We know that

$$[F] = [M][L][T^{-2}] \qquad [v] = [L][T^{-1}] \qquad [R] = [L]$$

where the brackets are to be read as "the dimensions of." From its defining equation

---

[2] The drag force is quite complicated in this case. It involves the inertia of the air as well as its viscosity.

$$[\eta] = \frac{[F]}{[A]}\frac{[l]}{[v]} = [M][T^{-1}][L^{-1}]$$

Placing these dimensions in the proposed equation for F yields

$$[MLT^{-2}] = [MT^{-1}L^{-1}]^a[LT^{-1}]^b[L]^c$$

For the equation to be true, M must be raised to the same power on each side. Hence we see at once that

$$M = M^a \qquad \text{so} \qquad a = 1$$

Equating the L factors gives

$$L = L^{-a}L^bL^c$$

from which

$$1 = -1 + b + c \qquad \text{or} \qquad b + c = 2$$

Similarly for the T factors,

$$T^{-2} = T^{-a}T^{-b}$$

from which

$$-2 = -1 - b$$

and so we find $b = 1$. Placing this value in the equation relating b and c, $b + c = 2$, gives $c = 1$. We therefore conclude that the drag force on a sphere is given by

$$F = (\text{const})(\eta v R)$$

The difficult detailed theory for this problem was first carried out by Stokes, and he arrived at what is now known as **Stokes' law:**

The force F required to pull a sphere of radius R with speed v through a fluid of viscosity η is given by

$$F = 6\pi\eta R v \tag{8.13}$$

As you see, our dimensional-analysis approach yielded the proper form of this relation. However, Stokes' law is found to be in error at high speeds where turbulent flow is important; extra terms must be added to this equation if it is to correspond with reality.

Illustration 8.11 A tiny water droplet has a radius of 0.010 cm. Describe its motion as it falls through the air from a high building.

Reasoning Because the air drag on the assumed sphere increases with speed, the drop accelerates rapidly at first under the overpowering force of gravity which pulls it downward. However, the upward-directed drag force on it increases as the drop's speed increases. The net force on the drop is

$$\text{Net force} = \text{weight} - \text{drag force}$$

As the drop's speed continues to increase, the drag force eventually approaches the weight in magnitude. Finally, when the magnitude of the drag force becomes equal to the weight, the net force acting on the drop is zero. From then on, the drop will fall with constant speed, a speed that we call the *terminal speed* (or *terminal velocity*).

Any object allowed to fall for a long distance through a fluid will reach a constant velocity called the **terminal velocity.**   **Definition**

To find the terminal velocity v_T in this case, we use the Stokes' law form for the drag force. Equating it to the weight of the drop, we have

$$mg = 6\pi\eta R v_T$$

Solving for v_T gives

$$v_T = \frac{mg}{6\pi\eta R}$$

But the mass m of the drop is ρV, which is $\rho(\frac{4}{3})(\pi R^3)$. Substitution gives

$$v_T = \frac{2gR^2\rho}{9\eta}$$

Using the given value for R of 1.0×10^{-4} m and $\eta = 19 \times 10^{-6}$ Pl from Table 8.3 gives

$$v_T = \frac{2(9.8 \text{ m/s}^2)(1.0 \times 10^{-4} \text{ m})^2(1000 \text{ kg/m}^3)}{9(19 \times 10^{-6} \text{ Pl})}$$

from which $v_T = 1.1$ m/s.

The value of v_T we have found in this case is small enough that Stokes' law applies. But for large raindrops, the terminal speed is high enough that turbulence occurs in the air through which they fall. As a result, the value of v_T found in that case by use of Stokes' law is larger than the observed value.

Illustration 8.12 A certain globular protein particle has a density of 1246 kg/m³. It falls through pure water ($\rho_w = 1000$ kg/m³) with a terminal speed (called its *sedimentation speed*) of 3.0 cm/h. Find the radius of the particle.

Reasoning Unlike the situation of the previous illustration, the falling particle experiences an appreciable buoyancy force in addition to the force of gravity and the Stokes' law force. At its terminal speed, the following force balance occurs:

$$\text{Weight} = \text{BF} + \text{Stokes' law force}$$

Calling the radius of the falling particle R and its density ρ_P, we have its weight as $\frac{4}{3}\pi R^3 \rho_P g$.

The buoyancy force is equal to the weight of the displaced fluid, water in this case:

$$\text{BF} = (\text{particle volume})(\text{density of water})(g)$$
$$= \frac{4}{3}\pi R^3 \rho_w g$$

Placing these values in the force balance equation gives, after rearranging,

$$\tfrac{4}{3}\pi R^3 g(\rho_P - \rho_w) = 6\pi \eta R v_T$$

Solving for R, we have

$$R = \sqrt{\frac{9\eta v_T}{2g(\rho_P - \rho_w)}}$$

Making use of the values given in the problem together with η for water from Table 8.3, namely, 8.0×10^{-4} Pl, yields $R = 3.5 \times 10^{-6}$ m.

8.16 Bernoulli's Equation for Liquids in Motion

As we have seen, all liquids have a characteristic viscosity. If the viscosity is large, a great deal of work is needed to push the liquid through a pipe. This energy is lost as the molecules rub against one another in the liquid. This lost energy appears eventually as heat energy.

Many liquids have such a small viscosity that their energy loss as a result of friction effects can be neglected, at least for certain purposes. When this is the case, an important, simple relation can be found for the pressure in the fluid. It is called **Bernoulli's equation** and was published by Daniel Bernoulli in 1738.

Consider the pipe system shown in Fig. 8.24. It is completely filled with

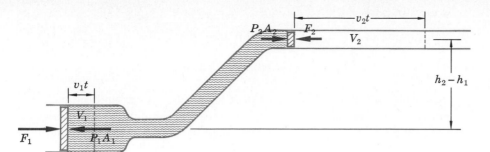

liquid between the two frictionless pistons. We shall say that the lower piston is being pushed to the right with speed v_1 and that the upper piston is moving to the right with speed v_2. The force on the lower piston F_1 is balanced by the force resulting from the pressure of the liquid P_1A_1, where A_1 is the area of the lower piston. (The forces on the piston must balance, or it would be accelerating, and we have already specified that the piston is moving with constant speed.) Similarly, at the top piston, $F_2 = P_2A_2$. Now in a time t the lower piston will move a distance v_1t, thereby displacing a volume of liquid $(v_1t)(A_1)$. However, if the liquid is incompressible, the upper piston must make way for an equal volume of liquid. Hence

$$(v_1t)(A_1) = (v_2t)(A_2)$$

Bernoulli asked what happens to the work done by piston 1. The work done by piston 1 is just $F_1(v_1t)$, or

$$\text{Input work} = P_1A_1v_1t$$

since $P_1A_1 = F_1$. Since piston 2 does an amount of work $F_2(v_2t)$, some of the input work is transformed there.

In addition, the liquid pressed to the right by piston 1 is essentially transferred to the upper tube. As a consequence, that liquid (with volume A_1v_1t) is given some PE. Moreover, since it will now be traveling with a different speed v_2, its KE will also be changed. Of course, some energy is lost in viscous-friction work, but we are assuming this to be small. We therefore have the following equation which tells us what happened to the energy input:

$$\text{Input work} = \text{output work} + \text{change in PE} + \text{change in KE}$$

or, by using the symbols of Fig. 8.24,

$$P_1A_1v_1t = P_2A_2v_2t + Mg(h_2 - h_1) + \tfrac{1}{2}Mv_2^2 - \tfrac{1}{2}Mv_1^2$$

But the volume of liquid involved is A_1v_1t, and its mass is found from the definition of density to be

233

$$M = \rho A_1 v_1 t = \rho A_2 v_2 t$$

Substitution of this in the above equation gives, after rearrangement,

Bernoulli's Equation

$$P_1 + \tfrac{1}{2}v_1^2\rho + gh_1\rho = P_2 + \tfrac{1}{2}v_2^2\rho + gh_2\rho \qquad (8.14)$$

This equation is **Bernoulli's equation.** Clearly, the pistons need not be present. Points 1 and 2 could be any two points in the liquid. All that is needed are surfaces in the liquid; these surfaces can be imaginary, and the computation will still be the same. *Notice, however, that the equation is applicable only if friction losses can be neglected.*

8.17 Torricelli's Theorem

Illustration 8.13 A simple application of Bernoulli's equation is shown in Fig. 8.25. Suppose that a large tank of fluid has two small spigots on it, as shown. Find the speed with which the water flows from the spigot at the right.

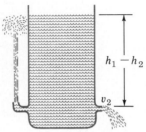

$h_1 - h_2$

v_2

FIGURE 8.25
Torricelli's theorem tells us
how fast the liquid is
moving as it flows out the
spigot.

   Reasoning Since the spigots are so small, the efflux speed v_2 will be much larger than the speed v_1 of the top surface of the water. We can therefore approximate v_1 as zero. Bernoulli's equation can then be written as

$$P_0 + gh_1\rho = P_0 + \tfrac{1}{2}v_2^2\rho + gh_2\rho$$

since $P_1 \approx P_2 =$ atmospheric pressure $= P_0$.
   Rearrangement of this equation gives

Torricelli's Theorem

$$v_2 = \sqrt{2g(h_1 - h_2)}$$

This is **Torricelli's theorem.** *Notice that the speed of the efflux liquid is the same as the speed of a ball that falls through a height $h_1 - h_2$.* This points out the fact that when a little liquid flows from the spigot, it is as though the same amount of liquid had been taken from the top of the tank and dropped to the spigot level. The top level of the tank is a little lower, and the PE lost has gone into KE of the efflux liquid. If the spigot had been pointed upward, as at the left in Fig. 8.25, this KE would allow the liquid to rise to the level shown before stopping. In practice, viscous-energy losses would alter the result somewhat.

8.18 Other Applications of Bernoulli's Equation

Illustration 8.14 Suppose that water flows through a pipe system like the one shown in Fig. 8.26. Clearly, the water must flow faster at B than it

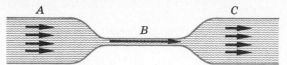

does at *A* or *C*. Assuming the flow speed at *A* to be 0.20 m/s and at *B* to be 2.0 m/s, compare the pressure at *B* with that at *A*.

Reasoning Applying Bernoulli's equation and noting that the average PE is the same at both places, we have

$$P_A + \tfrac{1}{2}v_A^2\rho = P_B + \tfrac{1}{2}v_B^2\rho$$

Substituting $v_A = 0.20$ m/s, $v_B = 2.0$ m/s, and $\rho = 1000$ kg/m³ gives $P_A - P_B = 1980$ N/m². Hence, *the fluid pressure within the constriction is much less than in the large pipes on either side.* This is probably opposite to what you would guess at first. However, it is true and has wide application. Aspirators, for example, obtain a partial vacuum by forcing water through a constriction where the pressure is greatly reduced.

It is easy to see in Fig. 8.26 that the pressure at *A* must be larger than at *B*. Since each little volume of fluid is accelerated as it moves from *A* to *B*, an unbalanced force toward the right must exist on it. To supply this unbalanced force, the pressure must decrease as one goes from *A* to *B*. Can you reverse this line of reasoning to show that the pressure at *C* is larger than at *B*?

This result—that *where the speed is high the pressure will be low*—affords an interpretation of such diverse facts as the lift on an airplane wing and the curve ball pitched by a good ballplayer. The flow around an airplane wing is illustrated in Fig. 8.27. In this case you will notice that the air is traveling faster on the upper side of the wing than on the lower. The pressure will be lower at the top of the wing, and the wing will be forced upward.

FIGURE 8.27
The airplane wing experiences a force from the low-velocity (high-pressure) region below the wing to the high-velocity (low-pressure) region above the wing.

8.19 Blood-Pressure Measurement

There are many other examples in which fluid flow is of importance and interest. We have space here for only one more, the measurement of blood pressure. The pressure of the blood in your arteries and veins varies widely as a function of both time and position in your body. As the heart throbs, the blood pressure on the exit side of it alternately rises and falls. These pressure fluctuations persist throughout the arterial system. But as the blood flows into smaller and smaller channels, friction effects and the elasticity of the channels themselves tend to even out the flow pattern. Finally, as the blood flows into the veins for the return trip to the heart, the flow is almost uniform.

The wide pressure fluctuations in the arteries is of importance in several

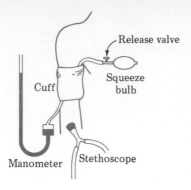

FIGURE 8.28
Apparatus used to measure
blood pressure.

respects, although only two are mentioned here: (1) Extremely high maximum pressures can lead to rupture of the channel walls through which the blood flows. Strokes are one evidence of such a rupture. (2) The magnitudes of the pressure peaks and valleys during the heart-beat cycle provide information concerning constrictions in the channels as well as other body factors which influence blood circulation.

Ordinary blood-pressure data give two numbers, the systolic pressure and the diastolic pressure. These are the pressure readings (*usually expressed in millimeters of mercury*) *at the peak* (*systolic*) *and low point* (*diastolic*) *of the blood-flow cycle.* To measure these values, use is made of the arrangement shown in Fig. 8.28.

An inflatable cuff is placed around the upper arm near the same level as the heart. The pressure exerted by the cuff when inflated is monitored using a mercury manometer. By pumping air into the cuff, the pressure can be made to exceed the peak pressure in the arm artery. As a result, flow of blood into the lower arm is cut off. A stethoscope (a listening device) placed on the artery below the cuff detects no sound since blood is not flowing through the artery.

The pressure in the cuff is slowly reduced by allowing it to deflate. Suddenly, at a well-defined pressure called the systolic pressure, one begins to hear the pulse beat in the stethoscope. At that point, the pressure is low enough in the cuff for the blood to surge past the cuff, through the artery, at the pressure peaks. This pressure reading for the cuff therefore gives the blood pressure at the peak of the heart-throb cycle. Actually, the sound one hears is the result of turbulent flow past the constricted artery at the cuff.

As the pressure is reduced further in the cuff, the blood flow becomes less turbulent. The sound heard through the stethoscope changes character, losing its sharpness. Eventually, the sound ceases. At that pressure, the diastolic pressure, the blood is able to flow past the cuff during all portions of the cycle. This reading therefore gives the lowest pressure during the pulse-beat cycle. In a normal young person, the systolic and diastolic pressures are about 120 and 80 mmHg, respectively (usually reported as 120/80). As one ages, these pressures often change, although very high systolic pressures (200 mmHg, for example) are almost always reason for serious concern. Other pressure variations usually require careful consideration before their implications can be ascertained.

Summary

Substances can be divided into two groups, fluids and solids. There are two types of fluids, namely, liquids and gases. The solids can also be subdivided into those which are crystalline and those which are not. In all these classifications, borderline examples exist. Glasses, for example, can be classed as solids or very viscous liquids.

The mass density ρ of a substance is the mass of substance per unit volume: $\rho = m/V$. Increases in temperature cause the density of most substances to decrease. However, water, in the range 0 to 4°C is an exception to this rule.

Hooke's law applies to many situations in which an elastic deformation occurs. It states that the distortion is proportional to the distorting force. When distortion exceeds the limit of elasticity, the system no longer returns to its undistorted form when the distorting force is removed. In terms of the stress and strain, Hooke's law can be stated as "stress is proportional to strain."

By definition, the ratio of stress to strain is a modulus of elasticity. For a tensile deformation, the modulus is called Young's modulus. In a shear-type distortion, the shear modulus applies. When distortion is caused by equal pressure on all sides, the ratio of pressure to $\Delta V/V$ is called the bulk modulus. The reciprocal of the bulk modulus is called the bulk compressibility.

The pressure increase due to a height h of fluid whose mass density is ρ is given by $P = \rho g h$. In a fluid, the pressure at a point acts equally in all directions. Moreover, if a confined liquid is subjected to a pressure, that pressure is transmitted to every point within the liquid. This is called Pascal's principle. It forms the basis for operation of the hydraulic press.

Archimedes' principle states that "a body partially or wholly immersed in a fluid is buoyed up by a force equal to the weight of the fluid which it displaces." The volume of an object can be found by use of this principle by comparing the apparent weights of the object in air and in a fluid of known density.

The pressure of the atmosphere can be measured by use of a barometer. Differences in pressure can be measured by use of a manometer. Standard atmospheric pressure is 1.013×10^5 Pa, which is the same as 14.7 lb/in^2 and 76 cmHg.

In laminar flow, the fluid follows a set pattern along definite flow lines called streamlines. When turbulent flow occurs, regions exist where stable flow lines are not followed.

Viscosity is a measure of the force required to cause a shearing-type flow in a fluid. It is represented by η, and its SI units are newton-seconds per square meter, denoted poiseuille (Pl). The flow rate of a viscous fluid through a cylindrical tube is given by Poiseuille's law. Viscosities can be compared by measuring the flow rates through the same tube. With proper precautions, the flow rates vary in inverse proportion to the viscosities.

When an object is pulled through a viscous fluid, its motion is opposed by a drag force. In the case of a sphere moving with low velocity through a fluid, the drag force is given by Stokes' law. An object falling through a fluid under the action of gravity may attain a terminal speed if its speed becomes large enough for the drag force to equal the force pulling the particle downward.

Bernoulli's equation describes the flow of fluids in which viscous effects are negligible. It can be used to obtain Torricelli's theorem and to explain the operation of aspirators, atomizers, and similar devices.

In measuring blood pressure, two pressure values are obtained. The highest pressure in the heartbeat cycle is the systolic pressure. The lowest pressure is the diastolic value. Normal values for a young person are about 120 and 80 mmHg for the systolic and diastolic values, respectively.

Minimum Learning Goals

Upon completion of this chapter, you should be able to do the following:

1 Place each of a series of given substances in the category(s) appropriate for it: gas, liquid, solid, fluid, crystalline, amorphous, glass. Give the distinguishing characteristics of each category.

2 Define weight density and mass density. Compute each for a substance when appropriate data are given. Use the density to compute the volume of a given mass or weight of substance. Give the relation between ρ and D.

3 Sketch an elongation-versus-force graph for a sub-

stance which obeys Hooke's law. Give the meanings (definitions) of stress and strain in the cases of tensile, shear, and bulk deformations. Define the modulus in each case. State Hooke's law in terms of stress and strain.

4 Relate bulk compressibility to bulk modulus, and give appropriate units for each.

5 Explain why the can in Fig. 8.10 collapses as the pump removes air from it.

6 Find the pressure due to a column of fluid whose density is known. Explain how a manometer can be used to measure pressure differences. Explain the principle of the mercury barometer. Find the force due to a fluid on a given area at a known depth in the fluid.

7 State Pascal's theorem and explain how use is made of it in the hydraulic press.

8 State Archimedes' principle; then use it to find the buoyant force on an object of known volume submerged in a fluid of known density.

9 Use Archimedes' principle to compute the density

of a substance whose volume is found by immersion in a fluid of known density.

10 State the value of standard atmospheric pressure in newtons per square meter, pounds per square inch, and centimeters of mercury.

11 Explain what is meant by laminar flow, streamlines, and turbulent flow.

12 List several given substances in order of increasing viscosity. Explain what is meant by viscosity and how it is related to friction energy losses. Give the common units in which viscosity is measured.

13 Given Poiseuille's law, identify each quantity in it, and be able to use it for simple calculations.

14 Define the term *drag force* and explain under what conditions Stokes' law can be used to compute it.

15 Given Bernoulli's equation, identify each quantity in it, and be able to use it for simple calculations. Derive Torricelli's theorem from it. Show how it predicts that pressure decreases as flow rate increases.

16 Explain what a doctor is doing when he or she measures blood pressure.

Important Terms and Phrases

You should be able to define or explain each of the following:

Gases, liquids, solids
Fluids
Crystalline and amorphous solids
Weight and mass densities
Hooke's law
Elastic, elastic limit
Stress, strain, and modulus
Tensile, shear, and bulk modulus
Compliance; bulk compressibility
$P = \rho g h$
Pascal's principle

Archimedes' principle; buoyant force
Barometer and manometer
Standard atmospheric pressure
Streamlines; laminar flow
Turbulent flow
Viscosity; poise, poiseuille units
Poiseuille's law
Drag force
Terminal velocity
Stokes' law
Bernoulli's equation
High speed implies low pressure
Systolic and diastolic pressure

Questions and Guesstimates

1 Suppose you are given a solid cube of metal. How could you determine the density of it?

2 How could you determine the density of the glass from which tiny glass spheres are made? Assume the spheres are too small to be weighed and measured with the equipment available. However, a considerable quantity of the material is available.

3 How could you determine the density of a liquid? Of a gas?

4 Compare the pressure at a depth of 5.0 m in a small lake with the pressure 5.0 m below the surface of a large lake. How does the force which a dam must withstand depend on the size of the lake which it dams?

5 Explain why "water seeks its own level."

6 How could you measure the shear modulus of a gelatin dessert? Why would the modulus depend on whether the gelatin had pieces of fruit in it?

Reinforced plastics such as fiberglass have glass fiber embedded in the plastic. What effect does the glass fiber have on the mechanical properties of the plastic? (Bone is reinforced by collagen fibers in this same way. As a result, the tensile modulus and strength of bone are higher than they would be otherwise.)

7 The viscosity of nearly all liquids decreases with temperature. As we shall learn in the next chapter, the KE of the molecules in the liquid increases with increasing temperature. How can this latter fact be used to justify the viscosity change with temperature?

8 Blood plasma consists of blood from which the platelets and other particulate matter have been removed. Whole blood has a viscosity of about 4 mPl, while blood plasma has a viscosity of about 1.5 mPl. Explain the reason for the difference.

9 A glass filled to the brim with water sits on a scale. A block of wood is gently placed in the water so it floats in the glass. Some of the water overflows and is wiped away, but at the end the glass is still filled to the brim. Compare the initial and final readings of the scale.

10 Explain how you can determine the density of an irregular object by use of Archimedes' principle. Consider two cases, when the material is (a) more dense and (b) less dense than the flotation fluid.

11 From your own experience in floating, estimate the density of your body. (E)

12 Explain the principle behind the operation of a siphon.

13 To make a baseball follow a curved path, the pitcher puts "spin" on the ball. Explain why a spinning ball should follow a curved path.

14 A Ping-Pong ball can be suspended in the air by blowing a jet of air just above it, as shown in Fig. P8.1. Explain.

15 Discuss the meaning of Bernoulli's equation when the liquid is not flowing.

FIGURE P8.1

Problems

1 A solid metal cube has an edge length of 2.00 cm and a mass of 21.60 g. What is the density of the metal?

2 A steel sphere has a radius of 0.500 cm and a mass of 4.15 g. Find the density of the steel.

3 Assuming the density of a typical human to be 1005 kg/m³, what is the volume of a 70-kg person?

4 About what mass of air is there in a box-shaped room that is $7.0 \times 5.0 \times 2.8$ m?

5 A 100.00-cm³ flask has a mass of 61.28 g when empty and a mass of 183.45 g when filled with a certain liquid. What is the density of the liquid?

6* A pycnometer is a small flask used for the density determination of liquids. In a certain case, the pycnometer has a mass of 20.000 g when empty and 22.000 g when filled with water at 20°C. After being dried and refilled with a benzene solution, its mass is 21.760 g. From these measurements and the density of water, find the volume of the flask and the density of the benzene solution. For very accurate measurements of density, the mass of air in the empty flask must be taken into account. What is the mass of the air in the pycnometer described in this problem?

7 In the (unlikely) event that a 60-kg woman could rest all her weight on the stiletto heel of one shoe, what would be (a) the stress on the floor beneath it; (b) the pressure (assume the heel to have a bottom area of 0.80 cm²)? (c) Compare your answer in part b to atmospheric pressure.

8* A clothesline is made of wire that has a cross-sectional area of 5.0 mm². What is the stress in the line when a 20.0-kg load is hung at the middle of the line, causing the line to sag 15° below the horizontal?

9 A 5.0-kg mass is hung from the end of a 0.70-mm-diameter steel wire which is 1.40 m long. How far will the wire stretch under this load?

10 A broken elastic is clamped at one end and held taut by a small load at its other. Its length is 22.0 cm, and its cross-sectional area is 0.050 cm². When a 1.50-kg mass is added to its end, the elastic stretches 0.62 cm. Find the average Young's modulus for the elastic in this range of stress.

11* A 35-kg stoplight hanging over the center of a street is held by two equal steel cables fastened to poles on either side of the street. If the cables make an angle of 20° with the horizontal, by what fraction are the cables stretched because of the weight of the light? Take the cross-sectional area of the cables to be 0.180 cm².

12* A copper wire with cross-sectional area 0.0030 cm² is to be used in lifting a 1.00-kg object. How fast can the object be accelerated if the wire is to stretch not more than 0.10 percent? Assume $Y = 11.5 \times 10^{10}$ N/m².

13* Show that the fractional increase in length under load of a wire consisting of length L_1 of wire 1 attached to L_2 of wire 2 is given by

$$\left(\frac{F}{\pi a^2 Y_1 Y_2}\right)\left(\frac{L_1 Y_2 + L_2 Y_1}{L_1 + L_2}\right)$$

provided that they both have the same radius a.

14* How long would a vertical wire have to be for the weight of the wire to cause a 1 percent strain in the uppermost portion of the wire? Express your answer in terms of the density and Young's modulus for the material.

15 A cube of gelatin has top dimensions of 4.0×4.0 cm and a height of 3.0 cm. A shearing force of 0.50 N applied to its upper surface causes the upper surface to displace 2.5 mm in the direction of the force. Find the shear modulus (in pascals) for the gelatin.

16 The shear modulus for a typical metal is about 5×10^{10} Pa. Suppose a shear force of 200 N is applied to the upper surface of a metal cube that is 3.0 cm on each edge. (*a*) How large a shear strain will it cause in the cube? (*b*) How far will the top surface be displaced?

17 (*a*) By about what fraction will the volume of a bar of steel change as the air around it is removed in a vacuum chamber? (*b*) How large an increase in pressure is needed to decrease a volume of benzene by 1 percent?

18 To what pressure must water be subjected if its density is to be increased by 0.01 percent?

19 Standard atmospheric pressure is 1.013×10^5 Pa. Find the force exerted by the atmosphere on the side of a can if the area of the side is 600 cm². Express your answer in both newtons and pounds.

20 What is the average pressure exerted by the floor on the foot of a 70-kg woman as she stands tiptoe on one foot? Assume the contact area between foot and floor to be 12 cm².

21 What is the pressure due to the water at a depth of 15.0 m in a freshwater lake? What is the total pressure at that depth, assuming atmospheric pressure is 1.00×10^5 Pa?

22 Standard atmospheric pressure is 1.013×10^5 Pa (14.7 lb/in²). (*a*) How high a column of water can be supported by this pressure? (*b*) To what maximum height can a vacuum-type water pump lift water in a well when the atmospheric pressure has the standard value?

23* Tire pressure gauges read what is called *gauge pressure,* the difference between the pressure being measured and atmospheric pressure. A 1200-kg automobile has its tires inflated to a gauge pressure of 180 kPa (26 lb/in²). How large an area for each tire is in contact with the pavement?

24* A bicycle tire is inflated to a gauge pressure (see Prob. 23) of 300 kPa. With what force must you push your thumb in on the tire to indent the tire? Assume the contact area of your thumb with the tire to be 1.5 cm². Give your answer in both newtons and pounds.

25* By sucking on one side of a mercury manometer, a person is able to make one side 74 mm lower than the other. (This practice is not recommended since mercury vapor is poisonous.) Assuming atmospheric pressure to be 1.013×10^5 N/m², what is the pressure of air in the person's mouth when doing this experiment? Express your answer in both centimeters of mercury and newtons per square meter.

26* A U tube (much like the manometer shown in Fig. 8.19) has water in one side and oil in the other. The oil column on one side is 62 cm high and balances a column of water which is 54 cm high on the other side. What is the density of the oil?

27 Hydraulic stamping machines exert tremendous forces on a sheet of metal to form it into the desired shape. Suppose the input force is 1000 N on a piston which has a diameter of 1.40 cm. The output force of the hydraulic press is exerted on a piston which has a diameter of 35 cm. How large a force does the press exert on the sheet being formed?

28 The plunger on a certain hypodermic-needle system has a cross-sectional area of 0.79 cm². How large a force must be applied to the plunger if liquid in the needle is to move into a vein where the pressure is 20 kPa above atmospheric pressure?

29* Water is confined to a very strong container by means of a piston with a 0.50-cm² cross-sectional area. How large a force is required on the piston to increase the density of the confined water by 0.010 percent?

30* Find the difference in density between the water

at the top and bottom of a freshwater lake that is 15.0 m deep. Assume that the only factor influencing the density is the pressure of the water.

31* A metal cylinder has a volume of 2.15 cm³. Find the buoyant force on it when it is completely submerged in a liquid that has a density of 800 kg/m³.

32* A metal cube, when submerged in water, is found to have an apparent mass that is 1.63 g less than its true mass. What is the volume of the cube?

33* When a solid object is weighed in air, its mass is found to be 2.371 g. But when it is completely immersed in water at 20.0°C, its apparent mass is found to be 1.863 g. What is the specific gravity of the material that composes the object?

34* A solid object is weighed in air and found to have a mass of 3.087 g. When completely submerged in an oil that has a density of 845 kg/m³, the object's apparent mass is 2.165 g. What is the specific gravity of the material of the object?

35* An irregular piece of metal has a mass of 10.00 g and an apparent mass of 8.00 g when submerged in water. Find the volume and density of the metal.

36* If the piece of metal described in Prob. 35 has an apparent mass of 8.50 g when fully immersed in a particular oil, what is the density of the oil?

37* If a woman floats with 95 percent of her body beneath water, what is her density? Use $\rho_w = 1000$ kg/m³.

38* A block of foam plastic has a density of 820 kg/m³. What fraction of the block will be above water as the block floats in water? Use $\rho_w = 1000$ kg/m³.

39* The density of ice is 917 kg/m³, and the approximate density of the seawater in which it floats is 1025 kg/m³. What fraction of an iceberg is beneath the water surface?

40* What is the minimum volume of a block of material ($\rho = 820$ kg/m³) if it is to hold a 75-kg man entirely above the surface of the water when he stands on the block?

41** When a beaker partly filled with water is placed on an accurate scale, the scale reads 20.00 g.. If a piece of wood having a density of 800 kg/m³ and a volume of 2.0 cm³ is floated on the water in the beaker, how much will the scale read?

42** When a beaker partly filled with water is placed on an accurate scale, the scale reads 20.00 g. If a piece of metal with density 3000 kg/m³ and volume 1.50 cm³ is suspended by a thin string so that the metal is submerged in the water but does not rest on the bottom of the beaker, how much does the scale read?

43** As shown in Fig. P8.2, a piece of wood "weighs" 10.00 g in air. When a heavy piece of metal is suspended below it, the metal being submerged in water, the "weight" of wood in air plus the "weight" of the metal in water is 14.00 g. The "weight" when both wood and metal are submerged in water is 2.00 g. Find the volume and the density of the wood. Use $\rho_w = 1000$ kg/m³.

44** A 2000-cm³ block of foam plastic ($\rho = 450$ kg/m³) is to be weighted down with aluminum so that it will just sink in water. What mass of aluminum is required if the aluminum is to be hung from the block?

45 A liquid that has essentially the same density as water requires 213 s to flow through a capillary viscometer. Water at 30°C requires 127 s to flow through the same viscometer. What is the viscosity of the liquid?

46 A pressure differential of 20 cmHg between the two ends of a long capillary tube causes 50 cm³ of water

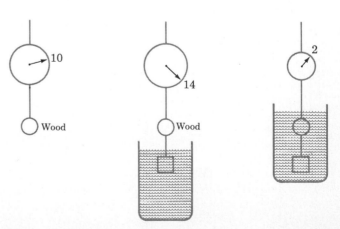

FIGURE P8.2

to flow through the tube in 87 s. How much fluid with a viscosity of 2.3 mPl would flow through the tube in the same time under the same pressure?

47 What was the diameter of the tube in Prob. 46 if its length was 15 cm?

48 By what factor will the quantity of fluid flowing through a capillary tube change if the tube is changed to one having twice the length and one-third the diameter of the original tube?

49 How large a horizontal force is needed to pull a 1.40-cm-diameter sphere through a fluid with a horizontal speed of 3.0 cm/min if the viscosity of the fluid is 6.3 Pl?

50** A tiny glass sphere has a radius of 0.50 mm and a density of 2600 kg/m³. It is let fall through a vat of oil ($\rho = 950$ kg/m³, $\eta = 210$ mPl). Find (a) the buoyant force on the sphere and (b) the gravitational force on the sphere. (c) What will be the terminal speed of the sphere as it falls through the oil?

51** A spherical balloon has a diameter of 20 cm and a mass of 1.405 g. The gas filling it has a density of 0.944 kg/m³ while the density of the surrounding air is 1.290 kg/m³. (a) Will the balloon rise or fall in the air? (b) What will be its terminal speed?

52** Certain colloidal particles in a fluid fall at an average rate of 2.3×10^{-4} cm/s under the action of gravity alone. What would be their sedimentation velocity in a centrifuge if the rotation rate were 5000 rev/min and they were at a radius of 6.0 cm from the rotational axis?

53 The pipe near the lower end of a large water-storage tank springs a small leak, and a stream of water shoots from it. The top of the water in the tank is 15 m above the point of the leak. (a) With what speed does the water gush from the hole? (b) If the hole has an area of 0.060 cm², how much water flows out in 1 s?

54* Water is flowing smoothly through a closed pipe system. At one point the speed of the water is 3.0 m/s, while at another point 3.0 m higher the speed is 4.0 m/s. (a) If the pressure is 80 kPa at the lower point, what is the pressure at the upper point? (b) What would the pressure at the upper point be if the water were to stop flowing and the pressure at the lower point were 60 kPa?

55* A fire hose is to shoot water straight upward a distance of 20 m. What is the minimum required gauge pressure of the water in the hose line?

56* A certain hypodermic needle has an inside diameter of 0.40 mm and a length of 5.0 cm. It is operated by a plunger which has an end area of 3.0 cm². How large a force must be applied to the plunger so that water flows from the needle at a rate of 0.20 cm³/s? Assume the needle is squirting into the air.

57* An airplane wing is designed so that when the speed of the air across the top of the wing is 450 m/s, the speed of air below the wing is 410 m/s. What is the pressure difference between the top and bottom of the wing?

58 The boundary between laminar and turbulent flow can be estimated by use of a number called the Reynolds number, given by

$$\text{Reynolds number} = \frac{Q\rho}{\pi \eta r}$$

for flow rate Q through a cylindrical pipe of radius r. The density of the fluid is ρ, and its viscosity is η. Turbulent flow sets in if the Reynolds number exceeds about 1000. Find the approximate maximum laminar-flow rate for benzene through a 0.50-cm-diameter tube. Express your answer in cubic centimeters per second, and use 1000 as the critical Reynolds number.

9

Temperature, Motion, and the Gas Law

With this chapter we begin the study of thermal energy and its effects. The definition of temperature is given, and the nature of heat is investigated. We conclude, after a discussion of the gas law, that temperature is a measure of molecular KE and that thermal energy is KE of random molecular motion of the gas. These concepts are needed as a foundation for the chapter which follows, the study of thermal properties of materials and the equivalence of heat and other forms of energy.

9.1 Early Ideas about Heat

Ever since the discovery of fire by early peoples, the nature of heat has been a source of interest. We know that the wonderful properties of fire caused ancient peoples to deify it. Even when the spiritual maturity of most of the human race had reached a level such that fire was no longer considered a god, the feeling persisted that heat and fire had mystical qualities.

As recently as the eighteenth century, people believed that the heat that warmed them as they stood near a fire was actually a material quantity that flowed to them through space. They called this fluid **caloric.** When a hot piece of metal was cooled by water, caloric was believed to flow out of the metal into the water. When a burning piece of wood was torn apart by the flames, caloric was allowed to escape and flow to other bodies, or so the people of the 1780s believed. Each piece of material contained more or less caloric, depending on its temperature.

It was not until the 1790s that this prevalent concept of a heat fluid was effectively challenged. The most conclusive blow to the caloric theory was dealt by Benjamin Thompson (who was awarded the title Count Rumford, by which he is widely known). Rumford was an American who joined the British forces during the American Revolution. He became an armament expert and served for many years as an official in the government of Bavaria.[1] During that time he performed a number of scientific experiments. The experiments in which we are interested here were conducted in a cannon workshop of the military arsenal at Munich.

Rumford noticed that as cannon barrels were being bored, a tremendous amount of heat was given off. Since the metal chips cut off by the drill had lost caloric, i.e., the heat given off in the drilling process, the chips should not be the same as the original metal, which had not lost caloric. In spite of this reasoning, Rumford was unable to find any difference between the chips and the original metal in their ability to hold or give off heat. Seeking to investigate this effect further, he tried a very dull drill. In spite of the fact that it was unable to cut the metal, this borer also generated heat as it was rotated while rubbing against the metal. In fact, the drill generated enough heat to boil water in a cavity of the metal. No matter how long the borer had previously been used on a piece of metal, it readily generated heat each time it was started once again. From this, Rumford concluded that the supply of heat was inexhaustible. He further concluded that the heat came not from the metal, but from the rotation of the drill. As a result, in 1798 Rumford was led to discard the caloric concept of heat. Instead, he postulated that the motion of the drill transmitted motion to the particles within the metal and heat was actually this motion. As long as the drill kept transmitting motion to the metal by frictional forces, the motion and heat within the metal would keep on increasing. This is essentially the picture of heat that is accepted by physicists today. We return to this discussion of the nature of heat, but first let us examine the concept of temperature.

9.2 Thermometers

Long before Rumford's discovery, practical means had been found for measuring how hot an object was. Qualitatively, we all know what is meant by hot and cold. We also know that when a hot object is placed in contact with a cold one, the hot object will cool while the cold object warms. To measure the "hotness" of an object, some method must be found to place a number on this property. We call this number the *temperature* of the object. Over the years, many number scales have been used to assign temperatures, but only a few temperature scales have survived.

To define a scale for temperature, we need a thermometer—a device used to measure the temperature. Nearly all thermometers make use of the

---

[1] Rumford was an interesting person, and several biographies of him have been written. Your library probably has one available.

fact that most liquids, solids, and gases expand as they are heated. The ordinary mercury thermometer, sketched in Fig. 9.1, uses the expansion of liquid mercury with temperature as a measure of temperature. As shown, a bulb filled with mercury is attached to a capillary tube. As the bulb is heated, the mercury expands and rises higher in the capillary. Hence, the height of the mercury in the capillary can be used as a measure of the temperature of the bulb.

Three common methods for assigning numerical values to temperatures by use of a thermometer are shown in Fig. 9.2. If the device of Fig 9.1 is placed in a vessel containing water and melting ice, the mercury will reach a certain level in the capillary. A mark is made at that level. Now the device is placed in boiling water (at an atmospheric pressure of 760 mmHg), and the capillary is marked at the new higher level of the mercury. There are now three common ways of placing numbers along the capillary tube so as to read temperature. These give rise to three different temperature scales. They are illustrated in Fig. 9.2, where the two scale marks are shown, together with three separate scales which can be superposed on them.

It should be noticed that both *the Celsius* (formerly called centigrade) *and Kelvin scales split the region from freezing to boiling of water into 100 parts called* **degrees.** *The Fahrenheit scale splits this region into 180 degrees. A Celsius degree is larger than a Fahrenheit degree by a factor of $\frac{180}{100}$, or $\frac{9}{5}$.*

The Celsius scale is used throughout the whole world. The United States will officially accept it when the country converts to the metric system. Although once widely used in the English-speaking countries, the

FIGURE 9.1
As the mercury in the glass container is heated, it expands and rises in the capillary. The height at which it stands is a measure of the temperature.

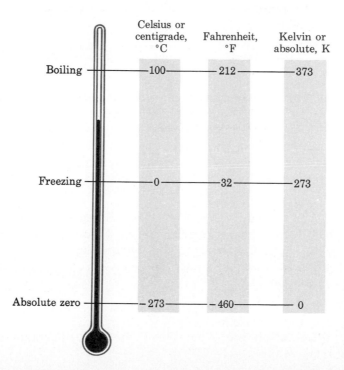

FIGURE 9.2
The boiling and freezing points of water can be used to illustrate the interrelations among the three usual temperature scales.

Fahrenheit scale is seldom used outside the United States today. The Kelvin scale makes use of Celsius-size degrees, but its zero point is taken differently. It is often used by scientists.

Mercury thermometers will not function below $-39°C$, because mercury freezes at this temperature. This difficulty is surmounted by using alcohol or pentane or some similar low-freezing-point liquid at low temperatures. In addition, no two liquids expand in exactly the same way over the whole temperature range. A better way of subdividing the scales is needed. In recognition of this fact, a completely new definition of the temperature scales was adopted by international agreement. This definition makes use of a gas-filled thermometer, and we shall learn more about it after we describe how gases respond to temperature changes.

Although the thermometers we have been discussing are most common, others also exist. *Any physical quantity which varies in a repeatable way with temperature can, in principle, be used as a temperature-measuring device.* Practical considerations determine which type of device should be used in a given circumstance. Perhaps one of the most useful of these alternative devices is the *thermistor,* whose electrical resistance changes with temperature. This resistance change is converted electrically to a temperature reading on a meter scale. The thermistor itself can be very tiny and can be connected to the readout system by thin wires. Its response time is of the order of a second, and so readings are almost instantaneous. This device is in widespread use in both medicine and industry.

Illustration 9.1 The temperature of a room is $77°F$. What is the Celsius temperature in the room?

Reasoning We always solve such problems in the following way:

1 Find out how many degrees the temperature is above or below the freezing point of water ($32°F$ or $0°C$).
2 Use the fact that a Fahrenheit degree is equivalent to $\frac{5}{9}$ Celsius degree to convert from one type of degree to the other.
3 Find the temperature on the new scale by adding to (or subtracting from) $32°F$ or $0°C$.

In this case we had $77°F$. This temperature is $77 - 32 = 45$ Fahrenheit degrees above freezing. But 45 Fahrenheit degrees is equivalent to $(\frac{5}{9})(45) = 25$ Celsius degrees. Therefore the temperature is 25 Celsius degrees above freezing. Since freezing is $0°C$, we see that the temperature in question is $25°C$.

Illustration 9.2 What is the Fahrenheit temperature on a day when the temperature is $-10°C$?

Reasoning   The temperature is 10 Celsius degrees below freezing. Since a Fahrenheit degree is $\frac{5}{9}$ as large as a Celsius degree, 10 Celsius degrees are equivalent to $(\frac{9}{5})(10) = 18$ Fahrenheit degrees. The required temperature is 18 Fahrenheit degrees below freezing. Since freezing is 32°F, the temperature in question is $32 - 18 = 14$°F.

9.3 Gas Law

In their laboratory work in science, many students perform simple experiments that illustrate the behavior of gases. The law that governs the behavior of air and many other gases is simple and easily found by experiment. Three variables enter into the law: pressure, temperature, and number of molecules per unit volume. Let us see what facts experiment tells us about the relation among these variables.

When a gas that is confined to a fixed volume is heated, its pressure increases. A graph showing the measured pressure versus temperature looks much like that shown in Fig. 9.3. Different gases and different initial pressures yield similar graphs. Provided the gases are not near the conditions at which they liquefy, the graph *is a straight line.* Moreover, *the intercept of the line is always* $-273.15°C$.

Another simple experiment is to measure the volume of a gas as a function of temperature but subject to a constant pressure. *The measured volume-versus-temperature graph is* typified by Fig. 9.4. Again, *a straight-*

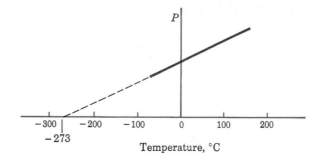

FIGURE 9.3
When a gas in a closed container is heated, the pressure of the gas varies linearly with the temperature. On the absolute scale, the equation of the line is $P = (\text{const})(T)$.

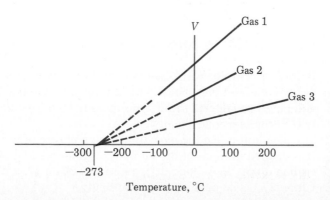

FIGURE 9.4
A gas maintained at constant pressure contracts as it is cooled. On the absolute temperature scale, the equation of the line for an ideal gas is $V = (\text{const})(T)$.

line relation is found *with an intercept at* −273.15°C. Although different gases and different volumes give different straight lines (provided the gas is not near liquefaction), this intercept is always the same.

Notice in both Figs. 9.3 and 9.4 that *the temperature-axis intercept is* −273.15°C. *This is taken as the zero point on the absolute (or Kelvin) temperature scale.* Let us represent absolute temperatures by the symbol *T*. Then the data of these two graphs can be represented by

$$PV = (\text{const})(T)$$

To check this against Fig. 9.3, recall that *V* is constant in that case. Then

$$P = \frac{\text{const}}{V} T = (\text{const})(T)$$

which is the equation of a straight line on the *P*-versus-*T* graph. When *T* = 0, that is, at *t* = −273.15°C, *P* = 0. As we see, the equation and Fig. 9.3 agree. Similarly, for *P* = constant, the equation becomes

$$V = \frac{\text{const}}{P} T = (\text{const})(T)$$

This is the equation of a straight line with intercept at $T = 0$ (or $t = -273.15$°C). It agrees with the data shown in Fig. 9.4. We conclude that *a gas which is far removed from conditions under which it liquefies obeys the equation*

$$PV = (\text{const})(T) \tag{9.1}$$

It is important to recognize that T is the absolute (Kelvin) temperature of the gas.

A gas that obeys Eq. (9.1) is defined to be an **ideal gas.** As we shall see, the constant in Eq. (9.1) has a definite, simply stated value. Before proceeding to discuss this value, we first introduce two terms often used to express it.

9.4 The Mole and Avogadro's Number

Before we continue with discussion of the gas law, let us take a few moments to review a few terms that will be of importance in the discussion. Although you have encountered these terms in your chemistry courses, it is imperative that you understand them well.

The unit called the *mole* (mol) is a measure of the number of particles. It is defined in terms of the number of atoms in 12 g of a particular type of carbon atom, the isotope designated as carbon 12. (We discuss isotopes in Chap. 27.) The definition is as follows:

A **mole** *of a substance contains as many particles as there are atoms in*   **Definition**
12 grams of carbon 12.

We most often use the kilomole (kmol), the quantity of substance that contains as many particles as there are atoms in 12 kg of carbon 12.

Next let us define a quantity called Avogadro's number:

Avogadro's number N_A *is the number of particles in one mole of*   **Definition**
a substance.

Its measured value is found to be 6.022×10^{23} particles per mole. Because we most often use kilograms and kilomoles in discussions, *we usually take* N_A *to be* 6.022×10^{26} *particles per kilomole.*

Notice that the terms *mole* and *Avogadro's number* apply to any type of particle. For example, 1 mol of sugar contains N_A sugar molecules, while 1 mol of baseballs contains N_A baseballs. As you see, the mole is a measure not of mass, but of number of entities.

A related term that we should also review is *molecular,* or *atomic, mass* (often called molecular or atomic weight).

The **molecular** (*or* **atomic**) **mass** *M of a substance is the mass in kilo-*   **Definition**
grams of one kilomole of the substance.

Because 1 kmol of carbon 12 is equivalent to 12 kg, M for carbon 12 is 12 kg/kmol, or its equivalent, 12 g/mol. Similarly, the value of M for the hydrogen atom is 1 kg/kmol, and M for the nitrogen molecule N_2 is 28 kg/kmol. In the same way, M for benzene (C_6H_6) is 78 kg/kmol. The atomic masses of the elements are given in Appendix 4.

Illustration 9.3   The atomic mass of the copper-63 isotope is 62.93 kg/kmol. Find the mass of an atom of copper 63.

Reasoning   We know that 62.93 kg of copper atoms contains N_A atoms. Therefore the mass per atom is

$$\text{Mass per atom} = \frac{M}{N_A} = \frac{62.93 \text{ kg/kmol}}{6.02 \times 10^{26} \text{ atoms/kmol}}$$
$$= 1.045 \times 10^{-25} \text{ kg/atom}$$

This same method can be used to find the mass of any atom or molecule for which M is known.

9.5 Gas-Law Constant

Let us now return to the law obeyed by ideal gases,

$$PV = (\text{const})(T)$$

where P is the pressure of the ideal gas in volume V at absolute temperature T. By means of the experiments described in Sec. 9.3, the proportionality constant can be evaluated. It is found to have the value nR, where n is the number of kilomoles of gas in the volume V and

$$R = 8314 \text{ J/(kmol)(K)} = 8.314 \text{ J/(mol)(K)}$$

Here R is called the *gas constant*.
    We can therefore write the **ideal-gas law** *as*

Ideal-Gas Law

$$PV = nRT \qquad\qquad (9.2)$$

It is very important in using the law that the proper units for the various quantities be used. The temperature must always be expressed in the Kelvin scale. If n is the number of kilomoles in V, then R must be 8314 J/(kmol)(K). If for some reason you insist on using n in moles, then R must be taken as 8.314 J/(mol)(K).

Illustration 9.4   Standard atmospheric pressure and temperature are 1.01×10^5 Pa and 0°C. Find the volume which 1 kmol of N_2 occupies under these conditions. The molecular mass of N_2 is 28 kg/kmol.

Reasoning   To use the gas law, we note in this case that

$$P = 1.01 \times 10^5 \text{ Pa}$$

$$T = 0 + 273.15 \text{ K} = 273.15 \text{ K}$$

where we read this latter value as "273.15 kelvins." Since we are concerned with the volume occupied by 1 kmol, $n = 1$ kmol. Then

$$PV = nRT$$

becomes

$$(1.01 \times 10^5 \text{ N/m}^2)(V) = (1 \text{ kmol})\left[8314 \frac{\text{J}}{\text{(kmol)(K)}} \right](273 \text{ K})$$

From this we find

$$V = 22.4 \text{ m}^3/\text{kmol}$$

You should be able to generalize this example to show the following:

One kilomole of ideal gas occupies a volume of 22.4 m³ under standard conditions.

This is sometimes a convenient fact to remember.

9.6 Molecular Basis for the Gas Law

We have now considered the pressure of gases from two separate viewpoints. In Chap. 5 we discussed how gas molecules exert pressure on the walls of a container by collisions with the wall. We found that the pressure was related to the KE of the gas molecules. Our result was, in slightly different symbols,

$$P = (\tfrac{2}{3})(\nu_0)(\tfrac{1}{2}m_0v^2) \tag{5.6}$$

In this equation, we assume there are ν_0 molecules per cubic meter, each molecule has a mass m_0, and its average translational KE is $\frac{1}{2}m_0v^2$. Notice that the temperature of the gas does not appear in this equation.

On the other hand, the ideal-gas law, obtained by experiment, states that

$$PV = nRT \qquad \text{or} \qquad P = \frac{nRT}{V}$$

By comparing this equation with Eq. (5.6), we see that

$$\frac{nRT}{V} = (\tfrac{2}{3})(\nu_0)(\tfrac{1}{2}m_0v^2)$$

or, after rearranging,

$$T = \frac{2\nu_0 V}{3nR}\left(\frac{1}{2}m_0v^2\right)$$

But $\nu_0 V$ is simply the total number of molecules in the volume V. Moreover, nN_A is also the total number of molecules in the volume V. Therefore

$$\nu_0 V = nN_A$$

and the expression for T becomes

$$T = \frac{2N_A}{3R}\left(\frac{1}{2}m_0v^2\right) \tag{9.3}$$

Absolute Temperature

We therefore find that T is directly proportional to the average translational KE of the gas molecules. We conclude that *absolute temperature is a measure of the translational KE of the gas molecules.* This fact has great importance, as we shall soon learn.

Illustration 9.5 Estimate the average speed of the nitrogen molecules in the air under standard conditions.

Reasoning   Under standard conditions, the temperature is 0°C, or 273.15 K. From Eq. (9.3) we have

$$v^2 = \frac{3RT}{N_A m_0}$$

To find m_0 for nitrogen molecules, we note that the molecular mass of N_2 is 28 kg/kmol. Therefore,

$$m_0 = \frac{28 \text{ kg/kmol}}{N_A}$$

Substitution yields

$$v^2 = \frac{3RT}{28 \text{ kg/kmol}}$$

Since $R = 8314$ J/(kmol)(K) and $T = 273$ K, we find

$$v^2 = \frac{(3)(8314)(273)}{28} \frac{\text{m}^2}{\text{s}^2}$$

or

$$v = 493 \text{ m/s}$$

The nitrogen molecules of the air have an average speed of about 500 m/s under standard conditions.

9.7 Use of the Gas Law

The ideal-gas law is found to apply to all gases provided the following two conditions are satisfied:

1   The volume occupied by the molecules themselves, i.e., the volume of a molecule multiplied by the number of molecules, must be a negligible fraction of the volume in which the gas is enclosed.
2   The translational KE of the molecules must be large compared with the energy needed to separate two gas molecules from each other.

This latter restriction has to do with the fact that uncharged atoms and molecules usually attract one another. As a result, two molecules tend to stick together. For the ideal-gas law to apply, the attraction energy must be small compared with the average translational KE of a molecule.

But we know that the average translational KE of a molecule is related to the absolute temperature through Eq. (9.3). We can therefore restate the second condition in the following way:

2′   The temperature must be high enough to ensure that only a negligible number of molecules stick together.

Of course, when the molecules stick together, the gas must condense to a liquid. This then allows us to state the second condition in still another way:

2″ The gas must be far removed from conditions under which it liquefies.

Condition 2 is normally satisfied for gases such as air, oxygen, nitrogen, helium, hydrogen, and all the gases difficult to liquefy. Condition 1 is satisfied for all gases whose density is comparable with, or smaller than, that of air.

An interesting feature of Eq. (9.3) (and Figs. 9.3 and 9.4) can be seen if you notice what happens at absolute zero ($T = 0$ K). Since T is proportional to the translational KE of the molecules, when $T = 0$, the KE should also be zero. The molecules would no longer be moving, and so they would exert no pressure. This is why, in Fig. 9.3, we see that $P = 0$ when $T = 0$. Can you explain why, as shown in Fig. 9.4, $V = 0$ when $T = 0$?

But, of course, no gas can be cooled to absolute zero without condensing. Conditions 1 and 2 become invalid before we come close to $T = 0$, and so the ideal-gas law no longer applies. In fact, additional queer behavior is observed near absolute zero because classical physics must be modified at very low temperatures. At these temperatures, quantum effects become important and lead to such phenomena as superfluidity (zero-viscosity fluids) and superconductivity (zero-resistance electrical conductors). In spite of these restrictions on the ideal-gas law, the law has a wide range of application.

The ideal-gas law is fundamental to the presently agreed upon definition of the various temperature scales. In 1954 the following method for defining the temperature scales was accepted. It involves the behavior of an ideal gas confined to a fixed volume. This gas can be used as a thermometer, called a constant-volume gas thermometer. The gas obeys the ideal-gas law, which becomes, in the case of V constant,

$$P = (\text{const})(T)$$

We agree to calibrate this standard thermometer by measuring the pressure of the gas at a certain temperature called the triple point of water. *The **triple point of water** is that unique temperature at which water, ice, and* **Definition** *water vapor can coexist.* This temperature is arbitrarily assigned the value 273.16 K.[2] But now P can be measured at this temperature, and so P and T are both known for the gas thermometer. Therefore the constant in the relation $P = (\text{const})(T)$ can be evaluated for this thermometer.

Now that the constant is known, the gas thermometer can be used to measure absolute temperature in terms of the pressure of the gas within it. Of course, a gas thermometer is seldom used because it is inconvenient. But, when needed, it can be used to calibrate any other type of thermometer one wishes to use.

The ideal-gas law is widely used for computations involving one or the

---

[2]This value was chosen because it brings the new scale into agreement with the old scale.

other of the following situations: (1) we wish to compute P, V, n, or T provided all but one of these quantities are known or (2) knowing P, V, n, and T under one set of conditions, we can compute the parameters under other conditions. These uses are illustrated in the following examples.

Illustration 9.6 An empty oil drum is closed at a temperature of 20°C. It is then set out in the sun, where it heats up to 60°C. If the original pressure was 1.0 atm, what is the final pressure in the drum?

Reasoning Write Eq. (9.2) twice,

$$P_1V = nRT_1 \qquad P_2V = nRT_2$$

where V is the constant volume of the drum and n is the number of kilomoles of gas in the drum. Dividing one equation by the other yields

$$\frac{P_1}{P_2} = \frac{T_1}{T_2}$$

where T_1 and T_2 are absolute temperatures. We had $P_1 = 1$ atm, $T_1 = 20 + 273 = 293$ K, and $T_2 = 333$ K. Therefore

$$P_2 = (1.0 \text{ atm}) \left(\frac{333 \text{ K}}{293 \text{ K}} \right) = 1.14 \text{ atm}$$

Notice that absolute temperatures must be used. The pressure units are not important as long as they are the same.

Illustration 9.7 The gas in the piston of a diesel engine is originally at a temperature of 27°C and a pressure of 74 cmHg when it is suddenly compressed. If the final pressure of the gas is 3700 cmHg and the temperature has been increased to 547°C, what is the final volume of the gas in terms of the original volume?

Reasoning Write the gas law twice,

$$P_1V_1 = nRT_1 \qquad P_2V_2 = nRT_2$$

Dividing yields

$$\frac{P_1V_1}{P_2V_2} = \frac{T_1}{T_2}$$

Substituting gives

$$\frac{V_1}{V_2} \frac{74}{3700} = \frac{273 + 27}{273 + 547}$$

from which

$$V_2 = 0.058 \, V_1$$

Illustration 9.8 A car tire is filled to a pressure of 32 lb/in² on a cold day when the temperature is $-3°C$. What is the pressure in the tire when the temperature rises to $+47°C$? (Assume the volume of the tire to remain constant.)

Reasoning Once again,

$$\frac{P_1 V_1}{P_2 V_2} = \frac{T_1}{T_2}$$

But $V_1 = V_2$, $T_1 = 270$ K, and $T_2 = 320$ K. *Notice that the pressure read by a tire gauge* (*called gauge pressure*) *is the excess pressure within the tire.* (The gauge reads zero when the pressure is atmospheric pressure.) Assuming atmospheric pressure to be 14.7 lb/in², we have $P_1 = 46.7$ lb/in². Hence

$$P_2 = \frac{P_1 T_2}{T_1} = (46.7 \text{ lb/in}^2)\left(\frac{320 \text{ K}}{270 \text{ K}}\right) = 55.4 \text{ lb/in}^2$$

and the gauge would read 40.7 lb/in².

Illustration 9.9 A tank of oxygen gas at 0°C has a volume of 4 liters, and the pressure is 50 atm. What mass of oxygen is in the tank?

Reasoning We can solve this in two different ways.

Method 1: Let us first find what volume this gas would occupy at standard temperature and pressure:

$$\frac{P_1 V_1}{P_2 V_2} = \frac{T_1}{T_2} = 1$$
$$(50 \text{ atm})(4 \text{ liters}) = (1 \text{ atm})(V_2)$$
$$V_2 = 200 \text{ liters} = 0.20 \text{ m}^3$$

But 22.4 m³ will hold 1 kmol of O_2 under these conditions. Since the atomic mass of oxygen is 16 and the molecular mass is 32, the mass of oxygen in the tank is

$$\text{Mass} = \left(\frac{0.2 \text{ m}^3}{22.4 \text{ m}^3/\text{kmol}}\right)(32 \text{ kg/kmol})$$
$$= 0.285 \text{ kg} = 285 \text{ g}$$

Method 2: Working directly from the gas law, Eq. (9.2),

$$PV = nRT$$

we substitute $(50)(1.01 \times 10^5 \, \text{N/m}^2)$ for P and $4 \times 10^{-3} \, \text{m}^3$ for V. Using $R = 8314 \, \text{J/(kmol)(K)}$ and $T = 273 \, \text{K}$, we find

$$n = 8.9 \times 10^{-3} \, \text{kmol}$$

But 1 kmol of oxygen is 32 kg, and so the mass in the container is simply

$$\text{Mass} = (8.9 \times 10^{-3} \, \text{kmol})(32 \, \text{kg/kmol}) = 0.285 \, \text{kg}$$

Illustration 9.10 What is the mass of a molecule of carbon dioxide (CO_2)?

Reasoning The atomic mass of carbon is 12 and of oxygen is 16. Hence the molecular mass of CO_2 is 44. Since 44 kg of CO_2 will contain Avogadro's number of molecules, 6.02×10^{26}, the mass of a molecule will be

$$m_0 = \frac{44 \, \text{kg}}{6.02 \times 10^{26}} = 7.31 \times 10^{-26} \, \text{kg}$$

Illustration 9.11 Show that the constant-temperature bulk modulus E of an ideal gas is the pressure P of the gas.

Reasoning We recall that a substance subjected to a pressure increase ΔP will undergo a small fractional volume decrease $\Delta V/V$ that is related to the bulk modulus E by

$$E = \frac{\Delta P}{\Delta V/V}$$

In the case of an ideal gas at constant temperature before compression,

$$PV = nRT$$

and after compression at constant temperature

$$(P + \Delta P)(V - \Delta V) = nRT$$

Equating these two expressions for (nRT) and simplifying gives

$$PV = PV + V \Delta P - P \Delta V - \Delta P \Delta V$$

or, after a little algebra,

$$0 = \frac{\Delta P}{P} - \frac{\Delta V}{V} - \frac{\Delta P}{P} \frac{\Delta V}{V}$$

from which

$$\frac{\Delta V}{V} = \frac{\Delta P}{P}\left(1 - \frac{\Delta V}{V}\right)$$

But we are concerned with only very small fractional changes in volume, and so $\Delta V/V$ is much smaller than unity. As a result, we can neglect it in the above relation and then find that

$$\frac{\Delta V}{V} = \frac{\Delta P}{P}$$

Substituting this for $\Delta V/V$ in the expression for E gives

$$E = \frac{\Delta P}{\Delta P/P} = P$$

and so we find that the bulk modulus of an ideal gas is equal to the pressure of the gas in a compression carried out at constant temperature.

9.8 Variation of Molecular Speeds in Gases

As we saw in the previous sections, temperature is a measure of the average molecular KE. In particular, from Eq. (9.3) we find that the average translational KE of a gas molecule is simply

$$\text{KE of translation} = \tfrac{1}{2}m_0 v^2 = \frac{3}{2}\frac{R}{N_A}T = \tfrac{3}{2}kT \tag{9.4}$$

where the constant k is called *Boltzmann's constant* and is given by

$$k = \frac{R}{N_A} = 1.3807 \times 10^{-23} \text{ J/K}$$

Boltzmann's constant is named after Ludwig Boltzmann, a pioneer in molecular calculations. A short biography of him is given on page 258.

Boltzmann, together with James Clerk Maxwell (who is famous also for his work in electricity), and a few other highly creative scientists of the late 1800s devised the model for gases that we have been using. Although their work was much more sophisticated than the approach we have used, they still (correctly) pictured gas molecules to be like balls darting here and there. This basic model is called the *kinetic theory of gases,* and despite much opposition at first, the theory became highly developed many years before our experimental prowess became capable of confirming it.

Long before the experimental skills of scientists had progressed enough to enable them to measure the speeds of gas molecules directly, Maxwell

Ludwig Boltzmann (1844–1906)

Boltzmann was born in Vienna and graduated from its university in 1866. Thereafter he held professorships at Graz, Munich, Leipzig, and Vienna. The molecular picture of gases is a monument to his life's work. His mathematical description of the kinetic theory of matter is recognized still today as an achievement of the highest order. But the brilliance of his work was not fully recognized by many scientists of his time. The molecular picture of matter on which it was based was then under serious attack by several leaders in the scientific community. His opposition held that atoms and molecules were imaginary constructs and, as such, had no place in a philosophically sound science. Lacking experimental proof of his molecular models, Boltzmann was unable to win over his critics. In his later years, Boltzmann became increasingly subject to depression and ultimately killed himself. Less than three years after his death, researchers achieved the first of a series of direct experimental proofs of his molecular theories.

developed a theory showing how the speeds of gas molecules should vary. He found that a wide distribution of speeds should exist with the speeds ranging from zero to extremely large values.

Maxwell's prediction for molecular speeds is most easily seen from a graph. Suppose we select a molecule in a gas and measure its speed at the instant of selection. The probability (or chance) that it will have any given speed is shown in Fig. 9.5. For the purposes of this example, the gas is assumed to be nitrogen at the indicated temperatures. We see that the molecules in nitrogen gas at 460 K (187°C) are most likely to have a speed near 500 m/s. However, a vast range of speeds are possible, and we should not be surprised to find the molecule selected to have any speed between about 50 and 1500 m/s. The curves do not actually reach zero even when infinite

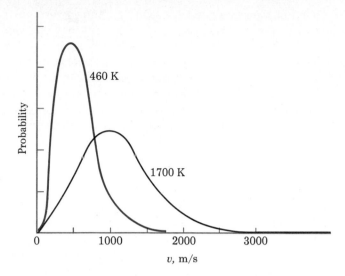

FIGURE 9.5
The chance that a nitrogen molecule in a gas at the temperature indicated will have a certain speed is proportional to the height of the curve at that particular speed.

speeds are reached; but, as we see, the chance of finding a molecule with such a high speed is extremely small—in fact, it is entirely negligible.

It was not until 1926 that Maxwell's predictions (made in 1860) were susceptible to test by experiment. When the test was made, excellent agreement was found with the results of Maxwell's theory. Not only does experiment confirm that the average translational kinetic energy of a gas molecule is $\frac{3}{2}kT$, but also the predicted distribution of speeds is found.

9.9 Brownian Motion

The molecular concepts we have used in discussing the ideal-gas law have very wide applicability. As we have just seen, Maxwell was able to predict the variation of speeds within a gas by use of these concepts. Many other facets of gas behavior have been successfully treated by using this model of an ideal gas. Indeed, these results of the kinetic theory of gases have been amply confirmed by many different experiments. In this and the two following sections, we discuss extensions of this model to systems we would not ordinarily think of as gases.

Consider tiny smoke or dust particles in the air or, perhaps, colloidal particles such as those that make milk white or impure water slightly hazy. Each of these tiny particles acts like a "gas molecule" and has a translational KE given by Eq. (9.4), namely, $\frac{3}{2}kT$. If their KE is much larger than their gravitational PE, the particles will not settle to the bottom of the container. They will float through the air (in the case of dust) or through the liquid (for colloidal particles) just as though they were molecules of an ordinary gas.

A botanist, Robert Brown, reported in 1827 an interesting type of motion (now called *Brownian motion*) that he had observed colloidal pollen particles to undergo. Examining through a microscope pollen particles sus-

FIGURE 9.6
Tiny particles suspended in a fluid undergo a zigzag motion called Brownian motion.

pended in water, he noticed that they follow a zigzag path[3] much like that shown in Fig. 9.6. Let us see qualitatively why Brownian motion occurs.

At first you might think that the zigzag motion is caused by collision of the colloidal particle with individual molecules of the surrounding fluid. But when you consider that the particle is so large that hundreds of molecules of the fluid touch it at the same time, you can see that the individual forces exerted by the molecules are too small to cause the observed deflections. Instead, the deflections are caused by the resultant force exerted on the particle by many surrounding molecules. Although most often the resultant force on the particle is near zero, this is not always the case. Now and then, by chance, the resultant of these forces is large enough to deflect the particle.

We can better understand this effect by an analogy. Suppose you throw 100 coins and count those which come up heads as +1 and those that come up tails as −1. Usually the number of heads and tails is about the same. So the usual resultant is close to zero. But occasionally, by chance, many more heads come up than tails. In that case the resultant is large. As we see, random effects that usually balance out occasionally do not do so.

The Brownian motion effect is important in several respects. It occurs in all colloidal systems. But, from a fundamental standpoint, it provides a visible means for observing the molecular properties of an ideal gas. By means of the Brownian motion of colloidal particles, it was possible to obtain the first direct confirmation of the kinetic theory. In addition, one of the earliest measurements of Avogadro's number was made by means of an experiment involving the motion of colloidal particles.

9.10 Osmotic Pressure

Using the ideas of the kinetic theory of gases, we can draw an important conclusion concerning a mixture of two or more gases. The conclusion was first stated by Dalton and is known as **Dalton's law of partial pressures:**

Dalton's Law of Partial Pressures

The total pressure due to a mixture of ideal gases is equal to the sum of the pressures each gas would exert if alone.

For example, suppose certain amounts of N_2, H_2, and O_2 gas are placed in a container of volume V. Suppose further that this same amount of N_2 would exert a pressure P_N if placed in the same container by itself. The similarly determined pressures for H_2 and O_2 are P_H and P_O. Then, Dalton's law tells us that the total pressure due to the mixture of these gases will be

---

[3] Brown initially believed the motion of the particles was a result of the "life" carried by the genetic material of the pollen. He later abandoned this idea when he found that even dust particles from powdered stone showed the same effect. The detailed theory of this effect was given by Albert Einstein in 1905, the same year that he developed the theory of relativity. He was 26 years old at the time.

$$P = P_N + P_H + P_O$$

As we see, each gas is unaffected by the presence of the other gas as far as pressure is concerned. This is not surprising since the ideal-gas law applies at any gas pressure for ideal gases. It tells us that $P \propto n$, the number of moles of gas in the container. The law contains nothing which depends on the particular type of molecule involved: 1 mol of N_2 plus 1 mol of H_2 in the container should give the same pressure as 2 mol of either N_2 or H_2. Indeed, the ideal-gas law predicts Dalton's law of partial pressures.

But we have also seen that the kinetic theory applies to colloidal particle motion. We expect, then, that the ideal-gas law should apply to both molecular and colloidal solutions. We might therefore guess that molecules dissolved to form a dilute solution should act as an ideal gas. This turns out to be true and gives rise to what is known as *osmotic pressure*. Let us now examine the very useful concept of osmotic pressure.

When you were in grade school, you may have demonstrated the effects of osmotic pressure by means of a hollowed out carrot and colored sugar water. The colored sugar water is sealed inside the hollowed out carrot, but is free to rise up in a capillary tube inserted through the top of the carrot. When the carrot is partly submerged in a beaker of pure water, a curious behavior is noted. Some of the water slowly passes from the beaker of water through the wall of the carrot and enters the sugar water. This causes the sugar water to rise up in the capillary tube. When the flow stops, the sugar water in the capillary tube stands much higher than the water in the beaker. Because the water stands at a greater height inside than outside the carrot, we conclude that the final pressure inside the carrot is much larger than the water pressure in the beaker. Amazingly, the water has flowed from a region of low pressure, the beaker, to a region of high pressure, inside the carrot. The pressure difference between the beaker and the interior of the carrot is called the osmotic pressure. Let us now investigate the cause of osmotic pressure.

We describe here a more precise version of the experiment. It is used to measure molecular masses of proteins, polymer molecules, and other rather large molecules. The experiment makes use of an osmometer, a device which is described now.[4] An osmometer consists of two chambers separated by a very thin, semipermeable membrane. A diagram of a simple osmometer is shown in Fig. 9.7. In the grade-school experiment, one chamber is the hollow space inside the carrot and the other is the beaker in which the carrot is immersed. The carrot wall acts as the semipermeable membrane. Modern osmometers make use of very thin sheets of plastic that swell but do not dissolve in the solvent from which the solution to be measured is made. The membrane material must be such that the solute molecules do not pass through the membrane while the solvent molecules do pass through.[5] Such a membrane is said to be semipermeable.

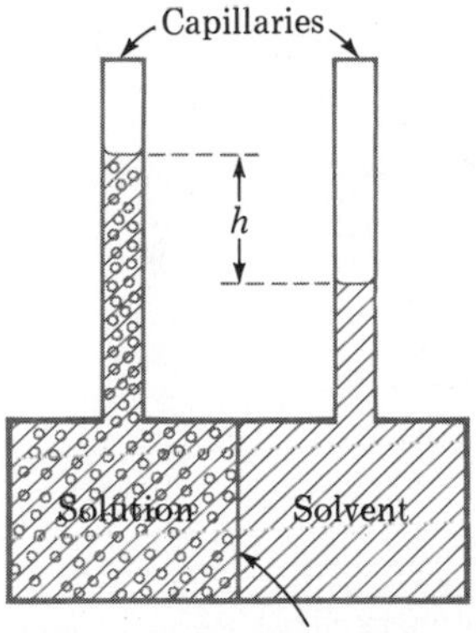

FIGURE 9.7
The osmotic pressure of the solute in the very dilute solution is $\rho g h$, where ρ is the density of the solution or solvent (the two densities are essentially the same).

---

[4] Pronounced os-mom'-i-ter.

[5] A solution is made by dissolving a solid in a liquid, e.g., sugar in water. The solid is called the **solute**. The liquid is called the **solvent**.

When the device in Fig. 9.7 is filled as shown, the levels in the two chambers adjust and eventually reach equilbrium heights. Solvent molecules diffuse through the membrane (either out of or into the solution) until there is the same solvent pressure on each side of the membrane. Notice carefully that the solute molecules do not pass through the semipermeable membrane. The system reaches equilibrium only when the solvent pressures are the same on the two sides. Let us call the pressure due to the solvent molecules at the membrane P_s.

But the total pressure at the left of the membrane is due to the solvent plus the solute "gas" molecules. Let us denote this solute "gas" pressure by π. Then, Dalton's law tells us that the total pressure on the left is

$$\text{Pressure on left} = \pi + P_s$$

However, the pressure on the right is due to the pure solvent only. It, too, is P_s once equilibrium is established. Therefore

$$\text{Pressure on right} = P_s$$

We now see that there is a difference in total pressure on the two sides of the membrane. *The difference in pressure between the two sides of the membrane is called the* **osmotic pressure.** It is

$$\text{Osmotic pressure} = \pi + P_s - P_s = \pi$$

In other words, *the osmotic pressure is simply the pressure of the solute "gas" in the solution.* We can measure it in the osmometer by measuring h, the difference in levels. Then, since the osmotic pressure is equal to the difference in pressure, we have $\pi = \rho g h$, where ρ is the density of the very dilute solution. (In practice, the solution and solvent have negligibly different densities.)

Let us now return to the ideal-gas law. The pressure due to the solute "gas" will be given by the gas law provided the solution is very dilute. We can then write

$$\pi V = nRT$$

where n is the number of kilomoles of solute in volume V. Or if we represent the molecular mass (or weight) of the solute by M, we have

$$n = \frac{m}{M}$$

where m is the mass of solute in the volume V.

Substituting this value for n gives, after division by V,

$$\pi = \frac{m}{V} \frac{RT}{M}$$

Definition

But m/V is the concentration of the solution (in kilograms per cubic meter), and we represent it by c. Therefore

$$\pi = \text{osmotic pressure} = \frac{cRT}{M} \qquad (9.5)$$

Osmotic Pressure of Dilute Solution

As we see, *the osmotic pressure is largest for molecules of small molecular weight.* However, for very small molecules, it is difficult to find membranes which will pass the solvent but not the solute.

We can make use of an osmometer and Eq. (9.5) to determine the molecular weight M of a solute molecule. Solving for M, we find

$$M = \frac{c}{\pi} RT$$

If we measure the osmotic pressure π of a solution of known concentration c, the solute M can be found. *In practice, care must be taken to make c small enough for the solute molecules to act as an ideal gas.* It is common practice to use osmometers to measure the molecular masses of molecules in the range from about 5000 to several million.

Illustration 9.12 One of the protein molecules found in blood plasma is albumin. A water solution containing 0.20 g of albumin per 100 cm³ has an osmotic pressure of 0.74 cm of water at 27°C. From these data, find the molecular mass of the albumin molecule.

Reasoning We can make use of Eq. (9.5). Since π is 0.74 cm of water,

$$\pi = \rho g h$$

becomes

$$\pi = (1000 \text{ kg/m}^3)(9.8 \text{ m/s}^2)(0.74 \times 10^{-2} \text{ m}) = 72.5 \text{ Pa}$$

Then, from the gas law,

$$M = \frac{c}{\pi} RT$$

But

$$c = \frac{0.20 \text{ g}}{100 \text{ cm}^3} = 2.0 \text{ kg/m}^3$$

and

$$T = 273 + 27 = 300 \text{ K}$$

Therefore, after substituting these values and 8314 J/(kmol)(K) for R, we find

$$M = 69,000 \text{ kg/kmol}$$

for the molecular mass of albumin.

Summary

Thermal energy is related to the KE of a substance. The hotter a substance, the more KE its molecules have. Thermometers are used to measure temperature. Three common temperature scales exist. The Celsius (°C) scale has the freezing point of water at 0°C and the boiling point at 100°C. The Kelvin (or absolute) scale (K) has these two points at 273.15 and 373.15 K. The Fahrenheit (°F) scale takes these two points to be 32 and 212°F. One Celsius degree (or kelvin) is equivalent to $\frac{9}{5}$ Fahrenheit degrees.

Many gases obey the ideal-gas law and are called ideal gases. The law is $PV = nRT$, where n kilomoles of gas is contained in a volume V at a pressure P. The temperature T must be measured on the Kelvin (absolute) scale. The symbol R represents the gas constant and has a value 8314 J/(kmol)(K).

A kilomole (kmol) of substance is a mass of substance whose mass in kilograms is equal numerically to the atomic or molecular mass (or weight) of the substance. In a more general sense, it is the mass of substance (in kilograms) that contains 6.02×10^{26} particles of the substance. The number 6.02×10^{26} particles per kilomole is called Avogadro's number and is represented by N_A. People often use the mole which is $\frac{1}{1000}$ kmol.

The kinetic theory of gases can be combined with the ideal-gas law to show that absolute (Kelvin) temperature is a measure of the translational KE of the gas molecules. Precisely, $T = [2N_A/(3R)](\frac{1}{2}m_0 v^2)$, where $\frac{1}{2}m_0 v^2$ is the average translational KE of a gas molecule. Often, the ratio R/N_A is replaced by k, Boltzmann's constant. Then one has $\frac{1}{2}m_0 v^2 = \frac{3}{2}kT$. The average speed of a nitrogen molecule in the air is about 500 m/s, but the molecules have a wide distribution of speeds at any instant.

A gas will behave as an ideal gas provided it satisfies two general restrictions. The gas must be far from the conditions under which it will liquefy. The actual volume of the molecules must be small in comparison to the volume available to the gas.

The ideal-gas law is used to define the Kelvin temperature scale. Absolute zero (0 K) and the triple point of water (273.16 K) are used as fixed points for defining the scale.

Tiny dust, smoke, and colloidal particles undergo a zigzag motion called Brownian motion. These particles, if dilute enough, often obey the ideal-gas law. The average translational KE of each particle is $\frac{3}{2}kT$.

Dalton's law of partial pressures states that the total pressure due to a mixture of ideal gases is equal to the sum of the pressures each gas would exert if alone.

The solute molecules in a dilute solution can be assigned a pressure called the osmotic pressure π. If dilute enough, the solute gas obeys the ideal-gas law. In that case, the osmotic pressure $\pi = cRT/M$, where c is the concentration of the solute and M is its molecular mass. This fact is basic to the measurement of molecular masses by use of an osmometer.

Minimum Learning Goals

Upon completion of this chapter, you should be able to do the following:

1 Define the three common temperature scales and locate the following points on each: absolute zero, freezing point, boiling point of water.

2 Change a temperature on one scale to the other two common scales.

3 Draw graphs showing the temperature variation of P and V for an ideal gas. Locate absolute zero on each graph.

4 Write the ideal-gas law and define each quantity in it. Find P, V, n, or T for an ideal gas when all but one of these quantities are given.

5 Define kilomole and mole. Explain the relation

between the kilomole and Avogadro's number. State the value of Avogadro's number.

6 Use Avogadro's number to find the mass of a molecule (or atom) when the molecular mass of the substance is given.

7 State under what conditions a gas is likely to behave as an ideal gas.

8 Explain the molecular meaning of absolute temperature by reference to the KE of the molecules of an ideal gas.

9 Compute the average translational KE of the molecules of an ideal gas when the gas temperature is given. Sketch a graph showing the distribution of speeds of gas molecules at two different temperatures.

10 Solve problems involving the gas law similar to those in Illustrations 9.6 to 9.10. Distinguish between gauge pressure and absolute pressure.

11 Explain what is meant by Brownian motion and why it occurs.

12 State Dalton's law of partial pressures. Use it to find the resultant pressure due to the mixture of several ideal gases.

13 Describe an osmometer and in the description point out what is meant by osmotic pressure. Explain how osmotic pressure can be used to determine molecular masses.

Important Terms and Phrases

You should be able to define or explain each of the following:

Temperature scales: Celsius, Kelvin, Fahrenheit
Ideal gas
Mole, kilomole
Avogadro's number N_A
$PV = nRT$

Gauge versus absolute pressure
Absolute zero
KE of translation $= \frac{3}{2}kT$
Kinetic theory of gases
Brownian motion
Osmotic pressure

Questions and Guesstimates

1 Suppose you were the only survivor of a plane crash on a tiny island in the South Pacific. While waiting for rescue, you decide to construct a thermometer. How might you accomplish this?

2 Sketch a graph showing the variation of the pressure of an ideal gas as a function of $1/V$ as the gas is compressed slowly at constant temperature.

3 We can imagine the molecules of an ideal gas to act as tiny balls in continual motion. An ideal gas of colloidal-size particles can also exist. But glass beads and pool balls do not act as an ideal gas. Where (in size) does the dividing line come, and to what is it due?

4 Hydrogen and oxygen gas are sealed off at atmospheric pressure in a very strong glass container containing two electrodes. A spark from the electrodes ignites the gases so that the reaction $2H_2 + O_2 \rightarrow 2H_2O$ results. Will the pressure in the tube be changed after the temperature has come back to its original value (200°C)? Explain. What if the final temperatue is 20°C?

5 In deriving the gas law we assumed perfectly elastic collisions with the wall. This is not a correct assumption, but it makes no difference as long as the walls are at the same temperature as the gas. Explain why.

6 Compare the gravitational PE of a nitrogen molecule which is 1 m above the ground with its KE when the temperature is (a) 0°C and (b) −270°C.

7 Although the air is mostly composed of N_2 molecules, some O_2 is present, of course. Do both kinds of molecules travel with the same average speed? What is the exact relation between the speeds?

8 Justify the fact that water molecules form a liquid under standard conditions even though H_2 and O_2 gases do not condense until much lower temperatures are reached.

9 The composition of the atmosphere changes with altitude. The percentage of hydrogen molecules in the air increases, and the percentage of nitrogen molecules decreases, as one goes farther above the earth. Why?

10 In order to escape from the earth, a rocket must be shot out from it with at least a speed of 11,200 m/s. Explain why only a tiny amount of hydrogen exists in the atmosphere even though billions of years ago there may have been more hydrogen than nitrogen in it.

11 The constitution of the atmosphere changes as we go to extreme heights above the earth. Why should there be a greater proportion of N_2 at high altitudes than at low altitudes compared with the O_2 content?

12 In outer space there is about one molecule per cubic centimeter of volume. Try to devise a method for measuring the pressure of such a thin gas.

13 Making use of Fig. 9.5, estimate the fraction of the nitrogen molecules having speeds in excess of 1250 m/s at 460 and 1700 K.

14 Estimate how large (in volume) a helium-filled balloon must be if it is to be able to lift 500 kg. To start at least, assume the mass of the balloon to be included in the 500 kg. (E)

Problems

Take atmospheric pressure to be 1.01×10^5 Pa ($= 14.7$ lb/in^2 $= 76$ cmHg $= 1.00$ atm).

1 What will Celsius and Kelvin thermometers read when the temperature is (a) 77°F and (b) −31°F?

2 Normal body temperature is 98.6°F. What temperature is this on the Celsius scale? On the Kelvin scale?

3 According to the handbook, the melting point of mercury is −38.9°C, and its boiling point is 357°C. Change these to (a) Fahrenheit temperatures and (b) Kelvin temperatures.

4 Methanol freezes at −98°C and boils at 65°C. Change these to (a) Fahrenheit temperatures and (b) Kelvin temperatures.

5* At what temperature do Fahrenheit and Celsius thermometers read the same numerical value?

6* An absolute temperature scale called the *Rankine scale* uses Fahrenheit-size degrees and has its zero at 0 K. On it, the freezing point of water is 492°R. Obtain a general relation between a temperature on the Rankine scale T_R and a Fahrenheit temperature t_F.

7 Find the mass of an average chlorine atom. See Appendix 4 for a table of average atomic masses.

8 What is the mass of an average oxygen atom? See Appendix 4 for a table of average atomic masses.

9* Find the mass in kilograms of an NH_3 molecule.

10* The chemical formula for a typical polystyrene molecule is C_nH_{n+2}, where n is of the order of tens of thousands. Find the mass of a polystyrene molecule for which $n = 10,000$.

11 How many molecules are there in 10.0 g of water? (The formula for a water molecule is H_2O.)

12 Find the number of molecules in 5.0 g of benzene (C_6H_6).

13* The molecular weight of a typical polyethylene molecule in a piece of polyethylene is about 25,000 kg/kmol. (a) Find the mass of this typical molecule. The density of polyethylene is close to 0.95 g/cm^3. How many of these molecules exist in (b) 1 g of polyethylene and (c) in 1 cm^3?

14* Benzene has a density of 879 kg/m^3 and a molecular weight of 78 kg/mol. Find (a) the mass of a benzene molecule and (b) the number of benzene molecules in 1 cm^3.

15 A test tube filled with air is sealed off at STP.[6] It is then heated to 300°C. What, then, is the air pressure in the tube in newtons per square meter, in pascals, in atmospheres, and in centimeters of mercury?

16 A sealed test tube contains nitrogen gas at STP. What will be the pressure in the tube after it is cooled to −45°C? Express your answer in newtons per square meter, in pascals, in atmospheres, and in centimeters of mercury.

17 A sealed tank with volume of 3.0 ft^3 has an ordinary pressure gauge on it which reads 100 lb/in^2. The temperature of the tank is then increased from 27°C to an unknown higher temperature. (a) If the gauge now reads 140 lb/in^2, what is the new temperature? (b) What was the final pressure in the tank in atmospheres?

18 A tank of compressed oxygen has a volume of 20,000 cm^3 at a gauge pressure of 1500 lb/in^2. What volume will this gas occupy at atmospheric pressure and the same temperature?

19 Air at 27°C and atmospheric pressure in a chamber is suddenly compressed to a volume one-twentieth

---

[6]STP stands for "standard temperature and pressure." These are 0°C and 101 kPa.

as large as the original and a pressure of 30 atm. What is the temperature of the gas?

20 A gas at an absolute pressure of $1000 \, lb/in^2$ and $27°C$ is suddenly expanded into a chamber having 10 times the volume. What is the new pressure of the gas after a few minutes if its temperature is found to be $-3°C$ at that time?

21 The gauge pressure for an automobile tire is $24 \, lb/in^2$, or 165 kPa, and the volume of the tire is V_0. What volume of air under standard atmospheric pressure and the same temperature was used in filling the tire?

22 A car tire is filled to a gauge pressure of $24 \, lb/in^2$ (1.65×10^5 Pa) when the temperature is $20°C$. After running at high speed, the tire temperature rises to $60°C$. Find the new gauge pressure within the tire, assuming the tire's volume did not change.

23 In a diesel engine, the cylinder compresses air from approximately standard pressure and temperature to about one-sixteenth the original volume and a pressure of about 50 atm. What is the temperature of the compressed air?

24 One way to cool a gas is to let it expand. Typically, a gas at $27°C$ and a pressure of 40 atm might be expanded to atmospheric pressure and a volume 33 times larger. Find the new temperature of the gas.

25** An air bubble of volume V_0 is released by a fish at a depth h in a lake. The bubble rises to the surface. Assuming constant temperature and atmospheric pressure is P_a above the lake, what is the volume of the bubble just before it touches the surface? The density of the water is ρ.

26** An air bubble at the bottom of a lake 10.0 m deep has a volume of $0.23 \, cm^3$. Find its volume just before it reaches the surface. Assume the temperature at the bottom to be $7°C$ and at the top $17°C$.

27* What mass of air is there in a room that is $3 \times 8 \times 9$ m? Consider the air to be at 1 atm and $20°C$; take the average molecular mass of an air molecule to be 28 kg/kmol.

28* A $3000\text{-}cm^3$ tank contains oxygen gas ($M = 32 \, kg/kmol$) at a temperature of $20°C$ and a gauge pressure of 300 kPa. What is the mass of the oxygen in the tank?

29* A piece of dry ice, CO_2, is placed in a test tube, which is then sealed off. If the mass of dry ice is 0.48 g and the sealed test tube has a volume of $20 \, cm^3$, what is the final pressure of the CO_2 in the tube if all the CO_2 vaporizes and reaches thermal equilibrium with the surroundings at $27°C$? (The molecular weight of CO_2 is 44.)

30* When a $20\text{-}cm^3$ test tube was sealed off at very low temperatures, a few droplets of liquid nitrogen were condensed in the tube from the air (the boiling point of nitrogen is $-210°C$). What will the nitrogen pressure in the tube be when the tube is warmed to $27°C$ if the droplets' mass is 0.14 g?

31** The temperature at the sun's center is estimated to be about 14×10^6 K, while the density there is about $100 \, g/cm^3$. Assuming the protons which constitute most of the sun's core to act like a perfect gas even at these very high densities, find the pressure of the proton gas (in atmospheres) at the sun's core. (Note: $m_p = 1.67 \times 10^{-27}$ kg.)

32* Find the density that water vapor (steam) would have at a pressure of 1 atm and a temperature of $100°C$ if it could be considered an ideal gas. For comparison, the actual density is $0.598 \, kg/m^3$. Try to justify any difference you may note.

33 Find the average speed of a hydrogen molecule ($M = 2.0 \, kg/kmol$) in hydrogen gas at $20°C$.

34 At what temperature will the molecules of an ideal gas have an average speed that is 5 times as large as their speed at $20°C$?

35* The escape velocity for a projectile on earth is about 11.2 km/s. (a) At what temperature would hydrogen molecules have an average speed equal to this speed? (b) Repeat for a nitrogen molecule.

36* The temperature of outer space is about 3 K and consists mainly of a gas of single hydrogen atoms. On the average, there is about one such atom per cubic centimeter. (a) Find the pressure of this gas in outer space; express your answer in atmospheres. (b) Find the average KE of one of the hydrogen atoms. (c) What speed will an atom of this energy have? (The atomic weight of a hydrogen atom is 1 kg/kmol.)

37* The temperature of the interior of the sun is approximately the same as that in a hydrogen bomb, about 10^7 K. Find the average translational KE of a proton (a hydrogen nucleus) at this temperature, and from it compute its average speed. (The atomic mass of a hydrogen atom is 1 kg/kmol, and, to good approximation, this same atomic mass applies to the proton.)

38* Show that the pressure of an ideal gas can be written as $P = \frac{1}{3}\rho v^2$.

39* For many purposes, a pressure of 1×10^{-6} mmHg is considered to be a reasonably good vacuum. (a) At this pressure, find the number of kilomoles of nitrogen in $1 \, m^3$. (b) How many nitrogen molecules exist in $1 \, cm^3$ at this pressure? Assume the air to be pure nitrogen at room temperature. (Nitrogen actually composes about 78 percent of air.)

40* The actual volume of a nitrogen molecule is about 8×10^{-30} m³. Find the fraction of the volume actually occupied by nitrogen molecules in nitrogen gas under STP.

41* Hydrogen gas is to be burned in oxygen to obtain 9.0 g of water. What volume of hydrogen under STP is needed?

42* What volume of hydrogen at STP is needed if it is all to react with chlorine to obtain 1.00 kg of HCl? (Note: $M_{Cl} = 35.5$ kg/kmol.)

43* A closed cylinder has an insulated, movable piston separating the cylinder into two parts. If equal masses of the same gas are on each side of the cylinder, show that the piston will position so that $V_2 = V_1(T_2/T_1)$, where V_1 and T_1 are the volume and temperature on one side of the piston and V_2 and T_2 apply to the other side.

44* The density of dry air at STP is 1.29 kg/m³. Its composition is (by weight) 78% N_2, 21% O_2, 0.9% argon, plus traces of CO_2, helium, and other gases. (a) Find the pressure predicted by the gas law for each of the three major components. (b) Use Dalton's law to find the air pressure under STP. (The molecular mass of argon gas, 40 kg/kmol, is equal to the atomic mass of argon since the gas is monatomic.)

45* On a day when the air temperature is 20°C and the relative humidity is 80 percent, there are 13.7 grams of water vapor per cubic meter of air. As-suming the water vapor to act as an ideal gas, what fraction of the atmospheric pressure (100 kPa) is due to the water vapor?

46* Find the osmotic pressure of a 2 g per 100 cm³ water solution of sugar. The molecular mass of sugar is 342 kg/kmol. Assume the temperature to be 27°C. Express your answer in centimeters of water.

47** A solution is made by dissolving 0.250 g of polystyrene in benzene so as to make 100 cm³ of solution. The osmotic pressure of the solution is measured at 20°C and is found to be 1.67 cm of benzene. Find the molecular weight of the polystyrene. (The density of benzene at 20°C is 879 kg/m³.)

48*** A thin sheet of plastic separates two containers, as shown in Fig. P9.1. Volume 1 (V_1) contains H_2 gas at STP, while V_2 contains N_2 gas at STP. It is also known that $V_2 = 2V_1$. A hole is now made in the plastic. What is the final pressure after equilibrium is achieved? Repeat if the H_2 pressure is 240 cmHg, while the N_2 pressure is 60 cmHg.

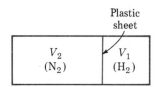

FIGURE P9.1

10

Thermal Properties of Matter

When a substance cools, it loses thermal energy. By what mechanism did the energy leave the substance? What effect does this loss have on the atoms of the material? How much thermal energy did the material lose? The answers to these questions are discussed in this chapter. In addition, we investigate such topics as the boiling and freezing of liquids and the mechanisms by which heat is transferred. The knowledge gained from our study of these topics will help us to better understand the nature of heat as well as the behavior of atoms and molecules in matter.

10.1 Thermal Energy, Heat Energy, and Internal Energy

We saw in the last chapter that the Kelvin temperature of an ideal gas has an easily understood meaning: *The average translational KE of a gas molecule is $\frac{3}{2}kT$, where k is Boltzmann's constant.* From this we conclude that *the average KE of a molecule in a hot gas is larger than in a cold gas. Most types of gas molecules,* however, *possess energy in addition to translational KE.* For example, a diatomic molecule such as N_2 or O_2 has KE of rotation as well as of translation. One such molecule is shown schematically in Fig. 10.1. For many purposes, the molecule can be thought of as two balls connected by a spring. The spring, which represents the chemical bond between the two atoms, is often distorted in collisions between molecules. As a result, the

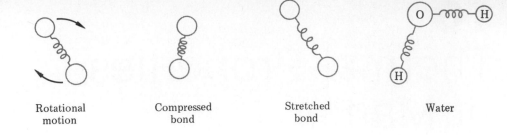

Rotational
motion

Compressed
bond

Stretched
bond

Water

FIGURE 10.1
A gas molecule has
rotational energy in addition
to KE of translation. It also
has energy associated with
the springlike bond between
its atoms.

springlike bond is compressed or stretched. Energy is therefore stored in the distorted bond between the atoms composing the molecule. As we see, even a simple diatomic gas possesses a great deal of energy in addition to its translational KE. A similar situation occurs for the water molecule shown in the figure.

The situation becomes even more complicated as we proceed to substances with more complex molecules. They possess several springlike bonds per molecule, and so the molecules can store even more energy in their distorted bonds. An ideal gas composed of even such complicated molecules conforms to the kinetic theory. As such, the translational KE of its molecules is still given by $\frac{3}{2}kT$, but each molecule has considerable energy in addition.

As the gas molecules collide with one another, the energy of the gas is distributed among translational KE, rotational KE, and energy associated with the "springs" between atoms. We know that the translational KE increases in proportion to T, the absolute temperature of the gas. The rotational and vibrational (spring-associated) energies also increase as T increases. In other words, *T is a measure of the total energy of the gas molecules.* Indeed, it is shown in courses in statistical mechanics that the total energy is proportional to T provided the temperature is not too low.[1]

These same ideas can be carried over to liquids and solids. A liquid is much like a highly compressed gas. Even though its molecules are too close together for it to behave as an ideal gas, it is still true that T is a measure of its molecular energy. A piece of solid can be likened to a very complex molecule. Although its complexity is immense, still there is energy associated with the springlike bonds between its atoms. This energy, too, increases with an increase in T.

Definition *In all forms of matter, the energy associated with the random motion of the molecules and atoms increases with T. This energy associated with the random motion of the atoms and molecules is often called* **thermal energy.** We should point out at once that *thermal energy* is a colloquial term. It has no precise scientific definition. When we wish to be quantitative, we use the terms *heat energy* and *internal energy.*

---

[1] This restriction (that the temperature not be too low) has profound meaning. Departures from the kinetic theory at low temperatures presented a great puzzle for physicists. It was not until the ideas of quantum mechanics were developed that the reasons for these discrepancies became known.

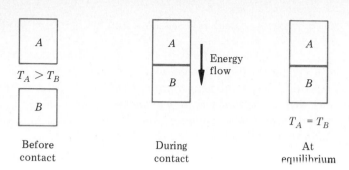

FIGURE 10.2
Heat energy flows from a
hot object to a cooler one
until their temperatures
become equal.

*The **internal energy** U of a substance is the total of all kinds of energy possessed by the atoms and other particles constituting the substance.* As we see, U includes the kinetic, potential, chemical, electrical, nuclear, and all other forms of energy possessed by the molecules of a substance.

    Internal Energy

To define the quantity we call heat energy, we must first consider what happens when two substances at different temperatures are placed in contact. The situation is shown in Fig. 10.2. If the temperature of object A is higher than that of object B, and if the two objects are composed of the same substance, then the average energy of a single molecule in A is larger than that in B. As a result, when the objects are placed in contact, the molecules in A lose energy to those in B, and A is thereby cooled and B warmed.

*The energy which is transferred (or flows) from a high-temperature object to a lower-temperature object because of the temperature difference is called **heat energy**. It is represented by the symbol ΔQ.* Experiment shows that the flow of heat energy will cease when the temperatures of the two objects become equal. This is true even if the objects are made of different substances. We conclude that

    Definition

Heat energy flows spontaneously from hot objects to cooler objects, but not vice versa.

    Heat Energy

In the next section we define the practical units by means of which heat energy is measured.

10.2 Units of Heat Energy

Historically, the use of heat energy far preceded our understanding of the nature of heat. Consequently, the measurement of heat energy was done in a purely practical way, and the units chosen for it were selected because of their utility. Since water is involved in most uses of heat, it is not surprising that the quantity and units of heat energy were chosen in terms of an experiment involving water.

*The two common units of heat energy are the **calorie** (cal), and the **British thermal unit** (Btu), defined today as follows:*

    Heat Units

$$1 \text{ cal} = 4.184 \text{ J} \quad \text{and} \quad 1 \text{ Btu} = 1054 \text{ J}$$

These definitions conform to the *original definitions* of these units, which were stated in the following way: *One calorie is the amount of heat energy required to raise the temperature of one gram of water from 16.5 to 17.5°C; one British thermal unit is the amount of heat energy required to raise the temperature of one pound of water from 63 to 64°F.* From either of these definitions we find

$$1 \text{ Btu} = 252 \text{ cal}$$

Nutritionists use still another unit for heat energy. It is also called the calorie, but more properly it should be called the kilocalorie (or large calorie) since

$$1 \text{ nutritionist's calorie} = 1000 \text{ cal}$$

For example, when health scientists tell us our daily diet should contain about 2000 cal of food energy, they are actually speaking in terms of 2000 kcal.

10.3 Specific Heat Capacity

To increase the temperature of an object, we must increase the thermal energy of its molecules. We can do this by letting heat flow into the object from a hotter object. Similarly, if we wish to cool an object, we can allow heat energy to flow from the object to a still cooler object. To be able to describe such processes as cooling and heating quantitatively, we must know how much energy is required to change the temperature of an object. *The quantity of heat which must flow into or out of unit mass of a substance to change its* **Definition** *temperature by one degree is called the* **specific heat capacity** *of the substance.*

We represent the specific heat capacity by c. Its definition can be written as an equation. When a quantity ΔQ of heat flows into a mass m of substance, its temperature will increase by an amount ΔT. Then, by definition,

Specific Heat Capacity
$$\text{Specific heat capacity} = c = \frac{\Delta Q}{m\,\Delta T}$$

After clearing fractions we have

$$\Delta Q = cm\,\Delta T \tag{10.1}$$

The units of c are typically calories per gram per Celsius degree. A somewhat similar definition for c is used in the British units system, but in that case, 1 lb of substance is considered and the temperature change is meas-

TABLE 10.1
SPECIFIC HEAT CAPACITIES,
cal/(g)(°C) or Btu/(lb)(°F)

Water	1.000	Aluminum	0.21
Human body	0.83	Glass	0.1–0.2
Ethanol	0.55	Iron	0.11
Paraffin	0.51	Copper	0.093
Ice	0.50	Mercury	0.033
Steam	0.46	Lead	0.031

ured in Fahrenheit degrees. The appropriate units are Btu per pound per Fahrenheit degree.

Because of the different complexities of substances, each substance has its own unique specific heat capacity. Moreover, the heat requirements for a 1-degree temperature change vary slightly with temperature. This variation is usually slight in a limited temperature range, and so it is often ignored. Values for c appropriate to many substances are given in Table 10.1. Notice that the values apply to both calories per gram per Celsius degree and Btu per pound per Fahrenheit degree. From the definition of c, can you show why these two values should be equal?

Illustration 10.1   How much heat is given off as 20 g of water cools from 90 to 30°C?

Reasoning   The specific heat capacity of water is 1.00 cal/(g)(°C). In this case, $\Delta T = -60°C$, and so, from Eq. (10.1),

$$\Delta Q = cm\,\Delta T = [1.00\ \text{cal/(g)(°C)}](20\ \text{g})(-60°C) = -1200\ \text{cal}$$

The negative sign tells us this is a loss of heat.

Illustration 10.2   A thermos jug contains 300 g of coffee (essentially water) at 90°C. Into it is poured 50 g of milk (also essentially water) at 15°C. What is the final temperature of the coffee?

Reasoning   The thermos jug is well insulated, and so we assume that no heat flows from the coffee to it. Then the heat lost by the coffee will go into warming the milk. The law of conservation of energy tells us that

Heat lost by coffee = heat gained by milk

Using Eq. (10.1), we can write this equation as

$$(cm|\Delta T|)_{\text{coffee}} = (cm|\Delta T|)_{\text{milk}}$$

In each case, $c = 1.00$ cal/(g)(°C). The mass of the coffee is 300 g, and that of the milk is 50 g. If we call t the final temperature of the coffee-milk solution, then $|\Delta T|$ for the coffee is $90° - t$ while for the milk it is $t - 15°$. Be very careful in situations such as this. We want both sides of the equation to be positive. Therefore ΔT, which is negative for the coffee, is replaced by $|\Delta T|$. This requires that we use $90° - t$ rather than the usual expression for ΔT, which is $t - 90°$ in this case. The above equation then becomes

$$(300 \text{ g})(90°C - t) = (50 \text{ g})(t - 15°C)$$

from which

$$t = 79.3°C$$

The coffee was cooled 10.7°C by the milk.

Illustration 10.3 An insulated aluminum container "weighing" 20 g contains 150 g of water at 20°C. A 30-g piece of metal is heated to 100°C and then dropped into the water. The final temperature of the water, can, and metal is 25°C. Find the specific heat capacity of the metal.

Reasoning Notice in this case that both the container and the water gain heat. The law of conservation of energy allows us to write

Heat gained by can + heat gained by water = heat lost by metal

These quantities are

$$\text{Heat gain of can} = [0.21 \text{ cal/(g)(°C)}](20 \text{ g})(5°C)$$

$$\text{Heat gain of water} = [1.00 \text{ cal/(g)(°C)}](150 \text{ g})(5°C)$$

$$\text{Heat loss of metal} = c_x(30 \text{ g})(75°C)$$

Putting these quantities in the above equation yields $c_x = 0.34$ cal/(g)(°C).

10.4 c_V and c_P for Gases

Values for the specific heat capacity of several gases are given in Table 10.2. *Two values for the specific heat capacity of gases, c_P and c_V, are listed.* The subscripts on these quantities tell us whether the pressure or the volume of the gas was maintained constant during measurement. You will notice from the table that c_V, *the specific heat at constant volume, is less than c_P, the specific heat at constant pressure.* The reason is easily understood from the following considerations.

When a gas is heated in a constant-volume container, the heat energy

TABLE 10.2
SPECIFIC HEAT CAPACITIES
OF GASES AT 15°C, cal/(g)(°C)
or Btu/(lb)(°F)

Gas	c_V	c_P	$\gamma = c_P/c_V$
He	0.75	1.25	1.66
Ar	0.075	0.125	1.67
O_2	0.155	0.218	1.40
N_2	0.177	0.248	1.40
CO_2	0.153	0.199	1.30
H_2O (200°C)	0.359	0.471	1.31
CH_4	0.405	0.53	1.31

added to the gas must all appear as additional energy of the gas molecules. It is therefore all used to increase the temperature of the gas. But this is not the case if the gas can expand while being heated. To see this, refer to Fig. 10.3. The container is closed on top by a movable piston. The weighted piston exerts an unchanging pressure on the gas. If heat energy is added to the gas, the gas will expand by lifting the piston. However, the pressure of the gas, equal to the pressure applied by the piston, remains unchanged. But in lifting the piston, the gas does work. This work is done at the expense of the heat energy which flows into the gas. Therefore, not all the heat energy added to the gas is used to heat the gas. Some of it is used to do work as the piston is lifted by the gas.

We see from this that *more heat energy will be needed to raise the temperature of a gas under constant pressure than under constant volume. In the first case, some of the heat energy is used in doing work. Because of this, c_P is larger than c_V. The difference in these two values is equal to the work done on the piston as unit mass of the gas is heated 1 degree under constant pressure.* In a special note on page 278 we evaluate $c_P - c_V$ for an ideal gas and find it to be R/M, where R is the gas constant and M is the molecular weight of the gas.

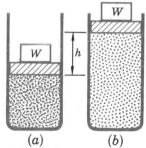

FIGURE 10.3
When the gas in the chamber is heated, the piston rises, thereby maintaining constant pressure. The heat energy must do work in lifting the piston as well as in heating the gas.

10.5 Heat of Vaporization and the Boiling of Liquids

As we have seen, when heat energy is added to a substance, the internal energy of the substance is increased. This increase in internal energy is usually accompanied by an increase in temperature. However, *a temperature increase does not always occur even though the internal energy of the substance is increased. At the temperature where a phase change occurs in the substance, the temperature remains constant until the phase change is completed.* For example, at the temperature where a crystalline substance melts or where a liquid boils, a phase change is occurring. In the one case, a crystalline phase is changing to a liquid phase. In the other, a liquid phase is changing to a vapor phase. *Even though heat must be added to accomplish the phase change, no temperature increase occurs for the substance.*

When a substance changes from one phase to another, its internal energy is changed. Water molecules in steam have more internal energy than they had when condensed. When the molecules are frozen into ice, they have even less internal energy than in the liquid phase. Let us now investigate what happens when a substance changes phase. We begin our discussion by considering the phase change from liquid to vapor.

A liquid placed in an open dish will slowly evaporate. This is a result of the fact that not all the molecules of the liquid have the same energy. It was pointed out in the last chapter that *some molecules of a gas have energy far in excess of the average, and others have very little. This is also true in a liquid. The very-highest-energy molecules, if at the surface of the liquid, may actually escape from the other molecules and leave the liquid. This is the process of* **evaporation.**

The molecules that evaporate are the highest-energy molecules. They carry energy away from the liquid, and so evaporation leads to a decrease in the average internal energy of the remaining molecules. Since the temperature of the liquid is merely a measure of this internal energy, it is clear that evaporation will lead to a cooling of the liquid. This process actually forms the basis for many cooling systems.

Definition

The energy required to tear unit mass of molecules loose from each other to change them from the liquid phase to the vapor phase is called the **heat of vaporizaton** *of the liquid.* Since the molecules of a liquid are less tightly bound together at a high than at a low temperature, the energy required to tear a molecule loose decreases with increasing temperature. For example, the heat of vaporization of water is 590 cal/g at 10°C and 539 cal/g at 100°C. That is, 539 cal of heat energy are required to tear 1 g of water molecules apart at 100°C. The reverse process also occurs. *When a vapor condenses to liquid, the heat of vaporization is given off in the process.* Typical values of heats of vaporization are given in Table 10.3.

Heat of Vaporization

Most often, the heat of vaporization is quoted for the normal boiling temperature of the liquid. It is common experience that liquids evaporate most readily when this temperature is reached. In fact, bubbles form within

TABLE 10.3
HEATS OF VAPORIZATION AND FUSION

Substance	Melting temperature, °C	Boiling temperature, °C	Heat of vaporization, cal/g	Heat of fusion, cal/g
Lead	327			5.9
Water	0	100	539	80
Mercury	−39	357	65	2.8
Ethanol	−114	78	204	25
Nitrogen	−210	−196	48	6.1
Oxygen	−219	−183	51	3.3

the liquid, and we use this as a common means for telling when a liquid has reached its boiling temperature. Let us now examine the meaning of this temperature in terms of the vaporization of the molecules.

Suppose that a liquid is placed in a closed container from which the air has been removed. Some of the liquid molecules will evaporate into the space above the liquid, as shown in Fig. 10.4. Of course, the reverse process is possible. From time to time, a vapor molecule will hit the surface of the liquid and stick to the liquid. As the number of molecules in the vapor increases, *a condition will be reached at which the number of molecules leaving the liquid will equal the number returning to the liquid from the vapor.* The number of molecules in the vapor will therefore remain constant at this point if the temperature of the system is not changed. *The vapor under these conditions is said to be* **saturated.** *We define the pressure of the molecules in the vapor under this condition to be the* **vapor pressure** *of the liquid. The vapor pressure will increase as the temperature is raised,* since molecules will evaporate more readily at higher temperatures. This variation of vapor pressure with temperature for water is shown in Table 10.4. Notice that even ice has an appreciable vapor pressure.

Once in a while, within the body of a liquid, a group of molecules will attain enough energy to tear apart the liquid and form a small bubble. A much exaggerated case is shown in Fig. 10.5. At low temperatures the vapor pressure within the bubble will be much smaller than the pressure of the atmosphere above the liquid. Hence, the bubble will collapse under the external air pressure before it has had a chance to grow to observable size. However, as the temperature of the liquid is raised, a temperature will be reached where the vapor pressure within the bubble equals the air pressure on the top of the liquid. The bubble will no longer collapse. Instead, it will grow and rise to the surface of the liquid. This process, happening at many places within the liquid, gives rise to the phenomenon of boiling.

Clearly, *a liquid boils at a temperature where the vapor pressure of the liquid just equals the external pressure on the liquid.* In science and industry, a vacuum is often applied when liquids are distilled. The reduced external pressure allows the liquid to boil at a lower temperature and decreases the possibility of chemical reaction within it.

FIGURE 10.4
When the vapor is saturated, equal numbers of molecules of the liquid evaporate from the surface and condense from the vapor in a given time.

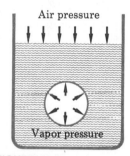

FIGURE 10.5
The boiling temperature is the temperature at which the vapor pressure in the bubble equals the external pressure on the liquid. (The size of the bubble is exaggerated.)

TABLE 10.4
VAPOR PRESSURE OF WATER AND ICE

Temperature, °C	Vapor pressure, mmHg	Temperature, °C	Vapor pressure, mmHg
−90	0.000070	90	526
−50	0.030	94	611
−10	1.95	99	733
0	4.58	100	760
10	9.21	110	1,075
30	31.8	150	3,570
60	149.4	200	11,650

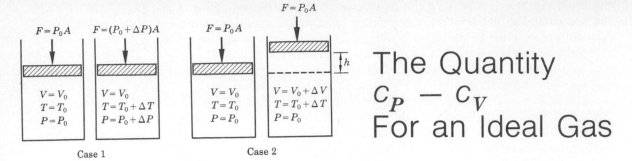

$F = P_0A$ $F = (P_0 + \Delta P)A$ $F = P_0A$ $F = P_0A$

$V = V_0$ $V = V_0$ $V = V_0$ $V = V_0 + \Delta V$
$T = T_0$ $T = T_0 + \Delta T$ $T = T_0$ $T = T_0 + \Delta T$
$P = P_0$ $P = P_0 + \Delta P$ $P = P_0$ $P = P_0$

Case 1   Case 2

The quantity of heat needed to raise the temperature of a gas an amount ΔT depends on how the gas is heated. Two extreme cases are important: (1) the gas is heated in a container having constant volume, or (2) the gas is heated in a container the volume of which is changed in such a way as to maintain the pressure constant. These two alternatives are illustrated in the figure. Let us examine these two situations to find the heat energy needed to raise the temperature the same amount ΔT in each case.

When heat energy is supplied to an ideal gas, energy may be given to the molecules as internal energy. In some instances, part of the heat energy is used doing work against forces used to confine the gas. If the volume of the gas is held constant, as in case 1 shown in the figure, no work is done against the external force holding the piston in place, since no movement of the piston occurs. Therefore, all the heat energy ΔQ_V given to the gas was used to raise the internal energy of the molecules. Calling the specific heat capacity in this case c_V, we have

$$\Delta Q_V = mc_V \Delta T$$

where m is the mass of gas in the container and the subscript V indicates that the volume is held constant.

In case 2, a larger amount of heat energy will be needed to raise the temperature by the same amount ΔT. This follows from the fact that the KE of the molecules must be raised by the same amount as in case 1, since the temperature rises the same amount in each case. But, in addition, an amount of work Fh is done against the force on the piston as the piston lifts through the distance h. We therefore have

$$\Delta Q_P = \text{increase in energy of molecules} + \text{work done against piston}$$

where the subscript P indicates that the pressure is maintained constant. Since the increase in energy of the molecules is the same in both cases, it is $mc_V \Delta T$. Replacing ΔQ_P by $mc_P \Delta T$ and substituting for the quantities on the right-hand side of the above equation, we find

$$mc_P \Delta T = mc_V \Delta T + Fh$$

However, the force on the piston F is merely the pressure times the area A of the piston, and so $Fh = P_0 Ah$. But Ah is the increase in volume of the container ΔV caused by the lifting of the piston, and so the above equation may be rewritten as

$$(c_P - c_V)m\,\Delta T = P_0\,\Delta V$$

This relation may be simplified by using Eq. (9.2), the ideal-gas law, which was, for the initial condition in case 2,

$$P_0 V_0 = nRT_0$$

After heating, in case 2, the equation becomes

$$P_0(V_0 + \Delta V) = nR(T_0 + \Delta T)$$

Subtraction of the first of these two equations from the second yields

$$P_0\,\Delta V = nR\,\Delta T$$

This value for $P_0\,\Delta V$ may be substituted in the above equation involving $c_P - c_V$ to give

$$(c_P - c_V)m\,\Delta T = nR\,\Delta T$$

Upon canceling like factors we find

$$c_P - c_V = \frac{nR}{m}$$

But n, the number of kilogram moles of gas, is related to the molecular weight M and the mass of gas m in the container by $n = m/M$. Therefore we can replace n/m by $1/M$ to obtain

$$c_P - c_V = \frac{R}{M}$$

As you will see in Prob. 24 at the end of this chapter, this theoretical result agrees well with the data in Table 10.2.

As the vapor of a boiling liquid escapes, the highly energetic molecules in the vapor carry energy away from the liquid. If the liquid is to continue boiling, heat must be continuously supplied to compensate for this energy loss. *If one tries to heat a liquid above its boiling temperature at the particular external pressure to which it is subjected, it will only boil more vigorously. Its temperature will not rise further until it has all evaporated.* An exception occurs if bubbles fail to form, in which case the liquid will superheat. Since bubbles first form most easily on impurities such as dust particles or air bubbles, it is not uncommon for a very clean, gas-free liquid to superheat. When it does begin to bubble, it may do so with nearly explosive force.

The boiling points and heats of vaporization for liquids are usually given for the boiling point under standard pressure, 760 mmHg. Under these conditions, the heat of vaporization of water is 539 cal/g, and the boiling point is 100°C, of course. It is interesting to notice that atmospheric pressure in the Rocky Mountain region is much lower than 760 mmHg at the more-than-mile-high altitudes of such cities as Denver and Cheyenne. At these altitudes pressure is near 600 mmHg, and water boils at about 94°C. Why is a pressure cooker a near necessity under such conditions?

10.6 Heat of Fusion and Melting

Ice crystals melt at 0°C under standard pressure (760 mmHg). Before melting, the water molecules are ordered in a crystalline lattice. They are held in place by rather strong intermolecular forces. To melt the crystal, one must tear the molecules out of this tight arrangement and cause them to become disordered. This process requires energy, which is usually supplied by heat.

We therefore find that when a crystalline material is heated, it begins to melt at a certain temperature. As heat is slowly added to the crystal-liquid mixture, the temperature remains constant until all the crystals have melted. The substance has a definite melting temperature, and a definite amount of heat must be furnished to melt the crystals at this temperature.

Heat of Fusion *The* **heat of fusion** *is the amount of heat energy required to melt unit mass of the crystalline material. This same amount of heat must be given off to the surroundings when unit mass of the liquid crystallizes.* The heat of fusion of water to ice is 80 cal/g. Values for other materials are listed in Table 10.3. It is seen that the hydrogen-bonded materials, water and ethanol, have much higher heats of fusion and vaporization than the others. Why?

A few other effects should be mentioned in connection with melting phenomena. First, some plastics such as polyethylene are partly crystalline. That is, only a portion of the solid is in an ordered state, with the remainder being liquid. The two parts are intimately mixed, and so the material has intermediate physical properties. These crystalline regions differ in degree of perfection. The less perfect crystals melt at a lower temperature than the larger, more perfect crystals. Hence such materials exhibit a melting range, often 20 or more Celsius degrees wide. The quoted melting point for these materials is usually taken as the temperature at which all the crystallinity has disappeared.

Other plastics, such as polystyrene, are clear, hard, glassy materials. These materials are merely very viscous liquids, as explained in Chap. 8. The molecules are not well ordered in the solid. Although they soften at a fairly definite temperature, no heat of fusion is involved in this process as there would be for the melting of a crystalline material.

Freezing points of liquids can be altered somewhat by applying large pressures to the systems. Materials that contract upon freezing have their melting points raised by increased pressure. Most materials behave in this way. A few materials, such as water, expand when they freeze. Increased pressure decreases the freezing point of such substances. The pressure of an ice skater's skate on the ice can cause the ice below it to melt. In this case, the skater is actually skating on ice lubicated with a thin film of water.

10.7 Calorimetry

A large number of situations involving heat interchange can be elucidated by the application of the following simple equation:

$$\text{Heat lost} = \text{heat gained}$$

In general, such situations involve the interchange of heat within an insulated vessel, a calorimeter, which effectively isolates the system from its surroundings. *The above equation is simply a statement of the law of conservation of energy.* We illustrate its use by several examples.

To handle these examples, we also make use of the following facts:

1 When a mass m of substance is heated (or cooled) through a temperature ΔT, it gains (or loses) a quantity of heat given by Eq. (10.1), namely,

$$mc\,\Delta T$$

(We assume that no phase change occurred in this temperature range.)
2 When a mass m of substance is melted (or crystallized), it gains (or loses) an amount of heat

$$mH_f$$

where H_f is the heat of fusion.
3 When a mass m of substance is vaporized (or condensed), it gains (or loses) an amount of heat

$$mH_v$$

where H_v is the heat of vaporization.

Illustration 10.4 A 100-g piece of lead is heated to 100°C and then dropped into a cavity in a large block of ice at 0°C. How much ice will melt?

Reasoning Heat lost by lead = heat used to melt ice, so

$$m_{Pb}c_{Pb}\,\Delta T = m_w H_{fw}$$
$$(100\text{ g})[0.031\text{ cal/(g)(°C)}](100 - 0°C) = (m_w)(80\text{ cal/g})$$
$$m_w = 3.9\text{ g}$$

The various constants were obtained from Tables 10.1 and 10.3.

Illustration 10.5 If 20 g of ice in a 10.0-g copper calorimeter can is originally at −30°C, how much steam at 100°C must be condensed in this can if the ice is to be changed to water and heated to 40°C?

Reasoning

$$\text{Heat lost by steam} = \text{heat gained by ice and can}$$

First the steam must be condensed to water at 100°C, and then the water must be cooled to the final temperature of 40°C. Hence

$$\begin{aligned}\text{Heat lost} &= mH_v + mc\,\Delta T\\ &= (m)(540) + (m)(1)(100 - 40) = 600m\end{aligned}$$

The copper can only needs to be heated from −30 to +40°C. On the other hand, the ice must first warm to 0°C, melt to water, and then be heated to 40°C. Therefore

$$\begin{aligned}\text{Heat gained} &= (m_{\text{Cu}}c_{\text{Cu}})(70) + (m_w c_{\text{ice}})(30) + m_w H_{fw} + (m_w c_w)(40)\\ &= (10)(0.093)(70) + (20)(0.50)(30) + (20)(80) + (20)(1)(40)\\ &= 65 + 300 + 1600 + 800\\ &= 2770 \text{ cal}\end{aligned}$$

The units have been omitted from these equations to conserve space. You should supply them and check the units of the answer. Equating heat gained to heat lost yields the mass of steam required,

$$m = \tfrac{2770}{600} = 4.6 \text{ g}$$

Notice how little steam is required to accomplish this. It is clear that the steam lost most of its heat during the process of condensing, namely, 2500 of the total 2770 cal. Can you see from this why, under certain conditions, hot steam can cause much more severe burns than hot water?

Illustration 10.6 A 10.0-g lead bullet is traveling at 100 m/s when it strikes and embeds itself in a wooden block. By about how much will its temperature rise on impact? For this rough computation, assume all the kinetic energy of the bullet changes to heat energy in the bullet.

Reasoning The heat energy gained by the bullet will equal the KE loss of the bullet:

$$\begin{aligned}\text{KE of bullet} &= \tfrac{1}{2}mv^2\\ &= (\tfrac{1}{2})(0.0100 \text{ kg})(100 \text{ m/s})^2 = 50 \text{ J} = 12.0 \text{ cal}\end{aligned}$$

$$\text{Heat gained} = m_{\text{Pb}} c_{\text{Pb}} \Delta T$$
$$12 \text{ cal} = (10 \text{ g})[0.031 \text{ cal}/(\text{g})(°C)](\Delta T)$$
$$\Delta T = 39°C$$

Hence, if the original temperature of the bullet were 20°C, its final temperature would be 59°C. If the bullet had been traveling at 600 m/s, we would find ΔT obtained in the above way to be 36 times as large, or the final temperature would be about 1430°C. Of course, the bullet would melt before this temperature was reached, and so the above computation would no longer be correct. How could the computation be carried out in this latter case?

Illustration 10.7   When nutritionists state that 1 kg of bread has a food value of 2600 cal, they mean that if the dried bread is burned in pure oxygen, it will give off 2600 kcal of heat energy. (Basically, the body generates heat from food in a somewhat similar chemical reaction.) Estimate how much heat energy a human body gives off each day.

   Reasoning   Depending on the person, his or her nutritional calorie intake each day is in the range of 2000 to 3000 "calories." Since these are actually kilocalories, the process of metabolism within the person's body will generate on the order of 2×10^6 cal of heat. Because the body temperature remains nearly constant, the body must lose this energy as it is generated. The methods by which this energy is generated and released within the body are complex and are discussed in texts on biochemistry and biophysics. The air we breathe out and the evaporation of perspiration from the skin are well-known mechanisms for cooling the body, but others are important as well.

10.8 Thermal Expansion

As we have seen, the temperature of a substance is a measure of the internal energy of its molecules. *As the temperature of a liquid or solid is raised, the molecules,* having greater energy, *will generally vibrate through larger distances. This increased amplitude of vibration of a given molecule will force its neighbor molecules to remain at a greater average distance from it. Hence, the solid or liquid will expand.* Although there are some notable exceptions to this rule over small temperature ranges (for example, water contracts in going from 0 to 4°C),[2] it is generally true that substances expand with increasing temperature provided that a phase change does not occur.

---

[2] In water, hydrogen bonding binds the molecules into groups of several molecules each in a definite structure even above the melting point of ice. As the temperature increases, these groups break up, causing a new, more compact arrangement of the molecules.

Clearly, the thermal expansion of the metal in a building or a bridge can be a matter of considerable practical importance. If provision were not made for thermal expansion, railway tracks and concrete highways would buckle under the action of the hot summer sun. You may know of situations where faulty design has caused difficulty in this respect. Many of us have lived or worked in buildings where lengthening steam pipes in the heating system give rise to noticeable effects. For these reasons, and many more, it is necessary to know exactly how a material expands with temperature. To this end, a constant of linear thermal expansion α (alpha) and a constant of volume thermal expansion γ (gamma) are defined and tabulated.

Definition   *The* **coefficient of linear thermal expansion** α *is defined to be the increase in length per unit length of a material for a temperature change of one degree.* Written as an equation, this definition is

$$\alpha = \frac{\Delta L / L}{\Delta T}$$

In other words, if a bar of length L expands in length an amount ΔL when the temperature is raised by an amount ΔT, the value of α is specified by the above equation. Notice that the length units cancel, and so the units of α are reciprocal degrees, that is, $°C^{-1}$ or $°F^{-1}$. A few typical values for α are given in Table 10.5.

As an example, a brass rod with length L will lengthen an amount ΔL, given by the definition of α to be

TABLE 10.5
COEFFICIENTS OF THERMAL
EXPANSION PER CELSIUS
DEGREE AT 20°C

Substance	α,* $\times 10^6$	γ,* $\times 10^6$
Diamond	1.2	3.5
Glass (heat-resistant)	~3	~9
Glass (soft)	~9	~27
Iron	12	36
Brick and concrete	~10	~30
Brass	19	57
Aluminum	25	75
Mercury		182
Rubber	~80	~240
Glycerin		500
Gasoline		~950
Methanol		1200
Benzene		1240
Acetone		1490

*This notation at the top of the column means that all values of α and γ have been multiplied by 10^6. Therefore, α for iron is 12×10^{-6} $°C^{-1}$.

$$\Delta L = \alpha L \, \Delta T \qquad\qquad (10.2)$$

for a temperature change ΔT. For a rod of length $L = 1.00$ m and a temperature change of $50°C$, we have (see Table 10.5 for α)

$$\Delta L = (19 \times 10^{-6}°C^{-1})(1.00 \text{ m})(50°C) = 0.00095 \text{ m}$$

Since this change in length is really very small, the value of L used to determine ΔL is not sufficiently temperature-dependent to cause worry about its temperature of measurement. Actually α varies somewhat with temperature, and for very precise work one should use the value appropriate to a given temperature range. In actual practice, however, this complication is very seldom of importance.

Thermal expansion of volumes is also of importance. In a manner analogous to the way we defined the linear-expansion coefficient, we define a volume-expansion coefficient. *The* **coefficient of volume thermal expansion** γ *is the relative change in volume per unit change in temperature,* or, as an equation,

$$\gamma = \frac{\Delta V / V}{\Delta T}$$

which yields at once

$$\Delta V = \gamma V \, \Delta T \qquad\qquad (10.3)$$

The units of γ are reciprocal degrees. As an example of its use, suppose that 100 cm^3 of benzene at $20°C$ is heated to $25°C$. According to Eq. (10.3), its volume will change by an amount (see Table 10.5 for γ)

$$\Delta V = (1.24 \times 10^{-3}°C^{-1})(100 \text{ cm}^3)(5°C) = 0.62 \text{ cm}^3$$

This is a 0.6 percent change in volume and is an appreciable change in V for many purposes. It is necessary therefore to stipulate the temperature at which V should be measured if the γ coefficients in Table 10.5 are to apply. The values given there are for $T = 20°C$. Of course, for small temperature changes not too far removed from $20°C$, we can compute ΔV to fairly good precision using V measured anywhere in this small temperature range. We examine this point further in an example.

An examination of Table 10.5 shows that the linear-expansion coefficient for solids is essentially one-third the volume-expansion coefficient. This is a general rule for most solids that are isotropic, i.e., the same in all directions. An example at the end of this chapter asks you to show why this rule follows from the definitions of α and γ.

Illustration 10.8 A slab of concrete in a highway is 20 m long. How much longer will it be at $35°C$ than it is at $-15°C$?

Reasoning We have

$$\Delta L = \alpha L \, \Delta T$$

From our tabulated values, $\alpha \approx 10 \times 10^{-6}°C^{-1}$ and so

$$\Delta L = (10^{-5}°C^{-1})(20 \text{ m})(50°C) = 0.010 \text{ m}$$

An expansion crack of this magnitude would be required, or the concrete might buckle.

Illustration 10.9 Late one evening an automobile owner had his gasoline tank filled. The temperature of the gasoline was 68°F. The tank held 16 gal. He then parked the car. By the time he returned the next day, the hot sun had warmed the gasoline to 131°F. How much gas had overflowed from the tank?

Reasoning Let us first change the temperatures to the Celsius scale. They are 20 and 55°C. Since the initial temperature was 20°C, no difficulty arises in connection with the γ values listed in Table 10.5. From the definition of γ

$$\Delta V = \gamma V \, \Delta T = (0.95 \times 10^{-3})(16)(35) = 0.53 \text{ gal}$$

Even if the initial temperature were 55°C, the error involved in using the volume at that temperature to compute ΔV would amount to only about 3 percent, since ΔV is only about 3 percent of the total volume. However, if the initial volume had been specified at a temperature T other than 20°C and high precision were needed, we would first have to find the volume at 20°C by use of the relation

$$\Delta V = \gamma V_{20}(T - 20)$$
where
$$\Delta V = V - V_{20}$$

and V is the measured volume at the temperature in question. We have neglected the expansion of the gasoline tank itself. Is this a serious error?

10.9 Transfer of Heat: Conduction

As we have seen, much of the thermal energy in a substance is KE of the molecules of the material. *Where heat energy is transmitted from the hot end of a metal rod to its cold end,* as shown in Fig. 10.6, the process of heat transmission is quite obvious. The molecules at the warm end are vibrating with high energy. They collide with the cooler molecules on their right and give them more energy. These in turn hit the slower molecules on the right

FIGURE 10.6
Heat moves from the hot
end of the rod to the cool
end by conduction.

side of them, giving them more energy. Hence, *the heat is transmitted down
the rod by means of molecular collisions. This mechanism of heat transfer is
called* **conduction.**

Heat Conduction

The rapidity with which heat energy flows down a bar depends on the
material of which the bar is made. We all know that metal conducts heat
better than wood or glass. A rather general relation states that *good electrical conductors are good heat conductors.* In particular, metals are good conductors of heat because the valence electrons in the metal move about freely
and, in so doing, carry kinetic energy (heat) along with them.

To express the flow of heat in mathematical terms, we imagine the
experiment illustrated in Fig. 10.7. A bar having cross-sectional area A and
length L is connected between two constant-temperature devices, as shown,
with $T_1 > T_2$. Heat energy will flow down the bar. Let us now ask how
much heat ΔQ will flow down the bar in time t. (Assume the sides of the bar
to be insulated so that heat cannot flow radially from it.)

One would suppose that, the larger A was, the more heat would flow.
Indeed, experiment shows that $\Delta Q \propto A$. Also, if $T_1 = T_2$, no heat would
flow. As a first guess, one might say that $\Delta Q \propto T_1 - T_2$. If the bar were
very long, the heat flow would probably be decreased and so one might
suppose that $\Delta Q \propto 1/L$. Of course, the longer the time t, the greater the
amount of heat that will flow. All these guesses are confirmed by experiment, which shows that

$$\Delta Q = \frac{\lambda(T_1 - T_2)At}{L} \tag{10.4}$$

where λ *(lambda) is a proportionality constant called the* **heat conductivity** *for
the material* of the bar. Clearly, λ *will be large for a good conductor* and
small for a poor conductor. Some typical values are given in Table 10.6.
Notice that the units we must use in Eq. (10.4) are stated in the table of heat
conductivities.

TABLE 10.6
HEAT CONDUCTIVITIES,
cal/(cm)(s)(°C)

Material	λ
Silver	1.0
Copper	1.0
Aluminum	0.50
Brass	0.20
Glass	$\sim 20 \times 10^{-4}$
Brick	20×10^{-4}
Concrete	16×10^{-4}
Asbestos paper	$\sim 5 \times 10^{-4}$
Rubber	$\sim 5 \times 10^{-4}$
Wood	$\sim 3 \times 10^{-4}$
Glass wool	1×10^{-4}
Down	0.6×10^{-4}

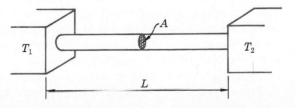

FIGURE 10.7
The rate of heat flow down
the rod is proportional to
$T_1 - T_2$ and to A and is
inversely proportional to L.

287

Illustration 10.10 A 1.00-m-long brass bar with a 2.0-cm² cross section is placed with one end in boiling water and the other end on a cake of ice. How much ice will be melted by the heat conducted down the bar in 10 min?

Reasoning First we find the heat conducted down the bar in 10 min. In proper units we have $A = 2.0$ cm², $L = 100$ cm, $T_1 - T_2 = 100°C$, $t = 600$ s, and $\lambda = 0.20$ cal/(cm)(s)(°C). Substitution in Eq. (10.4) gives

$$\Delta Q = 240 \text{ cal}$$

But 80 cal is needed to melt 1 g of ice. Hence, $240/80 = 3.0$ g of ice would be melted in this time.

10.10 Transfer of Heat: Convection

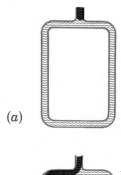

(a)

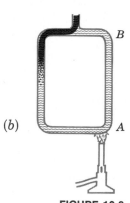

(b)

B

A

FIGURE 10.8
When heat is applied to the liquid in the tube, the dye in the liquid shows that the liquid circulates counterclockwise. Heat is carried along by the liquid, giving rise to heat convection.

A simple experiment devised to illustrate convection is shown in Fig. 10.8. When the glass tube illustrated is filled with water, a little colored dye placed near the neck remains nearly motionless in the position shown in part *a*. However, if the tube is heated as shown in part *b*, the liquid begins to flow counterclockwise around the tube, carrying the dye with it as illustrated.

The reason for this motion is quite simple. A heated liquid or gas expands, and so the water, in the lower right corner of the tube at *A*, expands when heated. It is now lighter than the rest of the liquid. The heavier column of liquid, on the left, will no longer be supported by the lighter column, on the right. It falls, pushing the water along in the tube. Hence the liquid on the right flows upward. It cools as it moves along, and by the time it reaches the left side, it is cooler and denser than when it was at point *A*. In summary, heated liquid at *A* will rise to *B*. In so doing, it will carry heat with it, and so *heat is moved from A to B by the actual motion of the liquid from A to B. This means of heat transfer is called* **convection.**

Conduction does not involve the motion of the molecules over large distances. Heat is transferred from molecule to molecule by collision. In convection, however, the molecules of the transferring material actually move along with the heat. *Only liquids and gases can transfer heat by convection, since only in these materials can the molecules move over large distances.*

Many homes are heated by air-convection methods. Even in the case of heating systems without fans, the circulatory movement of the air is appreciable. For example, to a person standing over a hot-air register above an air furnace, the rush of hot air from the register is often quite noticeable. Proper design of such convection systems must allow the cool air to return to the furnace much as the cool liquid circulates back to point *A* in Fig. 10.8*b*. Clearly, the purpose of the cold-air registers in such heating systems is to return the cool air to the furnace.

Weather phenomena are partly the result of convective air currents: thermal air-circulatory currents near the edges of mountain ranges are particularly interesting. Quite large effects are noticed at various fixed times of day as the cool air from the mountains flows down and causes the warm air on the nearby plains to rise. The Gulf Stream and the Japan current are other interesting examples of large-scale transfer of heat by convection.

10.11 Transfer of Heat: Radiation

We know that the sun warms the earth. It is, in fact, our major source of heat. We can easily see that the heat from the sun is not transferred to us by either conduction or convection. There are essentially no molecules in the vast reaches of space between us and the sun. Hence vibrational transfer by conduction and circulatory transfer by convection are impossible. We have here a case of *heat transfer through vacuum*, i.e., through nothing. This method of heat transfer *is called* **radiation.**

Heat Radiation

For many years, even to the first decade of this century, most scientists could not conceive of heat and light transfer from the sun through *nothing*. They therefore postulated that all space was filled by a "luminiferous ether." It was not until Einstein showed (in 1905) that such a concept as a mechanical ether was neither useful nor verifiable that it was eventually discarded. However, even before that time, the concept of an ether had run into formidable conceptual difficulties. We no longer consider the ether concept either necessary or convenient. As we shall see in our study of electromagnetic radiation, an understanding of radiation can be achieved without it.

As we have intimated, light radiation and electromagnetic (radio) radiation are closely related to heat radiation. We see in later chapters that these phenomena are essentially all the same.

10.12 Laws of Cooling

Newton showed that a convenient and simple law is obeyed in the cooling of bodies which are not too hot. He found by experiment that if the temperature of a body is T_1 and the temperature of its surroundings is T_0, the heat ΔQ lost by the body in time t is

$$\Delta Q = (\text{const})(T_1 - T_0)(t) \tag{10.5}$$

Newton's Law of Cooling

This is called **Newton's law of cooling.** We now know it is only approximately true and then only if $T_1 - T_0$ is not too large. It represents the combined effects of conduction, convection, and radiation.

In heat loss by radiation alone, it was first shown experimentally and later by theory that a hot object should radiate heat according to

$$\Delta Q = (\text{const})(T_1^4 - T_0^4)(t) \tag{10.6}$$

This is known as **Stefan's law.** The temperatures involved are absolute temperatures, T_1 being that of the hot object and T_0 being that of its surroundings. *For very small temperature differences, Stefan's law reduces to Newton's law of cooling.* Can you show this? In doing so, recall that $(x^4 - y^4) = (x^2 - y^2)(x^2 + y^2)$.

Furthermore, it was shown that a blackbody (a body that does not reflect light) will radiate heat much better than a highly reflective body. As a general rule, *a good heat absorber is a good heat radiator.*

10.13 Humidity

Everyone knows that on a day when the humidity is high, the air contains a great deal of water vapor. Humidity is a measure of the water content of the air. Precisely, *the* **relative humidity** (RH) *is defined to be the ratio of the mass of water vapor per unit volume in the air to the mass per unit volume required to produce saturation at the same temperature.* As pointed out earlier in this chapter, when a saturated vapor is in contact with a liquid, identical numbers of molecules leave and return to the surface of the liquid in a given time. Hence, at saturation, no net evaporation occurs. If the vapor is more than saturated, i.e., **supersaturated,** droplets will condense from the vapor and fog or rain will result.

Since the pressure of an ideal gas is proportional to the number of molecules in it, the definition of relative humidity is often couched in terms of pressures rather than masses. Water vapor is nearly an ideal gas, and so the two definitions are nearly identical. In equation form

$$\text{Relative humidity} = \frac{m}{m_s} \approx \frac{P}{P_s} \tag{10.7}$$

where m and P are the mass per unit volume and pressure of the water vapor in the air and m_s and P_s are the respective values for saturated vapor. Some data for saturated water vapor at various temperatures are given in Table 10.7.

According to Table 10.7, saturated air at 20°C contains 17.1 g/m³ of water. Suppose that the air actually did contain 17.1 g/m³ of water vapor. If the air were at any temperature above 20°C, it could contain still more vapor. However, if this amount of water vapor were present and the air cooled below 20°C (as it would, perhaps, after the sun went down), it would become supersaturated as soon as the temperature dropped below 20°C. At that temperature and below, water droplets would begin to fall out of the air in the form of fog, dew, or rain. *We term the temperature at which the air just* *becomes saturated the* **dew point.**

The dew point of the air is a useful quantity. Suppose that the temperature of the air on a certain day is 32°C. On that day the meteorologist at the weather bureau cooled some of the air down until fog or dew began to settle

TABLE 10.7
PROPERTIES OF SATURATED
WATER VAPOR

T, °C	T, °F	m, g/m³	P, cmHg
−8	+17.6	2.74	0.23
−4	24.8	3.66	0.33
0	32.0	4.84	0.46
4	39.2	6.33	0.61
8	46.4	8.21	0.80
12	53.6	10.57	1.05
16	60.8	13.50	1.36
20	68.0	17.12	1.75
24	75.2	21.54	2.23
28	82.4	26.93	2.82
32	89.6	33.45	3.55
36	96.8	41.82	4.44

out of it. Suppose that she found the dew point to be 16°C. She now knows from Table 10.7 that the air contains 13.50 g/m³ of water vapor, since this is the value for saturated vapor at 16°C. However, since the actual air temperature is 32°C, she knows that saturated air at this temperature holds 33.45 g/m³ of water. From this she computes the relative humidity to be

$$\text{RH} = \frac{m}{m_s} = \frac{13.50}{33.45} \approx 0.40$$

We usually multiply this answer by 100 and say that the relative humidity is 40 percent.

The relative humidity can be measured in other ways besides determining the dew point. One common method, the **wet-bulb–dry-bulb method,** makes use of the fact that liquids cause a cooling effect when they evaporate (because the heat of vaporization is abstracted) and that when the vapor is saturated, no evaporation occurs. Hence, if the reading of a dry thermometer is compared with that for a thermometer having a wet cloth around its bulb, the wet-bulb thermometer will usually read cooler than the dry-bulb. The difference in temperatures is a direct measure of relative humidity; the lower the relative humidity, the greater the difference. Tables have been compiled relating relative humidity to this temperature difference, so that the relative humidity can be determined by reading the two thermometers.

High relative humidity in summer often causes us discomfort. This results from the fact that we perspire when we are hot and the evaporation of the perspiration cools us. However, if the relative humidity is 100 percent, there will be no evaporation and therefore no cooling. That is why we do not feel the heat nearly so much in a hot, dry climate as in a hot, moist climate.

Summary

When the temperature of an object is raised, its internal energy is increased. The internal energy U of the object is the total energy resident in the object. It includes the kinetic, vibrational, chemical, nuclear, and all other forms of energy possessed by the particles composing the object.

The energy that is transferred from a high-temperature object to a lower-temperature object because of the temperatue difference is called heat energy. We represent it as ΔQ. Heat energy flows spontaneously from hot objects to cold objects, but not vice versa.

The two commonly used units of heat energy are the calorie and the Btu. They are related to the joule through 1 cal = 4.184 J and 1 Btu = 1054 J. Moreover, 1 Btu = 252 cal. Nutritionists make use of the kilocalorie and call it a calorie.

By definition, the specific heat capacity c of a substance is the heat energy needed to raise the temperature of unit mass of the substance by one degree. As an equation, $\Delta Q = cm\,\Delta T$, where ΔQ is the heat needed to raise the temperature of a mass m by an amount ΔT.

In dealing with gases, two specific heats are usually given. When the temperature change occurs at constant volume, the specific heat capacity is c_V. At constant pressure, the value is c_P. And c_P is always larger than c_V since heat must be furnished to do work in expanding the volume in the case of constant pressure.

The energy needed to vaporize unit mass of substance is called the heat of vaporization H_v. An equal amount of energy is given off when unit mass of the vapor is condensed. In a similar manner, the energy needed to melt unit mass of crystalline substance is called the heat of fusion H_f. An equal amount of heat is given off when unit mass crystallizes from the molten state.

When a liquid is maintained at constant temperature in contact with its vapor, the equilibrium pressure of the vapor is called the saturated vapor pressure. Its value increases with temperature. As the temperature of a liquid is increased, its saturated vapor pressure increases until, at high enough temperature, it equals the external pressure on the liquid. At that temperature, vapor bubbles form and grow in the liquid. The liquid is then said to boil.

If an object of original length L is subjected to a temperature change ΔT, the length will change by an amount ΔL given by $\Delta L = \alpha L\,\Delta T$. The quantity α is the coefficient for linear thermal expansion. Similarly, an object of volume V undergoes a volume change ΔV when a temperature change ΔT takes place. A coefficient for volume thermal expansion γ is defined by $\Delta V = \gamma V\,\Delta T$.

Heat can be transferred by three mechanisms. In conduction, the energy is passed from atom to atom (or other particle) by collision. No long-range motion of the particles take place. In convection, high-energy molecules flow as a current and carry heat energy along with them. The particles move over large distances, distances comparable to that over which the heat is transported. In radiation, energy is transmitted through vacuum. This is the mode of transport of energy in a light or radio-wave beam.

Relative humidity is the ratio of the mass of water vapor per unit volume in the air to the mass per unit volume required to produce saturation at the same temperature. Air can hold more water per unit volume at high than at low temperature. If air is cooled far enough, a temperature is reached at which the air is saturated. This temperature is called the dew point. At temperatures below this, some form of precipitation will occur.

Minimum Learning Goals

Upon completion of this chapter, you should be able to do the following:

1 Distinguish among the terms thermal energy, internal energy, and heat energy by explaining the meaning of each.

2 State the direction of heat flow when two objects of known temperature are placed in contact.

3 Give the meaning of the units calories and Btu (a) by reference to their values in joules and (b) by reference to the heat energy needed to heat water. Explain the relation of the calorie to the nutritionist's calorie.

4 Define the specific heat capacity both in words and by equation. Give its value for water.

5 Use the equation $\Delta Q = cm\,\Delta T$ to solve simple problems involving the heating and cooling of objects.

6 Explain why c should be larger for a substance

composed of complex molecules than it is for a monatomic gas.

7 Explain why c_V differs from c_P for a gas.

8 Define the following terms and relate them to one another: saturated vapor pressure, boiling point, heat of vaporization, evaporation, cooling due to evaporation.

9 Tell qualitatively how the temperature of a crystalline substance changes as a function of time as it is slowly heated to the melting point, melted, heated further, and then vaporized. In so doing, point out the effects of heat of fusion and heat of vaporization.

10 Describe what happens to the boiling point of a liquid as the external pressure on it is changed.

11 Solve simple problems in calorimetry such as those in Illustrations 10.4 to 10.7.

12 Compute the thermal expansion of a rod or a volume when the appropriate thermal-expansion coefficient is given. Or, given the results of an expansion experiment, compute the coefficient.

13 Determine how much heat flows through a slab of material when the temperatures of the two faces of the slab are given. Assume the heat conductivity of the material is known.

14 List the three processes by which heat is transferred and explain each.

15 Define the terms relative humidity and dew point. Calculate the RH when the dew point (or water-vapor concentration) and temperature of the air are given. Assume a table such as Table 10.7 is available to you. Explain why the RH is an important factor in cooling the body by evaporation or perspiration.

Important Terms and Phrases

You should be able to define or explain each of the following:

Internal energy U
Heat energy ΔQ
Specific heat capacity c
$\Delta Q = cm\,\Delta T$
Calorie, Btu, nutritionist's calorie
c_P is larger than c_V
Saturated vapor pressure

Evaporation and boiling
Heats of vaporization and fusion
Boiling under reduced pressure
$\Delta L/L = \alpha\,\Delta T$ and $\Delta V/V = \gamma\,\Delta T$
Heat conduction, convection, radiation
Newton's law of cooling
Stefan's law
Relative humidity
Dew point
Supersaturated vapor

Questions and Guesstimates

1 Although argon gas molecules (Ar) and oxygen molecules (O_2) have roughly the same molecular mass, it takes about twice as much energy to heat oxygen as argon. By reference to their internal energies, explain why this result is not unexpected.

2 In an experiment, a student is given a thermos jug containing an unknown substance at temperature T_1. A quantity ΔQ of heat is added by adding hot water. After equilibrium is again established, the temperature is still T_1. The student concludes that the specific heat of the material in the thermos jug is infinite. Explain why the experiment implies $c = \infty$. What is the probable explanation of these experimental results?

3 Can heat be added to something without its temperature changing? What if the "something" is a gas? A liquid? A solid? Explain.

4 Explain why c_P is larger than c_V for a gas.

5 A particular type of wax melts at 60°C. Describe an experiment by which you could determine the heat of fusion of the wax.

6 How would you go about computing the heat needed to raise the temperature of a mixture of known weights of two gases by 10°C?

7 It is possible to make water boil furiously by cooling a flask of water which has been sealed off when boiling at 100°C. Explain why.

8 Liquid chlorine will boil at 30°C if it is subjected to a pressure of 8.60 atm. Where should one look to find data of this sort? Be specific.

9 In what ways would the earth be different if water contracted as it froze?

10 Why should a gas cool more when it expands an amount ΔV by pushing out a piston than when it expands into a vacuum chamber of size ΔV?

11 Why does a piece of steel feel colder than a piece of wood at the same temperature?

12 What would happen if the earth were to be covered by a dense layer of smog so that the rays from the sun could not reach its surface?

13 Temperature fluctuations are much less pronounced on land close to large bodies of water than they are in the central regions of large land masses. Explain why.

14 Estimate how much the temperature of a human body would rise in 1 day if it retained the approximately 2000 large calories (kilocalories) acquired in food in one day. The value for c is about 0.83 cal/(g)(°C) for a person. (E)

15 It is well known that a room filled with people becomes very warm unless it is properly ventilated. Assuming that a person gives off heat equivalent to the person's food energy in a steady way throughout the day, estimate how much the temperature of your classroom would rise in 1 h if there were no heat loss out of the room. (E)

16 About how much water would have to evaporate from the skin of an average-size person to cool his or her body by 1 Celsius degree? How does this fit in with what you have heard about the effect of perspiration on the body? [$c_{body} \approx$ 0.83 cal/(g)(°C).] (E)

17 If ice is subjected to high pressure, its melting point is decreased below 0°C. To a rough approximation, the melting temperature decreases by about 5 Celsius degrees for each additional 6.0×10^7 Pa of applied pressure. Estimate the melting temperature of ice beneath an ice skater's skate. (E)

Problems

1 How much heat (in calories) is needed to change the temperature of 500 g of water from 10 to 80°C?

2 How much heat (in calories) is lost by 250 g of water at 95°C as it is cooled to 25°C?

3 (a) If 20 g of copper is cooled from 80 to 30°C, how many calories of heat are given off? (b) How many Btu is this?

4 (a) How many calories of heat are needed to change the temperature of 10.0 g of lead from 20 to 100°C? (b) How many Btu is this?

5* The *energy efficiency ratio* (EER) of a refrigerator or an air conditioner is defined as

$$\text{EER} = \frac{\text{heat removed (Btu) per hour}}{\text{electricity input (W)}}$$

Find the ratio of the energy removed to the energy input for a refrigerator that has an EER of 10, a very high value.

6* How much heat (in calories) does a refrigeration system remove each second if it requires an electrical input power of 300 W and has an EER = 7.0? (See Prob. 5.)

7* An electric heater supplies 1800 W of power in the form of heat to a tank of water. How long will it take to heat the 200 kg of water in the tank from 10 to 70°C? Assume heat losses to the surroundings to be negligible.

8* Cool water at 9.0°C enters a hot-water heater from which warm water at a temperature of 80°C is drawn at an average rate of 300 g/min. How much average electrical power (in watts) does the heater consume in order to provide hot water at this rate? Assume there is negligible heat loss to the surroundings.

9* A 70-kg person consumes about 2500 nutritionist's calories, that is, 2.5×10^6 cal, of food per day. If all this food energy were changed to heat and none of the heat escaped, what would be the temperature rise of the person's body?

10* Fifty people are sitting in a rectangular room which has dimensions 3.5 by 10 by 10 m. On the average, each person consumes 2500 nutritionist's calories per day and loses one-twenty-fourth of this amount of heat each hour. By how much do they raise the temperature of the air in the room in 1 h? For simplicity, assume the air to be N_2 at a density of 1.29 kg/m³. Neglect the volume occupied by the people. Assume the air to be stagnant with no heat loss to the walls or windows of the room and to be maintained at constant volume.

11 How much heat (in calories) must be removed from 30 g of water at 0°C to change it to solid ice?

12 How many calories are required to melt a 45-g ice cube at 0°C?

13 How many calories of heat are needed to change 30 g of ice at −5°C to water at 20°C?

14 How much heat must be removed from 200 g of water at 15°C to change it to ice at −10°C?

15 How much ice at 0°C is needed to cool 250 g of water at 25°C to 0°C?

16* An 18-g ice cube (at 0°C) is dropped into a glass containing 150 g of soda at 25°C. If negligible heat exchange occurs with the glass, what will be the final temperature of the soda after the ice melts?

17* Molten lead at 327°C is poured into a hole in a block of ice at 0°C. How much ice will be melted by 40 g of lead? Assume temperature equilibrium with the block is achieved.

18* Thirty grams of solid mercury at its freezing point (−39°C) is dropped into an ice-water mixture at 0°C. After equilibrium is again achieved, the mercury–ice water mixture is at 0°C. How much additional ice is produced by the addition of the mercury?

19 How much heat is given off as 50 g of gaseous ethanol at its boiling point is condensed to liquid at this same temperature?

20 How much heat is required to change 8.0 g of liquid nitrogen at its boiling point to vapor at this same temperature?

21* How much perspiration must evaporate from a 5.0-kg baby to reduce its temperature by 2°C? The heat of vaporization for water at body temperature is about 580 cal/g.

22* The average energy reaching us from the sun each second is 0.134 J/cm². Most of this energy is absorbed as it passes through the earth's atmosphere. Assume that 0.10 percent of it strikes the surface of a lake and is used to evaporate the water. How much water would evaporate from 1 m² in 1 h? Use $H_v = 590$ cal/g.

23* When a volatile liquid such as alcohol or ether evaporates from your skin, noticeable cooling occurs. Suppose 0.020 g of dichloroethane ($H_v \approx$ 85 cal/g) evaporates from a 1-cm² area of your skin and cools a surface layer 0.035 cm thick. (a) How much will the temperature of the skin be lowered? Assume c for the skin to be 0.75 cal/(g)(°C) and its density to be 0.95 g/cm³. Neglect the fact that the skin may undergo a phase change. Also, consider the value of c to be negligibly small for dichloroethane. (b) Why does the temperature of the skin not decrease as much as you calculate?

24* Earlier in this chapter it was shown that $c_P − c_V = R/M$ for an ideal gas. Compute the quantity $(c_P − c_V)(M)$ for each of the real gases in Table 10.2 from the data in the table. For comparison, $R = 8314$ J/(kmol)(K). The values of M for the gases (in the order listed in the table) are 4.0, 40, 32, 28, 44, 18, and 16 kg/kmol.

25* Assuming air to be composed of 78% N_2 and 22% O_2 by weight, what should be the value of c_V for air?

26** Gas is confined in a vertical cylindrical container by a piston having a weight of 10 N much as in the situation shown in Fig. 10.3. When the system is at 20°C, the piston rests at a certain height in the cylinder. After heating to 100°C, the piston has risen 20 cm. How much more heat is required to heat the gas in the container from 20 to 100°C under constant pressure than under constant volume? Assume the gas to be ideal.

27* If 100 g of lead shot at 100°C is dropped into 50 g of water at 20°C contained in a copper calorimeter can of mass 50 g, what is the resulting temperature?

28* To 100 g of water contained in a 50-g copper calorimeter at 35°C is added 20 g of ice at −10°C. What is the final temperature?

29* How much steam at 100°C is needed to change 40 g of ice at −10°C to water at 30°C if the ice is in a 50-g copper calorimeter?

30* A 70-g calorimeter can [$c = 0.20$ cal/(g)(°C)] contains 400 g of water and 100 g of ice at equilibrium. To this is added a 300-g piece of metal [$c = 0.10$ cal/(g)(°C)] which has been heated to a high temperature. The final temperature is 10°C. What was the original temperature of the metal?

31* A copper calorimeter has a water equivalent of 5.9 g. That is, in heat exchanges, the calorimeter behaves like 5.9 g of water. It contains 40 g of oil at 50.0°C. When 100 g of lead at 30.0°C is added to this, the final temperature is 48.0°C. What is the specific heat capacity of the oil?

32* Benzene boils at about 80°C. Benzene vapor at 80°C is bubbled into a calorimeter, the water equivalent of which is 20 g (see Prob. 31), containing 100 g of oil [$c = 0.50$ cal/(g)(°C)] at 20°C. The final temperature after 7.0 g of benzene has been condensed is 30°C. What is the value of the heat of vaporization of benzene? For benzene $c = 0.40$ cal/(g)(°C).

33 When 150 g of ice at 0°C is added to 200 g of water in a 100-g aluminum cup, with the cup and water at 30°C, what is the resulting temperature?

34* Assuming that the total heat of vaporization of water, 539 cal, can be used to supply the energy needed to tear 1 g of water molecules apart, how much energy is needed per molecule for this purpose? Find the ratio of this energy to kT at the boiling point.

35 (a) How much KE does a 20-g lead bullet have when it is moving at 50 m/s? (b) How many calories is this? (c) If all the energy could be used to heat the bullet, how much would the temperature of the bullet change?

36 (a) A 0.050-g water droplet falls from a height of 2.0 m. What is its KE after falling that far if air drag is negligible? (b) How many calories is this? (c) Assuming all its KE to be changed to heat in the water as the drop splatters on the floor, how much does the temperature of the water increase?

37* In an extruder used to make synthetic fibers, a piston applies a pressure of 70 MPa to the molten plastic and forces it through a tiny nozzle at a rate of 0.00100 cm^3/s. Assume all the energy loss occurs as friction work in the plastic as it goes through the nozzle. (a) How much friction work is done each second? (b) How much is the temperature of the plastic raised as it is forced through the nozzle? [Assume $c_{plastic} = 0.20$ cal/(g)(°C) and $\rho = 1000$ kg/m^3.]

38* A 60-kg girl running at 5.0 m/s while playing basketball falls to the floor and skids along on her leg until she stops. How many calories of heat are generated between her leg and the floor? Assuming all this heat energy is confined to a volume of 2.0 cm^3 of her flesh, what will be the temperature change of the flesh? Assume $c = 1.0$ cal/(g)(°C) and $\rho = 950$ kg/m^3 for flesh.

39* A weight attached to a string which passes over a pulley causes a paddle wheel to turn in a water container as the weight falls. The paddle is constructed so that the weight falls slowly, thereby acquiring negligible kinetic energy. Its decrease in potential energy is therefore converted entirely into friction work or heat energy in the water. Suppose the 200-g mass fell 2.0 m. How much would it heat the 100 g of water in the container if no heat were lost to the paddle or container? (This is the basic idea behind Joule's experiment for determining the mechanical equivalent of heat, i.e., the relation between the joule and the calorie.)

40* A mass m of lead shot rests at the bottom of a closed cardboard cylinder L units long, as shown in Fig. P10.1. When the cylinder is quickly inverted by rotating it about its center, the lead shot falls through the length of the tube. Show that the rise in temperature of the lead after n reversals is given by

$$\Delta T = \frac{ngL}{c_{Pb}J}$$

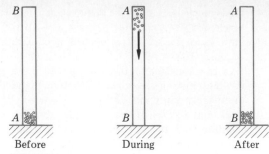

FIGURE P10.1

where J is the mechanical equivalent of heat, i.e., the number of joules equivalent to 1 cal, and g is the acceleration due to gravity.

41** A certain 6-g bullet melts at 300°C and has a specific heat capacity of 0.20 cal/(g)(°C) and a heat of fusion of 15 cal/g. (a) How much heat is needed to melt the bullet if it is originally at 0°C? (b) What is the slowest speed at which the bullet can travel if it is to just melt when suddenly stopped?

42** A 2.2-g lead bullet is moving at 150 m/s when it strikes a bag of sand and is brought to rest. (a) Assuming that all the frictional work was transferred to thermal energy in the bullet, what is the rise in temperature of the bullet as it is brought to rest? (b) Repeat if the bullet lodges in a 50-g block of wood that is free to move.

43 A 75-cm-long iron bar is heated from 20 to 300°C. By how much does its length change?

44 A brass sphere has a diameter of exactly 3.0 cm at 20°C. What will be its diameter at −80°C?

45 A certain straight roadway is made of concrete slabs, placed end to end. Each slab is 25 m long. How large an expansion gap should be left between slabs at 20°C if the slabs are to just touch at 45°C?

46 The marks on an aluminum measuring tape were made when the tape was at 25°C. What is the percentage error due to contraction if the rule is used at −5°C?

47 A salad-dressing jar has an aluminum, screw-type lid. At 20°C, the lid fits so tight that it will not screw off. By what factor will the lid's diameter expand if it is heated to 85°C under the hot-water faucet? Why should you not heat the can lid any longer than necessary?

48 It is common machine-shop practice to "shrink-fit" cylindrical rods into holes in wheels, blocks, and plates. Suppose a 2.000-cm-diameter rod is to be fitted into a 1.985-cm-diameter hole in a brass block. How much must the block be heated so that the cool rod will fit into the hole?

49 A heat-resistant glass flask is calibrated to hold exactly 100 cm³ at 20°C. How much will it hold at 30°C? *Hint:* The hollow flask will expand as though it were actually a solid volume.

50 A 100-cm³ volumetric flask is filled with benzene at 30°C. How much more benzene must be added to have it full after it is cooled to 20°C? (Ignore the expansion of the flask.)

51** An iron beam 10 m long has its two ends embedded in substantial concrete pillars. If the structure was made at 5°C, what force will be exerted by the beam on the pillars when its temperature is 30°C? The cross-sectional area of the beam is 60 cm².

52** A flask of mercury is sealed off at 20°C and is completely filled with the mercury. Find the pressure within the flask at 100°C. Ignore the expansion of the glass. (Assume $k_{Hg} = 4.0 \times 10^{-10}$ m²/N.)

53** Show that the coefficient for volume thermal expansion in the case of an ideal gas is $1/T$ provided the expansion is done at constant pressure.

54** A uniform, solid brass sphere of radius b_0 and mass M is set spinning with angular speed ω_0 about a diameter. If its temperature is now increased from 20 to 80°C without disturbing the sphere, what will be its new (a) angular speed and (b) rotational kinetic energy?

55** When a substance is heated and expands, it does work against the atmosphere. This work contributes a part of, but not all, the difference between c_P and c_V for the substance. Compute the difference between c_P and c_V that results from this cause in the case of benzene (density of benzene = 879 kg/m³).

56** A cube of original edge length L_0 has a volume $(L_0 + \Delta L)^3$ after its temperature is increased by ΔT. Use this fact to show that the volume thermal-expansion coefficient for the material of the cube is 3 times larger than its linear thermal-expansion coefficient.

57 Consider a glass window of area 1 m² and thickness 0.50 cm. If a temperature difference of 20°C exists between one side and the other, how many calories would flow through the window each second? Why is this result *not* applicable to a house window on a day when the temperature difference between inside and outside is 20°C?

58 If the two ends of a brass rod 0.50 m long with a 0.50-cm radius are maintained at 100 and 20°C,

how much heat flows down the rod in 1 min? Assume the sides of the rod are insulated.

59 An asbestos sheet 2.0 mm thick is used as a spacer between two brass plates, one at 100°C and the other at 20°C. How much heat flows through 40 cm² of area from one plate to the other in 1 h?

60 Deep bore holes into the earth show that the temperature increases about 1°C for each 30 m of depth. Assuming that the earth's crust has a thermal conductivity of about 2×10^{-3} (cal/s)/(°C)(cm), how much heat flows out through the surface of the earth each second for each 1 m² of surface area?

61* A brass pipe 10 cm in diameter having a 0.25-cm wall thickness carries steam at 100°C through a vat of circulating water at 20°C. How much heat is lost per meter of pipe in 1 s?

62** A 0.50-cm thick sheet of brass has sealed to one face a 0.50-cm-thick rubber sheet. The other side of the brass sheet is connected to a bath maintained at 20°C. The other rubber surface is attached to a circulating bath at 80°C. Find the temperature at the rubber-brass junction.

63** Two brass plates, each 0.50 cm thick, have a rubber spacer sheet between them which is 0.10 cm thick. The outer side of one brass plate is kept at 0°C, while the outer side of the other is 100°C. Find the temperatures of the two sides of the rubber spacer.

64* What thickness of wood has the same insulating ability as 8 cm of brick?

65** Commercial R values for insulation are defined as follows for a sheet of material:

$$R \text{ value} = \frac{\text{temperature difference (°F)}}{\text{heat-transfer rate (Btu/h)}}$$
$$\text{per square foot}$$

What is the R value for a ¼-in-thick glass windowpane?

66** Wall insulation for houses is often specified as approximately R-10. That is to say, the wall's R value (see Prob. 65) is 10. How thick would a solid wood wall have to be to give this R value?

67** Two sheets of insulation have R values of R_1 and R_2, respectively. Show that the R value for the combination of one sheet on top of the other is $R_1 + R_2$.

68 On a certain day the temperature is 24°C, and the humidity is 80 percent. At what temperature will fog form in the air?

69 On a particular day it is found that water condenses on a water glass when the water temperature is 16°C or lower. The room temperature is 20°C. Find (a) the amount of water in unit volume of the air and (b) the relative humidity.

70 If the air temperature is 32°C and the relative humidity is 80 percent, how cold must a soft drink bottle be before moisture will condense on it?

71 A person who wears glasses has been outside for some time on a day when the temperature is −4°C. How low must the RH be in a house which the person enters if the glasses are not to fog up? Assume the house temperature is 20°C.

72* In a closed room with volume 500 m³, the temperature is 24°C and the humidity is 95 percent. A dehumidifier is turned on to reduce the humidity. Assuming unchanged temperature, what mass of water will be removed in reducing the humidity to 40 percent?

73* On a day when the outside air is at 10°C and the relative humidity is 70 percent, the air is used to ventilate a house maintained at 24°C. What will be the RH in the house? Take account of the fact that the air expands as it is heated.

11
Thermodynamics

Long before the nature of atoms and molecules was known, a powerful way of discussing heat, work, and internal energy had been found. It involves the description of matter in terms of gross, overall properties such as pressure, temperature, volume, and heat flow. This way of discussing the behavior of objects and substances is called **thermodynamics.** Today, even though we understand quite well how atoms and molecules behave, thermodynamics is widely used in all branches of science. This chapter introduces this important and useful area of study.

11.1 State Variables

In thermodynamics, we most frequently discuss the behavior of a definite group of molecules which we call a **system.** A typical system may be the gas molecules in a container, or those in a solution, or even such a complex system as the molecules in a rubber band. For any very meaningful thermodynamic discussion, the system must be well specified. Only then can we give an unambiguous description of it.

To describe the system, we use quantities that apply to the whole system or to some well-defined portion of the system. Typical measurable quantities are pressure, temperature, and volume. In thermodynamics we also use such quantities as internal energy, heat, work, and a quantity we shall encounter later called entropy. As the condition of a system changes, these quantities may change. It is important that we know which quantities are suitable in representing the exact condition of the system. Let us now see what they are.

When a container of gas reaches equilibrium, the gas has a definite temperature, pressure, and volume. The ideal-gas law reflects this fact since it tells us that $PV = nRT$. We see that, *given any two of these variables, the other can be calculated and is therefore known. This particular situation, where the gas (the system) has specified values of P, V, and T, is called a* **state** *of the system. Whenever the gas is returned to these same values of P, V, and T, its state will be the same.* Even though each individual molecule within the system may not be doing exactly the same thing whenever the system is brought to this state, the system as a whole still appears the same in macroscopic measurements.[1]

Definition

Certain features of a system are always the same when the system is in a given state. The variables that describe these features are called **state variables.** For example, P, V, and T are state variables for a system that consists of a gas. No matter how the system reaches a particular state, the state itself is characterized by the same values for pressure, volume, and temperature. For example, a system consisting of the gas in a car tire can be specified by the values of P, V, and T for the gas. No matter what the history of the tire, a given state of the gas within the tire always has the same values for P, V, and T, its state variables.

State Variable

We have seen that the internal energy of a substance consists of all the energy of every type resident in the substance. Since energy is conserved, we can keep a balance sheet of it. A system's internal energy is a well-defined quantity. If you tell me how much internal energy a system has today, I can (in principle, at least) tell you how much internal energy it has at a later time. All I need to do is measure the energy changes that the system undergoes in the meantime. Then, like a good accountant, I can total the energy balance sheet and tell you how much internal energy the system still possesses. *Internal energy is a state variable.*

But heat is *not* a state variable. There is no way that we can evaluate how much heat a system contains. Even though we can compute how much heat must flow into a block of metal, for example, to cause a given temperature change, this same temperature change could have been caused by the action of friction without heat flow into the block. We see that an accountant cannot associate unambiguously a certain amount of heat with a given state of the system. The same change in states can be achieved by adding different quantities of heat. *Heat is not a state variable.*

11.2 First Law of Thermodynamics

The law of conservation of energy is one of the two fundamental pedestals on which all thermodynamics is based. Indeed, those who first developed

---

[1] **Macroscopic** is used to denote measurements involving the average effect of billions or more molecules.

the field of thermodynamics were also the earliest investigators of energy conservation. *When the law of conservation of energy is stated in a general way for a system, the statement is called the first law of thermodynamics.* Let us now see how the statement is made.

A system in a given state has a definite amount of internal energy U. This energy can be of any and all kinds including kinetic, potential, chemical, and nuclear energy. The system's internal energy can be changed in only two general ways: (1) heat energy can flow into or out of the system, and (2) the system can do work of some kind against external forces. To summarize this fact in equation form, we define

ΔU = change in internal energy of system[2]

ΔQ = heat that flows *into* system

ΔW = work done *by* system

Then we have

$$\Delta U = \Delta Q - \Delta W \qquad (11.1)$$

First Law of Thermodynamics

This equation is a statement of the **first law of thermodynamics.** *It is also a statement of the law of energy conservation.*

In using the first law, we must be very careful about plus and minus signs. It is a balance-sheet equation, and so debits and credits must be noted with care. By convention, ΔW is the work done *by* the system. It therefore decreases U and hence appears with a negative sign in Eq. (11.1). On the other hand, ΔQ is the heat that flows *into* the system. It therefore increases U and carries a positive sign in Eq. (11.1).

The first law is applicable to all systems, no matter how complex. For example, consider your body as the system. (In order for the system to remain simple, you will not be allowed to eat or excrete.) It loses internal energy as the day goes on. The food you have previously eaten is part of your internal energy. This and other forms of internal energy are used up as the day progresses. Much of the energy is lost as heat flows from your body to the surroundings. A smaller amount is lost as your body, the system, does work on the objects of the world about you. We can write the first law for your body in the following way:

(Change in internal energy) = ΔQ − (work done by body)

Notice in this case that ΔQ will be a negative quantity because heat flows from the system, your body, rather than into it. Hence we see that ΔU is negative for the system, and so your body loses internal energy.

---

[2] As usual, an increase in U is a positive change while a decrease is a negative change.

11.3 Work Done during Volume Change

In many practical applications of thermodynamics, the work done by the system is carried out by means of a volume change. It is therefore necessary to know how work and volume change are related. For simplicity, let us consider the system shown in Fig. 11.1. We see there a gas confined to a cylinder by a piston. Let us consider how much work the gas does as it expands and pushes the piston slightly upward.

If we keep the displacement of the piston small, the pressure of the gas will not change much. The force exerted on the piston by the gas is given by the definition of pressure to be

$$P = \frac{F}{A} \quad \text{or} \quad F = PA$$

where A is the end area of the piston. In the slight displacement Δy, the work done against the piston by the gas is

$$\Delta W = (\text{force})(\text{distance})(\cos \theta)$$
$$= (PA)(\Delta y)(1) = P(A\,\Delta y)$$

where $\cos \theta$ is unity because the force is in the same direction as the displacement. But $A\,\Delta y$ is simply the increase in volume of the gas ΔV. We therefore find

$$\Delta W = P\,\Delta V \tag{11.2}$$

Because the pressure that the piston exerts on the system is the same as the pressure that the gas exerts on the piston, we can summarize our result in the following way:

The work that a system does in an expansion ΔV against a constant pressure P is simply $P\,\Delta V$.

If the system contracts, then ΔV is negative, and so the work done by the system is negative. In that case, the surroundings have actually done work on the system.

Often it is important to notice that the expansion work of a system can

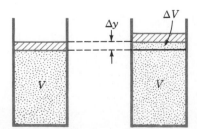

FIGURE 11.1
If the piston has an area A, then $\Delta V = A\,\Delta y$.

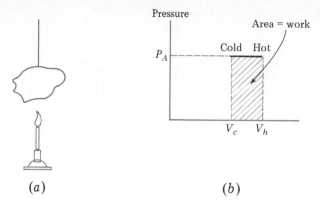

FIGURE 11.2
The work done by the
object as it expands against
the atmosphere equals the
area under the P-V curve.
(a) The object expands as it
is heated. (b) P-V diagram
for the experiment in part a.

be represented by an area on a particular type of graph called a *P-V* diagram. As a very simple example of such a diagram, consider the case of the thermal expansion of a solid metal object as shown in Fig. 11.2. When the object is heated, it expands. The pressure to which it is subjected is that of the atmosphere, P_A. Of course, P_A remains constant during the expansion. We can represent this expansion as shown in the *P-V* graph in part *b* of the figure.

As we see, the system starts at the point labeled "Cold" in the graph. The object has a volume V_c, and the pressure is P_A. During expansion, the system follows the graph line shown. Let us now find the work done by the system. It is, from Eq. (11.2),

$$\Delta W = P \Delta V = P_A(V_h - V_c)$$

But the quantity P_A is the height of the shaded rectangle in Fig. 11.2*b* while $V_h - V_c$ is its width. Therefore, $P_A(V_h - V_c)$ is simply the shaded area. We therefore conclude that

The expansion work done by a system is equal to the area under its *P-V* curve.

Work Related to Area

Although this has been derived for an especially simple case, it is true in general. To see this, refer to Fig. 11.3. We see there a gas confined by a piston. The pressure on the piston can be changed by adding or removing weights. The pressure and volume can also be changed by heating or cooling the gas. Suppose the conditions are varied in such a way that the volume changes from V_A to V_B by the path shown in part *b* of the figure.

Consider the most heavily shaded region in which the pressure is P, approximately, and the volume change is ΔV. The work done in this small part of the total expansion is $P \Delta V$. But this is equal to the area of the heavily shaded rectangle. Similarly, the whole expansion from A to B can be thought of as a series of tiny expansions. The work done during each is an area. The total work done is the sum of all these areas, and so it is the area under the curve from A to B. As we see, *the expansion work is always equal to the area under the P-V curve.*

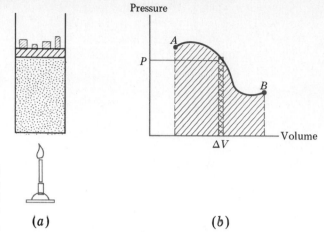

FIGURE 11.3
The work done by the
system in going from *A* to *B*
is equal to the area under
the curve.

(a) *(b)*

Illustration 11.1 A cube of brass 15 cm on each side is heated from 20 to 300°C. How much work is done by the cube as it expands?

Reasoning The cube expands against atmospheric pressure, about 1×10^5 Pa. According to the thermal-expansion equation,

$$\Delta V = V\gamma \, \Delta T$$

In our case $V = (0.15)^3$ m³, $\Delta T = 280°$C, and $\gamma = 5.7 \times 10^{-5}°$C^{-1}. Substitution gives $\Delta V = 5.4 \times 10^{-5}$ m³.

Now we can make use of

$$\Delta W = P \Delta V$$

to obtain

$$\Delta W = 5.4 \text{ J}$$

It is of interest to note that the heat energy needed to raise the temperature of this much brass 280 Celsius degrees is about 10^5 cal. Therefore, only about 10^{-5} of the total heat energy furnished is used in doing expansion work.

11.4 Typical Processes in Gases

As we have just seen, the *P-V* curve for a system undergoing change is related to the work done by the system. In drawing such a graph, we assume that the change is slow enough for the pressure and temperature to be uniform throughout the whole system at any instant. Let us now examine several important ways in which a system composed of a gas can undergo

change. *An* **isothermal process** *is one during which the temperature remains*
constant. Since the temperature of an ideal gas is a measure of its internal
energy, an isothermal process is a constant-internal-energy process. For an
ideal gas, then, we see that $\Delta U = 0$ during an isothermal process. The first
law then tells us that

$$\Delta Q = \Delta U + \Delta W$$

and becomes

$$\Delta Q = \Delta W \qquad \text{isothermal, ideal gas}$$

To illustrate the behavior of an ideal gas in an isothermal change,
refer to the situation shown in Fig. 11.4a. We see there a container of gas in
good thermal contact with a heat reservoir. The heat reservoir might be an
oven, a cooling bath, or any other constant-temperature device. It will
maintain the container of gas at constant temperature provided the piston
closing the container is not moved too rapidly.

Suppose weights are slowly added to the piston. The pressure on the
gas will slowly increase, and the volume will decrease. The P-V diagram for
this isothermal process is given in Fig. 11.4b. Its form is given by the ideal-
gas law

$$PV = nRT$$

which becomes

$$P = \frac{\text{const}}{V}$$

in this case because T is maintained constant. This is the equation for the
graph line shown in Fig. 11.4b, the P-V diagram for the isothermal compression.

Suppose now that the gas has been compressed from point A to point B
on the graph. If the force on the piston is slowly reduced, the above relation

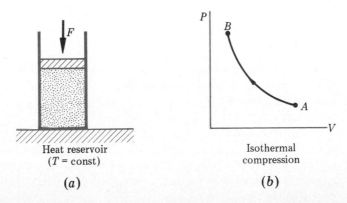

Heat reservoir
(T = const)

(a)

Isothermal
compression

(b)

FIGURE 11.4
The P-V diagram for an
isothermal compression.

305

Reversible Process

between P and V will still apply. The system will follow the same graph line as it moves from state B back to state A. Such a process is said to be *reversible.* Notice that *in a* **reversible process,** *the state variables acquire the same values at all stages of the process independent of the direction in which the process is being carried out.* Not all processes are reversible. For example, *a process which has appreciable friction losses cannot be reversible.* Why?

Adiabatic Process

An **adiabatic process** *is one in which no heat is lost or gained by the system during the process.* For example, if the system is well insulated from the surroundings, heat transfer is often negligible. Or if the process is carried out extremely rapidly (such as a very sudden compression of a gas), no appreciable heat will flow into or out of the system in that short time. The process will be adiabatic.

For an adiabatic process, the first law

$$\Delta U = \Delta Q - \Delta W$$

becomes

$$\Delta U = -\Delta W \qquad \text{adiabatic}$$

This relation is not restricted to an ideal gas. It tells us that *if a system does adiabatic work, the internal energy of the system must decrease.* The work is done at the expense of internal energy. But if adiabatic work is done on the system, the internal energy will be increased. The following illustrations show two practical uses of these processes. First, however, we examine the adiabatic behavior of an ideal gas in more detail.

In the case of an ideal gas, an adiabatic process is not described in terms of $PV = nRT$ alone. The difficulty is that all three state variables (P, V, and T) change during the process. To find how each changes, we need another equation. It can be found by noticing that the work done on the gas goes completely into increased *internal* energy. This increase in internal energy causes a temperature change in the system. But this same temperature change could have been carried out by adding heat to the system. Hence, a relation among heat, temperature change, and work can be found even for an adiabatic process. For an ideal gas, this line of thought leads to the following result: *if an ideal gas undergoes an adiabatic change from P_1, V_1, T_1 to P_2, V_2, T_2, then*

$$P_1 V_1^{\gamma} = P_2 V_2^{\gamma} \qquad (11.3)$$

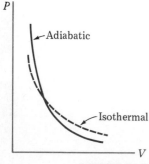

FIGURE 11.5
Comparison of the adiabatic and isothermal changes for an ideal gas.

where $\gamma = c_P/c_V$ for the gas. You will recall from Chap. 10 that c_P is the specific heat at constant pressure while c_V is the specific heat at constant volume. Because $c_P > c_V$, we know that $\gamma > 1$. Typical values for c_P/c_V were listed in Table 10.2.

The P-V diagram graph line for an adiabatic change is given by Eq. (11.3). A typical case is shown in Fig. 11.5. For comparison, the graph line for an isothermal change is shown by the dashed curve. As you might expect, the adiabatic graph line is steeper than the isothermal line at comparable values of P and V.

Illustration 11.2 In a cylinder of a diesel engine, the fuel vapor is ignited by suddenly compressing the gas in the cylinder. Suppose the gas volume is compressed to one-fifteenth its original value. Its initial pressure is about atmospheric, 1×10^5 Pa. If its initial temperature is 27°C, find its final temperature. Assume the gas in the cylinder to be mostly air or nitrogen.

Reasoning This is essentially an adiabatic process since the gas in the cylinder is compressed so rapidly. When we assume the gas to be N_2 and to act as an ideal gas, Table 10.2 gives $\gamma = 1.40$. We know that

$$P_1 V_1{}^\gamma = P_2 V_2{}^\gamma \qquad \text{or} \qquad \frac{P_1}{P_2} = \left(\frac{V_2}{V_1}\right)^\gamma$$

But we are interested in the variation in temperature, not pressure. The ideal-gas law tells us that

$$P_1 V_1 = nRT_1 \qquad \text{and} \qquad P_2 V_2 = nRT_2$$

Dividing one equation by the other gives, after simplifying,

$$\frac{P_1}{P_2} = \frac{T_1}{T_2}\frac{V_2}{V_1}$$

Equating this value of P_1/P_2 to the one found above, namely $(V_2/V_1)^\gamma$, gives, after each side is divided by V_2/V_1,

$$\frac{T_1}{T_2} = \left(\frac{V_2}{V_1}\right)^{\gamma-1}$$

In our case, $V_2/V_1 = \frac{1}{15}$, and so

$$T_2 = T_1(\tfrac{1}{15})^{-0.40}$$

We must use absolute temperatures for T, and so $T_1 = 300$ K. Taking logarithms of both sides of the equation then gives

$$\log T_2 = \log 300 - 0.40(\log 1 - \log 15)$$
$$= 2.477 - 0.40(0 - 1.176)$$
$$= 2.947$$

Then, taking antilogs, we find $T_2 = 886$ K, which is 613°C. Notice how very hot the gas has become because of the adiabatic compression.

Illustration 11.3 As shown in Fig. 11.6a, a container is sectioned into two parts with gas at high pressure in one part and vacuum in the other, much larger part. A small hole is opened in the connecting wall so the gas

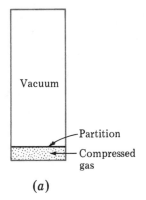

Vacuum

Partition

Compressed gas

(a)

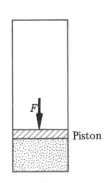

F

Piston

(b)

FIGURE 11.6
(a) A hole is made in the partition so the gas can expand. (b) The expanding gas raises the piston. Under which condition would an ideal gas be cooled most?

expands into the vacuum chamber. Describe the temperature change of the gas, assuming the process to be adiabatic.

Throttling Process

Reasoning *This type of process, in which a fluid expands through a small opening or porous disk, is called a* **throttling process.** Since it is adiabatic, we can write, from the first law,

$$\Delta U = -\Delta W$$

where ΔW is the work done by the fluid in the process.

We are asked to deal with the special case of a gas. Suppose first that the gas is ideal. In the expansion shown in Fig. 11.6a, the ideal gas does no net work as it expands into the vacuum. This follows because the external pressure resisting the expansion is zero, so $P\,\Delta V = 0$. Since the gas does no work, the fact that $\Delta U = -\Delta W$ tells us that the internal energy of the gas does not change. Since for an ideal gas $T \sim U$, we see that T does not change. Therefore, if the expanding gas is ideal, the temperature remains unchanged.

However, the result will be different if the fluid is not an ideal gas. Suppose the compressed material is actually a liquid, e.g., butane, which vaporizes as it expands into the vacuum. Then energy must be furnished to the molecules to tear the liquid apart, i.e., to supply the heat of vaporization. Since the process is adiabatic, the required energy must come from the internal energy already resident in the fluid. As a consequence, the KE of the molecules is decreased during the expansion process. The temperature of the vapor will therefore be lower than that of the original liquid. (In a sense, this is very much like cooling by evaporation.) You can see that many possibilities exist between the ideal gas, where no cooling occurs, and the volatile liquid, where a great deal of cooling occurs.

In certain cases, even an ideal gas will cool upon adiabatic expansion. For example, suppose the partition is replaced by a movable piston, as shown in Fig. 11.6b. Then the gas will do work as it expands against the piston. This work will lead to a decrease in the internal energy of the gas, and the gas will cool.

11.5 Cyclic Processes and Energy Conversion

The subject of thermodynamics was greatly advanced during the onset of the industrial revolution. In order to develop the machines and engines of this new era, it was necessary to achieve an understanding of their operation and their relation to the basic laws of nature. In particular, it became important to know how to improve the efficiency of these new tools of society. All heat engines, as well as refrigeration systems, are similar in one respect: They perform the same operation (or cycle) over and over. As we shall see, the efficiency of such a device is limited by a law of nature which makes it very difficult to design engines with near 100 percent efficiency.

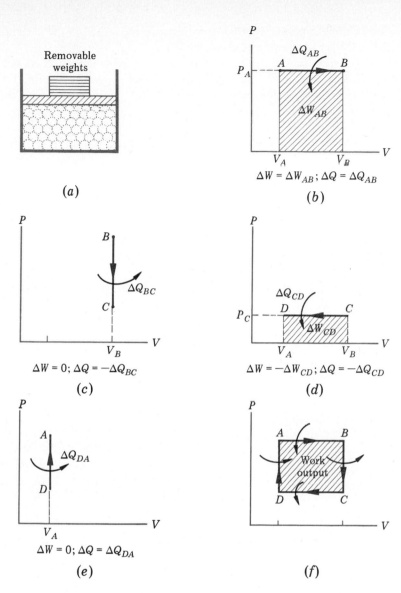

$$\Delta W = \Delta W_{AB};\ \Delta Q = \Delta Q_{AB}$$

(a)

(b)

$$\Delta W = 0;\ \Delta Q = -\Delta Q_{BC}$$

(c)

$$\Delta W = -\Delta W_{CD};\ \Delta Q = -\Delta Q_{CD}$$

(d)

$$\Delta W = 0;\ \Delta Q = \Delta Q_{DA}$$

(e)

(f)

FIGURE 11.7
A simple thermodynamic
cycle. The net work done
equals the area enclosed.

To begin our discussion, consider the system shown in Fig. 11.7a. Suppose the system has the P, V values given by point A of the P-V diagram in part b of the figure. Let us now slowly heat the gas. The volume will then increase from A to B. During this process, the pressure remains constant since the weights on the piston are not changed. Further, heat ΔQ_{AB} is added to the system as indicated. Also, as the gas expands, it does work ΔW_{AB} equal to the shaded area shown.

Next, let us slowly cool the system while simultaneously removing weights from the piston so as to keep the volume constant. This is part BC of the cycle shown in Fig. 11.7c. Since the piston does not move, zero work is done in this portion of the cycle. (Notice that the area beneath this portion of the graph line is zero.) But heat is removed from the system, and so $\Delta Q = -\Delta Q_{BC}$, as indicated.

The next step of the cycle is shown in *d*. Now the gas is cooled further at constant pressure. The gas contracts, and so ΔV is negative. For this reason, the system does negative work equal to $-\Delta W_{CD}$, where this is the shaded area shown. Heat is lost by the system in this step, and so $\Delta Q = -\Delta Q_{CD}$.

Finally, the cycle is completed by the process shown in *e*. Here the gas is heated while the volume remains unchanged by adding weights to the piston. The pressure then increases from *D* to *A*. No work is done by the system during this process. But heat is added to the system, and so $\Delta Q = \Delta Q_{DA}$.

The complete cycle is shown in Fig. 11.7*f*. Notice that the net work done by the system is simply area $\Delta W_{AB} - \Delta W_{CD}$. This is the shaded area in part *f* of the figure. In this case, as well as in all others,

Work/Cycle Is Area Enclosed

The net output work done during a thermodynamic cycle is the area enclosed by the cycle on its *P-V* diagram.

This is an important result because it allows us to find the output work of an engine directly from its *P-V* diagram.

Illustration 11.4 Figure 11.8 is an idealized *P-V* diagram for one cylinder of an ordinary gasoline engine. (This idealized cycle is called the **Otto cycle**.) The air-filled cylinder at *C* is compressed adiabatically to *D*. Gasoline vapor is ignited in the cylinder during portion *DA,* and so heat is added rapidly to the gas with negligible change in volume. The piston then moves out as the hot gas expands adiabatically in portion *AB.* Then, during portion *BC,* heat is released to the exhaust, and the original process repeats. Assuming the shaded area to be one-third of the total area of rectangle *AECFA,* find the net output work of this cylinder per cycle.

Reasoning The net output work is equal to the area enclosed by the cycle. Let us first find the rectangular area *AECFA*. It is

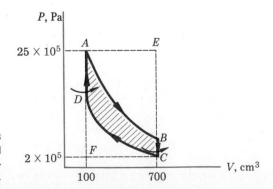

FIGURE 11.8
The Otto cycle, an idealized cycle for the internal-combustion engine.

Area $AECFA = (23 \times 10^5 \text{ N/m}^2)(600 \times 10^{-6} \text{ m}^3) = 1380$ J

But the shaded area, equal to the net output work, is one-third this. Therefore, the net output work per cycle is 460 J.

11.6 Efficiency of an Engine

As the world becomes more concerned about its energy resources, we seek ways of decreasing energy use. One of the most obvious ways is to improve the efficiency of the engines used in our cars and in other vehicles. But we are severely limited in our quest for engine efficiency. The first law of thermodynamics places a limit on the efficiency attainable. Let us see what this limit is.

We define the efficiency of an engine in the same way as for any machine:

$$\text{Efficiency} = \frac{\text{net output work/cycle}}{\text{input energy/cycle}}$$

The net output work is given by the area enclosed by the engine's P-V cycle. But we do not use that value for our computation. Instead, we compute the efficiency from a direct consideration of the first law of thermodynamics.

Consider any engine which burns fuel for its operation. (An electric motor is actually part of the larger "engine" which includes the power-plant generators.) The burning gasoline, for example, in the internal-combustion engine furnishes heat energy to the system. This constitutes the input energy to the engine. We call it ΔQ_{in} and assume it to be taken for one cycle of the engine.

Some of this input heat is always lost from the engine in a nonproductive way. For example, the hot gases of the car's exhaust carry unused energy away from the engine. In addition, frictional energy losses may also be important. We denote the total of these exhaust processes ΔQ_{ex}.

Now we invoke the first law, which is really the energy-conservation law. The net output work done by the engine is just the difference between the input and exhaust energies. (This is often represented in diagrammatic form, as shown in Fig. 11.9.) We then have

$$\text{Net output work/cycle} = \Delta Q_{\text{in}} - \Delta Q_{\text{ex}}$$

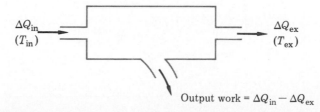

ΔQ_{in}
(T_{in})

ΔQ_{ex}
(T_{ex})

Output work $= \Delta Q_{\text{in}} - \Delta Q_{\text{ex}}$

FIGURE 11.9
The output work of the engine is always less than the Input by an amount equal to the exhaust heat.

But the input energy per cycle is simply ΔQ_{in}. We then find from the definition of efficiency that

$$\text{eff} = \frac{\Delta Q_{in} - \Delta Q_{ex}}{\Delta Q_{in}}$$

or
$$\text{eff} = 1 - \frac{\Delta Q_{ex}}{\Delta Q_{in}} \tag{11.4}$$

Notice that Eq. (11.4) tells us *the efficiency is always less than unity (or 100 percent) by the amount* $\Delta Q_{ex}/\Delta Q_{in}$. You might then think that all we have to do is design an engine that loses very little heat in exhaust processes. But this is not as simple as it sounds. The trouble lies in the fact that *work is done by thermal energy only when the heat flows from a hot object to a cooler one. Thermal energy does work only when a heat flow exists.*

As a guess, you might think that the work that a given quantity of heat can do would depend on the temperature difference through which it flows. Then the higher the temperature difference between the engine's intake and exhaust, the more efficient the engine should be. This turns out to be true in a general way. But the temperature difference of importance turns out to be the difference between the highest and lowest temperature portions of the engine's cycle.

For most real engines, the *P-V* cycle is quite complex. As a result, the engine efficiency depends on exactly how the engine is constructed. Even so, in 1824 Sadi Carnot (pronounced car-no) proved from theory that one particular engine (and cycle) has the highest possible efficiency. This cycle, *Carnot Cycle* called the **Carnot cycle,** is shown in Fig. 11.10.

Carnot showed that the ratio $\Delta Q_{ex}/\Delta Q_{in}$ in Eq. (11.4) is equal to the ratio of the respective temperatures, T_{ex}/T_{in}. *For the Carnot cycle,* then, we have

$$\text{eff} = 1 - \frac{T_{ex}}{T_{in}} \tag{11.5}$$

Maximum Efficiency Moreover, *Carnot proved that the efficiency given by Eq. (11.5) is the highest*

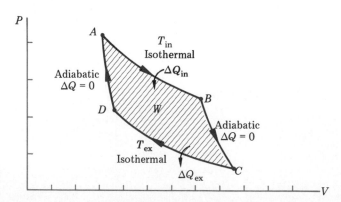

FIGURE 11.10
The Carnot cycle.

Engine	Approximate efficiency, %
Steam engine	15
Steam turbine	35
Gasoline engine	30
Diesel engine	40

possible efficiency for any engine operating between the two temperatures T_{ex} *and* T_{in}. Usually the exhuast temperature will be no lower than atmospheric temperature, about 300 K. Therefore, to achieve high efficiency, the engine fluid must have a very high temperature during the hottest part of the cycle. Ultimately, this temperature is limited by the material used to construct the engine. In practice, however, other limiting factors make this ultimate consideration unimportant. Typical efficiencies for modern engines are given in Table 11.1. Note that the values are maxima and only approximate.

Illustration 11.5 Estimate the maximum possible efficiency for a steam engine.

Reasoning In a steam engine, the hot gas used to drive the piston is steam. Since water boils at 373 K, the pressure of steam goes up rapidly at higher temperatures. As a result, the upper temperature is limited by the strength of the steam boiler. Let us suppose the entering steam has a temperature of $T_{in} = 453$ K (which is 180°C). The exhaust temperature cannot easily be less than ambient. Let us therefore take $T_{ex} = 300$ K (which is 27°C). Then

$$\text{eff} = 1 - \frac{T_{ex}}{T_{in}} = 1 - \frac{300}{453} = 0.34 \text{ or } 34\%$$

In practice, the efficiency might be only one-third this. Our computation assumed the Carnot cycle, and, of course, the actual cycle will be less efficient.

11.7 Heat Pumps; Refrigerators

Engines use heat energy to do work. Refrigerators and heat pumps do just the opposite; they use work to transfer heat from a cold region to a warmer one. The energy diagram of a typical compression-type refrigerator is given in Fig. 11.11. Notice that work is used to lift heat energy from a cold region (the inside of the refrigerator) to a warm region outside the refrigerator.

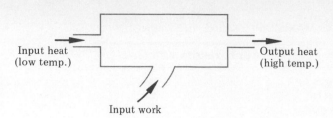

FIGURE 11.11
A heat pump, or
refrigerator.

Input heat
(low temp.)

Output heat
(high temp.)

Input work

Heat would flow in the reverse direction by itself. Work from outside is required to maintain a flow in the direction shown. The first law tells us that

$$\text{Work input} + \text{heat input} = \text{heat output}$$

Many refrigeration systems operate by means of a compressor acting on a fluid. Freon gas is frequently used as the fluid.[3] It can be liquefied at room temperature by application of moderate pressure. Let us refer to Fig. 11.12 to see the processes used.

The compressor at the top of the diagram is used to compress the Freon gas. Because of the work done on the Freon by the compressor, the gas is heated as it is compressed. This hot, highly compressed gas is then cooled by circulating it through coils over which a fan blows or air can circulate. (When a refrigerator or an air conditioner is running, you can feel hot air being blown from it in the uncooled region. This air is carrying heat away from the coils at the right of the diagram.) As the Freon cools somewhat at this high pressure, it condenses to a liquid. This liquid then enters a jet system, which acts as a throttling valve.

You will recall that, in a throttling process, a fluid is suddenly expanded into a region of low pressure. As a result, in this case, the Freon is greatly cooled as it emerges from the jet as a low-pressure gas. This cold gas is then circulated through coils in the cooling compartment of the refrigerator. Or, in an air-conditioning unit, air is blown into the room across these coils. The cold gas warms somewhat as it cools the region near the cooling coils. It then flows back to the compressor, where the whole process is repeated.

As we see, *work done on the compressor by outside agents* (such as an electric motor) *is used to transport heat from a cold region to a warm region. The heat exhausted is larger than the heat taken up. The heat equivalent of*

[3]Freon is a trade name for dichlorodifluoromethane and similar compounds.

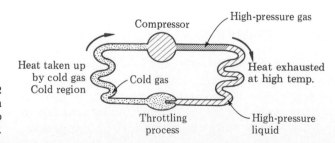

FIGURE 11.12
Some details of a
compressor-type heat pump
or refrigerator.

Compressor

High-pressure gas

Heat taken up
by cold gas
Cold region

Cold gas

Heat exhausted
at high temp.

Throttling
process

High-pressure
liquid

the work done by the compressor is added to the heat transferred from the cold to the warm region.

Refrigeration units are often rated in tons. This unit is a purely practical unit. A 1-ton refrigeration unit is able to change 1 ton of water at $0°C$ to ice at the same temperature in a running time of 1 day. Or, in more basic units,

$$1 \text{ ton of refrigeration} = 12,000 \text{ Btu/h} = 3513 \text{ J/s} = 840 \text{ cal/s}$$

As we see, a 1-ton refrigeration unit should be able to freeze about 10 g of water per second. In practice, the heat conductivity of water is so low that it would be difficult to achieve such a high rate of heat transport.

A heat pump can be used to heat a building. Since it pumps heat from a cold region to a warm region, it can pump heat from the cold outdoors to the interior of a home. In mild climates, many home refrigeration systems are of the reverse-action type. During winter, the refrigeration unit is effectively reversed so that the unit heats instead of cools. Hot air is blown into the house by extraction of heat from outside. But in very cold climates, heat pumps are usually augmented by oil or gas furnaces or by direct electrical heating for the coldest days of the year.

As we have seen, heat pumps transport heat in a direction opposite to the direction in which the heat would normally flow by itself. An outside energy source must do work to accomplish this. The natural tendency in the universe is for heat to flow from hot to cold. We have already seen, in terms of molecules, why this should be. The fact that a preferred direction exists for natural energy flow is basic to the world in which we exist. This effect is summarized by the second law of thermodynamics. In the following sections, we examine this law and its consequences.

11.8 Second Law of Thermodynamics

Someone once remarked about the universe that "left to themselves, things go from good to bad to worse." In a very crude sense, this summarizes the second law of thermodynamics. As we have seen, the first law is a statement of energy conservation, but it has nothing to say about the course of events in the universe. Energy is conserved when a stone falls and has its gravitational PE changed to KE. As the stone strikes the ground and comes to rest, its KE is changed to thermal energy. However, a stone resting on the ground never changes the thermal energy in and near it to KE and goes shooting up into the air. The first law does not rule out such a possibility since this reverse process also conserves energy. But the process does not occur.

There are many other processes in the universe that are not ruled out by the first law but that do not occur spontaneously. For example, heat flows from hot to cold, but not from cold to hot. Water evaporates from a saucer, but the vapor in the air does not by itself recondense into the saucer. A dead body decays and turns to dust, but the elements of the earth do not spontaneously form the body in the reverse process. *Nature has a preferred direction for the course of spontaneous events. The second law tells us what that direction is.*

During the history of thermodynamics, the *second law* has been stated in several fully equivalent ways. One of the earliest statements simply summarizes the fact that heat flows naturally from hot to cold:

Second Law of Thermodynamics

Heat flows spontaneously from a hotter to a colder object, but not vice versa. Because of this, it is impossible for a cyclical system to transfer heat from a lower-temperature body to a higher-temperature body indefinitely unless external work is done on the system.

This statement agrees with what we have found out about engines and refrigerators. To operate a refrigerator, an input energy source is required. If heat is to be transferred from a cold region to a warmer region, external input work is needed. And, in the case of an engine, a difference in temperature must exist. The efficiency of the engine is given by $1 - T_{ex}/T_{in}$ and becomes zero if the two temperatures are equal.

As we see, *the second law tells us it is impossible to make use of thermal energy unless the energy can flow to a region of lower temperature.* For example, the waters of the ocean have a huge amount of thermal energy. But we cannot use this energy unless a cooler place is found to which it can flow. As a result, for all usual purposes, the thermal energy resident in the oceans is of no use to us. In the next section, we explore the basic reason for this lack of usefulness.

11.9 Order versus Disorder

As any gambler knows, the odds are best for an event happening if the event can occur in many different ways. To illustrate this fact, let us consider a game in which five identical coins are tossed onto a table after being well shaken. Only six events can result from such a toss:

Event	Number of heads up	Number of tails up
1	0	5
2	1	4
3	2	3
4	3	2
5	4	1
6	5	0

At first guess you might think that each of these events is equally likely to occur. But that is not correct. The reason is that there is only one way in which event 1 or event 6 can occur. For event 1 to occur, all coins without exception must come up tails. For event 6, all coins must come up heads.

However, for event 2 to occur, there are actually five ways in which it could happen. If we call the five coins A, B, C, D, and E, these ways are as follows:

	Configuration				
Way	Coin A	B	C	D	E
1	II	T	T	T	T
2	T	H	T	T	T
3	T	T	H	T	T
4	T	T	T	H	T
5	T	T	T	T	H

Because there are 5 times as many ways that event 2 can happen, event 2 is 5 times as likely to occur as event 1. Event 5 can also happen in five different ways. As a result, events 2 and 5 are equally likely to occur. And both these events are 5 times as likely as events 1 or 6.

In the same way, we can show that events 3 and 4 are equally likely and each can occur in 10 different ways. Therefore, 3 and 4 are twice as likely to happen as events 2 and 5, while events 3 and 4 are 10 times as likely to happen as events 1 and 6. If you were a gambler, it is obvious which events you should lay your money on if no odds are given.

We can extend this to a situation in which more coins are involved. Suppose 100 coins rather than 5 are tossed. Then, as before, there is only one way in which all the coins can come up heads (or tails). But the number of ways in which other combinations can occur becomes almost unbelievably large. The results are shown in Fig. 11.13. Notice that the number of ways in which 50 heads can come up is about 10×10^{28}. As you can see, the odds against all heads or all tails coming up are so small as to be negligible.

Indeed, the total number of ways for all combinations of heads and tails is about 1×10^{30}. Therefore, the chance that all coins would come up heads

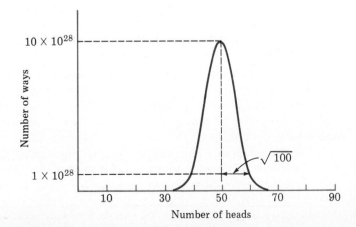

FIGURE 11.13
The number of ways in which the indicated number of heads come up when 100 coins are tossed.

is 1 in 10^{30}. If you throw the coins once each 10 s for 10^{22} yr, your chance of all heads coming up once is about 10 percent. For all practical purposes, there is no chance at all of all heads or all tails occurring. As we see from Fig. 11.13, the only really likely occurrence is for nearly equal numbers of heads and tails to occur.

If we consider 10^6 coins instead of 10^2, the situation becomes even more striking. We can summarize all such results in a very simple way. Notice in Fig. 11.13 that the graph line decreases to about one-tenth of its maximum value at the following two numbers of heads: 40 and 60. To give an estimate of the width of the peak, we could say it extends from $50 - 10$ to $50 + 10$. In other words, if you throw 100 coins, the numer of heads that should come up is about 50 ± 10. The more general result is as follows:

If N coins are tossed, the expected number of heads will be about[4]

$$\frac{N}{2} \pm \sqrt{N}$$

In the case of 10^6 coins, we should expect $500,000 \pm 1000$ heads to come up. Notice how very precise this estimate is. It says the expected number of heads lies between 501,000 and 499,000, a very narrow range indeed. As you can see, when the number of coins becomes very large, the percentage deviations one will find from the average are very small.

This example with the coins is typical of our universe in general. When things are left to happen by themselves, they occur by chance. As a result, the probability laws applicable to tossed coins apply to these other situations as well. For example, suppose you have a box containing gas molecules, as shown in Fig. 11.14. In the air, there are about 3×10^{19} molecules per cubic centimeter. Let us say the box has 10^{20} molecules in it. We now ask: What are the chances that the molecules will all bunch up in one half of the box?

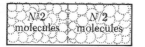

FIGURE 11.14
What is the likelihood that all the molecules will appear in one side of the box?

From our results with the coins, we can easily clarify this situation. To make the situations similar, call a molecule in the left side of the box a "heads" molecule. Molecules on the right will be "tails" molecules. Our general result from above tells us that the number of "heads" will be about $\frac{1}{2}N \pm \sqrt{N}$. In this case, the number of heads is about

$$50,000,000,000,000,000,000 \pm 10,000,000,000$$

Notice how small the expected deviation is. It is only $10^{10}/(5 \times 10^{19})$, or about 1 part in 5 billion. For all practical purposes, the number of molecules in the two halves of the box will be the same. And, of course, there is really no chance at all that all the molecules will spontaneously move into one side of the box.

These considerations have fundamental importance for all spontaneous processes. Reasoning from them, we can predict that thermal motion (or

---

[4] Precisely, if N is large, 95.3 percent of the time the number of heads will lie in this range.

other random-type disturbances) causes systems to change from order to disorder. As a crude example, consider the case of 100 coins again. Suppose we carefully arrange all with heads up. They then have a high degree of order. Now let us give them a type of motion similar to random thermal motion by shaking them. They quickly disorder and never return to their original state of order.

Similarly with the gas molecules in the box of Fig. 11.14, we can give the system order by placing all the molecules in one end of the box. But if we allow them to adjust with spontaneous thermal motion, they become disordered and fill the whole box. Never will they again, spontaneously return to their original ordered state.

Basic to this discussion are the concepts of order and disorder. We can give a simple method for comparing the disorder of two states. If a state can occur in only one way, it is a highly ordered state. In such a state, each molecule (or other particle) must be placed in a single exact way. In a disordered state, however, there are many possible ways for achieving the state. With these facts in mind, we can relate disorder and probability (or number of ways) of achieving a state. That state which has the highest disorder is the most probable state; it can occur in the largest number of ways. For example, the probability of N coins all coming up heads is very small. This is a state of very low disorder. As we have seen, systems left to themselves move toward states of high disorder.

There are many examples we could give that illustrate behavior of this type. We conclude from them that *in a system composed of many molecules,*

If a system is allowed to undergo spontaneous change, it will change in such a way that its disorder will increase or, at best, not decrease.

Second Law: Alternate Form

This law of nature, applicable to large numbers of molecules, is an alternate form of the second law of thermodynamics. In the next section we shall see yet another way in which the law can be stated.

11.10 Entropy

The implications of order and disorder in a system can be approached in two quite different ways. Both approaches use a quantity called *entropy.* The earliest use of entropy occurred in the mid-1800s by R. Clausius. Because the concept of atoms was still quite speculative at that time, Clausius followed conventional procedures and described the behavior of systems in terms of their overall properties, namely, P, V, T, and U. To describe the consequences of the fact that heat flows preferentially from hot to cold, he found it convenient to define a quantity that he called entropy in the following way.

Suppose an amount of heat ΔQ is added to a system whose temperature is maintained at a value T. Then the attendant entropy change of the system, denoted by ΔS, is defined to be

$$\Delta S = \frac{\Delta Q}{T} \qquad (11.6)$$

As we see, ΔS is positive (the system gains entropy) if heat flows into the system. Two cautions are necessary in the use of this relation. First, T is the absolute temperature. Second, the heat must be added to the system in a reversible way; that is, the accompanying change of the system must be reversible. By definition, *a reversible change is one that occurs in identical ways in both the forward and backward direction.* For example, in changing reversibly from state A to state B, the state variables for the system take on definite values at each step during the change. In the reverse process, going from B to A, the system's state variables repeat these steps in reverse order.

Clausius was able to show that *entropy is a state variable.* Under identical conditions, a system has the same entropy. In this respect, it is like P, V, T, and U. Because entropy is a state variable, it proves to be an important, useful quantity in the description of thermodynamic processes.

As the understanding of atoms and molecules was advanced by such men as Boltzmann and Maxwell, it became apparent that Clausius' entropy could be given a molecular meaning. About a quarter of a century after Clausius had introduced the concept, Boltzmann showed that entropy could be given an equivalent definition in terms of atoms and molecules. His definition related entropy to molecular disorder and is as follows.

Suppose a system can achieve a given state in Ω ways. (Ω is capital omega.) For a system of 100 coins, $\Omega = 1$ for the all-heads configuration while, for equal heads and tails, $\Omega \approx 10^{29}$. Then it turns out that *the entropy of the system is*

Entropy Related to Disorder

$$S = k \ln \Omega \qquad (11.7)$$

where ln stands for the natural logarithm and k is Boltzmann's constant.

Notice what Eq. (11.7) tells us. If a state of the system can occur in only one way, then $\Omega = 1$. But the logarithm of 1 is zero. So the entropy of such a highly unlikely state is zero. However, if a state can occur in many ways, Ω will be large. The entropy of a highly probable state is therefore large.

Let us now recall that a highly ordered state has a low number of ways in which it can occur while a disordered state can occur in many ways. Equation (11.7) then tells us that *entropy is a measure of disorder. The more disordered the state of a system, the larger its entropy will be.* As we see, *entropy is a state variable which measures disorder.* Because of this, we can restate the *second law of thermodynamics* in still another way:

Second Law in Terms of Entropy

If an isolated system undergoes change, it will change in such a way that its entropy will increase or at best remain constant.

The following illustration shows how the change in entropy in a process can be computed.

Illustration 11.6 By how much does its entropy change as a 20-g ice cube melts at 0°C? Assume heat can flow into the system of the ice cube, but otherwise it is an isolated system.

Reasoning From Eq. (11.6) we have

$$\Delta S = \frac{\Delta Q}{T}$$

Notice that T is the absolute temperature. In our particular case, the temperature $T = 273$ K. Since 80 cal of heat must be added for each gram of ice melted, we have

$$\Delta Q = (20 \text{ g})(80 \text{ cal/g})(4.184 \text{ J/cal}) \approx 6700 \text{ J}$$

Then $\qquad \Delta S = \dfrac{6700 \text{ J}}{273 \text{ K}} = 24.5 \text{ J/K}$

Notice that the entropy of the ice increases as it changes to water. The increase is a measure of the increase in disorder of the H_2O molecules.

11.11 Heat Death of the Universe

All spontaneous changes occur in such a way as to increase the disorder of the universe. This is simply a statement of the second law as applied to the universe as a whole. In Chap. 28 you will learn that, to the best knowledge we now have, the universe was once highly compressed. Although its exact compression is not known with certainty, it is likely that the portion of the universe we can see with our largest telescopes was compressed into a region with a diameter perhaps 10 times that of the sun. At the time of that original compressed state, the beginning of time, the universe was an incredibly hot cauldron of energy. During the billions of years that have since elapsed, the universe has expanded adiabatically at speeds approaching the ultimate speed, the speed of light.

The laws of thermodynamics apply to this process, which has taken perhaps 15 billion years. During that time, heat energy has continuously flowed from hot to cold regions. As it did so, the disorder and entropy of the universe have increased. The thermal energy—indeed, all energy—has become less useful on the average. The temperature of the original fireball has fallen continuously. There are still local hot spots, the positions of the sun and the stars. But most of the universe has long since cooled far below the temperatures to which we are accustomed. The gas in the vast reaches of space has an average temperature of about 3 K.

We on earth are fortunate. Our nearby sun still floods us with energy.

Heat flows to our earth by radiation from the hot sun. We use this radiant energy to grow plants. These plants are then used as sources of energy for us and the other creatures on earth. As we pointed out previously, except for nuclear fuels, the sun is the ultimate source of the energy we use. Notice that the sun's usefulness to us is the result of its very high temperature. As it cools, it will radiate less energy. Over a few billion years, the earth will slowly lose its major energy source.

During all this time, the entropy of the universe will be increasing. The sun and other hot objects will be losing entropy at the rate of $\Delta Q/T_h$, where ΔQ is the heat lost in unit time by the hot object at temperature T_h. Although the cooler portions of the universe will receive this energy, they will be at a lower temperature T_c. As a result, they will gain an entropy $\Delta Q/T_c$ which is larger than $\Delta Q/T_h$. *The entropy and disorder of the universe will continue to increase as hot objects cool and cold objects warm.*

We can picture in our minds a time when everything in the universe has reached the same temperature. Then no heat flow can occur. The disorder of the universe (and its entropy) will have reached a maximum value. Even though all the original energy the universe once had is still present, the energy will be useless. No plants will grow since there will be no hot object to light them. No engines can function since there will be no cooler place to which heat can be exhausted. No life will exist anywhere in the universe. *The universe will have undergone what is known as its* **heat death.**

Fortunately for us, this situation will not occur for billions of years. Indeed, we are not certain that it will ever happen. As you will learn in Chap. 28, there is a possibility that the universe will contract and, once again, become a flaming fireball. But that is another story, and we postpone discussion of it until the last chapter in this text.

Summary

Thermodynamics is the branch of science in which the behavior of matter is described in terms of its state variables, chiefly pressure, volume, temperature, internal energy, and entropy. A state variable is any quantity that depends only on the macroscopic condition of a physical system. Under identical macroscopic conditions, each of the state variables always has the same unique value. Heat and work are not state variables.

The first law of thermodynamics is a statement of the law of conservation of energy. It can be stated in equation form as follows: If heat ΔQ flows *into* a system while the *system does work* ΔW, these two quantities are related to the change in internal energy of the system by

$$\Delta U = \Delta Q - \Delta W$$

When a system subjected to an external pressure P increases its volume by ΔV, it does work equal to $P\,\Delta V$. Therefore $\Delta W = P\,\Delta V$. On a P-V diagram for a system, the work done by the system during a change is equal to the area under the graph line which represents the change. If the system undergoes a cyclic change, the net work done by the system during the cycle is the area enclosed by the cycle on the P-V diagram.

An isothermal process is one carried out without a change in temperature of the system. In an adiabatic change, no heat flow occurs into or out of the system.

For an adiabatic process involving a system composed of an ideal gas, the following equation applies:

$$P_1 V_1^\gamma = P_2 V_2^\gamma$$

where $\gamma = c_P/c_V$ for the gas.

In a throttling process, a fluid is allowed to expand rapidly and adiabatically through a jet or porous disk. Such a process often leads to cooling of the fluid during expansion.

The efficiency of an engine is defined by

$$\text{eff} = \frac{\text{output work}}{\text{input energy}}$$

For a heat engine, this becomes

$$\text{eff} = 1 - \frac{\Delta Q_{ex}}{\Delta Q_{in}}$$

where ΔQ_{ex} is the heat exhausted by the engine in one cycle while ΔQ_{in} is the heat furnished to the engine on each cycle. The Carnot engine has the highest possible efficiency. In it, the engine operates between a high intake temperature T_{in} and a low exhaust temperature T_{ex}. For it

$$\text{eff} = 1 - \frac{T_{ex}}{T_{in}}$$

No engine can be 100 percent efficient.

A refrigerator or heat pump transfers heat from a cold reservoir to a warm reservoir. To do this, energy must be supplied to the device. The first law requires that the heat exhausted to the warm reservoir equal the sum of the input work and the heat removed from the cold reservoir.

The second law of thermodynamics tells us the direction in which spontaneous change will occur in a system. It can be stated in three equivalent ways:

1 Heat flows spontaneously from a hotter to a cooler object, but not vice versa.
2 If a system is allowed to undergo spontaneous change, it will change in such a way that its disorder will increase or remain constant.
3 If an isolated system undergoes change, it will change in such a way that its entropy will increase or remain constant.

The entropy change ΔS of a system can be given in the following ways: if heat ΔQ flows into a system at temperature T, then

$$\Delta S = \frac{\Delta Q}{T}$$

and if a system can occur in the same state in Ω ways, then the entropy of that state is given by

$$S = k \ln \Omega$$

where k is Boltzmann's constant. The entropy measures the disorder of a system. In the universe as a whole, entropy is constantly increasing.

Minimum Learning Goals

Upon completion of this chapter, you should be able to do the following:

1 Explain what is meant by a state variable. Give examples of quantities which are state variables and those which are not.
2 State the first law of thermodynamics.
3 Compute the work done by a system during a given volume change against a known pressure. Use a *P-V* diagram to compute work done by a system during an expansion (or a contraction) for which the graph line is given.
4 Define isothermal change and adiabatic change. Apply the first law to each in such a way as to describe the behavior of ΔQ, ΔU, and ΔW.
5 Explain why a gas warms when compressed adiabatically.
6 Describe what is meant by a throttling process, and explain why a fluid is often cooled by such a process.
7 Compute the work done per cycle by an engine when its *P-V* cycle diagram is given.
8 Draw on a *P-V* diagram the cycle for a system

when the processes carried out during the cycle are told to you in words.

9   Write down the efficiency of a heat engine in terms of work or heat, or, for a Carnot engine, in terms of temperature.

10 Explain why the Carnot engine and its cycle are important.

11 Describe the basic ideas of operation for a compressor-type refrigeration unit or heat pump.

12 Give several examples of physical systems which, when left to themselves, become more disordered.

Explain why the reverse process is not observed in nature when large numbers of molecules are involved.

13 Use the relation $\Delta S = \Delta Q/T$ to compute the entropy change of a simple system under isothermal conditions.

14 State the second law of thermodynamics in three ways: (a) heat flow; (b) order and disorder; (c) entropy.

15 Explain what is meant by the heat death of the universe.

Important Terms and Phrases

You should be able to define or explain each of the following:

State of a system
State variable
First law of thermodynamics: $\Delta U = \Delta Q - \Delta W$
Expansion work: $\Delta W = P\,\Delta V$
P-V diagram
Isothermal process
Adiabatic process
$P_1V_1^{\gamma} = P_2V_2^{\gamma}$

Throttling process
Reversible process
Thermodynamic cycle; area in P-V diagram
eff $= 1 - \Delta Q_{\text{ex}}/\Delta Q_{\text{in}}$
Carnot engine; eff $= 1 - T_{\text{ex}}/T_{\text{in}}$
Heat pump; refrigerator
Second law of thermodynamics
Entropy; $\Delta S = \Delta Q/T;\ S = k \ln \Omega$
Heat death of universe

Questions and Guesstimates

1   After heat energy has flowed into a system, where may this energy be found in the internal energy of the system? Give concrete examples of as many of these situations as you can.

2   The relation $\Delta U = \Delta Q - \Delta W$ is not always equivalent to the relation $\Delta U = \Delta Q - P\,\Delta V$. Give an example where the latter relation does not apply even though the former does.

3   In each of the following processes, point out what is meant by each quantity in the equation $\Delta U = \Delta Q - \Delta W$: an ice cube slowly melts in ice water; ice heats from -30 to $-10°C$; steam in a closed boiler cools from 120 to 110°C; solid CO_2 (dry ice) sublimates in dry air; a bottle of soda freezes and cracks the bottle.

4   Some people term the following statements the zeroth law of thermodynamics: Two objects in thermal equilibrium with each other must be at the

same temperature; two objects, each in thermal equilibrium with a third object, are in thermal equilibrium with each other. In the case of the objects being ideal gases, justify these statements.

5   A container of ideal gas is to be compressed to half its original volume. Under which condition would the work done on the gas be larger, isothermal or adiabatic?

6   Consider the simple heat engine shown in Fig. P11.1. The heated liquid on the right expands and is lifted by the cooler liquid on the left. As a result, the liquid circulates counterclockwise in the tube. As it does so, it rotates the paddle wheel, which is then coupled to external devices to do output work. Explain what factors affect the efficiency of this engine. What should be done to make it have its highest efficiency?

7   Body temperature is 98.6°F. Even so, heat is car-

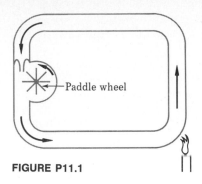

FIGURE P11.1

ried away from the body to the environment even when the surrounding temperature is higher than this. Doesn't this contradict the fact that heat flows only from hot to cold? Similarly, a watermelon can be cooled by wrapping it in a wet cloth even on a very hot day. How can you reconcile this with the second law?

8 Can a system absorb heat energy without a change in its internal energy? Can its internal energy be increased without adding heat to it?

9 A baby produces a highly ordered structure as it grows. Each molecule is a carefully structured entity, and the molecules are assembled in a highly ordered way. Don't this growth and its accompanying high degree of order contradict the second law? Ultimately, what is the energy source for the child? Repeat this question for the case of a growing plant.

10 Each of a pair of dice has six sides labeled 1 to 6. When the pair is tossed, what is the ratio of the chance that the up sides will total 2 to the chance that they will total 3? What if the total is to be 4 instead of 3?

11 A child wishing to cool off the kitchen of a home opens the refrigerator door and leaves it open. Will this work? Answer the question from both short- and long-term considerations. Would the situation be any different if an old-fashioned ice box were used rather than a refrigerator?

12 Two cylinders sit side by side and are enclosed by a movable piston. They are identical in all respects except that one contains oxygen (O_2) and the other contains helium (He). Both are now compressed adiabatically to one-fifth their original volume. Which gas will show the largest temperature rise?

Problems

1 What is the change in internal energy of 200 g of nitrogen as it is heated from 10 to 30°C at constant volume? For nitrogen, $c_V = 0.177$ cal/(g)(°C).

2 (a) By how much does its internal energy change as 15 g of ice at 0°C melts to water at 0°C? Ignore the fact that the ice is less dense than water. (b) If you were to include the contraction effect, would your answer be larger or smaller?

3 An ideal gas in a cylinder is slowly compressed to one-third its original volume. During this process, the temperature of the gas remains constant, and the work done in compression is 45 J. (a) By how much did the internal energy of the gas change in the process? (b) How much heat flowed into the gas?

4 An ideal gas in a cylinder is compressed adiabatically to one-third its original volume. During the process, 45 J of work is done on the gas by the compressing agent. (a) By how much did the internal energy of the gas change in the process? (b) How much heat flowed into the gas?

5* Five grams of helium gas is heated from −30 to 20°C. Find its change in internal energy and the work done by the gas if the heating occurs (a) at constant volume and (b) at constant pressure. For helium, $c_V = 0.75$ cal/(g)(°C) and $c_P = 1.25$ cal/(g)(°C).

6* A 10-g mass of ideal gas is compressed adiabatically in such a way that the gas heats 30°C when 200 J of work is done on it by the compressing force. (a) How much did the internal energy of the gas change during the compression? (b) How much heat will flow from the compressed gas as it cools to its original temperature? (c) What is c_V for the gas?

7 A gas is carried around the cycle shown in Fig. P11.2. (a) Find the work done by the gas in parts AB, BC, CD, and DA of the cycle. (b) What is the total work done by the gas during the cycle?

8 Figure P11.3 shows a thermodynamic cycle for an ideal gas. (a) What is the total work done by the gas during the cycle? (b) How much heat (net) was added to the gas during the completion of the cycle?

9* The ideal gas whose thermodynamic cycle is shown in Fig. P11.3 is at 20°C at point A. Find its temperature at (a) point B, (b) point C.

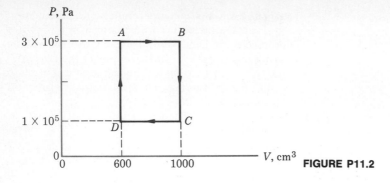

P, Pa

3 × 10⁵ region with points A and B at top

1 × 10⁵ region with points D and C at bottom

V, cm^3 **FIGURE P11.2**

600   1000

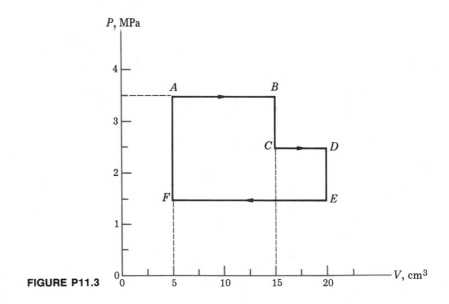

P, MPa

FIGURE P11.3

V, cm^3

10* Figure P11.2 shows the thermodynamic cycle for an ideal gas. The temperature of the gas at point D is 20°C. (a) Find the temperature at points A and B. (b) In carrying through the cycle, in which portions is heat added to the gas?

11* Suppose the thermodynamic cycle shown in Fig. P11.3 is run backward, i.e., from A to F, E, D, C, B, and back to A. (a) How much work is done by the gas in portion FE of the cycle? In portion DC? (b) What net amount of heat flows from the gas as it completes one cycle?

12* The thermodynamic cycle of Fig. P11.2 is run backward (i.e., from A to D, C, B, and back to A). (a) How much work does the gas do in sections DC and BA of the cycle? (b) How much total heat is given off by the gas during the cycle?

13** Assume the gas subjected to the thermodynamic cycle shown in Fig. P11.2 is nitrogen. (a) How

many kilomoles of gas is confined in the system if the temperature at point D is 20°C? (b) What was the temperature at point A? At point B? (c) How much heat was added to the gas in portion AB of the cycle? In portion DA? (d) What is the efficiency of this cycle?

14** Assume the gas subjected to the thermodynamic cycle in Fig. 11.2 is 1 mmol (millimole) of helium. (a) What is the temperature of the gas at point A? At point B? At point D? (b) How much heat was added to the gas in portion DA of the cycle? In portion AB? (c) What is the efficiency of this cycle?

15* For air, $c_V = 0.177$ cal/(g)(°C). Suppose air is confined to a cylinder by a movable piston under a constant pressure of 3.0 atm. How much heat must be added to the gas if its temperature is to be changed from 27 to 300°C? The mass of air in the cylinder is 20 g, and its original volume is

5860 cm³. *Hint:* Notice that $mc_V \Delta T$ is the internal energy one must add to the gas to change its temperature by ΔT.

16** A 16,000-cm³ cylinder is closed at one end by a piston and contains 20 g of air at 30°C. The piston is suddenly pushed in so as to change the gas volume to 1600 cm³. The compression is adiabatic, and the final temperature of the gas is 500°C. How much work was done in compressing the gas? For air, $c_V = 0.177$ cal/(g)(°C).

17** Thirty grams of highly compressed air is confined to a cylinder by a piston. Its volume is 2400 cm³, its pressure is 10×10^5 Pa, and its temperature is 35°C. The gas is now expanded adiabatically until its volume is 24,000 cm³. During the process, 4100 J of work is done by the gas. What is the final temperature of the gas? Assume $c_V = 0.177$ cal/(g)(°C).

18* Helium gas at 20°C and 1-atm pressure is adiabatically compressed to one-fourth its original volume. What is (*a*) its final pressure, (*b*) its final temperature?

19* Nitrogen gas is to be adiabatically compressed in such a way that its temperature is to rise from 20 to 500°C. To what fraction of its original volume must the gas be compressed?

20* Nitrogen gas is to be expanded adiabatically from its original pressure of 3.0×10^6 Pa and a temperature of 27.0°C to such a volume that its temperature is -15°C. By what factor must its volume be expanded?

21 In modern, high-pressure steam-turbine engines, the steam is heated to about 600°C and exhausted at close to 90°C. What is the highest possible efficiency of any engine which operates between these two temperatures?

22 Temperature differences between the surface water and bottom water in a large lake might be 5°C. Assuming the surface water to be at 20°C, what is the highest efficiency a heat engine could have if it operates between these two temperatures?

23* When gasoline is burned, it gives off 11,000 cal/g, called its heat of combustion. A certain car uses 9.5 kg of gasoline per hour and has an efficiency of 25 percent. What horsepower does the car develop?

24* A 1500-kg car is to accelerate from rest to 8.0 m/s in 7.0 s. (*a*) What is the minimum horsepower that the engine must deliver to the car if all friction losses are ignored? (*b*) By using this value and assuming the car uses its fuel with an efficiency of 20 percent, how much gasoline will the car burn in the

7.0 s? Gasoline has a heat of combustion of about 11,000 cal/g; that is, it furnishes this much heat energy for each gram burned.

25* A moderate-sized nuclear power plant might have an output of about 7.0×10^8 W. (*a*) Assuming the plant to have an overall efficiency of 30 percent, how much energy is consumed by the plant each second? (*b*) How much heat does it exhaust to the cooling system each second? (*c*) Repeat parts *a* and *b* for a fossil-fueled plant having the same output and efficiency.

26* A certain heat engine has an efficiency of 25 percent and an output of 400 hp. (*a*) How much input energy does it require per second? (*b*) How much heat does it give off each second?

27* A certain refrigerator requires $\frac{3}{4}$ hp for its operation. It is capable of removing 600 cal/s from the refrigerator. How much heat does it provide each second to the room in which it sits?

28* An air conditioner that requires 0.50 hp for its operation exhausts 500 cal of heat to the outdoors each second. How many calories does it take each second from the room that it is cooling?

29 Refrigerators and heat pumps are often rated by a *coefficient of performance* (COP), defined by

$$\text{COP} = \frac{\text{heat } Q_1 \text{ removed from cold chamber}}{\text{input work}}$$

Show that this may be rewritten as

$$\text{COP} = \frac{Q_1}{Q_2 - Q_1}$$

where Q_2 is the heat exhausted to the surroundings.

30 Refer to Prob. 29. Suppose a refrigerator has a COP of 5. (*a*) How much energy must be supplied to operate the refrigerator if it is to remove 2000 cal from its cooling chamber? (*b*) What horsepower rating must the refrigerator have to accomplish this in 60 s?

31 You are given three pennies which can be tossed to come up heads or tails. If each penny is labeled so we can keep track of it, in how many different ways can the pennies arrange themselves (heads or tails) when they are tossed? What is the chance that all three will come up heads?

32 A pair of dice is thrown. (*a*) Is the sum of the two up sides of the dice more likely to be 3 than it is 5? (*b*) In how many ways can the sum be 3? (*c*) In how many ways can the sum be 5? (*d*) What is the most likely sum for the dice?

33* When N labeled coins are tossed, 2^N different combinations of heads and tails are possible. How many combinations are possible for (a) three coins, (b) five coins, (c) 50 coins? *Hint:* If you don't have a calculator or don't know how to use logarithms, try a method based on the idea that $a^{5q} = a^q \cdot a^q \cdot a^q \cdot a^q \cdot a^q$.

34* (a) Using the explanation given in Prob. 33, what is the chance that 10 noninteracting ants will all end up in the same particular half of a box? (b) Repeat for the case where all but one end up in the same particular half.

35 What is the change in entropy of 30 g of water at 0°C as it is changed to ice at 0°C?

36 What is the change in entropy of 2.0 g of water at 100°C as it is changed to steam at 100°C?

37 About what is the change in entropy of 15 g of water as it is cooled from 20 to 18°C? (Assume T to be nearly constant.)

38* Heat is slowly added to 2.0 m³ of nitrogen at 20°C and a constant pressure of 300 kPa. During the heating, the volume increases by 500 cm³. By about how much did the entropy of the gas change during the process? The mass of nitrogen involved is 6.9 kg (how can we calculate this?), and for it $c_P = 0.25$ cal/(g)(°C).

39* Five coins can each come up heads or tails. (a) What is the entropy of the configuration where all coins are heads? (b) For the configuration where all but one are heads?

40** Show that the bulk modulus E for an adiabatic change in an ideal gas is γP, where $\gamma = c_P/c_V$ and P is the pressure of the gas. *Hint:* Refer to Illustration 9.11. You may want to make use of the mathematical approximation $(1 \pm x)^a = 1 \pm ax$ for $x \ll 1$.

12
Vibratory Motion

In this chapter we return to the mechanics of motion in order to prepare for a study of waves, e.g., waves on a violin string, sound waves in air, and electromagnetic waves traveling through vacuum. As you continue your study of physics, you will find that the motion of waves is at least as important as the motion of material objects. It is therefore important that we understand their behavior. We begin this study by discussing the motion of the material objects that generate mechanical waves. Chapter 13 is devoted to a study of certain types of waves generated by the vibratory motions discussed here.

12.1 Vibrating Systems

Any elastic body or system can be made to vibrate. A few common examples are given in Fig. 12.1. The first three shown are rather simple cases. If the system is displaced to the indicated position and released, the elastic restoring force will cause the system to vibrate.

The fourth example in Fig. 12.1, the drumhead, undergoes a much more complex form of vibration if it is elastically distorted by a blow. Although the fifth system, a pendulum, has no elastic restoring force, we know that it, too, will vibrate. All these systems, together with *all* other vibrating objects, have one thing in common. *If any one of these systems is displaced from its equilibrium position, a restoring force is set up which acts to return the system to its equilibrium position.*

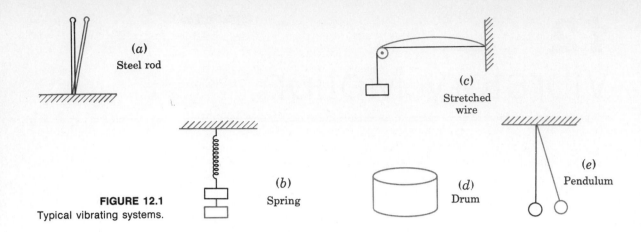

(a) Steel rod

(b) Spring

(c) Stretched wire

(d) Drum

(e) Pendulum

FIGURE 12.1
Typical vibrating systems.

12.2 Vibrating Spring

One of the simplest vibrating systems is a spring with a mass attached at one end. To understand the motion of the mass-spring system, we must recognize that energy can be stored in a spring. A compressed spring can be used to accelerate an object; for example, the compressed spring of a popgun can accelerate a projectile and send it shooting from the gun. Similarly, a stretched spring is also capable of doing work; for example, a stretched rubber band, a springlike device, can shoot a paper wad. As we see, energy can be stored in distorted, springlike systems. We compute the magnitude of this potential energy in a later section.

Let us now examine the motion of the mass-spring system shown in Fig. 12.2. We shall see that the motion of this device is typical of all simple vibrating motions. In order to remove all complicating features, we assume the mass in Fig. 12.2 slides with negligible friction on the surface.

The system is shown in its equilibrium position in part *a*. There is no

FIGURE 12.2
(a) The mass *m* is in equilibrium. (b) When the spring is compressed, PE is stored in the spring. Upon release of the system, this PE is changed to KE and finally back to PE when the mass reaches the position shown in (c). Notice that the spring furnishes a restoring force *kx*.

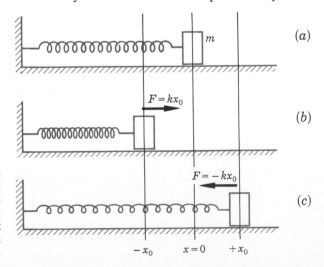

(a)

$F = kx_0$

(b)

$F = -kx_0$

(c)

$-x_0$ $x = 0$ $+x_0$

horizontal force acting on the mass in this position. (The pull of gravity down is balanced by the push of the table up, and so the net vertical force on the mass is always zero.)

Suppose that the spring is now compressed by moving the mass to the position shown in part *b* of the figure. During the process of compressing the spring, we did work on it, and so we have stored PE in the spring. The compressed spring exerts a force on the mass, tending to drive it back to the $x = 0$ position. If the mass is now allowed to move freely, the spring will keep accelerating it to the right until the position $x = 0$ is reached. The mass will now be moving to the right quite swiftly, and the spring will have lost all the PE stored in it when it was compressed. It is clear that the PE stored in the spring has been given to the mass and now appears as the KE of the moving mass.

Of course, the mass will not stop at $x = 0$, since its KE must be lost doing work before the mass can come to rest. As the mass proceeds to the right of $x = 0$, it begins to stretch the spring and to store energy in it. By the time the mass reaches the position shown in part *c* of Fig. 12.2, it has lost all its KE doing work against the spring. The KE has been changed completely to PE in the stretched spring.

The spring will now accelerate the mass to the left. By the time the mass reaches the $x = 0$ position, all the energy will be in the form of KE once again. The mass will again compress the spring to the position $x = -x_0$, at which point all the KE will be changed to PE stored in the compressed spring. Now the process will repeat itself, and the mass will vibrate back and forth between $x = +x_0$ and $x = -x_0$ forever if there are no friction losses. *This is typical of vibrating systems. Notice that as the mass oscillates back and forth, the energy oscillates back and forth between KE and PE, but the total energy must remain constant.* Why?

Several terms are used to describe vibratory motion:

1 **Amplitude of vibration** is the distance x_0 in Fig. 12.2. It is the maximum displacement of the mass from its equilibrium position.
2 **Period of vibration** τ is the time taken to make one complete oscillation. It is the total time taken for the mass starting at $-x_0$ to move to $+x_0$ and return to $-x_0$ once again. (τ is Greek tau.)
3 **Frequency of vibration** f is the number of complete oscillations the mass makes in unit time. For example, if the period of vibration is 0.10 s, the frequency is 10 cycles per second (cps), which is $1/\tau$. The cps unit is frequently referred to as **hertz** (Hz), named after Heinrich Hertz, who *Hertz Unit* first demonstrated radio-type waves. Both terms are used interchangeably in this book. (Often ν, Greek nu, is used in place of f.) By noticing that when $f = 10$ Hz, $\tau = 0.10$ s and when $f = 100$ Hz, $\tau = 0.01$ s, you can readily infer the following reciprocal relation between frequency and period:

$$\tau = \frac{1}{f} \tag{12.1}$$

4   The **displacement** x is the directed distance from the equilibrium position to the mass at any time during the vibration.

12.3 Simple Harmonic Motion

To discuss the vibrating spring in a quantitative manner, we must decide how the force exerted on the mass varies. *If a spring is not stretched or compressed too far, it obeys Hooke's law,* which we first encountered in Chap. 8. You will recall that in a Hooke's law system, a distortion of the system produces a restoring force proportional to the distortion. For the case shown in Fig. 12.3, this means that the force F with which the distorted spring pulls on the mass is given by

$$F = -kx \qquad (12.2)$$

The negative sign shows the force to be in such a direction as to decrease the distortion. When x is positive, the force on the mass is in the negative direction and is tending to bring the mass back to its equilibrium position. In which direction is the force on m when x is negative?

All systems that obey Hooke's law and in which friction is small undergo similar motion, called **simple harmonic motion.** We use this fact later when the motion found for the spring-mass system is generalized to other systems. In all these cases, the constant k in Eq. (12.2) is called the **spring constant.** Since it is numerically equal to F/x, it measures the force needed to distort the system a unit distance. For a stiff spring, k is large; it is small for an easily deformed system.

To compute the energy stored in an elastic system that obeys Hooke's law, let us refer to Fig. 12.4. We see there a graph of the stretching force on a spring versus the distance x that the spring is stretched by the force. Because the spring, or other elastic system, is assumed to follow Hooke's law, the graph line is a straight line whose equation is[1]

$$\text{Stretching force} = kx$$

---

[1]In Eq. (12.2) there was a negative sign. There is none here because we are discussing the stretching force, not the oppositely directed restoring force.

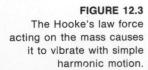

FIGURE 12.3
The Hooke's law force acting on the mass causes it to vibrate with simple harmonic motion.

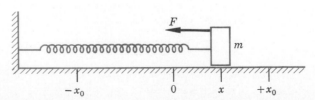

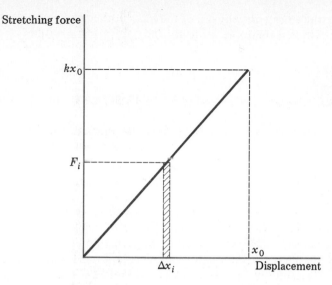

FIGURE 12.4
The work done in stretching an elastic system is equal to the area under its graph line.

Notice that at the maximum displacement x_0, the stretching force is kx_0, as shown on the graph.

We can show quite easily that the work done in stretching the spring from $x = 0$ to $x = x_0$ is simply the area under the graph line. To do that, we examine the shaded rectangle under the graph line. Its area is $F_i \Delta x_i$, where F_i is the stretching force that prevails during the small increase in distortion Δx_i. But because $W = F_s \Delta s$, this is also the work done by the stretching force during this small increase in displacement. Thus we can conclude that the work done in stretching a spring is equal to the sum of many rectangular areas such as these. Hence

> The work done in stretching (or compressing) an elastic element is equal to the area under its F-versus-x graph line.

You should be able to extend this discussion to confirm the "compression" portion of this statement.

Since the area of a right triangle is just half its base times its height, we see from Fig. 12.4 that the area under the graph line is $(\frac{1}{2}x_0)(kx_0)$. However, this equals the work done in stretching the spring, and so it is equal to the PE stored in the spring. We therefore conclude that the PE stored in a spring with constant k that has been stretched (or compressed) a distance x_0 is given by

$$\text{Spring PE} = \tfrac{1}{2}kx_0^2 \qquad (12.3)$$

Now that we know how much elastic energy is stored in a spring (or other Hooke's law system), we can use the law of energy conservation to learn much about the vibration of the system shown in Fig. 12.3. Because we are assuming friction losses to be negligible, the sum of the PE stored in the spring and the KE of the mass at the spring's end must remain constant. We can therefore write for the spring-mass system

$$PE + KE = \tfrac{1}{2}kx_0^2$$

where x_0 is the maximum distortion of the spring. Substituting gives

$$\tfrac{1}{2}kx^2 + \tfrac{1}{2}mv^2 = \tfrac{1}{2}kx_0^2 \tag{12.4}$$

where m and v pertain to the mass at the end of the essentially massless spring. Simplifying, we have

$$mv^2 = k(x_0^2 - x^2)$$

Hence, if the amplitude of vibration x_0 and the spring constant are known, the velocity of the mass can be computed for any position or displacement by using Eq. (12.4), a statement of the law of conservation of energy applied to this situation.

Of course, the mass is continuously changing its velocity. It travels fastest at $x = 0$, when all the energy is in the form of KE. Let us now observe the character of the acceleration of the mass. This is easily done. At any displacement x, the unbalanced force acting on the mass is a result of the tension in the spring. According to Eq. (12.2), that force is just $-kx$. Using Newton's law, $F = ma$, we have at once for the mass

$$-kx = ma$$

or the acceleration of the mass is just

$$a = -\frac{k}{m}x \tag{12.5}$$

Both Eqs. (12.4) and (12.5) are typical forms for bodies undergoing simple harmonic motion.

Illustration 12.1 A particular spring stretches 20 cm when a 500-g mass is hung from it. Suppose a 2.0-kg mass is attached to the spring, and it is displaced 40 cm from its equilibrium position and released. Find (a) the maximum speed of the mass, (b) the maximum acceleration of the mass, and (c) the speed and acceleration when $x = 10$ cm.

Reasoning Let us first find k, the spring constant. Since 0.50 kg weighs 4.90 N, we have $k = F/x = (4.90 \text{ N})/(0.20 \text{ m}) = 24.5 \text{ N/m}$.

(a) The mass will be traveling fastest when it is at its center position, $x = 0$. Let us find the maximum PE. It is

$$(PE)_{\text{max}} = \tfrac{1}{2}kx_0^2 = \tfrac{1}{2}(24.5 \text{ N/m})(0.16 \text{ m}^2) = 1.96 \text{ J}$$

But the maximum KE is equal to the maximum PE, and so we have

$$\tfrac{1}{2}mv^2_{\text{max}} = 1.96 \text{ J}$$

This is the KE when the mass is at its center position. Using $m = 2.0$ kg, we find

$$v_{\text{max}} = 1.40 \text{ m/s}$$

(b) The maximum acceleration occurs when the force kx is greatest, i.e., when $x = x_0$. In that case

$$a_{\text{max}} = \frac{F_{\text{max}}}{m} = \frac{kx_0}{m} = \frac{(24.5 \text{ N/m})(0.40 \text{ m})}{2.0 \text{ kg}} = 4.9 \text{ m/s}^2$$

(c) At $x = 0.10$ m, we have

$$a = \frac{F}{m} = \frac{(24.5 \text{ N/m})(0.10 \text{ m})}{2.0 \text{ kg}} = 1.22 \text{ m/s}^2$$

Also, since

$$\tfrac{1}{2}kx_0^2 = \tfrac{1}{2}mv^2 + \tfrac{1}{2}kx^2$$

we find

$$1.96 \text{ J} = \tfrac{1}{2}(2.0 \text{ kg})(v^2) + \tfrac{1}{2}(24.5 \text{ N/m})(0.10 \text{ m})^2$$

from which

$$v = 1.36 \text{ m/s}$$

12.4 Frequency of Vibration

We have already computed nearly everything concerning harmonic motion. In the last section we showed how the speed and acceleration of the vibrating object can be found. It remains to find the frequency of vibration.

There are several ways of finding the oscillation frequency f. One of the simplest makes use of the similarity between a ball moving along a circular path and a mass vibrating at the end of a spring. The situation is shown in Fig. 12.5. As the ball moves around the circle with constant speed v_0, its shadow on the plane below moves back and forth. The motion of the shadow turns out to be simple harmonic. Indeed, if v_0 is properly chosen, the shadow will move in unison with the vibrating mass at the end of the spring shown there. Let us now prove these statements and then use our results to find the frequency of simple harmonic motion.

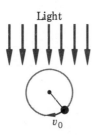

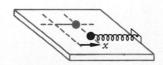

FIGURE 12.5
As the ball rotates around the circle with speed v_0, its shadow moves back and forth in coincidence with the mass at the end of the spring.

The mass at the end of a spring does not move with constant acceleration. Equation (12.5) says that $a \sim x$, and so the acceleration is changing
continuously. For this reason, simple harmonic motion is not too simple in
spite of its name. But by use of the analogy between circular and simple
harmonic motion just mentioned, we can easily arrive at useful results. To
do so, let us refer to Fig. 12.6.

We have learned that a ball moving on a circular path with constant
speed v_0 has a centripetal acceleration given by v_0^2/r. This centripetal acceleration is shown in part b of the figure. But there, since we choose the radius
of the circle equal to x_0, the acceleration is v_0^2/x_0. As the ball moves along
the circle, its shadow (shown in part a) will oscillate back and forth along the
x axis. We claim that the oscillation of the shadow is actually a simple
harmonic motion. To prove this, consider the x-directed acceleration of the
shadow.

The x coordinate of the shadow in part a is the same as the coordinate
labeled x in part b. But in part b, the x acceleration of the ball (and the
shadow) is simply the x component of the centripetal acceleration. By examination of part b, you should convince yourself that the x component of
the acceleration is

$$\text{Acceleration of shadow} = -\frac{v_0^2}{x_0}\cos\theta$$

The minus sign tells us the acceleration is in the $-x$ direction for the situation shown. In addition, we see from part b that $\cos\theta = x/x_0$, and so

$$\text{Acceleration of shadow} = -\frac{v_0^2}{x_0^2}x$$

Notice that this acceleration is very similar to Eq. (12.5) for the acceleration in simple harmonic motion. There we had

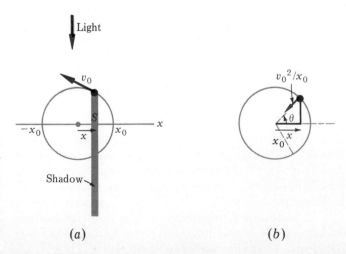

FIGURE 12.6
The ball moves around the
circle with constant speed
v_0. The x displacement of
its shadow moves with
simple harmonic motion
of amplitude x_0. For it,
$v_0 = x_0\sqrt{k/m}$.

(a)

(b)

$$\text{Acceleration in simple harmonic motion} = -\frac{k}{m}x$$

We can therefore conclude that the motion of the shadow is the same as the motion of a mass at the end of a spring. To make the motions identical, the radius of the *reference circle* must be the same as the amplitude of motion. And the speed of the ball in the circle must be chosen so that

Reference Circle

$$\frac{v_0^2}{x_0^2} = \frac{k}{m} \quad \text{or} \quad v_0 = x_0\sqrt{\frac{k}{m}} \tag{12.6}$$

You might notice in passing that Eq. (12.6) is a special case of Eq. (12.4). Let us now use this similarity between circular and simple harmonic motion to obtain an expression for the frequency of the motion.

If we wish to compute the time taken for the vibrating mass on the spring to make one complete vibration, it will be equivalent to computing the time that the mass on the circle takes to go around the circle once. The time for one complete trip around the circle is the period τ. If the radius of the circle is r, we find

$$\tau = \frac{2\pi r}{v_0} = 2\pi\frac{x_0}{v_0}$$

However, since we know from Eq. (12.6) that

$$v_0 = x_0\sqrt{\frac{k}{m}}$$

the expression for the period becomes

$$\tau = 2\pi\sqrt{\frac{m}{k}} \tag{12.7}$$

Period of Vibration

The frequency of vibration f is just $1/\tau$. Since $F = ma = -kx$ was used in deriving this relation, the units of k are newtons per meter or pounds per foot.

Equation (12.7) for the period of vibration of a mass m in simple harmonic motion is a general one. Any system having a mass m vibrating under a Hooke's law force with force constant k will have a period of vibration given by Eq. (12.7). We shall see that it also applies to systems far different from simple springs.

Illustration 12.2 Find the period and frequency of vibration for the system of Illustration 12.1.

Reasoning We knew in Illustration 12.1 that

$$k = 24.5 \text{ N/m} \quad \text{and} \quad m = 2.0 \text{ kg}$$

Applying Eq. (12.7) gives

$$\tau = 2\pi\sqrt{\frac{2.0 \text{ kg}}{24.5 \text{ N/m}}} \approx 1.8 \text{ s}$$

Also, $\qquad\qquad\qquad f = 1/\tau = 0.56 \text{ Hz}$

12.5 Sinusoidal Vibration: Simple Harmonic Motion

We shall now see that a simple mathematical equation can be written for a mass vibrating with simple harmonic motion. As we have seen, the vibrating shadow in Fig. 12.6 undergoes simple harmonic motion. The value of x, the displacement, is easily found from part b of the figure. We see that

$$x = x_0 \cos\theta$$

But θ is constantly changing as the ball moves around the reference circle. Since our motion equations state that

$$\theta = \omega t$$

we have at once that

$$x = x_0 \cos\omega t \qquad\qquad (12.8)$$

where the angular speed ω is measured in radians per second. If we wish to express our answer in terms of f, the frequency in revolutions per second (or, in this case, vibrations per second), we have

$$\omega = 2\pi f \qquad\qquad (12.9)$$

since 1 rev is equivalent to 2π rad.

Equation (12.8) states that the mass vibrating with simple harmonic motion will do so as the cosine of ωt. To see exactly what this type of vibration entails, consider the experiment illustrated in Fig. 12.7.

A weight is suspended from a spring as shown. If the object is raised to y_0 and released, it will undergo simple harmonic motion with amplitude y_0. Behind the vibrating object we place a sheet of paper on which the object marks its position as it vibrates up and down. The paper will be pulled out to the left at constant speed, and a trace of the motion of the object will be plotted on it. This is shown in the lower part of the figure.

Let us count time from the instant the object was released. This is the end of the trace on the left. We take this point as $t = 0$. The position of the

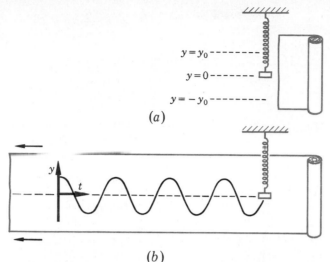

(a)

(b)

FIGURE 12.7
The vibrating mass traces a cosine curve as a function of time.

object at the instant shown in the figure occurs at some later time. Hence, this can be considered a plot of the displacement of the object, y, as a function of the time. According to Eqs. (12.8) and (12.9), the equation of this trace is

$$y = y_0 \cos \omega t = y_0 \cos 2\pi f t = y_0 \cos \frac{2\pi t}{\tau} \qquad (12.10)$$

This general type of trace is called a **sinusoidal curve,** and the motion that causes it is called **sinusoidal motion.** Clearly, *simple harmonic motion and sinusoidal motion are the same.*

12.6 Simple Pendulum

We know that a simple pendulum like that shown in Fig. 12.8 vibrates with apparent simple harmonic motion. If the restoring force on the body is proportional to the displacement, Hooke's law applies and the vibration will be simple harmonic motion, as we saw previously. Let us now examine the restoring force for the pendulum.

The pendulum ball will move along the arc of the circle shown. Since the tension in the pendulum cord always pulls perpendicularly to this arc, the tension T neither speeds nor slows the motion. It is the weight mg of the pendulum ball that causes the motion. However, only F, the component of the weight tangential to the circular arc is effective. We have from Fig. 12.8

$$F = -mg \cos \theta$$

where the minus sign is used to show that F is a restoring force. But we can also see from the figure that $\cos \theta = x/L$, and so we have

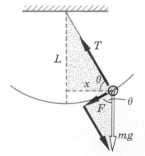

FIGURE 12.8
The two θ's are equal. Notice also that if the swing is not too large, the component of F along x (which is the x-direction restoring force) is nearly equal to F itself.

$$F = -\frac{mg}{L}x \qquad (12.11)$$

This is similar in form to Hooke's law,

$$F = -kx$$

except in one detail. In Hooke's law, the displacement x should be in the direction opposite to the force. We see from Fig. 12.8 that this is not true in this case; F and x are not exactly in line. However, if the angle of swing of the pendulum is very small, F and x will be almost exactly in line. There-fore, *the pendulum* does *approximate Hooke's law if it is not swinging too widely. To that approximation, the pendulum will undergo simple harmonic motion.* From the comparison of Eq. (12.11) with Hooke's law, we see that the spring constant k for the pendulum is just

$$k = \frac{mg}{L}$$

When this value for k is placed in the general formula for the period of a body in simple harmonic motion, Eq. (12.7) gives

Period of Pendulum

$$\tau = 2\pi\sqrt{\frac{L}{g}} \qquad (12.12)$$

Notice that *the period of a simple pendulum does not depend on the mass of the bob.* It depends on only the length of the pendulum and the accelera-tion due to gravity g. This offers a precise means for measuring g in a simple experiment. If a pendulum of known length is timed so that its period of vibration is known, g can be computed at once from Eq. (12.12). With proper precautions, this method for determining g is extremely accurate.

12.7 Forced Vibrations

In any real vibrating system, there is always some loss of energy to friction forces. As a result, a vibrating pendulum or mass at the end of a spring vi-brates with constantly decreasing amplitude as time goes on. This fact is illus-trated in Fig. 12.9. Part *a* shows the vibration we would find in the ideal case with no friction. It is the situation we discussed in previous sections. A more realistic situation is shown in *b*, where the vibration is fairly strongly influenced by friction forces. *We say such a system is* **damped** *and the vibra-tion* **damps down** *fairly quickly.*

When the friction forces are very large, the system does not vibrate at all; instead, it simply returns slowly to its equilibrium position, as shown in Fig. 12.9*c*. Such a system is said to be **overdamped.** This situation will exist

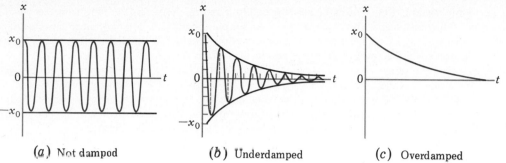

(a) Not damped (b) Underdamped (c) Overdamped

FIGURE 12.9
The free vibration of a system depends on the extent of energy losses within it.

if the mass at the end of a spring is immersed in a very viscous liquid. The mass does not vibrate at all in such a case. *When the friction forces are just large enough to cause no vibration of the system, we say the system is* **critically damped.**

If any real system is to vibrate for an extended time, energy must be added continually to replace the energy lost doing friction work. For example, to keep a child swinging at constant amplitude in a swing, you must push the swing from time to time to add energy to the system. This is typical of all vibrating systems. An outside agent must feed energy into the system if the vibration is not to damp down.

Everyone knows that there is a right and a wrong way to push a swing if it is to swing high. You must push with the motion of the swing and not against it. Only in that way can energy be added effectively to the swing. In fact, if you push against the motion, the vibration can be brought to a stop since the vibrating object must then do work on the pushing agent. These simple facts have importance in all forced, or **driven,** vibrating systems.

In a driven system, the vibration is usually sustained by a repetitive force acting on the system. This force has a frequency f, which may or may not be the same as the natural frequency of vibration of the system f_0. When $f = f_0$, the driving agent will be most effective in adding energy to the system. At all other frequencies, the driving force will not be quite in step with the motion of the system, and so its action will be less effective in adding energy to it. The variation in the amplitude of a vibrating system with the frequency of the applied force is shown in Fig. 12.10. Notice, as we said above, that *the driving force is most effective when its frequency f equals the natural frequency f_0 of the system. In that case we say the force is in* **resonance** *with the system.* More is said about f_0, the **resonance frequency** of the system, in the next chapter, where we discuss the resonance vibration of strings.

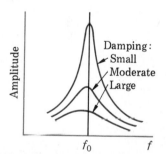

FIGURE 12.10
Response of vibrating systems to a driving force of variable frequency f.

Summary

In a system which vibrates, energy is constantly interchanged between kinetic and potential. The sum of these two energies is a constant provided dissipative factors are absent. The energy stored in a spring is $\frac{1}{2}kx^2$, where k is the spring constant.

A system subjected to a Hooke's law restoring force

undergoes simple harmonic motion. For a spring-mass system, the acceleration of the mass is given by $F = ma$ to be $a = -(k/m)(x)$. The velocity of the mass at any displacement x can be found from the energy-balance equation, namely, $\frac{1}{2}kx^2 + \frac{1}{2}mv^2 = \frac{1}{2}kx_0^2$.

When a ball moves along a circle with constant speed, its shadow executes simple harmonic motion. We can use the ball on a reference circle to compute the period of the simple harmonic motion. For a mass at the end of a spring, the period is found to be $2\pi\sqrt{m/k}$.

A simple pendulum approximates simple harmonic motion provided its angle of swing is small. Its period is given by $2\pi\sqrt{L/g}$.

The amplitude of simple harmonic motion depends on the frequency of the driving force. When the driving force has the same frequency as the system's natural frequency, resonance is achieved. Then the amplitude of vibration is maximum.

Minimum Learning Goals

Upon completion of this chapter, you should be able to do the following:

1 Point out the meaning of the following words for a mass vibrating at the end of a spring: amplitude, displacement, frequency, period.
2 Explain how the kinetic, potential, and total energies are related as a system vibrates.
3 Compute the energy stored in a spring in terms of the spring constant and the amount that the spring is stretched or compressed.
4 Use the conservation of energy to justify the relation $\frac{1}{2}kx_0^2 = \frac{1}{2}mv^2 + \frac{1}{2}kx^2$. Make use of this relation to find the speed of the mass at the end of a spring.
5 Use $F = ma$ to show that the acceleration of the mass at the end of a spring is given by $a = -(k/m)(x)$.

6 State the importance of Hooke's law for simple harmonic motion.
7 Explain how motion in a reference circle is related to simple harmonic motion.
8 Find the natural frequency and period of vibration for a mass at the end of a spring.
9 Explain why simple harmonic motion is called sinusoidal motion. Give the equation for a sinusoidal-type curve and explain the quantities in it.
10 Point out what causes the restoring force in the case of a simple pendulum, and explain why the motion is only approximately simple harmonic motion. Give the equation for the period of the motion.
11 Explain the meaning of the following terms in relation to vibration: damped, overdamped, critically damped, resonance. Sketch the graph relating vibration amplitude to frequency for a driven vibration.

Important Terms and Phrases

You should be able to define or explain each of the following:

Frequency
Period
$f = 1/\tau$
Amplitude
Displacement
Hertz unit (Hz)
Spring constant k
Simple harmonic motion

$PE = \frac{1}{2}kx^2$
Reference circle
$\tau = 2\pi\sqrt{m/k}$
Sinusoidal motion
$x = x_0 \cos 2\pi f t$
$\tau = 2\pi\sqrt{L/g}$
Critical damping
Overdamping
Underdamping
Resonance frequency

Questions and Guesstimates

1 What basic condition is necessary if a frictionless system is to undergo simple harmonic motion?
2 A tiny sphere (perhaps a ball bearing) is rolled on a mixing bowl as shown in Fig. P12.1. Explain why vibratory motion occurs in (a) but not in (b). What causes the restoring force in (a)? What does the

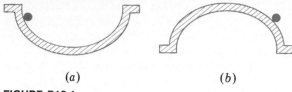

(a)   (b)

FIGURE P12.1

energy-conservation law tell us about (*a*) if friction forces are negligible?

3 Does a perfectly elastic ball bouncing on a hard floor undergo simple harmonic motion? Is there any similarity at all? Explain your answer.

4 Do all vibrating bodies without friction losses make periodic complete interchanges between the KE and PE? Explain.

5 A common method used to free a stuck car is to rock it back and forth, with proper shifting of the gears alternately to forward and to reverse many times over. Discuss the physical basis for this method, paying particular attention to energy transfer.

6 Two equal weights hang together at the end of the same spring, and the system is set vibrating. What happens to the amplitude, frequency, and maximum speed of the end of the spring if one of the weights falls off when (*a*) the spring is at its largest extension and (*b*) the mass is passing through the center position?

7 Will a pendulum clock, properly adjusted at sea level, keep good time in the mile-high city of Denver?

8 Various portions of your body have characteristic vibration frequencies: a freely swinging arm or leg could be cited. Discuss how these natural frequencies influence how you walk or run. Estimate the frequencies.

9 Discuss the influence of design characteristics on the suitability and performance of a diving board or trampoline.

10 When a car's wheels are out of balance, the car is likely to vibrate strongly at a certain speed. Explain why.

11 Suppose a pendulum bob consists of a hollow sphere. Would the behavior of the pendulum change if the sphere were filled with water? If one-fourth were filled?

12 Estimate the natural frequency for up-and-down vibration of a car on its springs. Under what conditions could this natural frequency become of importance? (E)

13 A glass tube (inner diameter = 1.0 cm) is bent to form a U tube with each side tube 40 cm long and the bottom of the U about 10 cm long (see Fig. P12.2). Estimate the natural up-and-down frequency of vibration for the liquid column when the U is filled to a height of 30 cm with (*a*) water and (*b*) mercury. (E)

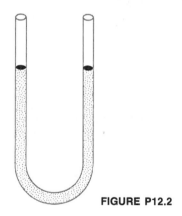

FIGURE P12.2

Problems

1 As seen in Fig. P12.3, a mass vibrates between two limits, 10.0 and 24.0 cm above the floor. It reaches the lowest point 20 times a minute. What are (*a*) the frequency, (*b*) the period, and (*c*) the amplitude of vibration for this system?

2 A pendulum swings through a side-to-side distance of 20 cm. It takes 230 s for the pendulum to reach the starting point, one side, for the 100th time after its release. What are the (*a*) period, (*b*) frequency, and (*c*) amplitude of the pendulum motion?

3 When a 200-g load is added to the lower end of a

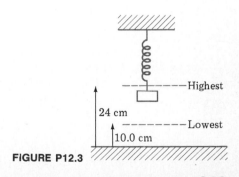

FIGURE P12.3

vertical, 15-cm-long spring, the spring stretches 18.0 cm. What is the spring constant for this spring?

4 A certain spring scale used for measuring food portions (similar to a postage scale) is designed so that the platform of the scale depresses 0.50 cm when a 40-g load is placed on the platform. What is the spring constant for the scale's spring?

5 A 20-g mass hangs at the end of a vertical spring. When an additional 30 g is added to the spring, the spring stretches 15.0 cm more. (a) What is the spring constant for the spring? (b) What is the natural period of vibration for the system after the additional 30 g is added?

6 A 2.0-kg mass is hung at the end of a vertical steel wire which has a length of 1.50 m and a cross-sectional area of 8.0×10^{-3} cm^2. (a) How much does the wire stretch when the load is added to it? (b) What is the natural frequency for vertical vibration of the mass at the end of the wire? (Young's modulus for the wire is 19×10^{10} N/m^2.)

7* Two children notice that they can make a car vibrate up and down by periodically pushing downward on it. The car vibrates through eight complete cycles in 13.0 s. (a) Assuming the car to have a mass of 1800 kg, find the spring constant for its suspension system. (b) If this value is correct, about how much should the car lower as a 70-kg person enters the car and sits in the front seat?

8* When a 1.50-kg suit is hung on the center point of a clothes line that already has an identical suit hanging there, the center of the line sags an additional 12.0 cm. The addition of a third identical suit causes the line to sag 12.0 cm more. With what frequency do you expect the suits hanging on the line to vibrate up and down when all three suits are hanging at its center?

9 How much energy is stored in a spring whose constant is 120 N/m when the spring is (a) stretched 5.0 cm, (b) compressed 2.0 cm?

10 A certain Hooke's law spring stretches 20 cm when a load of 0.35 N is added to it. How much energy is stored in the spring when it is compressed 5.0 cm?

11 A vertical spring is compressed 1.20 cm when a 500-g mass is set on its upper end. Assume the spring obeys Hooke's law. How much potential energy is stored in the spring by setting the mass on it?

12 The spring on a popgun obeys Hooke's law and requires a force of 220 N to compress it 15.0 cm to its cocked position. How much energy is stored in the cocked spring?

13* A 300-g object is attached to the end of a spring ($k = 80$ N/m). The spring is then stretched 2.5 cm from its equilibrium position and released. Find (a) the speed of the mass as it passes through the equilibrium position and (b) its acceleration just as it is released.

14* A 200-g mass hangs at the end of a spring ($k = 6.0$ N/m). The mass is now pulled down 3.0 cm from its equilibrium position and released. What are (a) the initial acceleration of the mass and (b) its speed as it passes through the equilibrium position?

15* A spring-mass system slides on a horizontal, frictionless surface (much like Fig. 12.3). The mass is 50 g. A horizontal force of 0.70 N applied to it stretches the spring 4.0 cm. (a) What is the acceleration of the mass when the system is released? (b) What is the speed of the mass as it passes through its equilibrium position?

16* The force constant for a spring in a popgun is 1000 N/m, and the spring is compressed 8.0 cm when the gun is cocked. If the pellet shot by the gun has a mass of 15.0 g, what is the maximum speed with which the pellet could leave the gun?

17* As shown in Fig. P12.4, a rotating wheel drives a piston. For the situation shown there, find the following for the motion of the piston: (a) frequency, (b) amplitude, (c) maximum speed, (d) speed when the connecting rod is horizontal.

18* For the piston shown in Fig. P12.4, write the piston's equation of motion in the form $x = x_0 \cos \omega t$ with x_0 and ω given numerical values.

19** For the situation shown in Fig. P12.4, find the magnitude of the acceleration of the piston (a) when the connecting rod is horizontal, (b) when the radius to the pivot point is vertical, (c) when the radius to the pivot point makes an angle θ to the horizontal.

20** A piston undergoes vertical simple harmonic motion with amplitude of 8.0 cm and frequency f. A washer sits freely on top of the piston. At low piston frequencies, the washer moves up and down with the piston. However, at very high frequencies,

FIGURE P12.4   3.0 rev/s

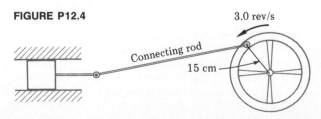

the washer momentarily floats above the piston as the piston starts its downward motion. (*a*) What is the maximum acceleration of the piston when the washer begins to separate from it? (*b*) What is the lowest frequency at which separation will occur?

21* A 2.0-kg mass oscillates with simple harmonic motion at the end of a spring. The amplitude of motion is 30 cm, and the spring has a constant of 400 N/m. Find the speed of the mass when its displacement is (*a*) 30 cm, (*b*) 0 cm, (*c*) 15 cm.

22* A spring whose constant is 125 N/m has a 500-g mass at its end. The mass is displaced 3.0 cm from its equilibrium position and released. Find (*a*) the frequency of vibration; the speed of the mass when its displacement is (*b*) 0 cm, (*c*) 2.0 cm. (*d*) What is the acceleration of the mass when its displacement is 2.0 cm?

23* A 400-g mass at the end of a spring vibrates with 3.0-cm amplitude and a frequency of 0.50 Hz. (*a*) What is the maximum speed of the mass? (*b*) What is its maximum acceleration? (*c*) What is its speed when it is 1.0 cm from the end of its swing?

24** Show that the following statements are true for a spring that obeys Hooke's law in the region of concern. (*a*) A force *F* applied to the spring causes the spring to stretch (or compress) the same amount whether or not the spring is subject to a prior additional stretching force. (*b*) The frequency of vibration of a mass at the end of the spring is the same for the spring hanging vertically as it is for the horizontal spring with the mass supported on a frictionless surface.

25 What is the period of vibration for a 70-cm-long pendulum? What is its frequency of vibration?

26 A certain simple pendulum has a period of 1.000 s. (*a*) How long is the pendulum? (*b*) If the pendulum bob is a sphere, is the length measured to the top or to the center of the sphere?

27* The length of a certain simple pendulum is 25 cm. (*a*) With what frequency does it vibrate? (*b*) If it is released at an angle of 37° to the vertical, how fast is the bob going when it passes through the center point?

28* A large clock is controlled by a pendulum. If the clock is taken to the moon, where objects weigh only about one-sixth their weight on earth, how long (in hours) will it take the clock to tick out 1 h on its dial?

29* A steel clothesline is strung between two large posts. When a 2.0-kg object is hung from the center point of the line, the line sags 13 cm. It sags

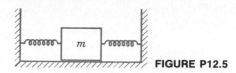

FIGURE P12.5

26 cm under a 4-kg mass. If only the 2.0-kg object is hanging on it and the object is then pulled down 10 cm more and released, find the (*a*) frequency of vibration and (*b*) the speed of the object as it passes the equilibrium position.

30* Two identical springs hold a mass *m* on a frictionless table, as shown in Fig. P12.5. Prove that the frequency of vibration of this device is

$$\frac{1}{2\pi}\sqrt{\frac{2g}{d}}$$

if each spring separately stretches *d* units under the weight of a mass *m*.

31** A certain spring has a constant of 15 N/m. The spring is mounted vertically in a tube; a 20-g mass is set on top of the spring and allowed to come to rest. We refer to this position as the zero position. By means of a lever, the mass is now pushed 10.0 cm farther down the tube. How high above the zero position will the mass fly when the spring is now released? (Equate the PE in the spring to the PE of the mass. What assumption does this make about the spring?)

32** A compressed spring with a mass attached to its end is immersed in a container of water at 20.00°C. The heat capacity of the container, spring, mass, and water is 60 (cal)(°C⁻¹). After the spring is released, it vibrates back and forth with decreasing amplitude as a result of friction (or viscous) forces imposed by the liquid. When the system stops vibrating, the temperature is found to be 20.100°C. (*a*) How much energy was stored in the spring? (*b*) If the spring was compressed 8.0 cm, what is the constant of the spring?

33* Show that the maximum speed of a pendulum bob is given by

$$v = \sqrt{2gL(1 - \cos\theta)}$$

if the pendulum starts to swing from an angle of *θ* to the vertical.

34* A pendulum is drawn aside to a certain angle and released. When the bob passes the center point, the tension in the string is twice the weight of the

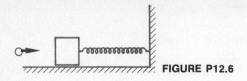

FIGURE P12.6

bob. Show that the original displacement angle was 60°.

35** A 5.0-g bullet is shot with a speed of 1000 cm/s into the 95-g block shown in Fig. P12.6 and becomes embedded in the block. If the block was originally at rest and if the spring has a constant of 100 N/m, with what amplitude will the spring vibrate after the collision? Ignore friction forces between the block and the table.

36* If a 2.0-kg mass vibrates at the end of a spring according to the relation $y = 0.20 \cos 9.42t$ centimeters, find (a) the amplitude of vibration, (b) the frequency of vibration, (c) the spring constant, and (d) the maximum speed of the mass. It is assumed that t is in seconds.

13
Wave Motion

The vibrating objects discussed in the last chapter are capable of generating waves. But many other types of waves can be generated by use of varied wave sources. Whether these waves are in air, in metal, or even in vacuum, they have many features in common. In this chapter we discuss those principles that are basic to all wave motion. In later chapters we apply these principles to sound, light, and other types of waves.

13.1 Importance of Wave Motion

Nearly every vibrating body sends out a wave. We are all aware of the fact that the musical tones given off by a piano, a guitar, or a violin are the result of waves sent out into the air by the vibration of a stretched string. The clarinet also sends out sound waves through the air, but in this case the source of the vibration is the reed blown by the player. In a trumpet, the wave is generated by the vibration of the trumpeter's lips in the mouthpiece of the instrument. There are many other sources of musical sounds—the drum, the triangle, cymbals. Each source consists of a vibrating body which acts in such a way as to generate waves in the air.

Have you ever stopped to think when you are watching TV that everything you see and hear was transmitted to you from the station by waves? The station generated electromagnetic waves by vibrating electrical charge in an antenna. Your TV set responded to these waves and regenerated light and sound waves, which then traveled across the room to you.

If you were to break a bone, the doctor would almost certainly use x-rays, another type of wave, to determine the nature of the fracture. Strained muscles can be treated with another form of wave motion—heat, or infrared, radiation.

Many other types of waves are commonplace to us, but there are yet other wave phenomena that are not so noticeable. For example, we shall see in Chap. 25 that, to a certain extent, even a baseball may be considered to have wavelike properties.

In this chapter we learn the general rules about how waves behave. We find that there are two general types of waves into which all mechanical waves can be divided. Within each of these groups there are many different forms of waves. A considerable portion of physics is devoted to the study of their behavior.

13.2 Waves on a String: Transverse Waves

Suppose that a string is tied at one end to a solid wall and at the other to a piece of spring steel, as shown in Fig. 13.1. If the end of the flexible steel rod is struck from below so as to give it an upward velocity, as shown in part *b* of the figure, the end of the rod will vibrate back and forth, as shown in the succeeding parts of Fig. 13.1. We assume it to undergo essentially simple harmonic motion.

Notice, in part *b*, that the string is pulled upward along with the end of the rod. Its velocity at various places is in the direction of the arrows shown. We see that the string near the rod tends to pull the adjacent string on the right in an upward direction. As the string on the right begins to respond to this pull, it in turn pulls the string adjacent to it upward, and so on. Continuing down through parts *c* to *g* of Fig. 13.1, we see that the disturbance moves continually toward the right. Let us call the velocity with which this disturbance travels along the string v. Later in this section we shall see what its value is in terms of the tension and mass of the string.

While the initial disturbance is traveling to the right along the string for the reasons discussed above, the bar continues to vibrate. In so doing, it sends a continually changing disturbance down the string, as shown in Fig. 13.1. Since the source vibrates with simple harmonic motion, the wave along the string has a sinusoidal shape.

It will be noticed that only the wave travels to the right in Fig. 13.1. The little pieces of the string simply move up and down. *A wave such as this, where the wave travels in a direction perpendicular to the direction of motion of the particles, is called a* **transverse wave.** This is easily remembered if you recognize that *trans* means "across" and that the particles actually travel across the direction of propagation of the wave.

Definition

You may well ask: If the particles do not move down the string, what does move in the direction of propagation of the wave? *It is characteristic of waves that they carry energy.* In this particular case, the string had no KE or

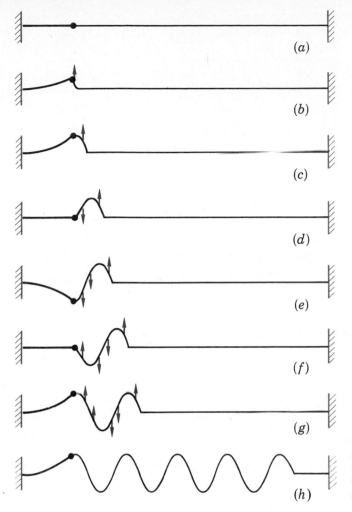

(a)

(b)

(c)

(d)

(e)

(f)

(g)

(h)

FIGURE 13.1
The vibrating rod sends a
wave down the elastic
string. Since the motion of
the ball is simple harmonic,
the wave on the string is
sinusoidal.

PE at the start. However, it is clear that in those portions of the wave where
the string is moving swiftly, KE is present. This, together with PE, was
carried down the string from the vibrating bar. Hence, energy is carried
along with the wave in its direction of propagation. The energy travels
along with the speed of the wave v.

The terms used in speaking of waves are easily understood by reference
to Fig. 13.2. Points A and B on the wave are called the **crests** of the wave,
while points D and E are called **troughs.** One complete cycle, the portion of
a wave generated in one complete vibration of the source, is from crest to
crest, i.e., from A to B, or from trough to trough, i.e., from D to E. The
horizontal distance between either of these two sets of points is called the
wavelength of the wave and is represented by the Greek letter lambda (λ).
The amplitude of the wave is designated by y_0.

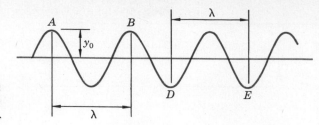

FIGURE 13.2
The amplitude of the wave
is y_0, and its wavelength is λ.

There is an important relation among the wavelength λ, the speed of the wave v, and frequency f of any type of wave. To see what it is, let us refer to Fig. 13.3. Notice that the waves travel down the string with speed v as the source sends out an additional wave. It takes a time τ, equal to the period of vibration, for the source to send one additional wavelength down the string. During that time, the waves move a distance λ. Therefore, the motion equation $x = \bar{v}t$ becomes, in this case,

$$\lambda = v\tau$$

Or if we replace τ by $1/f$, this becomes

A Fundamental Relation
for Waves

$$\lambda = \frac{v}{f} \tag{13.1}$$

This extremely important relation is true for all periodic waves. We make frequent use of it.

The speed of a wave on a string is given by a particularly simple relation, which we state without derivation. If the tension in the string is F and if the mass of a length L of the string is m, then the speed of the wave along the string is

$$v = \sqrt{\frac{F}{m/L}} \tag{13.2}$$

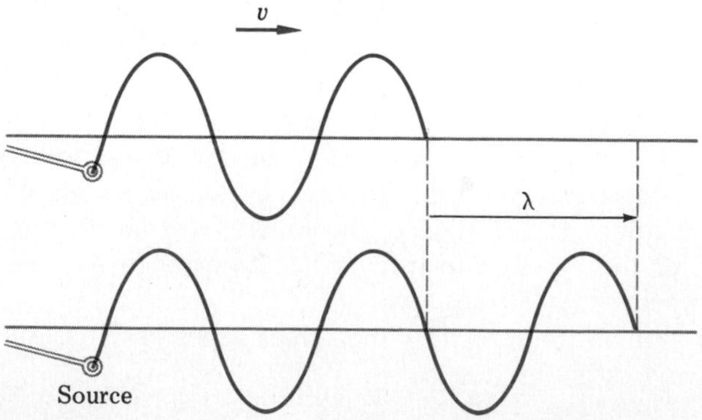

FIGURE 13.3
As the source sends out a
complete wave, the wave
moves a distance λ along
the string. Therefore, the
wave travels a distance λ in
a time τ.

Source

We can easily justify the qualitative dependence of speed on tension and mass per unit length. The tension in the string is responsible for the force that accelerates a piece of the string sideways as the pulse passes through the region. The larger the tension, the larger will be the acceleration, and so the motion of the pulse is swift if the tension is large. On the other hand, the more massive the string, the more inertia the string will have. The mass per unit length therefore affects the speed with which the pulse moves. A massive string has large inertia, and the speed of a pulse on it will be small.

As another check on the relation for v, let us use dimensional analysis to examine it. Using brackets to signify the words *the dimensions of,* we know that

$$[v] = [LT^{-1}] \qquad [F] = [MLT^{-2}]$$

By using these quantities together with Eq. (13.2), the dimensional equation becomes

$$[LT^{-1}] = \left[\sqrt{\frac{MLT^{-2}}{ML^{-1}}} \right]$$

or

$$[LT^{-1}] = \left[\sqrt{L^2 T^{-2}} \right]$$

which checks out correctly.

Illustration 13.1   A certain guitar string has a mass of 2.0 g and a length of 60 cm. What must the tension in the string be if the speed of a wave on it is to be 300 m/s?

Reasoning   From Eq. (13.2) we have $F = (m/L)(v^2)$, and since $v = 300$ m/s, $m = 0.0020$ kg, and $L = 0.60$ m, substitution gives F to be 300 N. Notice that this tension is really quite large since it is equivalent to the weight of about 30 kg pulling on the string.

13.3 Reflection of a Wave

Until now we have not worried about what happens to a wave when it hits the rigid support on the right. Since the wave represents energy traveling to the right, the energy must either be absorbed by the support or be reflected backward. In practice, some of the energy will be absorbed by the connection at the wall. However, if the support is quite solid (or if the end is completely free), a large fraction of the energy will be reflected.[1] Hence a reflected wave will travel to the left on the string at the same time the original wave is moving to the right. For simplicity, we assume that no energy is lost upon reflection.

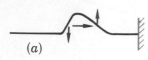

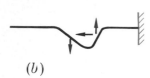

(a)

(b)

FIGURE 13.4
A pulse on a string is inverted when it is reflected from a fixed end.

To study this effect, let us consider only a single wave crest propagating along the string, as illustrated in Fig. 13.4a. When this wave reaches the wall, it cannot pull the wall upward in the same way as it would have pulled more string up. Because the wall does not "give," the force at that point is larger than would have been the case if the string had not ended at that point. The wall exerts a downward pull on the string. This pull accelerates the string downward to such an extent that the string's momentum carries it below the zero line. The result is that the pulse is turned upside down as it hits the wall, and the reflected wave appears as shown in part b of Fig. 13.4. If the string had been completely free to move up and down at that end, the wave would not have been turned over, although it would still have been reflected (because the energy in the wave could not just disappear at the end of the string!). In summary, *a pulse is inverted by reflection at a fixed end. It is reflected, but not inverted, by a free end.*

Next let us consider what happens when a reflected pulse traveling backward along the string meets a second pulse going forward on the string. Suppose two rectangular pulses going in opposite directions meet as shown in Fig. 13.5. The original pulses are shown by dashed lines in the region of overlap. It is found from experience that the string will displace as shown by the full line in this region. We see that the string undergoes the vector sum of the individual wave displacements. This is true for all wave systems as long as the displacement is a linear function of the force causing the displacement. All the waves we deal with in this text conform to the **principle of superposition:**

Superposition Principle

A point subjected to two or more waves simultaneously displaces an amount equal to the vector sum of the individual disturbances.

---

[1] No work can be done at a fixed end, since the distance moved in the direction of the applied force is zero. Why can no energy loss occur, i.e., work be done, at a perfectly free end?

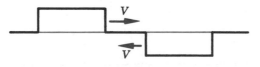

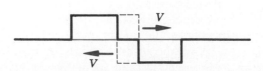

FIGURE 13.5
According to the principle of superposition, two waves on the same string will add as shown when they meet.

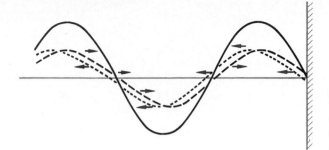

FIGURE 13.6
The incident and reflected waves add to produce a single wave having a node at the fixed end.

We are now ready to see what happens when a sinusoidal wave traveling down a string is reflected. You will have to use your imagination a little here, since we shall be drawing pictures of two waves on the same string. These waves, representing displacements of the particles of the string, must be added to obtain the true behavior of the string.

In Fig. 13.6 we show the two waves traveling to right and left as the dashed curves.[2] The true displacement of the string is obtained by adding the two dashed curves, and the result is shown as the solid curve in the figure. You should add the dashed curves at a few points to check the displacement of the string. Notice that the two waves exactly cancel each other at the wall, and so the string is not displaced at that point. Furthermore, you can easily see that as time goes on and the waves continue to move, they will always be matched so that they cancel at the wall. This must be true, of course, since the string must remain at rest at that point.

Look, now, at any of the other places where the two waves cancel in Fig. 13.6. You will notice that as the waves travel along, they will always cancel at these points as well. Hence we have the important result that *the forward and the reflected wave interfere with each other so as to keep certain points along the string motionless at all times. These points of zero motion are called* **nodes.**

It is instructive to observe what will happen at the points on the string midway between the nodes. These are the places of large displacement in Fig. 13.6. It is left as an exercise for you to show that as the two waves travel along, these intermediate points oscillate back and forth between very large positive and very large negative displacements. This is illustrated in Fig. 13.7. *These points of maximum motion are called* **antinodes.**

As shown in Fig. 13.7, the string will oscillate back and forth between the two positions shown. The nodes at points P, Q, R, S, and T represent points on the string which remain at rest. No motion of the string occurs at these points. On the other hand, at points A, B, C, and D, the antinodes, the string vibrates back and forth with large amplitude. By comparing Fig. 13.7 with Fig. 13.2, we see that *the distance between adjacent antinodes is $\lambda/2$. The distance between adjacent nodes is also $\lambda/2$.*

Definition

Definition

Distance between Nodes $= \lambda/2$

---

[2] At first you might not recognize that the reflected wave, the dashed one, has been inverted. Notice, however, that if it had not been reflected, it would be a continuation of the dashed curve and would lie below the axis near the end.

13.4 Resonance

We saw in Chap. 12 that a vibrating object can be made to vibrate very strongly if the force producing the vibration has the proper frequency. If the force pushes on the object with a frequency equal to the natural vibration frequency, the force will be in resonance with the object and vibration with very large amplitude will result. Let us review what happens when a force is in resonance with a vibrating system.

As a common example, we know that we do not really have to push very hard on a child in a swing for the child to swing quite high. A small periodic push when the swing is in the proper position will soon build up a large amplitude of oscillation. We can determine why this is true by utilizing some of the concepts we learned about work.

When someone pushes on the swing in the direction in which it is moving, work is done on the swing. The energy thus furnished causes the child to rise somewhat on the next oscillation. When this process is continued over and over, the child swings higher and higher as energy is put into the system. The limiting amplitude of oscillation occurs when the energy fed in by the person pushing is exactly equal to the frictional energy losses during one complete oscillation.

Any vibrating system can be excited to wide oscillations by a periodic driving force, as in the case discussed above. Notice, though, that the driving force must be applied at just the proper time if it is to feed energy into the system. (If the driving force pushes on the system when it is moving against the force, the system will do work on the driving mechanism and will lose, rather than gain, energy.) *In the case in which the driving force is just in step with the vibrating system so that the amplitude becomes quite large, the*

Conditions for Resonance

driving force and the system are said to be in **resonance.** *Two conditions must be satisfied for resonance: (1) The applied force should have the* **same frequency** *as the free vibration frequency of the system. (2) The force must be applied* **in phase** *with the vibration.* That is, the force and the vibration must be in step.

Many examples of resonance could be cited. The child in the swing is one. Grandfather clocks have a pendulum that is excited by a mechanism operating in resonance with the pendulum. All musical instruments make use of resonance to produce audible sounds; we shall discuss this later.

Not all resonances are desirable. It is quite common for a rattle in an automobile to become particularly noisy at a certain speed. The rattling device, a vibrating object of some sort, is in resonance with the vibratory motion of the car at that speed. This type of disturbance becomes very serious in cars whose wheels are badly out of balance. The thumping action of the wheel against the roadway varies in frequency with the speed of the automobile. It is not surprising that various portions of the car begin severe vibrations when the proper resonance frequency is maintained.

In the next section we examine the resonance motions of a vibrating string. We see that the string will resonate to driving forces of more than one frequency. Although the computations carried out for the string are very simple examples, it will be seen that much more complicated vibratory systems show the same general features of resonance.

13.5 Resonance Motion of a String: Standing Waves

Let us now consider what happens when a vibrator sends waves down an actual string, as shown in Fig. 13.8. If we perform the experiment indicated, interesting results are found. The string shown is attached to the vibrator at one end and is held rigid by the massive pulley at the other. However, the length of the string can be increased by pulling the tuning fork farther to the left, as shown in succeeding parts of the figure.

For most string lengths, the string remains nearly motionless. The vibrator sends tiny wave pulses down the string, and these are reflected by the

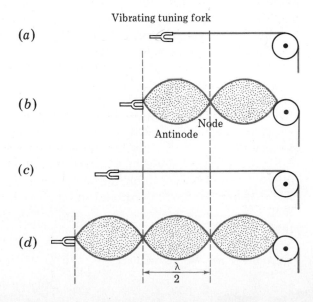

FIGURE 13.8
At what other lengths will the string resonate?

pulley. As the wave pulse goes down the string, it is analogous to a child on a swing moving away from the person pushing the swing. Like the child on a swing, the pulse is reflected and returns to the pusher—the vibrator in this case. Usually the vibrator will not be moving in the precise way needed to reflect the pulse once again and add energy to it. Like a person pushing improperly on the swing, the push will tend to cancel the motion rather than reinforce it. For this reason, the tiny pulses sent down the string by the vibrator do not usually reinforce. As a result, the string remains nearly motionless, as shown in part a.

However, a startling result is seen if the string is slowly lengthened. *At some very definite length, the string begins to vibrate widely,* as shown in part b. *The string vibrates back and forth between the limits shown. Certain points of the string remain motionless, and these are the nodes. Other points, the antinodes, vibrate most widely. We call this pattern consisting of nodes and*

Standing Wave *antinodes a* **standing wave** *on the string.*

It is easy to understand what is happening to cause the standing wave. In this very special situation, the pulse reflected back to the vibrator arrives just in time to be reinforced by the vibrator. It is reflected back down the string by the vibrator, but because the vibrator is exactly in step (or in phase) with it, the pulse is augmented by the vibrator. Consequently, the vibrator keeps adding energy to the pulses as they move back and forth along the string. The wave pulses therefore become very large.

But we have already seen in Sec. 13.3 that the pulses moving to right and left along the string combine to form nodes and antinodes. Therefore the string vibrates in a definite pattern of nodes and antinodes. Moreover, because the sum of these pulses is so very large, the pattern is large and easily visible. As we see, *the large standing wave is a result of the fact that the vibrator pushes in resonance with the wave pulses on the string.*

If now the string in Fig. 13.8b is lengthened further, the reflected pulses will reach the vibrator slightly later than before. They will no longer be in phase with the vibrator, and so resonance will no longer occur. As seen in part c of the figure, the string will now remain nearly motionless.

But if the string is lengthened still further, resonance will suddenly occur at the elongation shown in part d. Once again the reflected pulses will be in phase with the vibrator when they reach it, and again large pulses will build up on the string and a large, visible standing wave will result. As we see, the string will resonate to the vibrator at very special conditions. Let us now see what these conditions are in a quantitative form.

You will recall that the distance between adjacent nodes is $\lambda/2$. If we examine Fig. 13.8, we see that the vibrator is very close to the position of a node since it cannot move far. Therefore the two ends of the resonating string are at nodes. As a result, the string can resonate only if it is $\lambda/2$ long,

Resonance Conditions or $2(\lambda/2)$ long, or $3(\lambda/2)$ long, and so on. Indeed, *in general we can state that a string fastened firmly at its two ends will resonate only if it is a whole number of half wavelengths long.* For example, in parts a, b, c, and d of Fig. 13.9, the length L of the string is equal to $\lambda/2$, $2(\lambda/2)$, $3(\lambda/2)$, and $4(\lambda/2)$. In general, then, for resonance,

$$L = n\frac{\lambda}{2} \qquad \text{where } n = 1, 2, 3, \ldots . \qquad (13.3)$$

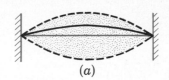

(a)

Since the wavelength is related to the frequency by Eq, (13.1), we see at once that a string of fixed length will resonate to only certain very special frequencies. In modern terminology, *the resonant frequencies of the string are* **quantized,** *meaning that the resonance frequencies f_n are separated by frequency gaps; quantum jumps in frequency exist between the resonance frequencies.* We have

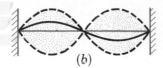

(b)

$$f_n = \frac{v}{\lambda_n} = \frac{v}{2L/n} = n\frac{v}{2L} = nf_1$$

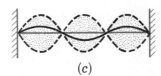

(c)

Illustration 13.2 The speed of a wave on a particular string is 24 m/s. If the string is 6.0 m long, to what driving frequencies will it resonate? Draw a picture of the string for each resonant frequency.

Reasoning The possible resonance wavelengths are given by Eq. (13.3). We have, for $L = 6.0$ m,

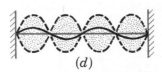

(d)

FIGURE 13.9
Resonant modes of motion in a string must have nodes at the two fixed ends. The dashed lines show the outer limits of the vibration, while the solid line is the position of the string at one instant.

$$\lambda = \frac{2L}{n} \qquad \lambda_1 = 12 \text{ m} \qquad \lambda_2 = 6 \text{ m}$$

$$\lambda_3 = 4 \text{ m} \qquad \lambda_n = \frac{12}{n} \text{ meters}$$

Now we can make use of Eq. (13.1) to find the frequency.

$$f = \frac{v}{\lambda} \qquad f_1 = \tfrac{24}{12} = 2 \text{ Hz} \qquad f_2 = \tfrac{24}{6} = 4 \text{ Hz}$$

$$f_3 = \tfrac{24}{4} = 6 \text{ Hz} \qquad f_n = \frac{24}{12/n} = 2n \text{ hertz}$$

The modes of vibration, i.e., the various standing waves, of the string are as shown in parts *a*, *b*, and *c* of Fig. 13.9 when the frequency is f_1, f_2, and f_3.

13.6 Other Transverse Waves

We have spent a great deal of space discussing waves on a string because the principles that apply there apply to many other vibrating systems. For example, if a metal bar clamped at its center is struck at its end, as shown in Fig. 13.10, the bar will vibrate. The mode of vibration of the bar is indicated in part *b* of Fig. 13.10.

The center of the bar must be a node, because it is tightly clamped in

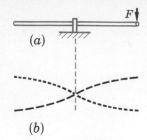

(a)

(b)

FIGURE 13.10
A transverse standing wave is set up in a bar when the bar is struck as shown.

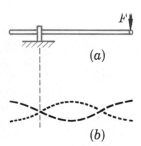

(a)

(b)

FIGURE 13.11
The position of the clamp must be a node.

place. Since the ends of the bar are not held rigidly, we expect antinodes near them. If we assume the ends to be antinodes, the length of the bar L is one-half wavelength. This follows from the fact that the distance between two successive antinodes is $\lambda/2$. Knowing that $\lambda = 2L$, we could measure the frequency of vibration of the bar and use Eq. (13.1) to compute the speed of a transverse wave in it.

If the bar had been clamped as in Fig. 13.11, at a distance $L/4$ from its end, the vibration would have appeared as in part b of that figure. Once again, the ends would approximate antinodes, and the clamp point would be a node. In this case

$$\lambda = L$$

while, in the case of Fig. 13.10,

$$\lambda = 2L$$

Since the frequency of vibration is given by the relation

$$f = \frac{v}{\lambda}$$

the respective frequencies of vibration for the bar are

Fig. 13.10: $\qquad\qquad f = \dfrac{v}{2L}$

Fig. 13.11: $\qquad\qquad f = \dfrac{v}{L}$

Hence, the bar in Fig. 13.11 would be vibrating with twice the frequency of the bar in Fig. 13.10. If f and L were measured, v could be computed in each case, as was pointed out above.[3]

Many other forms of transverse waves are common. Waves on the surface of water are approximately transverse, since the water particles move vertically up and down while the wave travels horizontally across the water. We speak more about these waves when we begin our study of light waves, since there are great similarities between the two.

When a drumhead is struck, transverse waves travel across its surface. Similar waves are set up in a cymbal or, for that matter, any flat sheet of material that is made to vibrate. These waves are more complicated than those on a string, for the wave can travel in any direction in the plane, while in the string it could travel in one direction only. The nodal points in such cases form nodal lines on the plane, lines along which the plate does not

---

[3] More detailed computations show that when the finite rigidity and inertia effects are considered, these conclusions are only approximate. In practice, the node in Fig. 13.11 occurs at $0.224L$ rather than at $0.250L$.

vibrate. We do not discuss this behavior here, but later in our study of light we encounter a similar situation.

Finally, perhaps the most important type of transverse wave is the electromagnetic wave. This type of wave is discussed after we have studied the subject of electricity. It will be found there that radio, radar, heat, infrared, light, and ultraviolet waves, as well as x-rays, are all forms of electromagnetic waves. Clearly, from this list, electromagnetic waves are of great practical importance to us.

13.7 Longitudinal Waves

An interesting experiment can be done with a very long spring placed on a smooth tabletop and tied at one end. One possible form of this experiment (not the best from a practical standpoint) is illustrated in Fig. 13.12. The spring at equilibrium on the tabletop is shown in part a. If it is suddenly compressed as in part b, the loops near the end will be compressed before the rest of the spring experiences the disturbance. The compressed loops will exert a force on the loops to the right of them, and so the compression will travel down the spring as indicated. At the fixed end, the compressional energy is reflected; thus the compression is reversed and ends up traveling to the left, as shown in part d.

This type of wave is not a transverse wave, since the particles of the spring actually vibrate back and forth in the direction in which the wave is

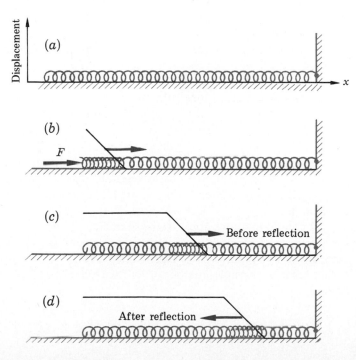

FIGURE 13.12
A longitudinal pulse travels down the spring and is reflected by the wall. The black curves are graphs of the displacement of the spring.

propagated, along the spring. *A compressional wave such as this, where the motion of the particles is along the direction of wave propagation, is called a* **Definition** **longitudinal wave.**

A difficulty arises when we attempt to plot a graph of a compressional wave. In a transverse wave, the particle motion was up and down, say, and the displacement on the graph was plotted up and down also. Here we would encounter difficulty plotting the spring displacement in the same direction as that in which it occurs, since distance along the wave and displacement both lie in the horizontal direction. We therefore still plot the magnitude of the displacement vertically, as shown by the black curves in Fig. 13.12. We should always be careful to remember that *no motion occurs in this direction in a longitudinal wave.* The plot is made in this way merely as a matter of convenience.

13.8 Standing Compressional Waves on a Spring

It is clear that *the longitudinal wave on a spring has many features in common with a transverse wave on a string.* If a compressional wave is sent down a spring, the wave and its energy are usually reflected at the end of the spring. This reflected wave can interfere with the later waves being sent down the spring from the source. *If the proper relation is maintained between the frequency of the driving source oscillating the end of the spring and the various parameters of the spring, resonance will occur.* It is this feature of the spring system that we now investigate.

(a)

(b)

As with resonance on a string, *the position of the driving source will usually be close to a node* (a point of zero motion), since the spring itself at resonance will move much more than the driving source. Also, *if the other end of the spring is fixed solidly to a wall or some other similar object, that end must also be a node. The resonance motion of the spring must then appear as shown in the graphs of Fig. 13.13.*

(c)

Remember that although the displacement of the loops of the spring along the spring is plotted vertically, the motion actually occurs in a horizontal direction. For example, point *A* in Fig. 13.13*a* means that the center point of the spring vibrates back and forth horizontally, with an amplitude indicated by the dashed waves. In the mode of motion represented in Fig. 13.13*b,* the center point does not move at all. The center point can be either an antinode or a node, depending on the frequency that is causing the spring to resonate.

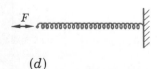

(d)

FIGURE 13.13
Standing waves result from longitudinal vibration of the spring.

The same relations apply to the vibration along the spring as we found for a wave on a string. From Fig. 13.13, we see that the distance between adjacent nodes is still $\lambda/2$. Also, at resonance, in this particular case, the spring must be a whole number of half wavelengths long. That is, at resonance

$$n\frac{\lambda}{2} = L \qquad \text{where } n = 1, 2, \ldots$$

This relation, when combined with the relation between wavelength and frequency, $\lambda = v/f$, tells us at once that the spring resonance frequencies will be

$$f = n\frac{v}{2L} \qquad \text{where } n = 1, 2, \ldots$$

Illustration 13.3 A spring 300 cm long is found to resonate in three segments (nodes at both ends) when the driving frequency is 20 Hz. What is the speed of the wave in the spring?

Reasoning The spring is vibrating as shown in Fig. 13.13c. It is seen that in this case

$$3\frac{\lambda}{2} = L$$

or $$\lambda = 200 \text{ cm}$$

Knowing that $\lambda = v/f$ and since $f = 20$ Hz, we have

$$v = (20 \text{ s}^{-1})(200 \text{ cm}) = 4000 \text{ cm/s}$$

We could, of course, have obtained this result by simply substituting in the relation

$$f = n\frac{v}{2L}$$

given above, using $n = 3$. However, most physicists prefer not to memorize a different relation for each case. They ordinarily use the number of half wavelengths on the total spring to find λ and then the relation $f = v/\lambda$ to find the unknown. As a matter of fact, almost all the situations in resonance that we encounter can be described by the use of this relation and an examination of the resonant system. It is not necessary to memorize an equation for each case.

13.9 Compressional Waves on a Bar

We have already discussed the transverse vibrations of a bar. It is possible to set up longitudinal vibrations in such a bar as well. There are many possible ways of doing this. One simple method is to strike the end of the bar, as shown in Fig. 13.14a.

The elastic bar acts like a very stiff spring. The blow on the end of the bar sends a compressional wave down the bar. This wave is quite complex

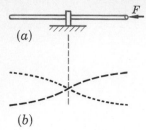

(a)

(b)

FIGURE 13.14
When the bar is struck as shown, a longitudinal standing wave is set up in it.

TABLE 13.1
SPEED OF
COMPRESSIONAL
WAVES

Material	Speed, m/s
Air (0°C)	331
Water (15°C)	1447
Copper	3500
Glass	4000–5500
Wrought iron	4900–5100
Steel	5000

in form and is actually a large group of waves of various frequencies. The bar will resonate to only certain frequencies, and hence it selects the proper frequency to which it will resonate.

In this particular case, the wave in the bar must have a node at its center and antinodes at its ends. The lowest resonant-frequency mode of motion is shown in Fig. 13.14b. Remember that the motion of the bar is longitudinal even though the graph of the motion looks like the similar case for transverse vibration. Clearly in this case the bar is one-half wavelength long. With the length L of the bar known, the wavelength is then also known. If the frequency of vibration of the bar is known as well, the velocity of compressional waves in the material can be computed by using $\lambda = v/f$. (A simple way to measure the frequency of vibration of the bar is to compare the sound given off by the bar with the sound given off by tuning forks of known frequency. More accurate methods are discussed later.) Some typical values for the speed of compressional waves in various materials are given in Table 13.1.

Other resonance vibrations of the bar in Fig. 13.14 are also possible. Any vibration that will have a node at the center of the bar and antinodes at the ends will be allowed. You should examine this case to see what the next higher resonant frequency of the bar would be. It will have three nodes along the bar. (Why is the mode of motion with two nodes along the bar not allowed in this case?) Since friction losses between molecular planes within the vibrating bar are more serious at the higher frequencies, these modes of motion usually die out rather rapidly; i.e., their energy is lost to heat, and the bar ceases to vibrate.

The speed of compression waves in various materials can be related to the density and elasticity of the material. One can show that the speed in extended solids and liquids is given by

$$v = \sqrt{\frac{E}{\rho}} \tag{13.4}$$

where ρ is the mass density of the material and E is the bulk modulus, defined in Eq. (8.8). Notice that consistent units must always be used in this relation. In the case of thin rods, the elastic bulk modulus E is replaced by Young's modulus Y.

It seems reasonable that the speed should increase with increased modulus of the material because a pulse should be transmitted most swiftly down the stiffest spring. A stiff spring has very little give, and so an impulse is felt almost at once by the entire spring. The density factor in the expression for v represents the inertia of the material. As we would expect, large inertia (large ρ) gives rise to slowly moving pulses.

We can also check the relation for v by subjecting it to dimensional analysis. Because modulus has the dimensions of force per unit area, we have

$$[E] = [\text{force}/\text{area}] = [(MLT^{-2})/L^2]$$
$$= [ML^{-1}T^{-2}]$$

Also

$$[\rho] = [ML^{-3}]$$

Using these results to perform a dimensional check on Eq. (13.4), we have

$$[v] = \left[\sqrt{\frac{ML^{-1}T^{-2}}{ML^{-3}}}\right]$$
$$= [L/T]$$

which is correct. Therefore Eq. (13.4) is dimensionally correct.

Illustration 13.4   If the thin metal bar of Fig. 13.14 is 0.925 m long and resonates as shown with a frequency of 2700 Hz, find Young's modulus of the metal. The density of the metal is 7.86 g/cm³.

   Reasoning   We need first to determine v, so that use can be made of Eq. (13.4). From Fig. 13.14, we see that 0.925 m is equivalent to $\lambda/2$. Hence $\lambda = 1.85$ m. Using $v = \lambda f$ gives

$$v = (1.85\ \text{m})(2700\ \text{s}^{-1}) = 5000\ \text{m/s}$$

Then from Eq. (13.4)

$$(5.00)(10^3\ \text{m/s}) = \sqrt{\frac{Y}{7.86 \times 10^3\ \text{kg/m}^3}}$$

or
$$Y \approx 2.0 \times 10^{11}\ \text{N/m}^2$$

Examination of the Young's moduli given in Table 8.2 indicates that this bar might well have been made of steel.

Summary

An oscillator undergoing simple harmonic motion can send a sinusoidal wave down a string. The distance between two adjacent crests of the wave is the wavelength of the wave λ. It is related to the vibration frequency f and the speed of the wave through $\lambda = v/f$. This is a general relation that applies to all periodic waves.

   The principle of superposition of waves states that a particle subjected to two or more wave disturbances responds to the vector sum of the individual disturbances.

   Under special circumstances, a string will resonate to a periodic driving force. For a string held rig-idly at both ends, resonance occurs if the string's length is $n(\lambda/2)$, where n can be any integer. The string then vibrates with a definite, steady pattern called a standing wave. In the pattern, certain points of the string remain motionless and are called nodes. Other points vibrate most widely and are called antinodes. The distance between adjacent nodes (or antinodes) is always $\lambda/2$ in a standing wave.

   When the particles subjected to a wave disturbance move perpendicularly to the direction of propagation, the wave is said to be transverse. Typical transverse waves are waves on a string and electromagnetic waves.

But if the particles move along the direction of propagation of the wave, the wave is said to be a compressional or longitudinal wave. Compression waves on a spring or in gases, liquids, and solids are waves of this type.

Minimum Learning Goals

Upon completion of this chapter you should be able to do the following:

1 Sketch a sinusoidal wave on a string and point out the following features of it: crest, trough, wavelength, amplitude.
2 Show by means of a sketch what happens to a wave pulse on a string as it strikes a fixed end and a free end.
3 State the relation among λ, v, and f for any wave.
4 Sketch several standing-wave forms for a string solidly held at its two ends. Point out the positions of nodes and antinodes. Use the number of segments in the standing wave to state the relation between L and $\lambda/2$ for each pattern sketched. Then compute f or v for the string provided L and either f or v are given.
5 Explain the difference between transverse and longitudinal waves and give examples of each.
6 Draw the compressional standing-wave resonance forms for the following different situations: (a) spring held rigidly at both ends; (b) rod held rigidly at one point with that point being either at an end or $0.5L$ or $0.25L$ from one end.
7 Compute the resonance frequency for each situation listed in item 6 provided sufficient data are given.

Important Terms and Phrases

You should be able to define or explain each of the following:

Wavelength
Crest; trough
$\lambda = v\tau = v/f$

Transverse wave; longitudinal wave
Resonance; standing wave
Node; antinode
In phase
A segment is $\lambda/2$ long

Questions and Guesstimates

1 The two idealized pulses shown in Fig. P13.1 are moving down the string at 20 m/s. Sketch how the string would look after 0.40 s. Repeat for 0.20 s.
2 Give qualitative physical arguments to justify the fact that the speed of transverse waves on strings should increase as the tension increases, but should decrease as the mass per unit length increases.
3 Some stringed musical instruments have strings made of catgut with a thin wire wound around it. What is the purpose of the wire winding?
4 A string fixed at both ends is vibrating with four antinodes, i.e., in four segments. Can one touch the string with a knifeblade without disturbing its vibration? Explain.
5 A variable-frequency oscillator sends waves down a string (of length L) whose ends may be considered nodal positions. At a distance of $L/5$ from one end, a tiny clamp holds the string nearly motionless although it still allows wave energy to pass by. Describe the standing waves one will notice on the string as the frequency of the oscillator is slowly increased from a very low value.
6 The little pieces of a string near the antinodes have a great deal of KE. Since the parts of the string near the nodes have very little KE, how did the energy get from the source to the antinodes?
7 At resonance in a string, the reflected wave cancels the vibration of the incident wave at the nodes. Was energy destroyed? What happened to it?
8 Two identical vibrators are attached at opposite

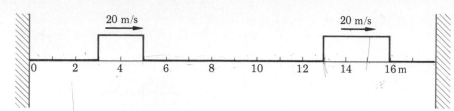

FIGURE P13.1

ends of a stretched string. They are adjusted so that when used one at a time, they will cause the string to resonate under transverse motion. When both vibrators are vibrating, will the string resonate? If so, under what conditions?

9 All common metals expand when heated. Try to devise a method for monitoring the temperature of a wire by a vibration-resonance technique. Steel wire lengthens about 0.001 percent for each degree change in temperature. Do you think the vibration method is feasible?

10 Is it possible for two identical waves traveling in the same direction down a string to give rise to a standing wave?

11 If you watch people trying to carry a large pan full of water, you will see that some are quite successful at it while for others, who are equally careful, the water sloshes badly. What makes the difference?

12 A steel guitar string is tuned to 330 Hz. Make an order-of-magnitude estimate of how much the frequency of the string changes when its temperature is lowered 20°C. (E)

Problems

1 Consider the wave pulses shown in Fig. P13.1. They are traveling along the string with speed 20 m/s. How long after the instant shown will the pulses once again look the way they are shown?

2 The wave shown in Fig. P13.2 is traveling to the right with a speed of 2.00 m/s. How long after the instant shown will the crest on the left have moved to the position of the crest on the right?

3 The wave shown in Fig. P13.2 travels toward the right with a speed of 100 cm/s. (a) How many wave crests go by point A each second? (b) What is the wavelength of the wave? (c) What is the frequency of the wave as computed from $\lambda = v/f$? (d) Compare the answers to parts a and c.

4 For the wave of Fig. P13.2, it is noticed that 25 wave crests pass point A in a second. Find the following for the wave: (a) wavelength; (b) frequency; (c) speed.

5 A certain radio station sends out radio waves whose frequency is 750 kHz. All radio waves travel with a speed of 3.0×10^8 m/s. How far apart are the crests of the radio wave from this station?

6 The radio-type waves sent out from a certain TV station have a frequency of 50 MHz (megahertz). These waves, like all electromagnetic waves, travel with a speed of 3.0×10^8 m/s. How far apart are the crests of the waves sent out by this station?

7 A string 2.0 m long has a mass of 4.0 g. It is stretched horizontally by running one end over a pulley and attaching a 1.0-kg mass to it. Find the speed of a transverse wave in the string.

8 How large a weight must be hung on the end of a thread 200 cm long if the speed of transverse waves in it is to be 4000 cm/s and if 100 cm of thread "weighs" 0.50 g? The string is horizontal.

9* An electric wire that has a mass of 5.0 kg for a 20-m length is strung between two poles 20 m apart. What must be the tension in the wire if it is to take 0.80 s for a pulse to travel along the wire from one pole to the other?

10* A clothesline is stretched between two poles that are 12.0 m apart. This length of line has a mass of 1.50 kg. If the tension in the line is 200 N, how

FIGURE P13.2

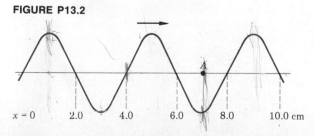

$x = 0$ 2.0   4.0   6.0   8.0   10.0 cm

long will it take for a pulse to travel from one pole to the other?

11* Refer to Fig. P13.1. What will the disturbance on the string look like when the center of the pulse shown at the right is coincident (after reflection) with the center of the other pulse?

12* What will the disturbance on the string of Fig. P13.1 look like when the midpoint of the wider pulse reaches the right-hand end of the string?

13 A standing wave is set up on a string by an oscillator that has a frequency of 120 Hz. The distance between nodes in the pattern is 43 cm. Find (a) the wavelength of the waves and (b) their speed.

14 A string 180 cm long is set into vibration so that a standing wave consisting of three segments exists on it. A variable-frequency strobe light is used to determine the frequency with which the string vibrates, and it is found to be 82 Hz. Find (a) the wavelength of the waves and (b) their speed.

15 As a strong wind blows over a telephone wire stretched between two poles 15 m apart, the wire hums in the wind. The hum has a frequency of 20 Hz, and so it is reasonable to assume that the wire is vibrating with this frequency. Assuming that the two poles are at nodes for the vibrating wire, what are (a) the probable wavelength and (b) the speed of the wave produced on the wire by the wind?

16 When a certain wire stretched between two points 60 cm apart is plucked near its center, it gives off the same tone as a 250-Hz tuning fork. Find the wavelength and the speed of the wave in the wire. *Hint:* Since the wire is plucked near its center, it will be vibrating with one antinode and with its ends as nodes.

17 A wire 80 cm long vibrates with four nodes, two of them at the two ends. Find (a) the wavelength and (b) the speed of the wave in the wire if the wire is being vibrated at 500 Hz.

18 A 60-cm-long string vibrates as a standing wave with four segments. The speed of the waves on the string is 18.0 m/s. Find (a) the wavelength and (b) the frequency of the waves.

19* A certain string held at its two ends will resonate to several frequencies, the lowest of which is 100 Hz. What are the next three higher frequencies to which it resonates?

20* A certain string resonates in three segments for a frequency of 60 Hz. Give four other frequencies to which it will resonate.

21* A vibrating string has one of its resonance frequencies at 350 Hz while the next higher resonance frequency is 420 Hz. What is its lowest resonance frequency?

22* The lowest resonance frequency for a string is 200 Hz. (a) If now the tension in the string is doubled, what will be its new fundamental resonance frequency? (b) If instead the tension is decreased by 10 percent, what will be the resonance frequency?

23* A wire held firmly at both ends vibrates in three segments to a frequency of f_1 and four segments to a frequency f_2. What is the ratio f_1/f_2?

24* Two wires are exactly the same length and are under the same tension. However, the lowest resonance frequency of one is only half that of the other. (a) Which wire has the largest mass per unit length? (b) By what factor is it bigger?

25* Show that the resonance frequencies of a string of length L and mass m can be given by

$$f_n = \tfrac{1}{2}n\sqrt{\frac{T}{mL}}$$

where T is the tension in the string.

26* A string of length L and mass m is subjected to a tension T such that it resonates in segments of length $L/3$. Show that T is given by $4mLf^2/9$, where f is the frequency with which the string is being vibrated.

27* An iron bar clamped at its center is set into longitudinal vibration by pulling outward on its end. (a) If the bar is 100 cm long, find the lowest frequency at which it will resonate. (b) Repeat if it is clamped at one end. ($v = 5 \times 10^3$ m/s.)

28* An iron bar 200 cm long is clamped 50 cm from one end. Give the lowest resonance frequency for longitudinal waves in the bar. ($v = 5 \times 10^3$ m/s.)

29* A coil spring is stretched to a length of 4.00 m and set into longitudinal vibration by a vibrator at one end. When the driving frequency is 2.0 Hz, the spring vibrates with five antinodes along its length. What is the speed of compressional waves on the spring? Assume the two ends are nodes.

30* A steel bar 1.00 m long is clamped at one end. What are the first three resonance frequencies for compressional waves? ($v = 5000$ m/s.)

31* A 40-cm-long brass rod is dropped one end first onto a hard floor, but it is caught before it topples over. According to a bystander who claims the gift of perfect pitch, the rod emits a tone of frequency 3000 Hz. If he is right, what is the speed of sound in brass?

32* A metal bar clamped at its center resonates in its simplest form to longitudinal waves of frequency 4000 Hz. (*a*) What will be its lowest resonance frequency when the clamp is moved to its end? (*b*) What will then be its two higher resonance frequencies?

33** A coil spring 4.0 m long lies on a smooth table, with one end fixed and the other end free. The speed of compressional waves in it is 800 cm/s. Give several frequencies with which one could keep hitting the end of the spring if resonating compressional waves are to be sent down the spring.

34** In Fig. P13.3 is shown a steel wire bent into a loop with a radius of 20 cm. The loop is held rigidly at *P* and vibrated transversely back and forth by an oscillator at *A*. Assuming both *P* and *A* to be nodes, find the first three resonance frequencies for the loop. ($v = 0.40$ m/s in the steel wire.)

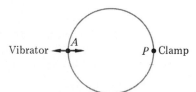

Vibrator   *A*   *P* Clamp

FIGURE P13.3

14
Sound

The concepts involving wave motion discussed in the last chapter are now applied to one particular form of wave motion, sound. Not only is a study of sound important in its own right, but also it affords us a valuable means for consolidating our knowledge of wave motion in general. We shall find that many of the ideas discussed in connection with sound are also important in our study of light and other types of wave motion.

14.1 Origin of Sound

Sound is usually defined to be any compressional disturbance traveling through a material in such a way that it is capable of setting the human eardrum into motion, thereby giving rise to the sensation of hearing. Notice that sound waves, because they are compressional waves in a material, require a substance in which to travel. Sound waves cannot travel through vacuum because there is nothing there to transmit the wave compressions. A common example is to show that the ringing of a bell cannot be heard if the bell is in a vacuum chamber. The bell itself is vibrating, but there is no material around the bell that can transmit the vibration.

The source of sound waves must be capable of sending out a compressional wave into the substance near the source. Whether it is a vibrating string on the sound box of a violin or an exploding firecracker, the sound source generates vibrations in the air or other sound-carrying material. Although sound waves, being compressional waves, can be transmitted through almost any substance, solid or fluid, we are usually concerned with sound waves in air.

14.2 Sound Waves in Air

Let us now consider the action of a loudspeaker when it is used to generate simple sounds. A simple loudspeaker consists of a cone-shaped sheet of flexible material, with a diaphragm that can be oscillated back and forth, as shown in Fig. 14.1. The oscillatory driving force F is applied at the center of the diaphragm.

When the diaphragm moves to the right, it compresses the air in front of it and a compressional wave travels out through the air. An instant later, the diaphragm is moving to the left, leaving a region of decreased air pressure in its wake, a so-called **rarefaction.** This disturbance, too, travels out from the loudspeaker. Hence a series of pressure waves travel out from the loudspeaker. They consist of alternate regions of high pressure (the compressions) and low pressure (the rarefactions). This situation is illustrated in Fig. 14.2.

The compressions of the air sent out by the loudspeaker as the speaker diaphragm moved toward the right are shown as A, B, and C in that figure. When the diaphragm was pulled to the left, a region of low pressure was left behind, and the rarefactions at P, Q, and R were sent out. A plot of the pressure in the air along this sound wave is given in the lower portion of Fig. 14.2. Notice that the center line is not at zero pressure, but is instead at a value equivalent to the average atmospheric pressure. The rarefactions are regions of somewhat reduced pressure. It is assumed in drawing the figure that the diaphragm was oscillating sinusoidally. Even for very loud sounds, the actual pressure variations are only about 0.01 percent of atmospheric pressure.

It is sometimes convenient to talk about the displacement of the air molecules by the sound wave rather than about the pressure in the wave. Certainly the air particles will be moved back and forth in much the same way as the diaphragm moves back and forth. The motion of the air is back and forth along the direction of propagation of the wave, and so this is a longitudinal wave. Moreover, since the air will vibrate in the same way as the diaphragm vibrates, a plot of the particles' displacements will appear much like the graph shown in Fig. 14.2. However, the two graphs would be displaced by one-fourth cycle from each other. Why?

We often draw sound waves propagating from a source in the manner

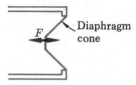

FIGURE 14.1
The flexible loudspeaker diaphragm vibrates back and forth to send out compressions in the air.

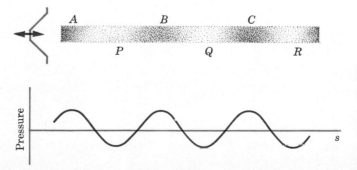

FIGURE 14.2
The sound wave sent out by the loudspeaker consists of alternate high- and low-pressure regions in the air. In practice, P changes by only about 0.01 percent or less.

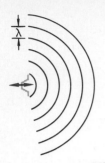

pictured in Fig. 14.3. The solid lines represent a particular point on the waves as they spread out and travel through the air. Often these lines (called **wavefronts**) are drawn at the crests of the waves. It is seen that *the distance between these wavefronts is the distance* λ *from crest to crest on the wave.* Even though most loudspeakers focus the sound waves somewhat in a particular direction, the waves do spread out in many directions from the source, much as is indicated in Fig. 14.3. The waves travel through the air at the speed of sound.

FIGURE 14.3
Crests of waves (wavefronts) spreading out from the loudspeaker are one wavelength apart.

14.3 Speed of Sound

Sound waves are compressional waves in matter. We saw in the last chapter that the speed of such waves in fluids is given by Eq. (13.4).

$$v = \sqrt{\frac{E}{\rho}} \qquad (13.4)$$

You will recall that ρ is the density of the fluid, and the bulk modulus E is given by

$$E = \frac{\Delta P}{\Delta V/V}$$

where ΔV is the change in the volume V caused by a pressure change ΔP. Notice that fluids that are difficult to compress (large E) will have large v. *Sound travels fastest in materials that are least compressible and that have low densities.* We have already listed the speed of sound for various materials in Table 13.1.

Of particular practical importance is the speed of sound in air. At 0°C, it is 331 m/s, that is, 1086 ft/s. We can gain a great deal of insight into the speed of sound in air by applying Eq. (13.4) to the case of an ideal gas. If you refer to Illustration 9.11 and Prob. 40 at the end of Chap. 11, you will find that we have already computed E, the bulk modulus, for an ideal gas. In the case of an isothermal compression, $E = P$, where P is the gas pressure. But in an adiabatic compression, $E = \gamma P$, where $\gamma = c_P/c_V$, the specific heat ratio. It turns out that the compressions in sound waves in air occur so rapidly, and air is such a poor heat conductor, that the compressions in them are essentially adiabatic. Therefore $E = \gamma P$ for the present purpose.

We thus find that the speed of sound in an ideal gas is given by

$$v = \sqrt{\frac{\gamma P}{\rho}}$$

But the pressure of the gas is given by the ideal-gas law, $PV = (m/M)(RT)$. Therefore

$$P = \frac{m}{V}\frac{RT}{M} = \rho\frac{RT}{M}$$

Substituting this in the velocity equation yields

$$v = \sqrt{\frac{\gamma RT}{M}} \qquad (14.1)$$

This is a very interesting result because it tells us that *the speed of sound in an ideal gas is independent of the gas pressure or density.* It depends only on the temperature of the gas, together with the values of γ and M. In the case of air near room temperature, the speed of sound increases by about 0.60 m/s for each 1°C rise in temperature.

Illustration 14.1 Find the speed of sound in helium gas at 27°C.

Reasoning Helium is a monatomic gas with $M = 4.0$ kg/kmol. Its value for γ is given in Table 10.2 and is 1.66. Therefore we have

$$v = \sqrt{\frac{\gamma RT}{M}} = \sqrt{\frac{(1.66)(8314\,\text{J/kmol} \cdot \text{K})(300\,\text{K})}{4.0\,\text{kg/kmol}}}$$
$$= 1020\ \text{m/s}$$

Notice that, because of the low molecular weight of helium, the speed of sound is much larger in helium than in air.

14.4 Intensity and Loudness of Sounds

We saw in the last chapter that the vibrator that sends a wave down a string also sends energy with the wave. Indeed, all waves carry energy along with them. Sound waves are no exception. For example, the loudspeaker in Fig. 14.3 sends out sound-wave energy. The energy flows in the direction of propagation of the wave.

Suppose a sound wave is traveling in the propagation direction shown in Fig. 14.4. We define the intensity of the wave in terms of the energy carried by the wave. To be precise, we erect a unit area perpendicular to the direction of propagation as shown. We then define the intensity of the wave to be the energy carried per second through this unit area by the wave. Since power is energy per second, **sound intensity** *is the power passing through a unit area erected perpendicular to the direction of propagation of the wave. The units of sound intensity are typically watts per square meter.* Representative sound intensities are listed in Table 14.1. Notice what a wide range of sound intensities the ear can hear. The ear is a truly remarkable measuring device.

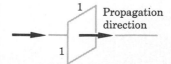

FIGURE 14.4
Sound intensity is measured as the amount of energy flowing through a unit area per second. The area must be perpendicular to the direction of propagation, as shown.

TABLE 14.1
APPROXIMATE SOUND INTENSITIES

Type of sound	Intensity, W/m^2	Intensity level, dB
Pain-producing	1	120
Jackhammer or riveter*	10^{-2}	100
Busy street traffic*	10^{-5}	70
Ordinary conversation*	10^{-6}	60
Average whisper*	10^{-10}	20
Rustle of leaves*	10^{-11}	10
Barely audible sound	10^{-12}	0

*For a person near the source of the sound.

As you can see by perusing Table 14.1, the *loudness* of a sound does not correspond directly to its intensity. Although the sound of ordinary conversation is certainly louder than the sound of a whisper, it is certainly not 10,000 times as loud even though the intensities differ by this factor. Loudness is a qualitative property that is different from intensity. However, there is a way that we can specify the approximate loudness of sounds.

To correspond more closely with the way the ear judges loudness of sound, an intensity-level scale, called the *decibel scale,* is defined. The intensity level (or sound level) measured in decibels (dB) is defined by

$$\text{Intensity level (dB)} = 10 \log \frac{I}{I_0} \qquad (14.2)$$

where I is the intensity of the sound under consideration and I_0 is a reference intensity usually taken to be 10^{-12} W/m^2. Table 14.2 relates various intensities to the corresponding intensity levels.

To interpret the relation of the decibel scale to loudness, the following fact is often useful. If the loudness of a sound is to be doubled, its intensity

TABLE 14.2
THE DECIBEL* SCALE

Intensity, W/m^2	Intensity level, dB
10^{-12}	0
10^{-11}	10
10^{-10}	20
10^{-9}	30
$\vdots$	$\vdots$
10^{-1}	110
1	120
10	130

*1 B (bel) = 10 dB and is named after Alexander Graham Bell, the inventor of the telephone.

level must be increased by about 10 dB. For example, we see that the intensity levels of the average whisper (20 dB) and of ordinary conversation (60 dB) differ by 40 dB. In increasing by 40 dB, the sound level has increased by 10 dB four times. Thus the loudness has undergone four doublings. It has been increased by $2^4 = 16$ times. Notice also that the intensity level of barely audible sound is zero. This is a result of the fact that the reference intensity I_0 is taken equal to 10^{-12} W/m², the least intensity the ear can detect.

Illustration 14.2 Find the sound level in decibels of a sound wave that has an intensity of 10^{-5} W/m².

Reasoning From Eq. (14.2) we have, after replacing I by 10^{-5} W/m² and I_0 by 10^{-12} W/m²,

$$\text{Sound intensity level (dB)} = 10 \log \frac{10^{-5}}{10^{-12}}$$
$$= 10 \log 10^7 = (10)(7) = 70 \text{ dB}$$

14.5 Frequency Response of the Ear

People vary in their ability to hear sounds. We all know persons whose hearing has been in some way impaired. The sensitivity of their ears has decreased considerably below that of a normal person. However, most people agree fairly well on the intensity of a sound that is just audible and also on how loud a sound must be before it causes pain. We can therefore set average limits of audibility for the normal human ear. The lower limit is the intensity of a just-audible sound, while the upper limit is a sound so intense that it hurts the ear.

The response of the ear to sound is also dependent on the frequency of the sound. The ear is more sensitive to some frequencies than to others. *Most people cannot hear compressional waves in the air which have frequencies higher than about 20,000 Hz. Waves of higher frequency than this are called* **ultrasonic waves**—meaning "beyond" or "above" sound in the sense of frequency. Similarly, we are not able to hear sounds below a certain frequency, about 20 Hz. *The ear is most sensitive near 3000 Hz.* At frequencies other than this, the sound must be made more intense before it is audible. This variation of sensitivity of the ear with frequency is shown in Fig. 14.5.

Frequencies of Audible Sound

The lower curve in Fig. 14.5, labeled "Threshold of hearing," shows the minimum audible intensity level as a function of frequency. For example, a sound wave with a frequency of 1000 Hz cannot be heard unless it has an intensity level of about 5 dB. But at a frequency of 100 Hz, the sound level must be about 30 dB if the sound is to be audible. Of course, near the

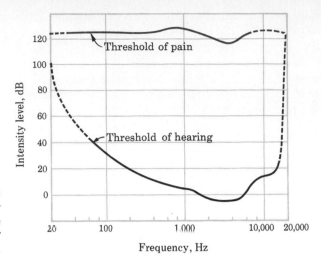

FIGURE 14.5
The normal ear can hear
sounds which have
intensities above the lower
curve in the figure.

frequency limits of audible sound, the intensity must be very large for a sound to be heard.

The upper curve in Fig. 14.5, labeled "Threshold of pain," shows how intense a sound must be to produce pain. It is not very frequency-dependent and shows that an intensity level of about 120 dB is painful. Such high sound levels can cause permanent damage to the ear. Indeed, long-term exposure to sounds in the range down to about 90 dB can cause hearing impairment. Of course, loss of hearing can be caused by other factors as well.

There are some people who have a rather strange impairment of hearing, often unknown to themselves; they are unable to hear sound frequencies above perhaps 6000 Hz. Since most of the sounds we hear consist, partly at least, of frequencies below this, these people are still able to hear sounds that are audible to other people. However, the *quality* of the sounds they hear will be quite unlike that of sounds heard by a normal person. Quality and pitch of sound are complex, rather subjective properties of sounds. They are discussed in the next section.

14.6 Pitch and Quality of Sound

If a high-quality loudspeaker is driven by an electrical oscillator that generates a sine-wave voltage, the sound wave given off by the loudspeaker will be an almost pure sine wave having the frequency being generated by the oscillator. Anyone who is not tone-deaf will be able to compare the pitch of this sound with another sound. If the frequency of the oscillator driving the loudspeaker diaphragm is increased, the listener will agree at once that the new sound has a higher pitch than the first tone. In such cases pitch and

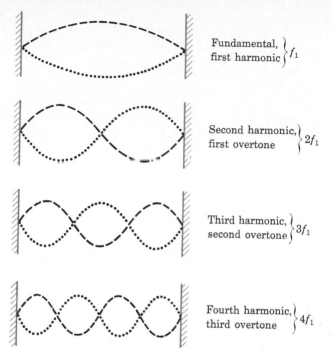

Fundamental, first harmonic $\}f_1$

Second harmonic, first overtone $\}2f_1$

Third harmonic, second overtone $\}3f_1$

Fourth harmonic, third overtone $\}4f_1$

FIGURE 14.6
The four simplest modes of motion for standing waves on a string.

frequency are nearly synonymous. But pitch, our qualitative conception of whether a tone is high like that of a soprano or low like that of a bass singer, is somewhat dependent on loudness and other factors.

Single-frequency sound waves are not very common, however. When a violin string is plucked or bowed, for example, the sound wave given off will not be a pure sine wave. This is readily apparent to anyone who has compared the tone obtained from a violin by an expert with that obtained by a beginner. In one case the tone will be full and melodious, whereas the beginner may obtain rather rasping sounds on the same string. We say that the quality of the tone is different in the two cases.

As we saw in Chap. 13, a string may resonate in more than one way. Typical, simple vibration patterns are shown in Fig. 14.6. These modes of vibration are named as shown in the figure. Since the wavelengths in the cases shown are in ratio $1 : \frac{1}{2} : \frac{1}{3} : \frac{1}{4}$ and since $f = v/\lambda$, the vibration frequencies are in the ratio $1 : 2 : 3 : 4$, as indicated on the figure.

It is very difficult, however, to cause a string to vibrate exactly as shown in any single pattern of Fig. 14.6. Instead, *if the string is bowed near one end, as is usually the case, it vibrates in several ways at once, causing several harmonics to occur at the same time. To find the resulting vibration, it is necessary to add the waves for the various harmonics excited.* Of course, because each harmonic will be excited to a different intensity, the proper intensity must be used for each harmonic when they are added.

A typical example for a vibrating violin string is shown in Fig. 14.7.

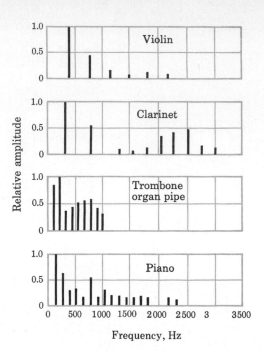

The amplitudes of vibration of the various harmonics are indicated by the lengths of the vertical bars. In this case, all but the first two harmonics are relatively weak. But, clearly, the tone heard by the ear will be different from either the first or the second harmonic alone.

Also shown in Fig. 14.7 are similar diagrams for the sounds of various other instruments. The piano string shows many more harmonics than the violin string. This is probably the result of the way in which the strings are set into vibration. The violinist pulls a bow slowly across the string, whereas the piano string is excited by a hammer blow. Moreover, the soundboard construction is of importance.

The quality of a sound or tone is dependent on the number and type of harmonics occurring in the sound. If all sounds were pure sine waves, much of the variety of sound would be lost. The tone of all human voices would be the same, and a voice could be recognized only by a characteristic frequency or inflection. Much of the beauty of music would be lost if the qualities of all sounds were the same.

In a complex sound such as that shown for the piano or even the clarinet, *the pitch of the sound is not always easily defined. It can no longer be taken as identical to the frequency of the sound, since the sound contains several nearly equal intensity waves of various frequencies.* In such cases, it is difficult to match tones accurately. For this reason it is quite common for an inexperienced singer to sing a tone that is twice the frequency of the fundamental tone of a violin and still be unaware that the tones are not the same. Moreover, many listeners would be unable to recognize that the singer was not sounding the desired tone.

14.7 Interference of Sound Waves

Suppose we have a pipe system such as that shown in Fig. 14.8. A pure sine wave is sent into the pipe at the left by a loudspeaker. The sound splits, half the sound intensity going up through section A and the remaining half through the lower section. Each pipe carries half the sound, and this sound is a wave motion in the air, a series of compressions and rarefactions.

Eventually the two waves are reunited at the outlet on the right, at D, where a sound detector such as the ear or a microphone is placed. It is observed that the sound emitted at D can be made loud or very faint, depending on the position of the sliding pipe EAF. Moreover, as the pipe at A is slowly pulled upward, the sound intensity at D becomes alternately large and small. We now investigate the reasons for this interference phenomenon.

When a compression of the air is caused by a rightward movement of the loudspeaker diaphragm, a region of high pressure starts into the pipe at C. The region of high pressure, the compression, causes compressions to move in both pipes, toward A and toward B. We say that the original compression in the entrance pipe at C has split into two equal parts and that one part went up toward A, the other down toward B. Since the compression, represented by the filled circles, propagates through the pipe with the speed of sound, the compressions will reach point D simultaneously *provided that L_A, the pipe length through A, is the same as L_B through B.* They will then reunite at D, giving the original compression, and this will then exit from the pipe system at D. This situation is shown in Fig. 14.8a.

Of course, the loudspeaker is sending out a pure sound, a sinusoidal wave, consisting of alternate compressions and rarefactions. However, if $L_A = L_B$, the two portions of the original compression will always meet at D. The same is true for the two halves of the rarefactions. Hence, compressions and rarefactions identical to the originals will exist at point D, and the

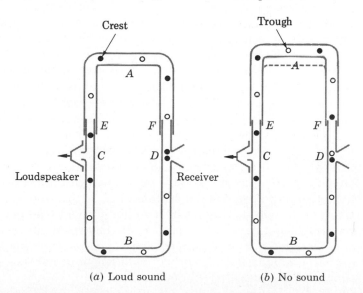

(a) Loud sound   (b) No sound

FIGURE 14.8
A sound wave from the loudspeaker is split into two parts. When they are reunited at D, either a loud or a weak sound results, depending on the path lengths traveled by the two parts.

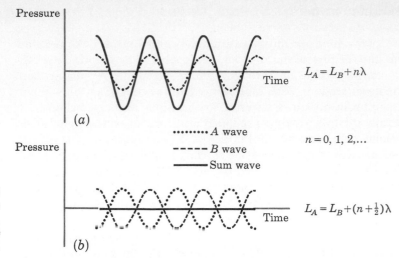

FIGURE 14.9
Waves *A* and *B* can
reinforce or cancel
depending on their relative
phase.

sound will be loud. Examine Fig. 14.8*a* to convince yourself that as time progresses, the *A* and *B* waves will reinforce each other when they meet at the receiver.

We can represent the situation that prevails at the receiver in Fig. 14.8*a* by means of the graph shown in Fig. 14.9*a*. Plotted there are the *A* wave (dots), the *B* wave (dashes), and the sum of the *A* and *B* waves (color) at the receiver. Notice that, as we have already concluded, crest falls on crest and trough falls on trough so as to cause a large-amplitude wave (loud sound) at the receiver.

Let us now return to Fig. 14.8 and look at the situation in part *b*. Notice that path *A* has been lengthened by sliding the *A* pipe away from the source and receiver. As you can see, path *A* is now one-half wavelength longer than path *B*. Consequently, when the *A* and *B* waves rejoin at the receiver, a crest of the *B* wave joins a trough of the *A* wave. Furthermore, as time goes on, crests always meet troughs when the two waves meet. Because a crest is canceled by a trough, we conclude that the two waves in Fig. 14.8*b* will rejoin so as to cancel each other at the receiver. No sound is heard at the receiver in this case. This cancellation of waves is also shown by means of the graph in Fig. 14.9*b*. We are therefore led to the following conclusion: *If two identical waves are displaced one-half wavelength relative to each other, the waves will cancel.*

This conclusion can be stated in a more compact way by using the terminology of the reference circle described in Chap. 12. You will recall that the vertical displacement of a particle traveling with constant speed around a circle varies sinusoidally. As the particle completes one full circle, it executes one complete vertical vibration. If the vibrating particle were acting as a source for a wave on a string, it would send out one complete wavelength as it traversed the circle once. In view of this, we often say that *a path length that is one wavelength long is equivalent to a phase of 360°*, one rotation on the circle. Similarly, a path length of one-half wavelength is equivalent to a phase of $\frac{1}{2} \times 360° = 180°$.

We extend this terminology to path length differences. For example, when one wave is half a wavelength behind another wave, we say the waves are a half wavelength, or 180°, out of phase with each other. Hence a retardation of $\frac{1}{2}\lambda$ is equivalent to a phase difference of 180°. Similarly, 90° is equivalent to $\frac{1}{4}\lambda$, and so on. Referring to Figs. 14.8 and 14.9, we see that the waves are in phase in *a* but are 180° out of phase in *b*.

Returning now to the apparatus of Fig. 14.8*b*, we see that if L_A is increased still further by pulling the pipe at *A* upward an additional amount, the *A* wave will be held back still more. When L_A is one whole wavelength longer than L_B, a crest of the *A* wave will again reach *D* at the same time as a crest of the *B* wave. Although these crests did not start together at point *C* (the *A* crest occurred one compression prior to the *B* crest), this is of no concern when they reach point *D*. The wave will appear once again as in Fig. 14.9*a*, and so sound of the original intensity will be produced at *D*.

Similarly, if L_A is increased until its length is $1\frac{1}{2}$ wavelengths longer than L_B, the situation in Fig. 14.9*b* will arise once again. No sound will be heard at *D*. Moreover, it is clear that no sound will be heard at *D* whenever $L_A = L_B + (n + \frac{1}{2})\lambda$, where *n* can be any integer, including zero. *When two waves exactly cancel each other* in this way, *we say that there is complete* **destructive interference.** Moreover, reinforcement, which we call **constructive interference,** of the waves occurs whenever $L_A = L_B + n\lambda$, where *n* is any integer, including zero. Of course, if L_B is greater than L_A, these interferences will also occur.

Conditions for Destructive and Constructive Interference

It is not necessary to have a pipe system such as this to obtain interference. We need only obtain two waves *which are exactly the same in their frequency and shape.* If these waves are combined after traveling different distances they will interfere with each other. The following illustration furnishes another example of interference. We see later in this text that the interference of light waves is of very great importance.

Illustration 14.3 Two identical sound sources send identical waves toward each other, as shown in Fig. 14.10. They both send out wave crests at the same time, and the sound waves have $\lambda = 70$ cm. Loud sound is heard at the midpoint, point *P*. But as one proceeds away from *P*, the sound intensity decreases to near zero at point *Q*. As one proceeds still farther beyond *Q*, the intensity again increases. How far is point *Q* from point *P*?

Reasoning It is easy to see why the sound is large at *P*. This point is equidistant from the two sources, and so crests from both *A* and *B* will arrive there at the same time. The waves from the two sources will therefore reinforce at *P*.

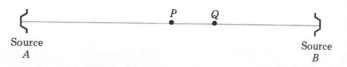

FIGURE 14.10
A loud sound is heard at *P*, while a very weak sound is heard at *Q*. Explain how this can be.

But as we move farther toward Q, the A wave has to travel farther than the B wave does to reach us. To reach some point such as Q, the wave from A will have to travel one-half wavelength farther than the wave from B does. In that case the two waves will be $\frac{1}{2}\lambda$, or 180°, out of phase, and they will cancel each other. Therefore very little sound will be heard at Q.

Knowing that the distance from A to Q is $AP + PQ$ and that the distance from B to Q is $BP - PQ$, we have, since $AP = BP$,

$$\text{Path difference for two waves} = AP + PQ - (BP - PQ)$$
$$= 2PQ$$

But we have already concluded that because Q is a point of near-zero sound, the path length difference is $\frac{1}{2}\lambda$. Therefore, we have

$$2PQ = \tfrac{1}{2}\lambda \qquad \text{or} \qquad PQ = \tfrac{1}{4}\lambda$$

Since λ is given as 70 cm, we see that $PQ = 17.5$ cm.

14.8 Beats

When people tune a string on a piano, they do not merely listen to see if the tone of the string is the same as that of the standard tuning fork used for comparison. Instead, they use a much more precise way of judging the accuracy to which the string is adjusted. They listen to *beats* between the sounds of the two vibrating objects. This is a very sensitive method for obtaining agreement of frequency and is widely used for that purpose. We now explain the phenomonon of beats between the sounds given off by two nearly tuned vibrating bodies.

Suppose that two vibrating bodies, A and B, vibrate with slightly different frequencies. For example, they might be two loudspeakers as shown in Fig. 14.11. Consider what happens if source A is vibrating with frequency 1000 cps and source B with $f = 999$ cps. At time $t = 0$, shown at top left in the figure, we say that the loudspeakers are in phase; i.e., both are sending out a compression as shown. If the ear is equidistant from the loudspeakers, the compressions will arrive at the ear together, and a large compression will be heard.

As time goes on, loudspeaker B, vibrating at a slightly slower frequency than A, will begin to fall behind. After $\frac{1}{2}$ s, loudspeaker A will have vibrated 500.00 times and will just be sending out a compression, as shown for $t = \frac{1}{2}$ s in the top part of Fig. 14.11. On the other hand, loudspeaker B will have vibrated only 499.50 times and will be exactly one-half cycle behind loudspeaker A. It will be sending out a rarefaction, as shown. The compression from A will now reach the ear at the same time as the rarefaction from B, and they will exactly cancel each other. Hence, at this instant, no sound will be heard.

As time continues, loudspeaker B will fall still further behind A. After

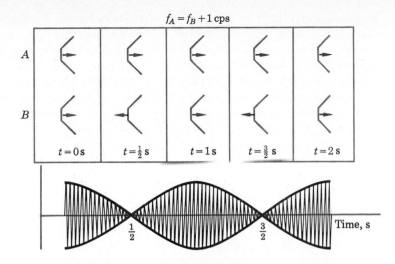

$f_A = f_B + 1\,\text{cps}$

A

B

$t=0\,\text{s}$ $t=\frac{1}{2}\,\text{s}$ $t=1\,\text{s}$ $t=\frac{3}{2}\,\text{s}$ $t=2\,\text{s}$

$\frac{1}{2}$ $\frac{3}{2}$ Time, s

FIGURE 14.11
Beats result when two
similar sources vibrate with
slightly different frequencies.

1 s, B will have vibrated exactly 999 times, while A has vibrated exactly 1000 times. Source B has now fallen exactly one cycle behind source A. Hence once again they will both be sending out compressions together, and a loud sound will be heard.

This process will continue, as shown in the succeeding portions of the upper part of Fig. 14.11. At times of 0, 1, 2, 3, . . . s, the sources will be in phase and a loud sound will be heard. But at times of $\frac{1}{2}$, $1\frac{1}{2}$, $2\frac{1}{2}$, . . . s, nothing will be heard, since the sources are 180° out of phase. Hence the ear will hear a series of sound pulses, or beats, one beat each second.

An attempt to show graphically what the beating sound wave is like is shown in the lower part of Fig. 14.11. We show there the combined wave from the two sources,[1] together with its envelope, as a function of time. As you see, the combined wave has maximum intensity at $t = 0$, 1, 2 s, as we discussed previously. As this combined wave strikes the ear, a beat is heard each time that a maximum in the sound envelope reaches the ear.

Notice that the frequencies differ by exactly 1 cps in this case, and exactly one beat is heard each second. You should carry through the above analysis for the case in which the sources differ by, say, 3 cps. In this case there will be 3 beats per second. In fact, *the number of beats per second equals the difference in frequencies of the two sources.* The piano tuner tries to adjust the tension in a piano string until the beats between the sound from the string and the sound from the tuning fork have become extremely far apart in time.

Sometimes the beat frequency between two sound waves gives rise to an audible sound. This occurs when the beat frequency is in the audible range. In that case, even though the beats are too close together in time to

---

[1]It is shown in more advanced courses that the combined wave is much like the wave shown. It is a nearly sinusoidal wave that has the same frequency as the average frequency of the component waves.

be heard individually, they make their presence known by a sound wave that has the same frequency as the beat frequency.

This phenomenon of beats occurs in any type of vibration that is a combination of vibrations from two sources. It may therefore be used to compare frequencies of vibrations other than sound. As in the case illustrated here, *it is a sensitive means for comparing frequencies.* In general, *two waves of any single type will give rise to beats when the two waves are combined, provided the waves differ in frequency. The beat frequency is the difference between the frequencies of the two waves.*

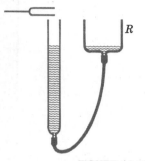

FIGURE 14.12
When the water is at just the right height in the tube, resonance will occur.

14.9 Resonance of Air Columns

If you hold a vibrating tuning fork over the open end of a glass tube partly filled with water, the sound of the tuning fork can be greatly amplified under certain conditions. While the fork is held as shown in Fig. 14.12, the reservoir R is raised so that the water level will rise in the glass tube. At a certain height of the water, the air in the tube will resonate loudly to the sound being sent into it by the tuning fork. In fact, there are usually several heights at which the tube will resonate.

The situation here is much like the case of the vibrating string. A wave is sent down through the air in the tube (or through the string); it is reflected at the end and is once again reflected when it reaches the position of the vibrating source. If the tube (or string) is just the proper length, the reflected wave will be reinforced by the vibrating source as it travels down the tube a second time. Hence, the vibrating source builds up a large motion much in the way a swing can be pushed very high with a series of small pushes. This whole general idea of resonance is discussed fully in Chap. 13. We can apply the results found there to the case of a vibrating air column.

The tube of Fig. 14.12 will have a node at the closed end and an antinode near the open end. This follows because the air molecules cannot move at the closed end, for the water at that end will not allow them to move downward. At the open end, the molecules can easily move out into the open space, and so at that point there will be maximum motion, i.e., an antinode.[2]

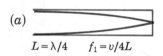

(a)

$L = \lambda/4 \quad f_1 = v/4L$

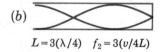

(b)

$L = 3(\lambda/4) \quad f_2 = 3(v/4L)$

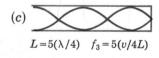

(c)

$L = 5(\lambda/4) \quad f_3 = 5(v/4L)$

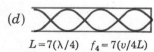

(d)

$L = 7(\lambda/4) \quad f_4 = 7(v/4L)$

FIGURE 14.13
Simple modes of vibration for a resonating pipe closed at one end.

We see, then, the tube will resonate only to waves that will fit into the tube with a node at one end and an antinode at the other. Several of these waves are shown schematically in Fig. 14.13. Remember that the distance between two nodes or two antinodes is $\lambda/2$. Hence, the distance from a node to an antinode is $\lambda/4$. If we call the length of the pipe L, the length in Fig. 14.13a is from a node to an antinode, or $L = \lambda/4$. In part *b*, the tube is three node-to-antinode lengths long; hence $L = 3(\lambda/4)$, and so on.

The frequencies for the resonances shown in Fig. 14.13 can be found from the fact that $f = v/\lambda$. Upon using the values found for λ in terms of L

---

[2]The antinode is not precisely at the end of the pipe. However, for pipes with radii much smaller than λ, this complication can usually be ignored.

as shown in Fig. 14.13, the resonance frequencies shown are easily computed. Notice that the first overtone f_2 is just $3f_1$. The second overtone f_3 is $5f_1$, and so on. The customary terminology is to call the frequency $2f_1$ the second harmonic, $3f_1$ the third harmonic, etc. In this case, the tube resonates only to the odd harmonic frequencies. Since the frequency of a given tuning fork is usually known, resonances in a tube such as the one shown in Fig. 14.12 can be used to measure v, the velocity of sound.

A pipe or tube need not be closed at one end to resonate. For example, you can use a piece of glass tube as a whistle by blowing across either of its ends. The simplest possible resonances of a tube that is open at both ends are shown in Fig. 14.14. In each case the ends of the tube are at antinodes because the air is free to move there. The resonant frequencies are computed in the usual way by using the fact that $f = v/\lambda$, with λ being determined as indicated in the figure.

When you blow across the end of a tube, this very complex process sends waves of many frequencies down the tube. Of this multitude of frequencies, the tube will resonate to only one or two. For this reason, the resonating tube usually gives off a loud sound of a single frequency. However, if you try hard enough, you can often cause the tube to resonate to two frequencies at the same time and thereby give off two tones simultaneously.

Many musical instruments make use of resonating air columns. The flute and piccolo are basically tubes whose length can be changed by means of holes along the tube. A clarinet is similar, but in that case the sound waves are generated by the vibration of a reed in the mouthpiece. More complex tube resonance systems are seen in the trumpet, trombone, and tuba, for example. These instruments elicit various resonance tones by changing the length of the resonating pipe. In addition, the sound waves in these instruments are excited by the vibration of the player's lips in the mouthpiece.

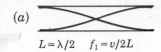
(a)
$L = \lambda/2 \quad f_1 = v/2L$

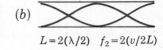
(b)
$L = 2(\lambda/2) \quad f_2 = 2(v/2L)$

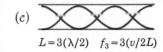
(c)
$L = 3(\lambda/2) \quad f_3 = 3(v/2L)$

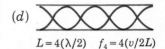
(d)
$L = 4(\lambda/2) \quad f_4 = 4(v/2L)$

FIGURE 14.14
Simple modes of vibration for a resonating pipe open at both ends.

14.10 Doppler Effect

Let us now turn to an entirely different aspect of waves of all kinds and of sound waves in particular, a phenomenon referred to as the *Doppler effect*.[3] You have certainly experienced this effect, although you may have been unaware of it. If you are standing near the edge of the road when a fast-moving car approaches you, the sound of the car or of its horn behaves in a peculiar way. The pitch of the sound is higher when the car is approaching you than it is after the car passes you and is receding. You can even notice this same effect when an airplane flies by you high overhead. A similar effect occurs with light and electromagnetic waves, as we see later. Indeed, the speed trap's radar uses this effect to measure speeds. We now investigate this very important aspect of wave behavior.

Moving Source: To see how the Doppler effect originates, let us consider

---

[3]Named after Christian Johann Doppler, who was the first to explain the effect clearly.

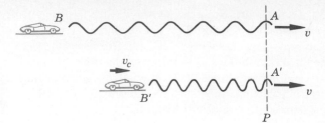

FIGURE 14.15
The Doppler effect results if
a sound source is moving.

the sinusoidal sound waves emitted by the identical horns of the two cars shown in Fig. 14.15. The car in the upper part of the diagram is standing still. Its horn vibrates with frequency f (period $\tau = 1/f$) and emits a sound whose wavelength is λ. As usual, $\lambda = v/f$, where v is the speed of sound in air. If we were to stand at point A, the number of wave crests that would pass us in 1 s could be found in the following way. The wave travels a distance $v\,\Delta t$ to the right in a time Δt. Hence a length of wave equal to $v\,\Delta t$ will pass us at point A during that time. In this length there are $(v\,\Delta t)/\lambda$ wavelengths, and so $(v\,\Delta t)/\lambda$ wave crests will pass us at point A in time Δt. But the frequency of a wave at A is equal to the number of wave crests that pass A in a time $\Delta t = 1$ s. We therefore find that the frequency of the wave is v/λ, which is simply the frequency of vibration of the horn. This confirms are usual supposition that the frequency of the sound we hear is the same as the frequency of the source that emits the sound.

However, such is not the case if the sound source is moving with velocity v_c, as shown in the lower part of Fig. 14.15 and also in Fig. 14.16. Notice that because the source is moving in the same direction as the waves, the wavelength of the waves is shortened. Let us call the wavelength in the lower part of Fig. 14.15 λ'. Then because the car moves ahead a distance $v_c\tau$ during the time taken for the horn to send out one complete wavelength, the distance between crests is shortened by that amount, $v_c\tau$. Therefore

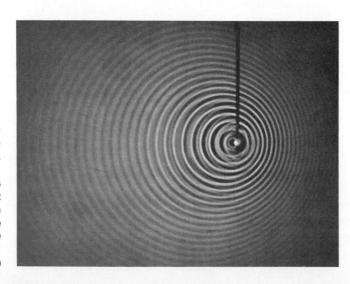

FIGURE 14.16
The vertical rod vibrates so
its lower end acts as a
water wave source.
Because the source is
moving toward the right, it
causes the wavelengths in
that direction to be
shortened. (*Education
Development Center,
Newton, Mass.*)

$$\lambda' = \lambda - v_c \tau$$

But $\lambda = v/f$ and $\tau = 1/f$, and so this becomes

$$\lambda' = \frac{v}{f} - \frac{v_c}{f}$$

We would like to express λ' in terms of the frequency f' of sound that an observer at point A' would hear. To do that, we notice that the wave travels a distance v in 1 s and that the wave crests are λ' apart. As a result, v/λ' wave crests will pass the listener each second, and so the frequency of sound heard will be $f' = v/\lambda'$. Therefore $\lambda' = v/f'$. Replacing λ' by this value, we have

$$\frac{v}{f'} = \frac{v}{f} - \frac{v_c}{f} = \frac{1}{f}(v - v_c)$$

This expression can be simplified by multiplying through the equation by f'/v, to give

$$1 = \frac{f'}{f}\left(1 - \frac{v_c}{v}\right)$$

from which

$$\frac{f'}{f} = \frac{1}{1 - v_c/v} \tag{14.3}$$

As we expected, f' is larger than f. Indeed, if the speed of the car v_c is equal to the speed of sound v, the denominator becomes zero, and so f' becomes infinite. A glance at Fig. 14.15 shows us that this is reasonable because when $v_c = v$, the distance between crests is zero and all the wave crests will be piled up on one another. We see in the next section that this results in a shock wave.

In Fig. 14.16 we see the water waves sent out by a moving wave source. The Doppler effect is shown clearly. Notice that not only is the wavelength shortened in the forward direction, but also it is lengthened in the reverse direction. An observer to the rear of a moving source notices that the frequency of the wave is decreased. To find f'/f in this case, we can repeat our derivation of Eq. (14.3), using the substitution $\lambda' = \lambda + v_c\tau$. The end result is Eq. (14.3) with the minus sign replaced by a positive sign.

Illustration 14.4 A car is moving at 20 m/s along a straight road with its 500-Hz horn sounding. You are standing at the side of the road. What frequency will you hear as the car is (*a*) approaching and (*b*) receding from you?

Reasoning We know that the speed of sound v is about 340 m/s, $v_c = 20$ m/s, and $f = 500$ Hz. Using Eq. (14.3), we have the following:
(*a*) car approaching

$$f' = f\frac{1}{1 - v_c/v} = \frac{500 \text{ Hz}}{1 - \frac{20}{340}} = 531 \text{ Hz}$$

(*b*) car receding

$$f' = f\frac{1}{1 + v_c/v} = \frac{500 \text{ Hz}}{1 + \frac{20}{340}} = 472 \text{ Hz}$$

Moving Observer: Even a stationary source can give rise to the Doppler effect if the observer is moving. To see why that is true, refer to Fig. 14.17. In the top part of the figure, both the source and the receiver of the wave are at rest. As usual, the frequency of the wave detected by the receiver is equal to the frequency of the source, $f' = f$.

In the lower part of the figure, however, we see at once that the receiver moving toward the source will encounter more wave crests per second than the stationary receiver does. To quantify this, we note that for the stationary receiver, all the crests in a length v will be encountered in 1 s. The frequency heard by the stationary observer is therefore

$$f = \frac{v}{\lambda}$$

But the moving observer encounters all these crests plus the number of crests in the length v_R through which the observer moves toward the source in 1 s. Hence, the frequency detected by the receiver is

$$f' = \frac{v}{\lambda} + \frac{v_R}{\lambda}$$

Using the fact that $1/\lambda$ is f/v, we can revise this equation to read

$$f' = f\left(1 + \frac{v_R}{v}\right) \tag{14.4}$$

Source

$v_R = 0$

Stationary receiver:
$f' = f$

$v_R > 0$

FIGURE 14.17
A Doppler shift occurs also when the receiver is moving.

Moving receiver:
$f' = f(1 + v_R/v)$

As we expected, an observer moving toward the source hears a larger frequency.

A similar expression applies if the observer is receding from the source. But in that case the plus sign in Eq. (14.4) is replaced by a minus sign. As we would expect, the observed frequency is lower for a receding observer.

General Case: We can combine Eqs. (14.3) and (14.4) into a single equation to cover all possible cases for motion along a line. The result is

$$f' = f\frac{1 \pm v_R/v}{1 \mp v_c/v} \qquad (14.5)$$

where the upper signs are to be used if the source-observer distance is decreasing and the lower signs are to be used if the source and observer are moving apart.

In many cases the ratios v_R/v and v_c/v are small compared to unity. Then use can be made of the approximations

$$(1 + x)(1 + y) \cong 1 + x + y \qquad \text{and} \qquad (1 + x)^{-1} \cong 1 - x$$

to obtain

$$f' \cong f\left(1 \pm \frac{v_R}{v} \pm \frac{v_c}{v}\right) \qquad (14.6)$$

Illustration 14.5 Using radar, a stationary police officer is monitoring car speeds. The radar sends out a wave ($v = 3 \times 10^8$ m/s and $f = 1 \times 10^{10}$ Hz) that strikes an automobile and is reflected back to the police officer. What is the relation between the frequency f' of the wave reaching the officer and the speed v_a of the automobile?

Reasoning The frequency of the radar wave will undergo two types of change. First, according to Eq. (14.6), the moving automobile approaching the source will detect a radar wave of frequency[4]

$$f_a = f\left(1 + \frac{v_R}{v}\right) = f\left(1 + \frac{v_a}{3 \times 10^8 \text{ m/s}}\right)$$

The radar waves reflected to the officer will have this frequency.

The officer sees the moving automobile as a radar source with source frequency f_a. But the frequency f' reaching the officer is changed because of the motion of the automobile. From Eq. (14.6), we have

---

[4] Electromagnetic waves obey a somewhat different Doppler-effect equation than sound waves because of effects first pointed out clearly by Einstein in his theory of relativity. However, the difference is very small when the wave speed is large compared to the source and observer speeds.

$$f' = f_a \left(1 + \frac{v_a}{3 \times 10^8 \text{ m/s}} \right)$$

By using the value found for f_a above, this becomes

$$f' = f \left(1 + \frac{v_a}{3 \times 10^8 \text{ m/s}} \right)^2 \cong f \left(1 + \frac{2v_a}{3 \times 10^8 \text{ m/s}} \right)$$

where we drop the squared term in v_a because $v_a \ll 3 \times 10^8$ m/s.
This can be simplified to yield

$$v_a = \frac{f' - f}{f}(1.5 \times 10^8 \text{ m/s})$$

Because v_a will be of order 20 m/s, it is apparent that $(f' - f)/f$ is a very small number. The police radar detector measures $f' - f$ by noticing that the beat frequency between f' and f is equal to this frequency difference. By mixing the emitted and reflected radar signals, a beat frequency is observed. An internal computer allows the radar detector to analyze the data and display the result for v_a.

14.11 Shock Waves and Sonic Booms

Let us return now to the situation shown in Figs. 14.15 and 14.16. We pointed out previously that the wave crests in front of the source all pile on top of one another if the source is moving at the speed of the waves themselves. In such a situation, all the energy of the sound waves is compressed into a very small region in front of the sound source. Consequently, this very concentrated region of sound energy, a *shock wave,* causes a very intense sound, akin to an explosion, as it passes an observer. Basically this is the origin of the sonic boom that accompanies supersonic aircraft.

When an airplane moves through the air with a speed near that of sound, the noise and air disturbances originating from the plane are built up into a shock wave, which, as we have just seen, is a region of very dense sound energy. The exact shape of the shock wave depends on the speed of the airplane. In general, it covers the surface of a cone, as illustrated in Fig. 14.18. The angle θ of the cone depends on the ratio of the speed of the plane v_p to the speed of sound v in the following way:

$$\sin \theta = \frac{v}{v_p}$$

As the plane's speed becomes larger with respect to v, the angle of the cone decreases.

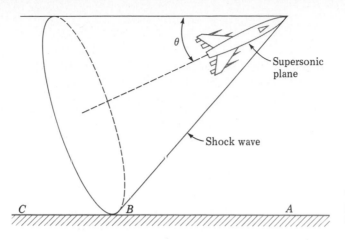

FIGURE 14.18
The sonic boom has already hit point *C* and is moving through point *B* toward *A*.

The familiar sonic boom is the result of the conical shock-wave surface passing over the earth. For example, in Fig. 14.18, the sonic boom will soon strike point *A*; it is currently striking point *B*; it has already hit point *C*. Since the shock wave is a region of very concentrated sound energy, it can cause severe damage when it strikes something.

A flash photograph of the shock wave generated by a projectile in air is shown in Fig. 14.19. Can you show that the speed of the projectile is about Mach 2.1, that is, 2.1 times the speed of sound?

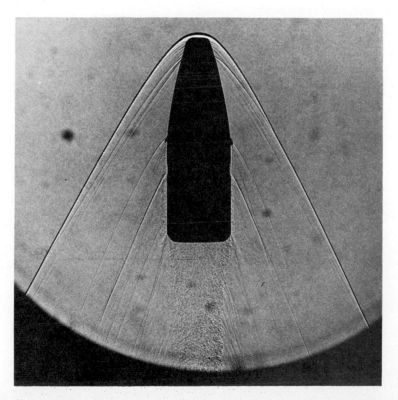

FIGURE 14.19
Shock waves generated by a projectile shooting through air. (*Hewlett-Packard Co.*)

Summary

Sound waves are compressional waves in a fluid or solid. When they strike the eardrum, they give rise to the sensation of sound. The sound source sends out compressions and rarefactions through the material near it. In air at 0°C, sound travels with a speed of 331 m/s. For each 1°C rise in temperature, the speed in air increases by about 0.60 m/s.

The intensity of a wave is the power carried by the wave through a unit area erected perpendicular to the wave's direction of propagation. Typical audible sounds cover an intensity range from about 10^{-12} to about 1 W/m^2. Since the ear is a logarithmic detector, it is customary to measure sound intensities on a logarithmic scale. The sound level in decibels is related to I, the intensity in watts per square meter, by $10 \log (I/10^{-12})$.

Most persons can hear sounds with frequencies in the range 20 to 20,000 Hz. Sound waves with frequency higher than 20,000 Hz are called ultrasonic waves. The normal ear is most sensitive to sounds close to 3000 Hz.

Most sounds consist of a mixture of several waves of different frequencies. The quality of a sound depends on the exact mixture of these waves. For this reason different musical instruments, for example, sound different even though they are playing the same fundamental frequency.

Waves of the same frequency can be combined to cause interference effects. If the waves are in phase, they interfere constructively and give a loud sound. If they are $\frac{1}{2}\lambda$ (or 180°) out of phase, they cancel each other. When two waves of slightly different frequencies interfere with each other, they give rise to beats. The number of beats per second is equal to the frequency difference.

Sound waves can resonate in pipes and tubes. A motion node exists at a closed end while a motion antinode exists near an open end.

The frequency of sound reaching the ear may change because of the motion of the source and/or the motion of the observer. This is called the Doppler effect. If the source and observer are approaching each other, the frequency of sound is increased. If they are receding, the frequency is lowered.

Shock waves originate when sound sources travel faster than the speed of sound. They consist of very intense wave disturbances and carry large amounts of sound energy in a limited region. When the shock wave strikes an observer, it gives rise to what is called a sonic boom.

Minimum Learning Goals

Upon completion of this chapter, you should be able to do the following:

1 Explain what is meant by a sound wave, and point out why sound cannot travel through vacuum.
2 Recall the speed of sound in air at 0°C, and compute its value at temperatures near room temperature.
3 Define the intensity of sound. State why the decibel scale is advantageous, and given the sound intensity in watts per square meter, find the intensity level in decibels.
4 Sketch a graph similar to Fig. 14.5 and interpret it in words.
5 Explain the concept of quality of sound and point out why it is different from frequency.
6 Combine two waves of the same frequency and amplitude but of different phase so as to obtain destructive interference and/or constructive interference.
7 Use the phenomenon of beats to find the difference in frequency between two sound sources.
8 Find the resonance frequencies of sound in given pipes.
9 Explain what is meant by the Doppler effect, and compute the frequency shift noticed for an approaching sound source.
10 Explain how a shock wave originates and why it gives rise to a sonic boom.

Important Terms and Phrases

You should be able to define or explain each of the following:

Sound wave
Wavefront
Sound intensity
Intensity level in decibels
The ear is a logarithmic device
Quality of sound

Ultrasonic waves
In phase and out of phase
Phase difference of 0 and 180°
Beats
Overtones versus harmonics
Doppler effect
Shock wave and sonic boom
Mach number

Questions and Guesstimates

1 Describe clearly why a bell ringing inside a vacuum chamber cannot be heard on the outside.

2 Would you expect a sound heard underwater to have the same frequency as when heard in air if the sources vibrate identically? Explain.

3 When a deep-voiced man inhales helium and then speaks, his voice sounds high-pitched. Why? (v_{He} = 1020 m/s)

4 Suppose that a few pipes of a pipe organ were mounted close to a hot steam pipe. Would this affect the performance of the organ? Explain.

5 A siren can be made by drilling equally spaced holes on a circle concentric to the axis of a solid metal plate or disk. When the disk is rotated while a jet of air is blowing against it near the holes, a sirenlike tone is given off. Explain how this gives the sensation of sound to the ear, and state what factors influence the pitch and quality of the tone.

6 It has been claimed that a certain singer could shatter a wineglass by singing a particular note. Could this be true? Explain.

7 When a firecracker explodes in a large room such as a gymnasium, the sound persists for some time and then dies out. Explain what happens to the sound energy given off by the explosion.

8 The reverberation time of a room is the time taken for a sound, for example, the tone from an organ pipe, to die out after the sound source is shut off. Explain why the reverberation time is shorter when the windows are open. What other means can be used to control the reverberation time?

9 There is on the market a device which uses intense ultrasonic waves in water to wash dirt loose from cloth and other objects. Explain how this works, and list its advantages and disadvantages.

10 One can make a soft drink bottle give off a sound by blowing with one's lips placed properly at the top of the bottle. Estimate the frequencies to which the bottle will resonate. Check your estimate by using tuning forks or a calibrated oscillator and loudspeaker. (E)

11 In underwater investigations, it is common to use **sonar** to "see" underwater. Pulses of sound are sent out through the water, and the pulses reflected to the sending point are detected. Explain how one can make inferences about the surroundings by such a method. Bats use a form of sonar for their flights through pitch-black caves. How does this work?

12 Suppose on some distant planet there exist humanoids whose hearing mechanisms are designed as follows. From the outside, their heads look like our own. However, a 1-cm-diameter, hard-surfaced cylindrical hole passes through the head from ear to ear. At the midpoint of the channel, a thin, circular membrane acts like a drumhead, separating the two halves of the channel. These beings experience the sensation of sound when this drumhead vibrates. What can you infer about their hearing abilities and the ways they will communicate orally with one another?

13 In the adult human ear, the ear canal, the hollow tube leading from the outer ear to the eardrum, has a length of about 2.5 cm. How does the resonance frequency of such a tube correspond with the sensitivity curve of the ear?

14 Two loudspeakers are to be connected to a hi-fi set. The directions say: "Set the speakers side by side and connect their wires to the speaker terminals. Listen to the sound. Reverse the wires connected to the terminals for one speaker. Listen to the sound. Connect the speaker in the way that gives maximum sound." Explain the physical reasons for these instructions.

Problems

Unless otherwise stated, use 340 m/s for the speed of sound in air.

1 The sound of a lightning flash is heard 6.0 s after the flash. Assuming light to travel much more swiftly than the sound, how far away was the lightning?

2 A mountain climber notices that it takes 4.0 s for her voice to be echoed by a distant mountain face. How far away is the mountain face?

3 What is the wavelength of the sound waves given off by (a) a 220-Hz tuning fork, (b) a 1000-Hz tuning fork?

4 What frequency must a sound source have if the wavelength of its sound is to be (a) 3.0 m, (b) 0.30 cm?

5 The speed of sound in water is 1500 m/s. From this fact, what is the bulk compressibility of water?

6 The bulk modulus of material A is only one-third that for material B, although the densities of the two materials are about the same. How will the speed of sound in A compare to that in B?

7 From the fact that the molecular weight of oxygen molecules is 32 kg/kmol, find the speed of sound in oxygen at 0°C.

8 Find the speed of sound in gaseous carbon dioxide (CO_2) at 200°C. The molecular weight of CO_2 is 44 kg/kmol.

9* A certain gas mixture is composed of two diatomic gases (molecular weights M_1 and M_2). The ratio of the masses of the two gases in a given volume is $m_2/m_1 = r$. Show that the speed of sound in the gas mixture is as follows if the gases are ideal:

$$v = \sqrt{\frac{1.40RT}{M_1 M_2} \frac{M_2 + rM_1}{1 + r}}$$

10* Show that the temperature variation of the velocity of sound near 0°C for an ideal gas can be written as

$$v = v_0\left(1 + \frac{t}{546}\right)$$

where t is the Celsius temperature. By how much should the speed of sound in air change for a 1°C temperature change according to this expression? *Hint:* For x small compared to unity, the approximation $(1 + x)^n = 1 + nx$ is valid.

11 By what percentage will the speed of sound in air change if the air temperature is raised from 0 to 16.5°C?

12 By how much must the temperature of air near 0°C be changed to cause the speed of sound in it to change by 1 percent?

13 A certain hi-fi set is stated to consume power at a rate of 20 W. It has two loudspeakers, each of which has an 80-cm² area from which the sound comes. If 0.020 W of sound power comes from each speaker, what is the sound intensity at the speaker? With what efficiency does the set convert electrical energy to sound energy?

14 A certain loudspeaker has a circular opening with a diameter of 15 cm. Assume that the sound it emits is uniform and outward through this entire opening. If the sound intensity at the opening is 10^{-4} W/m², how much power is being radiated as sound by the loudspeaker?

15* A beam of sound has an intensity of 2×10^{-5} W/m². What is the intensity level in decibels?

16* What is the intensity level in decibels for a sound that has an intensity of 0.50 W/m²?

17* (a) What is the intensity of a 60-dB sound? (b) If the sound level is 60 dB close to a speaker that has an area of 120 cm², how much sound energy comes from the speaker each second? (c) What is the acoustic power output of the speaker?

18* Five typists in a room give rise to an average sound intensity level of 60 dB. What would be the sound intensity level in the room if an additional three typists, each generating the same amount of noise, were added to the room?

19* About how many times more intense will the normal ear judge a sound of 10^{-6} W/m² to be than one of 10^{-9} W/m²?

20* Assume the sound intensity is 10^{-4} W/m² at 2.0 m from a source, and the intensity decreases inversely as the square of the distance from the source.

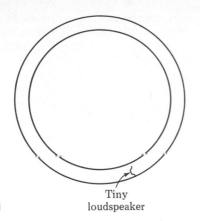

FIGURE P14.1

Tiny
loudspeaker

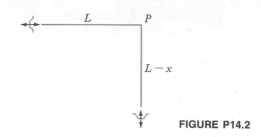

FIGURE P14.2

(a) What is the intensity at 20.0 m from the source?
(b) What is the ratio of the sound intensity levels at the two places?

21** A jet plane roars past an airfield at a height of 500 m causing a 100-dB sound on the ground below it. Assuming the sound to radiate uniformly in all directions, what was the sound intensity level at 50 m from the plane?

22** When the decibel level of a certain sound is tripled, its intensity also triples. What was the original intensity of the sound?

23* As shown in Fig. P14.1, a tiny loudspeaker is placed in an air-filled tube which is bent into a circle. If the circumference of the circle is 12.0 m, what are the three lowest frequencies of vibration of the loudspeaker for which intense sound will exist in the pipe?

24* The tiny loudspeaker in Fig. P14.1 sends sound through the air-filled interior of a hollow tube that is bent into a circle. The tube resonates to the following frequencies of the loudspeaker: 110, 220, 330, and 440 Hz, as well as to higher frequencies. What is the circumference of the circle into which the tube is bent?

25 Consider the harmonics shown for the violin string in Fig. 14.7. Assume that the fundamental vibration has a frequency of 380 Hz. From a consideration of the resonances of a string held rigidly at its two ends, what should be its next five resonance frequencies? Are these about the same as those shown in the figure?

26 Repeat Prob. 25 for the piano string shown in Fig. 14.7. In this case assume the fundamental frequency to be 130 Hz.

27* The two small loudspeakers shown in Fig. P14.2 send out identical sound waves. When the distance $x = 0$, loud sound is heard at point P. Suppose

now x is increased until $x = 40$ cm. Starting at low frequencies, the frequency of both sources is slowly increased; the frequencies of the two sources are still equal. (a) At what frequencies will loud sound be heard at P? (b) At what frequencies will minimum sound be heard at P?

28* The two loudspeakers shown in Fig. P14.2 send out identical sound waves of wavelength 34 cm. The speakers vibrate in phase. (a) Give at least three values of x for which loudness will be heard at point P. (b) Repeat for minimum sound heard at P.

29** Two tiny sound sources face each other and send out identical, in-phase waves. One source is at $x = 0$, and the other is at $x = 3.0$ m. If the wavelength of the waves sent out by the sources is 80 cm, at what points along the line from $x = 0$ to $x = 3.0$ m will a detector register minimum sound?

30** Assume the same situation described in Prob. 29 but with variable-frequency sources. Starting at low frequency, the source frequency is increased slowly. At what frequencies will minimum sound be heard at the point $x = 1.0$ m?

31 Two violins are slightly out of tune. When they both bow the supposedly same note, a beat frequency of 0.70 Hz is noticed. If the frequency emitted by one violin is f_0, what are the possible frequencies emitted by the other?

32 Two pianos sound the same note, but the vibration from one is 131.50 Hz while that of the other is 132.00 Hz. What will be the beat frequency between their two tones?

33 An unknown frequency is compared with a standard, 2200-Hz tuning fork and found to give rise to 2.0 beats per second. (a) What are the possible values of the unknown? (b) How could you determine the true value if you had a wad of chewing gum handy?

34 A violinist compares the tone given off by her violin string with the corresponding note on a piano and notices a beat every 3 s between the two sounds. (a) By how many vibrations per second do

393

the two instruments differ? (*b*) How could the violinist decide which instrument was higher in pitch?

35 A pipe open at one end and closed at the other is 0.75 m long. To what three lowest frequencies of sound will it resonate? Draw the wave within the tube for each frequency.

36 To what three lowest frequencies will a pipe open at both ends resonate? The pipe is 0.90 m long. Draw the wave within the tube at each frequency.

37 In an experiment like that shown in Fig. 14.12, resonances are found to occur when the water is 31.00 and 41.00 cm high in the tube. If no resonances occur in between, find the frequency of the tuning fork.

38 A man wishes to find how far down the water level is in the iron pipe leading into an old well. Being blessed with perfect pitch, he merely hums musical sounds at the mouth of the pipe and notices that the lowest-frequency resonance is 80 Hz. About how far from the top is the water level?

39 The Lincoln Tunnel under the Hudson River in New York is about 2600 m long. To what sound frequencies will it resonate? What practical importance, if any, do you think this has?

40* A certain organ pipe resonates to a frequency of 500 Hz when its temperature is 20°C. It, at 20°C, and another identical pipe near a heater at 30°C are sounded in unison. What will be the beat frequency between their two sounds?

41 How fast must a car be coming toward you if its horn is to appear to be 10 percent higher in frequency than when the car is standing still?

42 A sound source vibrates at 100 Hz and is receding from an observer with a speed of 18 m/s. If the speed of sound is 332 m/s, what frequency does the observer hear?

43* The speed of sound in hydrogen is about 1270 m/s. If a pipe which resonates in its fundamental frequency to 800 Hz in air is filled with hydrogen, what will be its fundamental resonance frequency?

44* A tube open at one end and closed at the other is 40 cm long. It will resonate under the action of a 210-Hz sound source provided the source is receding from it. How fast must the source be moving? (Use $v = 320$ m/s.)

45* Two identical sound sources vibrate with a frequency of 150 Hz. Suppose that along a straight line a detector is placed at $x = 0$, one source is placed at $x = 20$ m, and the other source is at $x = 30$ m and is moving toward larger x values at 8.0 m/s. What will be the beat frequency heard by the detector?

46* The device shown in Fig. P14.3, Kundt's tube, is sometimes used to measure the speed of sound in metal rods. When the rod R is stroked lengthwise with a rosined cloth, it resonates with a longitudinal motion such that the disk end is an antinode and the support is a node. The vibration of the disk D at the end of the rod sends sound waves of the same frequency down the glass tube. When the tube is adjusted to a position of resonance, such as the position shown, cork dust in the tube indicates clearly the position of nodes and antinodes in the air. If the separation between antinodes is found to be 10.0 cm and the rod is 100.0 cm long, find the speed of sound in the metal rod.

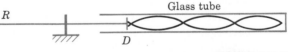

FIGURE P14.3

15
Electric Forces and Fields

We begin our study of electricity with electrostatics, the physics of electric charges at rest. Electric charges exert forces on one another, and these forces are the primary topic of this chapter. The force that one charge exerts on another is conveniently expressed in terms of electric fields, and so the electric field concept is also important to us. As we shall see, electrical concepts are closely related to the principles we learned in mechanics. This will become even more apparent in the next chapter, in which the concept of electric potential is related to work.

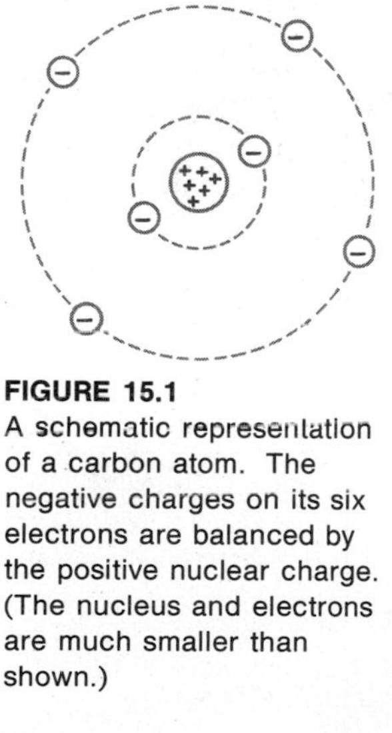

FIGURE 15.1
A schematic representation of a carbon atom. The negative charges on its six electrons are balanced by the positive nuclear charge. (The nucleus and electrons are much smaller than shown.)

15.1 Atoms as the Source of Charge[1]

An atom is composed of a tiny positively charged nucleus around which there are floating negatively charged particles called **electrons.** This is illustrated in a schematic way for an atom of carbon in Fig. 15.1. You will recall from courses in chemistry that all atoms are electrically neutral. That is, the quantity of positive charge on the nucleus exactly equals the total charge of the electrons about the nucleus. In the case of the carbon atom shown, if $-e$

[1]The material in this chapter and the next is basic to the entire study of electricity and the many facets of nature which involve electrical effects. It is therefore imperative that this material be thoroughly understood before passing on to the following chapters.

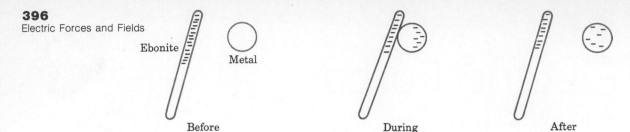

Ebonite   Metal

Before   During   After

FIGURE 15.2

When the negatively
charged ebonite rod
touches the uncharged
metal ball, electrons are
conducted off the rod onto
the ball.

is the charge on each electron, the charge on the nucleus will be exactly $+6e$. We postpone a detailed discussion of the atom to a later chapter and merely make use of its electrical constitution here.

It appears that the universe as a whole is nearly, if not completely, neutral from an electrical standpoint. The earth has very little, if any, excess of either positive or negative charge. For nearly all practical purposes, the earth can be considered to have zero net charge. The vast majority of the charges on and in the earth reside as constituents of atoms. When free negative or positive charges are found, they are usually assumed to have come from the tearing apart of an atom.

Actually, it is not difficult at all to remove an electron from an atom—under certain circumstances. For example, if a bar of ebonite (hard rubber) is rubbed with animal fur, some of the electrons from the atoms in the material of the fur are rubbed off onto the ebonite rod. (The reason for this charge transfer is not simply explained. It is covered in courses dealing with solid-state physics.) Hence, the ebonite rod acquires a net excess of electrons. When the rod is touched to a metal object, some of the excess electrons transfer to the metal, as illustrated in Fig. 15.2.

Similarly, if a glass rod is rubbed with silk, some of the electrons leave the atoms in the glass and give rise to an excess of electrons on the silk. Of course, now the glass rod has an excess of positive charges. If the glass rod is touched to a neutral metal ball, electrons will leave some of the atoms of the metal and replace those electrons lost by the atoms in the glass. As a result, the metal ball acquires a net positive charge. Many other materials give rise to a separation of charge when rubbed together. The ones we have described were originally used to define positive and negative charge before the existence of the electron was even known.

15.2 Forces between Charges

Now that we know how to obtain charged bodies, it is possible to examine the forces between charges. One of the simplest ways of doing this is to make use of very light metallized balls called pith balls. The balls can be charged by touching them with a glass or an ebonite rod. If the balls are suspended by light threads, four interesting experiments can be performed. They are illustrated in Fig. 15.3a to d.

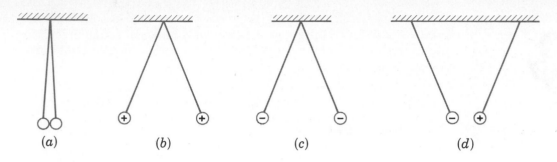

(a)   (b)   (c)   (d)

From the experimental results shown in Fig. 15.3 we can conclude the following:

FIGURE 15.3
Charged pith balls show that like charges repel one another while unlike charges attract one another.

1 Like charges repel one another; i.e., two positive charges will repel each other, as will two negative charges.
2 Unlike charges attract one another; i.e., positive charges attract negative charges, and vice versa.
3 The magnitude of electrical forces between two charged bodies most often exceeds the gravitational attraction between the bodies. (How do the experiments illustrated show this?)

15.3 Insulators and Conductors

Although all materials are made of atoms, and even though all atoms are made of electrons and nuclei, we are well aware that the electrical properties of substances vary widely. *There are two basic groups into which all substances can be divided according to electrical properties. They are the conductors and the nonconductors, or insulators.* Materials intermediate between these two groups are called *semiconductors* and are sometimes classified as a separate group.

In insulators, the electrons of any given atom are bound tightly to that atom and cannot wander through the material. Hence, even if an excess of charge is placed near the end of a rod made from an insulator, the electrons in the adjacent atoms cannot move under the attraction or repulsion of the nearby excess charge. They are strongly bound to their atoms and cannot get free, and so no large motion of charge occurs in the rod.

Conductors behave quite differently. In these substances the electrons farthest from the nucleus, called **valence electrons,** are so close to neighboring atoms that it is difficult to determine exactly which electron belongs to which atom. Under these conditions, the valence electrons in the atoms move about through the metal and may be looked on, for some purposes at least, as an electron gas contained within the confines of the piece of metal. Even in metals, though, most of the electrons are tightly held in the atom, and only the electrons near the outside of the atom are at all free to move through the solid. Usually only one, two, or three valence electrons are associated with each atom.

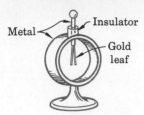

Metal — Insulator
        Gold
        leaf

The precise formulation of the motion of charges in solids is a very complex problem. It can be handled well only by use of quantum mechanics. In fact, the exact behavior of electrons in solids is still an active field for research today.

15.4 Electroscope

The **electroscope** is a simple device used for detecting and measuring charges of small magnitude. It is shown in Fig. 15.4. A metal rod from which are suspended two very thin leaves of gold foil is held inside a metal case by the aid of an insulator, which keeps the rod from touching the case. Two faces of the case are covered with glass, so that the disposition of the leaves can be seen.

Suppose that some negative charge is placed on the central rod by touching to it a charged piece of ebonite. The charge is confined entirely to the rod and leaves, since they are insulated from other objects. Because like charges repel one another, the negative charges will distribute themselves more or less uniformly over the rod and gold leaves. However, the gold leaves, being free to bend and being repelled by the charges on one another, take up the position shown in Fig. 15.5a.

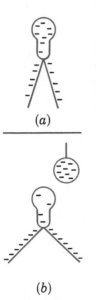

(a)

(b)

If a negatively charged ball is now brought close to the ball of the electroscope, as shown in part b of Fig. 15.5, many of the negative charges on the upper part of the electroscope will be repelled down the rod, causing the leaves to separate further. An opposite effect is observed if a positively charged ball is brought close to the electroscope (Fig. 15.5c). In addition, a ball having no net charge does not disturb the electroscope to any great extent. With this apparatus the sign of a charge and its rough magnitude can be determined. You should convince yourself that a similar procedure can be followed if the electroscope is charged positively.

15.5 Charging by Conduction and by Induction

(c)

There are two general ways for placing charge on a metal object by use of a second object which is already charged. As a concrete example, consider the ways in which you might charge a metal ball by use of a negatively charged ebonite rod. One way is to touch the ball to the rod. On contact, some of the excess negative charge on the rod will move off onto the ball. This process, shown in Fig. 15.6, is called **charging by conduction.**

The same rod can be used in a different way to charge the ball. This is shown in Fig. 15.7. In this process, **charging by induction,** the rod is not touched to the ball at all. Notice that when the rod is brought close, some of the electrons in the metal are repelled to the right side of the ball, leaving a positive charge on the left side. Since no charge has been added to or subtracted from the ball, it is still neutral, of course. Now, suppose that the

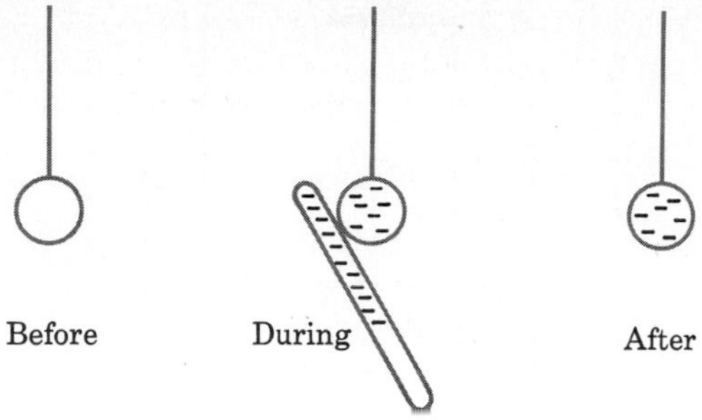

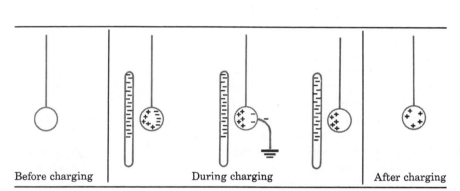

Before   During   After

FIGURE 15.6
A metal ball can be charged by conduction. Note that the charge is shared, and therefore the ball and the rod have like charges.

Before charging   During charging   After charging

FIGURE 15.7
The method for giving a metal sphere a positive charge by induction. Note that the rod and ball have unlike charges.

ball is touched by another object such as one's finger. (We say in that case that the ball has been **grounded,** and the symbol ⊣⊢ is used to show this.[2]) The negative charges can get still farther away from the negative rod if they flow to the object touching the ball. Once this has happened, the ball will no longer be neutral, since it has lost some of its negative charge. After the object touching the ball has been removed, the negative rod can be removed and the ball will have a positive charge. (Why must the touching object be removed *before* the rod?)

Upon comparing Figs. 15.6 and 15.7, you can see that an ebonite rod can be used to charge a metal object *negatively* by *conduction,* but that it charges the same object *positively* by *induction.* You may find it interesting to work out the similar diagrams using a positively charged glass rod. The charges are reversed in that case.

15.6 Faraday's Ice-Pail Experiment

In 1843 Michael Faraday first carried out a simple, but highly instructive experiment. He attached a metal ice pail to an electroscope, as shown in Fig. 15.8a. A metal ball suspended from a thread was given a positive

---

[2] Usually, to ground an object, we attach it to a water pipe or some other object entering the earth. Any excess charge can then flow to the ground, i.e., the earth.

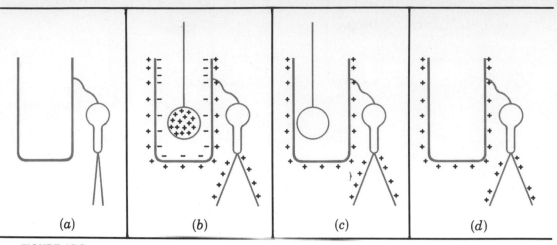

FIGURE 15.8
Faraday's ice-pail
experiment.

charge. The charged ball was then lowered into the pail (without touching it), at which time the leaves of the electroscope were found to diverge. Moreover, the ball could be moved from place to place within the pail, and the leaves remained in the same divergent position. Only when the ball was removed did the leaves return to their original position.

Faraday further noticed that if the charged metal ball was touched to the inside of the pail, the electroscope leaves remained in the divergent position they had maintained before the ball touched the pail. Now, however, when the ball was removed from the pail, the leaves remained divergent. When the ball was brought close to a second electroscope, it was found to be no longer charged. Apparently, on touching the inside of the pail, the excess charge on the ball had been completely neutralized. Since the leaves of the electroscope attached to the outside of the pail did not move when the ball touched the pail, Faraday concluded that the inner surface of the pail had just enough charge on it to neutralize the ball.

From these experiments we can draw the following conclusions, illustrated in Fig. 15.8*b* to *d*.

1   A charged metal object suspended inside a neutral metal container induces an equal and opposite charge on the inside of the container.
2   When the charged metal object is touched to the inside of the container, the induced charge exactly neutralizes the excess charge on the object.
3   If a charged object is placed within a metal container, an equal charge of the same sign will be forced to the outer surface of the container.
4   When a path for conduction is provided, all the charge on a metal object will reside on its outer surface.

These are important facts concerning electric charges on metal. We interpret them more fully after we examine Coulomb's law and the concept of electric fields.

15.7 Coulomb's Law

The mathematical law by which like charges repel and unlike charges attract was formulated in 1785 by Charles Augustin de Coulomb (1736–1806) and is called **Coulomb's law.** By means of a very sensitive balance, similar to that used by Cavendish in his study of gravitation, Coulomb was able to measure accurately the force between two small charged balls. A diagram of the situation is shown in Fig. 15.9. Two small balls, much smaller than shown, with distance r between centers carry charges $+q_1$ and $-q_2$. After a number of experiments Coulomb concluded that the force on sphere 1 varied in proportion to the product of their charges and inversely as the square of their distance apart. In symbols,

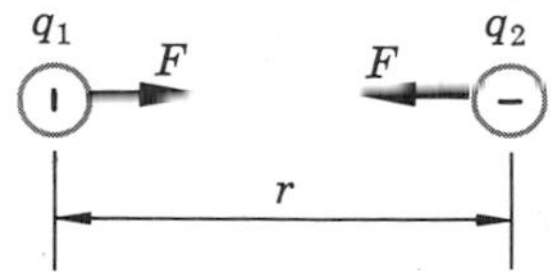

$$F \propto \frac{q_1 q_2}{r^2}$$

or

$$F = (\text{const}) \left(\frac{q_1 q_2}{r^2} \right) \tag{15.1}$$

FIGURE 15.9
The two unequal, unlike charges attract each other with equal force.

and the force was in the direction shown in Fig. 15.9. According to Newton's law of action and reaction, the force on sphere 2 was identical in magnitude but oppositely directed. Notice that Coulomb's law applies only to point charges. If the charges extend over a large region, the distance r between them is not easily defined.

To make the above statement of Coulomb's law into an equation, we must first decide on a unit in which to measure charge. The definition of the SI unit of charge is made in terms of electrical current, as we shall see in Sec. 21.5. For now we simply state that the SI charge unit is the *coulomb* (C). By using this unit for q_1 and q_2, Coulomb's law can be written as

$$F = k \frac{q_1 q_2}{r^2} \tag{15.2}$$

Coulomb's Law

where F is in newtons and r is in meters.[3] The constant of proportionality k is determined by experiment and is found to have the value $8.9874 \times 10^9 \text{ N} \cdot \text{m}^2/\text{C}^2$ when the experiment is performed in vacuum (or air, to good approximation). We shall usually take k to be $9.0 \times 10^9 \text{ N} \cdot \text{m}^2/\text{C}^2$.

The coulomb of charge is a very large quantity in comparison to the charge on an electron. If we call the electron's charge $-e$, the experimentally determined value is found to be

$$e = 1.60219 \times 10^{-19} \text{ C}$$

Charge Quantum

---

[3]The constant k is often written as $1/4\pi\varepsilon_0$, where ε_0, the *permittivity of free space*, has an experimentally determined value $8.85 \times 10^{-12} \text{ C}^2/\text{N} \cdot \text{m}^2$.

The proton has a charge $+e$. In fact, all the charged particles thus far discovered have charge $\pm e$ or whole-number multiples of this value. We therefore conclude that *there appears to be a basic unit of charge, the* **quantum of charge,** *of magnitude e.* However, we are still in doubt on this point. Recent theories of the fundamental particles such as the proton indicate that these particles contain still smaller charges. The particles that carry these charges, believed to be $e/3$ and $2e/3$, are called **quarks.** Despite an intensive search for them, isolated quarks have thus far eluded our detection. More is said about quarks in Sec. 27.16.

Illustration 15.1 A copper penny has a mass of about 3 g and contains about 3×10^{22} copper atoms. Suppose two pennies are 2.0 m apart and carry equal charges q. (a) How large must q be if the force of repulsion on one due to the other is equal to the weight of a penny? (b) How many electrons must be removed from a penny to give it this charge? (c) What fraction of the atoms have lost electrons in such a case?

Reasoning (a) The situation is shown in Fig. 15.10. Each penny has a mass of 3 g, or 3×10^{-3} kg. The weight of a penny is $mg = 0.0294$ N. Coulomb's law, $F = kq_1q_2/r^2$, gives the repulsion force on one of the pennies as

$$F = (9 \times 10^9 \text{ N} \cdot \text{m}^2/\text{C}^2)\frac{q^2}{(2.0 \text{ m})^2}$$

Solving for q when $F = 0.0294$ N gives $q = 3.6 \times 10^{-6}$ C.

(b) Since each electron removed from a penny leaves an unbalanced charge of 1.60×10^{-19} C behind, the number of electrons removed must be

$$\text{Number of electrons} = \frac{\text{charge}}{\text{charge/electron}} = \frac{3.6 \times 10^{-6} \text{ C}}{1.60 \times 10^{-19} \text{ C}} = 2.3 \times 10^{13}$$

(c) There are about 3×10^{22} atoms in the penny. So the fraction that have lost a single electron is

$$\text{Fraction} = \frac{2.3 \times 10^{13}}{3 \times 10^{22}} = 7.7 \times 10^{-10}$$

Notice how very small a fraction of the electrons must be removed to give an object a sizable charge. Notice also that charges as small as 10^{-6} C give rise to easily measurable forces between ordinary-sized objects.

FIGURE 15.10
As shown in Illustration 15.1, only a tiny fraction of the electrons need be removed from a penny to give rise to large electrical forces.

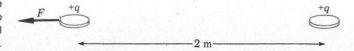

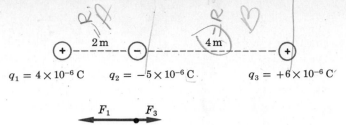

$q_1 = 4 \times 10^{-6}$ C $q_2 = -5 \times 10^{-6}$ C $q_3 = +6 \times 10^{-6}$ C'

F_1 F_3

FIGURE 15.11
The central charge is attracted to q_1 by force F_1 and to q_3 by force F_3.

Illustration 15.2 Find the force on the center charge in Fig. 15.11.

Reasoning Clearly, charge q_2 will be attracted to q_1 by force F_1. It is also attracted to charge q_3 by F_3. Using Coulomb's law, we have

$$F_1 = (9 \times 10^9 \text{ N} \cdot \text{m}^2/\text{C}^2) \frac{(4 \times 10^{-6} \text{ C})(5 \times 10^{-6} \text{ C})}{(2 \text{ m})^2} = 0.0450 \text{ N}$$

$$F_3 = (9 \times 10^9 \text{ N} \cdot \text{m}^2/\text{C}^2) \frac{(5 \times 10^{-6} \text{ C})(6 \times 10^{-6} \text{ C})}{(4 \text{ m})^2} = 0.0169 \text{ N}$$

The net force on q_2 is to the left and is

$$F = F_1 - F_3 = 0.0281 \text{ N}$$

Notice that the signs of the charges are not employed in the above use of Coulomb's law, since these are implicit in the direction of the forces. This enables us to compute the magnitude of the force by using Coulomb's law and indicate by a vector drawing which direction is positive or negative.

Illustration 15.3 Find the resultant force on the $+10$-μC charge of Fig. 15.12.

Reasoning First we construct the vector diagram for the forces on the charge in question. Using Coulomb's law, $F = kq_1q_2/r^2$, we then find

$$F_1 = 6 \text{ N} \qquad \text{and} \qquad F_2 = 18 \text{ N}$$

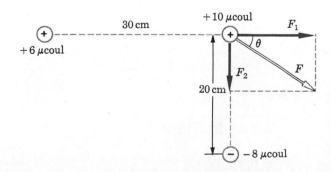

FIGURE 15.12
To find the resultant force on the $+10$-μC charge, we must add the two separate forces exerted by the other two charges on it.

403

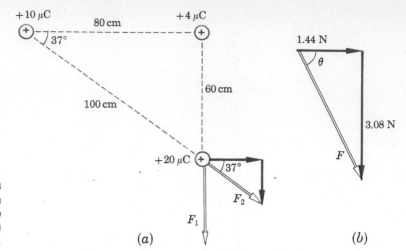

FIGURE 15.13
The vector forces acting on
the 20-μC charge produce
the resultant F shown in
part b.

As shown in the figure, these two forces act at right angles. We can therefore use the pythagorean theorem to find the resultant force F,

$$F = \sqrt{F_1^2 + F_2^2} = 19 \text{ N} \qquad \text{and} \qquad \tan \theta = \frac{F_2}{F_1} = 3.00$$

from which

$$\theta = 71.5°$$

Illustration 5.4 Find the force on the $+20$-μC charge in Fig. 15.13.

Reasoning From Coulomb's law, we find

$$F_1 = 2.0 \text{ N} \qquad \text{and} \qquad F_2 = 1.8 \text{ N}$$

Resolving F_2 into components, we have $F_{2x} = F_2 \cos 37° = 1.44$ N and $F_{2y} = 1.08$ N. The total y force on the charge is $F_y = 2.0 + 1.08 = 3.08$ N downward, while the total x force is 1.44 N to the right. They are shown in Fig. 15.13b. Clearly, the resultant force is

$$F = \sqrt{(1.44)^2 + (3.08)^2} = 3.4 \text{ N} \qquad \text{and} \qquad \tan \theta = \frac{3.08}{1.44} = 2.14$$

from which $\theta \approx 65°$ in the direction shown.

15.8 Electric Field

We find it convenient to discuss electrical forces in terms of a concept called the **electric field.** It serves much the same purpose in electricity as the concept of a gravitational field serves in mechanics. Before discussing this new concept in detail, let us first review the more familiar situation of the gravitational field.

It is common for people to say that a stone when released falls to the earth because of the downward force of the earth's gravitational field. By this they mean that the stone experiences a gravitational force and that this force causes it to accelerate. When we go far away from the earth out into space, the earth's gravitational attraction for objects becomes quite small. We say that the earth's gravitational field is very weak at large distances from the earth. A gravitational field is said to exist in a region of space where a gravitational force exists on a body. If the gravitational field is strong, the gravitational force on a given mass is strong.

It is convenient to draw pictures of gravitational fields. The gravitational field of the earth is shown in Fig. 15.14. We interpret this picture in the following way: If an object is placed at point A in the figure, it will experience a force directed toward the center of the earth. The lines of force drawn in Fig. 15.14 show the direction of the earth's gravitational field, i.e., the direction of the gravitational force on an object.

Actually, of course, Fig. 15.14 should be drawn in three dimensions, with lines of force directed from all sides in toward the center of the earth. Not only do the lines of force represent the direction of the force, but also they indicate the relative magnitude of the force. You can see this in Fig. 15.14 by noticing that the field lines are closer together near the earth where the force is strong than farther from the earth where the force is weaker. We return to this feature of the field lines after we define the electric field.

We define the electric field in a manner analogous to the gravitational field. In making the definition we refer to what is called a test charge. *A test charge is a fictitious charge endowed with a very special quality: the test charge exerts no forces on nearby charges* and therefore does not disturb the charges in its vicinity. In practice, we can approximate such a fictitious charge by using a tiny object that contains a very small charge. It will therefore disturb the charges by only a negligible amount.

The direction of the electric field at a point is the same as the direction of the force on a positive test charge placed at that point.

For example, suppose a positive test charge is placed at point A in Fig. 15.15a. It will experience a radial attractive force, as shown by the arrow at

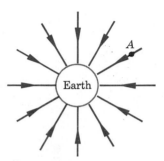

FIGURE 15.14
The gravitational field of the earth is directed radially inward and becomes stronger as one approaches the earth.

Definition

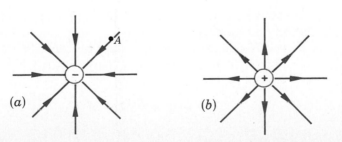

(a)   (b)

FIGURE 15.15
The electric field is directed radially inward toward a small negative charge and radially outward from a small positive charge.

A. Similarly, the force on the positive test charge will be directed radially inward no matter where it is placed in the neighborhood of the central negative charge. We therefore surmise that the electric field is directed as shown by the arrows; the electric field near a negative charge is directed radially into the charge.

We can obtain the electric field near a positive charge in the same way, and this is done in part *b* of the figure. The positive test charge is repelled radially outward by the central positive charge. Hence the electric field near a positive charge is directed radially away from the charge.

The directed lines we have drawn in Fig. 15.15 to show the direction of the electric field are called electric field lines. As we have just seen, electric field lines originate in and come out of positive charges, whereas they are directed toward and end on negative charges.

We wish also to define the electric field strength in a quantitative way. As stated above, the direction of the field at a point is the direction of the force on a positive test charge placed at the point in question. *The strength of the electric field is taken to be the electric force* **F** *on the tiny test charge divided by the magnitude of the test charge* q_t. *By using the symbol* **E** *for electric field strength,* this definition may be written in equation form as

$$\mathbf{E} = \frac{\mathbf{F}}{q_t} \tag{15.3a}$$

The units of **E** are clearly newtons per coulomb. Because **E** is a force per unit charge, we frequently state that the electric field strength **E** is the force per unit positive test charge. However, we should always realize that in order to measure the field, we would use a charge much smaller than 1 C so as not to disturb the other charges present.

Finally, we should point out that *the relative strength of the field can be estimated by examining the field line diagram.* For example, in Fig. 15.15 the field lines are closest together near the charges. The force on a unit positive test charge (the electric field strength) would also be largest close to the charges. In this case at least, the electric field strength is largest where the field lines are closest together. It is shown in more advanced courses that this is always true, and so we often estimate the field strength in a region by noticing the closeness of the field lines in that region.

Illustration 15.5 Find the electric field strength at a distance 50 cm from a positive charge of 10^{-4} C.

Reasoning We wish to find the field at point *A* in Fig. 15.16. Since it is the force on a unit positive charge, its direction is to the right, as shown by the arrow. To find the magnitude of the electric field, we make use of Coulomb's law (mentally placing a test charge q_t at *A*). We have

$$F = (9 \times 10^9 \text{ N} \cdot \text{m}^2/\text{C}^2)\left(\frac{q_1 q_2}{r^2}\right)$$

$q = 10^{-4}$ C

FIGURE 15.16
To find the electric field *E* at point *A*, we must compute the force which a unit positive charge would experience if placed at that point.

In our case, $q_1 = 10^{-4}$ C, $q_2 = q_t$, and $r = 0.50$ m. Since $F/q_t = E$, we have at once

$$E = (9 \times 10^9)\frac{10^{-4}}{(0.50)^2} = 3.6 \times 10^6 \text{ N/C}$$

You should show that the answer really does have these units.

Illustration 15.6 Find the electric field strength at point B in Fig. 15.17.

Reasoning Mentally we place a positive test charge q_t at point B and compute the force on it, using Coulomb's law. Making use of the vectors shown, we have

$$F_1 = (9 \times 10^9 \text{ N} \cdot \text{m}^2/\text{C}^2)\left(\frac{3.6 \times 10^{-6} \text{ C}}{0.0036 \text{ m}^2}\right)(q_t) = 9 \times 10^6 \, q_t \text{ N}$$

so
$$E_1 = \frac{F_1}{q_t} = 9 \times 10^6 \text{ N/C}$$

Similarly,

$$E_2 = 4.5 \times 10^6 \text{ N/C}$$

From this, after taking components,

$$E_y = 9 \times 10^6 - 2.7 \times 10^6 = 6.3 \times 10^6 \text{ N/C}$$
$$E_x = 3.6 \times 10^6 \text{ N/C}$$

and
$$E^2 = E_x^2 + E_y^2$$
$$E = 7.3 \times 10^6 \text{ N/C}$$

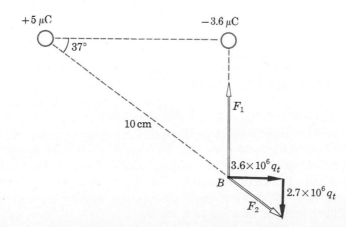

FIGURE 15.17
To find the field at B, we must find the forces F_1 and F_2 exerted on a positive test charge q_t placed at that point.

15.9 Electric Field in Various Systems

A great deal of insight into a problem can be obtained by examining the pertinent electric-field drawing. To illustrate this point, let us examine several electric fields and the charge distributions giving rise to them. In doing so we recall that we defined the electric field as the force per unit positive test charge. Lines of force must always be drawn out of positive charges and into negative charges.

Consider the electric field about two equal charges, one positive, the other negative, as shown in Fig. 15.18. The electric field will be similar about each, except that it will be directed into the negative charge and out of the positive charge. This field is shown in Fig. 15.18. You should examine several points in the figure to convince yourself that a positive test charge actually would experience a force in the directions indicated by the force lines in the figure. To see how this is done, consider point A of Fig. 15.18. A positive test charge at point A will be repelled by the positive charge and attracted by the negative charge. The attractive force equals the repulsive force because the test charge is as close to the positive charge as it is to the negative charge. The resultant of these two forces is tangent to the line of force at point A.

The field in the neighborhood of two like charges is shown in Fig. 15.19. Try to justify the way the lines are drawn. Where, according to the diagram, is the field strongest? Why should the field be weak midway between the charges?

Suppose that a positive, *uniformly* charged ball is held above a large metal plate. If the metal plate is attached to the ground, charge may run on or off it to the essentially inexhaustible capacity of the earth. The situation is shown in Fig. 15.20. We see at once that the electrons in the metal plate will be attracted by the positive charge. Although they cannot leave the plate, they congregate on the plate just below the positive charge. Negative charge will flow onto the plate from the ground in order to replace the electrons induced to take up the position near the center of the plate.

The positive test charge used to determine the direction of the field will be pulled and pushed down toward the plate in the central region. You can readily verify that the lines of force shown in Fig. 15.20 actually do represent the force on a positive test charge. Similarly, the density of the lines indicates the field strength. As one would expect, the field is strongest close to the positive charge.

Two rather subtle but important features of Fig. 15.20 should be pointed out. First, the lines of force do not penetrate into the metal plate. All stop at the surface. This would indicate that the electric field is zero within the metal; i.e., there is no force on a charge in the metal. This is readily understood from the following reasoning: We are considering only electrostatic conditions, i.e., no continuing flow of charge. If there were a field within the metal, it would exert a force on the free electrons in the metal. The electrons would then move, and a current would flow. But no currents are flowing under electrostatic conditions, and hence the electric field must be zero inside a metal.

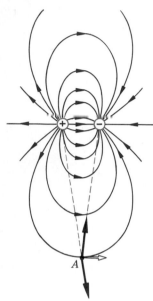

FIGURE 15.18
The electric field lines originate on the positive charge and end on the negative charge.

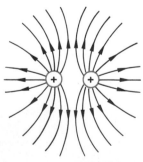

FIGURE 15.19
Notice how the lines of force about two like charges seem to repel one another. Why must this be so?

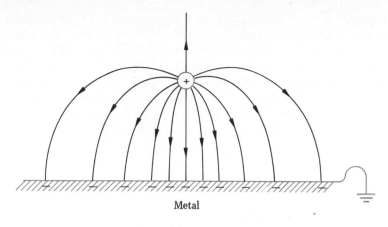

FIGURE 15.20
The positive charge attracts
negative charges to the top
of the metal plate. Why are
the field lines perpendicular
to the plate at its surface?

Metal

The second point has to do with the fact that the lines of force are perpendicular to the metal surface. First we point out that the lines of force represent the force on a positive charge. A negative charge would be forced in the opposite direction. In Fig. 15.20, the negatively charged electrons at the surface of the metal experience a force trying to pull them upward. This is the result of the attraction by the positive charge suspended above the metal. However, the electrons are held within the body of the metal and cannot come loose under this force.

Suppose that the force were not exactly perpendicular to the metal. There would be a component of the force along the surface of the metal. Since the electrons of the metal can move in this direction, they would flow under the action of this force and a current would flow. However, since this is an electrostatic situation, no current is flowing. Hence, there can be no component of the field parallel to the metal surface. We conclude that the field is directed perpendicular to the metal surface.

These last facts are used again later. They are quite general in any electrostatic problem. For this reason we state them once again.

Under electrostatic conditions:

1   The electric field within a metal is zero.
2   The electric field just outside the surface of a metal is perpendicular to the surface.

Since these facts were proved by reference to the movement of free electrons within the metal, the proofs will not hold true for insulators, which contain no free electrons. Hence the above rules apply only to metals and not to insulators.

Illustration 15.7   Show by means of a diagram using lines of force that a charge suspended within a metal cavity induces an equal and opposite charge on the interior surface.

Reasoning   Let us suppose the metal object to be a hollow metal

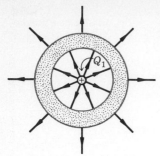

sphere as shown in cross section in Fig. 15.21. Lines of force come out of the positive charge Q_1 suspended in the cavity. Since the field within the metal must be zero in electrostatic situations such as this, the lines from Q_1 must stop when they hit the metal surface. They must therefore be as shown. However, this means the inner surface must possess a negative charge since lines of force go into and terminate only on negative charges. Moreover, since the number of lines originating on Q_1 is equal to the number terminating on the negative charge, we infer that the charge on the inner surface is $-Q_1$.

If the metal object was neutral before Q_1 was placed in the cavity, and if no charge was allowed to run on or off the object, the object must still have no net charge. Therefore, a charge $+Q_1$, must exist on the outer portions of the object. Can you prove, using lines of force, that the charge must be on the outer surface as indicated? Can you show that the results obtained here are independent of the shape of the object and cavity?

15.10 Parallel Metal Plates

The electric field between two oppositely charged metal plates is of particular importance in electricity, as we shall see as our studies progress. We show a typical situation in Fig. 15.22a. The charges are supplied to the plates by connecting them to a battery as shown. (We discuss batteries in more detail in the next chapter.) As indicated, the battery gives one plate a positive charge and the other a negative charge. Because the charges attract one another, they reside for the most part on the inner surfaces of the plates. This is shown clearly in the schematic diagram in b.

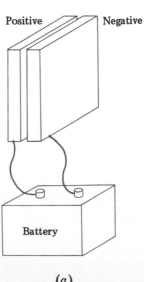

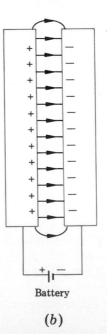

(a)

(b)

We are interested in the electric field in the region between the plates. A positive test charge placed in this region will experience a force from left to right because of the repulsion of the positive plate and the attraction of the negative plate. The electric field lines, shown in black, are therefore directed as indicated. Although some "fringing" of the lines occurs near the edges, the field lines are mostly directed straight across the gap between the plates. Moreover, as we see in the next chapter, the field is of constant strength throughout the region between the plates. This is indicated in the diagram by the equal spacing of the field lines.

Suppose a particle with positive charge q and mass m is placed between the plates. It will experience a force to the right because of the field $\mathbf{E}$. Since $\mathbf{E}$ is defined to be the force per unit positive test charge, our particle will experience a force q times larger, namely,

$$\mathbf{F} = q\mathbf{E} \qquad (15.3b)$$

toward the right-hand plate. Let us now investigate this situation further with a concrete example.

Illustration 15.8 The gap between the plates in Fig. 15.22 is 0.20 cm wide, and the constant field in the gap is 6000 N/C. A proton ($q = e$ and $m = 1.67 \times 10^{-27}$ kg) is released at the positive plate. What will be its speed just before it strikes the negative plate? Assume vacuum between the plates.

Reasoning The force on the proton will be

$$F = qE = (1.60 \times 10^{-19}\,\text{C})(6000\,\text{N/C}) = 9.6 \times 10^{-16}\,\text{N}$$

It will therefore experience an acceleration

$$a = \frac{F}{m} = \frac{9.6 \times 10^{-16}\,\text{N}}{1.67 \times 10^{-27}\,\text{kg}} = 5.7 \times 10^{11}\,\text{m/s}^2$$

To find its speed after it falls through the gap, a distance of 2.0×10^{-3} m, we use

$$v^2 = v_0^2 + 2ax = 0 + 2(5.7 \times 10^{11}\,\text{m/s}^2)(2.0 \times 10^{-3}\,\text{m})$$

from which $v = 48{,}000$ m/s.

15.11 Field of a Sphere

The electric field due to a point charge is given by Coulomb's law. We have that the force on a test charge q_t is

$$F = k \frac{q q_t}{r^2}$$

and so *the electric field due to the point charge q is*

$$E = \frac{F}{q_t} = k \frac{q}{r^2} \tag{15.4}$$

FIGURE 15.23
The field lines for a uniformly charged sphere act as though they come from a point charge at its center.

It is possible to show that this same relation gives the field *outside* a uniformly charged sphere which has a total charge q. In that case, r is the distance from the center of the sphere to the point in question. To show this, consider the sphere shown in Fig. 15.23 as viewed from a large distance. Because r is much greater than the dimensions of the sphere, the sphere will act as a point charge for points so far away from it. Hence Eq. (15.4) will certainly give the field at large distances from the sphere.

Notice, however, that the field lines radiating from a uniformly charged sphere will be identical to those radiating from a like charge at the sphere's center, provided we consider the region outside the sphere. Therefore the fields due to a point charge q and due to a uniformly charged sphere having a total charge q must be the same. For that reason, Eq. (15.4) must apply to a uniformly charged sphere as well as to a point charge. It is often convenient to use this fact.

Illustration 15.9 Sparking occurs through air if the electric field exceeds about 3×10^6 N/C. (We call this the *electric strength* of air.) About how much charge can a 10.0-cm-diameter metal sphere hold before sparking will occur?

Reasoning We know that Eq. (15.4) applies to the sphere where q is the charge on the sphere. At the sphere's surface, E cannot exceed 3×10^6 N/C, and $r = 0.050$ m in this case. Equation (15.4) therefore becomes

$$3 \times 10^6 \text{ N/C} = (9 \times 10^9 \text{ N} \cdot \text{m}^2/\text{C}^2) \frac{q}{(0.050 \text{ m})^2}$$

from which $q = 8.3 \times 10^{-7}$ C. As we see, a sphere this size can hold a charge of about 1 microcoulomb (μC) in air.

Summary

There are two kinds of charges, designated plus and minus. The electron carries a charge $-e = -1.60 \times 10^{-19}$ C, whereas the charge on a proton is $e = 1.60 \times 10^{-19}$ C. Like charges repel one another, and unlike charges attract one another.

The force on a point charge q_1 at a distance r from a point charge q_2 is given by Coulomb's law to be

$$F = k \frac{q_1 q_2}{r^2}$$

where q_1 and q_2 are measured in coulombs and $k \approx 9 \times 10^9$ in SI units. Also, $k = 1/4\pi\varepsilon_0$, where $\varepsilon_0 = 8.85 \times 10^{-12}\,C^2/(N \cdot m^2)$ is the permittivity of free space.

The electric field strength **E** at a point is the force per unit positive test charge placed at that point. Its units are newtons per coulomb. Electric fields are represented by field lines which are tangent to **E**. These lines emerge from positive charges and end on negative charges. The field is strongest where the lines are closest together.

Under electrostatic conditions, the field in the body of a metal is zero. The field lines end on the metal surface and are perpendicular to the surface. All the excess charge on a metal object resides on the surface of the object.

Minimum Learning Goals

Upon completion of this chapter, you should be able to do the following:

1 State what the two kinds of charge are and which kind exists on the proton and on the electron. Describe the force which one charge exerts on another charge.
2 Distinguish among insulators, conductors, and semiconductors.
3 Explain how an object can be charged by conduction and by induction. Describe qualitatively how charges on a given metal object redistribute when a charged object is brought nearby.

4 State the conclusions that can be drawn from the Faraday ice-pail experiment.
5 Use Coulomb's law to find the force on a given charge in the vicinity of other point charges.
6 Explain what information can be gleaned from an electric-field diagram. In so doing, define E, explain how to draw field lines, and tell how the lines behave near point charges and near and in metals.
7 Use the relation $F = qE$ in simple situations.
8 Sketch the electric field lines in the vicinity of simple charged objects.

Important Terms and Phrases

You should be able to define or explain each of the following:

$e = 1.602 \times 10^{-19}\,C$
Insulator versus conductor; semiconductor
Electroscope

Electrical ground
Induced charge
Faraday ice-pail experiment
Coulomb's law
Excess charges are a small fraction of the total charge
Electric field and E

Questions and Guesstimates

1 A plastic comb or ruler can be given an electrostatic charge by rubbing it vigorously with a dry cloth. How could you determine the sign of the charge on the comb or ruler?
2 A plastic comb or pencil charged by rubbing it with a dry cloth will attract tiny bits of paper. Explain why. (*Hint:* The usually slightly moist paper conducts electricity slightly.) Why does a phonograph record wiped with a dry cloth usually attract dust and lint?
3 Sometimes your whole body becomes charged by walking across a deep carpet or by sliding across a

plastic car seat. If you then extend your finger toward a metal door knob or handle, a spark will jump from it. Explain.
4 How can you charge a metal object positively using a positively charged plastic comb? Using a negatively charged plastic pen?
5 You have probably noticed that clothes dried in a rotating drier often cling together. What causes this? Will it happen if all the clothes are of the same material? What is done to decrease the problem?
6 Two equal-magnitude positive point charges are a

distance D apart. Where can you place a third charge so that the resultant force on it is zero? Is it in stable equilibrium there?

7 A positive point charge and a much larger negative point charge are a distance D apart. Is there anyplace where a third point charge can be placed where the resultant force on it would be zero?

8 A tiny ball with charge Q is suspended between two very large parallel metal plates. The plates are grounded. Sketch the electric field between the plates. What can you infer about the induced charges on the plates?

9 As shown in cross section in Fig. P15.1, a metal sphere has a cavity inside it. Within the cavity is suspended a charge $+Q$. Sketch the induced charges on the metal and the electric field both inside and outside the sphere.

10 Why cannot properly drawn electric field lines cross?

11 When surrounded by the atmosphere, sparks will

FIGURE P15.1

jump from a metal object if the field at the metal surface exceeds about 3×10^6 N/C. Why can more charge be placed on a metal sphere if it is highly polished than if its surface is rough?

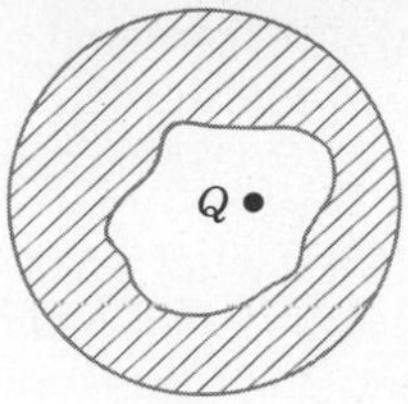

12 Sensitive apparatus is frequently shielded from unwanted electric fields by placing the apparatus inside a metal cavity. The "cavity" might consist of the interior of a metal can or the interior of a fine-mesh wire box that is grounded. Explain why the field of a charge placed outside such a shield does not affect the interior region.

Problems

1 Two point charges, $+3 \times 10^{-6}$ and -5×10^{-5} C, are 80 cm apart. (a) How large a force does the positive charge exert on the negative charge? What is its direction? (b) Repeat for the force on the positive charge.

2 Two point charges $+6 \times 10^{-4}$ and $+2 \times 10^{-5}$ C are 30 cm apart. Find the magnitude and direction of the force on (a) the 6×10^{-4} C charge and (b) the 2×10^{-5} C charge.

3 Two equal point charges exert a repulsive force of 0.60 N on each other when they are 40 cm apart. (a) Find the magnitude of each charge. (b) What can you conclude about the signs of the charges?

4 A point charge q at $x = 0$ exerts a force of 1.50 N on a charge $3q$ placed at $x = 50$ cm. How large is q?

5 A point charge 8×10^{-6} C is placed on the x axis at $x = 0$ while an unknown point charge is placed on the axis at $x = 50$ cm. The force on the unknown charge is 300 N in the $-x$ direction. Find the sign and magnitude of the unknown charge.

6* In the Bohr model of the hydrogen atom, an electron circles the nucleus (a proton) at a radius of 0.53×10^{-10} m. (a) How large a force does the proton exert on the electron? (b) This force supplies the centripetal force which holds the electron in orbit. How fast is the electron moving? (The

charge on each particle has magnitude e. The electron $m = 9.1 \times 10^{-31}$ kg.)

7 The following three point charges are placed on the x axis: $+4\,\mu$C at $x = 0$, $+2\,\mu$C at $x = 30$ cm, and $+6\,\mu$C at $x = 60$ cm. Find (a) the force on the 2-μC charge; (b) the force on the 6-μC charge.

8 Three point charges with values -2, $+4$, and $-3\,\mu$C are placed at $x = 0$, $x = 40$, and $x = 120$ cm, respectively. Find the force (a) on the -3-μC charge and (b) on the $+4$-μC charge.

9** Two point charges are placed on the x axis: a -3-μC at $x = 0$ and a $+4$-μC charge at $x = 80$ cm. At what point(s) in the vicinity of these two charges can a $+5$-μC charge be placed so that it will experience no resultant force?

10** Two point charges are place on the x axis, a $+6$-μC charge at $x = 0$ and a $+5$-μC one at $x = 100$ cm. At what point(s) in the vicinity of these two charges will the resultant force on a $+3$-μC charge be zero?

11* As shown in Fig. P15.2, four point charges are placed on the four corners of a square. If $q_1 = q_2 = q_3 = q_4 = 5\,\mu$C and $a = 30$ cm, find the magnitude and direction of the force on q_3.

12* All four charges in Fig. P15.2 are equal in magnitude. However, q_1 and q_2 are positive while q_3 and q_4 are negative. If $q_1 = 6\,\mu$C and $a = 20$ cm, find

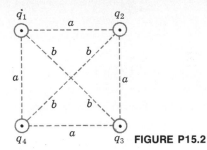

q_1 a q_2

a b b a

   b b

q_4 a q_3 **FIGURE P15.2**

the magnitude and direction of the force on q_4.

13** Two point charges q_1 and q_2 that are 90 cm apart exert a force of 0.050 N on each other. It is known that $q_1 + q_2 = 4.5\,\mu C$. Find the magnitudes of q_1 and q_2. Is the force attractive or repulsive?

14* Radium nuclei are radioactive and throw out alpha particles $(m = 4 \times 1.67 \times 10^{-27}\,kg,\ q = +2e)$. The nucleus left behind has a charge of $+86e$ and a very large mass. (a) Find the force exerted on the alpha particle by the nucleus when they are $5 \times 10^{-14}\,m$ apart. (b) What is the acceleration of the alpha particle at that instant?

15 In a certain region of space, a tiny pith ball with charge $-3.0 \times 10^{-12}\,C$ experiences a force of $5.0 \times 10^{-7}\,N$ in the $+x$ direction. What are the electric field magnitude and direction in this region?
ILL 15.5

16 The electric field in a certain region is directed eastward and has a magnitude of 4000 N/C. Find the magnitude and direction of the force experienced by a charge $-8.0 \times 10^{-5}\,C$ when in this region.

17* A tiny droplet has mass m and charge $+q$. It is to be supported against gravity by an electric field. Find the magnitude of E (in terms of q and m) as well as its direction.

18 Two charges are placed on the x axis, a $+6$-μC charge at $x = 60$ cm and a -3-μC one at $x = 0$. Find the electric field at (a) $x = 30$ cm and (b) $x = 100$ cm.

19* Find the electric field strength at the center of the square in Fig. P15.2 if (a) $q_1 = q_2 = q_3 = q_4$; (b) $q_1 = q_2 = 2\,\mu C$ and $q_3 = q_4 = -4\,\mu C$ and $b = 50$ cm.

20* If $q_1 = q_3 = -6.4\,\mu C$ and $q_2 = 2\,\mu C$ in Fig. P15.3, find the electric field strength at point A.

21* In a certain region the electric field is 3000 N/C in the $+x$ direction. An electron $(q = -e, m = 9.1 \times 10^{-31}\,kg)$ is released at $x = 0$. Find its position and speed after 1×10^{-8} s.

22 A 2.0-cm-diameter metal sphere carries a charge of -50 nC (nanocoulombs). Find the magnitude and direction of its electric field at the following distances from the center of the sphere: (a) 1.05 cm, (b) 100 cm.

23* Between collisions an air molecule in the atmosphere travels about 10^{-7} m. Suppose an ion $(q = e, m = 28 \times 1.66 \times 10^{-27}\,kg)$ is released in air in a region where $E = 3 \times 10^6$ N/C, the breakdown field for air. (a) What will be the acceleration of the ion? (b) How fast will it be moving after it has moved 10^{-8} m? (c) What will be its energy at that time? For comparison purposes, the energy a particle must have to ionize an air molecule on collision is about 2.4×10^{-18} J.

24** (a) In Fig. P15.3, if $q_1 = q_2 = +6.4\,\mu C$, how large must q_3 be if the field at A is to be directed along a line parallel to the line of the three charges? (b) How large is the field in that case?

25* Each aluminum atom contains 13 electrons, and the atom has $M = 27$ kg/kmol. (a) How many atoms are there in 2.0 g of aluminum? (b) How many electrons are there in 2.0 g of aluminum? (c) What fraction of the electrons must be removed from a 2.0 g aluminum sphere to give it a net charge of $3.0\,\mu C$?

26* Refer to the data concerning aluminum given in the previous problem. Suppose a small aluminum sphere has a mass of 0.50 g and that it is possible to remove 0.10 percent of its total electrons. Then what would be the electric field at a distance of 4.0 m from the charged sphere? For comparison purposes, sparking occurs in air if the field exceeds a value of about 3×10^6 N/C.

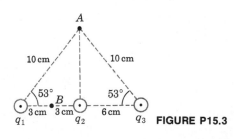

FIGURE P15.3

A

10 cm   10 cm

$53°$ $53°$

q_1 3 cm B 3 cm q_2 6 cm q_3

16
Electric Potential

In our study of mechanics we found the concepts of work and potential energy to be of great utility. Even though many situations were too involved to be solved in detail by using forces, the energy approach often allowed us to obtain useful results quickly. We see in this chapter that the concept of electric potential energy is extremely useful in applications of electricity. It is indispensable for an understanding of such widely diverse topics as electric circuits and nuclear accelerators.

16.1 Electric Potential Energy: Potential Difference

When discussing the movement of an object from place to place within a gravitational field, we used the concept of gravitational potential energy. It will be recalled that the difference in potential energy between two points was defined to be the work needed to move the object from one point to the other. Moreover, the path along which the object is moved from one point to the other point is of no consequence; the work done, and therefore the potential energy difference, is the same for all paths. The gravitational field is said to be a *conservative* field because it possesses this property. The work required to move an object from one point to another point against the gravitational force is independent of the path taken between the two points.

We have a similar situation in electricity. Consider the electric field between the two oppositely charged, parallel metal plates shown in Fig. 16.1. As explained in the last chapter, the field is uniform except near the edges of the plates. Assume that the region shown in the figure is not close to a plate edge. The electric field **E** in this region is constant and is directed from the positive to the negative plate.

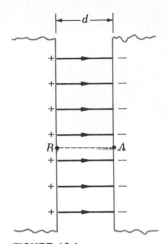

Suppose now that a unit positive test charge (an imaginary charge, you will recall) is placed between the plates. Because we defined $\mathbf{E} = \mathbf{F}/q_t$, and because we are taking $q_t = +1$ C, **F** and **E** have the same numerical value. The force on the unit positive test charge placed between the plates of Fig. 16.1 will be directed toward the right, and its numerical value is E. Work must be done to move the test charge from place to place within the field. Referring to Fig. 16.1, let us compute the work done in moving the unit positive test charge from A to B.

In order to move the unit positive charge from A to B in Fig. 16.1, we must pull on it with a force of magnitude E to balance the force exerted on it by the electric field. Hence the work done in carrying the test charge from A to B is (force per unit charge) × (distance), or Ed. We define the potential difference from A to B to be this work. In general,

FIGURE 16.1
The electric field between the two oppositely charged parallel plates is uniform.

> The **potential difference** *from point A to point B is the work required to carry a unit positive test charge from A to B; it is work per unit positive charge.*

Definition

The unit for potential difference is that of work per unit charge, which is joules per coulomb, and we denote this unit as *the volt* (V).

We have just seen that the work done in carrying a unit positive test charge from A to B in Fig. 16.1 is Ed. Using the symbol V_A for the potential at A and V_B for the potential at B, we can write

$$V_B - V_A = Ed \qquad \text{constant field} \qquad (16.1)$$

Notice that this relation applies only to the special case shown in Fig. 16.1, a situation where the electric field is constant. In other more complicated cases, we must return to the definition of potential difference:

> $V_B - V_A$ is the work one must do to carry a unit positive test charge from A to B.

16.2 Work and Electrical Potential Energy

Suppose we are to carry a positive particle of charge q and mass m from point A to point B in Fig. 16.1. How much work must we do to carry it from A to B? The answer to this question is almost obvious if we recall that $V_B - V_A$ is the work done in carrying a $+1$-C charge from A to B. A charge of $+3$ C, for instance, would require an amount of work 3 times as large, namely, $3(V_B - V_A)$. Or, in general,

To carry a charge q from A to B requires

$$\text{Work} = q(V_B - V_A) \qquad (16.2)$$

This work is stored in the system as electrical potential energy as we shall now see.

To move a positive particle of charge q and mass m from A to B requires work in the amount $q(V_B - V_A)$. But after the particle reaches B, suppose we now release it. The charge will be repelled by the positive plate and will be attracted by the negative plate. It will therefore accelerate and shoot back to point A. In so doing, it will acquire kinetic energy equal to the work we did originally in moving the charge from A to B. We therefore see that the work given by Eq. (16.2) is stored in the system and can be converted to kinetic energy. Because of its position in the electric field, the charged particle has electrical potential energy that can be converted to kinetic energy provided the particle is allowed to move in response to the electric field force.

Illustration 16.1 Suppose the potential difference between the two plates in Fig 16.1 is 12.0 V. What is the electric field between the two plates if their separation is 0.50 cm?

Reasoning We know that $V_B - V_A = 12.0$ V and $d = 5.0 \times 10^{-3}$ m. Using Eq. (16.1), we have

$$12.0 \text{ V} = E(5.0 \times 10^{-3} \text{ m})$$

from which $E = 2400$ V/m $= 2400$ N/C. Can you show that the units for E which we have been using, namely newtons per coulomb, are equivalent to volts per meter?

Illustration 16.2 (a) How much work is required to move a proton ($q = e, m = 1.67 \times 10^{-27}$ kg) from A to B in Fig. 16.1 if the potential difference between the plates is 45 V? (b) If the proton is released from point B, what will be its speed just before it strikes the plate at A?

Reasoning (a) We can use Eq. (16.2) to write

$$\text{Work} = e(45 \text{ V}) = (1.60 \times 10^{-19} \text{ C})(45 \text{ V}) = 7.2 \times 10^{-18} \text{ J}$$

The units are obtained by noticing that 1 V $= 1$ J/C.

(b) When the proton, a positive charge, is released from B, it will accelerate toward A. Its electrical PE will be changed to KE. But the PE is equal to the work done in moving the charge from A to B, and this was found in part a. Therefore,

$$\text{KE at } A = \text{PE at } B$$

$$\tfrac{1}{2}mv_A^2 = 7.2 \times 10^{-18} \text{ J}$$

Using $m = 1.67 \times 10^{-27}$ kg gives $v_A = 9.3 \times 10^4$ m/s.

Illustration 16.3   Find the speed of an electron ($q = -e$, $m = 9.1 \times 10^{-31}$ kg) at B if it is released from point A in Fig. 16.1. Assume the potential difference between the plates is 45 V.

Reasoning   Notice that, unlike the positive proton which falls from B to A, the negative electron falls from A to B because it is attracted by the positive plate and repelled by the negative plate. We have

$$\text{KE at } B = \text{PE at } A$$

$$\tfrac{1}{2}mv_B^2 = e(V_B - V_A)$$

or $\qquad \tfrac{1}{2}(9.1 \times 10^{-31} \text{ kg})(v_B^2) = (1.6 \times 10^{-19} \text{ C})(45 \text{ V})$

which gives $v_B = 4.0 \times 10^6$ m/s.

16.3 Equipotentials

Let us now look at other points besides A and B in the region between two charged plates. For example, we might ask for the potential difference between points M and N in Fig. 16.2. Because potential difference is simply work per unit charge, we must find the work required to move a unit positive test charge from M to N. To do this, we note that to hold the test charge in place, we must exert a force to the left on it. This force is needed to counterbalance the effect of the electric field on the test charge. But if we move the charge from M to N, our balancing force does no work since the direction of motion is perpendicular to the force. Indeed, we see that no work is needed to move the test charge in a direction perpendicular to the electric field. Therefore, no potential difference exists between points M and N in Fig. 16.2. In fact, it should be clear that all points on the dashed line through M and N are at the same potential. We call this line of constant potential an *equipotential line*. Moreover, the plane that lies through this line and is parallel to the plate is a constant-potential plane, an equipotential plane. *No work is done in moving a charge along an equipotential line or equipotential plane* since such motion is always perpendicular to the lines of force, i.e., the electric field. Conversely, *the lines of force are always perpendicular to the equipotential lines.*

There are an infinite number of equipotential lines and planes between the plates in Fig. 16.2. Can you show that no total work is done on a charge

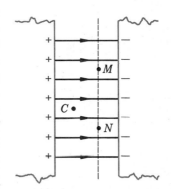

FIGURE 16.2
Points *M* and *N* are on an equipotential line.

in moving it along a path that begins and ends on the same equipotential line or plane? The same reasoning can be used to prove that the work done in carrying a charge from point M to a point such as C in Fig. 16.2 is independent of the path taken. We therefore conclude that *the static electric field is a conservative field. The electrical potential difference between two points is a constant independent of the path used for its computation.*

Before leaving our discussion of equipotentials, we should point out that metal objects are equipotential volumes under electrostatic conditions. For example, suppose you are concerned with a solid block of metal of any shape. We know that, under electrostatic conditions, the electric field everywhere within it is zero. As a result, no force will exist on a test charge within the metal, and so no work is required to move the test charge from place to place. Hence there can be no potential difference from place to place within the metal. The metal constitutes an equipotential volume.

Illustration 16.4 Sketch the equipotentials and electric field lines near a charged, solid metal object.

Reasoning Consider a typical charged object such as the one shown in cross section in Fig. 16.3. The metal object itself is an equipotential volume, and so its surface is an equipotential area. Because lines of force must be perpendicular to equipotential lines and areas, the field lines must radiate from the object as shown. But the equipotentials must be perpendicular to the field lines, and so they are as shown by the colored lines.

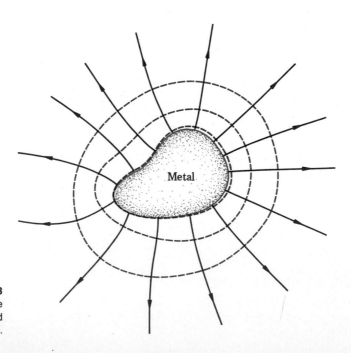

FIGURE 16.3
The equipotentials are perpendicular to the field lines.

16.4 Batteries as a Source of Potential Difference

One of the easiest ways to supply a potential difference to two points is by means of a battery. There are many kinds of batteries. Although most batteries are essentially chemical devices, other types of batteries are now becoming prominent. The ordinary lead-cell battery, the common automobile battery, makes use of a chemical reaction to supply energy. This is likewise true of the "dry cell," which is not dry inside in spite of its name. Perhaps you have heard of solar cells, which are used to supply electrical energy to devices in space. Solar batteries operate on quite different principles and transform light and heat energy directly into electrical energy. Other types of nonchemical batteries are in the process of development.

For most purposes in this book, we ignore the internal workings of the battery except to consider it a source of energy and potential difference. The basic property of all batteries is that they provide two terminals (actually metal posts on the battery box) that are at different potentials. *The symbol ordinarily used for a battery is* ⊣⊢ , *where the longer vertical line, marked plus, is the positive terminal of the battery and the shorter vertical line, marked minus, is the negative terminal.* Usually the plus and minus symbols are left off, and you are expected to know that the long side is positive. (In passing, we might note that the terminals on batteries are usually stamped plus and minus, or, in some cases, the positive terminal is merely painted red.)

A simple chemical battery (or cell) can be made by immersing two dissimilar metal rods in a dilute acid, as shown in Fig. 16.4. Most metals dissolve at least slightly in the acid. When the metal dissolves, each atom of the metal leaves at least one electron behind on the rod and enters the solution as a positive ion. The electrode (rod) from which the ion came is thus charged negatively. Finally, it becomes so negative that equal numbers of positive ions are attracted back, and the net number of ions leaving the electrode becomes zero. Since the electrode has an excess of negative charge, a potential difference exists between it and the solution.

In Fig. 16.4 the potential difference between the electrode and solution is greater for metal B than for metal A because B dissolves more extensively than A. Even though both electrodes are negative, metal B is at a lower potential than A. As a result, A is at a higher potential than B, and so A is called the positive terminal. The potential difference between electrode B and electrode A is called the *electromotive force* (emf) of the battery. We denote it by the symbol $\mathcal{E}$. Do not confuse it with the symbol E used for electric field. *The* **electromotive force** *(emf)* $\mathcal{E}$ *is the potential difference between the terminals of a battery when the battery is isolated electrically.*

Symbol for a Battery

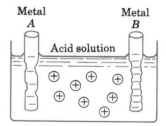

FIGURE 16.4
Positive ions go into solution, leaving electrons behind on the electrodes. In the case shown, metal B has lost more ions per unit of volume than metal A. Which electrode is at the higher potential?

16.5 Batteries as Energy Sources

We shall see as our study of electricity progresses that *the usual purpose of a battery is to supply energy to charges.* Let us now investigate a few instances

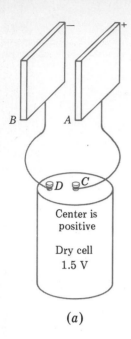

B   A

Center is positive

Dry cell
1.5 V

(a)

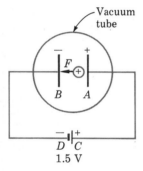

Vacuum tube

F

B   A

D   C
1.5 V

(b)

FIGURE 16.5
The potential difference from B to A is 1.5 V, the emf of the battery.

FIGURE 16.6
The electron gun shoots a beam of electrons toward the right. In a typical TV tube, the gun is considerably more complicated than the one shown. The beam travels through vacuum and eventually strikes a fluorescent screen, which would be far to the right.

which show how charges acquire energy from a battery. Consider first the situation shown in Fig. 16.5a. We see there a battery connected to two metal plates. The situation is also shown in the shorthand diagram in Fig. 16.5b. As indicated there, we assume the two plates to be in vacuum.

The battery places excess charge on the two plates. As indicated, plate A is positive since it is connected to the positive side of the battery. This transfer of charge to the plates requires only a tiny fraction of a second. After that, the situation is electrostatic. You will recall that under electrostatic conditions metal solids are equipotential volumes. For that reason, the wire from C to A and the plate at A are all at the same potential. We say that they are all at the same electrical level. Similarly, points D and B as well as plate B are all at identical levels. But point D is 1.5 V lower than point C. Therefore, plate B must be 1.5 V lower than plate A. The potential difference from B to A is 1.5 V, with A being at the higher level.

Suppose now that a positive charge is released from plate A as shown in Fig. 16.5b. It will be repelled by plate A and attracted by plate B. As a result, it will accelerate from plate A to plate B. We see, in effect, that the positive charge "falls" from the high-potential plate A to the low-potential plate B.

Positive charges fall from high potentials to lower potentials.

We have already discussed situations such as this in detail, and we concluded that

When a charge q falls through a potential difference $V_A - V_B$, it gains kinetic energy in the amount $q(V_A - V_B)$.

Let us now apply our knowledge to a more practical situation.

Many electronic devices give visual displays of electrical signals. The picture on a TV tube is a familiar example. To create the picture, an electron gun shoots a beam of electrons at the fluorescent material on the end of the tube. As the beam sweeps repeatedly across the tube face, it traces the picture one sees. A highly simplified electron gun is shown in Fig. 16.6.

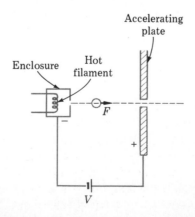

Accelerating plate

Enclosure   Hot filament

F

V

Electrons boil out of a white-hot filament within a metal enclosure. Many of these electrons wander out of the enclosure through a tiny hole as shown. They then "see" that the enclosure is charged negative and the plate on the right is positive. As a result, they accelerate toward the plate. By the time they reach the plate, they are moving at high speed, and some electrons shoot through the tiny hole in the plate. As a result, a pencillike beam of electrons shoots from this device. (In practice, the gun has a more complicated structure than shown.)

Illustration 16.5 In a typical TV set, the potential difference through which the charges accelerate is of the order of 20,000 V. Find the speed of the electrons which shoot from the gun.

Reasoning Notice in this case that the charge falls from minus to plus. The electron, being negative, behaves oppositely to a positive charge. In either case, however, the change in KE of the particle has a magnitude qV, where $V = 2 \times 10^4$ V. We have, since $q = -1.6 \times 10^{-19}$ C,

$$\text{KE gained} = |qV|$$
$$= (1.6 \times 10^{-19} \text{ C})(2 \times 10^4 \text{ V}) = 3.2 \times 10^{-15} \text{ J}$$

But the electron had negligible initial KE, and so its final $\text{KE} = \frac{1}{2}mv_f^2$ is equal to 3.2×10^{-15} J. The mass of an electron is 9.1×10^{-31} kg, and so we can write

$$\tfrac{1}{2}mv_f^2 = 3.2 \times 10^{-15} \text{ J}$$

and solve for v_f to give $v_f = 8.4 \times 10^7$ m/s.

This is an extremely high speed. The speed of light, the ultimate speed with which any particle can move, is only 30×10^7 m/s. Consequently, you might suspect that we would be in difficulty if we tried this computation for a much higher potential difference. You would be correct. When particle speeds close to the speed of light are encountered, we must resort to computational methods first explained by Einstein in his theory of relativity. We investigate this topic in Chap. 25. For now we are content to state that the result we obtained above is slightly in error. According to relativity, the correct result is $v_f = 8.1 \times 10^7$ m/s.

16.6 Electronvolt Energy Unit

In our studies of atomic physics, the energy of electrons and similar particles is needed frequently. In most cases, these energies are acquired by movement of the charged particle through an electrical potential difference. For example, to remove the single electron from a hydrogen atom, you must furnish it with an energy equivalent to that which the electron would acquire in falling through a potential difference of 13.6 V. In other words, the energy required to ionize the hydrogen atom, i.e., to tear its electron loose, is Vq, where V is 13.6 V and q is the electronic charge, 1.6×10^{-19} C. Upon evaluation, this turns out to be 2.18×10^{-18} J.

It is inconvenient to multiply V by q in stating the energy of a particle that has fallen through a potential difference. We therefore *define a new energy unit called the* **electronvolt** (eV). *By definition,*

The eV Unit

$$1 \text{ eV} = 1.60 \times 10^{-19} \text{ J}$$

where 1.60×10^{-19} C is the charge quantum e. Since the energy of a charge q which has fallen through a potential difference V is Vq joules,

$$\text{Energy (eV)} = \frac{(\text{potential difference})(\text{charge})}{\text{charge quantum}} \tag{16.3}$$

As typical examples, if an electron falls through a potential difference of 2 million volts (MV), its energy is 2×10^6 eV, or 2 MeV, where MeV stands for mega (or million) electronvolts. However, an α particle (which has a charge of 3.20×10^{-19} C, or two charge quanta) will have an energy of 4 MeV when it falls through a potential difference of 2×10^6 V. Of course, the electronvolt is not a proper SI unit, and so we must always change it to joules (by multiplying by 1.6×10^{-19}) before using it in fundamental formulas.

16.7 Absolute Potentials

So far, we have been concerned only with differences in potential because, as in gravitational potential, the choice of a position for zero PE is merely a matter of convenience. Gravitational PE can be measured with respect to any point we choose: the tabletop, the ground, the top of a building, or wherever. Similarly, in electrical-PE problems, the zero-PE location is a matter of choice. In electrical-circuit theory, a particular wire in the circuit may be attached to the ground, e.g., soldered to a water pipe. This point would usually be taken to have zero PE. However, a different zero for electrical potential is frequently taken.

When dealing with atoms and molecules, we frequently specify the zero of electrical PE in a different way. To illustrate this method of approach, let us refer to Fig. 16.7, where we see a charged sphere carrying charge $+Q$. This sphere might represent a proton, for example. It is customary in such cases to take the zero of electrical PE to be an infinite distance away from the charge. Let us now see the consequences of such a choice.

Consider the point B in Fig. 16.7, which is at a radius r from the center of the charge. *We define the* **absolute potential** *at B to be the work done in carrying a unit positive test charge from infinity up to B.* In effect, what we are doing is as follows. So far, we have discussed situations in terms of

Definition

Absolute Potential

potential differences, $V_B - V_A$. Now, however, we specify that point A is to be taken at infinity. Further, we specify that the potential at infinity is to be taken as zero so that $V_A = 0$. Then V_B becomes what we refer to as the absolute potential at B. Notice carefully that *when we speak of the absolute potential at a point, we are really speaking about the potential difference between infinity and that point.*

A particularly simple expression can be found for the absolute potential outside a single, isolated charge such as the one shown in Fig. 16.7. We need to compute the work done in carrying our unit test charge from infinity up to the point B, for example. This computation is not as simple as finding the potential difference between two parallel plates. In the parallel-plate case, E is constant, and so the force on the test charge is constant. Hence the potential difference was simply Ed.

But in this case, E becomes stronger as we come closer to B. At infinity, E is zero. Clearly, the value of E changes, and so does the force on our test charge as we bring it in from infinity. In spite of that fact, the work done in bringing the test charge from infinity to point B is still easily found by using calculus.

We only state the result here. *The absolute potential V, at point B of Fig. 16.7, due to the charge $+Q$; that is, the work needed to carry a unit positive charge from infinity up to point B, is*

$$V = k\frac{Q}{r}$$ (16.4)   *Point-Charge Potential*

or

$$V = (9 \times 10^9 \text{ N} \cdot \text{m}^2/\text{C}^2)\left(\frac{Q}{r}\right)$$

This equation, like Coulomb's law, *applies only to point charges and uniformly charged spheres.* In other cases, the distance r is not usually well defined.

Illustration 16.6 Compute the absolute potential at point B in Fig. 16.7 if $r = 50$ cm and $Q = 5 \times 10^{-6}$ C.

Reasoning From Eq. (16.4) we have at once

$$V = (9 \times 10^9 \text{ N} \cdot \text{m}^2/\text{C}^2)\left(\frac{5 \times 10^{-6} \text{ C}}{0.50 \text{ m}}\right) = 90,000 \text{ V}$$

Verify the units given with the answers.

Illustration 16.7 Compute the absolute potential at point B in the vicinity of the three charges shown in Fig. 16.8.

Reasoning When a unit positive test charge is carried from infinity up

FIGURE 16.7
We define the absolute potential at B to be the work done in carrying the unit positive test charge from infinity up to B.

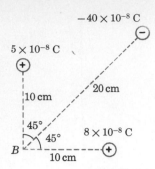

−40 × 10⁻⁸ C

5 × 10⁻⁸ C

10 cm

20 cm

45°

45°

8 × 10⁻⁸ C

B

10 cm

FIGURE 16.8
Find the absolute potential
at point B.

to point B, work is done against three separate forces. It is done against the repulsions of the two positive charges, and the negative charge actually tends to pull the positive test charge in toward B. Hence the work done consists of two positive amounts of work, done against the repulsions of the two positive charges, and a negative amount of work, the work done *on* the test charge by the attraction of the negative charge. These are each found by use of Eq. (16.4).

Due to $+5 \times 10^{-8}$: $\qquad V_1 = (9 \times 10^9)\left(\dfrac{5 \times 10^{-8}}{0.10}\right) = 4500 \text{ V}$

Due to $+8 \times 10^{-8}$: $\qquad V_2 = (9 \times 10^9)\left(\dfrac{8 \times 10^{-8}}{0.10}\right) = 7200 \text{ V}$

Due to -40×10^{-8}: $\qquad V_3 = -(9 \times 10^9)\left(\dfrac{40 \times 10^{-8}}{0.20}\right) = -18,000 \text{ V}$

Unlike forces and fields, work and potential are scalars, not vectors. Hence, being numbers without direction, merely plus or minus, they may be added directly. The result for the absolute potential at point B is $V_B = V_1 + V_2 + V_3$, so

$$V_B = -6300 \text{ V}$$

Since the result is negative, we conclude that the unit positive charge actually has less potential energy at B than it did at infinity. This must mean that the effect of the attraction of the negative charge for it actually was stronger than the repulsion of the two positive charges.

Illustration 16.8 In the Bohr model of the hydrogen atom, the pointlike electron ($q = -e$) moves in a circular orbit ($r = 0.053$ nm) with the positive nucleus ($q = +e$) at the center. (*a*) Find the absolute potential due to the nucleus at the electron's position. (*b*) How much energy is needed to pull the electron loose from the atom?

Reasoning (*a*) The absolute potential due to the nucleus at the position of the orbit is $V = kQ/r$ with $r = 5.3 \times 10^{-11}$ m and $Q = 1.60 \times 10^{-19}$ C. Using these values, we find that the answer is

$$V_{\text{orbit}} = 2.72 \text{ V}$$

(*b*) The potential due to the nucleus at a point very far away from the nucleus is zero because $r \to \infty$ in kQ/r. Therefore, if the atom's electron is to be removed from the orbit and carried to infinity (i.e., the atom is to be ionized), it must be carried through a potential difference

$$\Delta V = V_\infty - V_{\text{orbit}} = 0 - 2.72 \text{ V} = -2.72 \text{ V}$$

The negative sign tells us it is a potential drop from the orbit to infinity. However, because the electron is negatively charged, it is attracted by the nucleus, and so work must be done on it to pull it to infinity. The work done is

$$\text{Work} = q(\Delta V) = (1.60 \times 10^{-19} \text{ C})(2.72 \text{ V}) = 4.35 \times 10^{-19} \text{ J}$$

In practice, only half this amount of energy must be furnished from outside the atom to ionize it. The electron's KE in orbit furnishes the other half of the required energy.

16.8 Capacitors (Condensers)

We have made frequent reference to two oppositely charged metal plates. This is one form of a device of considerable practical importance for storage of electric charge and energy, as we see in later chapters. It is called a *capacitor* (or *condenser*). Such a device connected to a battery is shown in Fig. 16.9. The positive terminal of the battery places a positive charge on one plate, and the negative terminal charges the other plate negatively. These charges attract one another, and so they reside on the inner surfaces of the plates, as shown. Such a device is capable of storing charge. As shown in part *b*, *the symbol used for* a capacitor *is* —||— *or sometimes* —|(— .

We now are interested in how much charge Q resides on one of the plates, say the positive one. (Of course, since the system is electrically neutral, an equal negative charge exists on the other plate.) The amount of charge on the plate will depend on several factors. First, the larger the voltage of the battery, the more charge it will put on the capacitor. Hence, if V is the voltage difference across the capacitor,

$$Q \propto V$$

Of course, the larger the plate area, the more charge it should hold. In addition, if the plates are brought closer together, the positive plate will attract more negative charge from the battery to the negative plate. All these factors have to do with the geometry and construction details of the device. They are a constant of the device. Hence we can write

$$Q = CV \tag{16.5}$$

where C is a constant for a given capacitor. We call it the **capacitance** of the capacitor. When C is large, the charge on the capacitor will be large for a given voltage across it. Since Q is measured in coulombs and V in volts, the unit of capacitance is coulombs per volt, which is called the **farad** (F).

If one of the plates has an area A, and if the separation between the plates is d, it is possible to show that *the capacitance of a parallel-plate capacitor is given by*

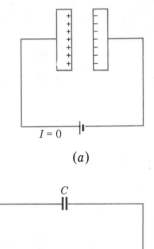

$I = 0$

(a)

C

(b)

FIGURE 16.9
Equal and opposite charges reside on the inner faces of the capacitor plate. Notice the symbol used for a capacitor in (b).

$$C = \frac{\varepsilon_0 A}{d} \qquad \text{parallel plates} \qquad (16.6)$$

where ε_0 (read "epsilon subzero") is called the **permittivity** of free space, 8.85×10^{-12} F/m. (The constant ε_0 is related to the Coulomb's law constant k through $\varepsilon_0 = 1/4\pi k$. In Chap. 21 we see that ε_0 plays an important part in the definition of electrical units.) As an example, suppose that the two plates are square, 10 cm on a side, and separated by 0.10 mm. Upon changing the length units to meters and substituting in Eq. (16.6), the capacitance is found to be

$$C = 8.85 \times 10^{-10} \text{ F} = 8.85 \times 10^{-4} \,\mu\text{F}$$

where $1 \,\mu\text{F}$ is equal to 10^{-6} F. Since capacitances are usually of the order of $1 \,\mu\text{F}$ or less, the latter unit is often used. However, farads must always be used in Eqs. (16.5) and (16.6).

In practice, most parallel-plate capacitors have an insulating sheet of material, called a **dielectric,** *placed between the plates.* This is done for two main reasons. First, the plates can be placed very close together with no fear that they will touch and permit the charges to flow. Many commercial capacitors are formed by taking two thin sheets of metal foil and laying one on top of the other, with a thin film of plastic between them to keep them from touching. The layered sheets are then rolled up into a tight cylinder and packaged for convenience. The device is essentially a parallel-plate capacitor, but it looks very different from the device shown in Fig. 16.9. Capacitors with a capacitance of $1 \,\mu\text{F}$, a common size, occupy a volume of about 1 cm^3 when made this way.

Dielectrics have a higher dielectric strength, i.e., spark over less readily, than air. For this reason, parallel plates with a dielectric between them can be used at higher voltage differences than air capacitors. Moreover, dielectrics increase the capacitance of a capacitor. *A capacitor with vacuum (or air to a good approximation) between its plates will have a capacitance given by Eq. (16.6). But if a dielectric is placed between the plates, the capacitance is* *Dielectric Constant* *found to be larger by a factor k_d, the* **dielectric constant** *or* **relative permittivity** *of the dielectric.* Typical dielectric constants are given in Table 16.1.

TABLE 16.1
DIELECTRIC CONSTANTS AT 20°C

Air	1.006	Ice (-5°C)	2.9
Paraffin	2.1	Mica	6
Petroleum oil	2.2	Acetone	27
Benzene	2.29	Methyl alcohol	31
Polystyrene	2.6	Water	81

HCl	HOH	CO	CH₃OH

HCl
(Hydrogen
chloride)

HOH
(Water)

CO
(Carbon
monoxide)

CH₃OH
(Methyl
alcohol)

FIGURE 16.10
Examples of dipolar
molecules.

16.9 Dielectrics

As pointed out in the previous section, when an insulating material (a dielectric) is placed between the plates of a capacitor, the capacitance of the device is increased. This fact is important in biophysics and chemical physics as well as from a purely technical viewpoint. From the change in capacitance caused by a given material, you can make inferences regarding the molecular behavior of the material itself. Let us now investigate this subject further.

Many molecules do not have their charges spread out uniformly. Although the molecule as a whole is neutral, the various atoms within it may have (at least partly) lost an electron to other atoms of the same molecule. A few typical examples are shown schematically in Fig. 16.10. Molecules of this type are called **dipolar molecules** since a dipole is by definition two equal and opposite charges separated by a very small distance. The **dipole moment** μ of such a molecule is equal to the magnitude of either charge multiplied by the distance of separation. In symbols, $\mu = Qd$.

In a typical molecule, the dipole moment is of the order of 1 D (debye[1]), which is equivalent to $\frac{1}{3} \times 10^{-29}$ C · m. For example, if the separated charge in the HCl molecule of Fig. 16.10 is taken to be the electronic charge, 1.6×10^{-19} C, then, since the dipole moment of this molecule is 1.03 D,

$$(1.6 \times 10^{-19} \text{ C})(d) = (1.03)(\tfrac{1}{3} \times 10^{-29}) \text{ C} \cdot \text{m}$$

where d is the charge separation. We then find $d \approx 0.5 \times 10^{-10}$ m. In many cases, d is smaller than the known separation of the atoms, and so the situation corresponds more to an unequal sharing of charge between the atoms. For HCl the value we have found for d is only about half the known distance between the atom centers.

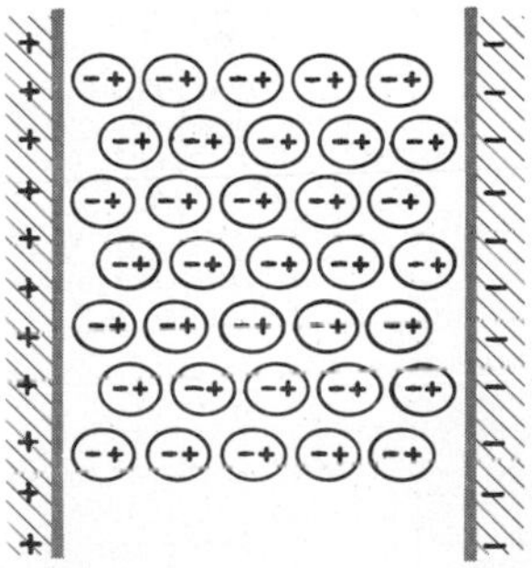

Suppose a dielectric composed of dipolar molecules is placed between two charged plates. The electric field between the plates will rotate the polar molecules so that their negative ends will be closest to the positive plate, and vice versa for the positive ends. This is shown schematically in Fig. 16.11.

FIGURE 16.11
When a polar dielectric is placed between charged plates, the molecules tend to align in the field. They thereby induce more charge flow onto the plates. The alignment is exaggerated in the figure.

---

[1] In honor of Peter Debye, a famous American physical chemist. He received the Nobel prize in chemistry in 1936 for his work with molecules and x-rays.

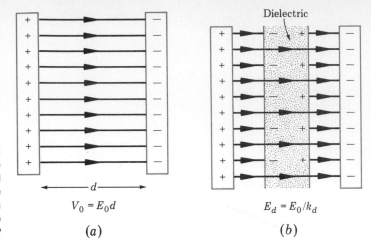

FIGURE 16.12
The dielectric reduces the field between the plates and thereby decreases the potential difference. Can you estimate k_d in the case shown?

$$V_0 = E_0 d$$

(a)

$$E_d = E_0/k_d$$

(b)

In practice, thermal motion keeps the alignment of the molecules from being even approximately perfect. Notice that, in effect, the negative ends of the molecules have caused a negative charge to be induced on the left side of the dielectric, while a positive charge has been induced on the opposite side. This causes a much reduced electric field within the region between the plates, as we shall now see.

In Fig. 16.12a, we show two charged metal plates. Let us say that the battery that charged them is no longer attached, so that charge cannot run on or off the plates. If the electric field between the plates is E_0 and the plate separation is d, the potential difference between the plates is simply $V_0 = E_0 d$. Moreover, the charge on the plates Q_0 is given by

$$Q_0 = C_0 V_0 \quad \text{or} \quad Q_0 = C_0 E_0 d \qquad (16.7)$$

Now let us suppose a dielectric sheet is placed between the plates as shown in Fig. 16.12b. Although we assume the dielectric to completely fill the space, we show separations at the edges to simplify the description of what is going on. First, we must notice that the charge on the metal plate is still Q_0 since the battery has been disconnected and no charge can flow on or off the plate. (The dielectric is an insulator.) Therefore the field in the airspace between the dielectric and plate is still E_0. However, as indicated previously, a negative charge layer is induced on the surface of the dielectric. Some of the field lines coming from the positive plate end on these negative induced charges. However, since there is less induced charge on the dielectric than the charge on the metal plate, some of the lines do not stop there but continue on through the dielectric. As we see, the field E_d inside the dielectric is smaller then E_0.

If, in actuality, the dielectric fills the whole region between the metal plates, the field everywhere between the plates will be E_d and the potential difference between the plates will be given by $V_d = E_d d$. Moreover, if we call C_d the capacitance of the parallel-plate device when dielectric is between the plates, we have

$$Q_0 = C_d V_d$$

since the charge Q_0 is still the same. After replacing V_d by $E_d d$, we find

$$Q_0 = C_d E_d d$$

Dividing this relation by Eq. (16.7) and rearranging gives

$$C_d = C_0 \frac{E_0}{E_d} \qquad (16.8)$$

In other words, *the capacitance has been increased in proportion to the ratio of the original electric field to the field within the dielectric.*

It will be recalled that *the dielectric constant k_d of the insulating material was defined equal to the ratio C_d/C_0.* We see from Eq. (16.8) that this means *a material having high dielectric constant most effectively reduces the electric field between the plates.* In other words, a high-dielectric-constant material is a material that can have charges induced easily on its surface. As a rule, the molecules with the largest dipole moment will have the highest dielectric constants since the induced charge layer will be largest for them. However, water and methyl alcohol are exceptions to this rule; in them the molecules are bound together more or less in "clumps" of molecules by so-called hydrogen bonds. These molecular clumps then act like molecules of much higher dipole moment and give rise to the very large dielectric constants listed for them in Table 16.1. Even nondipolar molecules such as benzene and methane have noticeable dielectric effects. These molecules are made dipolar by the distorting force of the electric field into which they are placed.

Effect of Dielectric on Forces

The dielectric constants of liquids are of extreme importance in many chemical and biological reactions because the dielectric liquid greatly changes the electrical forces between ions. As we saw, *the electric field in the dielectric between the plates of a capacitor is only $1/k_d$ as large as it would be without the dielectric.*

This same relation holds in more complicated cases. Two ions in solution exert forces on each other decreased by a factor $1/k_d$ by the presence of the solvent. Since water has a dielectric constant of 80, the force between ions in water is one-eightieth as large as it would be without the water. As a result, the Na^+ and Cl^- ions in NaCl, for example, are rather easily separated in water, and thermal motion is sufficient to cause solution of the NaCl. However, in benzene, $1/k_d = 1/2.3$, and so the forces between the ions are still too large to be overcome by thermal motion. Hence benzene does not dissolve NaCl. Many other similar situations exist in chemical and biological systems where the dielectric nature of the solvent is the controlling factor.

16.10 Energy Stored In a Capacitor

Suppose that the charged capacitor shown in Fig. 16.13 is discharged by connecting a wire from *a* to *b*. If the wire is quite thin, it will be found that

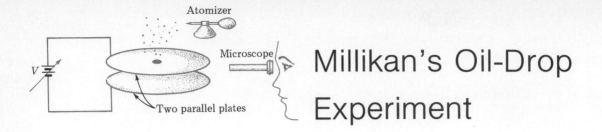

Millikan's Oil-Drop Experiment

The magnitude of the quantum of charge, the electronic charge e, was first measured accurately by Robert A. Millikan and his coworkers (1909–1913). Millikan's experiment has a simplicity and directness which establish without doubt that charge is quantized. His apparatus is illustrated schematically in the figure.

Basically, the apparatus consists of two parallel metal plates (separation ≈ 1 mm and area ≈ 100 cm^2) with a known variable potential difference between them. Because of the potential difference V, an electric field $E = V/d$ exists between the plates, where d is the plate separation. If now a charged oil droplet (mass m and charge q) finds itself between the plates, a vertical electric force Eq will act on the droplet. By adjusting the potential difference between the plates, this force can be made equal and opposite to the gravitational force mg on the droplet. When this condition is achieved, the resultant force on the droplet is zero, and the droplet will remain motionless. Then

$$mg = Eq$$

from which

$$q = \frac{mg}{E}$$

When the electric field E, the mass of the droplet m, and the acceleration due to gravity g are known, the charge q on the droplet can be found.

In practice, Millikan allowed very small oil droplets sprayed from an atomizer to fall through a small hole in the top plate. Many of these droplets contain excess charge generated by friction as the drops are produced in the atomizer. An intense light shining between the plates causes the droplets to sparkle, and so they can be observed by looking through the microscope, as indicated. Provision was also made for changing the charge on the droplet by subjecting the region between the plates to a weak x-ray beam when desired.

The electric field between the plates was calculated from the measured plate separation d and the potential difference between the plates V. To find the mass of the oil particle m, Millikan made measurements of the time taken for the droplet to fall a measured distance through the air in the absence of an electric field. Since the droplets were moving at constant speed—their terminal speed—through the distance chosen, the gravitational force mg was equal to the viscous-friction force between the droplet and the air. Assuming that the viscous force was proportional to the speed of the droplet v, he could write

$$mg \propto v$$

One of the major sources of error involved in the experiment was associated with the proportionality constant in this relation. To fair accuracy, however, one can use the so-called Stoke's law form for this viscous force and write

$$mg = 6\pi\eta av$$

where η is the viscosity of air and a is the radius of the droplet. Since the radius of the drop is related to its mass through the fact that the density of the oil ρ multiplied by the volume of the drop is equal to the drop's mass,

$$m = \tfrac{4}{3}\pi a^3 \rho$$

the mass of the drop can be found from a knowledge of ρ, η, and its terminal speed v.

As shown, when m and E are known, the charge on the droplet can be found. Millikan carried out thousands of measurements of this general type (slightly modified for higher accuracy). His result for q was always an integer multiple of 1.60×10^{-19} C. He therefore concluded that charge was quantized, the magnitude of the charge quantum being 1.60×10^{-19} C, which we designate by e. All charges thus far found in nature are integer multiples of this value. In particular, the charge on the electron is $-e$, while that on the proton is $+e$. The currently accepted value of e is 1.6022×10^{-19} C.

---

when the negative charge from the negative plate runs through it to neutralize the positive plate, the wire becomes hot. Since heat is energy, the charged capacitor, which produced the heat energy, must have had energy stored in it. We now compute how much energy it had.

Let us consider how much work is required to charge the capacitor. That work will then be equal to the energy stored in the capacitor. When the capacitor is uncharged, no work at all is required to carry a small quantity of positive charge from plate b to plate a. However, as we continue to carry charge across, the field between the plates will become sizable. We then need to pull the positive charges across. In fact, if we have charged the capacitor to the point where the potential difference between its plates is V' volts, the work required to carry a charge increment of ΔQ across would be $\Delta Q V'$.

Clearly, the first charge carried across when the capacitor is not charged requires no work, since V' is zero. On the other hand, the final increment of charge carried across from one plate to another is carried through the final

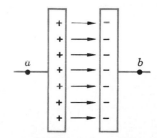

FIGURE 16.13
The energy stored in a capacitor is equal to the work done in charging it, $\tfrac{1}{2}QV$.

433

potential difference V. The total work done is easily found by using calculus. However, it appears reasonable that the total work done should be equivalent to the work that would be done in carrying the whole charge Q across the average potential difference during the charging process, $\frac{1}{2}V$. This turns out to be a correct assumption. We therefore have that *the energy stored in a capacitor with charge Q and potential difference V is*

Energy of Capacitor

$$\text{Energy} = \tfrac{1}{2}QV \qquad (16.9)$$

By use of the relation $Q = CV$, Eq. (16.9) can be written in the alternate forms

$$\text{Energy} = \tfrac{1}{2}QV = \tfrac{1}{2}CV^2 = \frac{Q^2}{2C}$$

Illustration 16.9   A 2-μF capacitor is charged across a 12-V battery. How much energy is stored in it?

Reasoning   We have

$$\text{Energy} = \tfrac{1}{2}CV^2 = (\tfrac{1}{2})(2 \times 10^{-6})(144) = 1.44 \times 10^{-4}\,\text{J}$$

You should carry the units through this equation to verify the unit of the answer. In so doing, note from $Q = CV$ that 1 F is equivalent to 1 C/V.

16.11 Energy Stored in an Electric Field

In the last section we saw that the energy stored in a charged capacitor is $\frac{1}{2}CV^2$, where V is the voltage difference across a capacitor having a capacitance C. Although it is not necessary to specify exactly how and where this energy is stored, it is sometimes convenient to think of the energy as being stored in the electric field between the capacitor plates. With this in mind, it would be well to express the equation for the stored energy in terms of the electric field E between the capacitor plates. This can be done by recalling that in the case of a parallel-plate capacitor $V = Ed$, where d is the separation of the plates.

We therefore have for the energy stored in a parallel-plate capacitor

$$\text{Energy} = \tfrac{1}{2}CV^2 = \tfrac{1}{2}CE^2d^2$$

But from Eq. (16.6) the capacitance of a parallel-plate capacitor with plate area A is given by

$$C = \frac{\varepsilon_0 A}{d}$$

provided that the capacitor has vacuum between its plates. If it is filled with
a dielectric having a constant k_d, this becomes

$$C = \frac{k_d \varepsilon_0 A}{d}$$

Substituting this value for C in the energy equation yields

$$\text{Energy} = (\tfrac{1}{2}\varepsilon_0 k_d E^2)(Ad)$$

But Ad is the volume of the space between the capacitor plates, in other
words, the volume in which the constant electric field E exists. Dividing both
sides of the equation by the volume gives us an expression for the energy per
unit volume, i.e., *the energy we picture to be stored in unit volume of the
region of space where the electric field is E.* We have

$$\text{Energy per unit volume} = \tfrac{1}{2}\varepsilon_0 k_d E^2 \qquad (16.10)$$

Notice that the energy stored in unit volume is proportional to the square of
the electric field strength. It is often convenient to use Eq. (16.10) for assigning energy to an electric field. Although this expression was derived for a
very special case, it is shown in more advanced texts that it has general
validity.

Summary

The static electric field is a conservative field. By definition, the potential difference from point A to point B is
the work done in carrying unit positive test charge from
A to B. Its units are joules per coulomb, or volts, and it
is represented by $V_B - V_A$. If the line connecting B and
A is along the direction of a constant field E, then the
potential difference has a magnitude Ed, where d is the
distance between A and B.

Equipotential lines, surfaces, and volumes consist
of points which are all at the same potential. The electric field lines always intersect equipotential lines and
surfaces at right angles.

Batteries act as sources of energy for electrical systems. They provide a potential difference through
which charges can be made to fall. The potential difference between the terminals of an electrically isolated
battery is called the emf ($\mathcal{E}$) of the battery.

When a charge q falls through a potential difference V, it acquires an energy qV. Positive charges, left
to themselves, fall from high potentials to low potentials. Negative charges fall in the reverse direction.

The potential difference between any point and
infinity is called the absolute potential at the point. In
this terminology, points at infinity are assigned a value
of zero electric potential. The absolute potential at a
distance r from the center of a point charge q is kq/r.

A capacitor is a device for storing charge. It carries
equal but opposite charges on its two plates If a potential V across a capacitor causes a charge Q on its plates,
the capacitance C of the capacitor is given by
$C = Q/V$. Its units are farads (F). The capacitance of a
parallel-plate capacitor with plate area A and plate sep-
aration d is $C = \varepsilon_0 A/d$, where ε_0 is the permittivity of
free space, 8.85×10^{-12} F/m.

Dielectrics usually reduce the electric field in which
they are placed. A parallel-plate capacitor with fixed
charge Q will have the electric field between its plates
reduced by a factor $1/k_d$ when a dielectric fills the region between the plates. The quantity k_d is called the
dielectric constant for the dielectric; k_d is about 80 for
water and on the order of 2 to 5 for most oils and similar
materials; k_d for vacuum is unity.

A charged capacitor has energy $\tfrac{1}{2}QV$ stored in it.
Energy can be assumed to be stored in an electric field.
In a dielectric with constant k_d where the electric field is
E, the energy stored per unit volume is $\tfrac{1}{2}\varepsilon_0 k_d E^2$.

Minimum Learning Goals

Upon completion of this chapter, you should be able to do the following:

1 Define what is meant by a test charge, and use it to define the electric potential difference between two points.

2 Compute the potential difference between two specified points in a given constant electric field. Show the equipotential surfaces and lines in simple situations involving specified charged objects.

3 Use the relation $W = (V_B - V_A)q$ in simple, specified situations.

4 Explain what is meant by the emf of a battery.

5 Explain in words what is meant by the absolute potential at a point. Give its value at a specified distance from a point charge. Find the absolute potential at a point due to several specified point charges near the point.

6 Find the change in KE of either a positive or negative known charge as it moves through a given potential difference. If its initial speed is given, find its final speed. (Neglect relativistic effects.)

7 Describe a simple electron gun.

8 Give the change in energy in electronvolts of a charge q that falls through a known potential difference. Convert energies in electronvolts to joules, and vice versa.

9 Draw a diagram of a parallel-plate capacitor and relate C, Q, and V for it. State the units of C.

10 Outline a method by which the dielectric constant of a liquid can be determined if an instrument for measuring capacitance is available.

11 Compute the energy stored in a given capacitor charged to a known potential difference.

Important Terms and Phrases

You should be able to define or explain each of the following:

Electric potential difference
The static electric field is conservative
Equipotentials
Field lines are perpendicular to equipotentials
Emf

Positive charge falls from plus to minus while negative charge does the reverse
Electronvolt
Absolute potential at a point
$Q = CV$; farad
Dielectric constant k_d
Energy $= \frac{1}{2}QV$

Questions and Guesstimates

1 Two points A and B are at the same potential. Does this necessarily mean that no work is done in carrying a positive test charge from one point to the other? Does it mean that no force will have to be exerted to carry the test charge from one point to the other? Explain.

2 Can two equipotential surfaces intersect? Explain.

3 The absolute potential midway between two equal but oppositely charged point charges is zero. Can you find an obvious path along which no work would be done in carrying a positive test charge from infinity up to this point? Explain.

4 Starting from the fact that a piece of metal which has no current flowing in it is an equipotential body, prove that the electric field inside a hollow piece of metal is zero.

5 Draw the electric field between a highly charged cloud and a lightning rod mounted on a building. Why must the rod be securely grounded?

6 If V is zero at a point, must E be zero there as well?

7 What can be said about E in a region where V is constant?

8 Prove that all points of a metal object are at the same potential under electrostatic conditions. Does this also apply within a hole inside the object? Does it matter if there is a charge suspended in the hole?

9 A parallel-plate capacitor has a fixed charge Q on its plates. The plates are now pulled farther apart. The puller must do work. Why? Does the potential difference change during the process? What happens to the work done by the puller?

10 Suppose a sheet of metal is used as the dielectric between the two plates of the capacitor. (It does not quite touch the plates.) Does it appear to have a high or low dielectric constant?

11 As an electron beam shoots down the length of a TV tube, about how far do the electrons fall because of the effect of gravity? Will a TV set operate properly if the set is laid on its back so the electron beam must shoot straight upward?

Problems

1 The potential difference from A to B is $+12.0$ V. How much work is required to carry a 3.0-μC charge (a) from A to B, (b) from B to A?

2 The potential difference from A to B is $+60$ V. How much work is needed to carry (a) a proton ($q = e$) from A to B, (b) an electron ($q = -e$) from A to B?

3 The potential difference between two parallel metal plates is 6.0 V, and the separation of the plates is 2.0 mm. (a) What is the electric field strength between the plates? (b) How large a force would a proton ($q = e$) experience if it were between the plates?

4 Two parallel metal plates are separated by a 0.50-cm gap in which the electric field is 3000 V/m. (a) What is the potential difference between the plates? (b) How large a force would a 4.0-μC charge experience if it were between the plates?

5 Two parallel metal plates are connected to a 6-V battery. The field between the plates is 300 V/m. (a) How far are the plates separated? (b) What would be the magnitude of the force on an electron between the plates? The charge on the electron is -1.6×10^{-19} C.

6* The electric field in a certain region is 5000 V/m and is directed outward along the $+x$ axis. What is the potential difference from A to B if (a) $x_A = 3.0$ m, $x_B = 5.0$ m, and both are on the x axis; (b) $x_A = 7.0$ m, $x_B = 2.0$ m, and both are on the x axis; (c) the two points are on the y axis with $y_A = 0$ and $y_B = 7.0$ m?

7* In a certain region of space, the electric field is directed in the $+y$ direction and has a magnitude of 4000 V/m. What is the potential difference from the coordinate origin to the following point? (a) $x = 0$, $y = 20$ cm, $z = 0$; (b) $x = 0$, $y = -30$ cm, $z = 0$; (c) $x = 0$, $y = 0$, $z = 15$ cm.

8* In Fig. P16.1 is shown a particle between two parallel, oppositely charged plates. Assume the particle to be an electron ($q = -e$, $m = 9.1 \times 10^{-31}$ kg). (a) How large must the electric field be between the plates if the particle is to remain mo-

FIGURE P16.1

tionless under the combined effects of gravity and the electric field? (b) Should the top plate be positive or negative? (c) If the distance between the plates is 2.00 mm, what is the required potential difference between the plates?

9* The particle between the plates in Fig. P16.1 has a mass of 4.0×10^{-12} g. It can be held stationary against the pull of gravity by making the potential difference between the two plates 245 V with the upper plate negative. The distance between the plates is 2.00 mm. Find (a) the electric field strength between the plates and (b) the sign and magnitude of the charge on the particle.

10** The parallel metal plates shown in Fig. P16.2 are separated by a distance of 10 cm, and the voltage difference between them is 28 V. A small ball of 0.60-g mass hangs by a thread from the upper plate. What is the tension in the thread if the ball carries a charge of 20 μC? Two answers are possible. Find both.

11 An electron ($m = 9.1 \times 10^{-31}$ kg, $q = -e$) initially at rest falls from the negative filament to the positive plate in a vacuum tube. The terminals of a 12.0-V battery are connected to the filament and plate. (a) What is the speed of the electron just before it hits the plate? (b) What is its kinetic energy in electronvolts at that time?

12 It is desired to accelerate a proton from rest to a speed of 5.0×10^6 m/s. (a) Through how large a

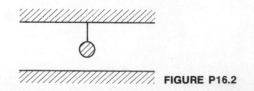

FIGURE P16.2

potential difference must the proton fall? (*b*) What will be its kinetic energy in electronvolts? ($m_p = 1.67 \times 10^{-27}$ kg and $q = e$.)

13* In a certain region the electric field is 3000 V/m and is directed out along the $+x$ axis. If a proton is released at the coordinate origin, what will be its speed as it passes $x = 50$ cm?

14* When a proton is released at the coordinate origin, it shoots out along the y axis. Its speed as it passes the point $y = 30$ cm on the axis is 5.0×10^5 m/s. What is the potential difference from the origin to the point $x = 0$, $y = 30$ cm, $z = 0$?

15* Two parallel plates have a potential difference of 50 V. (*a*) A proton is shot from the negative plate to the positive plate with an initial energy of 80 eV. What is its kinetic energy just before it strikes the opposite plate? (*b*) Repeat if it is shot from the positive plate toward the negative plate.

16* A proton is shot with a KE of 5000 eV from a negative plate toward a positive plate. The potential difference between the two plates is 1500 V. (*a*) How much KE (in electronvolts) will the proton lose as it shoots to the positive plate? (*b*) What will be its KE (in electronvolts) just before it hits the plate? (*c*) Repeat for an α particle. (The charge on an α particle, a helium nucleus, is $+2e$.)

17* An electron moving with a speed of 5.0×10^6 m/s is further accelerated through a potential difference of 20 V. What will its new speed be?

18* A proton moving with a speed of 3.0×10^7 m/s is to be slowed to 1.2×10^7 m/s. How large a potential difference must it rise through in order to slow it this much?

19 Two charges $+6$ and -3 μC, are 60 cm apart. Find the absolute potential at a point halfway between them. How much work is required to carry a 3-μC charge from infinity up to that point?

20 What is the absolute potential at the center of the square in Fig. P16.3 (*a*) if $q_1 = q_2 = q_3 = q_4 =$

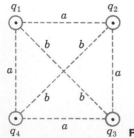

q_1 a q_2

a b b a

b b

q_4 a q_3 **FIGURE P16.3**

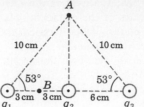

FIGURE P16.4

2μC and (*b*) if $q_1 = q_2 = q_3 = 2 \mu$C and $q_4 = -6 \mu$C? Take $b = 30$ cm.

21 In Fig. P16.4, if $q_1 = +2 \mu$C, $q_2 = -8 \mu$C, and $q_3 = +4 \mu$C, find the absolute potential at point A.

22 See Fig. P16.4. (*a*) How large is the potential difference between point A and B if $q_1 = +3 \mu$C, $q_2 = -6 \mu$C, and $q_3 = -9 \mu$C? (*b*) What are V_A and V_B?

23** A 30-cm-diameter metal sphere hangs from a thread in a very large room, so its surroundings are essentially at infinity. If the electric field at its surface is to be equal to the breakdown strength of air, 3×10^6 V/m, what must be the absolute potential of the sphere?

24** Assume a sodium ion ($q = +e$) to have a radius such that a chloride ion ($q = -e$) can approach to a center-to-center distance of 3×10^{-10} m. How much work is required to separate the two ions completely if they are originally as close together as possible and if the ions are (*a*) in vacuum and (*b*) in water? (*c*) Compare the energy in (*a*) to the average thermal energy ($3kT/2$) of an ion.

25 A typical capacitor used in a radio might have a capacitance of 2.0×10^{-8} F. How much charge exists on the capacitor when it is connected across a 9.0-V battery?

26 Two identical metal plates are placed parallel to each other, one opposite the other. The gap between them is 0.50 mm, and the area of each is 400 cm². (*a*) Find the capacitance of the plates if air exists between them. (*b*) If the capacitor is connected to a 12.0-V battery, how much charge will exist on the capacitor? (*c*) Repeat parts *a* and *b* if the space between the plates if filled with polystyrene.

27 A certain capacitor has a capacitance of 1.750×10^{-9} F when it has air between its plates. When the capacitor is filled with acetone, what will be its capacitance?

28 How much energy is stored in a 2-μF capacitor which is charged to a potential difference of 300 V?

29* A certain parallel-plate capacitor can be adjusted so that the gap between the plates can be changed without otherwise disturbing the electrical system. In position A, the capacitance is 3.0×10^{-9} F, and in position B, it is 2.7×10^{-9} F. The capacitor is charged by a 12-V battery when in position A. The battery is then removed, and the capacitor is changed to position B without changing the charge on it. (*a*) How much charge is on the original capacitor? (*b*) What is the voltage across it in position B? (*c*) By how much does its stored energy change in going from A to B? (*d*) What minimum work must the person holding the plates have done to change the capacitor from A to B?

30** Repeat the previous problem if the battery had been left attached to the plates as the plate separation was changed.

17
Direct-Current Circuits

Most practical applications of electricity involve charges that are in motion. Our light bulbs are lighted by charges flowing through them. An electric motor is caused to run by the interaction of electric charges that move through the motor's coils. When we turn on a radio or TV, we allow charge to flow through the device so it may operate. In this chapter we discuss the flow of charge through direct-current (dc) circuits, circuits in which the flow is unidirectional.

17.1 Electric Current

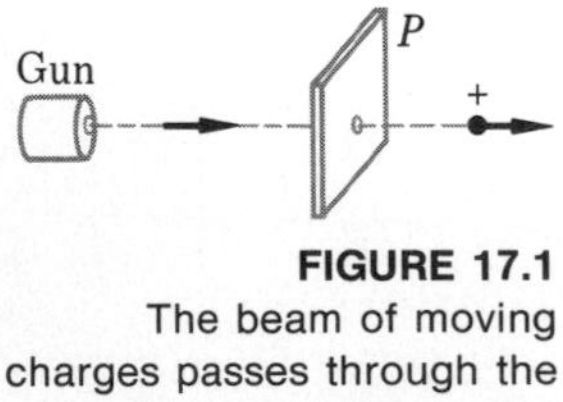

We begin our discussion of charges in motion by defining a quantity called electric current. Suppose we have a device much like the electron gun we discussed in Sec. 16.5, except that it shoots out positively charged particles instead of electrons, as shown in Fig. 17.1. As we see, a beam of positive particles shoots to the right from the gun. (It is assumed that the device is in vacuum so that the particles can move freely.) The beam shoots through a hole in the plate P.

    The beam passing through the hole constitutes a flow of charge, and we now wish to characterize the magnitude of the flow. We do so by defining a quantity called **electric current,** which we designate by the symbol I. The definition is as follows:

FIGURE 17.1
The beam of moving charges passes through the hole in the plate P. If a charge ΔQ passes through the hole in time Δt, the current is $\Delta Q/\Delta t$.

If a charge ΔQ is carried past a given point (plate P in this case) by a beam in a time Δt, then the current carried by the beam is

$$I = \frac{\Delta Q}{\Delta t} \tag{17.1}$$

The unit of current, one coulomb per second, is given the name **ampere** (A).

Notice that if the moving charge is positive, ΔQ and I will be positive. But if the moving charge is negative, then ΔQ and I are negative. For this reason,

A flow of negative charge results in a current in a direction opposite to the flow direction.

As you see, our definition favors positive charge over negative charge. The reason can be traced to Benjamin Franklin in the mid- and late 1700s. It was he who proposed that positively charged objects carry an excess of charge and that negative objects are deficient in charge. When the two types of object are placed in contact, charge runs from the positive object to the negative one, or so he believed. This view proved extremely useful and was adopted universally. Much later it was discovered that, in metals, negative charges move but positive charges do not. However, we still retain Franklin's original terminology. There is no great compulsion to change the terminology because a flow of positive charge in one direction produces the same overall effect as a flow of negative charge in the opposite direction, except in very specialized experiments. It is surprisingly difficult to determine by experiment which type of charge is moving.

Although beams of charge are quite common (in TV sets, for example), we are even more familiar with charge flow in wires. To see how current is defined in this case, refer to Fig. 17.2. In a metal, there are many free electrons. Under the action of an electric field, they move through the metal. Such a movement of charge is shown in the figure. However, in keeping with our emphasis on positive charge, we couch our definition in terms of positive charge. *If ΔQ is the quantity of charge passing through a given cross section of a wire* (such as A in the figure) *in time Δt, the current in the wire is*

$$I = \frac{\Delta Q}{\Delta t} \tag{17.2}$$

Current in a Wire

It is measured in amperes (A). As with the charge beam, the current is

FIGURE 17.2
The current in the wire in amperes is defined to be the quantity of positive charge in coulombs flowing through a cross section such as A in 1 s.

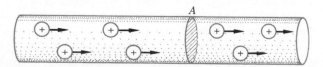

opposite in direction to the flow of negative charges. It should be noticed that I is the quantity of charge which passes a given point in a wire each second, provided the current is steady.

Illustration 17.1 A current of 3.2 A is flowing in a wire. How many electrons pass a given point on the wire in a second?

Reasoning If we knew the total charge passing the point in 1 s, we could divide that by the charge on the electron, 1.6×10^{-19} C, and thereby obtain the number of electrons. Since I is the number of coulombs passing a point on the wire in 1 s, we have at once that 3.2 C of charge flows past the point in 1 s. But each electron carries a charge 1.6×10^{-19} C, and so the number of electrons passing per second is

$$\text{Number} = \frac{3.2}{1.6 \times 10^{-19}} = 2 \times 10^{19} \text{ electrons}$$

This is a tremendously large number, of course. We therefore conclude that any attempt to observe the action of a single electron moving along the wire will be fruitless. There are cases where the effects of individual electrons can be observed, however, For example, the Geiger counter and Wilson cloud chamber are capable of detecting individual electrons. These devices are discussed in Chap. 27.

17.2 A Simple Electric Circuit

Let us begin our study of electric circuits by examining the simple circuit shown in Fig. 17.3. We see there a flashlight bulb connected to a battery. Current flows from the positive terminal of the battery (P), through the bulb's filament, and back into the negative terminal of the battery (N). (Of course, the *electrons* in the wire flow in the reverse direction.)

We know that the filament of the flashlight bulb glows white-hot in a situation like this. Clearly, the bulb is being furnished energy by the battery. The charge flow through the filament generates heat, and the filament is thereby made white-hot. Apparently the battery gives energy to the charges and causes them to flow through the filament. As they pass through, they generate heat in a manner somewhat analogous to the way heat is generated by friction in a mechanical system.

In the circuit of Fig. 17.3a, only the filament from A to B becomes hot. The current flowing through the other wires (the connecting wires from P to A and from B to N) causes negligible heating. This is a result of two factors: (1) the filament wire is much thinner than the others, and (2) the material of the filament offers more resistance to the flow of charges through it. We say that the filament has a high electrical resistance, much higher than the electrical resistance of the other wires. For many purposes, the heat energy

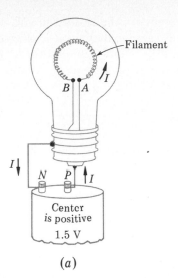

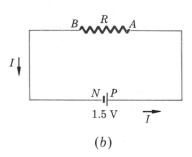

FIGURE 17.3
A simple electric circuit.
The filament of the flashlight
bulb glows white-hot as
current flows through it.

generated in the connecting wires is negligible. In such cases, the connecting wires can be assumed to have negligible resistance.

The circuit shown in Fig. 17.3a is customarily drawn as shown in part b. *Any wire in which a current generates appreciable heat is called a* **resistor.** In this case, the filament is a resistor. *We represent all resistors by the symbol* —⋀⋀⋀—. This is the symbol used in b to represent the filament. The essentially resistanceless connecting wires are shown by the lines running from the resistor to the battery. Note that the current flows out of the positive terminal of the battery and into its negative terminal. **Definition**

Certain features of the circuit shown in Fig. 17.3 can be clarified by comparing it to an analogous water system. For example, consider the two systems shown in Fig. 17.4. Part a is a water system consisting of a water pump which causes water to flow through a pipe system. The pipes and pump are completely filled with water. As a result, if the pump pushes a little water into the pipe at P, an equal quantity must flow back into the pump at N. Moreover, a similar quantity of water must pass point A and flow into the narrow pipe. Of course, this same quantity of water must exit from the narrow pipe at B. As we see, the quantity of water flowing past points A, B, N, and P each second must be the same. Since the quantity of water passing a point each second is the current, we see that I, the water current in this case, is the same at all points in the water circuit.

Similar considerations apply to the electric circuit shown in part b. The wires are "filled" with charges (electrons), and these act like the water molecules in the water pipes. They act like an incompressible fluid which flows like the water in the pipes of part a. (This fact is reflected in the terminology of electricians who sometimes refer to the "juice" in the wires.) We can conclude at once that the current I in the electrical circuit is the same at all points of this simple circuit.

There is another important point to notice about the circuits of Fig. 17.4. As we have said, when a little water leaves the pump at P, an equal

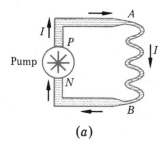

(a)

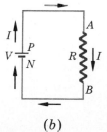

(b)

FIGURE 17.4
The water system in (a) is
analogous to the electrical
system in (b). Notice that
the pump and the battery
are the energy sources.

amount enters at N. The situation is similar in the electric circuit. When a little charge flows out of the battery at P, an equal amount of charge enters at N. This occurs almost instantaneously. Notice in particular that the charge that leaves P is not the same charge that enters at N. All the charges in the wire simply move along the wires a little. Equal amounts enter and leave the two ends.

It is of interest to examine the energy in these two circuits. Work must be done in the water circuit to push the water through the thin, curling pipe. This pipe offers much, much more resistance to the water flow than the larger pipes. Therefore, we can usually ignore the resistance effects of the larger pipes. The work done in pushing the water through the narrow pipe is lost doing viscous (or friction) work. An amount of heat equal to this work is generated in the resistance pipe. As we see, the energy supplied to the water by the pump appears as friction work in the resistance pipe.

A similar situation occurs in the electric circuit. The battery supplies energy to the charges and pushes them through the resistance wire. This energy is lost to heat as the charges pass through the resistor. As with the water circuit, usually only negligible heat is generated in the portions of the circuit from P to A and from B to N. These connecting wires offer negligible resistance to charge flow.

17.3 Ohm's Law

Let us now examine this simple circuit to obtain a quantitative relation among the current, the battery voltage, and the resistance effect of the wire. We redraw the circuit in Fig. 17.5. Notice that the part of the circuit from P to A is designated "high potential," while the part from B to N is "low potential." The reason is as follows.

Consider a positive charge at point P of the circuit. It is at the same potential as the positive terminal of the battery, namely, V volts above point N. As the charge moves along the resistanceless wire from P to A, it loses no energy since negligible heat is generated in this wire. Therefore, when it reaches point A, it is still at the same potential as it was at P. The whole portion of the circuit from P to A is at the same potential. We say that this portion of the circuit is all at the same electrical level.

However, point N is at a potential V volts lower than P. Since the whole circuit from N to B is at the same electrical level, we conclude that this portion of the circuit is V volts lower than the section from P to A. As we

FIGURE 17.5
Current always flows from high to low potential through a resistance. The emf of the resistanceless battery is $\mathcal{E}$.

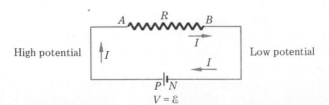

see, the potential difference from A to B is V volts. And since positive charges move from points of high potential to points of lower potential, it is not surprising that the current (a flow of positive charge) is directed from A to B. Indeed,

> Current always flows through a resistor from its high-potential end to its lower-potential end.

We are now experienced enough to accept as reasonable an important fact first discovered by Georg Simon Ohm (1787–1854). He found by experiment that the current that flows through a resistor is often directly proportional to the potential difference between its two ends. For example, in Fig. 17.5 he found that

$$I \propto V$$

where V is the potential difference between the two ends of the resistor. For example, if he doubled the battery voltage, the current would double.

The other factor that influences the current is the magnitude of the resistance effect of the resistor. We define this resistance effect in the following way: *If a potential difference V across a resistor causes a current I to flow through it, the resistor's* **resistance** *is*   **Definition**

$$R = \frac{V}{I} \quad \text{from which} \quad V = IR \qquad (17.3) \quad \textit{Ohm's Law}$$

The unit of resistance, one volt per ampere, is designated the ohm (Ω). This defining equation for resistance is often *called* **Ohm's law.** However, Ohm also implied that R does not change as the potential across the resistor is changed. This is often true, but not always, as we shall see later. In spite of this, Eq. (17.3) is always true provided we understand R to be defined by it.

Illustration 17.2 Suppose the light bulb in Fig. 17.3 draws a current of 0.25 A when connected across a 1.5-V battery as shown. What is the resistance of the bulb under these conditions?

Reasoning We can make direct use of Ohm's law, Eq. (17.3). The potential difference across the bulb V is 1.5 V. We are told that I through the bulb is 0.25 A. As a result,

$$V = IR$$

becomes

$$R = \frac{1.5 \text{ V}}{0.25 \text{ A}} = 6.0 \ \Omega$$

The resistance of the hot bulb is 6 Ω. We see later that the bulb's resistance is considerably lower when it is not white-hot.

17.4 Resistivity and Its Temperature Dependence

Not all materials are good conductors of electricity. There are even differences in the abilities of the various metals to conduct current. For this reason we need a quantity that will tell us exactly how good a conductor a material is. Or we could, if we wished, describe the material by stating how much resistance it offers to charge flow.

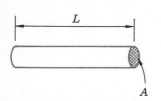

L

A

FIGURE 17.6
The resistance of a uniform wire varies directly as *L* and inversely to *A*.

Suppose that we have a cylindrical piece of wire like that shown in Fig. 17.6. The resistance of this length *L* of wire depends on the length as well as the cross-sectional area of the wire. Since a wire twice as long would be expected to have twice as much resistance, the resistance of the wire is proportional to *L*. Moreover, the larger the cross-sectional area *A*, the less the resistance should be. Therefore, *R* should vary inversely as *A*. In equation form,

$$R = \rho \frac{L}{A} \tag{17.4}$$

Resistivity

The proportionality constant ρ (Greek rho) *is called the* **resistivity** *of the* material from which the wire is made. If ρ is large, the material is a poor conductor. Solving for ρ, we find

$$\rho = R \frac{A}{L}$$

and so its units are ohm-meters or ohms times a length unit. Resistivities of various materials are listed in Table 17.1. Notice that copper and silver are the two best conductors listed. Most electrical wire is made from copper or aluminum.

Sometimes a quantity called the **conductivity** is used to describe the electrical properties of metals and other conductors. It is equivalent to $1/\rho$, the reciprocal of the resistivity.

It will be noticed that the resistivity is quoted in Table 17.1 for a particular temperature, 20°C. This is necessary because the resistance of wires changes rather markedly with temperature. Over a limited temperature range, the resistivity may be represented by an equation of the form

Variation of Resistance with
Temperature

$$\frac{\rho - \rho_{20}}{\rho_{20}} = \alpha_{20}(t - 20°) \tag{17.5}$$

or

$$\frac{R - R_{20}}{R_{20}} = \alpha_{20}(t - 20°)$$

where *t* is the Celsius temperature.

TABLE 17.1
RESISTIVITIES AND THEIR TEMPERATURE
COEFFICIENTS

447
17.4 Resistivity and Its
Temperature Dependence

Material	Resistivity ρ at 20°C, $\Omega \cdot m$	α at 20°C, °C^{-1}
Silver	1.6×10^{-8}	4.1×10^{-3}
Copper	1.7×10^{-8}	3.9×10^{-3}
Aluminum	2.8×10^{-8}	4.0×10^{-3}
Tungsten	5.6×10^{-8}	4.5×10^{-3}
Iron	10×10^{-9}	6.5×10^{-3}
Graphite (carbon)	3500×10^{-8}	-0.5×10^{-3}

In these equations, the resistivity and resistance at 20°C, ρ_{20} and R_{20}, are used as the reference values. The quotient is the fractional change in resistivity in the first equation and in resistance in the second. The temperature coefficient of resistance α is a measure of how much the resistance changes with temperature. If α is zero, the resistance is constant. Typical values for α are listed in Table 17.1.

It is seen that the value of α used in Eq. (17.5) must be measured at the same temperature as the reference temperature for the resistance. Sometimes reference temperatures other than 20°C are used, in which case Eq. (17.5) would be altered accordingly. The variation of the resistivity of copper with temperature is shown in Fig. 17.7. Notice that the changes in resistance with temperature are actually quite large.

If an equation such as Eq. (17.5) were strictly correct, the graph in Fig. 17.7 would be a straight line. Actually, the graph is slightly curved. Hence, Eq. (17.5) is only approximately correct for large temperature intervals.

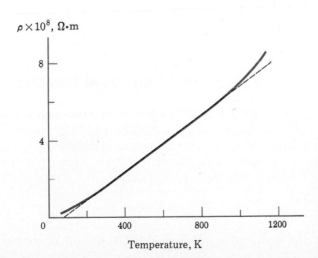

FIGURE 17.7
The actual resistivity of copper is shown as the solid curve. A straight line that approximates the room-temperature data is shown as the broken line.

Illustration 17.3   A rectangular bar of iron is 2 by 2 cm in cross section and 40 cm long. How large is its resistance?

Reasoning   We know that

$$R = \rho \frac{L}{A}$$

From Table 17.1 we find $\rho = 1.0 \times 10^{-7}\ \Omega \cdot m$, and so

$$R = (1 \times 10^{-7}) \left[\frac{0.40}{(0.02)(0.02)} \right] = 1 \times 10^{-4}\ \Omega \qquad \text{at } 20°C$$

You should carry through the units to verify the units of the answer.

Illustration 17.4   At a temperature of 520°C, what is the resistance of the bar in the previous illustration?

Reasoning   Making use of the data in Table 17.1 as well as the results of the previous example, we have

$$\frac{R - R_{20}}{R_{20}} = \alpha_{20}(t - 20)$$
$$\frac{R - 1 \times 10^{-4}}{1 \times 10^{-4}} = (6.5 \times 10^{-3})(520 - 20)$$

from which

$$R = 4.3 \times 10^{-4}\ \Omega$$

The resistance of the bar is 4.3 times higher at this temperature than it was at 20°C.

17.5 Power and Electrical Heating

When a battery causes current to flow through a resistor, as shown in Fig. 17.8, the battery is furnishing energy. The battery is continuously lifting charge uphill through the potential difference V, the battery voltage.

We know that the work done in carrying a charge ΔQ up through a potential difference V is given by

$$\text{Work done} = V \Delta Q$$

If the charge ΔQ is transported in a time Δt, the work done per second is

$$\text{Work/s} = \frac{V \Delta Q}{\Delta t}$$

But work per second is power, and so we have

$$\text{Power} = V \frac{\Delta Q}{\Delta t}$$

However, current is defined to be charge transported per second, that is, $\Delta Q/\Delta t$. Hence we find that the power furnished by the battery is

$$\text{Power} = VI \qquad (17.6)$$

Electric Power

This is a general relation for power delivered from a source of current I operating on a voltage V. In this instance, the power furnished by the battery is expended in the resistor. The same general reasoning can be used to show that Eq. (17.6) also is a statement of the power expended in any element through which a current *I* flows and across whose terminals the voltage is *V*. *In the case of the resistor,* the application of Ohm's law leads to other forms of the equation.

$$P = VI = I^2 R = \frac{V^2}{R} \qquad \text{resistor} \qquad (17.7)$$

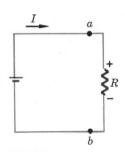

If we recall that volts are joules per coulomb and that current is coulombs per second, the units of power in this equation are joules per second, or watts.

FIGURE 17.8
The energy lost by the battery appears as heat in the resistor.

We are all aware of the fact that current flowing through a resistor produces heat in the resistance wire. The current through the filament of an incandescent light bulb heats the resistance element to a white heat. An electric stove has a resistance element as its heating unit, as do toasters, flatirons, electric clothes driers, portable heaters, and many other heating devices. A resistor converts the electrical energy furnished by the battery into heat energy.

The conversion of electrical energy into heat energy allows us to determine the mechanical equivalent of heat by a method different from the one discussed in Chap. 10. We need only measure how much heat is produced by a resistor *R* carrying a current *I*. This is conveniently done by placing the resistance wire in a calorimeter containing water. The electrical energy furnished to the water and calorimeter in time *t* is just *Pt,* or *I²Rt.* It is expressed in joules. The heat energy in calories is found from the temperature rise of the calorimeter and its contents. Then since the energies must be the same in either units, calories or joules, the conversion between them is easily found. The relation between the units must, of course, be the same as we found in considering the rotation of a paddle wheel in water:

$$1 \text{ cal} = 4.184 \text{ J}$$

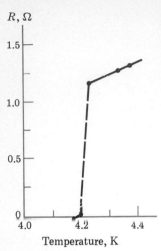

R, Ω

Data of H. Kamerlingh Onnes (1911) for the resistance of mercury near 4 K. The resistance at 0°C (273 K) was about 60 Ω.

Temperature, K

Superconductors

Ordinary conductors of electricity become better conductors as the temperature is lowered. This was already known in 1835 from the measurements of Heinrich Lenz. There was some speculation in later years that perhaps the resistance of a metal would actually be zero at absolute zero, i.e., at −273°C, or 0 K. However, no one was able to carry resistance measurements to very low temperatures until much later than 1835, because no means existed for cooling objects to such low temperatures.

The attainment of very low temperatures received considerable impetus in 1883, when Wroblewski and Olzewski succeeded in liquefying air. With liquid air as a cooling agent, it was possible to carry out experiments at the boiling point of liquid oxygen (−183°C) and liquid nitrogen (−196°C). Soon after, in 1898, James Dewar (1842–1923) succeeded in liquefying hydrogen, which has a boiling point of −253°C, or 20 K. There remained only one other gas which had not been liquefied, helium. The liquefaction of helium was finally achieved in 1908 by Heike Kamerlingh Onnes (1853–1926), and it was found that liquid helium boiled at 4.2 K.

Using his liquefied helium to achieve lower temperatures than had been possible before, Onnes set out to measure the resistance of metals at very low temperatures. The measurement technique was quite simple in principle, although quite difficult in practice because of the difficulty of maintaining the temperature constant. Since resistance R is defined by Ohm's law, Onnes needed to measure only the voltage drop V across a given cylinder of the metal when a definite current I flowed through it. Then, $R = V/I$. For his measurements, he used pure mercury as his metal. (It solidifies at −39°C, and so it was a solid.) When he carried out the requisite measurements in 1911, he obtained the astonishing data shown in the graph.

Although the metal was steadily approaching a very low resistance as the temperature was being lowered, at 4.2 K the resistance suddenly decreased to zero. The mercury had become what we now refer to as a **superconductor.** Subsequent experiments indicated that a current would flow essentially forever in a ring of the superconductor, even though no source of emf was maintained. As far as experiment has been able to tell, the resistance of the superconductor is essentially zero.

Onnes succeeded later in showing that lead, tin, and indium also become superconductors at 7.2, 3.7, and 3.4 K, respectively. Since that time, many other superconductors have been found. Most interesting has been the fact that certain alloys become superconductors at rather high temperatures. For example, the compound Nb_3Sn becomes superconducting at 17.9 K. Since no heat is generated when a current flows through a metal having no resistance, superconductors are now being used where this property is of great importance, e.g., in large electromagnets. The theoretical reasons for superconductivity in terms of atomic structure have been clarified only recently, especially since 1950.

Illustration 17.5 How much heat (in calories) does a 40-W light bulb generate in 20 min?

Reasoning A 40-W light bulb generates 40 J of heat per second. Therefore, in (20)(60) s it will develop

$$\text{Heat} = (20)(60)(40) = 48,000 \text{ J} = 11,500 \text{ cal} = 11.5 \text{ kcal}$$

Illustration 17.6 If electrical energy costs 10 cents per kilowatthour, how much does it cost to operate a 700-W coffee maker for 30 min?

Reasoning A kilowatthour (kWh) is the energy consumed in one hour by a one-kilowatt appliance. Hence

$$\text{Energy (kWh)} = [\text{power (kW)}][\text{time (h)}]$$
$$= (0.700 \text{ kW})(0.50 \text{ h}) = 0.350 \text{ kWh}$$

The cost will be 3.5 cents.

17.6 Kirchhoff's Point Rule

We are now prepared to begin a systematic study of electric circuits. An electric circuit is simply a path along which charge can move. Two basic rules apply to all electric circuits. They are very easy to understand and remember since they are almost obvious. To see what the first rule is, refer to Fig. 17.9.

We see there a situation where several wires meet at a junction point. Consider point A. The current flowing into the point is I, and the currents flowing out of the point are I_1, I_2, and I_3. What can we say about these

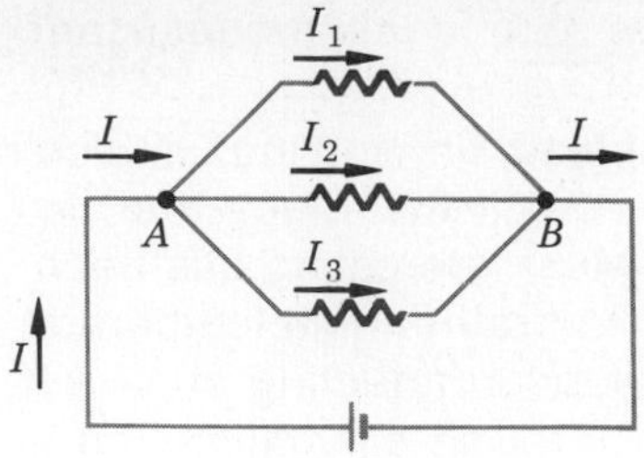

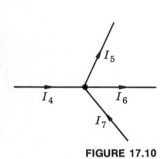

FIGURE 17.9
The point rule tells us that
$I = I_1 + I_2 + I_3$.

currents? We can easily answer this question if we think of the water-pipe analogy. The quantity I would represent a current of so many cubic centimeters of water flowing into point A each second. But if the pipes do not leak, this exact same amount of water must flow out of point A each second. In other words, since point A cannot store water, as much must flow into the point as flows out. We can therefore state that the current that flows into point A must equal the current that flows out of point A.

A similar situation applies to the flow of charge. The current coming into point A must equal the current which flows out of point A. Therefore, we can state that $I = I_1 + I_2 + I_3$. Moreover, this same rule must apply to any point in the circuit. At point B, for example,

$$\text{Current into } B = \text{current out of } B$$

FIGURE 17.10
According to the point rule,
$I_4 + I_7 = I_5 + I_6$.

becomes

$$I_1 + I_2 + I_3 = I$$

This is identical to the equation found for point A.

We can summarize this result in what is called **Kirchhoff's point rule:**

Point Rule

The sum of all the currents coming into a point must equal the sum of all the currents leaving the point.

We shall find that this simple rule is of very great importance. As another example of its use, notice that the rule, when applied to Fig. 17.10, gives $I_4 + I_7 = I_5 + I_6$.

17.7 Kirchhoff's Loop (or Circuit) Rule

We apply the second rule to circuits in which the current is steady or changing slowly.[1] In these cases, the electric field near the circuit is essentially electrostatic. Under such conditions, the electric field is a conservative field.

---

[1] As we shall see later, if the current is alternating in the microwave frequency range (10^9 Hz), induced emf's in the circuit loop become important.

By this we mean that the work done in carrying a positive test charge from one point to another is independent of the path followed. Let us see what this tells us about the electric circuit of Fig. 17.11.

Suppose we start at point A in the circuit and carry a positive test charge along it through points B, D, F, G, and back to A. How much total work have we done in carrying the test charge around the circuit and back to the starting point? Since this has been done in a conservative field, the answer is zero.

The situation here is much like a similar situation involving the gravitational field, another conservative field. Suppose you rise in the morning from your bed and return to it at night. The net amount of work you have done for the whole day on your body against the gravitational field is zero. Since the starting and ending points are the same, the gravitational PE of your body is unchanged. The total work done against gravity is zero.

In the case of the electrical circuit of Fig. 17.11, we see that the following is true. If we carry a positive test charge around the circuit and back to the starting point, zero net work is done. This must mean that the charge was carried through an equal amount of voltage rises and voltage drops. The net effect of all these voltage changes (rises taken as positive and drops as negative) is that their algebraic sum is zero. This fact is summarized in **Kirchhoff's loop** (*or circuit*) **rule:**

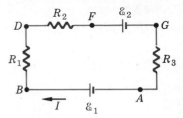

FIGURE 17.11
What does Kirchhoff's loop rule tell us about this circuit?

The algebraic sum of the voltage changes around a closed circuit must equal zero.

Loop Rule

As we see, the loop rule is intimately connected with potential rises and drops. For that reason, let us review what happens to the potential as we move across a resistor, a battery, and a capacitor.

Suppose we move from A to B across the resistor shown in Fig. 17.12. *We know that current always flows from high to low potential through a resistor.* Hence we know that the voltage change from A to B is a drop in potential. Therefore its sign will be negative. Ohm's law tells us its magnitude is IR. The voltage change in going from A to B is $-IR$.

Looking at the battery in Fig. 17.12, we recall that its symbol tells us the left side is positive. Therefore point A is $\mathcal{E}$ volts higher than point B. Going from A to B, the potential change is $-\mathcal{E}$.

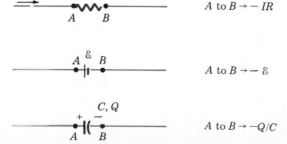

FIGURE 17.12
In each case shown, going from A to B is a voltage drop, a negative voltage change. In the reverse direction, the voltage change would be positive.

453

In the case of the capacitor, we must be told which plate is positively charged. According to the diagram, plate A is positive. It therefore is at the higher potential. Since the potential difference across a capacitor is given by $Q = CV$ to be Q/C, we see that the potential change in going from A to B is $-Q/C$.

The potential change in each of these three cases is negative when we go from A to B. If we were going from B to A, the change would be positive. Let us now use the loop rule in a few simple circuits before moving on to its more serious applications.

Illustration 17.7 Find the current that flows in the circuit of Fig. 17.13.

Reasoning Let us guess that the current will flow in the direction shown. (You might protest that this is wrong since the 12-V battery will certainly have more effect than the 3-V battery. But one of the nice things about Kirchhoff's rules is that even a poor guesser can use them, as we shall see.) We pick a point such as a and move around the circuit. The voltage changes are as follows, in volts:

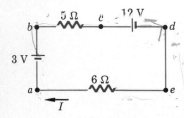

FIGURE 17.13
When we solve this circuit, how will our answers tell us that we have chosen I in the wrong direction?

$a \rightarrow b$: $+3$
$b \rightarrow c$: $-5I$
$c \rightarrow d$: -12
$d \rightarrow e$: 0
$e \rightarrow a$: $-6I$

Check them so that you are sure about the signs we have used. The sum of these voltage changes must be zero. Therefore

$$3 - 5I - 12 - 6I = 0$$

Solving for I, we find $I = -\frac{9}{11}$ A. The negative sign for I tells us we guessed its direction wrong. No harm is done. The current is $\frac{9}{11}$ A in a direction opposite to that shown.

Suppose we had circled the circuit in the reverse direction. Then our equation would have been

$$+6I + 12 + 5I - 3 = 0$$

from which $I = -\frac{9}{11}$ A, as before.

In solving this circuit, be sure you understand our choice of signs for the voltage changes. Also note that the current is the same at all points in the circuit. Why?

Illustration 17.8 Find the currents which flow in all the wires of the circuit of Fig. 17.14.

Reasoning We assign currents to all the wires and give each a symbol and direction. Once again, we waste little time trying to guess proper direction since our answer will indicate direction.

Starting at a, let us follow the loop $acda$ and write the loop rule. We have, in volts (be sure you understand the signs used),

$$-18I_2 - 9 + 0 = 0$$

from which we find at once that

$$I_2 = -0.50 \text{ A}$$

The current I_2 therefore flows in a direction opposite to that shown.

Now let us move around the loop $abcda$. We have

$$-6 + 12I_3 - 9 = 0$$

from which

$$I_3 = 1.25 \text{ A}$$

We could also write a loop equation for loop $abca$. But no new voltage changes would appear in it. Therefore, this equation would contain no new information and would be redundant.

Instead, we write the point rule for point c:

$$I_1 + I_2 = I_3$$

Substituting, we have, in amperes,

$$I_1 - 0.50 = 1.25$$

Notice that we carry along the signs we found for I_2 and I_3. Solving, we find $I_1 = 1.75$ A.

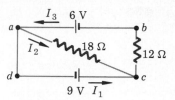

FIGURE 17.14
Find the currents in all three wires.

17.8 Resistors in Series and in Parallel

There are two configurations of resistor circuits that can be simplified very easily. If we recognize them, we can often greatly decrease the work needed to solve a problem. In the configuration shown in Fig. 17.15a, *the resistors are said to be in* **series**. *To move from point A to point B, only one path is possible. That path goes through all the resistors.*

If we connect this combination across a battery as in part *b*, we might guess it would act like a single resistor equal to the sum $R_1 + R_2 + R_3$. In what follows, we prove this guess to be correct. What we wish to do is find

Resistors in Series

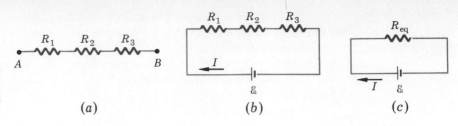

FIGURE 17.15
The three resistors are in
series. They are equivalent
to $R_{eq} = R_1 + R_2 + R_3$.

an equivalent resistor R_{eq} in part c of the figure that draws the same current as the combination in part b. To do this, we make use of the loop rule for the circuits of parts b and c.

Going clockwise around the circuit in part b, we have the following loop equation:

$$+\mathcal{E} - IR_1 - IR_2 - IR_3 = 0$$

This can be rewritten as

$$\frac{\mathcal{E}}{I} = R_1 + R_2 + R_3$$

Writing a similar loop equation for part c gives

$$+\mathcal{E} - IR_{eq} = 0$$

from which

$$\frac{\mathcal{E}}{I} = R_{eq}$$

From these two equations we see that

$$R_{eq} = R_1 + R_2 + R_3 \qquad \text{series} \tag{17.8}$$

In general, we can say

Equivalent Series Resistance

Several resistances in series are equivalent to a single resistor equal to their sum.

Resistors in Parallel

Another common configuration is shown in Fig. 17.16a. *These resistors are said to be in* **parallel**. *In a parallel configuration, one end of each resistor is connected to a point A. The other end of each resistor is connected to a point B. When a potential difference is placed across the combination,* as in part b of Fig. 17.16, *each resistor has this same potential difference across it.* We wish to find the equivalent resistor R_{eq} shown in c which can be used to replace the combination and still draw the same current.

Notice in Fig. 17.16b that the potential difference across each resistor is $\mathcal{E}$. Therefore, Ohm's law tells us that

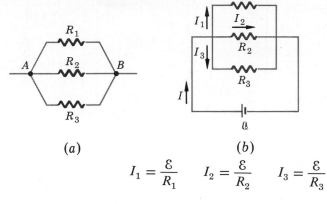

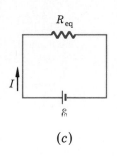

(a)   (b)   (c)

FIGURE 17.16
The three resistors are in
parallel. Their equivalent is
given by $1/R_{eq} = 1/R_1 +$
$1/R_2 + 1/R_3$.

$$I_1 = \frac{\mathcal{E}}{R_1} \qquad I_2 = \frac{\mathcal{E}}{R_2} \qquad I_3 = \frac{\mathcal{E}}{R_3}$$

But the point rule tells us that

$$I = I_1 + I_2 + I_3$$

and so

$$I = \frac{\mathcal{E}}{R_1} + \frac{\mathcal{E}}{R_2} + \frac{\mathcal{E}}{R_3}$$

If we divide both sides of this equation by $\mathcal{E}$, it becomes

$$\frac{I}{\mathcal{E}} = \frac{1}{R_1} + \frac{1}{R_2} + \frac{1}{R_3}$$

Now let us look at the circuit in part c. The loop tells us that

$$+\mathcal{E} - IR_{eq} = 0$$

from which

$$\frac{I}{\mathcal{E}} = \frac{1}{R_{eq}}$$

We can now equate these two expressions for $I/\mathcal{E}$ to give

$$\frac{1}{R_{eq}} = \frac{1}{R_1} + \frac{1}{R_2} + \frac{1}{R_3} \qquad \text{parallel} \qquad (17.9)$$

In general, we can say

For several resistors in parallel, the reciprocal of their equivalent resist-   *Equivalent Parallel Resistance*
ance is equal to the sum of the reciprocals of the resistances.

Let us now do a few practice examples to see more clearly how these equivalent resistors are obtained and used.

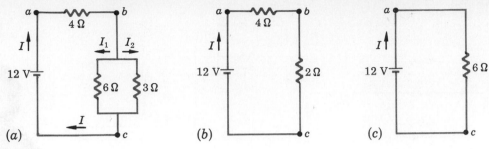

FIGURE 17.17
The parallel resistors
between b and c are
equivalent to $2\,\Omega$, as shown
in (b). The two series
resistors in (b) can be
combined, as in (c).

Illustration 17.9   Find the current which flows from the battery in Fig. 17.17a.

Reasoning   We could solve this by use of Kirchhoff's rules. However, it is usually simpler to combine obvious series and parallel resistors before writing the loop equations. The resistances may be reduced as shown in parts b and c of the figure. Let us first combine the two parallel resistors between points b and c. We have

$$\frac{1}{R_{bc}} = \frac{1}{6} + \frac{1}{3} = \frac{1}{6} + \frac{2}{6} = \frac{3}{6}$$

or
$$R_{bc} = 2\Omega$$

An equivalent circuit is now drawn in Fig. 17.17b. The parallel combination has been replaced by its equivalent resistance. We see that the 4- and 2-Ω resistors are connected in series between points a and c. Their equivalent is

$$R_{ac} = 4 + 2 = 6\Omega$$

A new equivalent circuit is now drawn, shown in part c. This is a situation to which Ohm's law can be applied. The voltage difference across the 6-Ω resistor is 12 V. Hence we have

$$I = \frac{V}{R} = \frac{12\ V}{6\ \Omega} = 2A$$

which is the result required.

Illustration 17.10   Find the current which flows from the battery in Fig. 17.18a.

Reasoning   Here, too, we have two parallel resistors, the 3-Ω and 6-Ω. They can be replaced by a 2-Ω resistor as in part b. There we see that the 2- and 4-Ω resistors are in series. So they are equivalent to a 6-Ω resistor. But this equivalent resistor is in parallel with the 6-Ω resistor. By using the

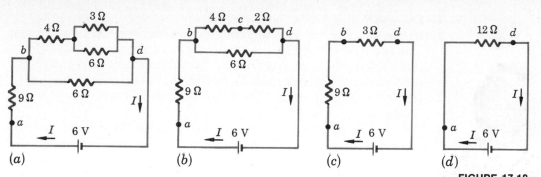

FIGURE 17.18
The complex circuit of (a)
can be reduced to the
simple equivalent circuit
shown in (d).

reciprocal formula, two 6-Ω resistors in parallel are equivalent to a 3-Ω resistor. This is drawn in part c. Finally, the circuit can be reduced to the form shown in part d. Usng Ohm's law for it, we have

$$I = \frac{V}{R} = \frac{6\ V}{12\ \Omega} = 0.50\ A$$

17.9 Capacitors in Series and in Parallel

We do not examine this situation in detail. Instead, we show the two configurations in Fig. 17.19a and b. The results for the equivalent capacitors are shown in each case. These results are obtained in problems at the end of the chapter. Notice that the relations are just opposite to those we found for resistors. *Parallel capacitors add directly while the reciprocals of series capacitances add.*

17.10 Solution of Circuit Problems

We now have at our disposal the tools needed to solve most direct-current (dc) circuit problems. Before using them in several examples, let us state a few facts we should remember in their use. Although each individaul problem will have its own peculiar features, *the following approach is most often useful.*

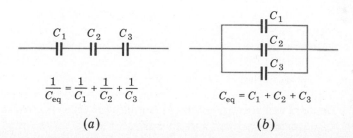

$$\frac{1}{C_{eq}} = \frac{1}{C_1} + \frac{1}{C_2} + \frac{1}{C_3}$$

(a)

$$C_{eq} = C_1 + C_2 + C_3$$

(b)

FIGURE 17.19
Reduction of series and parallel capacitances.
(a) series; (b) parallel.

1   Draw the circuit.

2   Assign a current and direction to each important wire. Be careful to use only one current designation for a given wire even though it may contain several elements. If a branch wire comes off from it, the currents on the two sides of the branch must usually be considered different.

3   Reduce series and parallel resistance systems whenever possible and convenient.

4   Write the loop equations for the remaining circuit. Remember that unless an equation contains a new voltage change, it will be redundant.

5   Write the point equations for the junction points. Unless a new current is used in the equation, it will be redundant.

6   Solve these equations for the unknowns.

Let us now proceed with some examples.

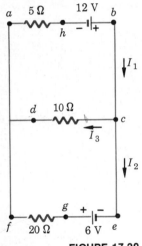

FIGURE 17.20
A circuit easily solved by Kirchhoff's rules.

Illustration 17.11   Find the currents in the wires of the circuit shown in Fig. 17.20.

Reasoning   Steps 1 and 2 of our procedure are already completed in the diagram. There are no simple resistance systems. (The resistors in *ab* and *ef* are *not* in simple parallel with the center resistor. Notice that the right-hand ends are connected to batteries, not to a common point.)

Let us now start at point *a* and write the loop equation for loop *abcda*. It is, in volts

$$-5I_1 + 12 - 10I_3 = 0 \qquad (1)$$

Similarly for loop *dcefd*

$$+10I_3 + 6 - 20I_2 = 0 \qquad (2)$$

We have three unknowns, I_1, I_2, and I_3, but only two equations. The loop equation *abcefa* includes no new voltage changes, and so we ignore it.

To obtain a third equation, we write the point equation for point *c*:

$$I_1 = I_3 + I_2 \qquad (3)$$

Substitution of this value for I_1 in Eq. (1) gives

$$-5I_3 - 5I_2 + 12 - 10I_3 = 0$$

which reduces to

$$-5I_2 - 15I_3 + 12 = 0 \qquad (4)$$

We now can solve (2) and (4) simultaneously.

Solving (2) for I_3, we find

$$I_3 = 2I_2 - 0.60 \qquad (5)$$

Substituting this in (4) gives

$$-5I_2 - 30I_2 + 9 + 12 = 0$$

from which

$$I_2 = 0.60 \text{ A}$$

Using this value in (5) gives

$$I_3 = 0.60 \text{ A}$$

Now, by use of (3) we find

$$I_1 = 1.20 \text{ A}$$

Illustration 17.12 For the circuit shown in Fig. 17.21, find I_1, I_2, and I_3.

Reasoning Choosing the currents to flow in the directions shown on the figure, and noticing that $I_1 + I_3 = I_2$, we can write the appropriate circuit or loop equations. Taking the loop through the 40- and 60-V batteries, starting at a, and going counterclockwise, we have

$$-40 - 10I_1 - 30I_2 + 60 = 0 \qquad \text{or} \qquad I_1 + 3I_2 = 2$$

Next from a through the 40- and 50-V batteries gives

$$-40 - 10I_1 + 15(I_2 - I_1) + 50 = 0$$

or
$$2.5I_1 - 1.5I_2 = 1$$

After multiplying the second equation by 2 and adding it to the first, we have

$$6I_1 = 4 \text{ A} \qquad \text{or} \qquad I_1 = \tfrac{2}{3} \text{ A}$$

Using this in the first equation gives

$$3I_2 = \tfrac{4}{3} \text{ A} \qquad \text{or} \qquad I_2 = \tfrac{4}{9} \text{ A}$$

And since $I_3 = I_2 - I_1$, we have

$$I_3 = -\tfrac{2}{9} \text{ A}$$

Clearly, I_3 is going in the opposite direction to that drawn on the diagram.

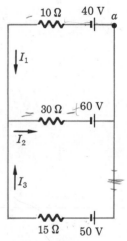

FIGURE 17.21
Find the three unknown
currents.

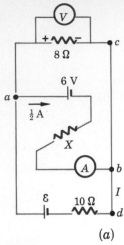

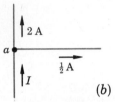

(a)

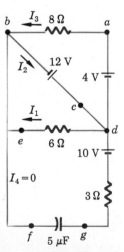

(b)

FIGURE 17.22
The ammeter and voltmeter readings are known. We wish to find I, X, and $\mathcal{E}$.

FIGURE 17.23
When the capacitor is fully charged, the current through the wire on the bottom is zero, and that portion of the circuit can be neglected.

Illustration 17.13 Find the values of $\mathcal{E}$, X, and I in Fig. 17.22. The ammeter reads 0.50 A, and the voltmeter reads 16 V. From the way in which the ammeter and voltmeter are connected, it is known that the polarity of the resistor is as shown and that the current in the center branch is in the direction shown. Assume the meters to be ideal; i.e., the ammeter resistance is zero, and the voltmeter resistance is infinite. This being true, the meters can be ignored.

Reasoning Since the voltmeter reads 16 V across an 8-Ω resistor, Ohm's law tells us that the current in the wire on the top is 2 A. From the polarity of the resistor at the top, the current is traveling to the right through the wire.

The currents at point a of the circuit must be as shown in Fig. 17.22b. From the point rule,

$$I = 2 + 0.50$$

and so

$$I = 2.5 \text{ A}$$

Writing the circuit equation for loop $abca$, and recalling that since an ideal ammeter has negligible resistance, the voltage change across it is zero, we have

$$-6 - \tfrac{1}{2}X + 16 = 0$$

from which

$$X = 20 \ \Omega$$

The circuit equation for loop $acbda$ gives

$$-16 - 25 + \mathcal{E} = 0$$

or

$$\mathcal{E} = 41 \text{ V}$$

Illustration 17.14 For the circuit shown in Fig. 17.23, fnd I_1, I_2, I_3, and the charge on the capacitor.

Reasoning Notice first that no current flows through wire fg once the capacitor is charged. Hence, $I_4 = 0$, and we can just ignore this wire when solving the rest of the circuit. At point d we find

$$I_2 = I_1 + I_3$$

Writing the loop equation for $dcbed$ gives

$$-12 + 6I_1 = 0$$

or
$$I_1 = 2 \text{ A}$$

Using the loop *dabcd* gives

$$+4 - 8I_3 + 12 = 0$$

or
$$I_3 = 2 \text{ A}$$

and
$$I_2 = I_1 + I_3 = 4 \text{ A}$$

If we knew the potential difference between points f and g, we could use the relation $Q = CV_{fg}$ to find Q. We can easily find V_{fg} by starting at point g and moving around the circuit *gdefg*, adding the voltage rises and falls. Doing this gives

$$0 + 10 - 6I_1 + V_{fg} = 0$$

or
$$V_{fg} = -10 + 12 = +2 \text{ V}$$

The positive sign shows we have gone uphill in potential in going from f to g, and so the g side of the capacitor is positive. We now have for the charge on the capacitor

$$Q = CV_{fg} = (5 \times 10^{-6})(2) = 1 \times 10^{-5} \text{ C}$$

17.11 Ammeters and Voltmeters

In Illustration 17.13 we saw a typical situation where an ammeter and a voltmeter are used in a circuit. Although we learn how these meters are constructed in Chap. 19, let us not postpone a discussion of how they are used.

An ammeter is used to measure the current that flows through a wire. We connect it directly in line with the wire as shown in Fig. 17.24a. Notice that the current to be measured flows through the meter. If the meter had much resistance, then the meter would disturb the circuit. Therefore an ideal ammeter should have zero resistance. The ammeters you will be using in laboratory will usually have a resistance of only a fraction of an ohm.

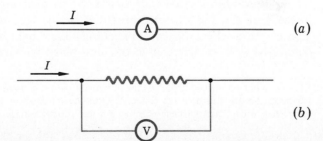

(a)

(b)

FIGURE 17.24
Why should an ammeter have little resistance while the resistance of a voltmeter should be large?

Voltmeters are used to measure potential difference. To measure the potential difference $V = IR$ across the resistor in Fig. 17.24b, the voltmeter terminals must be connected to the two ends of the resistor as shown. Ideally we want the voltmeter to leave the circuit undisturbed. This is possible only if the voltmeter resistance is very large. An ideal voltmeter has infinite resistance.

Students who confuse the ammeter and voltmeter in situations such as those shown in Fig. 17.24 are in serious danger to life and happiness because of the severe displeasure of their laboratory teacher. An ideal voltmeter has an infinite resistance. As such, no current passes through it when its terminals are connected to two points that differ appreciably in potential. But an ideal ammeter has zero resistance. If its terminals were inadvertently connected to two points of different potential, the current through the ammeter would be given by

$$I = \frac{V}{R} = \frac{\text{something}}{\text{zero}} \rightarrow \infty$$

This student error is accompanied by smoke issuing from the meter case, irreparable harm to the meter, and an antagonistic attitude on the part of the instructor. So beware.

17.12 House Circuits

We are all familiar with the ordinary electrical circuits that extend throughout our houses. The power company runs at least two wires to each house to provide a potential difference of about 120 V.[2] These lead-in wires to the house usually have a large diameter so that they can carry considerable current without heating up. (The larger the cross-sectional area of the wire, the less its resistance will be. Since heat generated is proportional to I^2R, the low resistance will ensure low heat dissipation.)

In most newer houses, the wires are capable of carrying about 20 A without undue heating. However, *to protect against too large a current, a fuse or circuit breaker is placed in series with the wire. Its purpose is to disconnect the wire* from the voltage source if greater than the allowed current is drawn from it. This procedure automatically disconnects any wire that is accidentally called on to carry more than the safe current.

A typical house circuit consists of two parallel wires strung through the house from the 120-V source provided by the lead-in wires to the house. This is shown schematically in Fig. 17.25. Each light bulb, appliance, etc. is connected with one terminal to the high-potential wire and the other to the

---

[2]The potential difference supplied by the power company is constantly reversing in a sinusoidal way. We discuss it in detail in Chap. 20. For the purposes of the present discussion, the alternating voltage has the same effect as a direct-current voltage.

low-potential wire. When the switch to that appliance is closed, current runs through the device from the plus to the minus wire of the power system. The low-potential wire is usually grounded.

Many 120-V appliances have a third prong on the power plug. This furnishes a connection between a ground wire and the metal frame of the appliance. If, by accident, the high-voltage wire touches the metal frame of the appliance, a direct connection to ground is made. The effect is the same as connecting the high- and low-voltage wires directly. A large current then flows through the high-voltage wire to ground. The fuse in the high-voltage wire will then blow. If the ground wire is absent, such a malfunction will leave the whole appliance "floating" at high potential. Anyone touching the metal frame will then suffer a shock.

Let us compute how much current is drawn by the 60-W bulb of Fig. 17.25 when it is turned on by closing the switch. Since power = VI, and since $P = 60$ W and $V = 120$ V in this case, we find the current through the bulb to be $I = 0.50$ A. Similarly, when turned on, the stove draws 10 A, the radio draws 0.167 A, and the 120-W bulb draws 1.0 A. If they are all turned on at once, a total of 11.667 A will pass through the fuse. Usually a house circuit would be fused for no less than 15 A, and so no danger exists in this case.

A house with a large number of electrical appliances requires more than one circuit. Most houses have several separate fused circuits similar to the one shown in Fig. 17.25. Each of these starts at the source furnished by the lead-in wires to the house and runs to various portions of the house.

It is interesting to compute the resistance of a light bulb. When the light bulb is cool, its resistance is not very large. However, when it is connected across the rated voltage, usually 120 V, its resistance element becomes white-hot. As discussed previously, its resistance will increase considerably when it heats up. When it is hot, it will operate at the wattage stamped on it. Suppose that we have a 60-W, 120-V bulb. We know that

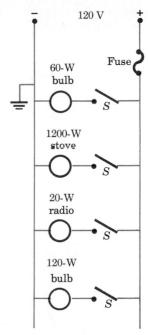

FIGURE 17.25
The household appliances shown act as resistors. Each is put into operation by closing the appropriate switch.

$$P = VI = \frac{V^2}{R}$$

or

$$60 = \frac{(120)^2}{R}$$

from which

$$R = 240 \ \Omega$$

where you should verify the units of the answer.

17.13 Electrical Safety

Since we use electrical apparatus daily, we should understand the elements of electrical safety. *Electricity can kill a person in two ways: It can cause the muscles of the heart and lungs (or other vital organs) to malfunction, or it can cause fatal burns.*

Even a small electric current can seriously disrupt cell functions in that portion of the body through which it flows. When the electric current is 0.001 A or higher, a person can feel the sensation of shock. At currents 10 times larger, 0.01 A, a person is unable to release the electric wire held in the hand because the current causes the hand muscles to contract violently. Currents larger than 0.02 A through the torso paralyze the respiratory muscles and stop breathing. Unless artificial respiration is started at once, the victim will suffocate. Of course, the victim must be freed from the voltage source before he or she can be touched safely; otherwise the rescuer, too, will be in great danger. A current of about 0.1 A passing through the region of the heart will shock the heart muscles into rapid, erratic contractions (ventricular fibrillation) so the heart can no longer function. Finally, currents of 1 A and higher through body tissue cause serious burns.

The important quantity to control in preventing injury is electric current. Voltage is important only because it can cause current to flow. Even though your body can be charged to a potential thousands of volts higher than the metal of an automobile by simply sliding across the car seat, you feel only a harmless shock as you touch the door handle. Your body cannot hold much charge on itself, and so the current flowing through your hand to the door handle is short-lived, and the effect on your body cells is negligible.

In some circumstances, the 120-V house circuit is almost certain to cause death. One of the two wires of the circuit is usually attached to the ground, so it is always at the same potential as the water pipes in a house. Suppose a person is soaking in a bathtub; his (or her) body is effectively connected to the ground through the water and piping. If his hand accidentally touches the high-potential wire of the house circuit (by touching an exposed wire on a radio or heater, for example), current will flow through his body to the ground. Because of the large, efficient contact his body makes with the ground, the resistance of his body circuit is low. Consequently the current flowing through his body is so large that he will be electrocuted.

Similar situations exist elsewhere. For example, if you accidentally touch an exposed wire while standing on the ground with wet feet, you are in far greater danger than if you are on a dry, insulating surface. The electrical circuit through your body to the ground has a much higher resistance if your feet are dry. Similarly, if you sustain an electrical shock by touching a bare wire or a faulty appliance, the shock is greater if your other hand is touching the faucet on the sink or is in the dishwater.

As you can see from these examples, *the danger from electrical shock can be eliminated by avoiding a current path through the body.* When the voltage is greater than about 50 V, avoid touching any exposed metal portion of the circuit. If a high-voltage wire must be touched, e.g., in case of a power-line accident when help is not immediately available, use a dry stick or some other substantial piece of insulating material to move it. When in doubt about safety, avoid all contacts or close approaches to metal or to the wet earth. *Above all, do not let your body become the connecting link between two points which are at widely different potentials.*

17.14 EMF and Terminal Potential of a Battery

Probably everyone has noticed at one time or another that the lights on an automobile dim when the motor is first started. The electrical starter used on an automobile draws considerable current from the battery. In so doing, it lowers the potential between the terminals of the battery, and the car lights dim. We now investigate this nonconstancy of the terminal potential difference of a battery.

As pointed out in the last chapter, the emf of a battery is generated by the chemical action within it. When no current is being drawn from the battery, the difference in potential between the battery terminals is equal to the emf of the battery. However, *a battery* is a very complex chemical device, and it actually *behaves like an emf in series with a resistor.* An equivalent circuit for a battery is shown in Fig. 17.26.

Notice that when no current is being drawn from the battery, there is no potential drop across the internal battery resistance r. Hence the potential difference between the terminals of the battery is equal to the emf. However, if the battery is connected across a resistor as shown in Fig. 17.27, a current I flows as shown and the terminal potential difference is $\mathcal{E} - Ir$.

For a good 12-V battery, the internal resistance would be only of the order of 0.01 Ω. If it were connected across a 3-Ω resistor, we would have

$$I = \frac{12}{3 + 0.01} \approx 4\,\text{A}$$

The terminal potential would be the voltage difference between points a and b, which is

$$\text{Terminal potential} = 12 - (4)(0.01) = 11.96\,\text{V}$$

In this case, the terminal potential is nearly equal to the emf.

However, as a battery becomes older, its internal resistance increases. If the resistance of the battery in Fig. 17.27, were 1.0 Ω, the current would be

$$I = \tfrac{12}{4} = 3.0\,\text{A}$$

and the terminal potential would be

$$\text{Terminal potential} = 12 - 3.0 = 9.0\,\text{V}$$

It should be clear that when a starter on a car draws 100 A from the battery, the terminal potential of even a new battery will decrease noticeably.

Illustration 17.15 What will be the terminal potential of each of the batteries in Fig. 17.28?

Reasoning The two batteries are opposing each other, and so the

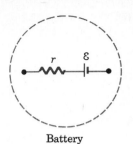

FIGURE 17.26
A battery acts as though it consisted of a pure emf and a resistor in series, as shown.

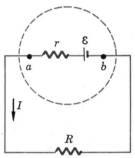

FIGURE 17.27
The terminal voltage *V* of the battery is $\mathcal{E} - Ir$.

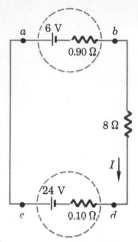

FIGURE 17.28
The 24-V battery is charging the 6-V battery. We find that the terminal potential of a discharging battery is less than its emf, while the reverse is true for a battery which is being charged.

effective driving emf in the circuit is 18 V. The current will flow in the direction shown and is

$$I = \tfrac{18}{9} = 2 \text{ A}$$

The potential difference from d to c is

$$V = -0.2 + 24 = 23.8 \text{ V}$$

Hence the terminal potential of the 24-V battery is less than its emf. For the terminal potential of the other battery from b to a we have

$$V = +1.8 + 6 = 7.8 \text{ V}$$

Notice that the terminal potential of this battery is larger than the emf. This is always the case when a battery is being charged as the 6-V battery is in this example.

17.15 Potentiometer

In measuring the potential difference between two points in a circuit, it is sometimes important that the voltage-measuring instrument draw no current, as in trying to measure the emf of a battery which has a high internal resistance. Or suppose that we wished to measure the potential difference between two points on a person's body, as is done in measuring brain waves. The body acts like a battery of extremely high internal resistance. Clearly, no current can be drawn during the measurement if the undisturbed emf is to be measured.

Any voltmeter draws some current from the points to which it is attached. Although electronic voltmeters draw only very small current from the voltage source they are measuring, ordinary laboratory voltmeters frequently have resistances of only a few hundred ohms. Hence, when connected across a few volts, they will draw appreciable current. We require *a simple device for measuring voltage which draws no current in the process. The* **potentiometer** *is such a device.*

Suppose you wish to measure an unknown battery of emf denoted as $\mathcal{E}_x$. If you could find a large group of calibrated batteries, you could connect each of the known batteries into a circuit such as the one shown in Fig. 17.29a. You could gently tap the switch to see whether the galvanometer G (a sensitive current meter) would deflect. When a known battery was found for which the galvanometer did not deflect, you would know that the emfs of the two batteries just balanced each other so that no current would flow. Hence you would know that the unknown battery had the same emf as this known battery. The measurement will have been made when no current was being drawn from the unknown, and so the internal resistance of the battery has no effect.

It is impractical to use a large group of known batteries, since they will not usually be available. However, the circuit of Fig. 17.29b serves the same purpose and, as we shall see, requires only one standard battery. One merely moves the sliding contact at B along the variable resistor until the voltage drop from A to B is equal to $\mathcal{E}_x$. In effect, the variable resistor provides a variable battery with terminals at A and B. (The part of the circuit below points a and b is frequently referred to as a **potential divider**.) When the voltage from A to B exactly equals $\mathcal{E}_x$, no current will flow in the upper circuit, and we know that

$$\mathcal{E}_x = I_0 R_x$$

If $\mathcal{E}_x$ is now replaced by a standard known battery $\mathcal{E}_s$, the potentiometer can again be balanced so that the galvanometer does not deflect. Then

$$\mathcal{E}_s = I_0 R_s$$

Dividing these two equations yields

$$\frac{\mathcal{E}_x}{\mathcal{E}_s} = \frac{R_x}{R_s}$$

In practice, the resistor is calibrated, and so R_x/R_s is known. With the value of the standard cell $\mathcal{E}_s$ known, the unknown emf is easily calculated. (Commonly, the resistor is merely a long, uniform resistance wire. Hence R will be proportional to the length of the wire from A to B. As a result, the ratio R_x/R_s can be replaced by the respective lengths of wire, L_x/L_s.)

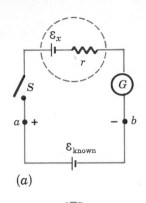
(a)

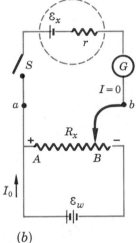
(b)

FIGURE 17.29
In (a), when no current flows with S closed, we know that $\mathcal{E}_x = \mathcal{E}_{\text{known}}$. To obtain a variable $\mathcal{E}_{\text{known}}$, the circuit below points a and b in (b) is used. The current through G is zero when the voltage drop from A to B equals $\mathcal{E}_x$. This device, a potentiometer, measures the emf of the unknown cell.

Illustration 17.16 When electric contacts are placed at various positions on someone's head, it is found that potential differences exist which fluctuate in characteristic ways. These potential differences can be recorded by an **electroencephalograph**. Typically these voltage differences are of the order of 5×10^{-4} V and fluctuate in times of the order of 0.10 s. Suppose the resistance of the head between two of these contacts is 10,000 Ω. How large a current can the voltage-recording device draw from the voltage being measured if the voltage read is to be in error by less than 1 percent? How large a resistance must the recording device have?

Reasoning The portion of the person's head between the two electrodes acts as a battery with emf $= V_0$ and internal resistance $= 10,000$ Ω. When a current I is being drawn, the internal resistance causes a voltage error $10,000I$. We wish

$$\frac{10,000I}{V_0} < 0.01$$

Therefore the recorder cannot draw a current in excess of $0.01 V_0/10,000$, or $10^{-6} V_0$ amperes. Since the current is caused to flow through the voltage recorder by the voltage source being measured V_0, we can use Ohm's law to find the resistance R_0 of the recorder. Thus

$$V = IR$$

becomes

$$V_0 = (10^{-6} V_0)(R_0)$$

from which the recorder resistance must be at least

$$R_0 = 10^6 \; \Omega$$

Many modern recorders use the potentiometer principle and, in effect, have infinite resistance when a measurement is made.

Summary

If a charge ΔQ flows past a given point in a time Δt, the current I is given by $I = \Delta Q/\Delta t$. Its units are amperes (A). Current is defined to have the direction of positive charge flow. If negative charges are moving, the current due to them is defined to flow in the opposite direction.

When a potential difference V exists between the two ends of a resistor, the current through the resistor is related to V by $V = IR$. The quantity R is the resistance of the resistor, and it is measured in ohms (Ω). Ohm's law not only states that $V = IR$, but also implies that R is a constant, independent of V and I. Current always flows through a resistor from its high-potential end to its low-potential end.

The resistivity ρ of a material measures the material's resistance to charge flow. In terms of it, the resistance of a wire L meters long with cross-sectional area A is given by $R = \rho(L/A)$.

The resistance of most materials changes with temperature. If α_{20} is the temperature coefficient of resistance based on a reference temperature of 20°C, then

$$\frac{R - R_{20}}{R_{20}} = \alpha_{20}(t - 20°)$$

where R is the resistance at t degrees if its resistance at 20°C was R_{20}. A similar relation exists with R replaced by ρ.

When a current I flows through a resistance R, the power expended in the resistor is $I^2 R$. This expended power appears as heat energy. If a current I flows through a potential drop V of any kind, the power expended by the current is VI.

Kirchhoff's point rule states that the sum of all the currents coming into a point must equal the sum of all the currents leaving the point. His loop rule states that the algebraic sum of the voltage changes around a closed circuit (or loop) is zero.

Resistances in series add directly; for parallel resistances, the reciprocals add.

Capacitances in parallel add directly; for series capacitances, the reciprocals add.

The primary safety rule in dealing with electricity is as follows: Never let your body become a connecting link between two points of widely different potential.

A battery acts like an emf $\mathcal{E}$ in series with an internal resistance r. The terminal potential difference of a battery differs from $\mathcal{E}$ by an amount Ir, where I is the current through the battery.

A potentiometer measures voltages without drawing current. As a result, it can be used to measure emf for sources that have appreciable internal resistance.

Minimum Learning Goals

Upon completion of this chapter, you should be able to do the following:

1. Make use of the relation $I = \Delta Q/\Delta t$ in simple situations. Define electric current in your own words.
2. Interpret a simple circuit diagram. State the potential difference between various pairs of points of the circuit.
3. State which end of a resistor is at the higher potential when the direction of current flow through the resistor is given.
4. State the equation form of Ohm's law and explain its meaning. Use it to define resistance and its units.
5. Compute the resistance of a given piece of wire provided the resistivity of the wire material is known.
6. Find the resistance of a wire at a given temperature when its resistance and temperature coefficient at some reference temperature are known.
7. Use the power equation $P = VI$ to find the power loss or gain in a resistor, battery, and capacitor under dc conditions.
8. State Kirchhoff's point rule and give examples of its application.
9. State Kirchhoff's loop rule and write the loop equation for a series circuit containing batteries and resistors.
10. Reduce a given set of series and parallel resistors to a single equivalent resistor.
11. Reduce a given set of series and parallel capacitors to a single equivalent capacitor.
12. Solve dc circuits that contain batteries, resistances, and capacitances by use of Kirchhoff's rules.
13. Sketch a typical house circuit and point out the various elements in it. Compute the current drawn by various portions of a house circuit when the appliances running from it are given.
14. Analyze a given practical situation from the viewpoint of safety.
15. Explain why the terminal potential of a battery is not always equal to the emf of the battery. Find the terminal potential if $\mathcal{E}$, I, and r are known.
16. Sketch the circuit diagram for a potentiometer, and explain the function of the device.

Important Terms and Phrases

You should be able to define or explain each of the following:

dc circuit
$I = \Delta Q/\Delta t$; ampere
Ohm's law: $V = IR$
Resistance
Resistivity

Temperature coefficient of resistance
Power $= I^2R = VI$
Kirchhoff's rules
Equivalent resistance (series and parallel)
Equivalent capacitance (series and parallel)
Battery-terminal potential difference

Questions and Guesstimates

1. Sometimes students insist that current is "used up" as it flows through a resistor. Arguing from the water analogy, how would you convince such a student that current is not lost in a resistor?
2. How do we know which end of a battery is at the higher potential, i.e., positive, in a schematic diagram of a circuit? Of a resistor?
3. Current and voltage are never lost in a circuit. What is lost as the current flows through the circuit?
4. Fluorescent light bulbs are usually much more efficient light emitters than incandescent bulbs. That is, for the same input energy, the fluorescent bulb gives off more light than the incandescent bulb does. Carefully touch each of the two bulbs after it has been lit a few minutes. Explain why the incandescent bulb is a less efficient light emitter.
5. In Fig. P17.1a, the pump lifts water to the upper reservoir at such a rate that the water level remains constant. The water slowly trickles out of the narrow tube into the lower reservoir. Point out the similarities between this water circuit and the electric circuit shown in part b.
6. A resistor is connected from point A to point B.

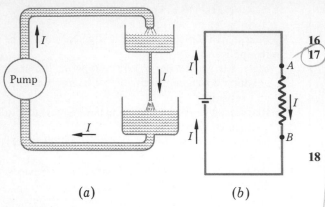

(a)   (b)

FIGURE P17.1

How does one tell which it is from A to B, a potential drop or a potential rise? Repeat for a battery and for a capacitor.

7 Explain the following statement: For series resistors, the equivalent is always larger than the largest; for parallel resistors, the equivalent is always smaller than the smallest. What is the similar statement for capacitors?

8 Why would a person connect two batteries in series? In parallel?

9 Batteries should never be connected in parallel unless they are nearly identical. Why?

10 Is the energy stored in three identical capacitors charged by a 12-V battery larger when the capacitors are in series or in parallel?

11 Does the energy stored in three capacitors in series depend on which one is in the middle? Explain.

12 Explain why the charges on several unlike capacitors charged in series are all identical while unlike capacitors in parallel have different charges.

13 Some time ago, car manufacturers changed from a 6-V electrical system to a 12-V system. Why?

14 Estimate the electric power consumption of a nearby city. How large a current would have to be supplied to the city at 220 V to supply this energy? Explain why very-high-voltage power transmission (that is, 10^5 V) is an advantage.

15 Explain why an ammeter must have a low resistance. What precautions must be taken in handling an ammeter?

16 Why must a voltmeter have a high resistance?

17 Using an ohmmeter (basically a battery in series with a very sensitive ammeter), measure your resistance from one hand to the other. A current of about 0.02 A through the midsection of one's body is sufficient to paralyze the breathing mechanism. About how large a voltage difference between your hands is needed to electrocute you?

18 If you grasp the two wires leading from the two plates of a charged capacitor, you may feel a shock. The effect is much greater for a 2-μF capacitor than for a 0.02-μF capacitor, even though both are charged to the same potential difference. Why?

19 Birds sit on high-tension wires all the time. Why aren't they electrocuted since sometimes the wires have gaps where there is no insulation on them?

20 If a current of only a small fraction of an ampere flows into one hand and out the other, the person will probably be electrocuted. If the current flows into one hand and out the elbow above the hand, the person can survive even if the current is large enough to burn the flesh. Explain.

21 Mothers frequently worry about their small children playing near electrical outlets. Discuss the various factors which determine how badly shocked the child could be. What would happen if a small child were to cut a lamp cord in two with a pair of noninsulated, wire-cutting pliers when the cord is plugged in? Is the child in any danger in this case?

22 Explain why it is much more dangerous to touch an exposed houselight circuit wire if you are in a damp basement than if you are on the second floor.

23 It is extremely dangerous to use a plug-in radio near a bathtub when you are taking a bath. Why? Does the same reasoning apply to a battery-operated radio?

24 For such purposes as electrocardiograms, brain-damage tests, lie detectors, etc., one measures the voltage differences between various portions of a person's body. Why can't a simple voltmeter be used for this purpose? What precautions must be taken?

Problems

Ignore internal resistances of batteries when the resistance is not given.

1 A current of 2.0 A is drawn from a 6-V battery for 20 min. (a) How much charge and (b) how many

472

electrons flow from the battery in this time?

2 A certain atom-smashing machine provides a current of 0.0010 A [= 1 milliampere (mA)] through a potential difference of 1 MV. (a) How much charge does it provide in $\frac{1}{2}$ h? (b) If the current consists of a stream of protons, how many protons are provided in $\frac{1}{2}$ h?

3 A 60-W light bulb draws a current of 0.50 A on 120 V. How many electrons pass through the bulb each minute?

4 A 900-W toaster draws a current of 7.5 A on 120 V. How many electrons pass through the toaster in 20 s?

5 A 40-W fluorescent bulb draws a current of $\frac{1}{3}$ A when connected across a potential difference of 120 V. What is the resistance of the bulb?

6 How much current flows from a 12-V battery when a 3.0-Ω resistor is connected across its terminals?

7 Find the resistance of a 20-m length of 3.0-mm-diameter aluminum wire at 20°C.

8 What is the resistance of a 50-cm length of 0.100-mm-radius silver wire at 20°C?

9 The resistance of a spool of insulated copper wire is measured by noticing that a 1.50-V battery causes a current of 0.60 A to flow through the entire length of wire. The diameter of the metal part of the wire is 0.100 cm. What length of wire is on the spool?

10 Number 10 copper wire has a diameter of 2.59 mm. The approximate safe current for this size wire is 30 A. (At higher currents, it becomes too hot.) (a) Find the resistance of a 5.0-m length of this wire at 20°C. (b) How large a potential difference will exist between the ends of a wire this long when it carries 30 A?

11 The filament of a certain light bulb is 8.0 cm long and has a diameter of 0.150 mm. When the bulb is connected across 120 V, its filament becomes white-hot and the bulb draws a current of 0.25 A. (a) What is the resistance of the bulb under these conditions? (b) What is the resistivity of the filament material at this temperature?

12 An ordinary 60-W tungsten-filament light bulb has a resistance of 240 Ω when lit. The temperature of the filament is about 2000°C under these conditions. If the α value in Table 17.1 is valid even for temperatures this high, what is the resistance of the bulb at 20°C?

13 At 20°C the resistance of a particular tungsten-filament lamp is 30 Ω. When it is lighted, the lamp has a resistance of 240 Ω. About what is the temperature of the filament in the hot lamp?

14** A graphite resistor is placed in series with a 10-Ω (at 20°C) iron resistor. How large should the resistance of the graphite resistor be in order for the combined resistance to be temperature-independent?

15** It is desired to make a 100-Ω resistor which is temperature-independent by using a carbon resistor in series with an iron resistor. What should the resistance of each be at 20°C?

16 (a) Find the equivalent resistance for the circuit of Fig. P17.2. (b) Find I if the value of $\mathcal{E}$ is 12.0 V.

17 In Fig. P17.3, the value of $\mathcal{E}$ is 12.0 V, and each resistor is 2.0 Ω. Find the equivalent resistance and I.

18* Find the equivalent resistance for the circuit in Fig. P17.4 when (a) switch S is open and (b) S is closed.

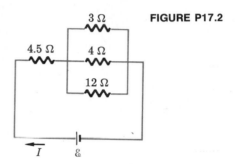

FIGURE P17.2

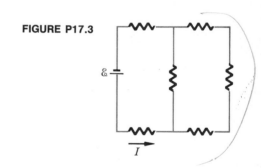

FIGURE P17.3

FIGURE P17.4

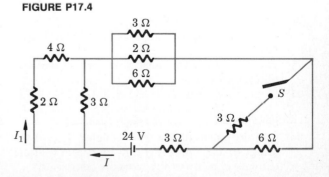

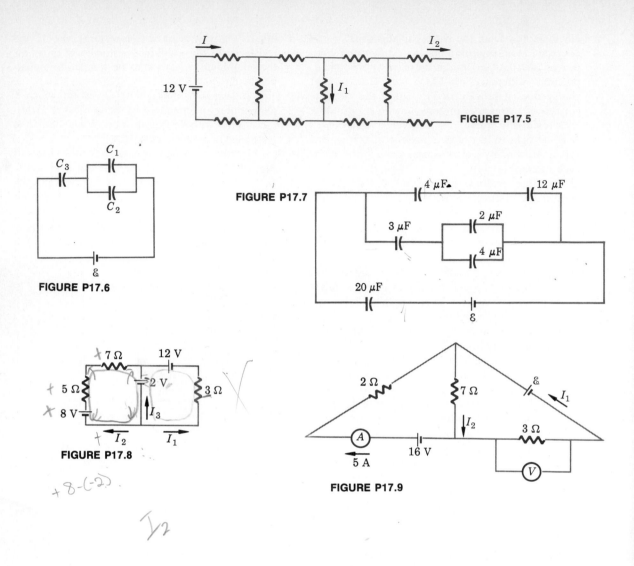

FIGURE P17.5

FIGURE P17.7

FIGURE P17.6

FIGURE P17.8

FIGURE P17.9

19* (a) Find the equivalent resistance for the circuit shown in Fig. P17.5 if all the resistors are 4 Ω. The circuit is shown in its entirety. (b) Find I. (c) What is I_2?

20 Each of the capacitors shown in Fig. P17.6 is 2.0 μF. Find the equivalent capacitance of the combination.

21 Find the equivalent capacitance for the combination shown in Fig. P17.7.

22* Find I_1, I_2, and I_3 in the circuit of Fig. P17.8.

23* In Fig. P17.9, the ammeter reads 5 A. Find I_1, I_2, ε, and the voltmeter reading.

24** In the circuit of Fig. P17.10, find I, I_1, and the charge on the 7-μF capacitor. *Hint:* Note that the currents in many of the wires are zero.

25* For the circuit shown in Fig. P17.11, find the (a) current through the batttery, (b) the current through the 12-Ω resistor, (c) the power loss in the 8-Ω resistor.

26* Refer to the circuit of Fig. P17.12. The voltmeter reads 5.0 V, and the ammeter reads 2.0 A with the current flowing in the direction indicated. Find (a) the value of R and (b) the value of ε. (Notice in writing the circuit equation that the voltage drop across R is 5 V.)

27* In Fig. P17.12, how large must ε be if the current through the 6-V battery is to be zero and R is 12 Ω?

28* In Fig. P17.12, what would (a) the ammeter and (b) the voltmeter read if ε were 20 V and R were 6.0 Ω?

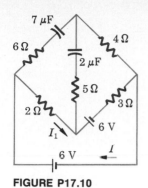

FIGURE P17.10

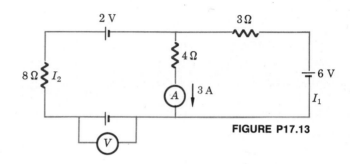

FIGURE P17.11

FIGURE P17.12

FIGURE P17.13

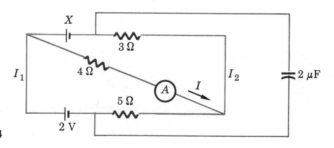

FIGURE P17.14

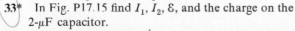

29 A particular 120-V house-light circuit has operating on it a 1200-W toaster, one 60-W lamp, and a 500-W soldering iron. The fuse in the circuit blows as soon as another 60-W bulb is turned on. About what was the rating of the fuse?

30 It is planned to operate a 1200-W dryer, a washer which requires 360 W, four 60-W bulbs, and a 40-W radio, all from the same 120-V line. For at least how large a current must this line be fused?

31* In Fig. P17.13, the ammeter reads 3.0 A. Find I_1, I_2, and the voltmeter reading.

32* The current through the ammeter in Fig. P17.14 is 3 A in the direction shown. (*a*) Find the unknown emf (X) and the two unknown currents. (*b*) What is the charge on the capacitor?

33* In Fig. P17.15 find I_1, I_2, $\mathcal{E}$, and the charge on the 2-μF capacitor.

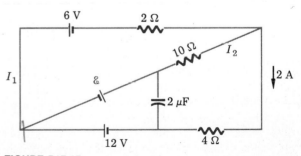

FIGURE P17.15

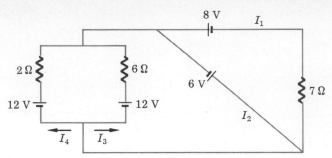

FIGURE P17.16

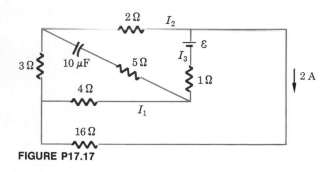

FIGURE P17.17

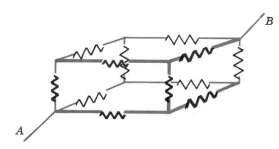

FIGURE P17.18

34* Find the currents I_1, I_2, I_3, and I_4 in the circuit shown in Fig. P17.16.

35* In Fig. P17.17 find I_1, I_2, I_3, $\mathcal{E}$, and the charge on the 10-μF capacitor.

36* Two capacitors, of 2 and 4 μF, are individually charged to a potential difference of 12 V by connecting them, one at a time, across a battery. After they are removed from the battery, they are connected, the positive plate of one to the positive plate of the other and the negative plate of one to the negative plate of the other. Find (a) the potential across each and (b) the resultant charge on each. *Hint:* Notice that the potential drop across both will be the same after they are connected.

37** Repeat Prob. 36 if the capacitors are connected with the positive plate of one connected to the negative plate of the other.

38** Each of the resistors shown in Fig. P17.18 has a value of 2 Ω. Find the equivalent resistance be-

tween A and B. *Hint:* From symmetry, many of the wires contain identical currents.

39* Prove the relation for parallel capacitances given in Fig. 17.19.

40* Prove the relation for series capacitors given in Fig. 17.19.

41 When a current of 3.0 A is drawn from a certain battery, its terminal voltage drops from its zero current value of 1.57 to 1.26 V. Find the internal resistance of the battery.

42 What is the maximum current that can be drawn from a battery that has an emf of 1.57 V and an internal resistance of 0.30 Ω?

43* The terminal potential of a battery is 11.40 V when a 20.0-Ω resistor is connected across its terminals and is 10.16 V when a 5.0-Ω resistor is connected across its terminals. Find the emf and internal resistance of the battery.

18
Magnetism

All of us, at one time or another, have performed simple experiments with magnets. We know that they have a magnetic field about them and that certain materials experience forces when placed in such a magnetic field. For example, the north pole of one magnet exerts a repelling force against the north pole of a second magnet placed nearby. Its south pole will attract north poles of other magnets. In addition, unmagnetized pieces of certain materials (iron is a common example) are highly attracted by both the south and north poles of magnets.

Although these are facts which many of us learned in grade school, the molecular explanation of these facts is not simple. The subject of magnetism today is still an area of active and exciting research. You will see in this chapter and the next that magnetism is important to us in many ways. As we discuss this subject, it will become apparent that magnets and their effects are only a small facet of magnetism.

18.1 Plotting Magnetic Fields

A magnetic field exists in the region about a magnet. As a start, we may describe a magnetic field in terms of its effect on a compass needle. As you probably know, *a compass needle is merely a small magnet. The end of the needle which points north as a result of the earth's magnetic field is denoted the* **north pole** *of the needlelike magnet.* Furthermore, *we define a magnetic field*

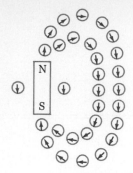

FIGURE 18.1
The direction of the
magnetic field in the vicinity
of the magnet can be found
by using a compass needle.

*in such a way that the north pole of the compass needle points in a direction
parallel to the field.* With this in mind, together with the fact that like poles
repel and unlike poles attract, it is a relatively simple matter to plot magnetic fields.

As a simple example, suppose that we wish to plot the magnetic field
about a bar magnet. A compass set in the vicinity of the magnet will align
itself in the direction of the magnetic field. This is illustrated in Fig. 18.1,
where several compasses are shown. You should satisfy yourself that the
north poles of the needles (the arrowheads) will line up, as shown, under the
repulsion from the north pole and attraction to the south pole of the bar
magnet. A "picture" of the magnetic field consists in drawing a series of
lines about the magnet in such a way that the lines show the direction in
which a compass needle would point. This is done for a bar magnet in Fig.
18.2*a*. The fields about other types of magnets are shown in parts *b* and *c*
of the same figure. Notice that *the field always points out from the north pole
and into the south pole.* Why? (Remember that the direction of the field at
a point is parallel to the direction in which the compass needle points.)

Many important features of a magnetic field can be ascertained from
pictures such as those of Fig. 18.2. In particular, *the relative density of the
lines in a given figure can be taken as a measure of the field strength,* much as
in the case of electric fields. As a result, *the magnetic field lines in a figure not
only picture the direction of the field, but also indicate its relative magnitude.*

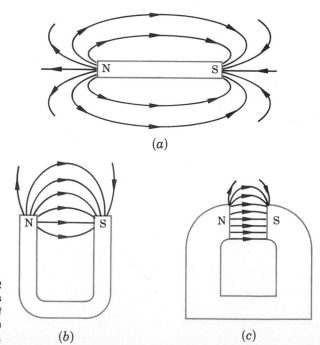

FIGURE 18.2
The magnetic field points
away from the north pole of
a magnet and to the south
pole.

18.2 Magnetic Fields of Currents

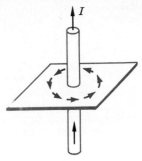

Magnets are not the only source of magnetic fields. In 1820, *Hans Christian Oersted discovered* that a current through a wire caused a nearby compass needle to deflect. This indicated *that a current in a wire is capable of generating a magnetic field.* We now know from many other types of experiments that this is indeed the case. Further, the magnetic field of a magnet is also the result of the motion of charges, as we shall show subsequently.

Oersted investigated the nature of the magnetic field about a long, straight wire which was carrying current. His experiment is illustrated in Fig. 18.3. The wire carries a current in the direction indicated. When a compass is placed near the wire, the needle lies with its length tangent to a circle concentric with the wire, and the inference is that a magnetic field exists in a circular form about the wire. As is to be expected, the strength of the field is greatest close to the wire. A three-dimensional representation of the magnetic field is shown in Fig. 18.4. In this, as well as later diagrams, the symbol · indicates an arrow coming toward the reader, while ✕ represents an arrow going away from the reader. The symbols are meant to suggest the point and the tail of the arrow.

There is a simple **right-hand rule** *for remembering the direction of the magnetic field about a wire.* If you grasp the wire in your right hand with the thumb pointing in the direction of the current, your fingers will circle the wire in the direction of the field. This is shown schematically in Fig. 18.5.

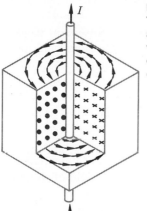

FIGURE 18.4
The magnetic field circles about the long, straight wire. Its magnitude decreases inversely with the distance from the wire.

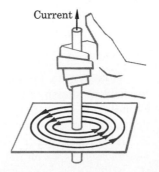

FIGURE 18.5
When you grasp the wire in your right hand with the thumb pointing in the direction of the current, your fingers circle the wire in the same fashion as the field.

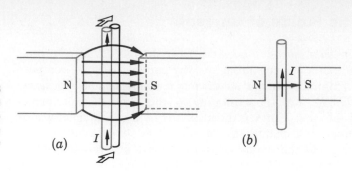

18.3 Force on a Current in a Magnetic Field

Thus far we have discussed only the qualitative features of the magnetic field. We now seek some means for precisely measuring it. This can be done by using the fact that *a wire carrying a current experiences a force when it lies in a magnetic field.* In a typical experiment to illustrate this phenomenon, a wire is placed between the poles of a magnet, as shown in Fig. 18.6a. When a current I is caused to flow through the wire in the direction indicated, the wire experiences a force tending to push it in the direction shown. If the direction of the current is reversed, the force on the wire also reverses. Redrawing the situation of Fig. 18.6a in two dimensions as shown in part *b,* we notice that *the line of the wire and a magnetic field line intersecting it determine a plane,* the plane of the page. (Remember, two intersecting straight lines determine a plane.) *The force on the wire is always perpendicular to this plane.* We come back to this rule shortly.

The magnitude of the force F on the wire depends on several factors. Resorting to experiment, we can isolate each factor in turn and determine precisely their interrelations. It is found that the force is directly proportional to both the current I in the wire and the length L of the wire in the magnetic field. When the current is doubled, the force is doubled. We *Magnetic Field Vector* **B** choose to *define the magnetic field in terms of a vector* **B,** *the direction of which at a given point is identical to the direction of the magnetic field line at that point. The magnitude of* **B** *is taken proportional to the force experienced by a wire that carries a current through the field.* We therefore have that the force on a wire of length L carrying current I in a magnetic field of strength B is given by

$$F \propto BIL$$

However, a complication exists. When the wire is directly in line with the field, as shown in Fig. 18.7a, the force on it is found to be zero. Further experiments show that the force on the wire is proportional to the component of **B** that is perpendicular to the wire. This component of **B**, namely, $B_\perp$, is shown in Fig. 18.7b. *We define our units of field strength B in such a way that the force on the wire is given by*

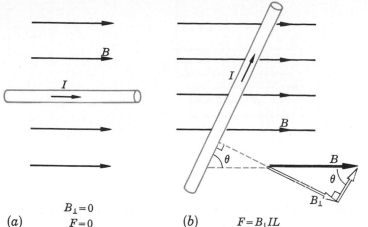

$$B_\perp = 0$$
$$F = 0$$

(a)

(b) $F = B_\perp IL$

FIGURE 18.7
The force on a current-
carrying wire is proportional
to the component of **B**
which is perpendicular to
the wire.

$$F = B_\perp IL \qquad (18.1) \qquad \textit{Magnetic Force on a Current}$$

Anyone who has a friend named Bill should have no trouble remembering
this equation. Notice that $B_\perp$ is merely $B \sin \theta$, where θ is the angle between
the wire and the field lines, as shown in Fig. 18.7*b*.

    The unit of B defined by Eq. (18.1) *is the tesla (T), also called the weber
per square meter (Wb/m²)*; this latter name will become reasonable to you
after we define magnetic flux later in this chapter. Substituting the usual
units for F, I, and L into Eq. (18.1), you can see that 1 T is equivalent to
1 N/(A)(m).

    We would like to call B the intensity of the magnetic field, but that
name has been preempted for another quantity. *Several names are used for
B, the more acceptable being* **flux density** *and* **magnetic induction.** *Colloqui-
ally, B is often referred to as the* **magnetic field strength.**

    Another unit used for B is the gauss (G). A gauss is much smaller than
a tesla. The relation is 10^4 G $= 1$ T. The gauss is not a member of the SI
family of units, and so it must first be converted to teslas before we can use it
in our equations.

Illustration 18.1 Find the force on a 300-cm length of the wire shown in
Fig. 18.7*b* if $\theta = 53°$, the current is 20 A, and B is 2.0 G.

    Reasoning The value of $B_\perp$ is $0.8B$. Hence, using Eq. (18.1), we have,
after changing all units to SI units,

$$F = B_\perp IL$$
$$= [(2 \times 10^{-4}\ T)(0.8)](20\ A)(3.0\ m) = 0.0096\ N$$

In the next section, we learn how to find the direction of the force.

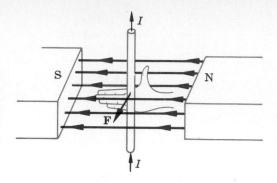

18.4 Extension of the Right-Hand Rule

In the previous section it was pointed out that the direction of the force experienced by a wire carrying a current in a magnetic field is perpendicular to the plane defined by the wire and the field. We now consider a simple, intuitive extension of the right-hand rule (mentioned in Sec. 18.2) which will allow us to state the direction of the force experienced by the wire. It is purely an intuitive aid for remembering the direction of the force. No real physical significance should be attached to the rule, since it is simply a memory device.

Right-Hand Rule   The rule is shown in Fig. 18.8. *Hold your right hand flat, and point your fingers along the lines of the magnetic field. Position your hand in such a way that your thumb points in the general direction in which the current is flowing in the wire. The force on the wire will then be found to be in the direction in which the palm of the hand would push.*

Let there be no confusion on this point. The line of the field vector **B** and the line of the wire together define a plane (the plane of the page in Figs. 18.6 and 18.7). The force on the wire is always perpendicular to this plane. Once you know this, a pure guess allows you a 50 percent chance of obtaining the proper direction for the force. It must be either into or out of a given side of the plane. To find which is the proper alternative, use the rule illustrated in Fig. 18.8. The direction of the force in Fig. 18.8 is toward you, out of the page. Using the same rule, you see that the direction of the force in Figs. 18.6 and 18.7*b* is into the page.

18.5 Forces on Moving Charges

A current is the result of the motion of charged particles, and a wire carrying a current experiences a force in a magnetic field. Is it really necessary that these particles be in a wire in order for them to experience a force from a magnetic field? It would appear that if the force were exerted directly on the charge carriers, they would experience a force even in open space in a region where there was a magnetic field. Indeed, this turns out to be the case.

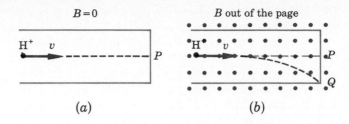

$B = 0$

B out of the page

H⁺ v - - - - - - - - - - - - • P

H⁺ v • • • • • • P

• Q

(a)

(b)

FIGURE 18.9
The proton follows an arc of a circle when deflected in a uniform magnetic field. The dots in (b) indicate **B** to be directed out of the page.

If a positively charged particle such as a hydrogen nucleus, i.e., a proton, is accelerated through a potential difference in an evacuated glass tube, it will travel down the tube and hit the end, as shown in Fig. 18.9a. The point of impact P can easily be seen as a bright spot on the glass if the end of the tube has been coated with a fluorescent material. (A television tube produces its picture when electrons hit the fluorescent screen at its viewing end. We consider the positively charged proton rather than the negatively charged electron because we defined current in terms of positive charge flow.)

Suppose, now, that the north pole of a magnet is placed behind the tube so that the field is coming out of the page as shown in part b. (As mentioned before, dots represent arrows coming toward you, and crosses represent receding arrows.) It is then found that the bright spot on the end of the tube will move to a position such as Q. The proton has been deflected by the magnetic field. Hence, we have direct experimental evidence for the fact that a charged particle moving in a magnetic field experiences a force.

Moreover, the force on the positively charged particle is perpendicular to both the direction of the field and the direction of motion of the particle. Examination of Fig. 18.9b will show that a current traveling to the right will also be deflected downward. This is not surprising in view of the fact that a conventional current is a flow of positive charge.

If the same experiment is performed by shooting electrons down the tube, it is found that the electrons are deflected upward rather than downward. The magnetic force on a moving, negatively charged particle is in the opposite direction from that on a positive particle. Hence, *no new rules need be learned to determine the direction of forces on moving charges in a magnetic field. We merely consider the moving charge to be a current in the direction of the motion and then use the extension of the right-hand rule to find the direction of the force on the current. If the particle is positive, it will be deflected in the same direction. If negative, it will be deflected in the opposite direction.*

Magnetic-Force Direction on Moving Charges

The magnitude of the force on a moving charge q is readily found by consideration of the equation for the force on a current,

$$F_{\text{wire}} = B_{\perp}IL$$

Consider the wire shown in Fig. 18.10. The current I is traveling to the right, and we can consider the current to be the motion of positively charged

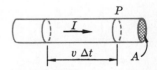

FIGURE 18.10
All the charge in the length $v\,\Delta t$ will pass through the area at P in time Δt.

particles. Each particle has a charge q and an average speed v. How large a force will each particle experience?

Suppose there are n free charges in a unit length of the wire. Because each charge moves to the right in Fig. 18.10 a distance $v \, \Delta t$ during a time Δt, all the charges in the length $v \, \Delta t$ shown in the figure will pass through the cross section at P during time Δt. In other words, during a time Δt, all the free charges in a length $v \, \Delta t$ of the wire will pass point P. In this length of wire, there are $n(v \, \Delta t)$ free charges, and so the charge passing P in time Δt is $\Delta Q = qn(v \, \Delta t)$. But current I is defined by $\Delta Q / \Delta t$, and so we find

$$I = \frac{qnv \, \Delta t}{\Delta t} = qnv$$

If we substitute this expression for I in $F_{\text{wire}} = B_\perp IL$, we have

$$F_{\text{wire}} = B_\perp qnvL = qvB_\perp(nL)$$

But nL is the number of free charges in length L of the wire. We can therefore find the force on each free charge by dividing F_{wire} by nL. Doing so, we find

$$F_{\text{charge}} = \frac{F_{\text{wire}}}{nL} = qvB_\perp \qquad (18.2)$$

Remember, this force is perpendicular to the direction of the motion as well as to the direction of B. Also remember that $B_\perp$ is the component of B perpendicular to v.

Illustration 18.2 A proton traveling at 10^5 m/s enters a region where there is a magnetic field with $B = 0.020$ T. Further, v and B are perpendicular. Describe the motion of the proton.

Reasoning The situation is as shown in Fig. 18.11. When the particle is at point M, the force is in the direction shown. Hence the proton will deflect upward. But when the particle reaches a point such as N, the force is still perpendicular to v and is now in the new direction shown. Clearly, the force F is going to force the particle around a circular path of radius r. *A charged particle shot perpendicular to a uniform magnetic field will follow a circular path.* Since the force is always perpendicular to v, *no work is done on the particle* because the motion is never in the direction of the force.

The magnetic force F furnishes the centripetal acceleration needed to hold the proton in a circular path. Our equation for this problem is therefore, from $F = ma$,

Magnetic force = (mass)(centripetal acceleration)

$$qvB_\perp = \frac{mv^2}{r}$$

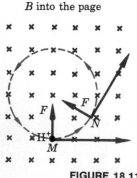

B into the page

FIGURE 18.11
The proton is deflected in a circular path by the uniform magnetic field.

Using the numbers given, we can solve for the radius of the path r. We have

$$r = \frac{mv}{B_\perp q} = \frac{(1.67 \times 10^{-27}\ \text{kg})(10^5\ \text{m/s})}{(0.020\ \text{Wb/m}^2)(1.60 \times 10^{-19}\ \text{C})} = 5.2 \times 10^{-2}\ \text{m}$$
$$= 5.2\ \text{cm}$$

Through how large a voltage had this proton fallen? What would have been its path if it had been an electron?

18.6 Determination of e/m

In the last section we discussed the motion of a charged particle in a magnetic field. One of the most important pieces of information concerning the nature of electrons can be obtained by observing their motion in a magnetic field. Let us illustrate the concepts of the last section by seeing how we can determine the ratio of the charge e to the mass m of the electron.

Suppose that an electron traveling with speed v enters a magnetic field, as pictured in Fig. 18.12. It will describe a circular path, as we have discussed previously. Notice that since the particle is negative, the circle in which the electron moves is not the same as that followed by the positive particle shown in Fig. 18.11. Since the magnetic force evB supplies the required centripetal force, we can write

$$evB = m\frac{v^2}{r}$$

or

$$\frac{e}{m} = \frac{v}{Br} \qquad (18.3)$$

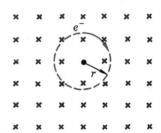

FIGURE 18.12
An electron, being negatively charged, is deflected in the direction opposite that in which a positive charge would be deflected.

We therefore have a means for determining the ratio e/m by experiment.

By using methods to be described subsequently, the strength of the magnetic field B can be measured, as can the radius of the circular path that the electron describes. If v were known, e/m could be computed. The path of the charged particles can be made visible in a number of ways. One of the simplest is to have air molecules present in the tube, but at a very low pressure. Some of the particles in the beam will collide with the air molecules. As we see in Chap. 26, light is given off as a result of such collisions, and so the circular path of the particle beam can be seen as a circular ring of light. The diameter of the ring can then be easily measured.

There are two basic ways of measuring v. One method simply makes use of the fact that a charged particle gains energy Vq when it falls through voltage V. Therefore, if we knew the voltage through which the electron was originally accelerated, we could write

$$Ve = \tfrac{1}{2}mv^2 \qquad \text{or} \qquad v = \sqrt{\frac{2Ve}{m}}$$

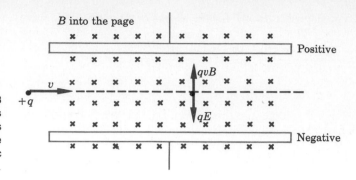

FIGURE 18.13
The velocity selector passes
undeflected those particles
for which the electric force
qE equals the magnetic
force qvB.

Substitution of this in Eq. (18.3) gives, after simplification,

$$\frac{e}{m} = \frac{2V}{B^2 r^2}$$

Notice that now we need know only the voltage through which the electron was accelerated, in addition to the value of B and r, in order to evaluate the ratio e/m for the electron. The value found is 1.7588×10^{11} C/kg.

A more precise method for measuring velocity makes use of a particle **velocity selector.** This device is shown schematically in Fig. 18.13. Two oppositely charged parallel plates produce a field E, which exerts a force qE downward on a positive particle, as shown. Also, the indicated magnetic field B exerts an upward force qvB on the same positive particle. If the voltage across the plates is adjusted properly, the magnitude of E will be such that

$$qvB = qE$$

and the particle will not deflect. Solving for v yields

$$v = \frac{E}{B}$$

To determine the speed v of the particle, we need only adjust the plate voltage V so that the particle travels undeflected through the crossed electric and magnetic fields. With $E = V/d$ known, d being the plate separation, the velocity of the particles can be computed from the above relation. Once v is known, an experiment in which the particle is deflected in a magnetic field alone will permit computation of q/m from Eq. (18.3). Can you show that the above discussion of the velocity selector, using positive charges, applies equally well to electrons?

18.7 Hall Effect

When current flows in a metal wire, the charge is being transported by the motion of the negatively charged electrons as they move along the wire.

This fact—that the charge carriers in metals are negatively charged—was not known until relatively late in the evolution of current electricity. As we have seen, the explanation of most experimental observations involving currents can be stated equally well in terms of either negative or positive charge carriers. It is for this reason that the situation in metals was not clarified until the qualitative features of electron behavior in metals was understood. There are very few experimental observations which depend on the sign of the charge carriers and can be used to determine their sign. Since in semiconductors the charge carriers need not be negative (as we point out in Chap. 20), an experimental method for determining their sign is needed. One of the very few methods available is discussed now.

In 1879, while reading Maxwell's classic text entitled *Electricity and Magnetism*, Edwin Herbert Hall noted a statement that he did not consider reasonable. Maxwell maintained (as did most others of the time) that even though a wire carrying a current experiences a force when placed in a magnetic field, the charges moving through the wire do not themselves experience this force. Hall, however, believed it reasonable that the force on the wire was probably the result of a force on the moving charges. He therefore designed an experiment to test his supposition. After several unsuccessful attempts, Hall finally made use of the apparatus shown in Fig. 18.14. Current from a battery is sent through a very thin, flat ribbon of gold foil. There is an appreciable voltage drop along the gold ribbon in the direction of current flow, since the thin foil has considerable resistance. However, points such as M and N on opposite sides of the ribbon are at the same potential, and so no voltage difference is recorded by a sensitive voltmeter connected between these two points. Hall noticed, however, that a steady voltage difference between points M and N occurred as soon as a magnetic field was applied perpendicular to the flat ribbon, as illustrated. He interpreted the results as follows.

Experiments on wires showed that in this situation the wire experiences

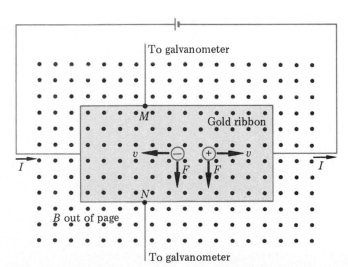

FIGURE 18.14
Apparatus used to measure the Hall effect.

a force in the direction of **F** in the figure (as you may easily verify by use of the right-hand rule). If this force were actually exerted on the charge carriers, either positive or negative, they, too, would be forced in the direction of **F**. This force would cause the charge carriers, the movable charges, to move toward the edge of the foil and thereby build up a charge on the edge of the ribbon. Thus points M and N no longer would be at the same potential. Since Hall had observed that the application of a magnetic field actually did cause a potential difference between M and N, he concluded that the force was exerted directly on the moving charges, in conformity with his ideas. Moreover, we notice that if the charge carriers are positive, point N should be positive with respect to M, since the positive charges are forced toward N. But if the charge carriers are negative, point N should be negatively charged. Here, then, we have a clear-cut experimental method for deciding the sign of charge carriers. In the case of Hall's gold foil, point N was found to be negative, and so his experiment confirmed the idea that electrons carry the current in metals. This type of experiment is widely used at present to discover the sign of the charge carriers in semiconductors. The Hall effect also forms the basis for a commercially produced device for measuring magnetic fields.

18.8 The Earth's Magnetic Field

Experiments such as those outlined in the previous section are complicated by the fact that the earth acts like a very large magnet. Its magnetic field is shown in Fig. 18.15a. Only the approximate location of the poles within the earth can be given, but they are indicated roughly on the figure. Notice, in particular, that the geographic poles, defined by the axis of rotation of the earth, do not coincide with the magnetic poles.

How do we locate the position of the magnetic pole on the north end of the earth? We use a compass, the north pole of which is, by definition, the end of the needle that points north. Therefore, *the pole at the north end of the earth must actually be the south pole of a magnet, because it attracts the*

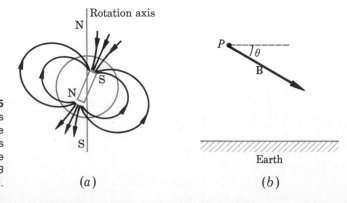

FIGURE 18.15
The earth's magnetic field is shown in (a). It can be specified at point P as shown in (b). The dip angle θ is the angle between B and the horizontal.

(a) (b)

north pole of the compass needle. This is, of course, purely the result of the fact that we choose to call the end of a magnet that is attracted to the north a north pole. Although this, the usual definition of north and south poles, causes some confusion, it should cause no difficulty for the careful student.

As seen from Fig. 18.15*a*, the earth's magnetic field is parallel to the surface of the earth only near the equator. Near the poles, the field is almost perpendicular to the earth. The quantities needed to specify the field of the earth at a given place are the direction and magnitude of the field. Usually the values given are the magnitude of B and the angle of dip. The dip angle is illustrated as the angle θ in Fig. 18.15*b*. The value of B is about 10^{-4} T, or 1 G, on the earth's surface.

18.9 Lines of Flux and Flux Density

In Fig. 18.16*a*, we have drawn the magnetic field between the poles of a U-shaped magnet such as that illustrated in part *b*. You should justify the way in which the field is drawn by recalling that a compass needle aligns itself parallel to the field lines and that the lines emanate from the north pole. As in the case of electric fields, the density of the lines is a measure of the field strength. We agree, for computational purposes at least, to draw the lines in the following way.

If the magnetic field has a magnitude B, we draw a number of lines equal to B through a unit area perpendicular to the field. Hence, in Fig. 18.16*a*, if the pole pieces had an area A, we would draw BA lines coming out of the upper pole and going into the lower pole (provided that we neglect the fringing effect). *The total number of lines going through an area is called the* **flux** *through the area and is represented by the Greek letter phi* (ϕ). We have, therefore,

$$\phi = B_\perp A \qquad (18.4) \qquad \textbf{Definition}$$

where the notation $B_\perp$ is used to remind us that we must employ the component of B perpendicular to the area.

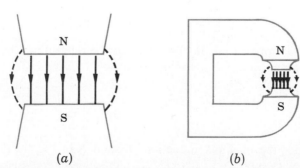

(a)   *(b)*

FIGURE 18.16
The magnetic field between the poles of the magnet shown in (*b*) is illustrated in (*a*). Why does the field fringe at the edge of the magnet?

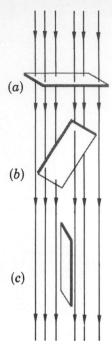

(a)

(b)

(c)

FIGURE 18.17
A maximum amount of flux goes through the area when the lines are positioned as shown in (a). No flux goes through it when the lines are oriented as shown in (c).

The reason for this insistence on perpendicularity in the definition of flux is easily seen from Fig. 18.17. In part *a*, the number of lines going through the area is just $\phi = BA$. However, in part *c*, all the lines skim past the area, and none go through it, so $\phi = 0$. The area is parallel to the lines. The intermediate case is most easily treated by considering the actual field to be composed of two component fields, one perpendicular to the surface and the other parallel to the surface. The parallel component contributes nothing to the flux through the area, as we saw in Fig. 18.17c. So the only flux through the area is the result of the perpendicular component of *B*, and it contributes a flux of $B_\perp A$, as stated in Eq. (18.4).

The SI unit for flux is the weber (Wb). It follows from Eq. (18.4) that the unit for *B* is that of ϕ/A, namely, webers per square meter. This unit is the same as the tesla, as stated earlier. *Because B is the flux per unit area, it is frequently called the* **flux density.**

18.10 Ampère's Law and the Computation of *B*

Although we have pointed out that a magnetic field is generated by a current or flow of charge, we have not yet presented a method for computing *B* from a knowledge of the current. In general, the computation of *B* outside a current loop or a more complicated system is a problem of great mathematical difficulty. However, in certain special cases the computation is rather simple, as we now show.

The right-hand rule has already provided us with a method for obtaining the qualitative features of the magnetic field outside a current-carrying wire. For example, we know that the magnetic field circles a long, straight wire in the manner shown in Fig. 18.18. Moreover, our intuition tells us that *B* should become smaller in magnitude as we move farther away from the wire. In order to obtain a quantitative value for *B*, however, we must make use of a relation first discovered by André Marie Ampère (1775–1836). It may be visualized in the following way.

It was known to Ampère through various experiments that the magnitude of the magnetic field strength *B* at a distance *r* from a long wire carrying a current *I* varied inversely as *r* and directly with *I*. In symbols,

$$B \propto \frac{I}{r}$$

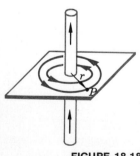

FIGURE 18.18
The magnetic field circles a long, straight wire which is carrying a current.

The proportionality constant can be determined by experiment and is denoted as $\mu_0/(2\pi)$, so that we have

$$B = \frac{\mu_0 I}{2\pi\, r}$$

for the magnetic induction at point *P* in Fig. 18.18.

The quantity μ_0 is called the **permeability of free space,** *and its exact value is* $4\pi \times 10^{-7}$ Wb/(A)(m), as we see in Sec. 21.5.

Looking at Fig. 18.18, we notice that the magnitude of B should be constant on a circle symmetric about the wire. This follows from the fact that there is no preferred point on such a circle: Since the wire is circular and is at the center of the circle, each point on the circle is indistinguishable from all other points on the circle. As a result, B should be the same at all points on the circle. If we now multiply B at radius r by the length of the circular path having that same radius (that is, $2\pi r$), we obtain, after using the experimental relation for B given above,

$$(B)(2\pi r) = \mu_0 I$$

Let us state this result in a somewhat different way. We chose a closed path (namely, a circle) that encircled a current I. The product of the length of this closed path and the magnitude of B on the path gave $\mu_0 I$ as the result. We are now tempted to ask whether this result cannot be generalized to a rectangular path instead of a circle. Or perhaps a simple answer might result if we used even such a complicated path as a many-sided polygon that circles a wire. Ampère was able to show that this was indeed the case. He showed that no matter how complicated the closed path which encircles a wire carrying current I, a simple result could be obtained. One need only break the path up into small-length vectors such that the magnetic-field-strength vector **B** was essentially constant on each given length. Then, if the product of each length times the component of **B** parallel to the length was taken, the sum of all such products was equal to $\mu_0 I$.

Ampère's discovery is embodied in what is now known as **Ampère's circuital law.** Before writing the law in mathematical form, let us apply it to the complicated path shown in Fig. 18.19. We have already split the path into a large number of small lengths $l_1, l_2, \ldots, l_{100}$, where we assume the number of lengths to be 100 for the purpose of discussion. Actually, we should probably use much smaller lengths so that **B** could not vary appreci-

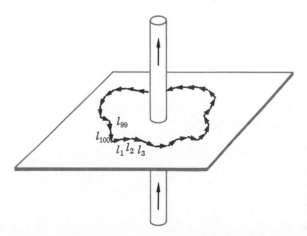

FIGURE 18.19
The closed path or loop formed by the vectors shown is used in connection with Ampère's law.

491

ably on the length. We must now take the length l_1 and multiply it by the component of **B** parallel to the length at that particular position. We denote this product as $(lB_{\parallel})_1$. Similarly at the second segment, the product is $(lB_{\parallel})_2$, and so on. Ampère's circuital law now states that

$$(lB_{\parallel})_1 + (lB_{\parallel})_2 + (lB_{\parallel})_3 + (lB_{\parallel})_4 + \cdots + (lB_{\parallel})_{100} = \mu_0 I$$

where $+ \cdots +$ represents all the other terms of the sum.

We may write the above equation in symbolic form in the following way:

Ampère's Circuital Law

$$\Sigma(lB_{\parallel})_n = \mu_0 I \qquad (18.5)$$

where the symbol $\Sigma(lB_{\parallel})_n$ means to take the sum of all the various $lB_{\parallel}$ products around the closed path. Equation (18.5) is the mathematical statement of Ampère's circuital law. It must be remembered that the elements of length l must be taken small enough for B to be sensibly constant on each element of length.

As an example of the use of this law, let us compute the flux density B outside a long, straight wire. Suppose that we wish to find B at point P in Fig. 18.20. We know from the right-hand rule that B will circle the wire in the direction of the arrows. Furthermore, the magnitude of B everywhere on the circle shown should be constant, since all points on the circle are symmetrical with respect to the wire. For the purpose of using Ampère's law, let us consider the circle of Fig. 18.20 to be composed of a large number, say 1000, of short segments or vectors extending around the circle. Writing down Eq. (18.5) gives

$$(lB_{\parallel})_1 + (lB_{\parallel})_2 + (lB_{\parallel})_3 + \cdots + (lB_{\parallel})_{1000} = \mu_0 I$$

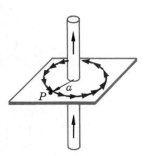

FIGURE 18.20
The vectors form a closed loop concentric to the wire, and B is constant on this path.

where again $+ \cdots +$ represents all the other terms of the sum not shown.

But we have already seen that B circles the wire in the same way that the little vectors do. In addition, B is constant at each place on the loop. Hence, all the $B_{\parallel}$ are equal, and their value is the same as the total B at point P, since B circles parallel to the loop. As a result, the above equation becomes, after we factor out $B_{\parallel}$ or B,

$$B(l_1 + l_2 + l_3 + \cdots + l_{1000}) = \mu_0 I$$

Now the sum of the lengths $l_1, l_2, \ldots$ is just the distance around the circle, $2\pi a$. Therefore we find

$$(B)(2\pi a) = \mu_0 I$$

$$B = \frac{\mu_0 I}{2\pi a} \qquad \text{for a long, straight wire} \qquad (18.6)$$

which is the result found by experiment, as mentioned earlier.

Illustration 18.3 Find B at a distance of 5 cm from a straight wire carrying a current of 20 A.

Reasoning Using Eq. (18.6), we have

$$B = \frac{\mu_0 I}{2\pi a} = \frac{(4\pi \times 10^{-7})(20)}{(2\pi)(0.05)}$$
$$= 0.80 \times 10^{-4} \text{ T} = 0.80 \text{ G}$$

Place the units in the above equation, and verify the units of the answer. The field at this distance from the wire is about as strong as the earth's field. With this in mind, can you think of any practical problem Oersted might have had in obtaining the form of the field about a wire?

18.11 Magnetic Field of a Circular Loop

Suppose that a long wire carrying a current I is bent into the form of a circle such as that shown in Fig. 18.21a. The field close to the wire must circle it, just as it did for the straight wire. However, the pattern at some distance from the wires is considerably distorted. This is shown more clearly in Fig. 18.21b, where the cross section of the loop is shown. The symbol · indicates a wire carrying a current out of the paper, while $\times$ represents a current going into the page. (Recall that the symbols arise from the pointed tip and the feathered tail of an arrow.)

It turns out that the computation of B for this case is extremely complicated except for a point on the axis of the loop. We cannot easily use Ampère's circuital law, because there is no obvious path in Fig. 18.21 on which B is constant or known. Other methods must be used in this case. They involve calculus, and so we treat this example in only a qualitative way.

If we had a number of loops of wire wound tightly together, the picture of the field would appear exactly as in Fig. 18.21. However, the loop shown would now be interpreted as being composed of several turns or loops of wire. If the current in the wire were I for both the single loop and the coil with N loops, then the field B for the coil would be N times as large as for the single loop.

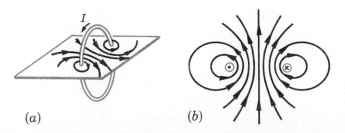

(a) (b)

FIGURE 18.21
Two views of the magnetic field about a current-carrying loop.

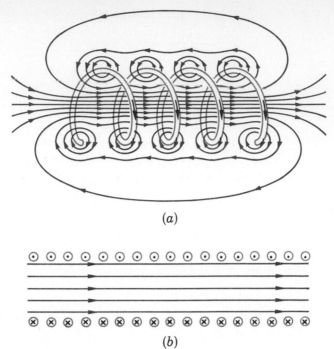

(a)

(b)

FIGURE 18.22
The fields are essentially
uniform inside the
solenoids.

(b)

18.12 Solenoid

A solenoid is a long, cylindrical coil of wire. To show the magnetic field within this device, a solenoid looser than ordinary is illustrated in Fig. 18.22a. The cross section of a more common, tightly wound solenoid is shown in Fig. 18.22b. Notice how the field within the solenoid is directed straight through it. In addition, the field is much stronger inside the tightly wound, long solenoid than it is on the outside. For many purposes, a long solenoid may be considered to have a field parallel to its axis within the solenoid. The flux lines outside the solenoid have fanned out so far in space that they are very widely separated, indicating that the field is negligibly weak outside the solenoid.

It will be possible for us to compute the value of B within a long solenoid by using Ampère's law. Consider the path, or loop, *abcda* in Fig. 18.23. In this case Ampère's law becomes

$$(lB_\parallel)_1 + (lB_\parallel)_2 + (lB_\parallel)_3 + (lB_\parallel)_4 = (\text{current encircled})(\mu_0)$$

Inside the solenoid B is parallel to l_1, and so

$$(lB_\parallel)_1 = l_1 B$$

We have already seen that B outside the solenoid is very small provided the solenoid is very long, and so we approximate it as zero. Thus

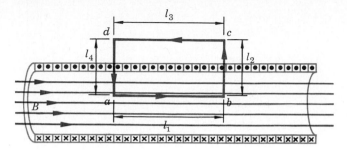

$$(lB_\parallel)_3 = 0$$

Moreover, B is perpendicular to l_2 and l_4 inside the solenoid, so

$$(lB_\parallel)_2 = (lB_\parallel)_4 = 0$$

Therefore, Ampère's law in this case reduces to

$$l_1 B = (\mu_0)(\text{current encircled})$$

Suppose there are n loops of wire on a unit length of the solenoid. Then on a length l_1 there will be nl_1 loops. Hence the number of loops encircled by our path is nl_1. (In Fig. 18.23 this number is actually 13.) Each loop carries a current I out of the area described by our path, and so the current encircled is just $nl_1 I$. Therefore,

$$l_1 B = \mu_0 nl_1 I$$

$$B = \mu_0 nI \qquad \text{inside a long solenoid} \qquad (18.7)$$

Notice that B is independent of the position within the solenoid, and so the field within a long solenoid is not only parallel to the axis but also uniform. This is true, of course, only for points far from the ends.

18.13 Toroid

As a final example, we consider a toroid. This device is just a solenoid bent into a circle so that it has no ends. A schematic diagram of a toroid is shown in Fig. 18.24. Of course, the toroid should have many more turns of wire on it than shown. *The magnetic field is confined almost entirely to the inside of the coil and circles the toroid as indicated.*

To compute B in this case, we take our closed loop to be the black circle shown. By symmetry, B will be constant everywhere on the circle, and it will be tangential to the circle. After we replace the circle by a series of vectors, remembering that B is constant, we can use Ampère's law to give

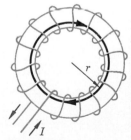

$$(lB_\parallel)_1 + (lB_\parallel)_2 + \cdots + (lB_\parallel)_n = (\mu_0)(\text{current encircled by our path})$$

Since B is constant,

$$B(l_1 + l_2 + \cdots + l_n) = (\mu_0)(\text{current encircled})$$

or
$$(B)(2\pi r) = (\mu_0)(\text{current encircled})$$

We see that the current encircled is NI, where N is the total number of turns of wire on the toroid. Therefore,

$$2\pi rB = \mu_0 NI$$

or
$$B = \frac{\mu_0 NI}{2\pi r} \qquad \text{toroid} \qquad (18.8)$$

Notice that in the case of the toroid, B is not uniform within the coil, since it varies with r. However, if the diameter of each loop is small compared with the diameter of the ring, r will not vary much from place to place within the toroid and so B will be nearly constant.

Illustration 18.4 A long solenoid has 50 turns of wire on it for each centimeter of its length. The current in the wire is 0.50 A. The diameter of the solenoid is 2.0 cm. Find B in the solenoid and the flux through the solenoid.

Reasoning To find B, we make use of Eq. (18.7):

$$B = \mu_0 nI = (4\pi \times 10^{-7})(50 \times 100)(0.50)$$
$$= 31.4 \times 10^{-4}\,\text{T} = 31.4\,\text{G}$$

where you should supply the units to the equations. The flux ϕ is the number of lines going through the solenoid. To find ϕ, we note that the number of lines going through a cross-sectional area A of the solenoid is $B_\perp A$. But B is perpendicular to the solenoid's cross-sectional area, and so

$$\phi = (31.4 \times 10^{-4}\,\text{Wb/m}^2)(\pi \times 10^{-4}\,\text{m}^2) \approx 10^{-6}\,\text{Wb}$$

where use has been made of the fact that $1\,\text{T} = 1\,\text{Wb/m}^2$.

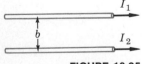

FIGURE 18.25
The magnetic field from the upper wire acts on I_2 so as to push the lower wire toward the upper wire.

Illustration 18.5 Consider the two parallel, long, straight wires shown in Fig. 18.25. Find the force on unit length of the lower one resulting from the current in the upper one.

Reasoning The current I_1 causes a magnetic field into the page at the position of I_2. Its magnitude is

$$B = \frac{\mu_0 I_1}{2\pi r} = \frac{\mu_0 I_1}{2\pi b}$$

But since the force on a wire is $B_\perp IL$, and since $I = I_2$ in this case, we have

$$F = \frac{\mu_0 I_1}{2\pi b} I_2 L$$

from which

$$\frac{F}{L} = \frac{\mu_0 I_1 I_2}{2\pi b}$$

Can you show that the force is in such a direction as to push the lower wire toward the upper one? Notice that when both wires carry the same current, $I_1 = I_2 = I$ can be equated to $[2\pi b F/(L\mu_0)]^{1/2}$. This provides a way of measuring currents in terms of force and length measurements. We see in Sec. 21.5 that this is the basic experiment used to define the ampere.

18.14 Ampère's Theory of Magnets

One cannot help noticing the similarity between the magnetic field of a bar magnet and that of a solenoid, as shown in Fig. 18.26a. Ampère noticed this too. In fact, he noticed that even the field of a single loop was exactly like what one would expect for a short magnet as shown in Fig. 18.26b. He theorized that all magnetic effects, even in bar magnets, are the result of circulating currents. We believe that his ideas are basically correct, although

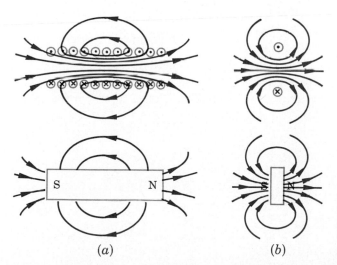

FIGURE 18.26
The magnetic field about a solenoid is very similar to that of a bar magnet having about the same size.

(a)   (b)

we still do not understand completely why some of the basic particles, the electron, for example, act as small magnets.

We see at once that if Ampère's theory is accepted, it is completely impossible to obtain a north pole by itself. The north pole is merely one side of a current loop (or of a group of current loops), and the other side is always present as a south pole (see Fig. 18.26). For this reason physicists today speak of the *pole fiction*. That is, there is no particle which can be considered to be a magnetic pole. Unlike free positive and negative charges, north and south poles can never be separated from each other completely. Magnetic monopoles do not exist.

Actually, of course, Ampère had no idea about the structure of atoms and molecules, even though he believed matter to be made up of atoms, and so he could not say how the current loops were present in a piece of magnetized iron. We know today that *each atom behaves as a little magnet. Atoms of some elements such as iron, cobalt, and nickel* (as well as gadolinium and dysprosium), *cooperate with each other in such a way as to act as quite strong little magnets.* We call materials made of atoms such as these **ferromagnetic materials.** Atoms of other materials act as much weaker magnets.

It is often convenient to think of the magnetic nature of the atom as being the result of motion of its electrons. If they moved on a circular path, they would constitute a circular current loop. In addition, the electron itself is known to act as a tiny bar magnet. We can picture the electron as a charge spinning about its axis and thereby constituting a current loop, and this effect is called the **spin** of the electron. We should point out, however, that most physicists frown on such detailed pictures of the atom, since there is reason to believe that these pictures cannot be proved experimentally. Moreover, this way of picturing the spin is not completely satisfactory in certain respects, and it cannot be reconciled with the quantitative results of experiment.

We know that if we place a group of tiny magnets close to one another, they will try to arrange themselves so that each south pole is close to a north pole. This is the result of the attractions of unlike poles and the repulsions of like poles. The lowest PE of the system is reached when the magnets are arranged somewhat as shown in Fig. 18.27a. Notice that the magnets arranged in this way are equivalent to a large magnet.

However, if the magnets are strongly agitated by shaking the paper or board on which they rest, they will break loose from one another and end up as shown in Fig. 18.27b. Notice now that the individual magnets tend to cancel one another. No longer do they act as a strong bar magnet.

FIGURE 18.27
(a) A magnetized piece of iron; (b) the disordered, unmagnetized iron shown schematically; (c) a more realistic picture of the domains.

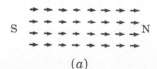

(a)

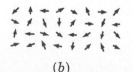

(b)

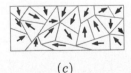

(c)

An analogous situation exists with atoms in a solid. Thermal vibrations tend to agitate the system and prevent the atoms from ordering themselves as in Fig. 18.27a. Only certain atomic magnets, iron, and the other ferromagnetic materials can preserve the pattern shown in part a at ordinary temperatures. Even these atoms, when heated enough, acquire enough thermal energy to break loose and disorient as in part b. The temperature at which this happens is quite definite for any type of atom and is called the **Curie temperature.** We should note, however, that the analogy is not complete. There are, in addition to the magnetic forces between the ferromagnetic atoms, other forces that are much more complex. These forces can be understood only in terms of quantum mechanics, and so we are unable to discuss them further here. They actually play a major role in the aligning of the atomic magnets.

Most materials have their atomic magnets, if any, randomly oriented as in Fig. 18.27b. The ferromagnetic materials, however, consist of little regions in which the atoms are all aligned as in Fig. 18.27a. Each of these oriented regions is called a **domain.** In an ordinary piece of iron pipe, each domain may contain as many as 10^{16} atoms and consist of a region a small fraction of a millimeter in linear dimension. However, the domains in an unmagnetized piece of iron will be randomly oriented, and so the situation will be much as in Fig. 18.27b, where now the arrows represent domains instead of atoms. A more realistic picture of the domains is given in part c. For a bar of iron to be magnetized, the domains within it must be lined up within the material. This can be done in the following way.

Suppose you start with the unmagnetized bar of iron shown in Fig. 18.28a. A solenoid with a current in it, such as shown in part b of the figure, possesses a rather weak magnetic field within itself, as we saw in Illustration 18.4. If now you place an iron core in the solenoid, the magnetic field B of the solenoid will exert forces on the domains. Those domains that are oriented along the field grow, while poorly oriented domains decrease in size. In effect, the domains orient under the action of the field, as shown in part c of Fig. 18.28. The iron is now a bar magnet, with north and south poles. In "soft" iron, the domains are easily oriented, but in "hard" iron the field must be made quite strong or the domains must be agitated by heat or mechanical means to allow them to turn into the direction of the field. (The designations *hard* and *soft* refer only to the magnetic properties, not to the physical hardness.) It is possible, however, to align the domains nearly perfectly and form a large bar magnet as shown in part c.

Once the domains have been aligned, the total flux density B consists of two parts. The original small field of the solenoid is still present. However, the field produced by the bar magnet is usually hundreds of times larger than the field of the solenoid alone. *The combination of a solenoid and a piece of very soft iron is called an* **electromagnet.**

If now the current in the solenoid is turned off, the domains in a soft bar of iron will return nearly to the original state shown in part a. Thermal motion causes them to disarrange. This is a desirable situation in the case of an electromagnet, because it makes it possible to turn it on or shut if off at will. A hard piece of iron, on the other hand, will retain most of its align-

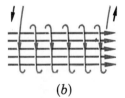

(a)

(b)

(c)

(d)

FIGURE 18.28
An unmagnetized piece of iron can be magnetized by using the field of a solenoid to line up the domains.

ment and will be a permanent bar magnet as shown in part *d* of Fig. 18.28.

Let us summarize what we have found about the magnetic properties of materials. Most materials, when placed in a magnetic field, change the field scarcely at all. But a very few substances, chiefly iron and its alloys, increase the magnitude of the magnetic field in which they are placed; often the field is strengthened by factors of hundreds. As we shall see, the ability of iron to greatly augment a magnetic field is of prime importance in many applications of magnetism.

18.15 Torque on a Current Loop

Many practical devices including motors and moving-coil meters make use of the torque that a current loop experiences when the loop is placed in a magnetic field. To see how such a torque originates, refer to Fig. 18.29*a*, where we show a current-carrying coil in a magnetic field. As indicated, the coil is mounted on an axle and can rotate. Using the right-hand rule, we find the forces on the various sides to be those illustrated. Notice that only the two forces F_h cause a torque about the axis of rotation. Even these two forces will cause no torque when the plane of the coil is perpendicular to the field of the magnet. Maximum torque occurs when the lines of B skim past the surface of the coil, i.e., when the field lines lie in the plane of the coil, for then the lever arm for F_h is maximum.

To obtain a quantitative expression for the torque on the coil, we note that each of the two forces F_h gives a torque

$$(F_h)(\text{lever arm})$$

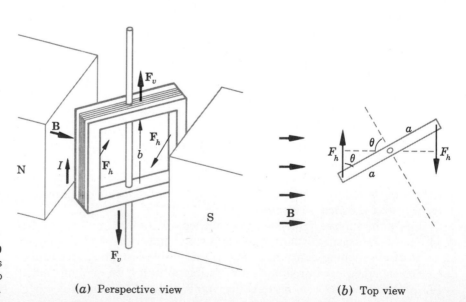

FIGURE 18.29
The magnetic field causes the current carrying coil to experience a torque.

(*a*) Perspective view   (*b*) Top view

From Fig. 18.29b we see the lever arm to be $a \sin \theta$, where θ is the angle so labeled. Therefore the torque on the coil is

$$\text{Torque} = 2F_h a \sin \theta$$

But F_h is simply the force on the vertical side of the coil. If the vertical side has a length b, and if the current is I, each vertical wire will contribute a force BIb to F_h. But there are N loops on the coil, so $F_h = NBIb$ and the torque becomes

$$\text{Torque} = (2ab)(NI)(B \sin \theta)$$

Notice that $2ab$ is simply the area of the coil. We can therefore write

$$\text{Torque} = (\text{area})(NI)(B \sin \theta) \qquad (18.9)$$

Although we have derived Eq. (18.9) for a very specially shaped coil, it turns out that the equation is true for all flat coils. Since NI is the current flowing around the coil, we see that the important features of the coil (aside from its orientation) are its area and the current in it. In view of this, it is customary to define a quantity called the **magnetic moment** of a current loop:

$$\mu = \text{magnetic moment} = (\text{area})(I)$$

There is a definite advantage to thinking of a current loop as a bar magnet characterized by its magnetic moment, as we now shall see.

We have already pointed out that a current loop has a magnetic field similar to that of a bar magnet. This is shown once again in Fig. 18.30. Notice that the coil acts like a short, fat bar magnet with north and south poles as indicated. This is also true for the single current loop in Fig. 18.30c. Moreover, when placed in a magnetic field, the coil and loop will experience a torque in the same direction as the torque on a bar magnet.

FIGURE 18.30
The coil in (a) acts as the bar magnet in (b). Notice how the direction of the magnetic moment μ is assigned in (c).

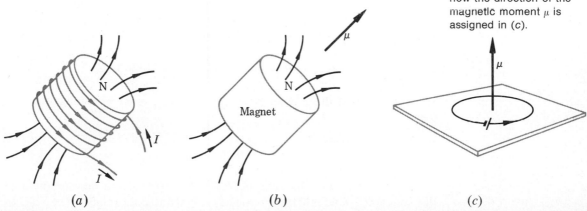

(a)   (b)   (c)

For example, if **B** is directed from left to right in Fig. 18.30, all three devices shown there will experience a clockwise torque.

We can obtain maximum usefulness for the concept of magnetic moment if we assign it direction. The direction assigned to μ is shown in Fig. 18.30*b* and *c*. Notice that μ is directed along the axis of the magnet, loop, or coil in such a way that it follows the central field line through the coil. As a result, the *magnetic-moment vector μ points out of the north pole of the equivalent magnet*. This has the following important consequence:

> When a current loop is placed in a magnetic field, it rotates so as to align its magnetic-moment vector with the magnetic-field vector.

You can appreciate why this is true by recalling that a compass needle is simply a bar magnet and that the field direction is defined to be the direction along which the needle aligns. *We shall find it convenient from time to time to think of a current loop as a magnet with magnetic moment μ.*

Illustration 18.6 Niels Bohr in 1913 postulated the following reasonably successful picture for the hydrogen atom. At its center is a proton; the proton charge is $+1.6 \times 10^{-19}$ C, and its mass is 1.67×10^{-27} kg. The diameter of the proton is of the order of 10^{-15} m. Circling the proton at a radius of 0.53×10^{-10} m is an electron whose charge is -1.6×10^{-19} C and whose mass is only $\frac{1}{1840}$ that of the proton. Bohr concluded that the electron circled the nucleus 6.6×10^{15} times each second. This motion of the charged electron around a circle is equivalent to a current in a loop of wire. Find the magnetic moment of the hydrogen atom resulting from the orbital motion of its electron.

Reasoning By definition, the magnetic moment μ is IA, where A is the area of the loop, namely,

$$A = \pi r^2 = (\pi)(0.53 \times 10^{-10})^2 \text{ m}^2 = 0.88 \times 10^{-20} \text{ m}^2$$

To find I, we note that the current in a loop is equal to the charge passing a given point in a second. For the present case, the electron circles the atom 6.6×10^{15} times each second, carrying a charge of 1.6×10^{-19} C past a given point that many times each second. Hence

$$I = (1.6 \times 10^{-19} \text{ C})(6.6 \times 10^{15} \text{ s}^{-1}) = 1.05 \times 10^{-3} \text{ A}$$

We therefore find that the atom acts as a small magnet with magnetic moment

$$\mu = IA = (1.05 \times 10^{-3} \text{ A})(0.88 \times 10^{-20} \text{ m}^2)$$
$$= 9.3 \times 10^{-24} \text{ A} \cdot \text{m}^2$$

When techniques became available to carry out the measurement of this magnetic moment, the predicted value was confirmed. However, in addition, the electron *itself* was found to act as a small magnet, and this was pictured as being the result of a spinning motion of the charged electron about an axis through its own center, as we mentioned previously.

18.16 Moving-Coil Meters

As we saw in the last section, a current-carrying coil in a magnetic field experiences a torque. Because the torque is proportional to the current in the coil, this effect can be used to measure current. In fact, most common electric meters are based on the movement of a coil in a magnetic field.

To see how this effect is utilized, let us refer to Fig. 18.31, where we see a current-carrying coil between the poles of a magnet. When current flows through the coil in the direction shown, the coil acts as a magnet with its north pole on the back side of the coil. (Check this and see.) We indicate this fact by the magnetic-moment vector μ shown.

Because the magnetic-moment vector tries to align with the field, the coil rotates so as to point μ toward the south pole of the permanent magnet. However, a coiled spring attached to the coil provides a torque to stop this rotation. As a net result, the coil rotates an amount proportional to the current in the coil. Therefore, the rotation of the coil, indicated by the pointer needle attached to it, can be used as a measure of the current that flows through the coil.

We call the device sketched in Fig. 18.31 a *meter movement*. In practice, the coil often has an iron core to intensify the field and the torque. Many galvanometers are simply a movement such as this placed in an appropriate case. For that reason, it is common to use the terms *meter movement* and *galvanometer* interchangeably.

The sensitivity of a meter movement, i.e., how large a deflection results from a given amount of current, depends on several factors. Of course, the stiffness of the restoring spring is of primary importance. If the instrument is to be rugged and portable, the spring must not be too delicate. The

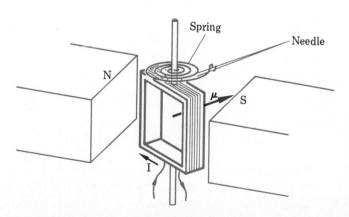

FIGURE 18.31
The current in the galvanometer coil causes it to rotate in the magnetic field provided by the permanent magnet.

sensitivity also depends on the number of turns of wire on the coil. If the number of turns is doubled, the torque on the coil is doubled as well.

A very sensitive galvanometer will give full-scale deflection if a current of only a fraction of a microampere (10^{-6} A, abbreviated μA) goes through it. Such a highly sensitive galvanometer must have a large number of turns of wire on its coil, and so its resistance might easily be 100 Ω. Even so, a voltage of 10^{-4} V across its terminals would cause a current of 10^{-6} A to flow through it. Hence, it could be used as a very sensitive voltmeter as well as an ammeter or galvanometer. The table-model galvanometers used widely for student work give a full-scale deflection for a current of about 1 mA (10^{-3} A). They have a resistance of about 20 Ω.

18.17 Ammeters

The core of an ammeter is a galvanometer movement. If it is to be a very sensitive ammeter, the ammeter may be a galvanometer without any modification. However, most movements will give full-scale deflection for a few milliamperes or less, and so they cannot be used directly to measure larger currents. Suppose, as a concrete example, that we wish to make an ammeter that will deflect full scale for a current of 2 A. The movement to be used has a resistance of 20 Ω and deflects full scale for a current of 3 mA. (This would be a 60-mV movement; that is, 60×10^{-3} V across its terminals would cause a current of 3 mA to flow through it, deflecting it full scale.)

In order to make the desired ammeter from this movement, we must find some way of allowing 2 A to flow through the meter while only 0.003 A flows through the movement. This can be done with the meter design shown in Fig. 18.32. The actual movement is shown as the circular device with resistance R_m.

When a current of 2 A flows into this meter through terminal *A,* we want only 0.003 A to flow through the movement, which has a resistance R_m. This means that *we must place a small resistance in parallel with the movement* so that 1.997 A will go through it. *We call this small resistor a* **shunt** resistance, and it is indicated as R_s in Fig. 18.32. To find how large R_s should be, we proceed as follows.

The movement will deflect full scale when there is a potential of 60×10^{-3} V across its terminals from *A* to *B*. But under these same conditions, the shunt resistor R_s must carry a current of 1.997 A. Using Ohm's law, we find

$$V = IR$$
$$60 \times 10^{-3} = 1.997 \, R_s$$
$$R_s \approx 0.030 \, \Omega$$

or

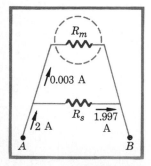

FIGURE 18.32
Only a small portion of the current goes through the movement of an ammeter. Most of it goes through the shunt resistor R_s.

Notice that *the shunt resistor is extremely small.* Often it is just a piece of copper wire. Moreover, the resistance of the ammeter will be less than 0.030 Ω, since R_s is in parallel with R_m. This is as it should be, since we do not wish the ammeter to disturb the circuit when it is connected in series to measure the current flowing. (What would happen if this ammeter were accidentally connected across a potential difference of 1 V?)

Many ammeters have more than one range. These meters contain several alternate shunts. The sensitivity of the meter is determined by which shunt is connected across the meter terminals.

18.18 Voltmeters

We can construct a voltmeter from the movement we used in the last section to make an ammeter. The movement deflected full scale for a current of 0.003 A and had a resistance of 20 Ω. Hence, the potential difference across its terminals at full-scale deflection is (0.003)(20), or 60 mV. Let us construct a 90-V voltmeter from this movement.

An appropriate circuit for this purpose is shown in Fig. 18.33. When the potential across the terminals is 90 V, we wish a current of 0.003 A to flow through the meter. To find the resistance R_x which must be placed in series with the movement, we apply Ohm's law to the circuit of Fig. 18.33. We have

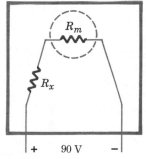

FIGURE 18.33
To make a voltmeter from a sensitive movement, we place a large resistor in series with the movement.

$$V = IR$$
$$90 = (0.003)(20 + R_x)$$

Solving for R_x yields

$$R_x = 29,980 \ \Omega$$

Clearly, this meter possesses a very high resistance. *Since we use a voltmeter by connecting it across a potential difference, we require it to have a high resistance* so that it will not disturb the circuit. The best voltmeters have a much higher resistance than the one we designed. Our meter is said to have a resistance of 29,980/90, or 333, Ω/V. This does *not* mean that the resistance varies depending on the scale reading. It is merely the ratio of the resistance of the meter to its maximum scale reading. A similar meter possessing a more sensitive movement would have a higher resistance. (Why?)

Summary

A compass needle is a tiny bar magnet. The end that points north is its north pole. Like poles of magnets repel, and unlike poles attract. Each magnet always has both a north and a south pole.

Magnetic fields exert forces on magnetic poles. A compass needle aligns itself tangentially to the magnetic field lines. These lines come out of north poles and enter into south poles of magnets.

A straight wire of length L that carries a current I in a magnetic field of strength **B** experiences a force given by $F = B_\perp IL$. In this expression, $B_\perp$ is the component of **B** perpendicular to the wire. The unit of B, the flux density (or magnetic induction), is the tesla (T), which is the same as webers per square meter. A plane is defined by the intersecting lines of B and L. The force is in a direction perpendicular to this plane and is given by a modification of the right-hand rule.

Moving charges also experience a force in a magnetic field. It is given by $F = qvB_{\perp}$, where $B_{\perp}$ is the component of $\mathbf{B}$ perpendicular to the velocity $\mathbf{v}$. The force direction is the same as on a current composed of the moving particles. A charge which moves with speed v perpendicular to a magnetic field $\mathbf{B}$ describes a circular path. The radius of the path is found by equating the magnetic force $qvB_{\perp}$ to the centripetal force.

We agree to draw B magnetic field lines through unit area perpendicular to the field. The flux ϕ through an area A is the number of lines which pass through it. It can be computed from $\phi = B_{\perp}A$.

Ampère's circuital law can be used to compute magnetic fields due to currents. The field outside a long, straight wire is $\mu_0 I/(2\pi r)$; the field in a long solenoid is $\mu_0 nI$.

Most materials do not change the magnetic field in which they are placed by much. The ferromagnetic materials, however, greatly increase magnetic fields in which they are placed and have regions of well-aligned atoms called domains. Each domain acts as a magnet. The domains can be aligned by placing the material in a strong magnetic field.

When a current-carrying coil is placed in a magnetic field $\mathbf{B}$, it experiences a torque. The magnitude of the torque on a flat coil of N loops and area A is $ANIB \sin \theta$, where θ is the angle between $\mathbf{B}$ and a perpendicular to the area. The magnetic moment of a current loop has a magnitude $\mu = IA$. It is a vector and has the direction of the loop's central magnetic field. A magnetic field tends to align the magnetic-moment vector along the field lines.

Ammeters and voltmeters make use of a meter movement, which consists of a coil in a magnetic field. The coil rotates in proportion to the current through it. Ammeters have a low-resistance shunt in parallel with the movement. Voltmeters have a high resistance in series with the movement.

Minimum Learning Goals

Upon completion of this chapter, you should be able to do the following:

1 Sketch the magnetic field in the vicinity of (a) various magnets, (b) a straight wire, (c) a loop of wire, (d) a solenoid, (e) a toroid.

2 Use a compass to determine the magnetic field lines in a given region.

3 Find the magnitude and direction of the force on a given straight-wire current in a specified magnetic field.

4 Use $F = B_{\perp}IL$ to find one of the quantities when the others are given.

5 Use $F = qvB_{\perp}$ to find one of the quantities when the others are given. Describe quantitatively the path followed by a particle of known q and v moving perpendicular to a given magnetic field.

6 Determine the flux through a given area when $\mathbf{B}$ in that region is known.

7 Compute the magnetic field for a long, straight wire and for a solenoid when sufficient data are given.

8 Given a list of common materials, select those which greatly alter the magnetic field into which they are placed.

9 Explain the meaning of a magnetic domain, and describe in terms of them what happens when a bar of ferromagnetic material is magnetized or demagnetized.

10 Explain why the Hall effect allows us to determine the sign of charge carriers.

11 State which way a current-carrying coil will turn when placed in a given position in a magnetic field. Compute the torque on the coil when sufficient data are given.

12 Point out where the effective north and south pole regions are for a current-carrying loop. Explain what is meant by the magnetic-moment vector for the current loop.

13 Explain the major features of a meter movement. Tell how it is used to make an ammeter or a voltmeter.

14 Compute the shunt resistance needed to make a given movement into an ammeter of stated range.

15 Compute the series resistance needed to make a given movement into a voltmeter of stated range.

Important Terms and Phrases

You should be able to define or explain each of the following:

Like poles repel, unlike poles attract
Magnetic field line
Right-hand rule
$F = B_\perp IL$; tesla, webers per square meter, gauss
Flux density, magnetic induction
Velocity selector
Solenoid
Toroid

Hall effect
Electromagnet
A south pole exists near the earth's north pole
Flux; $\phi = B_\perp A$
Ampère's circuital law
$B = \mu_0 I/(2\pi r)$; $B = \mu_0 nI$
Ampère's theory of magnets; poles come in pairs
Ferromagnetic materials; domains
Magnetic moment
Meter movement
Shunt resistor

Questions and Guesstimates

1 Sketch the magnetic field for two identical bar magnets placed as shown in Fig. P18.1a. Sketch the magnetic field for the situation shown in part *b* if the circular piece of metal is iron. Repeat if the circular piece is brass.

2 A magnet will attract an unmagnetized nail made of iron, and that nail will attract another. Explain why.

3 Suppose you have a strong bar magnet. How can you use it to determine which of a number of metal rods are magnetized?

4 A solenoid is viewed end-on in such a way that the solenoid current appears clockwise to the viewer. What is the direction of the field within the solenoid?

5 A straight wire viewed end-on carries current toward you. Describe its magnetic field.

6 A circular loop lies on a table. It carries a current in the clockwise direction as seen from above. At the center of the loop, a bar magnet sits vertically with its north pole on the table and its south pole straight above the north pole. Describe the forces on the loop caused by the magnet.

7 Two circular loops lie on a table and are concentric. The larger loop carries a current of 10 A counterclockwise, while the smaller carries a current of 5 A clockwise. Describe the forces on each loop.

8 As shown in Fig. P18.2, two high-voltage leads cause a beam of charged particles to shoot to the right through a partially evacuated tube. Their path is shown by a fluorescent screen placed along the length of the tube. When a horseshoe magnet is brought close, the beam deflects as shown. How could you determine the sign of the charge on the particles?

9 An electron is shot into a long solenoid at a small angle to the solenoid axis. Describe the motion of the electron.

10 A silver coin has a multitude of "free" valence electrons moving within it. When the coin is placed in a magnetic field, each moving charge experiences a force. But careful measurements show that the coin experiences no net force due to the constant field. Explain why.

11 Glib teachers sometimes say, "The earth's North Pole is a south pole, and vice versa." What does the teacher mean?

12 When a beam of electrons is shot into a certain region of space, the electrons travel in a straight line through the region. Can we conclude there is no electric field in the region? No magnetic field?

FIGURE P18.1

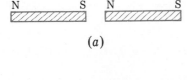

(a)

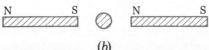

(b)

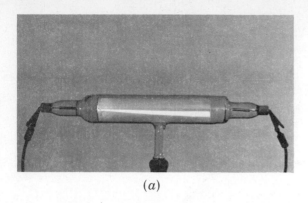

(a)

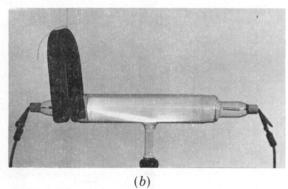

(b)

FIGURE P18.2

13 In a certain experiment, a beam of electrons is shot out along the positive x axis. It is found that the beam deflects toward positive y values in the xy plane. If this deflection is the result of a magnetic field, in what direction is the field? Repeat for an electric field.

14 It is found that a beam of charged particles is deflected as it passes through a certain region of space. By taking measurements on the beam, how could you determine which caused the deflection, a magnetic or an electric field?

15 A proton is shot from the coordinate origin out along the $+x$ axis. There is a uniform magnetic field in the $+y$ direction. (a) Describe the motion

of the proton, paying particular attention to the quadrants in which it travels. (b) Repeat for an electron. (c) Repeat if the proton velocity is such that $v_x = v_y$ and $v_z = 0$. (d) Repeat if $v_x = v_y = v_z \neq 0$.

16 Suppose you had no magnet or iron available when you landed on the moon. You could, however, do any other electrical experiment you wished. How would you determine the characteristics of the moon's magnetic field? It is assumed that, at the start, you do not know if an electric field exists on the moon.

17 We know that in a TV tube electrons are shot from one end of the tube to the other, where they strike the fluorescent screen. Suppose your little brother insists that his general science teacher says that protons are used, not electrons. How could you prove to him that he was wrong without dismantling the set?

18 Suppose you were given a material which was poorly conducting but still conducted enough to obtain a measurable current through it. How could you decide whether the current was being carried by positive or by negative charges or by both? Give as many ways as you can.

19 It is proposed to furnish the propulsion force to a spaceship in the following way. By use of a nuclear reactor or some other means, electricity is furnished. Large currents are sent through copper bars in the ship, and the forces exerted on these bars by the earth's magnetic field propel the ship. What objections do you see to such a plan?

20 Cosmic rays (i.e., charged particles coming to the earth from outer space) are unable to reach the surface of the earth unless they have very high energy. One reason is that they have to penetrate the earth's atmosphere. However, for particles coming toward the equator along a radius of the earth, magnetic effects are also important. Explain why, being careful to point out why particles can reach the poles of the earth without encountering this difficulty.

21 Give an order-of-magnitude estimate of the displacement of the electron beam on a TV screen as a result of the earth's magnetic field.

Problems

1 A power line parallel to the earth's surface carries a current of 20 A straight west. At that point, the earth's magnetic field is 0.80 G parallel to the earth's surface and directed straight north.

(a) Find the force due to the field on a 15-m length of wire. (b) What is its direction?

2 In a certain region, the magnetic field is 200 G in the $+x$ direction. Find the force on a 5.0-m-long wire that carries a 6.0-A current (a) in the $+x$ direction and (b) in the $+y$ direction.

3 A 2.0-cm length of wire carries a current of 5.0 A perpendicular to the magnetic field between the pole pieces of a magnet. The force on it is 0.080 N. What is the magnitude of the magnet's field?

4 A certain wire has a mass of 0.200 g per meter length. It is to carry a current perpendicular to a field of 0.40 T. How large must the current be if the magnetic force on the wire is to support the wire's weight?

5 In Miami, Florida, the earth's magnetic field is approximately straight north, but is directed at an angle of 57° below the horizontal. (The dip angle is 57°.) Its magnitude is about 0.51 G. Find (a) the magnitude and (b) the direction of the force it causes on a 15-m length of wire carrying a current of 20 A straight upward.

6 A particular power line is carrying a steady current of 50 A at 1000 V parallel to the earth and toward the north. (a) If the earth's magnetic field is 0.60 G at that point and the angle of dip is 53°, find the force on a 3-m length of wire. (b) What is the direction of the force?

7 How fast must a proton be moving if it is to follow a circular path with radius 2.0 cm in a magnetic field of 0.70 G? (You are given that $m_{proton} = 1.67 \times 10^{-27}$ kg.)

8 An electron having a speed of 10^7 cm/s enters a uniform magnetic field ($B = 0.020$ T) in a direction perpendicular to the field. Describe quantitatively the path taken by the electron.

9* A proton falls from rest through a potential difference of 3.34×10^5 V. It then enters a region perpendicular to a magnetic field, $B = 0.50$ T. How large is the radius of the circle in which the proton travels?

10* A proton is accelerated through an unknown potential difference, enters a region perpendicular to a magnetic field of 2.00×10^{-2} T, and describes a circle having a radius of 30.0 cm. What is the energy of the proton, expressed in electronvolts?

11* Describe quantitatively the path of an electron traveling at 10^7 m/s toward the west parallel to the earth, where $B = 0.5$ G and the angle of dip is 53°.

12* (a) A velocity selector having perpendicular electric and magnetic fields is to let protons with speed 3.0×10^6 m/s pass straight through. If

$B = 2.00$ G, what must be the magnitude of **E**? (b) Repeat for an electron beam.

13 A long wire carries a current of 20 A straight north. What are the magnitude and direction of the field due to the wire at a point 50 cm below the wire?

14 A long, straight wire carries a current of 12 A straight west along the floor in a house. (a) How large is **B** due to the wire at 2.0 m above the wire? (b) What is its direction?

15 Two long, parallel wires each carry a current of 5.0 A. They are 30 cm apart. Find the field due to them midway between them if the currents are (a) parallel and (b) antiparallel.

16 Find B midway between two long, straight, parallel wires (separation = 20 cm) if they carry currents of 10 and 20 A in (a) opposite directions and (b) the same direction.

17 A long solenoid has 5000 loops of wire on its 80-cm length. The solenoid diameter is 2.0 cm. Find B inside the solenoid when a current of 300 mA flows in it.

18 How large is B inside a long solenoid having 1000 turns on its 50-cm length if it carries a current of 3 A?

19 The radius of a narrow toroid is 10 cm, and it has 200 turns of wire on it. How large a current must flow in it if B is to be 0.50 T inside it?

20 If the toroid of Prob. 19 were wound on an iron core that increases B by a factor of 300, how large would the current have to be to produce $B = 0.50$ T?

21 (a) The magnetic field in a certain region is 300 G in the $+x$ direction. A 20 cm × 20 cm square of cardboard is held with its face perpendicular to the x axis. How much magnetic flux goes through the square? (b) Repeat if the x axis is parallel to a diagonal of the square.

22 A particular air-core solenoid is 2.0 m long and has 10,000 turns of wire wound on it. The diameter of its cross section is 3.0 cm. (a) If a current of 5.0 A flows in the coil, how large is the magnetic flux density in the central portion of the solenoid? (b) Repeat for the case where the solenoid is wound on an iron core which increases B by a factor of 50. (c) How much flux passes through the solenoid in part a?

23 A straight wire carrying a current of 10 A lies along the axis of a long solenoid in which the value of $B = 0.20$ Wb/m². How large is the force on a 1.0-cm length of the wire?

24 Two parallel, long, straight wires 20 cm apart carry currents of 20 and 10 A in opposite directions.

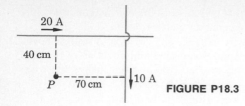

20 A

40 cm

P   70 cm   10 A

FIGURE P18.3

(*a*) How large is *B* resulting from the 10-A wire at the position of the other wire? (*b*) Find the magnitude and direction of the force on a 1-m length of the 20-A wire. (*c*) Repeat part *b* for the other wire.

25* A long, straight wire carries a current of 20 A along the axis of a solenoid in which the magnetic field (in the absence of the wire) is 5.0 G. Find the magnitude of the total magnetic field at a distance of 0.40 cm from the axis of the solenoid. The solenoid radius is 0.50 cm.

26* Find **B** at point *P* in Fig. P18.3.

27* For the situation shown in Fig. P18.3, where is the magnetic field zero?

28** A proton traveling at 10^5 m/s enters a region of space where there is a uniform magnetic field of 0.100 T. Its velocity makes an angle of 30° with the direction of **B**. Describe quantitatively the path of the proton. *Hint:* Split the velocity into two components, perpendicular and parallel to the field. The parallel component will not interact with the field. The perpendicular component alone will cause the particle to move in a circle.

29** The magnetic field in a long solenoid is 30 G. An electron is shot into the solenoid at an angle of 10° to the axis of the solenoid with a speed of 5×10^6 m/s. The electron follows a helical path. What are the radius and the pitch of the helix? (See the hint in Prob. 28.)

30* It is found that a particular beam of electrons travels in a straight line through crossed magnetic and electric fields in a velocity selector. The value of *B* is 0.050 T, and the plates are separated by 10 cm and have a voltage of 100 V across them. Find (*a*) the speed of the electrons and (*b*) the radius of the circle in which the electrons will travel when the plate voltage is reduced to zero.

31* A narrow, cylindrical beam of electrons is shot straight down a TV tube and causes a point of light on the screen. If the current carried by the beam is 0.10 mA, find (*a*) the number of electrons striking the screen each second and (*b*) the value of *B* at a radius of 2 cm from the beam. (*c*) As viewed by the

person watching TV, is *B* clockwise or counterclockwise about the beam as axis?

32* When a fast-moving particle such as an electron shoots through superheated liquid hydrogen, bubbles form along the path of the particle. In Fig. P18.4 one sees the paths of several particles in such a "bubble chamber." The paths are curved because of a magnetic field perpendicular to and into the page. Assume it to be 20 G. Is the particle that is leaving *A* and moving toward the right positive or negative? If you assume it to be an electron, approximately what is its speed? (The tracks are actual size. Assume them to be in the plane of the page.)

33* The particle which starts at *B* in Fig. P18.4 slows as it moves through the liquid hydrogen. As a result, it spirals inward. Assume the same data as in Prob. 32 and assume the particle to be an electron. Find its speed at point *C*.

34** Using Ampère's law, prove the following: The magnetic field within a nonferromagnetic wire of radius *a* that carries a current *I* spread uniformly across its cross section is $\mu_0 Ir/(2\pi a^2)$.

35** Using Ampere's law, prove the following: The magnetic field is zero inside a hollow copper pipe that carries a uniform current *I* lengthwise along the pipe.

36 A certain galvanometer movement has a resistance of 40 Ω and deflects full scale for a voltage of 100 mV across its terminals. How can it be made into a 2-A ammeter?

37 If a meter movement deflects full scale to a current of 0.010 A and has a resistance of 50 Ω, how can it be made into a 5-A ammeter?

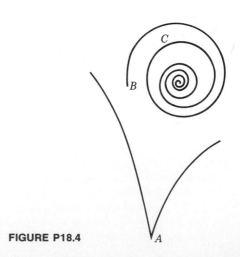

FIGURE P18.4

38 How can the meter movement described in Prob. 37 be made into a 20-V voltmeter?

39 How can the galvanometer of Prob. 36 be made into a 50-V voltmeter?

40* How can the meter movement of Prob. 37 be made into an ammeter having two ranges, 10 and 1.0 A?

41* How can the meter movement of Prob. 37 be made into a voltmeter having two ranges, 12 and 120 V?

42 A flat coil of wire has 30 loops of wire on it. Each loop has a radius of 2.0 cm. What is the magnetic moment of this coil when a current of 5.0 A flows through it?

43 The coil of Prob. 42 hangs in a magnetic field, as shown in Fig. P18.5. (*a*) Find the torque on it. (*b*) Will it tend to rotate so as to increase or to decrease θ? The angle $\theta = 120°$.

44* A circular coil with a 5.0-cm radius and with 20 loops of wire on it carries a current of 25 A. The coil lies on a horizontal tabletop. The current in it flows clockwise as seen from above. It is acted on by the earth's magnetic field, which at that place is directed north and downward with a dip angle of 55°. The value of B is 0.85 G. (*a*) Find the torque on the coil due to B. (*b*) Which part of the coil tries to rise from the tabletop, its north, east, south, or west side?

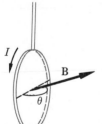

FIGURE P18.5

19
Electromagnetic Induction

The industrial revolution that transformed the world more than a century ago was based on three major scientific advances: the invention of the steam engine and the further development of heat engines through the use of thermodynamics; the discovery that forces can be produced by the interaction of currents with magnetic fields; and the discovery that currents can be produced by changing magnetic fields. We already discussed the first two advances. In this chapter we investigate the third.

19.1 Induced EMFs

The discovery that electric currents generate magnetic fields was made in 1820. As so often happens in science, this newly found facet of nature led to intense investigations of related phenomena. One avenue of experimentation was followed by those who attempted to answer the following question: If currents produce magnetic fields, is it not possible that magnetic fields produce currents? The answer to this question was shown to be affirmative about ten years later by Michael Faraday (1791–1867) in England and independently by Joseph Henry (1797–1878) in the United States.[1] Let us now discuss an experiment that shows this effect vividly.

---

[1] Henry's work, carried out in relative obscurity in Albany, New York, was published only in the United States, and few people knew of it. As a result, his experiments had very little influence on scientific progress at that time.

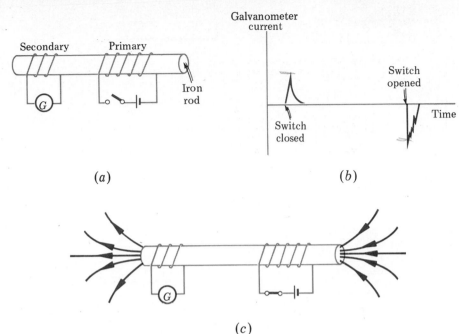

Secondary   Primary

Iron
rod

Switch
opened

Time

Switch
closed

(a)

(b)

FIGURE 19.1
An induced current exists in
the secondary only when
the primary current is
changing. The current
pulses are actually much
narrower than shown in (b).

(c)

The experiment makes use of the simple equipment shown in Fig. 19.1*a*. We see there two simple series circuits. One consists of a battery and switch in series with a long wire. The wire is coiled around an iron rod as shown. We call this coil the **primary coil** because it is attached to the battery. A second independent wire is also coiled around the rod. This coil is in series with a galvanometer, but has no battery in its circuit. It is called the **secondary coil.**

Since there is no battery in the secondary-coil circuit, one might guess that the current through it would always be zero. But a startling fact emerges if the switch in the primary circuit is suddenly either closed or opened. At that exact instant, the galvanometer suddenly deflects and then returns to zero. A current flows in the secondary-coil circuit for a short instant. It is as though the secondary circuit possessed a battery, a source of emf, for just an instant. We say that an **induced emf** exists in the secondary coil during that instant.

Figure 19.1*b* shows another feature of this induced current and emf. As we see there, the short-lived current flows in one direction when the switch is pushed closed. But the emf causes a current pulse to flow in the opposite direction when the switch is pulled open. This tells us that the direction of the induced emf depends on whether the current in the primary coil is increasing or decreasing.

A second, somewhat similar experiment is shown in Fig. 19.2. It involves a bar magnet and a coil in series with a galvanometer. When the magnet is stationary beside the coil, as in parts *a* and *c* of the figure, no current flows in the coil. However, if the magnet is moved relative to the

513

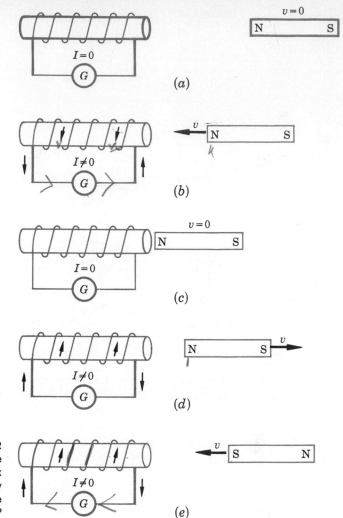

(a)

(b)

(c)

(d)

FIGURE 19.2
Current is induced in the
coil only when the flux
through it is changing. Why
does the current flow in the
directions shown?

(e)

coil, current flows in the coil as indicated in parts *b, d,* and *e.* As we see, an induced emf exists in the coil only when the magnet and coil are in relative motion. No induced emf exists when conditions are not changing.

Faraday concluded that an induced emf exists in a coil only if the flux (i.e., number of field lines) through the coil is changing. For example, refer to Fig. 19.1c. Since the field lines follow the iron rod as shown, a considerable flux passes through the secondary coil when the switch is closed. If the switch is now opened, this flux decreases to zero. During the time in which the change in flux is occurring, an induced emf exists in the coil. But no induced emf exists when the flux is not changing.

If we refer to Fig. 19.3, we can see that this same situation exists for the

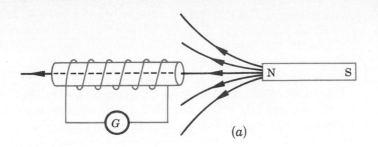

(a)

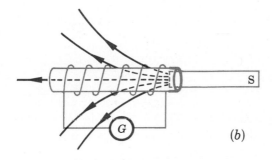

(b)

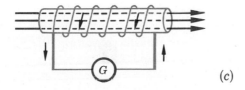

(c)

FIGURE 19.3
When the magnet is moved
from (a) to (b), the current
in the coil flows in the
direction shown in (c). Why?

coil and magnet. As shown there, the flux through the coil changes as the magnet's position is changed. During the time when the flux through the coil is changing, an induced emf exists in the coil. Let us now investigate this effect quantitatively.

The direction of the induced emf can be seen by reference to the experimental results in Figs. 19.1 to 19.3. For example, by reference to Fig. 19.3 we can learn how to predict the emf's direction. In parts *a* and *b*, the flux through the coil is caused to increase toward the left. But, as shown in part *c*, the flux generated by the induced current is in the opposite direction. The induced emf is in such a direction as to produce a current, the flux from which will tend to cancel the change in flux through the coil. If you examine Figs. 19.1 and 19.2 you will notice similar behavior. The induced current is in such a direction as to try to cancel the change in flux through the coil. This rule is embodied in **Lenz's law,** which may be stated as follows:

A change in flux through a loop of wire will induce an emf in the loop. The direction of the current produced by the induced emf will be such that the flux generated by the current will tend to counterbalance the original change in flux through the loop.[2]

In other words, if the flux through a coil is toward the *left* and *increasing,* the induced emf will be in such a direction as to produce flux through the coil toward the *right.* If the flux is toward the *left* and *decreasing,* the induced emf will try to put flux through the coil toward the left. *The induced emf exists only while the flux is changing.*

Faraday discovered the mathematical relation governing the induced emf in a coil. If a coil has N loops of wire, and if the flux changes by an amount $\Delta\phi = \phi_2 - \phi_1$ in time Δt, then the average induced emf during this time is

$$\text{emf} = -N\frac{\Delta\phi}{\Delta t} \tag{19.1}$$

This is called **Faraday's law.** The negative sign is purely a formality and reminds us that the induced emf is in such a direction as to oppose the change in flux.

Illustration 19.1 A solenoid having 100 loops of wire has a cross-sectional area of 4.0 cm². The coil is suddenly transferred from a region where there is no magnetic field to one where the field is 0.50 T directed down the length of the solenoid. If the transfer requires 0.020 s, how large is the average emf induced in the solenoid?

Reasoning Recall from our defining equation for flux, Eq. (18.4), that $\phi = B_\perp A$. Then

$$\phi_1 = 0 \quad \text{and} \quad \phi_2 = (0.50 \text{ T})(4.0 \times 10^{-4} \text{ m}^2) = 2.0 \times 10^{-4} \text{ Wb}$$

We therefore have that

$$\Delta\phi = \phi_2 - \phi_1 = 2.0 \times 10^{-4} \text{ Wb}$$

Faraday's law yields

$$\text{emf} = 100\frac{2.0 \times 10^{-4} \text{ Wb}}{2.0 \times 10^{-2} \text{ s}} = 1.0 \text{ V}$$

You should be able to show that since one weber is one newton-meter per ampere, the unit of the answer is volts.

[2] This is a consequence of the conservation of energy. If the flux generated by the induced current were in such a direction as to augment the change in flux, the induced current could continue to induce more current, without end.

19.2 Mutual Induction

Faraday's law for the induced emf in a coil applies to any method for changing the flux through a coil. Suppose that we have two coils placed side by side, as shown in Fig. 19.4. When the switch S is open, both coils have zero flux through them. We have called the coil in the battery circuit the primary coil and the other the secondary coil.

If the switch is suddenly closed, the primary coil will act as an electromagnet and will generate flux in the region near it. Some of the flux from the primary coil will go through the secondary coil. Hence, the flux through the secondary will change when S is suddenly closed. According to Faraday's law, an induced emf will be generated in the secondary for an instant as the current rises from zero to its final value in the primary. You should be able to show that the direction of the induced current through the resistor in Fig. 19.4 will be from B to A when S is just closed. The current will flow in the opposite direction just as the switch is opened.

The magnitude of the induced emf generated in the secondary will depend on many geometrical factors. Among these are the number of turns of wire on each coil, how close together the coils are, their cross-sectional area, and so on. (Why?) In addition, since the flux through the secondary will be proportional to the current in the primary, the induced emf in the secondary will be proportional to the rate of change of current in the primary, $\Delta I_p/\Delta t$. We therefore write the following equation for the induced emf in the secondary:

$$\text{emf}_{\text{sec}} = -M\frac{\Delta I_p}{\Delta t} \qquad (19.2)$$

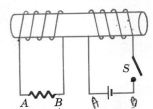

FIGURE 19.4
Why does the current flow from A to B through the resistor at the instant the switch S is just opened?

Mutual Inductance

As indicated above, *the proportionality constant M contains the effects of the geometry of the two coils. It is called the* **mutual inductance** *of the two coils. If the emf is in volts, I in amperes, and t in seconds, the unit M is defined to be the* **henry** *(H), or* $V \cdot s/A$.

Illustration 19.2 Two coils of wire wound on an iron core have a mutual inductance of 0.50 H. How large an average emf is generated in the secondary by the primary as the current in the primary is increased from 2.0 to 3.0 A in $\frac{1}{100}$ s?

Reasoning Let us first point out that when two coils are wound on a core of iron, the flux will be much larger than if they are wound on a nonmagnetic core. Hence, the value of M given here is much larger than one would ordinarily have for noniron-core coils.

Making use of Eq. (19.2) we have, using the proper SI units,

$$\text{emf} = 0.50\frac{1.0}{0.01} = 50 \text{ V}$$

AIP Niels Bohr Library

Michael Faraday (1791–1867)

Faraday, the son of an English blacksmith, received only a rudimentary education before finding work as a bookbinder's errand boy in 1804. Soon he became the bookbinder's apprentice, and this gave him the opportunity to read the books passing through his hands. Growing fascinated by science, he carried out such simple experiments as his meager salary allowed and attended public scientific lectures, the most inspiring of which was the series of lectures given by the great chemist Sir Humphry Davy. A few years later, in 1813, Faraday took the bold step of writing to Davy to request employment in his laboratory. Davy was impressed by the notes Faraday enclosed, which he had taken at the lecture series, and offered Faraday a job as a servant in his laboratory. From these humble beginnings, tutored by Davy, he advanced swiftly as a result of his obvious experimental skill and insight. In 1825, he was made director of the laboratory of The Royal Institution. His whole life was devoted to experimental investigations, particularly in electricity. He is recognized as one of the greatest experimental scientists of the nineteenth century.

Notice that this emf will exist for only an instant in the secondary. As soon as the current in the primary becomes steady, the flux will no longer be changing and there will be no induced emf.

19.3 Self-Inductance

Faraday's law tells us that any change in flux through a coil will induce an emf in the coil. This means that *when a current through a coil changes, the coil induces an emf in itself.* Therefore, if we consider the coil shown in Fig. 19.5, the flux in it will change from zero to some finite value when the switch is first closed. A flux will be generated by the current and will be directed toward the left through the coil. By Faraday's law, an emf will be induced in the coil and it will try to produce flux to the right through the coil. Hence, the induced emf must be opposed to the emf of the battery. If the switch is suddenly opened, the induced emf will be aiding, rather than opposing, the battery. (Can you show this?)

Here, too, the geometry of the coil as well as the core material will determine how large the induced emf will be. If $\Delta I/\Delta t$ is the rate of change of current through the coil, we can write for the average induced emf

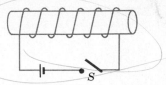

FIGURE 19.5
When the switch is first closed, the coil will induce an emf in itself. Will it aid or oppose the battery?

$$\text{emf} = -L\frac{\Delta I}{\Delta t} \qquad (19.3)$$

Self-Inductance

The constant of proportionality L is called the **self-inductance** *of the coil. It has the same units as mutual inductance, namely, henrys.*

Obviously, if the coil is wound on an iron core, the flux through it will be much greater than if no magnetic material was present. Hence, *if a large self-inductance is desired, the inductor should be wound on an iron core.* We return to the behavior of mutual and self-inductances in later sections. They are of particular importance in alternating-current (ac) circuits, where the current, and thus the flux, is changing continually.

Illustration 19.3 A certain solenoid has cross-sectional area *A*, length *D*, and *n* loops per unit length. What is its self-inductance?

Reasoning We know from Sec. 18.12 that the field in the solenoid is given by $\mu_0 nI$, where *I* is the current in the solenoid. Since the field is directed straight down the solenoid, the flux through each solenoid loop is

$$\phi = B_\perp A = \mu_0 nIA$$

Let us consider what happens as the current in the solenoid changes from *I* to $I + \Delta I$. The new flux will be an amount $\Delta\phi$ larger than before, and so

$$\phi + \Delta\phi = \mu_0 NA(I + \Delta I)$$

Subtracting the previous equation from this gives

$$\Delta\phi = \mu_0 nA\,\Delta I$$

But according to Faraday's law, the induced emf in the solenoid is

$$\text{emf} = -N\frac{\Delta\phi}{\Delta t} = -(nD)\mu_0 nA\frac{\Delta I}{\Delta t}$$

519

which can be compared to the defining equation for self-inductance, namely,

$$\text{emf} = -L\frac{\Delta I}{\Delta t}$$

From the comparison we see that the self-inductance

$$L = \mu_0 n^2 DA$$

This is the inductance we set out to compute. If the solenoid had possessed an iron core, the field within it, and therefore its inductance, would be hundreds of times larger than the value we have computed.

19.4 Inductance-Resistance Circuit

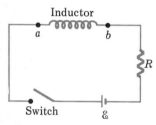

FIGURE 19.6

Why doesn't the current rise to a value $\mathcal{E}/R$ immediately after the switch is closed?

A self-inductance coil has some very interesting and useful properties in a circuit. They are explored in detail in Chap. 20. At present we are concerned with only one facet of the inductor's behavior—its ability to store energy.

Consider the circuit shown in Fig. 19.6. It consists of an inductance coil (represented by the symbol ⌇⌇⌇), a resistor, a battery, and a switch. If the inductor were not in the circuit, the current would rise quickly upon closing of the switch. The final current would be given by Ohm's law to be $\mathcal{E}/R$. However, as the current rises, flux is generated in the inductance coil. This induces an emf in the coil in such a direction as to oppose the increasing current. In other words, the inductor acts as a battery that has a polarity the reverse of that of the actual battery in the circuit.

The upshot of this is that the current is prevented by the coil from rising too rapidly. Despite that fact, the current continues to rise until it finally achieves its normal Ohm's law value, $\mathcal{E}/R$. This behavior is shown in Fig. 19.7. As we see there, *the current rises to a fraction 0.63 of its maximum value in a time L/R. This time, L/R, is called the* **time constant** *of the circuit.*

Definition

As you would expect, the current rise time is longest for a large inductance (because the larger the inductance, the larger the opposition to flow) and for a small resistance (because the smaller the resistance, the higher the final current).

It is important to notice that the inductor acts as a transitory battery of reverse polarity in the circuit. As the current rises, the current is, in effect, charging this emf source in the inductor. Let us now compute how much work is done against the opposing emf of the coil.

The emf induced in the coil is, from Eq. (19.3), $L(\Delta I/\Delta t)$. Therefore, when current flows through the coil, it carries charges through a potential difference $L(\Delta I/\Delta t)$. The work that the current does against the opposing effect of the inductor is, as it carries a charge ΔQ through the inductor,

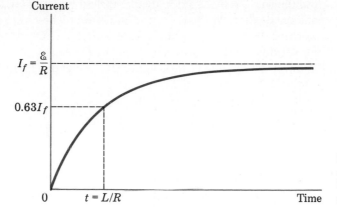

FIGURE 19.7
After the switch in Fig. 19.6
is closed, the current rises
as shown.

$$\Delta(\text{work}) = (\Delta Q)V = (\Delta Q)\left(L\frac{\Delta I}{\Delta t}\right)$$

This can be simplified if we notice that $\Delta Q/\Delta t$ is simply I. Then,

$$\Delta(\text{work}) = LI\,\Delta I$$

To summarize, this is the work done on the coil as the current with present value I is increased by an amount ΔI.

We wish to sum all these small quantities of work as the current in the circuit changes from zero to its final maximum value I_f. This sum is evaluated in one of the exercises at the end of this chapter. The result for the work expended on the coil while changing the current in it from $I = 0$ to $I = I_f$ is found to be

$$\text{Work} = \tfrac{1}{2}LI_f^2$$

This work supplied to the coil can be thought of as energy stored in the coil. Vivid evidence for this stored energy is shown when the switch in a circuit such as that of Fig. 19.6 is pulled open. If the inductance is large, a massive spark jumps across the gap of the switch. Moreover, very high voltage is induced in the coil as it tries unsuccessfully to oppose the loss of flux through itself. We therefore have found that

An inductor L through which a current I is flowing has stored in it an energy $\tfrac{1}{2}LI^2$.

19.5 Energy in Magnetic Fields

You will recall that we computed the energy stored in the electric field by considering the energy stored in a capacitor. (See Sec. 16.11.) Let us now find the energy stored in the magnetic field by considering the energy stored in an inductor. We assume the inductor to be a long solenoid. As was shown in the last chapter, the magnetic field is confined mainly to the core of the solenoid and has a uniform value, $B = \mu_0 nI$.

We computed the inductance of a solenoid in Illustration 19.3. The result was

$$L = \mu_0 n^2 DA$$

where D was the length of the solenoid and A was its cross-sectional area. Notice, however, that DA is simply the volume of the core region of the solenoid. It is the volume in which the magnetic field exists.

The energy stored in the solenoid is given by

$$\text{Energy} = \tfrac{1}{2}LI^2 = \tfrac{1}{2}\mu_0 n^2 I^2 DA$$

from which the energy per unit volume is

$$\frac{\text{Energy}}{\text{Volume}} = \tfrac{1}{2}\mu_0 n^2 I^2$$

But the field in the solenoid is $B = \mu_0 nI$, from which $I = B/(\mu_0 n)$. Substituting this value in the above equation gives

$$\frac{\text{Energy}}{\text{Volume}} = \tfrac{1}{2}\mu_0 n^2 \frac{B^2}{\mu_0^2 n^2}$$

$$\text{Energy per unit volume} = \frac{B^2}{2\mu_0} \tag{19.4}$$

for the energy density in a magnetic field B. This is to be compared with the value $\tfrac{1}{2}\varepsilon_0 E^2$ we found for the energy density in an electric field in vacuum. Of course, Eq. (19.4) applies only to magnetic fields in vacuum or to non-magnetic materials. Although this result was obtained for a solenoid, it applies to all other situations as well.

Illustration 19.4 A certain inductor has an inductance of 0.50 H and a resistance of 2.0 Ω. It is placed in series with a switch, a 12.0-V battery, and a 4.0 Ω resistor. Find the time constant of the circuit and the final energy stored in the inductor.

Reasoning The time constant is given by

$$\frac{L}{R} = \frac{0.50 \text{ H}}{(2.0 + 4.0)\,\Omega} = 0.083 \text{ s}$$

Although the rise of the current is delayed somewhat in this circuit, the delay is not large.

The final current in the circuit is $I_f = V/R = (12\ \text{V})/(6\ \Omega) = 2.0\ \text{A}$. Therefore, the energy stored in the inductor is

$$\text{Energy} = \tfrac{1}{2}LI^2 = \tfrac{1}{2}(0.50\ \text{H})(2\ \text{A})^2 = 1.0\ \text{J}$$

19.6 Motional EMFs

Induced emfs can occur in many ways. Until now we have been dealing mainly with flux changes through stationary coils and the attendant induced emfs. But sometimes the induced emf is the result of the motion of a wire through the magnetic field. In such cases it is often more convenient to use an approach that is not based directly on the concept of flux change through a loop.

Let us begin our discussion by reference to the simple experiment shown in Fig. 19.8. As shown there, a rod of approximate length d rolls with speed v along parallel wires that form a U-shaped loop from M through R and S back to N. Notice that the rod and wires form a loop ($PQRSP$) left of the rod. As the rod moves toward the right, the area of this loop increases.

Suppose a magnetic field **B** exists in this region and that it is directed out of the page as shown. As the rod moves along, the flux through the area of the loop increases *because the area increases*. Hence an emf is induced in the loop.

To find the emf, we note that in time Δt the rod will roll a distance $v\ \Delta t$. As a result, the loop area will increase by an amount $\Delta A = d(v\ \Delta t)$, the shaded area shown in the figure. The accompanying flux change is

$$\Delta\phi = B_\perp\ \Delta A = B_\perp dv\ \Delta t$$

Then, according to Faraday's law, the magnitude of the induced emf in the loop is

$$\text{Induced emf} = \frac{\Delta\phi}{\Delta t} = B_\perp vd$$

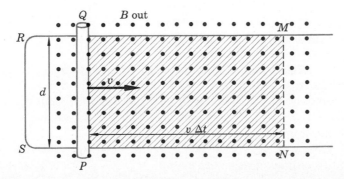

FIGURE 19.8
The induced emf can be treated as being the result of the change of flux through the loop.

Moreover, you should convince yourself that the induced emf is directed clockwise around the loop.

But there is another way we can approach this situation. Consider a positive charge q in the moving rod. This charge, because of its motion with speed v through the field **B,** experiences a force $qvB_{\perp}$. In this case, the total field is perpendicular to the velocity of the charge, and so $B = B_{\perp}$. We conclude that

$$F = \text{force on } q = qvB_{\perp}$$

Using the right-hand rule, you can see that the force on q is directed from Q to P along the rod.

However, from the definition of electric field as force per unit charge, we conclude that the charges moving with the rod experience an electric field directed from Q to P along the rod. Thus

$$E = \frac{F}{q} = vB$$

If we recall that the potential difference between two points equals the work done in carrying a unit test charge from one point to the other, then the potential difference from P to Q due to the electric field E is as follows:

$$V = Ed = B_{\perp}vd$$

Notice that this is exactly equal to the induced emf in the loop that we found using Faraday's law. Moreover, the electric field induced by the charge motion would cause current to flow clockwise in the loop, the same direction as we found from Faraday's law.

We can summarize these latter results in the following way:

A wire (or rod) of length d moving with velocity **v** perpendicular to both a field **B** and the wire itself has an induced emf along its length with

$$\text{Induced emf} = B_{\perp}vd \tag{19.5}$$

This is called a *motional emf.*

In the more general case where **B, v,** and the wire are not mutually perpendicular, the components of **B** and **v** that are perpendicular to each other and to the wire must be used.

This statement is often paraphrased in terms of lines of flux. When the rod in Fig. 19.8 moves so as to change the flux through the loop by an amount $\Delta\phi$, the rod cuts through $\Delta\phi$ flux lines. But the induced emf in the rod is simply $\Delta\phi/\Delta t$, the rate at which the rod is cutting the lines of flux. We can therefore state:

A moving wire has an induced emf within it equal to the rate at which the wire is cutting magnetic flux lines.

As we shall see, the concept of a motional emf is convenient in certain situations.

525
19.7 AC Generator

Illustration 19.5   A rod of 5.0 m length is held horizontal and with its axis in an east-west direction. It is allowed to fall straight down. What is the emf induced in it when its speed is 3.0 m/s if the earth's magnetic field is 0.60 G with a dip angle of 53°?

Reasoning   We use Eq. (19.5). Because $B_\perp = B \cos 53°$ and because **v** is already perpendicular to the rod, we have

$$\text{emf} = B_\perp vd = (B \cos 53°)vd$$
$$= (0.60 \times 10^{-4} \text{ T} \times 0.60)(3.0 \text{ m/s})(5.0 \text{ m}) = 5.4 \times 10^{-4} \text{ V}$$

Can you show that the same result would be obtained by finding the rate at which the rod is cutting flux lines?

19.7 AC Generator

A generator produces a voltage difference between two terminals by changing the flux through a coil. In theory, the flux could be changed either by moving a magnet with respect to the coil or by moving the coil with respect to a magnet. The latter procedure is more easily realized in practice and is the one ordinarily used.

The coil of wire shown schematically in Fig. 19.9 rotates about the axis AA'. A uniform magnetic field perpendicular to AA' is furnished by magnet poles as shown. Notice that one end of the coil is attached to the slip ring R and the other end to R'. These rings are fastened rigidly to the coil and rotate as a unit with the coil. Contact between the rotating slip rings and the stationary outside terminals is made by means of the brushes b and b' that

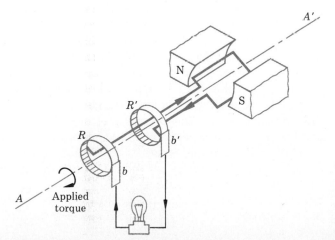

FIGURE 19.9
An alternating emf is produced between terminals b and b' as the coil rotates in an external magnetic field.

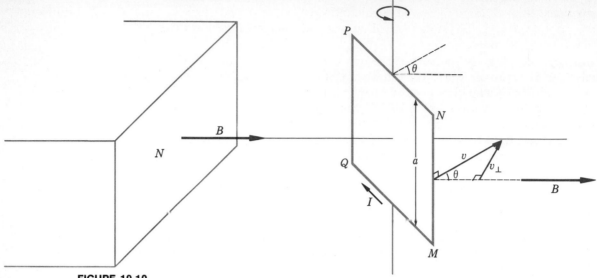

FIGURE 19.10
An induced emf is produced in the rotating coil.

slide along the rings. In a very simple motor, the brushes might be short ribbons of spring steel.

To see exactly how the induced emf between the terminals of the coil is generated, let us refer to Fig. 19.10. The coil is assumed to be rotating in the direction indicated. As you see, the coil is moving from a position in which the field lines were perpendicular to the plane of the coil to a position where the lines just skim past the coil. In other words, the flux through the coil toward the right is decreasing. Because of the change in flux through the coil, an emf will be induced in it.

Using Lenz's law, we see that the induced emf in the coil because of this changing flux will be directed from M to Q in the lower wire. Only then will the induced current flow as indicated, so as to replenish the flux through the coil and counteract the change.

Notice, however, what will happen once the coil has turned through 180° from the position shown. Everything in Fig. 19.10 will be unchanged except that points M and N will be interchanged with points Q and P. As a result, the induced emf will then be directed from Q to M, the reverse of what it was before. It is clear that the induced current in the coil will keep reversing as the coil continues to turn.

To analyze this situation quantitatively, we could find $\Delta\phi/\Delta t$ for the coil and use Faraday's law to compute the induced emf. When calculus is used, this approach is very convenient. However, to avoid use of calculus, we analyze the system in terms of motional emfs.

You will recall that when a wire cuts through the flux lines, an emf equal to Bvd is induced in it, where all three quantities must be mutually perpendicular. If we look at Fig. 19.10, we see that only sides MN and PQ of the loop cut through the flux lines. So only these two sides of the loop have induced emfs.

To compute the induced emf in side *MN,* we note that the component of **v** perpendicular to **B** must be used. From the figure, $v_\perp = v \sin \theta$. Because the wire *MN,* the field **B,** and $v_\perp$ are all mutually perpendicular, we find the induced emf in *MN* to be

$$\text{emf}_{MN} = B(v \sin \theta)a$$

Using the right-hand rule for the deflection of moving positive charges, you should satisfy yourself that the induced current will flow from *N* to *M* in side *MN* and from *Q* to *P* in side *PQ.* Therefore, an identical induced emf in *PQ* will augment the induced emf in *MN.* Hence we find that

$$\text{Induced emf in loop} = 2Bva \sin \theta$$

We can put this in a more convenient form if we note that v is the tangential speed of point *M* as it describes a circle about the axis of rotation. Calling the radius of this circle r (where $r = \frac{1}{2}\overline{MQ}$), we have

$$v = \omega r = 2\pi f r$$

where f is the frequency of rotation of the loop. Moreover, θ is simply the rotation angle of the loop, and it will increase continuously by the relation

$$\theta = \omega t = 2\pi f t$$

By making these substitutions, the induced emf becomes

$$\text{emf} = 2\pi f B(2ra) \sin 2\pi f t$$

But $2ra$ is simply the area A of the loop, and so we have as our final result

$$\text{emf} = 2\pi f A B \sin 2\pi f t \qquad (19.6)$$

If instead of a single loop we had a coil consisting of *N* loops, the emf would be *N* times as large.

As we see, *the induced emf in a rotating coil varies sinusoidally with time.* A graph of this behavior is given in Fig. 19.11. The induced emf (or voltage) has its maximum value when the sine is unity. Therefore the maximum emf is $2\pi f A B N$. It is reasonable that the induced voltage should be large for large f (flux is changing fastest), large A and B (the flux is large), and for a large number of loops on the coil.

Equation (19.6) is frequently written in the alternate form

$$v = v_0 \sin 2\pi f t$$

where v is the voltage at any instant and v_0 is the maximum voltage. Clearly, the voltage in the rotating coil varies sinusoidally and reverses its direction twice during each rotation cycle.

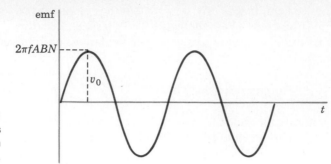

FIGURE 19.11
An alternating emf is
induced in a coil rotating in
a uniform magnetic field.

From the above considerations, it becomes evident that a coil of wire rotating in a magnetic field has an alternating emf generated between its terminals. If such a generator is used as the power source in the simple circuit shown in Fig. 19.12 the current through the resistor will reverse its direction $2f$ times per second. (Notice that the symbol for an alternating-voltage generator is ⊙.)

The ac generators used by power companies are usually more complex than the one discussed here, but their basic operation is the same. Mechanical energy to rotate the coil is usually furnished by steam turbines or by waterpower. Let us just briefly consider the conversion of energy in a system such as that shown in Fig. 19.12.

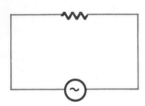

FIGURE 19.12
A simple ac circuit.

If the circuit is open so that no current can flow in the generator coil, very little force need be exerted to rotate the coil. However, as soon as current is drawn from the generator (the coil), the magnetic field will exert a force on the current-carrying wires of the generator, and these forces are in such a direction as to stop the coil from rotating. Hence, the mechanical energy fed into the generator is dependent on the current drawn from the generator—more current requires more mechanical energy.

At an instant when the voltage of the generator is V, the power being delivered to the resistor of Fig. 19.12 is VI. Clearly, if I is very small, the power consumed by the resistor is small and the mechanical energy needed to operate the generator is small. We therefore see that *the energy needed to operate the generator depends directly on the energy being drawn from it. The mechanical energy is transformed to electrical energy by means of the interactions between magnetic field and charge motion within the coil of the generator.*

Relation of Electrical and Mechanical Energy

19.8 Motors

A motor is essentially a generator run backward. As with a generator, *a simple motor consists of a coil rotating in the field produced by a magnet.* The simplified diagram of Fig. 19.13 shows the coil to be situated between the pole pieces of a magnet. In practice, this would probably be an electromagnet rather than a permanent magnet. Moreover, the coil itself is wound on a

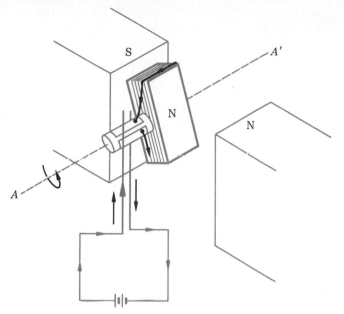

FIGURE 19.13
A simple dc motor. With the slip ring as shown, which way should the motor rotate?

soft iron core to intensify the magnetic field resulting from the current through it. The current-carrying coil itself acts as a bar magnet, as discussed earlier; its strength is increased more than a hundredfold by the iron core.

In the figure, the north pole of the coil is repelled by the north pole of the permanent magnet, and the coil will be made to rotate counterclockwise (as viewed from A to A') by the mutual repulsion of the poles. After rotating 180°, the north pole of the coil will be adjacent to the south pole of the permanent magnet and would be retarded from rotating away from that position. However, when it reaches that position, the sliding contacts on the split slip ring slide over the gap and the current flowing through the coil reverses. This in turn reverses the poles of the coil, and so once again the situation shown in Fig. 19.13 is achieved. Repulsion is maintained, and as a result rotation continues.

There are various modifications of such a motor. Most motors consist of several loops wound with their planes through AA' but at various angles to one another. Each loop has current flowing through it for only a small portion of a cycle during the time when its orientation to the field is right for obtaining maximum torque. Such a motor gives a much more uniform torque than one could obtain from a single loop.

Some motors use electromagnets, while others use permanent magnets to produce the magnetic field. The exact way in which the magnet coils and rotating coil (or armature) are connected differs from motor to motor. Some motors run on both ac and dc voltage, while others run on only one or the other.

Before leaving the subject of motors, we should point out that *the current through the motor is controlled chiefly by its* **counter emf,** an induced emf   *Counter (or Back) EMF* which is described now. (The resistance of a good motor is usually quite low.) When the coil (or armature) rotates in the field of the permanent magnet, an emf is induced in it. The induced emf is in such a direction as to oppose the emf which is sending current through the coil. For this reason it is called a *back* or *counter emf*. Since the resistance of a motor is usually quite small, the chief limitation on the current through it is caused by the counter emf. If the motor is overloaded, it will slow down and therefore draw more current from the source. (Why?) This increased current which flows through the overloaded motor on occasion may become large enough to burn it out.

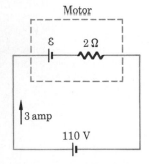

FIGURE 19.14
A motor acts as though it were a resistance in series with a counter emf $\mathcal{E}$.

Illustration 19.6 A particular permanent-magnet motor has a resistance of 2.0 Ω. It draws 3.0 A when operating normally on a 110-V line. How large is the counter emf it develops?

Reasoning The motor can be thought of as a battery in series with a resistance. Since the battery is to represent the counter emf, it must oppose the operating power source. The situation is shown in Fig. 19.14. Combining the batteries and writing Ohm's law give

$$110 \text{ V} - \mathcal{E} = (3 \text{ A})(2 \text{ Ω})$$
$$\mathcal{E} = 104 \text{ V}$$

19.9 Transformers

One of the most important applications of electromagnetic induction takes place in the transformer. These devices change (or transform) one ac voltage to another ac voltage. For example, in the typical TV set, a transformer changes the 120-V ac input voltage to the about 15,000 V needed to operate the picture tube. As another example, the common doorbell requires a voltage of about 9 V, and a transformer is used to obtain this voltage from the usual house-line voltage of about 120 V.

An idealized diagram of a typical transformer is shown in Fig. 19.15.

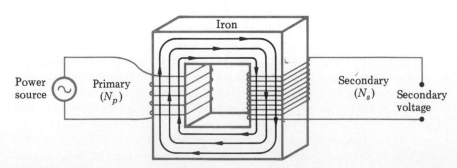

FIGURE 19.15
An iron core step-up transformer.

The transformer consists of an iron core on which are wound two coils, the primary (with N_p loops) and the secondary (N_s loops). As shown, the primary coil is connected to the ac power source, and the ac current in this coil sets up a changing magnetic field in the iron core. Because flux lines tend to follow iron, the lines circle through the secondary coil as indicated. Therefore the flux ϕ through the primary and secondary coils is the same.

The changing flux through the secondary coil gives rise to an induced emf in it given by

$$\text{Secondary voltage} = -N_s \frac{\Delta\phi}{\Delta t}$$

In most transformers, the resistance of the coils is negligible, and so the current-limiting factor in the primary is the back emf induced by the primary coil in itself. In other words, the induced emf in the primary is equal to the voltage of the power source. We can therefore write that

$$\text{Primary voltage} = -N_p \frac{\Delta\phi}{\Delta t}$$

where ϕ is the same flux that flows through the secondary coil.

Taking the ratio of these two induced voltages gives

$$\frac{\text{Secondary voltage}}{\text{Primary voltage}} = \frac{N_s}{N_p} \tag{19.7}$$

This is the transformer equation, and it tells us how the secondary voltage is related to the primary voltage. The two voltages are in the same ratio as the number of loops on the coils. A transformer that raises the input voltage ($N_s > N_p$) is called a *step-up transformer*. The reverse case is called a *step-down transformer*.

If the secondary circuit is not closed, current cannot flow in it. Hence, there is no power loss in the secondary coil when it is not in use. Moreover, we show in the next chapter that there is also no power loss in an inductor which has no resistance. This fact makes it possible for the power company to keep their transformers running throughout a city even when no one is using the electricity they are providing. The transformers themselves consume very little energy.

However, if current is drawn from the secondary, to run a heater, for example, energy is being consumed by the heater. This energy must be fed into the primary of the transformer so that it can be delivered to the secondary. Under these conditions, the loss in power at the secondary causes the primary to act as though it had resistance.

One of the most important uses of transformers has to do with power transmission. Many power companies provide power to cities that are perhaps 100 km from the generators. This proves to be quite a problem. Suppose that in a city of 100,000 people each person is using 120 W of power.

This would be the equivalent of one or two lighted light bulbs for each person. The power consumed is (120)(100,000) W, and at a voltage of 120 V we have

$$\text{Total power} = VI$$
$$(120)(100,000) = 120I$$
$$I = 100,000 \text{ A}$$

Since an ordinary house wire can safely carry only about 30 A without overheating, the power company would need the equivalent of about 3000 of these wires to carry power to the city. Although this is not impossible, the cost of the copper alone would be tremendous. The power companies get around this difficulty quite nicely by noticing that the important quantity is VI and not I alone. They therefore choose to transmit power over long distances at very high voltages. In the above example, if V had been 100,000 V, we would have had

$$(120)(100,000) = 100,000 \, I$$
or
$$I = 120 \text{ A}$$

For this reason the power companies use high-tension or high-voltage difference lines to transmit power over large distances.

Of course, they would not dare to have such high voltages wired directly to a house. The danger from electrocution and fire would be tremendous. Instead, the power companies use step-down transformers to convert these high voltages to the normal voltage used in houses in the United States, about 120 V.

Many houses also have 240-V lines. The large appliances such as ironers, dryers, stoves, etc., are sometimes run from 240-V lines rather than 120-V. This is for essentially the same reason that the power companies use high voltages. You should be able to explain why these large power-consuming devices are more profitably run on 240 than on 120 V.

19.10 Eddy Currents

In our discussion of transformers, we neglected one very important feature of their construction. If the core of the transformer were solid metal, the core would become extremely hot and considerable energy loss would occur. Let us now see why this would happen and what can be done to prevent it.

Consider the coil of wire wound on a metal core as shown in Fig. 19.16. When an ac source is used to drive current through the coil, an oscillating flux is set up through the metal. Because the flux through the dotted path shown in (*a*) keeps changing, an emf is induced around this path. This induced emf, similar to other emfs induced throughout the core, causes circular currents within the coil. We therefore see that

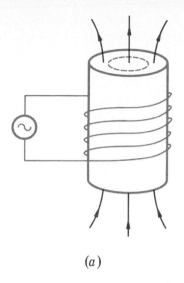

(a) (b)

FIGURE 19.16
In (a) the changing flux
induces eddy currents to
flow in the metal core. To
prevent these currents, the
core can be laminated as
in (b).

Induced currents, called **eddy currents,** flow in metal objects subjected to a changing magnetic field.

Although eddy currents are sometimes advantageous (as in the induction heating of metals), they are more often undesirable. In particular, they cause unwanted heating in transformers, motors, and generators. To eliminate eddy currents, the metal cores of these devices are *laminated*. That is, the core is slit into thin slices with the slices insulated from each other as shown in Fig. 19.16b. Because of the insulating barriers, current can no longer flow around paths such as the dotted circle shown in part a of the figure. As a result, unwanted heating of the metal is greatly reduced.

Summary

When the flux through a coil is changed, an emf is induced in the coil. Its magnitude is $N(\Delta\phi/\Delta t)$, where N is the number of loops on the coil. This is called Faraday's law. The direction of the current produced by the emf is given by Lenz's law. The induced emf is in such a direction that the flux generated by the current will tend to counterbalance the original change in flux through the loop.

If the current in one coil changes at a rate $\Delta I/\Delta t$, it can induce an emf in a nearby coil. The mutual inductance M of the two coils is defined by emf $= -M(\Delta I/\Delta t)$. When a current changes in a coil, an emf is induced in the same coil. This self-induced emf is given by $-L(\Delta I/\Delta t)$, where L is the self-inductance of the coil. Both M and L are measured in henrys (H).

An inductance coil retards the rise of current in a resistance-battery circuit. The time constant of such a circuit is L/R.

The energy stored in an inductor is $\frac{1}{2}LI^2$. This fact can be used to show that the energy stored in unit volume of a magnetic field is $B^2/(2\mu_0)$.

When a wire of length d moves with speed v perpendicular to a magnetic field **B**, an induced emf $= B_\perp vd$ is induced between the ends of the wire. Any time that a wire cuts through magnetic field lines, an induced emf will exist in it.

An ordinary alternating-voltage generator consists of at least one coil rotating in a magnetic field. The changing flux through the coil induces a sinusoidal emf in it. If the coil is rotating with frequency f, then the induced emf is given by $v = v_0 \sin 2\pi ft$.

Since motors consist of a coil rotating in a magnetic field, they have an emf induced in the coil. This is a back emf which opposes the voltage source that runs the motor. Under normal operation, the current through a motor is mostly limited by this back emf.

A transformer consists of a primary and secondary coil so arranged that the flux from one links the other. If we use an ac primary voltage, the ratio of the primary voltage to the secondary voltage equals N_p/N_s.

Minimum Learning Goals

Upon completion of this chapter, you should be able to do the following:

1 Give the direction of the induced emf in a coil caused by a nearby changing current or moving magnet. Describe qualitatively how the emf behaves as a function of time in simple experiments involving a nearby circuit or magnet. Relate your answers to Lenz's and Faraday's laws.

2 Make quantitative use of Faraday's law in simple situations.

3 Define mutual and self-inductance. Explain their qualitative features in terms of Faraday's law.

4 Sketch the current-versus-time graph for an L, R, battery circuit just after the circuit is closed. Locate the time constant on the graph.

5 Recall that energy is stored in an inductance and in a magnetic field.

6 Explain qualitatively why a wire cutting magnetic field lines should have an emf generated between its ends. Compute this emf in the case of a wire moving perpendicular to the field lines.

7 Sketch the details of a simple ac generator. Explain how it gives rise to an ac voltage. Sketch a graph of voltage versus time.

8 Explain the meaning of back emf for a motor and why it depends on the speed of the motor.

9 Explain what a transformer is and why it can give rise to voltage transformations.

10 Show how eddy currents arise and explain how laminating minimizes them.

Important Terms and Phrases

You should be able to define or explain each of the following:

Faraday's law; emf $= -N(\Delta\phi/\Delta t)$
Lenz's law
Mutual and self-inductance
Time constant $= L/R$

Energy $= \frac{1}{2}LI^2$
Motional emf $= B_\perp vd$
AC voltage; $v = v_0 \sin 2\pi ft$
Counter (or back) emf
Step-up and step-down transformers
Eddy current
Lamination

Questions and Guesstimates

1 An inventor claims that she has an electric generator that runs a motor which in turn keeps the generator running. Additional current from the generator is then used to light bulbs, etc. What do you think of this idea? Why?

2 The law of conservation of energy leads us to conclude that the self-induced emf in a coil is an opposing emf. Explain why. Does this also apply to the induced emf in a motor? How could a person devise a perpetual-motion machine if the induced emfs were not in the direction given by Lenz's law?

3 An overloaded motor frequently will blow a fuse before much damage is done. Explain what has happened.

4 Explain why a motor using permanent magnets will operate on direct but not on alternating current, while the same general motor using electromagnets will operate on both.

5 A long, straight wire carries a current along the top of a flat table. A rectangular loop of wire lies on the table as well. If the current in the long wire is shut off, in what direction will the induced current

flow in the loop? Draw a diagram for several positions of the loop relative to the wire, showing in what direction the induced current will flow in the loop.

6 How can we ascertain whether the earth is moving through a uniform magnetic field? Could we tell if the earth were moving through the field rather than just carrying the field along with it?

7 A copper ring lies on a table. There is a hole through the table at the center of the ring. If a bar magnet is held vertically by its south pole high above the table and is then released so that it drops through the hole, describe the forces that act on the magnet.

8 A very long copper pipe is oriented vertically. Describe the motion of a bar magnet dropped lengthwise down the pipe.

9 A 6-V battery is placed in series with a knife switch, a 2-Ω resistor, and a 0.5-H self-inductance. The time taken for the current to rise to about two-thirds its maximum value after the switch is first closed is $L/R = 0.25$ s. Estimate the induced emf in the inductor when (*a*) the switch is first closed and (*b*) the switch is suddenly opened. (E)

10 A closed wire loop experiences a rather large stopping force as it falls into a magnetic field. Justify this assertion by reference to the situation shown in Fig. P19.1. Will the same effect occur if a solid piece of metal, supported by a string, swings in a magnetic field? This general effect is referred to as *magnetic damping* of motion.

11 As shown in Fig. P19.2, a metal ring sits on the end of a solenoid and is held in place there. An alternating current (produced by a alternating emf) is sent through the solenoid. The ring becomes hot.

FIGURE P19.2

Why? A metal plate also becomes hot when held above the solenoid. Explain how eddy currents are induced in it and cause it to heat.

12 Two identical-size rings, one made of copper and the other of plastic, are placed in the same magnetic field. When the magnetic field changes, how do the induced emfs in the rings compare? The induced currents? The induced electric fields?

13 A metal plate separates point *P* from the bar magnet in Fig. P19.3. When the system is at rest, *P* experiences the full magnetic field from the magnet. But if the magnet is suddenly moved closer to *P*, then the field at *P* increases to its full value only after some time has passed. Why the delay? In fact, if the metal sheet is large and is a superconductor, the magnetic field at *P* does not change at all. Why? This general phenomenon is used to shield sensitive equipment from swiftly varying magnetic fields. Why will it not provide shielding from steady fields?

14 A copper ring is set as shown in Fig. P19.2 above a solenoid (which has an iron core to increase its field). When the current is turned on in the sole-

FIGURE P19.1

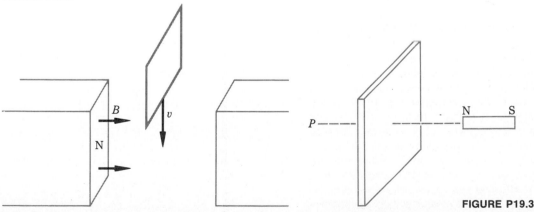

FIGURE P19.3

noid, the copper ring flies upward. Explain why. Be particularly careful about directions.

15 Motors do work on external objects. Explain clearly how this energy is transferred from the elec-

tric current to the rotating portion of the motor.

16 Electric generators transform mechanical work into electric energy. Explain how the energy is transferred.

Problems

1 A 20-loop coil lies on a tabletop in a region in which the magnetic field is vertically upward and has a value of 30 G. The field is reduced to zero in 0.50 s. If the radius of each loop is 7.0 cm, what is the average induced emf in the coil (a) while the field is changing and (b) before the field begins to change? (c) As viewed from above, is the induced emf in the coil clockwise or counterclockwise?

2 A 16-loop square coil has 20-cm-long sides. It rests flat against the north pole of a large electromagnet. The current in the electromagnet is slowly increased so that the magnetic field increases from zero to 0.40 T in 2.5 s. (a) Find the average induced emf in the coil while the current is being changed. (b) If you look toward the north pole, is the induced emf in the coil clockwise or counterclockwise?

3 At a certain place, the earth's magnetic field is 0.70 G and directed at an angle of 53° below the horizontal. (a) A man places his 2.0-cm-diameter wedding band flat on the table and slides it across the table with a speed of 50 cm/s. What is the average induced emf in the ring? (b) Repeat if the man rolls the band in a straight line across the table at this same speed.

4 As a laboratory experiment, a student attempts to measure the earth's magnetic field by connecting the two ends of a horizontal loop (area = 0.70 m²) to the terminals of a sensitive voltmeter. She then moves the loop parallel to the earth at 5.0 m/s. (a) If the vertical component of the earth's field there is 0.35 G, how large a voltage should the voltmeter read? (b) Suppose instead she flips the coil completely over in 1.50 s. What would be the average voltage induced in the loop?

5* A wire loop is being pulled out of a magnetic field at a constant rate as shown in Fig. P19.4. If the magnetic field is uniform (0.30 T) in the region shown and zero elsewhere, (a) what is the induced emf in the loop and (b) what is its direction?

6* Magnetic transducers are often used to monitor small vibrations. For example, the end of a vibrating bar is attached to a coil, which in turn vibrates in and out of a uniform magnetic field B, as shown

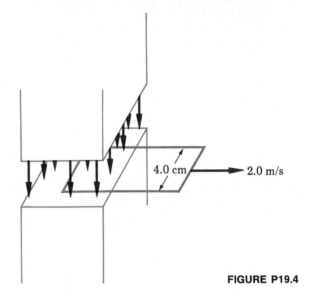

FIGURE P19.4

in Fig. P19.5. Show that the speed of the end of the bar, v, is related to the emf induced in the coil by emf = NBbv.

7 The average field produced within an iron-core toroid is 1.40 T. A secondary coil of 100 turns is wound on the toroid. If the cross-sectional area of the toroid is 0.50 cm², how large an average emf is

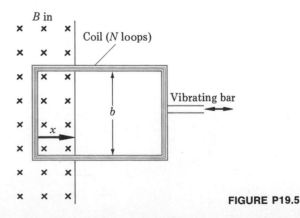

FIGURE P19.5

induced in the secondary coil if the current is stopped in the toroid in 0.010 s?

8   A bar magnet can be moved in or out of a 100-turn coil into which it fits snugly. It is found that an average emf of 0.30 V is induced in the coil when the magnet is suddenly brought up and inserted in it in 0.10 s. If the cross-sectional area of the magnet is 2.0 cm^2, find the value of B in it.

9*   Figure P19.6 shows a coil of wire (radius b) around a solenoid (radius a). The solenoid is actually much longer than shown. (a) If the magnetic field in the solenoid is changing at a rate of 0.020 T/s, find the induced emf in the outer coil. The outer coil has N loops on it while the solenoid has n loops per meter length. (b) If a bar of iron which intensifies the field by a factor of 300 is placed inside the solenoid so as to fill it, what will be the emf induced in the outer coil?

10*   The current in an air-core solenoid is increasing at a rate of 2.0 A/s. There are 10^6 turns of wire on the solenoid for each meter of its length, and its cross-sectional area is 2×10^{-4} m^2. A secondary coil of 10^4 turns is wound over the solenoid. How large an emf is induced in the secondary?

11**   Two coils are wound tightly on the same iron core. The cross-sectional area of both is about 4.0 cm^2. When a current of 3.0 A flows in the primary coil, $B = 0.20$ T. There are 100 turns on the secondary. (a) How large an emf is induced in the secondary if the current in the primary drops uniformly to zero in 0.050 s? (b) What is the mutual inductance of the coils?

12**   A long, iron-core solenoid with 1000 turns has a cross-sectional area of 4.0 cm^2. When a current of 2.0 A flows in it, $B = 0.50$ T. (a) How large an emf is induced in it if the current is turned off in 0.10 s? (b) How large is its self-inductance?

13   Find the self-inductance of a 30-cm-long solenoid that has a radius of 0.80 cm and a total of 500 loops of wire if it is (a) air-filled, (b) wood-filled, and (c) filled with iron that intensifies the field by a factor of 140.

14*   A straight solenoid with 500 loops of wire on it is constructed in such a way that its self-inductance is essentially zero. (a) How can such a coil be constructed? (b) What is the field inside the solenoid when the current through it is 3.0 A?

15*   A 30-mH self-inductance carries a current of 1.50 A. (a) How much energy is stored within the inductance? (b) If this energy is stored in a region that consists of 2.0 cm^3 of air, what is the average value of B in the region?

16*   A certain solenoid is wound uniformly on a wooden rod in such a way that the interior of the solenoid has a volume of 30 cm^3. When a current of 0.50 A flows in the solenoid, the field within it is 0.060 T. (a) Find the energy per unit volume of the magnetic field. (b) How much energy is stored in the inductor? (c) What is the value of the inductor?

17**   We saw in the text that a change ΔI in the current through an inductor increases the energy stored in it by $\Delta W = LI\,\Delta I$. Sketch a graph of I versus I. (It is a straight line at an angle of 45°.) Notice that $I\,\Delta I$ is the area of a vertical rectangle whose height is I and whose width is ΔI. Use this fact to show that an inductor, in which a current I_f flows, has an energy $\frac{1}{2}LI_f^2$ stored in it.

18*   An inductor L is in series with a resistor R and a battery. The steady current in the circuit is I_0. Without interrupting the current, the battery is removed from the circuit. The current in the remaining circuit decreases to zero in time T. (a) How much energy was originally stored in the inductor? (b) How much heat is dissipated in the resistor during the time T taken for the circuit to fall to zero? (c) What average current I_a flowing through the resistor for time T will produce this amount of heat? (d) How many time constants long would T have to be if $I_a = \frac{1}{2}I_0$?

19   A metal airplane flies parallel to the ground and toward the west at 200 m/s. (a) If the downward vertical component of the earth's field is 0.80 G, what is the potential difference between the tips of the wings, which are 25 m apart? (b) Which wing tip is positive, the north or the south? (c) Can this voltage be measured? (d) If so, how?

20   An engineer decides to light the lights in a train station by utilizing the emf induced in the axles of the trains running on the tracks. (a) If you assume the vertical component of the earth's field to be 0.80×10^{-4} T and the tracks to be 1.5 m apart, how

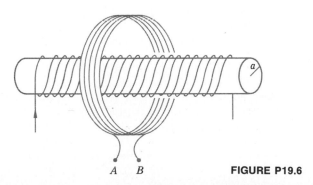

A   B   **FIGURE P19.6**

537

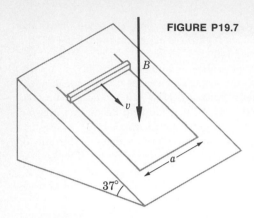

FIGURE P19.7

B

v

a

37°

large an emf is produced between the tracks by a train traveling 30 m/s? (b) Could this voltage be utilized on the moving train? (c) Could it be utilized in the train station at the end of the tracks? Explain your answers to (b) and (c).

21* The metal rod shown in Fig. P19.7 slides down the incline as indicated. It is part of the rectangular loop circuit shown. A vertical magnetic field B is present. (a) Find the induced emf in the rod when its speed is v. (b) If the resistance of the loop is R, what is the current in the loop? (c) Is the current clockwise or counterclockwise? (d) How large is the force that acts on the rod because of the current in the magnetic field? (e) Does this force tend to speed or stop the rod?

22* Refer to Fig. P19.4. Suppose the speed with which the loop is being pulled is v. The width of the loop is d, and the magnetic field has a constant value B between the poles and zero elsewhere. (a) Find the induced emf in the loop. (b) If the resistance of the loop is R, what is the current in it at the instant shown? (c) With how large a force must the loop be pulled to keep its speed constant?

23** For the situation shown in Fig. P19.7 and described in Prob. 21, find the terminal speed of the bar as it slides without friction down the incline. The mass of the bar is m, and the resistance of the loop is nearly zero except for the resistance of the bar which is R.

24** The square loop of resistance R shown in Fig. P19.1 has an edge length a and a mass m. Assuming the magnetic field to be B between the poles and zero elsewhere, find the terminal speed of the loop as it enters the region between the poles. Assume it is in about the position shown when its terminal speed is reached and that it is less wide than indicated.

25 A generator coil rotates in a magnetic field B_0 with

frequency f_0. Its peak emf is v_0. What would be its peak emf (a) if its rotation speed were tripled or (b) if the magnetic field were decreased to 0.40 of its original value?

26 The peak emf produced by a generator, which consists of a single coil rotating in a magnetic field, is v_0. What will be the peak emf if the speed of rotation is cut to one-third its original value and the magnetic field strength is doubled?

27 A 140-loop coil that has an area of 80 cm² rotates at a frequency of 2.5 rotations each second in a magnetic field B. If the maximum induced emf in the coil is 4.0 V, what is the value of B?

28 If the 200-loop coil in a generator has an area of 500 cm² and rotates in a field with $B = 0.60$ T, how fast must the coil be rotating in order to generate a maximum voltage of 150 V?

29 The coil of a motor has a resistance of 5.0 Ω. When the motor is turning at rated speed, it draws a current of 2.0 A from 120 V. (a) How large is the counter emf of the motor? (b) How much current would it draw if the coil were stopped from rotating?

30 Very large motors take nearly a minute to get up to speed after they are turned on. One such motor has a resistance of 0.50 Ω and normally draws 8.0 A on 120 V. (a) What resistance (the starting resistance) must be placed in series with the motor if it is not to draw more than 30 A when first turned on? (This resistance is later removed, of course.) (b) What is the back emf of this motor when operating at its normal speed?

31 A certain transformer used in a radio changes the 120-V line voltage to 9.0 V. (a) What is the turns ratio N_p/N_s for this transformer? (b) By mistake, it is connected into the circuit backward. About what output voltage does it deliver before everything burns out?

32 A neon sign transformer is designed to change 120 V ac to 15,000 V ac. (a) What is the turns ratio N_p/N_s for this transformer? (b) If the transformer were connected up backward (120 V to the secondary), what voltage would appear across the primary?

33* The resistance of no. 10 copper wire is about 5.2×10^{-3} Ω/m. It can carry a current of only about 30 A without overheating. By using wires of this size, it is desired to deliver 20 MW to a city 40 km from a generating station. What fraction of the power sent from the station is lost along the transmission lines if the transmission voltage is (a) 200 V and (b) 200,000 V? Assume the 30-A restriction is not exceeded.

20
Alternating Currents and Electronics

In the past few chapters, we have been concerned mainly with direct currents, i.e., currents which flow continuously in one direction. We saw in the last chapter, however, that a voltage source of alternating polarity is obtained by rotating a coil in a magnetic field. An alternating-voltage source such as this gives rise to alternating currents, and these, too, are of great importance. We see in this chapter how such currents behave when sent through resistances, capacitances, and inductances.

20.1 Charging and Discharging a Capacitor

Let us begin our study of varying current circuits by examining the simple circuit shown in Fig. 20.1a. Suppose the switch is open initially, and no charge exists on the capacitor, whose value we represent by C. We wish to know what will happen when the switch is suddenly closed.

The battery will try to send current around the circuit in a clockwise direction. Since there is initially no charge on the capacitor, the current i will be limited by only the resistor R. Therefore, just after the switch is closed (at $t = 0$), the current will be $i_0 = V_0/R$, as shown in part b. But as times goes on, the capacitor will become charged. The current to the capacitor will decrease. The current must drop to zero when the capacitor is fully charged. The exact way the current behaves in this circuit is shown in Fig. 20.1b.

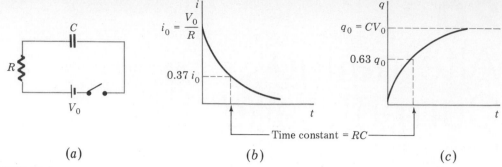

(a) (b) (c)

FIGURE 20.1
The time constant RC is a convenient measure of the time taken for a capacitor to charge or discharge.

The curve followed by the current in part b is called an **exponential decay curve.**[1] Analysis of the circuit shown in part a shows that the current drops to a value of $0.3679i_0$ in a time equal to the product of R and C. (It is an interesting problem in unit manipulation to show that ohm-farads are equivalent to seconds.) *We call the product RC the* **time constant** *of this circuit. It is the time in seconds required for the current to decrease to about 0.37 times its initial value.*

As the current flows in the circuit, the capacitor charges. When fully charged, the capacitor's charge is $q_0 = CV_0$. The charge q on it as a function of time is shown in part c of the figure. Notice that here the time constant measures the time in seconds taken for the capacitor to become about two-thirds charged. As we see, *the time constant is a rough measure of the time taken to charge a capacitor.*

If a charged capacitor C is connected directly across a resistance R, the capacitor will discharge through it. If we assume the initial potential difference between the capacitor terminals to be V_0, the current flowing from the capacitor as it discharges will follow the curve shown in Fig. 20.1b. The capacitor discharge current behaves the same as the charging current. It turns out that *the capacitor is about two-thirds discharged in one time constant.* Here, too, *the time constant RC is a rough measure of the time required for the process.*

Illustration 20.1   In most TV sets, a capacitor is charged to a potential difference of about 20,000 V. As a safety measure, a resistor is connected across its terminals so it will discharge after the set has been turned off. Suppose this so-called bleeder resistor is $10^6\ \Omega$ and $C = 10\ \mu\text{F}$. About how long must you wait after turning off the set before it is safe to touch the capacitor?

---

[1] The equation for the curve of Fig. 20.1b is

$$i = i_0 e^{-t/(RC)}$$

where e is the base for natural logarithms, $e \cong 2.718$, and $e^{-1} = 1/e \cong 0.3679$. A function of form e^x or e^{-x} is called an exponential function.

Reasoning The time constant for this RC circuit is RC. Therefore,

$$\text{Time constant} = (10^6 \ \Omega)(10^{-5} \ \text{F}) = 10 \ \text{s}$$

As a guess, we might say that it would be safe to touch the capacitor after 10 time constants have passed. However, we can be more precise if we know the following fact about the exponential decay curve of Fig. 20.1b. The value plotted on such a curve decreases by a factor of 0.37 during each time constant. As a result, in the figure we see that $i = 0.37i_0$ at $t = RC$. At $t = 2RC$, $i = (0.37)(0.37)i_0$. At $t = 3RC$, $i = (0.37)^3 i_0$, and so on. After 10 time constants, $i = i_0(0.37)^{10}$, which is $4.5 \times 10^{-5} i_0$. At the time $t = 10RC$, the current and charge have been reduced to 4.5×10^{-5} times their initial values. As we see, the current is extremely small after 10 time constants have passed; the capacitor is essentially discharged.

20.2 AC Quantities; RMS Values

Perhaps the most widely encountered type of varying current circuit is what we refer to as ac (for alternating-current) circuits. You will recall that a coil which rotates in a magnetic field gives rise to a sinusoidal emf. This type of emf causes an alternating current such as that shown in Fig 20.2. This is the type of voltage and current that power companies furnish to their customers. *All sinusoidal currents and voltages have an average value of zero over one or more complete cycles.* Even if the ac current flowing through a wire causes the wire to become white-hot (as in an incandescent light bulb), the average current in the wire is zero. This fact may be seen quite easily as we now show.

Average Value of AC Quantities Is Zero

The alternating current in Fig. 20.2 is positive as much as it is negative. To find its average value, we must add its values at various times and divide our sum by the number of values added. Clearly, there will be as many negative as positive values for the current. When added, they will give zero, and so the average current is zero. A similar line of reasoning shows that the average value of an alternating voltage is likewise zero.

Except for such applications as electroplating and battery charging, *an*

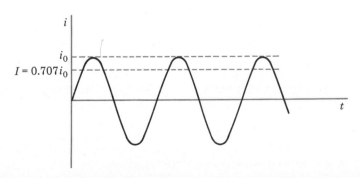

FIGURE 20.2
For an alternating current, the effective, or rms, current is $i_0 / \sqrt{2} = 0.707 i_0$.

alternating current can do useful work even though its average value is zero. For example, in using electricity to produce heat in a resistor, as in an electric stove, the fact that the current is reversing periodically is of no importance. We are interested in only the heat developed, and, as we have seen before, this is determined by the electrical power delivered to the resistor, $i^2 R$. (Notice that the current is squared in this expression, and so whether it is positive or negative is of no importance.)

Actually, then, *we are more interested in the average power delivered to a resistor than we are in the average current through it.* We therefore wish to find the average value of $i^2 R$. But since R is constant, we really need to know the average value of i^2. The average value of i^2, the **mean-square current,** is the average of a group of positive numbers, since i^2 is always positive. For a sinusoidally varying current (or voltage), it is possible to show by use of calculus that the average value of i^2 (or v^2) is one-half its maximum value, $\frac{1}{2}i_0^2$. Or, after the square root is taken, the **root-mean-square** (rms) **current** is just

RMS Current and Voltage

$$I \equiv i_{\text{rms}} = \frac{i_0}{\sqrt{2}} = 0.707 i_0$$

where i_0 is the peak current shown in Fig. 20.2. *The rms voltage is*

$$V \equiv v_{\text{rms}} = 0.707 v_0$$

where v_0 is the maximum voltage during the sinusoidal cycle. The factor 0.707 is $1/\sqrt{2}$. *Frequently, the rms values are called* **effective** *values.*

Most ac voltmeters and ammeters read the effective voltage or current. From time to time you may see a meter calibrated to read the peak voltage v_0 or peak current i_0. Of course, most dc (direct- or steady-current) meters read average values, and so they will not deflect when connected to ac systems. From the way in which the effective, or rms, current is defined, the *power loss* in a resistor is merely $I^2 R$, where I is the rms value. Of course, in a dc system, the rms current, average current, and instantaneous current are all equal.[2]

AC Power Loss

20.3 Resistance Circuit

We introduce the subject of ac circuits by considering in turn three different circuit elements connected in series with an alternating-voltage source. First let us consider the simple resistance circuit shown in Fig. 20.3a. Ohm's law tells us at once that the voltage difference from A to B is just iR. When the voltage is a maximum, the current also will have its maximum value. When

---

[2] We always use V and I for rms values, the meter readings. The letters v and i are used to indicate the instantaneous values of voltage and current.

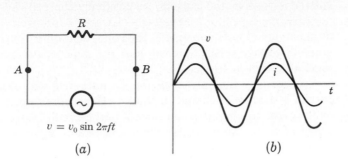

FIGURE 20.3
The current in a resistor is
in phase with the voltage
across the terminals.

the voltage is zero, the current will be zero as well. This behavior is shown graphically in Fig. 20.3*b*. We say that the current and voltage are *in phase* when they are thus in step.

As outlined in the previous section, the power loss in the resistor is I^2R. In this particular case, where only a resistance is present, $I = V/R$, and so the power loss could also be written as IV, where I and V are the rms meter readings. We see in the next sections that there is no average power loss in pure capacitors or inductors. *All power losses in simple ac circuits occur in resistors.*

20.4 Capacitance Circuit

Let us now consider the capacitance circuit shown in Fig 20.4*a*. We know that the potential from A to B is equal to the voltage of the ac source, $v_0 \sin 2\pi ft$. However, we recall that the potential across a capacitor is given by q/C. Hence we have

$$\frac{q}{C} = v_0 \sin 2\pi ft$$

Since C is a constant, we see that the charge on the capacitor will oscillate in value in the same way as the voltage of the source. When the

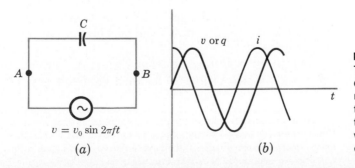

FIGURE 20.4
The voltage across a
capacitor reaches its
maximum $\frac{1}{4}$ cycle later than
the current. The charge on
the capacitor is in phase
with the voltage across the
capacitor.

voltage is positive, the capacitor is positively charged. When the voltage is negative, the capacitor is negatively charged. Moreover, the charge reaches its maximum at the same time as the voltage becomes maximum. Hence the graph of voltage versus time will look the same as a graph of charge versus time, as shown in Fig. 20.4b.

To see how the current in the circuit varies with time, we recall that current is the rate of change of charge. Hence, the rate of change of charge on the capacitor, $\Delta q/\Delta t$, is the current in the circuit. But $\Delta q/\Delta t$ is merely the slope[3] of the curve obtained when q is plotted against t. This curve was plotted in Fig. 20.4b, and so we need only plot its slope and we shall have a graph of the current in the circuit. This is shown as the black curve in Fig. 20.4b.

Notice now that the current is not in step with the voltage. While the current has its maximum at $t = 0$, the voltage does not reach its maximum until $\frac{1}{4}$ cycle later. We say that the *current leads the voltage* by 90° or $\frac{1}{4}$ cycle in such a capacitive circuit. We shall now see what this implies for the power dissipated in the circuit.

During the time interval from e to f in Fig. 20.5, the current is coming out of the positive terminal of the power source (since both v and i are positive), and so the source is furnishing power. However, during the portion fg of the cycle, current is actually going in the direction opposite to which the source would like it to go. The source is being charged rather than discharged. In the portion gh of the cycle, the source is once again furnishing energy, while in section hj it is being charged once more.

In this circuit it is apparent that the power source is being charged, i.e., energy is being fed back into it, as much as it is being discharged, i.e., furnishing energy to the circuit. Hence, the average power drawn from the source is zero. We see, therefore, that *the capacitor,* unlike a resistor, *does not consume energy.* Energy is stored within the capacitor by the source during part of the cycle, but the capacitor returns this energy to the source as it discharges. You may recall that we showed in Chap. 16 that a capacitor could store energy and that the amount stored was $\frac{1}{2}QV$.

It is also of interest to see how a capacitor behaves as the frequency of the source changes. At very low frequencies, very little current will flow in the circuit. When the frequency is zero, i.e., dc conditions, the current is always zero except at the first instant when the capacitor is being charged. If the source frequency is, say, 1 Hz, the capacitor will charge and discharge once each second, and so some current will flow, but not much.

However, if the frequency of the source is 1 million Hz, then the capacitor will charge and discharge 1 million times in 1 s. Since current is the rate of change of charge, the current will now be quite large because the charge is changing rapidly.

We therefore see that a capacitor keeps the current extremely small in a circuit such as that of Fig. 20.4a when the frequency is low. On the other

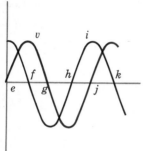

FIGURE 20.5
The capacitor is drawing energy from the voltage source in intervals *ef, gh,* and *jk,* while during intervals *fg* and *hj* it is pushing energy back into the voltage source.

---

[3] The **slope** of a curve is exactly what the word implies. It is the rate at which the curve is rising. If the curve is flat and horizontal, it is not rising at all and so its slope is zero. If the curve is rising rapidly, its slope is large. If the curve is decreasing, i.e., going down, it is, in effect, rising at a negative rate. Its slope is therefore negative.

hand, it does not impede the flow of current at high frequencies. *A capacitor acts much as a large resistance at low frequencies and as a small resistance at high frequencies,* but causes no power loss due to heating. We would like to speak of its "resistance," but this terminology would lead to confusion. We therefore designate its ability to impede the flow of current as its **reactance.** The reactance of a capacitor may be shown by means of calculus to be related to the rms current and voltage by an Ohm's law type of equation, namely,

$$V = IX_C \qquad\qquad (20.1)$$

where X_C, the reactance of the capacitor in ohms, is given by

$$X_C = \frac{1}{2\pi fC}$$

The expression for the reactance of the capacitor (the capacitive reactance) is reasonable. When the frequency f is small or when the capacitance C of the capacitor is small, X_C will be large. And as we surmised previously, X_C will be small when the frequency is high.

Illustration 20.2 Suppose in the circuit of Fig. 20.4 that $C = 0.4\ \mu F$ and that $v = 100 \sin 2\pi ft$ volts with $f = 20$ Hz. Find the rms current in the circuit. Repeat if $f = 2 \times 10^6$ Hz. (Ordinary power-line frequency is 60 Hz.)

Reasoning We know that $V = IX_C$, where V and I are the rms values. Since $V = 0.707v_0$, and since $v_0 = 100$ V, we see that $V = 70.7$ V. But $X_C = 1/(2\pi fC)$. Substituting $f = 20$ Hz and $C = 0.4 \times 10^{-6}$ F gives $X_C = 19,900\ \Omega$. Therefore,

$$I = \frac{V}{X_C} = 0.0036\ A$$

At a frequency of 2×10^6 Hz, the same procedure yields

$$X_C = 0.199\ \Omega \qquad \text{and} \qquad I = 355\ A$$

Notice that, as we have said before, *at high frequencies a capacitance impedes the current much less than at low frequencies.*

20.5 Inductance Circuit

The behavior of the simple self-inductance circuit shown in Fig. 20.6a can be analyzed in a manner similar to that used for the capacitance circuit.[4]

---

[4] Notice that the symbol ⌁ is used to represent an inductance coil.

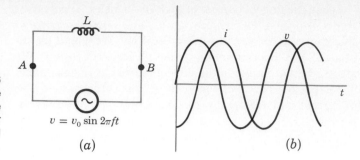

FIGURE 20.6
The voltage across the
inductance leads the
current through it by 90°, or
$\frac{1}{4}$ cycle. Notice the symbol
used for inductance.

First we notice that the voltage difference between points A and B is equal to the source voltage $v_0 \sin 2\pi ft$. However, it is also equal to the voltage induced in the inductor by the changing current and flux in the circuit. In the last chapter we found this voltage to be $L \Delta i/\Delta t$. Equating these two voltages yields

$$v_0 \sin 2\pi ft = L\frac{\Delta i}{\Delta t}$$

Since L is a constant, we see at once that the source voltage is proportional to the value of the rate of change of current in the circuit. But since the rate of change of current is the slope of the graph of current versus time, we have a way of finding one curve if the other is known. The voltage and current curves in the inductance circuit are shown in Fig. 20.6b. Notice that the voltage graph is indeed proportional to the value of the slope of the current graph.

Here, too, the current and voltage are 90° (or $\frac{1}{4}$ cycle) out of phase. In this case, though, the voltage is 90° ahead of the current. We say that *the voltage leads the current by 90°* in this case.

Once again, we can use the same reasoning as in the last section to show that *the inductor consumes no energy on the average*. Although the source stores energy in the inductor during part of the cycle, the inductor gives it back to the source in a later portion of the cycle. We showed in the last chapter that the energy stored in an inductor is $\frac{1}{2}Li^2$. You would do well to examine Fig. 20.6b and ascertain during which part of the cycle the source is losing energy and during which part energy is being returned to the source.

The general behavior of the inductance circuit as the source frequency is changed is also of interest. We know, of course, that the inductor will always try to counterbalance or impede the change in current. In fact, the induced emf in it is $-L \Delta i/\Delta t$ and is therefore proportional to the rate of change of the current. As a result, *when the current is changing very slowly, the inductor will not have much effect. However, at very high frequencies when the current is trying to change rapidly, the impeding effect of the inductor will be very large. We represent this impeding effect by the* **inductive reactance** X_L.

The inductive reactance X_L is related to the rms current through it and to the voltage across it by

$$V = IX_L \qquad\qquad (20.2)$$

where $X_L = 2\pi fL$ is the inductive reactance. As we expected, the reactance of the inductor is large at high frequencies and small at low frequencies. It is measured in ohms.

Notice that *capacitors and inductors behave oppositely as a function of frequency. The current-impeding effect of capacitors is large at low frequencies and small at high frequencies. The reverse is true for inductors.* Of course, *the impedance effect of a resistor is independent of frequency.*

Illustration 20.3 Suppose the inductance coil in Fig. 20.6 has a value of 15 mH. The source voltage, as read by an ac meter, is 40 V, and its frequency is 60 Hz. Find the current which flows through it. Repeat for a frequency of 6×10^5 Hz.

Reasoning We make use of the Ohm's law form $V = IX_L$. In the 60-Hz case, $V = 40$ V and

$$X_L = 2\pi fL = 5.65\ \Omega$$

Therefore, $I = 40$ V$/5.65\ \Omega = 7.1$ A.

At a frequency of 6×10^5 Hz, the value for X_L becomes $5.7 \times 10^4\ \Omega$. Then we find that $I = 7.1 \times 10^{-4}$ A. Notice how very much larger the inductor's impeding effect is at high frequencies than at low frequencies.

20.6 Combined *LCR* Circuit

Let us consider next the series *R, L, C* circuit shown in Fig. 20.7. As in any series circuit, the current through each element in it is the same; let us say that it is $i = i_0 \sin 2\pi ft$. A graph of this current is shown in Fig. 20.8a, and the graph is a sine curve, of course.

We have already discussed the voltage produced by a current through the three elements *R, L,* and *C*; the graphs shown in Figs. 20.3, 20.4, and 20.6 illustrate what we found. These graphs show that the voltage across the resistor v_R is in step (or *in phase*) with the current. But the curve for the voltage across the capacitance v_C has its maxima $\frac{1}{4}$ cycle to the right of the

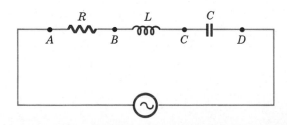

FIGURE 20.7
A series *RLC* circuit.

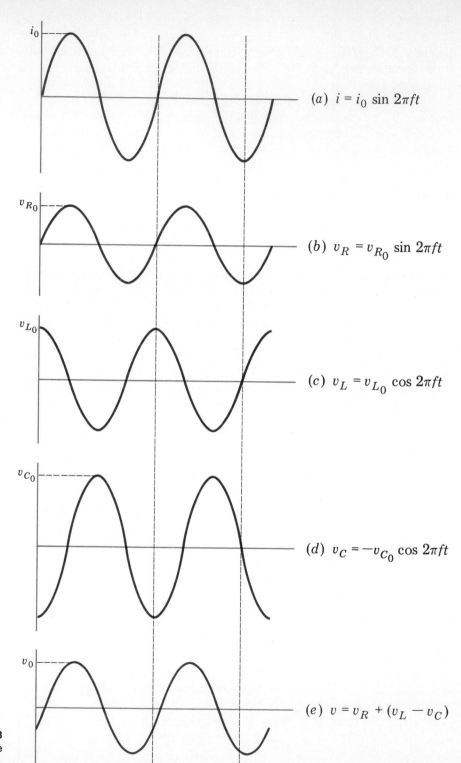

(a) $i = i_0 \sin 2\pi ft$

(b) $v_R = v_{R_0} \sin 2\pi ft$

(c) $v_L = v_{L_0} \cos 2\pi ft$

(d) $v_C = -v_{C_0} \cos 2\pi ft$

(e) $v = v_R + (v_L - v_C)$

FIGURE 20.8
Phase relations in the
circuit of Fig. 20.7.

maxima in the curve for the current. And the curve for the voltage across
the inductance v_L has its maxima $\frac{1}{4}$ cycle to the left of the maxima in the
curve for the current. With these facts in mind, the current curve is related
to the voltage curves as shown in parts *a* through *d* of Fig. 20.8.

Notice, as we have said, that the voltage across the resistor v_R is in
phase with the current. But the voltage across the inductor v_L is $\frac{1}{4}$ cycle out
of phase with the current. The voltage across the capacitor is $\frac{1}{2}$ cycle out of
phase with v_L and $\frac{1}{4}$ cycle out of phase with the current.

An ordinary ac voltmeter connected across the resistor in the series
circuit of Fig. 20.7 reads the rms voltage across the resistor, namely
$v_{R_0}/\sqrt{2} = V_R$. As shown in Fig. 20.8b, v_{R_0} is the amplitude of the resistor
voltage. Across the inductor, the meter reads $v_{L_0}/\sqrt{2} = V_L$, and across the
capacitor it reads $v_{C_0}/\sqrt{2} = V_C$.

But we notice a very strange feature of these voltages in Fig. 20.8: The
voltages v_L and v_C, across the inductor and capacitor, always oppose each
other. When v_L is positive, v_C is negative, and vice versa. If v_{L_0} and v_{C_0} were
the same, the two voltages would exactly cancel. Hence a voltmeter con-
nected across their combination, from *B* to *D* in Fig. 20.7, would read zero!
Clearly, *even though the voltmeter reads V_L across the inductor and V_C across
the capacitor, it will not read $V_L + V_C$ when connected across the combination.*

The situation is even more confusing when the voltage across the resistor is considered. Even though at *any instant* the *instantaneous* voltage
across the resistor-capacitor-inductor system (from *A* to *D* in Fig. 20.7) is
$v_R + v_L + v_C$, an rms voltmeter does not read $V_R + V_L + V_C$ when connected from *A* to *D*. Let us now see what it really reads.

If you add the v_R, v_L, and v_C curves in Fig. 20.8, you will obtain the
curve shown in Fig. 20.8e. It is a simple sinusoidal curve, as we can show by
use of a little trigonometry.

At any instant, the instantaneous voltage across the combination is

$$v = v_R + v_L + v_C$$

Because v_R is in step with the current, which we are taking to be
$i = i_0 \sin 2\pi ft$, we have

$$v_R = v_{R_0} \sin 2\pi ft = V_R \sqrt{2} \sin 2\pi ft$$

where use is made of the fact that $V_R = v_{R_0}/\sqrt{2}$. As we see in Fig. 20.8b, v_R
is indeed a sine curve, a curve that has a value of zero at the origin.

But the curve for v_L in Fig. 20.8c has its maximum at $t = 0$, and it is
therefore a cosine curve. Its equation is

$$v_L = v_{L_0} \cos 2\pi ft = V_L \sqrt{2} \cos 2\pi ft$$

Moreover, the curve for v_C is a cosine curve flopped over, a negative cosine
curve. Therefore

$$v_C = -v_{C_0} \cos 2\pi ft = -V_C \sqrt{2} \cos 2\pi ft$$

Now we can rewrite the *instantaneous* voltage v across the combination:

$$v = v_R + v_L + v_C = \sqrt{2}\,[V_R \sin 2\pi ft + (V_L - V_C) \cos 2\pi ft]$$

Notice, as we expected, that the voltages across the capacitance and inductor tend to cancel.

To proceed further, we need to use a relation from trigonometry. Those who remember much trigonometry will recall the relation

$$A \sin \alpha + B \cos \alpha = \sqrt{A^2 + B^2}\,\sin(\alpha + \theta)$$

where the phase angle θ is given by

$$\tan \theta = B/A$$

[We can check this by noting that when $B = 0$, $\tan \theta = 0$, and so $\theta = 0$. Then $\sqrt{A^2 + B^2}\,\sin(\alpha + \theta)$ becomes $A \sin \alpha$, as it should. Could you show that the formula is correct for $A = 0$?]

Using this trigonometric relation, we have

$$v/\sqrt{2} = \sqrt{V_R^2 + (V_L - V_C)^2}\,\sin(2\pi ft + \theta)$$

where $\tan \theta = (V_L - V_C)/V_R$. As we see, the voltage across the RLC combination is a sinusoidal voltage. Because an rms voltmeter will read its value to be $V = v_0/\sqrt{2}$, and because v_0 for it is simply $\sqrt{V_R^2 + (V_L - V_C)^2}$, we find that a voltmeter connected across the combination will read

$$V = \sqrt{V_R^2 + (V_L - V_C)^2}$$

Notice that the voltmeter does not read the simple sum of the individual meter voltages across each element. Because the voltages are not in phase (or in step), the addition process is more complicated.

We can obtain an Ohm's law form for the circuit of Fig. 20.7 in the following way. Notice that $V_R = IR$, $V_L = IX_L$, and $V_C = IX_C$, where I is the rms current through the series circuit. Using these values in the expression for V gives

$$V = \sqrt{I^2 R^2 + (IX_L - IX_C)^2} = I\sqrt{R^2 + (X_L - X_C)^2}$$

As we see, the quantity $\sqrt{R^2 + (X_L - X_C)^2}$ is the Ohm's law "resistance" factor for the series RLC circuit. We represent this factor, called the **impedance,** by Z. Our relation then becomes

$$V = IZ$$

for the simple series RLC circuit.

Let us summarize what we have found: *The equivalent Ohm's law form for the circuit shown in Fig. 20.7 or Fig. 20.9a is*

$$V = IZ \qquad\qquad (20.3)$$

where

$$Z = \sqrt{R^2 + (X_L - X_C)^2}$$

The quantity Z is called the **impedance** *of the circuit. Its units are ohms.* It is easily seen that Eq. (20.3) reduces to the forms given in the previous sections if all but one impeding element are zero.

You will notice that the equation for Z is reminiscent of the pythagorean theorem for a right triangle, namely,

$$\text{Hypotenuse} = \sqrt{(\text{side } 1)^2 + (\text{side } 2)^2}$$

With this in mind, we can represent Z by the vector diagram shown in Fig. 20.9b. Applying the pythagorean theorem to it, we see at once that $Z = \sqrt{R^2 + (X_L - X_C)^2}$.

Moreover, the angle θ in the diagram is given by

$$\tan \theta = \frac{X_L - X_C}{R}$$

This is the same expression we had previously for the phase angle between the current and the source voltage in the series RLC circuit. Therefore not only is the diagram in Fig. 20.9b a convenient representation for Z, but also it gives the phase angle for the circuit.

The power consumed by the circuit shown in Fig. 20.9 is I^2R. This follows from our previous discussions of the zero power loss in pure inductors and capacitors. Since the voltage drop across the resistor usually will not be the same as that of the source, it is clear that the power loss in the circuit cannot be equal to VI, as it was for dc circuits. To find the correct expression involving V, let us proceed as follows.

Notice in the triangle of Fig. 20.9b that

$$R = Z \cos \theta$$

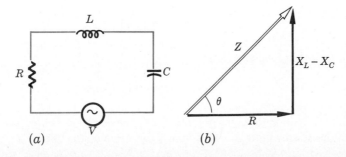

(a)   (b)

FIGURE 20.9
Since the impedance for the circuit shown in (a) is $Z = \sqrt{R^2 + (X_L - X_C)^2}$, we may think of Z as being the hypotenuse of the right triangle shown in (b). The angle θ is the phase difference between the current and the voltage.

If we multiply through this equation by I^2, the current in the circuit, we have

$$I^2R = (IZ)(I\cos\theta)$$

which is just the power loss in the circuit. But IZ is, from Eq. (20.3), the voltage V across the circuit. We can therefore write that *the power loss in the circuit is*

$$\text{Power} = I^2R = VI\cos\theta \qquad (20.4)$$

where θ is the phase angle between the current and the voltage. The factor $\cos\theta$ is often called the **power factor.** Notice that $\theta = 0$ in a pure resistance circuit. For a resistor, the power factor is unity. But for a pure capacitor or inductor, the power factor is zero. Why?

Illustration 20.4   Suppose in Fig. 20.9a that the voltage source has an rms value of 50 V and a frequency of 600 Hz. Suppose further that $R = 20\,\Omega$, $C = 10.0\,\mu\text{F}$, and $L = 4.0\,\text{mH}$. Find (a) the current in the circuit and (b) the voltmeter readings across R, C, and L individually.

Reasoning   We make use of $V = IZ$. Let us therefore find X_L and X_C at this frequency, 600 Hz. We have

$$X_L = 2\pi fL = 15.1\,\Omega \qquad \text{and} \qquad X_C = \frac{1}{2\pi fC} = 26.5\,\Omega$$

Then we find that

$$Z = \sqrt{(20)^2 + (15.1 - 26.5)^2} = 23.0\,\Omega$$

Now, using $I = V/Z$, we find $I = 2.17\,\text{A}$.

To find the voltage drop across R, we use $V_R = IR$ and note that $I = 2.17\,\text{A}$. Therefore,

$$V_R = 43.4\,\text{V}$$

The voltage drop across the inductance is given by $V_L = IX_L$ to be

$$V_L = (2.17)(15.1) = 32.8\,\text{V}$$

Similarly, we see that

$$V_C = IX_C = 57.5\,\text{V}$$

Notice that the potential difference across the capacitor is larger than the source voltage. This points out once again that *in ac circuits, the sum of* the voltmeter readings around a closed circuit is not zero; voltages don't add directly if the rms voltage readings are used. This fact is a result of the average character of the rms readings. They do not represent the instantaneous voltages, which can be either positive or negative. These instantaneous voltages do add directly. The rms voltages, however, are always positive by definition. Clearly, they cannot add to give zero. Kirchhoff's loop rule does not apply to them.

20.7 Electrical Resonance

Ac circuits that contain both capacitance and inductance show an important resonance phenomenon. To illustrate this fact, consider the series circuit shown in Fig. 20.10a. We know that the current in this circuit, which has no resistance, is given by

$$I = \frac{V}{Z} = \frac{V}{X_L - X_C}$$

Notice that *when $X_L = X_C$, the current in the circuit should become infinite.*

It is easy to obtain the condition $X_L - X_C = 0$ because X_L increases with frequency while X_C decreases with frequency. Figure 20.10b shows how these quantities vary for the C and L values given in this circuit. We see that the impedance becomes zero at $f = 4500\,\text{Hz}$ in this case. This fre-

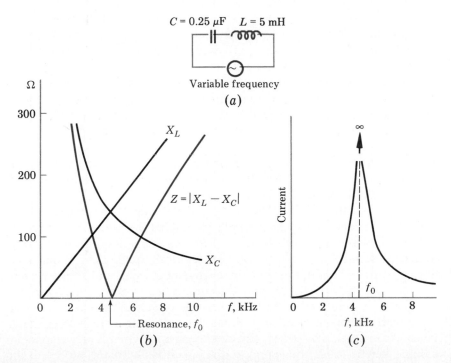

FIGURE 20.10
As the source frequency in (a) is changed, X_L and X_C change as shown in (b). The current in the circuit varies as shown in (c).

quency, *the frequency at which $X_L = X_C$, is called the* **resonance frequency** *of the circuit and we denote it by f_0.* Since $X_L = 2\pi f L$ and $X_C = 1/(2\pi f C)$, we have at resonance that

$$2\pi f_0 L = \frac{1}{2\pi f_0 C}$$

from which the resonance frequency is found to be

$$f_0 = \frac{1}{2\pi}\sqrt{\frac{1}{LC}} \tag{20.5}$$

Figure 20.10*c* shows how the current in this circuit varies as the oscillator frequency is changed. (Of course, the voltage of the oscillator must be kept the same for all frequencies.) As we see, *the current peaks sharply at the resonance frequency. In practical circuits, the peak would be finite rather than infinite because all wires have some resistance.* Even so, the effect is very dramatic and has important applications, as we see in the next chapter.

We can understand electrical resonance better if we recognize that it is much like mechanical resonance. You know that mechanical systems often have a natural frequency at which they vibrate. If pushed with this frequency, they will vibrate widely; they resonate.

A simple *LC* circuit also has a natural frequency of vibration. Let us explore this analogy between resonance in electrical and mechanical systems. For this purpose, consider the *LC* circuit and child on a swing shown in Fig. 20.11. Suppose that at the starting instant the current in the circuit is zero and that the child on the swing is at its highest position. The charge on the capacitor is q_0, and an energy $\frac{1}{2}(q_0^2/C)$ is stored in it. By analogy, the child on the swing possesses gravitational PE.

We know that in the electrical system the capacitor will begin to discharge through the inductor. The current will rise rather slowly because the inductor opposes any change in current. Similarly, the child on the swing will begin to pick up speed as the inertia of the swing system is overcome by the accelerating forces acting on it. Both the child and the capacitor lose their PE. Once the swing has reached the bottom of its path, all the PE has been changed to KE. Similarly with the circuit: once the capacitor has lost all its charge, the current is flowing strongly in the circuit and the original energy is now stored in the inductor. Its value is $\frac{1}{2}Li^2$. This situation is shown in part *b* of Fig. 20.11.

Of course, the child on the swing will not stop at the bottom of the path. The system's inertia will keep it moving until it comes to rest in the position shown in part *c* of Fig. 20.11. Now its energy is all PE once again. Much the same thing happens in the electrical circuit. The inductance, having inertia of a sort, opposes any change in current, and so the current does not stop at once. By the time the current finally stops, the capacitor is fully charged again, as in part *c*. These processes repeat over and over.

Clearly, then, the electrical circuit undergoes an energy interchange much like the child and swing. The child's energy alternates between potential and kinetic, while the energy in the circuit is alternately stored in the

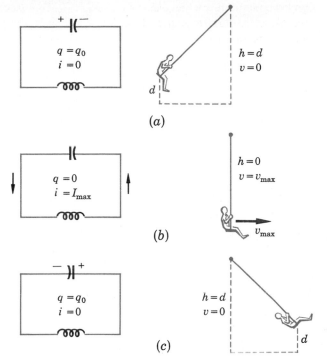

(a)

(b)

(c)

FIGURE 20.11
Just as the energy of the
swing continually oscillates
between potential and
kinetic, the energy of the
circuit is alternately stored
in the charged capacitor or
the current-carrying
inductor.

capacitor and inductor. Both systems would oscillate back and forth forever if there were no energy losses. In the case of the swing, friction losses eventually damp out the oscillation. In the electrical case, resistive effects cause some of the energy to be lost, and so the oscillation slowly damps down in amplitude.

The analogy can be extended further. Both the child and the circuit possess natural resonance frequencies for their motion. The swing system constitutes a pendulum; we have already computed its natural frequency of vibration in a previous chapter. In the case of the circuit, its natural resonance frequency is the resonant frequency computed in Eq. (20.5).

If we wish to cause the child to swing very high, we must push on the swing at just the proper time and with the same frequency as the resonant frequency of the swing. We have seen that a very large current could be built up in the LC circuit if the oscillator "pushed" on the circuit at its resonant frequency. Hence, even the resonance behavior of the two systems is quite similar. It is shown in the next chapter that the LC resonant circuit discussed here forms an integral part of any radio or TV receiver.

20.8 Thermionic Emission

Many of the applications of varying currents and voltages involve electronic circuits. There are two major classifications of electronic devices. One encompasses devices which use vacuum tubes, and the other encompasses electronic systems which use solid-state devices such as transistors. Of course, many electronic instruments use both vacuum tubes and solid-state devices. In this section we learn about thermionic emission, a phenomenon basic to all vacuum tubes.

To a first approximation we can consider the valence electrons in a metal to be free to move anywhere within the metal. They therefore behave in many respects as gas molecules in a container. In the case of a metal, the metal surface is the container. It is possible to learn a good deal about the behavior of the valence electrons in a metal by treating them as if they were an electron gas, i.e., a gas made up of electrons rather than molecules. We use the electron-gas approach to discuss the phenomenon of thermionic emission, the release of electrons from a white-hot metal.

We can easily compute the kinetic energy of the valence electrons within the metal. It will be recalled from Chap. 11 that the gas law can be written in two different forms,

$$PV = \tfrac{2}{3}\nu_0(\tfrac{1}{2}m_0v^2) \qquad \text{and} \qquad PV = \nu_0 kT$$

where P is the pressure of ν_0 gas molecules, each of mass m_0, confined to a volume V at an absolute temperature T. The quantity $\tfrac{1}{2}m_0v^2$ is the kinetic energy of the gas molecule, and k is Boltzmann's constant, 1.38×10^{-23} J/K. Equating the two expressions yields

$$\tfrac{1}{2}m_0v^2 = \tfrac{3}{2}kT$$

This equation says that the average translational KE of *any* ideal-gas molecule in *any* box is just $\tfrac{3}{2}kT$. It must also be true for the valence-electron gas in a block of metal provided that these electrons can really float freely within the metal. To that approximation we then have the important result that each valence electron in a metal has an average KE equal to $\tfrac{3}{2}kT$. Of course, some of the electrons will have more energy than this, and some will have less. But we can use this figure as a basis for discussion.[5]

Is there any possibility that the electrons can escape from the metal? In order to answer this question, we must consider what holds the electrons within the metal. A major portion of the force holding the electron to the metal is purely electrostatic in origin. Consider what would happen if the electron tried to get away from the metal block. As shown in Fig. 20.12, the negative electron just outside the surface of the metal will induce a positive charge on the surface. The positive charge in turn will exert an attractive force on the electron and will try to pull it back to the metal. The electron will not be able to escape unless it has enough KE to overcome this attraction.

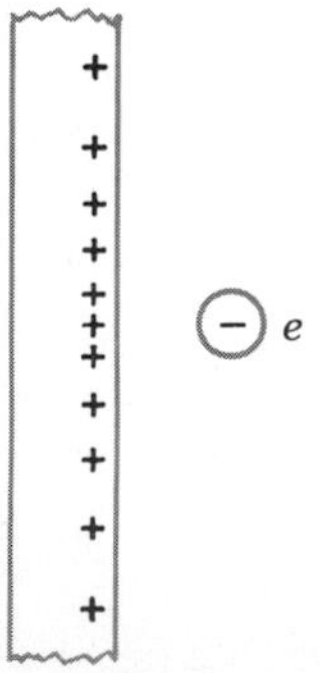

From these considerations we see that a certain amount of work is

FIGURE 20.12
The electron must overcome the attraction of the induced charges on the surface of the metal if it is to escape from the metal.

[5] At low temperatures and low electron energies, quantum effects become important, and the electron-gas approximation is badly in error. However, at high temperatures and high electron energies, this approximation is rather good. We are interested here in the high-energy electrons, and so the electron-gas approximation will be valid.

necessary to tear the electron away from the metal. Unless the electron possesses enough KE to do this work, it cannot escape. *The amount of energy needed to overcome the forces holding the electron within the metal and pull the electron loose is called the* **work function** *of the metal.* We should point out that the energy discussed above is only a portion of the work-function energy. The other energies involved are more difficult to compute, and we cannot discuss them here.

In view of the above discussion, we see that *an electron can escape from a metal provided that it has enough KE.* But since the average KE of an electron is proportional to the temperature of the metal, it is clear that the metal must be heated before the electrons can escape. *For most metals, no appreciable number of electrons can escape from the surface unless the metal is heated red-hot. This process, in which electrons in effect boil from a solid, is called* **thermionic emission.**

Definition

Work Function

20.9 The Diode and Rectification

The principle of thermionic emission is basic to the operation of a diode, the simplest type of vacuum tube. This tube consists of two basic elements enclosed in a glass vacuum tube T as illustrated in Fig. 20.13. A piece of hot metal, the cathode or filament, acts as a source of electrons. Sometimes the cathode is just a fine wire through which a current flows, heating it red-hot. This is the situation pictured. In practice, the cathode or filament voltage $\mathcal{E}_f$ is of the order of 5 V. Some diodes use an indirectly heated cathode, in which case $\mathcal{E}_f$ in Fig. 20.13 would be missing and a separate heating element, not connected electrically to C, would heat the cathode.

In any event, the cathode is heated enough for appreciable thermionic emission to occur. The purpose of the other element within the tube, an unheated metal plate P, is to collect the electrons emitted by the hot cathode. Since the tube is evacuated, the electrons from the filament move freely until they collide with the walls of the tube or with the plate. However, because of the voltage difference between the filament and the plate provided by the plate battery $\mathcal{E}_p$, the electrons emitted from the cathode are attracted by the plate. In most tubes, if $\mathcal{E}_p$ is made about 200 V or larger,

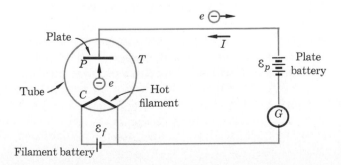

FIGURE 20.13
Electrons are emitted by the hot cathode C and travel through the vacuum tube T to the positive plate P.

557

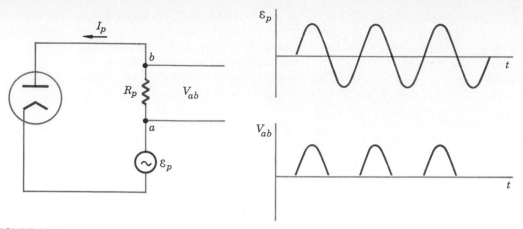

FIGURE 20.14
Although the voltage $\mathcal{E}_p$ is alternating as shown, the diode circuit provides a rectified voltage V_{ab} across the load resistor.

nearly all the electrons emitted by the cathode are collected by the plate.

The diode is most widely used for its rectifying action. As pointed out in the last chapter, it is more convenient in practice to produce and transmit alternating current than direct current. But for numerous purposes, for charging batteries, for electroplating, and especially in electronic devices, we need direct current. Direct current can be obtained from alternating current by using a diode in the circuit shown in Fig. 20.14. Here the plate battery has been replaced by the ac voltage we wish to rectify.

The voltage $\mathcal{E}_p$ alternates as shown in Fig. 20.14. When $\mathcal{E}_p$ is positive, the plate will be positive and the filament will be negative. Electrons emitted by the hot filament will flow to the plate. This gives rise to a plate current I_p. Since this plate current flows through R_p, the load, or plate, resistor, there will be a voltage difference between a and b of V_{ab}. However, when the voltage $\mathcal{E}_p$ reverses and makes the plate negative with respect to the filament, the electrons from the filament will be repelled by the plate and I_p will drop to zero. For this reason, V_{ab} will be zero when $\mathcal{E}_p$ is negative. The variation of V_{ab} is also shown in Fig. 20.14. It is seen that V_{ab} is never negative, and so we have succeeded in transforming the ac voltage $\mathcal{E}_p$ to a dc voltage V_{ab}.

20.10 Semiconductor Diode

As you are probably aware, more and more electronic instruments are being built using semiconducting devices, transistors, and diodes in place of vacuum tubes. These devices are small pieces of crystalline solid which require no heat for their operation. Since a filament as such is not required, the circuitry and power requirements are less than for ordinary vacuum-tube devices. In addition, they can be made very small, and so space can be saved by their use. For example, see the miniaturized circuit shown in Fig. 20.15.

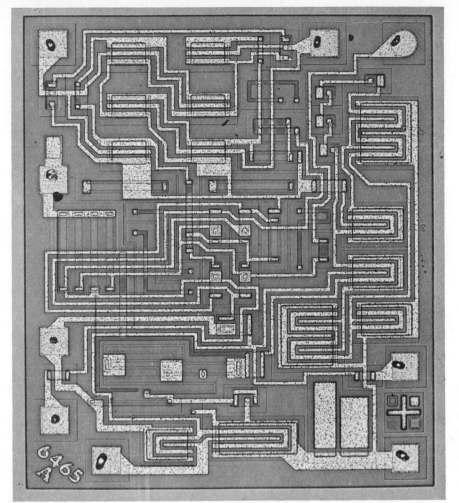

FIGURE 20.15
Although this circuit chip is less than 2 mm in largest dimension, it contains a complete operational amplifier circuit. A similar conventional circuit would occupy a volume about as large as a shoe box. (*RCA.*)

The atoms of silicon or germanium form the basis for many semiconducting devices. Pure silicon and germanium are nonconductors. Both these atoms have four outer, or valence, electrons. They must lose these four electrons or gain four other electrons in order to have complete electron shells. Hence, these atoms combine with other atoms in such a way that they can effectively gain or lose four electrons. In crystalline germanium and silicon, they do this by forming covalent bonds with their neighbors. That is, each atom shares its outer four electrons with four other atoms, while sharing four additional electrons from these four atoms, thereby satisfying all. This situation is shown schematically in Fig. 20.16, where each line between atoms represents a shared electron. Notice that each germanium atom has eight electrons about it, and so its outer shell is filled. This would also be the case for silicon.

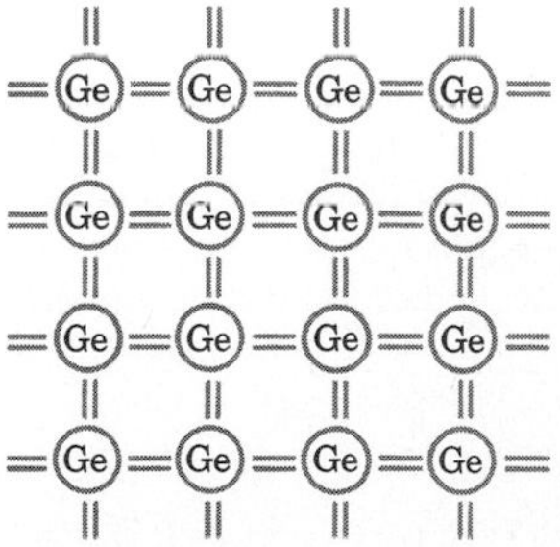

Suppose, now, that we purposely add an extremely small quantity of arsenic atoms as an impurity to the germanium from which we make a crystal. Arsenic is used since it is next to germanium in the periodic table

FIGURE 20.16
The covalent bonding allows each atom in the lattice to fill its outer shell. None of the electrons is free to move in the lattice.

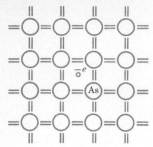

and has essentially the same size as a germanium atom. Therefore, an arsenic impurity atom fits into the lattice quite well and replaces a germanium atom, as shown in Fig. 20.17. Because arsenic is pentavalent rather than tetravalent, it has one too many electrons to fit in the lattice, with the result that the extra electron is not held very tightly by the arsenic atom and escapes easily. Once the electron is loose, it can travel through the crystal rather freely, much as a free electron in metal. Hence, the original nonconducting germanium (nonconducting because the electrons were all tied tightly in the lattice structure) has acquired a few free electrons, one from each arsenic impurity atom. The impure crystal will now be a conductor. Since the number of free electrons is small, it is a poor conductor or semiconductor.

This type of system may also be obtained by adding a few pentavalent impurity atoms to silicon. (Which ones would you expect to be best?) Such impure crystals have an excess of electrons, and these electrons act as charge carriers and carry current through the crystal. This type of crystal is called an **n-type semiconductor,** since *the charge carriers are negative.*

Another type of semiconductor can be made by adding a trivalent atom such as gallium to germanium. In this case, the impurity atom lacks one electron which is needed to make the lattice complete. *This electron vacancy,* shown by the oval at *a* in Fig. 20.18*a, is called a* **hole.** Now, however, the electron at *b* in Fig. 20.18*a* can easily slip over to fill the hole. This in turn leaves a hole at *b.* Another electron can slip into this hole, and so on. The hole becomes free from the original impurity atom and wanders about the crystal more or less freely. After a short time, the situation may be as shown in Fig. 20.18*b, c,* and *d.*

Notice that the region near the hole in Fig. 20.18*d* has an excess of positive charge and is not neutral. (Actually, it has a deficiency of one electron.) On the other hand, the gallium impurity atom now has one too many electrons and is not neutral either. However, the extra electron near the gallium atom is quite tightly held in the lattice, and so it is not free to move. The hole, though, is still free to move about. As it does so, the site of the excess positive charge moves with it. Hence the more or less free movement of the hole through the crystal is equivalent to a positive charge freely moving through the crystal. This type of semiconducting crystal is called a **p-type semiconductor,** since *positively charged holes carry the current.*

A semiconducting diode rectifier is made by combining an *n-* and a

(a)   (b)   (c)   (d)

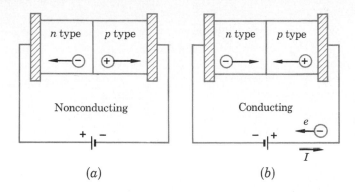

Nonconducting

(a)

Conducting

e

I

(b)

FIGURE 20.19
The diode conducts current
in (b), but does not pass
current when the voltage is
reversed, as shown in (a).

p-type semiconductor, as shown in Fig. 20.19a and b. (In practice, very small pieces of semiconducting crystal are placed between metal electrodes. The entire assembly is placed in an insulated container, with leads running to the two metal plates.) Originally, of course, each piece of semiconductor is electrically neutral. In addition, since the metal plates can receive or give up electrons at will, holes and excess electrons can travel relatively freely through the junction to the metal. This is not possible at the semiconductor junction, as we now see.

Suppose that one applies a voltage to the diode, as shown in Fig. 20.19a. The electrons and holes will be made to move in the directions shown. Since we are effectively trying to separate the negative from the positive charge, this will require considerable work. A steady current cannot flow in this situation, because it is possible to drain only a few of the electrons and holes from the semiconductors. As a result, the diode will not conduct current in the direction desired by the battery in part a. As shown in b, a reversed battery brings holes and electrons together, and so the diode then conducts.

20.11 Applications of Electronic Devices

In this section we present a few important applications of diodes and other simple electronic devices. Electronic technology is so complex that most users of electronic devices cannot be expected to master their detailed construction. Instead, we usually consider the complex of vacuum tubes, diodes, transistors, and other solid-state devices as units designed for specific purposes. For example, the inner workings of an electronic amplifier are usually quite complex, but the amplifier as a unit simply transforms a given input voltage to a much larger output voltage. For a good amplifier, the input and output voltage waves would be of exactly the same shape. As we see, the function and result of use of the amplifier unit are quite easily understood even though the inner workings are very complex.

In order to gain practice in the use of electronic devices as circuit units, we examine a few applications. As you will see, the effect of the device on

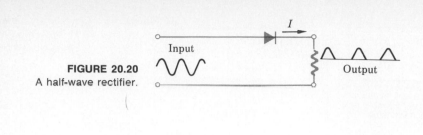

FIGURE 20.20
A half-wave rectifier.

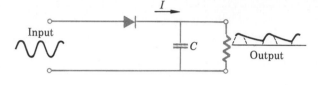

FIGURE 20.21
A filtered half-wave rectifier.

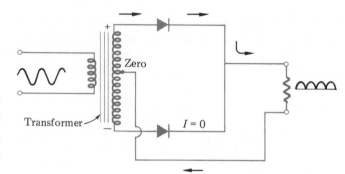

FIGURE 20.22
A full-wave rectifier. The center-tap transformer in essence provides two voltage sources which are one-half cycle out of phase.

an input signal must be known, but the internal workings of the device will not be of major importance for our discussion.

1 *Half-wave rectifier* We have already seen how this device operates. It is shown again in Fig. 20.20. The symbol for a diode is ▶︎|. It conducts in the direction of the arrow.

2 *Half-wave rectifier, filtered* As shown in Fig. 20.21, a capacitor C is placed across the output. If no current were being drawn from the output, the capacitor would become fully charged and maintain a constant dc voltage at the output. In practice, the output acts like a resistance of value R. If R is large enough, the time constant RC will be larger than the period of the voltage wave. Then the capacitor will drain only slightly during each voltage pulse, and the output will be nearly steady. (We specify the "steadiness" by the "ripple," the ratio of the voltage variation to the maximum voltage during the cycle.)

3 *Full-wave rectifier* At the instant shown in Fig. 20.22, the top end of the secondary is positive and the bottom is negative. One-half cycle later, the voltage reverses, but the lower diode will then conduct current to the right. As a result, current is furnished to the output on both halves of the cycle.

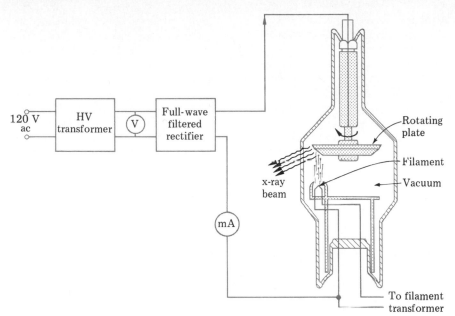

FIGURE 20.23
An x-ray tube circuit.

4 *An x-ray-tube circuit* Refer to Fig. 20.23. Inside the tube, electrons are accelerated through the high voltage between the filament and plate. Their impact with the plate generates x-rays by means of processes we discuss in Chap. 26. In high-output tubes, the electron beam heats the plate. To minimize this effect, the plate is rotated past the beam.

As you can see, we can convey information concerning an electronic circuit without showing the details of each device. We call a diagram like the left part of Fig. 20.23 a **block diagram.** Often in such a diagram only one of the two wires is shown. The second is often assumed to be grounded and is omitted.

There are many other interesting uses of electronics that we could describe. You may wish to pursue the subject further by referring to an elementary text on the subject.[6] Many such texts assume very little background and use only a minimum of mathematics.

Summary

When a dc source charges a capacitor C through a resistor R, the current decays exponentially. A characteristic time RC, called the time constant, is of importance. It measures the time taken for the capacitor to become about 63 percent charged (or discharged).

The average value of a sinusoidal-type current or

[6]See, for example, L. Temes, *Electronic Circuits for Technicians,* 2d ed., McGraw-Hill, New York, 1977 or J. Brophy, *Basic Electronics for Scientists,* 3d ed., McGraw-Hill, New York, 1977.

voltage is zero. Ordinary ac meters read $i_0/\sqrt{2} = 0.707i_0$ and $v_0/\sqrt{2}$, where i_0 and v_0 are the peak values. These meter readings are represented by I and V. They are called the rms or effective values.

When an ac voltage V is impressed across a resistor R, then $V = IR$ applies. For a capacitor C across which an ac voltage of frequency f is applied, $V = IX_C$, where $X_C = 1/(2\pi f C)$. In the case of an inductor L with an ac voltage across it, $V = IX_L$ with $X_L = 2\pi f L$. The quantities X_C and X_L are called the capacitive and inductive reactances.

In a series circuit containing an ac source together with inductance, capacitance, and resistance, the relation $V = IZ$ applies. In this relation, the circuit impedance Z is equal to $\sqrt{R^2 + (X_L - X_C)^2}$.

A pure inductor and a pure capacitor consume no average power. Power losses occur only in resistors. The power loss there is I^2R. In a series ac circuit containing L, C, and R, the power loss is $VI \cos \theta$, where V is the applied voltage. The quantity $\cos \theta$ is called the power factor. The angle θ is given by $\tan \theta = (X_L - X_C)/R$.

An LC circuit resonates at a frequency for which $X_L = X_C$. At the resonance frequency $f_0 = (1/2\pi)\sqrt{1/(LC)}$, the current in the circuit becomes very large. The less resistance in the circuit, the greater the current at resonance.

When substances are heated white hot, electrons "evaporate" from them in a process called thermionic emission. The most energetic electrons in the material have enough thermal energy to supply the work-function energy, the energy needed to tear an electron loose from the material.

Both thermionic and solid-state diodes conduct current in only one direction. As a result, they can be used to change ac voltages and currents to dc voltages and currents. Solid-state electronic devices use semiconducting materials. In n type materials, electrons constitute the current carriers. In p-type materials, the current carriers are electron vacancies called holes. They act as positive-charge carriers.

Minimum Learning Goals

Upon completion of this chapter, you should be able to do the following:

1 Sketch the current and charge curves for an RC circuit during charging. Define the time constant for the circuit and relate it to the curves. Explain the significance of the time constant for discharge of the capacitor through a resistor.

2 Sketch a typical ac voltage or current curve. On the sketch, show the peak, average, rms, and effective values. Relate the rms value to the peak value in a quantitative way.

3 State the Ohm's law form which applies to an ac voltage impressed on a resistor. Sketch the current and voltage curves on the same graph. Compute the average power loss in the resistor if sufficient data are given.

4 Explain why the impeding effect of a capacitor should be higher at low frequencies than at high. Use $V = IX_C$ in simple situations.

5 Sketch the current and voltage curves for a capacitor connected across an ac power source. State the average power loss in the capacitor.

6 Explain why the impeding effect of an inductor should be larger at high frequencies than at low. Use $V = IX_L$ in simple situations.

7 Sketch the current and voltage curves for an inductor connected across an ac source. State the average power loss in the inductor.

8 Use the relation $V = IZ$ for simple problems involving series RCL circuits.

9 By use of $V = IZ$, explain why a resonance frequency exists for an LC circuit. Show how to find the resonance frequency.

10 Give a qualitative explanation why thermionic emission depends on temperature.

11 Explain how a vacuum-tube diode operates and describe its use as a rectifier.

12 Describe how an n-type semiconductor can be made from silicon. Repeat for a p-type semiconductor.

13 Explain qualitatively why a p-n type of junction shows rectification. Give the symbol for a solid-state diode and show in which direction it passes current.

14 Draw a circuit for a half-wave rectifier and explain how its output can be smoothed. Draw a circuit for a full-wave rectifier and explain its principle of operation.

Important Terms and Phrases

You should be able to define or explain each of the following:

RC time constant
Ac voltage or current
Average, rms, effective values
$I = i_0/\sqrt{2}$; $V = v_0/\sqrt{2}$
$P = I^2R$; $P = VI \cos\theta$; power factor

Resistance; capacitive reactance; inductive reactance
Impedance
$X_L = X_C$ at resonance; resonance frequency
Thermionic emission
Vacuum-tube diode
p-type and *n*-type semiconductors
Solid-state diode
Half-wave and full-wave rectifiers

Questions and Guesstimates

1 You are given a 2-μF capacitor, a dry cell, and an extremely sensitive, versatile current-measuring device. How could you use these to measure the resistance of a resistor which is thought to be about $10^8 \, \Omega$? Could you do the measurement using an ordinary voltmeter in place of the current meter?

2 In some places, low-frequency ac voltage (considerably less than 60 Hz) is used. The electric lights operated on this voltage can be seen to flicker rapidly. Explain the cause of this flickering.

3 For which of the following uses would dc and ac voltage be equally acceptable: incandescent light bulbs, electric stove, electrolysis, TV set, fluorescent light, neon-sign transformer, battery charging, toaster, electric clock?

4 Draw an analogy between the vibration of a mass *m* on a spring and the oscillation of an *LC* circuit. What quantities in the mechanical system correspond to *L* and *C* in the electrical system? Explain.

5 Compare the equation for the resonance frequency of a mass vibrating at the end of a spring with the resonance equation for an *LC* circuit. What analogy can you draw between them?

6 A dc voltmeter is connected across the terminals of a variable-frequency oscillator. How would the meter behave as the frequency of the oscillating voltage is slowly increased from 0.01 to 100 Hz? Explain.

7 Why would it be unwise to use 1000-V ac lines in a home, even though this would be more economical from a wiring standpoint?

8 If by magic you had containers of positive and negative charge which you could ladle out in small portions, how could you use these charges to build up a large oscillation in an *LC* circuit? If you could ladle only a limited amount at a time, what effect would a slow increase in resistance of the circuit have?

9 The following statement was published in a daily newspaper. "A warning that home electrical appliances can cause fatal injuries has been sounded by City Health Director, J. R. Smith. His comments followed the death of an 18-year-old boy who accidentally electrocuted himself by inserting a fork in a toaster. Dr. Smith pointed out that even adults can be killed by such electrical shocks. Ordinary house current is 110 volts, but the voltage is increased if the current is grounded, he said." How should the last sentence have been worded?

10 The devices shown in Fig. P20.1 are called **filters.** When an ac voltage is put into the device, the output ac voltage depends on the frequency of the oscillating voltage. One of these devices lets the input voltage pass through undisturbed if the oscillation frequency is high. The other passes only low-frequency voltages. Explain which is which.

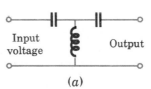

(*a*)

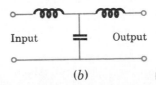

(*b*)   **FIGURE P20.1**

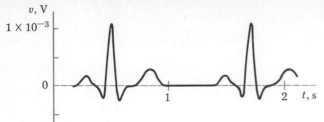

v, V

1×10^{-3}

0

-1×10^{-3}

1   2 t, s

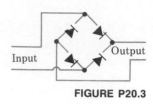

Input   Output

11 A typical electrocardiogram (ECG) graph is shown in Fig. P20.2. This is a graph of the voltage difference between the left leg and left arm. From the graph, estimate the average voltage, rms voltage, and the relation between peak voltage and rms voltage for this waveform. Why doesn't a simple galvanometer deflect in this way when attached to these two points on your body? (E)

12 All other things being equal, which type of material would be preferable for the filament of a vacuum tube, one with large work function or one with small work function?

13 The circuit shown in Fig. P20.3 is a full-wave rectifier (often referred to as a bridge-type rectifier). Examine it and explain why the current is rectified and flows during both halves of the cycle.

Problems

1 A series circuit consisting of a switch, a 200-Ω resistor, and a 0.50-μF capacitor is connected across a 12.0-V battery. (*a*) What is the time constant for this circuit? (*b*) After the switch is closed, about how long must elapse before the capacitor is about two-thirds charged?

2 How large a resistor should be placed in series with a 2.0-μF capacitor so that the time constant of the combination will be 5.0 s?

3 A series circuit consists of a 6.0-V battery, a 2×10^6 Ω resistor, a 4.0-μF capacitor, and an open switch. The capacitor is initially uncharged. The switch is now closed. (*a*) What is the time constant of the circuit? (*b*) About how long will it take for the capacitor to become two-thirds charged? (*c*) How much charge will flow onto it in the time calculated in part *b*? (*d*) About what was the average current which flowed into the capacitor during this interval?

4 Suppose you measure the resistance of your body between your two hands with an ohmmeter and find it to be 62,000 Ω. A 12.0-μF capacitor has been charged to 9.0 V and disconnected. You now grasp the two terminals of the capacitor with your two hands. (*a*) What is the time constant of the circuit involving your body and the capacitor? (*b*) About

what would be the potential difference across the capacitor after $\frac{3}{4}$ s? (*c*) What was the charge on the capacitor when the potential across it was 9 V? (*d*) About what was the average current which flowed through your body in this $\frac{3}{4}$ s?

5 An ordinary ac ammeter reads 2.00 A when connected in an ac circuit. How large does the current actually get in the circuit?

6 An ordinary ac voltmeter reads 120 V when connected across an ac house line. What is the maximum voltage between the lines?

7 A sinusoidal voltage has a peak value of 25 V. What is its rms value?

8 A sinusoidal current has a peak value of 40 A. What is its rms value?

9 When a sinusoidal current is sent through a 30-Ω resistor, the rms voltage across the resistor is 45 V. (*a*) What is the power loss in the resistor? (*b*) At what rate is heat (in calories per second) being generated in the resistor?

10 An rms ammeter in series with a 25-Ω resistor reads 4.0 A when the resistor is connected across a sinusoidal voltage. (*a*) What is the power loss in the resistor? (*b*) How much heat (in calories) is generated in the resistor in 1 min?

11 When a 200-Hz, 40-V rms power source is connected directly across a pure capacitor, the current drawn from the source is 5.0 mA. (a) What is the reactance of the capacitor? (b) What is the value of the capacitor?

12 A 300-Hz power source produces an rms current of 2.0 mA when connected directly across a pure capacitance that has a value of 8.0 μF. (a) What is the reactance of the capacitor under these conditions? (b) What is the rms voltage of the power source?

13* A current given by the relation $i = 5 \sin 360t$ amperes flows through a 20-Ω resistor. How much power does it dissipate in the resistor?

14* A voltage $v = 60 \cos 360t$ volts is impressed across a 20-Ω resistor. How much power is dissipated in the resistor?

15* An rms 120-V source having a frequency of 60 Hz is connected directly across a 10-μF capacitor. (a) How large is the rms current in the circuit? (b) How large is the peak current?

16* By what factor does the current in a capacitor circuit change if the frequency of the voltage source is made 10,000 times larger without changing the voltage? (The circuit contains only the capacitor and voltage source.)

17 What will be the reactance of an 8.0-mH inductance coil when used on a frequency of (a) 60 Hz and (b) 60 MHz?

18 If an inductance coil is to have a reactance of 20 Ω for a frequency of 3000 Hz, what must be its inductance? What will its reactance be at a frequency of 3.0 MHz?

19* An ac voltage source is connected directly across a resistanceless, 20-mH inductance coil. It causes a current of 3.0 A to flow in the coil when the voltage is 5.0 V. (a) What is the frequency of the source? (b) If the frequency is tripled, with the voltage remaining at 5.0 V, what will be the current in the coil?

20* An ac voltage source is connected directly across a resistanceless, 0.50-H inductance coil. How large must the voltage be to give a current of 2 A if the frequency is (a) 60 Hz and (b) 6×10^5 Hz?

21 A 100-V, 180/π-Hz voltage is connected across a 20-Ω resistor. (a) Find the current drawn from the voltage source. (b) Repeat for a frequency of 18,000/π Hz. (c) How much power is dissipated in each case?

22 Repeat Prob. 21 if the resistance is replaced by a 1.00-μF capacitor.

23 Repeat Prob. 21 if the resistance is replaced by a 0.10-H inductor.

24* A 0.10-H inductor having a resistance of 36 Ω is connected across a 120-V, 180/π-Hz source. How much current does it draw?

25* When connected across a 100-V, 60-Hz source, a 0.10-H coil draws a current of 2.0 A. What is the resistance of the coil?

26 (a) How large a capacitor must be connected in series with a 0.10-H coil if they are to resonate with a frequency of 60 Hz? (b) How large a coil would be needed to resonate at this same frequency with a 1-μF capacitor?

27 When connected in series with a 2.0-μF capacitor, a coil resonates fairly sharply to a frequency of 650 Hz. What is the inductance of the coil?

28* A capacitor and an inductor are connected in series across a 120-V, 60-Hz source. The inductor has an inductance of $\frac{1}{9}$ H, and its resistance is 800 Ω. The capacitor is 1.0 μF. (a) Find the current in the circuit. (b) Repeat for a frequency of 6000 Hz.

29** (a) How large an inductor must be connected in series with a 10-μF capacitor, a 20-Ω resistor, and a 100-V, 60-Hz source if the current is to be 4 A? (b) Repeat for a frequency of 6000 Hz.

30* The following elements are connected in series across a 100-V, 200/π-Hz voltage source: $R = 10.0$ Ω, $C = 2.50$ μF, $L = 2.50$ H. (a) What current is drawn from the source when they are all connected in series across it? (b) How much power is being dissipated by the circuit? (c) How large is the power factor?

31* Repeat Prob. 30 if $L = 5.00$ H.

32** A large coil of wire is wound on a cylindrical piece of wood. By using appropriate meters, it is found that a 10.0-V dc source produces a current of 0.50 A through it. A 70-V, 60-Hz ac source causes a current of 2.00 A to flow through it. What is the inductance of the coil?

33** An inductance coil draws a current of 0.60 A when connected across a 12-V battery. When connected across a 120-V, 60-Hz source, it draws 3 A. Find (a) the power drawn from the ac source and (b) the inductance of the coil.

34** A 60-mH, 10-Ω coil is connected across a 120-V, 60-Hz source. How much power does the coil dissipate?

35* The temperature of the white-hot filament in a light bulb or vacuum tube is of the order of 2500 K. (a) What is the average KE of a free electron in the filament? (b) Compare this with the work function of tungsten, about 4.5 eV.

36* Assuming the valence electrons in a metal to be free, what is the average speed of an electron in a white-hot piece of metal at 2500 K?

37** Assume one free electron per atom. (*a*) Find the pressure of the electron gas inside a block of copper at 300 K. (*b*) Compare this with standard atmospheric pressure. (The density of copper is 8.92 g/cm³, and its atomic mass is 63.5 kg/kmol.)

38* A half-wave rectifier system is connected to a 50,000-Ω resistor as its load. It is rectifying 60-Hz ac voltage. (*a*) What magnitude of filter capacitor must be used (as in Fig. 20.21) if the time constant of the filter-load system is to be 10 times as large as the "no-current" time of the rectifier? (*b*) Why is it difficult to obtain a smooth dc voltage from a 60-Hz rectifier system if large currents are to be drawn from it?

39* If, in Fig. 20.21, the output voltage has a maximum value of 12.0 and a minimum value of 9.6 V, what is the value of the ripple for the rectifier system?

21
Electromagnetic Waves

The nature of many types of waves is nearly obvious to everyone. A wave on a string consists of the vibration of the string; waves on the surface of water are the result of the motions of the surface of the water; sound waves are compressions and rarefactions in the air. But the nature of other waves is not as apparent. What vibrates in a radio wave? If light behaves as a wave, and it does, what is the nature of its vibrations? How are x-ray waves and heat radiation (a wave disturbance) carried through space? All these latter waves are electromagnetic waves. We learn about them in this chapter.

21.1 Maxwell's Equations

In our study of electricity we encountered a number of fundamental concepts. By 1865 it had become obvious that electric and magnetic phenomena are closely related and that a synthesis of these concepts could be achieved. A major step in this direction was made in 1865 by a 34-year-old Scottish physicist, James Clerk Maxwell.

Maxwell recognized that the major fundamental facts of electricity and magnetism could be summarized in terms of electric and magnetic fields. He stated his summary in four mathematical equations that may be described in words as follows:

1   Coulomb's law can be expressed in terms of electric field lines emanating from each point charge. The coulomb force consists of the action of this field on nearby charges.

2   Magnetic field lines circle back on themselves; they have no beginning or end.

3   A varying magnetic field induces an emf, which is equivalent to stating that it induces an electric field. This is simply Faraday's law.

4   Moving charges generate magnetic fields, and this fact is summarized in Ampère's circuital law.

Maxwell's four equations expressed in mathematical terms these four statements of fact.

Upon examining his equations in detail, Maxwell noticed that electric and magnetic fields played analogous roles. The differences that existed, except one, could be explained by recognizing that free individual charges can exist while free individual magnetic poles cannot; a south pole of a coil or magnet is always accompanied by a north pole. The one difference for which no excuse could be found was as follows. According to Faraday's law, changing magnetic fields induce electric fields, but no evidence existed for the analogous effect—that changing electric fields induce magnetic fields.

Despite the lack of supporting experimental evidence, *Maxwell postulated that changing electric fields generate magnetic fields.* He showed that if his postulate were true, a new term must be added to the equation that states Ampère's law.[1] This term turned out to be negligibly small when he applied Ampère's law to the experiments possible at that time. As a result, he could neither prove nor disprove the validity of the term.

It is now known that Maxwell's added term is correct, as we see later. His four final equations are called Maxwell's equations. They are as fundamental and as important to electricity theory as Newton's laws are to mechanics. *Maxwell's equations form the basis for all theoretical work in electromagnetism.*

In an attempt to find justification for the term he had added to Ampère's law, Maxwell applied his equations to various phenomena. He focused his attention on what would happen if charges were to oscillate rapidly back and forth so as to produce a swiftly changing electric (and magnetic) field. Under those conditions, the postulated generation of magnetic fields by changing electric fields should become most obvious.

Maxwell found that the application of his equations to the field gener-

---

[1] You will recall that Ampère's circuital law is $\Sigma(lB_\parallel)_n = \mu_0 I$. The new term Maxwell needed to add to the right-hand side was

$$\mu_0\varepsilon_0 \frac{\Delta\phi_E}{\Delta t}$$

where ε_0 is the permittivity of free space that occurs in the Coulomb law constant and ϕ_E is the electric field flux (i.e., number of field lines) that threads through the closed path used for the sum in the equation. Notice that the term is large only if the electric flux (and field) is changing rapidly.

ated by oscillating charges gave a very interesting result. They showed that,
under these conditions, electric and magnetic field waves should be sent out
into space by the oscillating charges. Moreover, the speed with which these
waves should travel turned out to be, according to the calculations, the
speed with which light waves had been measured to move (3×10^8 m/s).
Maxwell therefore made the bold supposition that he had discovered the
nature of light waves: they are waves that consist of electric and magnetic
field waves.

Because Maxwell's development is too mathematical for us to follow,
we approach the subject of electromagnetic waves somewhat differently. Of
course, since we have the benefit of hindsight, our approach will obscure the
genius necessary for his pioneering advance in this field. In fact, we just
show qualitatively why electromagnetic waves should exist, and we then use
a simplified calculation to show that their speed is the speed of light.

21.2 Generation of EM Waves

Let us begin our study of radio waves and electromagnetic (EM) waves in
general by considering the electric field near the charged rod or wire shown
in Fig. 21.1. When the wire is charged as shown, the electric field in the
surrounding area is as indicated. Of course, if the charges on the antenna
were reversed, the direction of the electric field would reverse. As we see,
the charged wire, the antenna, acts as a source for the electric field which
surrounds it.

Suppose now that the charge on the antenna were reversed continuously in a sinusoidal way with a frequency f. Then, of course, the field at
point A would reverse continuously with this same frequency. But there is
more to it than that. Just as a source of waves at the end of a string sends a
signal down the string, the source of the electric field sends an electric field
signal out into space. An attempt is made to show this in Fig. 21.2.

At point A the field is due to the charge on the antenna as it exists at the
instant shown. But the field at M is due to the previous time that the an-

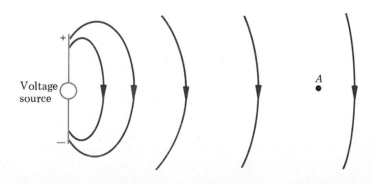

FIGURE 21.1
When the voltage source
keeps reversing, the electric
field at point A will
alternately point up and
down.

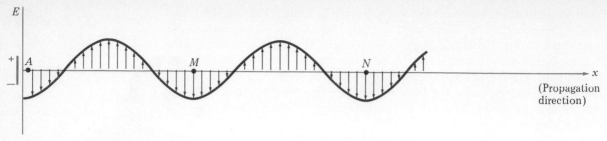

FIGURE 21.2
The antenna acts as a source for the electric field wave that travels away from the source.

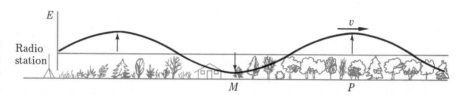

FIGURE 21.3
The electric field wave travels away from the antenna with speed *v*.

Radio station

tenna was charged positive on top. And at *N* the field is the signal sent out by the antenna when its top was positive, two oscillation cycles ago. Just as we found for the wave on a string, the electric field wave that is sent out by the antenna shows the previous history of the wave source. The electric field wave from a radio station blankets the surrounding region, as shown in Fig. 21.3.

But the electric field wave generated by the antenna travels out across the earth with a certain speed *v*. As a result, a person at point *P* on the earth will experience an oscillating electric field. The electric field wave shown in Fig. 21.3 is moving toward the right. When a crest passes point *P*, the field will be directed upward. When a trough passes point *P*, the field will be directed downward.

We see from this that each point on the earth experiences a vibrating electric field. The frequency of this vibration is equal to the vibration frequency of its source, the vibrating charge on the radio-station antenna. *The electric field wave sent out by the source obeys the same relation we found for all other waves. The wavelength of the wave* λ, *its frequency f, and its speed v are related by* $\lambda = v/f$. The frequency of the wave sent out by a radio station is of the order of 10^6 Hz. If we know the speed *v* of the electric field wave, we can find its wavelength. Before discussing the speed of these waves, though, let us consider for a moment another aspect of the signal sent out from the antenna.

In addition to the electric field wave, a magnetic field wave is sent out from the antenna. This wave is a result of the fact that when the charge on the antenna changes, a current flows in the antenna wire. As we have seen previously, this current will cause a magnetic field to circle the antenna as

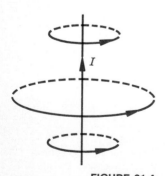

FIGURE 21.4
A magnetic field circles the antenna.

572

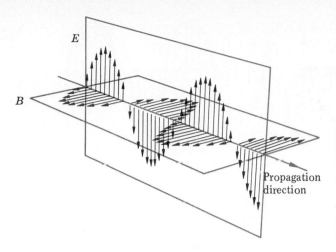

FIGURE 21.5
The oscillating charge
sends out a magnetic field
wave perpendicular to the
electric field wave. A
"snapshot" of this EM
wave along a line
perpendicular to the center
of the antenna is shown.

shown in Fig. 21.4. Notice that *the magnetic field is in a direction perpendicular to the electric field.* It, too, oscillates in direction with the same frequency as the radio station. We see, then, that *an observer some miles from the antenna experiences not only an oscillating electric field, but also a magnetic field perpendicular to the electric field which oscillates with it. This combined electric and magnetic field is what we call the station's electromagnetic (EM) radio wave.*

We represent this combined EM wave in Fig. 21.5. Notice that *the electric field* **E** *is perpendicular to the magnetic field* **B.** *Moreover, the two waves are in phase.* (This latter point is not obvious, but is the result of detailed computation.)

Illustration 21.1   The oldest radio station in the United States is station KDKA in Pittsburgh, which went on the air in 1920. It operates at a frequency of 1.02×10^6 Hz. What is the wavelength of its EM wave? Assume the speed of EM waves to be 3×10^8 m/s.

Reasoning   We know that for any wave $\lambda = v/f$. In our case $v = 3 \times 10^8$ m/s and $f = 1.02 \times 10^6$ Hz. Substitution gives $\lambda = 294$ m.

21.3 Reception of Radio Waves

As we have seen, a radio transmitter blankets the surrounding area with an electric field that is oscillating at a specified frequency. If you hold a straight piece of wire in this field, the electrons in the wire will move under the action of the oscillating electric field. Of course, the electric field will be very weak many miles from the transmitter, but a sensitive detecting system can still

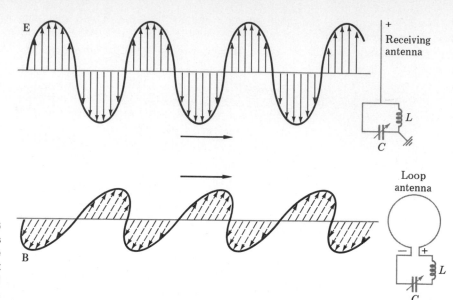

FIGURE 21.6
The radio's antenna acts as a voltage source since the radio wave passing by it induces potential differences in it.

measure the movement of the charges in the wire as a result of the field. A radio is a device constructed for this purpose.

Suppose that the electric field oscillates vertically, as shown in the upper part of Fig. 21.6. It will separate charges in the vertical radio antenna. For example, when the electric field points upward, the top of the antenna will be charged positive and the lower end negative. This, of course, produces a charge on the capacitor C. However, this charge will soon flow out of the capacitor, because $\frac{1}{2}$ cycle later the field will be pointing downward and the charge on the antenna and capacitor must be reversed. A half cycle later still, the antenna will charge the capacitor as it was originally. From this we see that the antenna is taking energy from the radio wave and giving it to the capacitor.

Frequency-Selection Method   *If we adjust the value of the capacitor so that the natural frequency of the LC circuit is the same as that of the radio station, the circuit will resonate under the driving action of the antenna. Consequently, the LC circuit will build up a large response to the action of radio waves to which it is tuned. Thus, it is possible to tune the radio to receive a particular station.* The resonant circuit will then show an oscillatory voltage across C which is nearly exactly proportional to the output voltage of the radio station. It will consist of a carrier wave modulated with the effect of the station's microphone system. However, the voltages across C will be extremely small except for the very nearest stations. For this reason most radios amplify this voltage by rather complex electronic circuits.

Most portable radios used today do not make use of the oscillating electric field. Instead, they receive the signal transmitted by the oscillating magnetic field portion of the EM wave. As shown in the lower part of Fig. 21.6, a loop antenna is used to detect the magnetic portion of the wave. As the wave passes by the loop, its changing flux induces an emf in the loop. This oscillating emf is used to drive the resonant circuit, as in the previous case. The loop antenna usually consists of a coil wound on a rod made of a ferromagnetic material, a ferrite. If you look inside a small portable radio, this rod is usually quite visible.

A loop antenna often shows directional properties. If the radio is oriented so that the loop is related to the field as shown in Fig. 21.6, reception is best. Why? But if it is rotated 90°, the reception will be minimum. This property of a loop antenna can be used to locate the source of the EM waves. How?

21.4 Speed of EM Waves

Now that we understand many of the qualitative features of EM waves, let us obtain an expression for their speed. We use a method that depends on a fact first pointed out clearly by Einstein in his theory of relativity, a fact that we discuss at greater length in Chap. 25. The fact is this: *Only relative motion can be determined*. An object can be said to be at rest relative to another object but not at rest in any absolute sense.

For example, as you read this, you are probably at rest relative to the earth. But the earth is in motion relative to the sun, and so, too, are you. Moreover, the sun is in motion in our galaxy, the Milky Way, and our galaxy is in motion relative to other galaxies in the universe. It makes sense to say that something is at rest relative to something else, but we cannot say which of two objects is at rest in any noncomparative way.

With this fact in mind, let us reconsider the force experienced by a charge q moving with speed v perpendicular to a magnetic field $B_\perp$. We found in Chap. 18 that the force experienced by the charge has a magnitude

$$F = qvB_\perp$$

However, who is to say that the charge is not at rest and the field is moving instead? After all, only relative motion is observable. Therefore our experiment can be interpreted in the following alternate way. A field **B** moving with speed v perpendicular to the field lines past a charge q exerts a force $F = qvB_\perp$ on the charge. But because the force per unit charge F/q is defined to be the electric field E, we can restate this as follows:

A magnetic field **B** moving with speed v perpendicular to the field lines generates an electric field $E = vB$ in the region through which it passes.

To illustrate this with a concrete example, consider the situation shown in Fig. 21.7. The magnet poles are moving with speed v as shown. They carry the magnetic field lines with them, and so we have a moving magnetic field **B** in this region. Our previously stated result tells us that at a point such as A, through which the field lines are moving, an electric field **E** exists

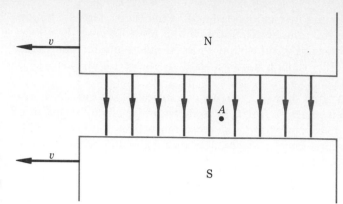

FIGURE 21.7
The magnetic field **B** moves past point *A* with speed *v*. It generates an electric field *Bv* directed into the page.

whose value is $E = vB$. Can you show that the direction of **E** is into the page?[2]

It seems from this line of reasoning that the magnetic field moving out from a radio antenna should generate an electric field in the region through which the field moves. At a given point, the electric field should be related to the speed v of the magnetic field wave and its magnitude B at that point and instant through $E = vB$. The question now arises of whether a moving electric field can generate a magnetic field. We now investigate that question, and we find that our investigation leads us to a very important result.

Consider the very long, uniformly charged wire shown in Fig. 21.8. The wire is moving to the right with speed v, and so the electric field lines from the wire are flying past point P with speed v. We know that the moving charged wire constitutes a current along the line of the wire. The magnitude of the current can be found by noticing how much charge passes point P each second.

Suppose the wire has a charge q per unit length. Since a length vt of the wire will fly past P in time t, we have

---

[2]*Hint:* Place a positive charge at *A* and remember that the motion is relative.

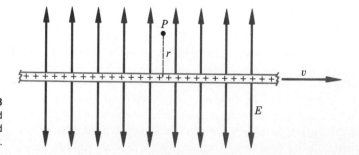

FIGURE 21.8
The moving charged rod carries the electric field lines past point *P*.

$$\text{Current} = \frac{\text{charge passing } P}{\text{time taken}} = \frac{qvt}{t} = qv$$

We see that the moving charged wire constitutes a current of magnitude qv.

But a current produces a magnetic field, and so the moving charged wire is surrounded by a magnetic field. (Can you show that it circles the wire and is out of the page above the wire?) We learned in Chap. 18 that the magnetic field due to a current I in a long, straight wire is given by $B = \mu_0 I/(2\pi r)$. Applying this to the present case, we find the magnetic field at point P to be given by

$$B = \frac{\mu_0 qv}{2\pi r} \tag{21.1}$$

We wish to relate this to the electric field outside the wire at point P.

The electric field outside a charged, long, straight wire is easily found by using Coulomb's law and calculus. We simply state the result here. The electric field at point P in Fig. 21.8 is

$$E = \frac{q}{2\pi\varepsilon_0 r} \tag{21.2}$$

where ε_0 is the permittivity of free space, $8.85 \times 10^{-12}\,\text{C}^2/(\text{N})(\text{m}^2)$.

Let us now eliminate q between Eq. (21.1) and (21.2). The result is

$$B = \varepsilon_0 \mu_0 vE \tag{21.3}$$

This is to be compared with the relation

$$B = \frac{1}{v}E \tag{21.4}$$

we obtained previously for the case of a moving magnetic field.

Although this was a very special situation in which a moving charged wire generated a magnetic field, it is a typical situation. Moving charges generate a magnetic field. But the moving charges have associated with them an electric field that travels along with them. *The magnetic field generated by the motion of the charges can equally well be attributed to the motion of the electric field.* We therefore are led to conclude that

An electric field **E** moving with speed v perpendicular to the field lines generates a magnetic field $B = \varepsilon_0 \mu_0 vE$ in the region through which it passes.

Let us now return to Fig. 21.5, where we see the electric and magnetic fields generated by an antenna. The fields are rushing out along the line of propagation with speed v. Consider the magnetic field as it flies past a point in space. It generates an electric field at that point. Similarly, the electric

James Clerk Maxwell (1831–1879)

Though born to wealth in Scotland, Maxwell had a somewhat spartan childhood. He was an undistinguished student until the age of 13, when his extraordinary intellectual abilities suddenly became apparent. He developed into a shy, quietly humorous man, sociable with friends but often reticent and withdrawn among strangers. After studying at Edinburgh and Cambridge, he held professorships at Aberdeen and London. When his father died, he retired to his family's estate in Scotland, where he researched and wrote his famous treatise on magnetism and electricity. In 1871, he became the first Cavendish Professor of Experimental Physics at Cambridge, and he proceeded to design the famed Cavendish Laboratory. Though he is best known for his profoundly original theoretical work concerning the behavior of gases and electromagnetic waves, his experimental and mathematical talents were extraordinary as well. His contributions to science have proved as original and important as those of Newton and Einstein.

field from the antenna also flies past the same point and generates a magnetic field there.

If you spend a little time considering the situation in Fig. 21.5, you can see that the electric field shown there is in the same direction as the electric field generated by the moving magnetic field. In addition, the magnetic field shown is in the same direction as the magnetic field generated by the moving electric field. We are therefore tempted to say that the electric and magnetic fields in an EM wave regenerate themselves as the wave travels out through space. Let us make this supposition and see where it leads us.

Suppose the electric and magnetic fields of an EM wave generate each other as the wave moves with speed v into space. Then both Eq. (21.3) and

Eq. (21.4) must apply to the wave. If so, B and E must be related in the same way in the two equations, and so the proportionality constants between E and B must be the same. Hence

$$\varepsilon_0 \mu_0 v = \frac{1}{v}$$

Solving this relation for v, the speed of the EM wave in vacuum, we find

$$v = \frac{1}{\sqrt{\mu_0 \varepsilon_0}} \qquad (21.5)$$

But $\mu_0 = 4\pi \times 10^{-7}$ in SI units, while in similar units $\varepsilon_0 = 8.85 \times 10^{-12}$. From them we obtain the result

$$v = 2.998 \times 10^8 \text{ m/s}$$

This is the speed that we compute for EM waves.

When Maxwell obtained this result, he was pleasantly surprised because this speed was well known to him. It is the speed with which light waves travel through vacuum. This was surprising at the time because no one had found any previous connection between light and electricity. Yet Maxwell had shown that the speed of light could be expressed in terms of the fundamental electrical constants ε_0 and μ_0. He was therefore led to the conclusion that *light waves are a form of electromagnetic wave.*

It was not until 1887, eight years after Maxwell's death, that EM waves were consciously generated and detected by electrical means. This feat was performed by Heinrich Hertz. Using an LC circuit, he generated a spark that jumped back and forth across a spark gap in a loop of wire. The wave generated from this "antenna" traveled to a second loop of wire, the "receiver," and induced an emf in it.

Subsequent experiments have fully confirmed Maxwell's supposition and the early experiments of Hertz. We now accept that

All EM waves travel through vacuum with the same speed, $c = 2.998 \times 10^8$ m/s, and light is one form of EM wave. In all EM waves, $E = cB$.

21.5 Definition of Electrical Units

In Maxwell's time, the electrical units were defined in a different way. We now base our definition of these units directly on his discovery concerning the speed of EM waves. Of course, our new definitions are chosen in such a way as to preserve as closely as possible the numerical results we are accustomed to in previous definitions. The SI electric units are defined in the following way.

The permeability of free space μ_0 *is arbitrarily defined to be* $4\pi \times 10^{-7}$ *$N \cdot s^2/C^2$. Using this value, we can define ε_0 in terms of the measured speed of light c.* Thus, since

$$\mu_0 = 4\pi \times 10^{-7} \, N \cdot s^2/C^2$$

and

$$c^2 = 1/\varepsilon_0 \mu_0 = (2.998 \times 10^8 \, \text{m/s})^2$$

we have

$$\varepsilon_0 = 8.85 \times 10^{-12} \, C^2/(N)(m^2)$$

In order to define the unit of current, the ampere, we use the fact that a wire carrying a current experiences a force when placed in a magnetic field. If we consider two long, parallel, straight wires through which the same current I flows, the wires experience forces because each is in the magnetic field of the other. When the separation of the wires is d, the magnetic field at one, because of the current of the other, is simply

$$B = \frac{\mu_0 I}{2\pi d}$$

as we found in Eq. (18.6). The field will be perpendicular to the second wire, and so the force on a length L of the wire will be

$$F = BIL$$

or, after substituting for B and dividing through by L,

$$\frac{F}{L} = \frac{\mu_0 I^2}{2\pi d} \qquad \text{or} \qquad I = \sqrt{\frac{2\pi d}{\mu_0} \frac{F}{L}}$$

Since the force on unit length of the wire, F/L, can be measured, together with the separation of the wires d, the current I can be evaluated in terms of known quantities. When F is measured in newtons and when μ_0 is given the value $4\pi \times 10^{-7} \, N \cdot s^2/C^2$, the current is in the unit we define to be the ampere. Hence, this equation defines the ampere directly in terms of force and length units which we have already defined.

The coulomb of charge is defined to be the charge carried through a cross section of a wire in one second when the current in the wire is one ampere. As a result, the definition of the coulomb is based directly on the same measurement used to define the ampere.

To define the unit of flux density B, we make use of the relation

$$F = B_\perp IL$$

If a 1-m length of wire carries a current of 1 A perpendicular to a magnetic field and the force on that wire is 1 N, then the value of the flux density B is, by definition, 1 T (or Wb/m^2), or $1 \, N \cdot s/(C)(m)$.

The other quantities used in electricity have already been defined in terms of the quantities defined above, together with force, length, and time units. We do not repeat them all here. However, it should be pointed out that we have succeeded in defining all the electrical units in terms of definite experiments involving the measurement of force, length, and time. As a result, anyone who is able to duplicate our units for these three basic quantities will be able to duplicate our electrical units as well.

21.6 Electromagnetic Wave Spectrum

During the more than 100 years since Maxwell's discovery, we have come to realize that there are many types of EM waves. Although all EM waves travel through vacuum with speed c, and even though they all consist of mutually perpendicular electric and magnetic fields, the waves vary widely in wavelength. It is the wavelength of an EM wave that distinguishes it from other types of EM waves.

The wavelength of light waves is exceedingly small. Light waves have wavelengths in the neighborhood of 0.0000005 m, or 5×10^{-7} m. The exact wavelength of light depends on its color, red light being near 6.5×10^{-7} m, while blue light is near 4.3×10^{-7} m.

Radio waves, on the other hand, are a few hundred meters long, while

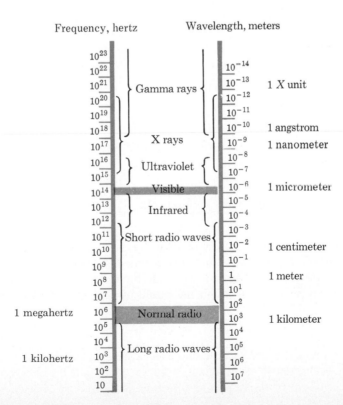

FIGURE 21.9

The electromagnetic spectrum.

radar waves may be as short as a centimeter. We now know that heat radiation consists of EM waves with wavelengths still shorter than radar waves but longer than light waves. As a matter of fact, short-wavelength heat radiation is nothing more than infrared light.

As you see, there is a vast spectrum (or range of wavelengths) of EM waves. This great span of EM wavelengths, extending from radio waves through x-rays and gamma rays, is shown in Fig. 21.9. Notice that EM radiation does not stop with blue light. There are electromagnetic waves still shorter. We are all familiar with the next shorter waves, ultraviolet radiation. Still shorter than these are x-rays. A type of radiation from radioactive substances, the gamma ray (γ ray), is essentially the same as x-ray radiation but still shorter in wavelength.

We see therefore that *EM waves differ widely depending on their wavelength.* The next few chapters are devoted to a very small range of wavelengths, ordinary light. In still later chapters we study other types of radiation in the EM spectrum.

21.7 Energy Carried by EM Waves

We have seen that EM waves consist of moving electric and magnetic fields. Because these fields contain energy, the EM wave carries energy forward with the wave.[3] Let us now compute how much energy is carried to a surface by an EM wave that is incident on the surface.

You will recall from Sec. 16.11 that the energy stored in unit volume of an electric field of magnitude E is $\frac{1}{2}\varepsilon_0 E^2$. Similarly, we showed in Sec. 19.5 that the energy stored in unit volume of magnetic field B is $B^2/(2\mu_0)$. With these facts in mind, let us look at the beam of EM radiation shown in Fig. 21.10; it carries energy through the plane indicated. The beam shown in the figure has an end area A and travels to the right with the speed of light c. Because the beam travels a distance ct during time t, a length ct of the beam moves through the plane indicated in this time. Hence the volume of the beam that travels through the plane in time t is $A(ct)$. This is the shaded volume in the figure. We therefore find that

$$\text{Volume through plane per second} = \frac{A(ct)}{t} = Ac$$

Let us take t short enough that ct is much smaller than the wavelength of the beam's radiation. Then both E and B will be substantially constant throughout the shaded volume in the figure. Therefore we can write that the energy carried through the plane by the beam volume Ac is

---

[3] For example, EM waves from the sun warm the earth and provide the energy for plant growth. Electromagnetic waves from a distant TV station carry the energy that activates your TV.

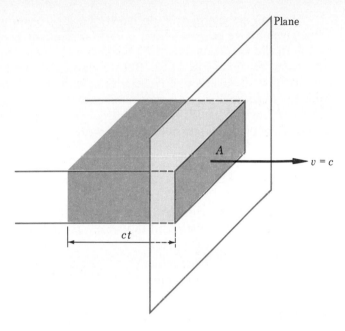

FIGURE 21.10
A volume $A(ct)$ of the beam
goes through the plane in
time t.

$$\begin{pmatrix}\text{Energy in} \\ \text{volume } Ac\end{pmatrix} = \begin{pmatrix}\text{magnetic field} \\ \text{energy density}\end{pmatrix}(\text{volume}) + \begin{pmatrix}\text{electric field} \\ \text{energy density}\end{pmatrix}(\text{volume})$$

or

$$\text{Energy in } Ac = \frac{B^2}{2\mu_0}Ac + \tfrac{1}{2}\varepsilon_0 E^2 Ac$$

To find the energy carried through unit area of the plane in unit time, we need only divide by the area A of the beam. We then have the following for the energy transported through unit area in unit time:

$$\begin{pmatrix}\text{Energy per unit} \\ \text{area per second}\end{pmatrix} = \frac{c}{2}\left(\frac{B^2}{\mu_0} + \varepsilon_0 E^2\right)$$

Or, because $B^2 = E^2/c^2 = E^2\varepsilon_0\mu_0$, this can be written as

$$\begin{pmatrix}\text{Energy per unit} \\ \text{area per second}\end{pmatrix} = \tfrac{1}{2}c\varepsilon_0(E^2 + E^2) = c\varepsilon_0 E^2$$

We can see in this last equation that the magnetic and electric field terms are equal in magnitude. Hence we conclude that

The electric field and the magnetic field of an EM wave transport equal energies.

The average energy transported by the EM beam is not equal to the value given above because E and B vary sinusoidally in a wave. We therefore need the average value of E^2 instead of the value that applies to only the shaded region of Fig. 21.10. But we learned in our study of alternating currents that the average of i^2, for example, was $i_0^2/2$, half the amplitude squared. Because E is also a sinusoidal quantity, the average value of E^2 is simply $\frac{1}{2}E_0^2$, where E_0 is the amplitude of the electric field wave.

In view of this discussion, we can write

$$\begin{pmatrix} \text{Power per} \\ \text{unit area} \end{pmatrix} = \begin{pmatrix} \text{average energy transported} \\ \text{per unit time per unit area} \end{pmatrix} = \tfrac{1}{2}c\varepsilon_0 E^2 \quad (21.6)$$

Or, if we choose, we can write this in terms of B_0, the amplitude of the magnetic field wave. We recall that $E = cB$, and so we have

$$\begin{pmatrix} \text{Power per} \\ \text{unit area} \end{pmatrix} = \tfrac{1}{2}c\varepsilon_0 c^2 B_0^2 = \frac{cB_0^2}{2\mu_0} \quad (21.7)$$

where use has been made of $c = 1/\sqrt{\varepsilon_0\mu_0}$. We therefore conclude that

The power transported through unit area by an EM wave incident perpendicular to the area is $\frac{1}{2}c\varepsilon_0 E_0^2 = cB_0^2/(2\mu_0)$.

Illustration 21.2 Suppose a 1000-W radio station sends its power uniformly in all directions from its antenna. (An actual station beams its signal.) Find E_0 and B_0 in its wave at a distance of 10 km (about 6 mi) from the antenna.

Reasoning The power sent out by the station is spread over the surface of a sphere with radius R and centered on the antenna. At 10 km, the power per unit area of the station's beam is

$$\frac{\text{Power}}{\text{Area}} = \frac{1000 \text{ W}}{4\pi R^2} = \frac{1000 \text{ W}}{4\pi(10{,}000 \text{ m})^2} = 8.0 \times 10^{-7} \text{ W/m}^2$$

We equate this to the expression given in Eq. (21.6) and get

$$\tfrac{1}{2}c\varepsilon_0 E_0^2 = 8.0 \times 10^{-7} \text{ W/m}^2$$

Using $c = 3 \times 10^8$ m/s and $\varepsilon_0 = 8.85 \times 10^{-12}$ C^2/(N)(m^2) gives $E_0 = 0.024$ N/C. Notice how very small the electric field is.

To find the magnetic field, we recall that $E_0 = cB_0$ and find

$$B_0 = E_0/c = 8.2 \times 10^{-11} \text{ T}$$

Even though the numerical value for B is much smaller than E, do not become confused by it; *the E and B fields transport equal energies.*

Summary

Maxwell summarized the behavior of electric and magnetic fields in four equations. Inspired by the symmetry between electricity and magnetism, he postulated that changing electric fields should generate magnetic fields. This postulate led him to predict the existence of electromagnetic waves, and he computed their speed to be the speed of light.

When an oscillator causes reversing charges to appear on the transmitting antenna of a radio station, an electric and magnetic field wave is sent out from the antenna. This electromagnetic (EM) wave consists of two mutually perpendicular waves, one electric, the other magnetic. The two waves are in phase. They travel through empty space with the speed of light, $c = 2.998 \times 10^8$ m/s.

Moving magnetic fields generate electric fields, and moving electric fields generate magnetic fields. In an EM wave, these two fields are related through $E = cB$, where c is the speed of light in vacuum.

There are many types of EM waves. They differ in wavelength. Radio waves have wavelengths of the order of hundreds of meters. At successively shorter wavelengths are radar, infrared (heat), visible light, ultraviolet light, x-rays, and γ rays. They cover a wavelength range down to less than 10^{-10} m.

Electrical units are defined in terms of the speed of light. By definition, $\mu_0 = 4\pi \times 10^{-7}$ N $\cdot$ s^2/C^2. Since the speed of light is given by $1/\sqrt{\varepsilon_0\mu_0}$, this fact is used to define ε_0. All other electrical units can then be defined in terms of the units of mass, length, and time.

Electromagnetic waves transport energy with a speed equal to the speed of the wave. Half the energy is carried by the magnetic field and half by the electric field. The power transported through unit area perpendicular to an EM wave beam is $\frac{1}{2}c\varepsilon_0 E_0^2$, where E_0, is the amplitude of the wave.

Minimum Learning Goals

Upon completion of this chapter, you should be able to do the following:

1 Describe Maxwell's equations in qualitative terms.
2 Explain qualitatively how EM waves are sent out from a radio-station antenna.
3 Sketch the **E** and **B** fields in an EM wave.
4 Describe the two ways in which radio waves can be detected by a radio. Explain the function of an LC circuit in the radio. Explain how a particular station is selected by the radio.
5 Give the speed of EM waves in vacuum.
6 Compute the wavelength of an EM wave when the frequency of its source is given.
7 Arrange a list of various types of EM waves in order of decreasing wavelength. State to which type of wave a given wavelength belongs.
8 Outline the method for definition of the following quantities: μ_0, ε_0, ampere, coulomb.

Important Terms and Phrases

You should be able to define or explain each of the following:

EM wave

EM wave spectrum
Maxwell's equations
Speed c of EM waves

Questions and Guesstimates

1 Some radio stations have their transmitting antenna vertical while others have theirs horizontal. Describe and compare the EM waves generated by these two types of antennas. In particular, how are **B** and **E** directed relative to the earth's surface?
2 Refer to the previous question. If you open up a

transistor radio, you can see how its coil antenna is mounted. How could you use the radio to tell whether a distant station's antenna is vertical or horizontal?

3 Electromagnetic waves from most of the radio stations in the world are passing through the region around you. How does a radio or TV set select the particular station you want to listen to? When you turn the dial on a radio, what is happening inside to select the various stations?

4 There are two types of radio and TV receiving antennas in use. One picks up the electric part of the EM wave, and the other picks up the magnetic. Examine a pocket transistor radio or a table radio and see which method is used. Is it possible to use both?

5 From time to time in the movies or on TV one sees the good guys trying to locate a clandestine radio transmitter by driving through the neighborhood with a device which has a slowly rotating coil on top. Explain how the device works.

6 It is claimed that in the vicinity of a very powerful radio-transmitting antenna, one can sometimes see sparks jumping along a wire farm fence. What do you think of this claim?

7 In microwave ovens, foods and utensils are sub-

jected to very-high-frequency radar (EM) waves. If a spoon is left in such an oven, it becomes very hot. What heats it? Can you explain the heating action in terms of the electric part of the wave? The magnetic? How are nonmetallic substances heated in the oven? Will a glass dish heat up in such an oven?

8 There is some doubt about the safety of human exposure to intense radio and microwaves. Why would one expect the danger to depend on the frequency of the waves? Which would you expect to present the most danger (if any), radio or microwaves?

9 Refer to Fig. 21.7. Find the direction of the electric field at A induced by the moving magnetic field.

10 Refer to Fig. 21.8. Find the direction of the magnetic field at P induced by the moving electric field.

11 Are the direction and phase of the magnetic part of the wave in Fig. 21.5 drawn properly if the magnetic field is generated by the moving electric field? Repeat for the electric field generated by the moving magnetic field.

12 Estimate the wavelength of the EM wave generated by the vibration of a positively charged ball suspended as a pendulum. Compare the wavelength to the diameter of the earth, 12,700 km.

Problems

1 Police radar uses EM waves with frequencies close to 1.0×10^{10} Hz. What is the wavelength of an EM wave that has this frequency?

2 Radio station WJR in Detroit operates on a frequency of 760 kc/s. (a) What is the wavelength of the wave it sends out? (b) How long does it take a wave crest sent out from the station to reach the moon, 3.84×10^8 m away?

3 As your heart beats, small potential differences are set up between various parts of your body. (These are measured by electrocardiograms and similar medical tests.) Suppose your heart beats 75 times each minute. (a) What is the frequency of the EM wave sent out by your body as a result of this consequent oscillating voltage and charge on your body? (b) What is the wavelength of the very faint EM wave your body sends out?

4 One way to provide heat to muscles and other portions of the body is by diathermy. In this, radar waves are sent into the body much as a microwave oven sends waves into the material to be heated.

The oscillating electric field of the wave causes dipolar molecules and ions to move back and forth, thereby generating friction-type heat. The oscillating magnetic field of the wave induces emf's which cause joule heating. Standard diathermy frequencies are 900 and 2560 MHz. (a) What are the wavelengths of EM radiation in air which arise from these frequencies? (b) Can you see any problem in using diathermy on the face of someone whose teeth have metal fillings?

5 An explosion occurs 5.0 km from an observer. How long after the observer sees the explosion will the sound from the explosion be heard? (Take the speed of sound to be 340 m/s.)

6 Galileo tried to measure the speed of light by stationing two people with lamps a distance D apart. Person A was to expose his lamp to person B and, when B saw it, she was to expose her lamp to A. The time between A exposing his lamp and his observation that B's lamp had been exposed was to be measured. How large would D have to be for this time to be 2.0 s?

7 The resonant circuit in the detection system of a radio is to be set to receive a station whose frequency is 1000 kHz. If the resonant circuit has $C = 800$ pF (that is, 800×10^{-12} F), what value must L have for the circuit?

8 Television channel 5 has a frequency of about 80 MHz. In order to be tuned to this frequency, what must be the value of the inductance in the tuning circuit if $C = 200$ pF?

9 A certain bar magnet has $B = 0.75$ T at its end. The magnet is given a speed of 5.0 m/s, so that it moves perpendicular to its length. (*a*) As its end passes close to a certain point, how large an electric field is induced at the point? (*b*) Is this an easily observed electric field?

10 Refer to Fig. 21.7. Assume $v = 5.0$ m/s and B between the pole pieces is 0.60 T. (*a*) How large is E at point A at the instant shown? (*b*) Would the electric field at A be easily observed? (*c*) What is the direction of **E** at point A?

11* Refer to Fig. 21.8. Suppose the electric field at point P due to the charge on the wire is 300 V/m and speed $v = 5.0$ m/s. (*a*) What value does the induced B field have at point P? (*b*) Is this value large enough that B could be measured easily? (*c*) What is the direction of B at point P?

12* The electric field between the plates of an air-filled, parallel-plate capacitor is 4000 V/m. Suppose the capacitor is moving in a direction parallel to the plates with a speed of 5.0 m/s. (*a*) What is the value of B at a point through which the electric field is moving? (*b*) Would this be a strong enough magnetic field to measure easily?

13 (*a*) What is the numerical value of the E/B ratio for an EM wave? (*b*) If B in an EM wave is to be 1.00 G, how large must the electric field be? (*c*) Is this an easily attainable electric field in air?

14 How large is the maximum magnetic field in an EM wave for which $E_0 = 2.0$ mV/m?

15 The pull-out, straight-wire antenna for a certain radio is 80 cm long. The antenna is oriented along the electric field line direction of an EM wave. How large a maximum voltage will be induced in the antenna if the amplitude of the radio wave is 1.00 mV/m? How large is the amplitude of the B field for this wave?

16 A wire 10 m long is oriented in the direction of the electric field vector in an electromagnetic (radio) wave. (*a*) If the voltage difference between the ends of the wire is to have a maximum value of 1.00×10^{-4} V, how large must the maximum electric field of the wave be? (*b*) What is the maximum value of the magnetic field in the same wave? (This would be a rather weak signal.)

17** A rectangular loop antenna measures 30×10 cm. It has 200 loops of wire on it. About how large an emf can be generated between the terminals of the loop by an electromagnetic wave of frequency 10^6 Hz which has a maximum electric field of 1.00×10^{-5} V/m? (*Hint:* As a rough approximation, use the change in flux per $\frac{1}{4}$ cycle. If you know calculus, use the fact that $B = B_0 \sin 2\pi ft$ and $\Delta B/\Delta t = dB/dt$.)

18** A certain coil has 200 loops of wire on it and has a cross-sectional area of 400 cm². It is placed in an EM wave from a nearby radio station that operates at a frequency of 8.0×10^5 Hz. By orienting the coil properly, a sinusoidal emf whose maximum value is 20×10^{-6} V is induced in the coil. About how large is the amplitude of (*a*) the magnetic field and (*b*) the electric field in the wave? (*Hint:* See the hint in Prob. 17.)

19* A typical laser used in student laboratory experiments has a cylindrical beam with a cross-sectional area of 10 mm². The power in its beam might be 0.25 mW. Assuming the beam to be composed of a single sinusoidal wave (not true), find the values of E_0 and B_0 for the wave.

20* A certain searchlight sends out 10,000 W of EM radiation by means of a beam that has a cross-sectional area of 0.80 m². Assuming the beam consists of a single sinusoidal EM wave (not true), find the values of E_0 and B_0 in the wave.

21 The negative electron circles the positive nucleus with a frequency of 3.6×10^{15} Hz in Bohr's picture of the hydrogen atom. If this system obeyed the laws studied in this chapter, it should act like an antenna radiating waves of this frequency. (*a*) Find the wavelength of the waves. (*b*) State in which portion of the EM spectrum they should be found.

22 Many dipolar gas molecules (such as carbon dioxide and hydrogen chloride) act as two oppositely charged balls connected by a spring. For example, the CN molecule has a natural resonance frequency for stretching vibrations that is about 3×10^{12} Hz. (*a*) Find the wavelength of EM radiation that these vibrations should cause the molecule to emit. (*b*) State in which portion of the EM spectrum these waves are found.

22
Properties of Light

In this and the next few chapters, we are concerned primarily with a very small portion of the entire electromagnetic spectrum. This portion is composed of the small range of wavelengths to which the eye is sensitive, the wavelengths referred to as **light.** However, even though our primary concern will be with light, much of what we learn is applicable to all EM radiation.

22.1 Concept of Light

Even in ancient times the properties of light were a source of wonder and a stimulus to experimentation. Its nature has always been a subject of great speculation. In Newton's time, scientific investigations into the properties and nature of light were conducted by nearly all the scientists of the day. Newton himself derived a great deal of his fame from his experiments with light.

In spite of this wide interest in light, its very nature remained in dispute even until the first decade of the present century. During Newton's time, and for many years later, there was disagreement whether a light beam was a stream of corpuscles or a wave of some sort. Newton was a great proponent of the corpuscular theory, and, because of his prestige, many others were inclined to this view as well. In 1670 Christian Huygens, Newton's contemporary, was able to explain many of the properties of light by considering it to be wavelike. Both these ideas concerning the nature of light had their supporters.

OPTICAL SPECTRA

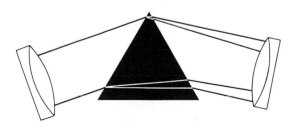

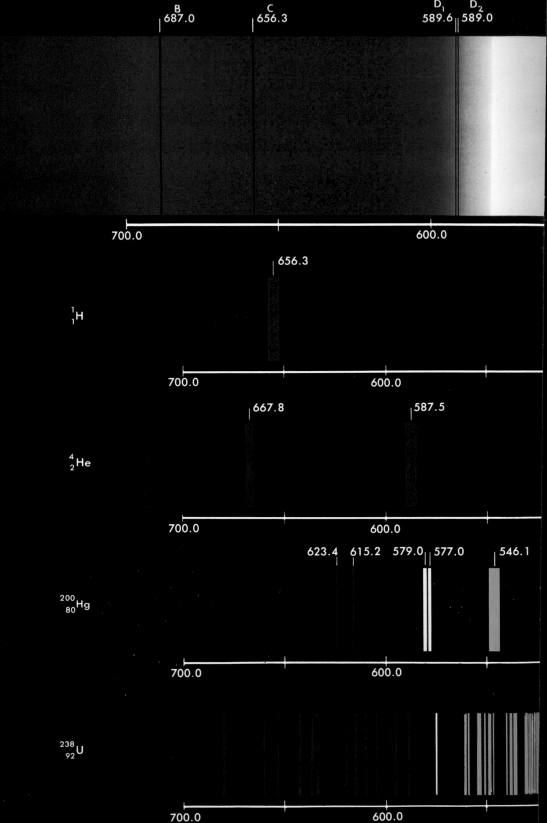

B
687.0

C
656.3

D₁
589.6

D₂
589.0

700.0   600.0

656.3

$^1_1$H

700.0   600.0

667.8   587.5

$^4_2$He

700.0   600.0

623.4 615.2 579.0 577.0   546.1

$^{200}_{80}$Hg

700.0   600.0

$^{238}_{92}$U

700.0   600.0

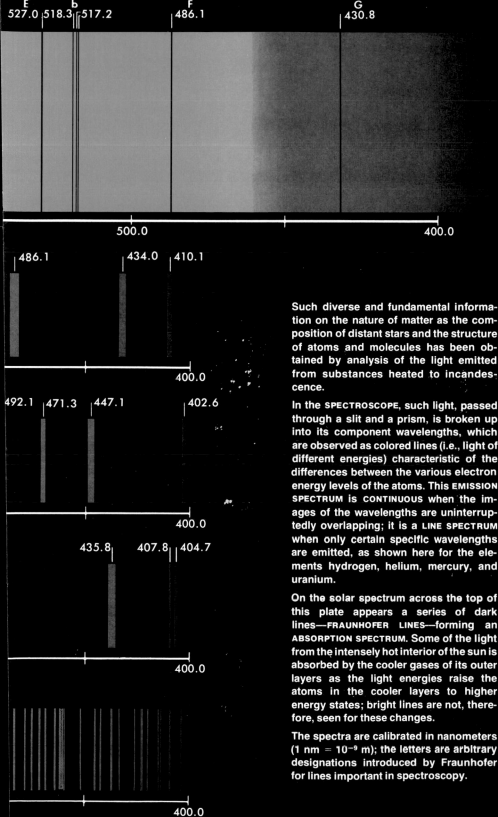

E
527.0 b 518.3 517.2   F 486.1   G 430.8

500.0   400.0

486.1   434.0   410.1

400.0

492.1   471.3   447.1   402.6

400.0

435.8   407.8   404.7

400.0

400.0

Such diverse and fundamental information on the nature of matter as the composition of distant stars and the structure of atoms and molecules has been obtained by analysis of the light emitted from substances heated to incandescence.

In the SPECTROSCOPE, such light, passed through a slit and a prism, is broken up into its component wavelengths, which are observed as colored lines (i.e., light of different energies) characteristic of the differences between the various electron energy levels of the atoms. This EMISSION SPECTRUM is CONTINUOUS when the images of the wavelengths are uninterruptedly overlapping; it is a LINE SPECTRUM when only certain specific wavelengths are emitted, as shown here for the elements hydrogen, helium, mercury, and uranium.

On the solar spectrum across the top of this plate appears a series of dark lines—FRAUNHOFER LINES—forming an ABSORPTION SPECTRUM. Some of the light from the intensely hot interior of the sun is absorbed by the cooler gases of its outer layers as the light energies raise the atoms in the cooler layers to higher energy states; bright lines are not, therefore, seen for these changes.

The spectra are calibrated in nanometers (1 nm = 10^{-9} m); the letters are arbitrary designations introduced by Fraunhofer for lines important in spectroscopy.

It was not until 1803, when Thomas Young (and a little later Augustin Fresnel) presented evidence to show that light beams could interfere with one another much as sound waves, that the wave theory became almost universally accepted. At about this time the speed of light in water was measured. The observed speed was less in water than in air. This contradicted the corpuscular theory and supported the wave theory. Hence, by 1865, when Maxwell found theoretically that EM waves should travel with the speed of light, the idea of light waves was fairly well accepted.

One would think, then, that by 1900 the nature of light was reasonably well understood. However, at that time people still knew very little about the emission of light by atoms. It was not until about 1913 that Bohr gave the first reasonably correct interpretation for the mechanism of light emission. His concepts were greatly modified, and it was not until about 1930 that the emission of light could be said to be well understood. In addition, Einstein showed in 1905 that at least one property of light, the photoelectric effect, which is discussed in Chap. 25, was best explained by considering light to act as quanta or particles. This concept has been expanded through the years until we now consider light to possess a sort of dual personality, part wavelike and part particlelike. More is said about these and other developments in later chapters.

It is clear that the subject of light has a long and varied scientific history. We expect that in years to come our understanding of the nature of light will continue to grow. For the next few chapters, however, it is sufficient to concentrate on the aspects of light evident from its EM character. Other characteristics of light involving its particle nature and dealing with its atomic origin are discussed later in this book.

The wavelengths of visible light waves can be measured by methods discussed in Chap. 24. These wavelengths turn out to lie in the range 400 to 700 mm. The position of the various colors on the wavelength scale is shown in Fig. 22.1 and also on the color plate insert. Other units frequently used to measure wavelengths are the angstrom (Å), where $1 \text{ Å} = 10^{-10}$ m, and the micron (μ), where $1 \mu = 10^{-6}$ m. When the complete spectrum of light shown in the upper part of the color plate enters the eye, as from a white-hot object, the light appears white and is called *white light*.

Light waves are electromagnetic in nature, consisting of an electric field perpendicular to, and in phase with, a magnetic field, as discussed in the last

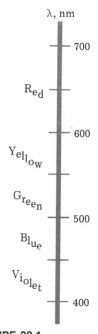

FIGURE 22.1
The correspondence between wavelengths and color shown here is only approximate. Colors such as blue-green and orange occupy the intermediate regions. (See the color plate.)

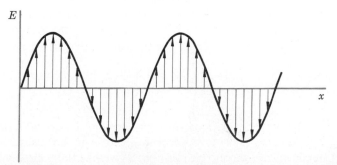

FIGURE 22.2
The electric field of the electromagnetic wave vibrates perpendicular to the direction of propagation. Hence the wave is transverse.

chapter. The electric field for a wave propagating in one direction in space is shown in Fig. 22.2. It is assumed that the wave is moving along the x axis to the right. Notice that the vibrating electric field is perpendicular to the x axis. Hence, *light waves are transverse waves,* since the vibration is perpendicular to the direction of propagation. As such, they have many properties in common with waves on a string or on the surface of water, since these, too, are transvese waves.

22.2 Speed of Light

There are several methods for determining the speed of light. We mention here only one of these general methods, the method which has been most widely used in high-precision work. It was used by Michelson in the early 1930s to measure the speed of light between two mountaintops in California. The same general method had been used previously over much smaller distances to measure the speed of light in materials other than air.

The method uses the apparatus shown in simplified form in Fig. 22.3. A beam of light from the source is reflected from one side of a cube M having mirrored surfaces on four sides. The beam of light is then reflected from mirror M' to the cube, where it is reflected again as shown. If the cube is at just the right position, the beam of light will enter an observer's eye in the position indicated. (Actually the observer looks through a telescope system placed where the eye is shown and sees an image of the source.)

Suppose, however, that the cube is rotating about an axis through its center, perpendicular to the page. When it is in the position indicated by the heavy lines in Fig. 22.3, the beam is reflected to mirror M', as shown. However, by the time the beam returns to the cube from M', the cube will perhaps have rotated to the dashed position and the light beam will not be reflected into the observer's eye. The cube must rotate through $\frac{1}{4}$ rev during the time needed for the beam to travel to M' and back if the beam is to be properly reflected, for only then will the cube reflect the beam into the eye of the observer.

The measurement technique is to speed the rotation of the cube until the reflected beam is properly reflected. At that speed of rotation, we know that the time taken for $\frac{1}{4}$ rotation of the cube is equal to the time the light takes to travel a distance $2D$. It is necessary only to know the speed of rotation of the cube and D in order to compute the speed of the light. The value at present accepted for the speed of light in vacuum is

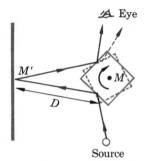

FIGURE 22.3
If the mirror is rotating at just the proper speed, the beam will be reflected into the eye of the observer. In practice the distance D is much larger than shown.

Speed of Light

$$c = 2.997\ 924\ 574 \times 10^8 \text{ m/s} \pm 1.2 \text{ m/s}$$

The speed of light in air is only about 0.03 percent less than this value.

Light travels fastest through vacuum. Its speed in other materials is always less than c. Moreover, the speed in materials other than vacuum depends on the wavelength of light as well as the constitution of the material.

TABLE 22.1
REFRACTIVE INDICES ($\lambda = 589$ nm)

Material	$c/v = n$	Material	$c/v = n$
Air*	1.0003	Crown glass	1.52
Water	1.33	Sodium chloride	1.53
Ethanol	1.36	Polystyrene	1.59
Acetone	1.36	Carbon disulfide	1.63
Fused quartz	1.46	Flint glass	1.66
Benzene	1.50	Methylene iodide	1.74
Lucite or Plexiglas	1.51	Diamond	2.42

*Normal temperature and pressure.

We list in Table 22.1 the ratio of the speed of light in vacuum to that in various materials. The wavelength used for the table is $\lambda = 589$ nm. This is the wavelength of the yellow light given off by a sodium-vapor lamp. This ratio is called the *index of refraction* of the material. More is said about it later.

22.3 Reflection of Light

When a stone is dropped into a large, still pond of water, a set of circular waves moves out from the point where the stone hit the water. We are all familiar with this situation, but let us look at it in some detail. The circular waves, or **wavefronts,** are shown in Fig. 22.4. They travel outward from the center in the directions shown by the arrows. These arrows, in the direction the wavefronts travel, are called **rays.** Notice that *the rays are always perpendicular to the wavefronts.* Hence we can specify the motion of a wave either by use of rays or by drawing the wave itself. Both methods are of value.

If we consider the case where we are far removed from the source of the wave, the wavefronts become nearly straight lines (or flat planes if we are dealing with three dimensions). This is illustrated in Fig. 22.5. As we can infer from that figure, at distances far removed from the source, the waves are plane waves and the rays are parallel. It is often convenient to use plane

FIGURE 22.4
The rays are perpendicular to the wavefronts and show the direction of motion of the wave.

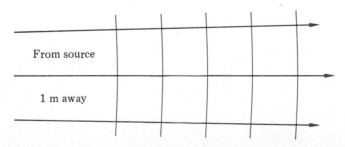

From source

1 m away

FIGURE 22.5
Rays from a distant source are nearly parallel. Notice that the wavefronts are nearly flat.

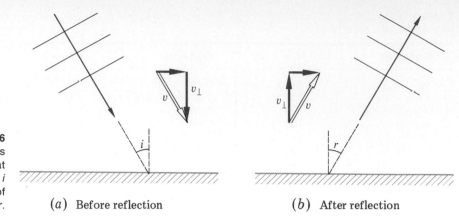

FIGURE 22.6
The incident wave is
reflected in such a way that
the angle of incidence i
equals the angle of
reflection r.

(a) Before reflection   (b) After reflection

waves in computations, and when the source is far removed, or when a suitable lens is used, the waves are very nearly of this sort.

Suppose that a series of plane water waves is incident on a flat wall, as shown in Fig. 22.6a. The velocity of the incoming wave can be thought of as being split into two components, one $v_\perp$ perpendicular to the wall, the other parallel to the wall. Upon striking the wall, $v_\perp$ will be reversed in direction, as shown in part b of the figure. As a result, the wave is reflected backwad from the surface as shown. Let us now find how the angle of incidence i, shown in part a, is related to the angle of reflection r, shown in part b.

As you can see from Fig. 22.6a, we have

$$\cos i = \frac{v_\perp}{v}$$

In Fig. 22.6b we see that

$$\cos r = \frac{v_\perp}{v}$$

Therefore, because their cosines are equal, the angle of incidence equals the angle of reflection. This is shown in Fig. 22.7a.

The fact that a water wave is reflected in such a way that the angle of incidence equals the angle of reflection is of general validity. We could have

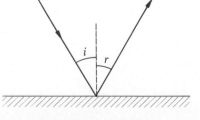

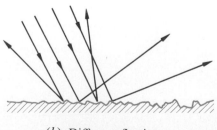

FIGURE 22.7
The angle of incidence
equals the angle of
reflection.

(a) Specular reflection   (b) Diffuse reflection

used the same reasoning to show that light waves would also be reflected in this way. Notice that the only basic assumption made was that the velocity component perpendicular to the surface was reversed upon reflection and the other component was unchanged. Our result is true for any type of wave for which this assumption is true. Measurements on light and other forms of electromagnetic radiation confirm our deduction. We may therefore formulate the following rule:

The angle of incidence equals the angle of reflection.

The type of reflection shown in Fig. 22.7a is called *specular reflection.* It occurs at a smooth, mirrorlike surface. Rougher surfaces such as paper, painted walls, and so on give rise to *diffuse reflection,* shown in Fig. 22.7b. For such surfaces, even though the law of reflection applies to individual rays, the nonsmooth surface causes the rays in a light beam to be reflected at various angles from the average plane of the surface.

22.4 Plane Mirror

Let us now apply what we have found about reflection to the important topic of image formation by mirrors. First we consider how a plane mirror forms an image.

Every day you look at yourself in a plane mirror. You see the image of your face in front of you. Actually, if you stop to examine exactly what you are doing, you perceive that the image of your face is behind the surface of the mirror. In fact, it appears to be about as far behind the mirror as your face is in front of the mirror. Let us now examine such a reflection in a plane mirror in order to understand clearly why the image is seen as it is.

Suppose that you place an object in front of a mirror, as shown in Fig. 22.8a. You wish to find where the eye shown in the figure will see the image of the object to be. When you see the tip of the object, your eye sees a ray of light which was emitted or reflected by the object tip. If you see the image of the object tip in the mirror, you really see light from the object tip which was reflected by the mirror, as shown in Fig. 22.8a. It is apparent from this

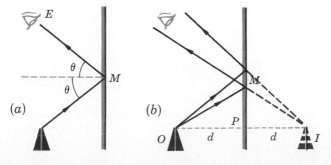

FIGURE 22.8
The image formed by the plane mirror is as far behind the mirror as the object is in front of it.

same figure that you will see the object as being off in the general direction from which the ray came. Hence you know that the image of the object tip will be somewhere along the line *EM* or its extension.

To see exactly where the image of the object will be, refer to Fig. 22.8*b*. The two reflected rays shown appear to the eye to have come from the single point *I*. It is at this point that the image of the object tip will appear to be. In fact, the same reasoning may be used to show that the total image of the object will appear as shown at *I*. From the geometry of the triangles *OMP* and *IMP,* it is clear that if the object is a distance *d* in front of the mirror, the image is a distance *d* behind the mirror.

Virtual (or Imaginary) Image

This type of image, one through which the observed rays do not actually pass, is called a **virtual** *or* **imaginary image.** In other words, the rays reaching the eye do not really come from the point where we see the image. There is no possibility whatsoever that a sheet of paper placed at *I* behind the mirror would have a lighted object appear on it. The mind merely imagines that the light comes from *I*. It is always true, of course, that the image of an actual object seen by reflection in a plane mirror is a virtual image. The image is always exactly as far behind the mirror as the object is in front.

22.5 Focus of a Concave Spherical Mirror

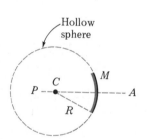

FIGURE 22.9
A spherical mirror *M* is a portion of a hollow sphere. The radius of the mirror is *R*, its center is point *C*, and its principal axis is line $\overline{PA}$.

Definition

Although flat, or plane, mirrors are used by all of us, spherical mirrors are not quite as common. However, makeup mirrors and shaving mirrors are portions of spheres as are surveillance mirrors in stores. A spherical mirror is actually a portion of the surface of a hollow sphere, as shown in Fig. 22.9. If the sphere reflects from its inner surface, it will reflect light as shown in Fig. 22.10*a*. Such a mirror is called a **concave spherical mirror.** In the event that the mirror reflects from its outside surface, it will reflect light as shown in Fig. 22.10*b*. This type of mirror is called a **convex spherical mirror.**

Notice that in drawing Fig. 22.10 we assumed that the light came from a distant source so that the rays are parallel and the wavefronts are straight or plane. (Recall that a water wave loses its curvature as it goes farther and farther from the source of the wave.) The figure indicates that the parallel rays traveling along the principal axis of the mirror (defined in the legend of Fig. 22.9) are all reflected to or from a point *F*. This is approximately correct, as we shall soon show. *This point to which the light from a distant object is reflected by a concave mirror is called the* **focus** *(or* **focal point**) *of the mirror.* If we were to reflect light from the sun by a concave mirror, the light would be reflected into a tiny spot (an image of the sun) very near the focus of the mirror. This follows because the light waves from a source as distant as the sun should be nearly plane by the time they reach the earth. The rays of light from the distant sun are so nearly parallel that we can consider them so.

Until now we have asked you to accept the above assertions without proof. However, it is a simple matter to demonstrate exactly how three special types of the infinite number of rays from a source will be reflected by

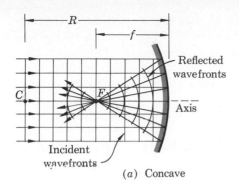

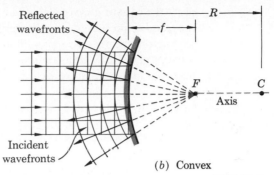

(a) Concave   (b) Convex

FIGURE 22.10
The reflected light from a
concave mirror is converged
to the focus F, while the
light reflected from a
convex mirror appears to
diverge from the focus.

a concave mirror. Once this is known, it becomes quite easy to ascertain the position of an image formed by such a mirror. We now consider how the concave mirror reflects these three important rays.

First, consider the ray shown in Fig. 22.11. It is parallel to the axis of the mirror, striking it at A. Let us draw a radius CA from the center of the sphere of which the mirror is a part. Then, since CA is a radius of the sphere, it is perpendicular to the mirror surface at A. The reflected ray makes the same angle to the perpendicular at A as the incident ray does, since the angle of incidence at any smooth reflecting surface equals the angle of reflection. Moreover, since the incoming ray is parallel to the axis CB, the angle at C must also be i, as indicated.

The triangle CFA is isosceles, and CF must therefore be equal to FA. But if the angle i is quite small, the length FA is nearly the same as FB. Hence any ray parallel to the axis and not too far from it will be reflected through a point F that is halfway between the mirror surface B and the center of the sphere of which the mirror is a part. In other words, if the mirror has a radius of curvature R, rays parallel to the axis of the mirror will be reflected through the point F, which is a distance $\overline{FB} = \frac{1}{2}R$ from the mirror. This point, *the point to which parallel rays are focused, is the focus of the mirror, and the length $\overline{FB}$ is called the* **focal length** *f of the mirror. We have $f = \frac{1}{2}R$.*

Focal Length f

Of course, it is not completely true that all rays parallel to the axis are focused exactly at the focal point. An approximation was made in arriving at this result. If the length AB in Fig. 22.11 is only a small fraction of the sphere diameter, the approximation is fairly good. However, you should

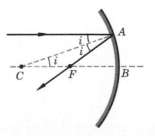

FIGURE 22.11
A ray parallel to the
principal axis of a concave
mirror is reflected through
the focal point.

595

draw the case of a ray reflected at an angle $i \approx 90°$ to see that, in that case, the ray will not be reflected through the focus. For this reason useful spherical mirrors are only a very small portion of a sphere in size. Larger spherical mirrors give fuzzy images. Parabolic mirrors do not suffer this disadvantage, but they are more expensive to make.

22.6 Three Reflected Rays and Image Formation

We now know that a ray parallel and near to the axis of a spherical mirror may be considered to reflect through the focal point. There are two other rays which are also easily traced. All three of the rays are shown in Fig. 22.12a to c.

Ray 1 in part *a* has already been discussed. Ray 2 in part *b* is just the reverse of ray 1, and the same geometric arguments used for ray 1 apply to ray 2 as well. Ray 3 in part *c* passes through the center of curvature of the mirror and strikes the mirror. Since it is traveling along a radius of the mirror surface, it hits the mirror perpendicularly. Hence it will be reflected straight back on itself, as shown. We therefore have the following rules *for concave mirrors:*

A ray parallel to the mirror axis is reflected through the focus.

A ray through the focus is reflected parallel to the mirror axis.

A ray through the center of curvature is reflected through the center of curvature.

These rules can now be used to locate the position of images.

Suppose that we wish to find the image formed by the mirror of the object *O* shown in Fig. 22.13. Let us say that the object is a light bulb. The bulb will emit light in all directions, but we know how to treat only three of the millions of rays we could draw coming from it. These three rays travel as shown in Fig. 22.13. They are exactly the rays described by our rules, and you should trace each in turn to see that it is properly drawn. Once the positions of *C* and *F* are known, only a straightedge is needed to draw these rays.

FIGURE 22.12
The three rays shown are easily drawn with a straightedge provided that the center *C* and the focus *F* of the mirror are known.

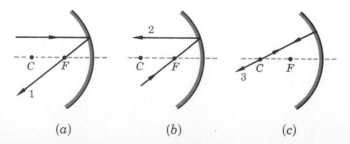

(a) (b) (c)

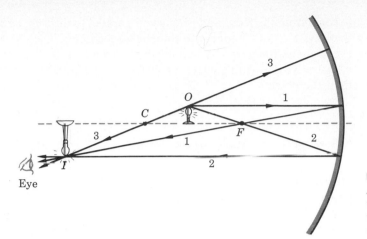

FIGURE 22.13
A real image *I* is formed
from the object *O*. Trace
the three rays from the
object.

If you now place your eye as shown, the three rays will appear to come from point *I*. You will actually see the light bulb to be at *I*, and this is called the **image** of the object at *O*. Moreover, since the rays of light actually do converge on point *I* and pass through it, a sheet of paper placed at *I* will show a lighted picture of the original bulb. This is a **real image**: *at a real image, the light actually passes through the point to reproduce the object.* Notice how this differs from the imaginary, or virtual, image found for the plane mirror.

Definition

Real Image

Suppose that the bulb was on the tip of the post represented in Fig. 22.13. Rays of light will emanate from it as well as from the bulb. (For example, light from the bulb might strike the post and be reflected from it. This light would appear to be emitted from the post, and it will act as if it were emitted rather than reflected light.) We could treat each little portion of the post as a new light source and find its image. You should trace a few rays to show yourself that the image of the post lies along the image shown at *I*. From now on, we locate only a single point of an image and immediately draw in the rest of the image in this way.

Consider the situation shown in Fig. 22.14. Once again, we draw our three rays. Now, however, ray 2 does not go through the focus on its way to the mirror, since the object is inside the focal point. However, it still appears

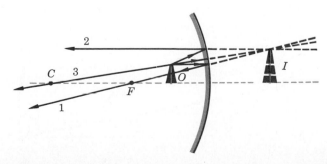

FIGURE 22.14
The three rays appear to
come from the virtual image
I. Notice especially rays 2
and 3.

to come from the focus and is reflected parallel to the axis, as always. Notice that rays 1, 2, and 3 appear to come from the image *I*, as shown. Hence, the eye sees the image to be behind the mirror in this case. Notice that the image is virtual (imaginary), erect (right side up), and magnified (larger than the object).

22.7 Mirror Equation

To derive a mathematical equation which describes the location of the image, let us refer to the example shown in Fig. 22.15. The ray *ABE* in part *a* is not a usual one. However, it is reflected in such a way that angle *ABH* equals angle *FBE*. For this reason, the shaded triangles in Fig. 22.15*a* are similar triangles. Taking the ratios of corresponding sides gives

$$\frac{O}{I} = \frac{p}{i}$$

In Fig. 22.15*b*, the shaded triangles are also similar. Hence

$$\frac{O}{I} = \frac{\overline{HF}}{\overline{FG}}$$

But $\overline{HF}$ is just $p - f$, and $\overline{FG}$ is nearly f. To this approximation we have

$$\frac{O}{I} = \frac{p - f}{f}$$

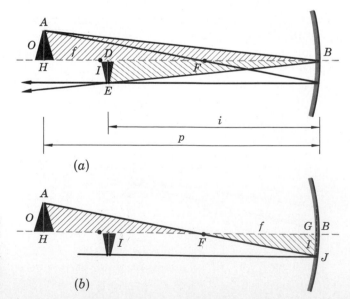

FIGURE 22.15
The shaded triangles are similar in each drawing. We assume that the distance $\overline{FG}$ in part *b* is negligibly different from $\overline{FB}$.

Equating this to the expression found in part *a* of the figure gives

$$\frac{p}{i} = \frac{p - f}{f}$$

After dividing this equation by p and rearranging, we have

$$\frac{1}{p} + \frac{1}{i} = \frac{1}{f} \qquad (22.1) \qquad \textit{Mirror Equation}$$

where $f = \frac{1}{2}R$ and R is the radius of curvature of the mirror.

Equation (22.1) is the mirror equation. It allows us to compute the distance i of the image from the mirror surface, provided that the object distance from the mirror surface p and the focal length f are known. To compute the relative heights of the object and image, we note that $O/I = p/i$, as found previously. *The* **magnification** *produced by the mirror is defined to be the ratio of the image height to the object height.* Therefore,

Definition

$$\text{Magnification} = \frac{I}{O} = \frac{i}{p} \qquad (22.2) \qquad \textit{Magnification}$$

Illustration 22.1 An object 2 cm high is placed 30 cm from a concave mirror having a radius of curvature of 10 cm. Find the position and size of the image.

Reasoning The mirror relation applies, with $p = 30\,\text{cm}$ and $f = \frac{10}{2} = 5\,\text{cm}$. Hence, with all distances in centimeters,

$$\frac{1}{30} + \frac{1}{i} = \frac{1}{5} \qquad \text{or} \qquad \frac{1}{i} = \frac{6}{30} - \frac{1}{30} = \frac{5}{30}$$

and $\qquad\qquad\qquad i = 6\,\text{cm}$

The image will be on the same side of the mirror as the object, i.e., on the silvered side, and since the light will actually be passing through it, the image is real. Its size is found from Eq. (22.2) to be

$$\frac{I}{2} = \frac{6}{30} \qquad \text{or} \qquad I = \tfrac{2}{5}\,\text{cm high}$$

It is always wise to check the algebraic solution by drawing the appropriate ray diagram.

Illustration 22.2 An object is placed 5.0 cm in front of a concave mirror having a 10-cm focal length. Find the location of the image.

Reasoning If we refer to Fig. 22.14, we see that this example should give an image on the wrong side of the mirror, i.e., the unsilvered back. Our answer should make this evident. Using the mirror equation, taking all distances in centimeters, we have

$$\frac{1}{5} + \frac{1}{i} = \frac{1}{10}$$

which gives

$$\frac{1}{i} = \frac{1}{10} - \frac{2}{10} = -\frac{1}{10} \qquad \text{or} \qquad i = -10 \text{ cm}$$

Notice that i is negative. Hence, when images are behind, i.e., on the back of, the mirror, the image distance is negative. This is not surprising in view of the fact that i was taken positive for the reverse case. Clearly, the image must be virtual in this event. An image behind the mirror is, of course, virtual. It does not appear on a screen placed there.

22.8 Convex Mirrors

A convex spherical mirror is a portion of a sphere which has a reflective coating on the outside surface. This surface is convex when viewed from outside the sphere, and hence the mirror is called a convex mirror. In Fig. 22.10 we illustrated the behavior of parallel rays reflected from such a mirror. They appear to diverge from a point behind the mirror. This point is said to be the **focal point** for the convex mirror. *Parallel rays incident on a convex mirror are reflected as though they came from the focal point.* To prove that the parallel rays behave in this way, we proceed in much the same way as we did for the concave mirror.

Focal Point of Convex Mirror

Referring to Fig. 22.16, we see from the law of reflection and the geometry involved that several angles are equal, as shown. The triangle AFC is isosceles so that $\overline{AF} = \overline{FC}$. If the length $\overline{AB}$ is small compared with the

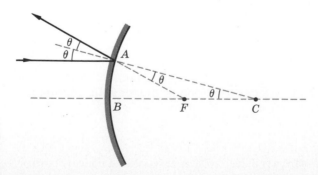

FIGURE 22.16
A ray parallel to the axis is reflected as though it came from the focal point.

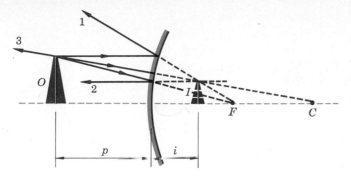

FIGURE 22.17
You should be able to draw
the three rays shown for
any situation involving a
convex mirror.

radius of curvature of the mirror, then $\overline{AF}$ is nearly equal to $\overline{BF}$. Consequently, $\overline{BF}$ nearly equals $\overline{FC}$, and so here, too, the focal point may be considered to be midway between the mirror and the center of curvature of the mirror.

We are therefore able to write rules for drawing three rays in the case of the *convex mirror:*

A ray parallel to the axis is reflected as though it came from the focal point.

A ray heading toward the focal point is reflected parallel to the axis.

A ray heading toward the center of curvature is reflected back on itself.

These three rays are illustrated in Fig. 22.17. You should trace them to see that they conform to the rules stated above. Notice that all three reflected rays appear to come from the image I behind the mirror. As you see, the image is virtual, upright, and diminished in size.

The algebraic relation used in locating the image for a convex mirror can be obtained by reference to Fig. 22.18a and b. You should be able to show that triangle ABH is similar to EBD in part a. In part b, triangle JFG is similar to triangle EFD. This being true, the following equations are found, as in the case of the concave mirror:

$$\frac{O}{I} = \frac{p}{i} \quad \text{and} \quad \frac{O}{I} = \frac{f}{f-i}$$

In writing these, the distance $\overline{BG}$ has been considered negligibly small.

Equating the two expressions, inverting, dividing by i, and rearranging yield

$$\frac{1}{p} - \frac{1}{i} = -\frac{1}{f}$$

Notice that, except for signs, this equation is the same as Eq. (22.1) for the concave mirror. The difference in signs alert us to the fact that the image in

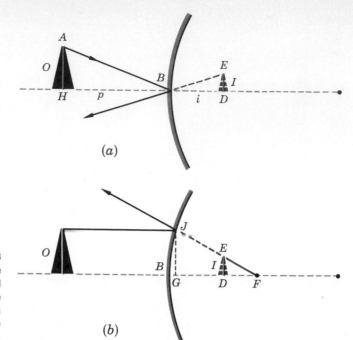

(a)

(b)

FIGURE 22.18
Triangles *HBA* and *DBE* are
similar, as are *DFE* and
GFJ. We make the
assumption that the length
$\overline{FG}$ is essentially the same
as $\overline{FB}$.

this case is behind the mirror rather than in front of it. Also, the negative
focal-length term is the result of the mirror being convex rather than concave.

Rather than remembering two mirror equations, we now set up rules
which will allow us to use Eq. (22.1) even for convex mirrors. If we agree
always to call image distances behind mirrors, i.e., virtual-image distances,
negative, the negative sign may be omitted from the i term in the convex-
mirror equation. Moreover, if we always say that the focal length of a
convex mirror is negative, the other negative sign may be omitted as well.
Hence for *all* mirrors we can write

Mirror Equation and Its Sign
Conventions

$$\frac{1}{p} + \frac{1}{i} = \frac{1}{f} \quad \text{mirrors} \tag{22.1}$$

where we agree that:

1 Object distances are positive if the object is on the reflecting side of
   the mirror, negative otherwise.
2 Image distances are positive if the image is on the reflecting side of
   the mirror, negative otherwise.
3 The focal length of a concave mirror is positive, and it is negative
   for a convex mirror.

In addition, we can use Eq. (22.2) for the magnification in either case without use of any negative signs whatsoever.

Illustration 22.3 A convex mirror with a 100-cm radius of curvature is used to reflect the light from an object placed 75 cm in front of the mirror. Find the location of the image and its relative size.

Reasoning Since the mirror is convex, $f = -R/2 = -50$ cm. Using the mirror equation and taking all distances in centimeters, we have

$$\frac{1}{75} + \frac{1}{i} = -\frac{1}{50} \quad \text{or} \quad i = -30 \text{ cm}$$

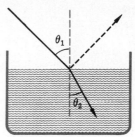

FIGURE 22.19
When a ray of light passes from an optically less dense material to an optically more dense material (air to water, for example), the ray is refracted toward the normal to the surface.

The negative sign tells us the image is behind the mirror, and it is, of course, virtual. Its size relative to the height of the object is, from Eq. (22.2),

$$\frac{I}{O} = \frac{i}{p} = \frac{30}{75} = 0.40$$

It would be wise to check this solution graphically to see that no mistake has been made.

22.9 Refraction of Light: Snell's Law

When a beam of light enters water from air, its path bends, as shown in Fig. 22.19. *This change in direction of a ray as it passes from one material to another is called* **refraction.** The angle θ_1 is, of course, the angle of incidence, and angle θ_2 is called the angle of refraction. Some of the light beam hitting the water surface is also reflected, as shown by the dashed ray in Fig. 22.19, but we ignore this in this section.

Refraction

In order to find a relation between θ_1 and θ_2, it is convenient to consider the motion of the wavefronts in a plane wave. The situation is shown in Fig. 22.20. We say that the light has a speed v_1 in the upper material and a speed v_2 in the lower material, with v_1 being greater than v_2. (If the upper material is air, $v_1 = c = 3 \times 10^8$ m/s.) You should be able to show that the angles labeled θ_1 and θ_2 are the same in the two parts of the figure. We assume that the wavefront ABC has moved to position $A'B'C'$ after a time t. Hence, we have in Fig. 22.20b

FIGURE 22.20
Since the wave travels more slowly in the lower material than it does in the upper material, θ_2 is smaller than θ_1.

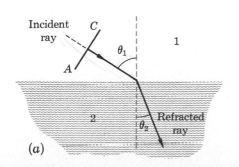

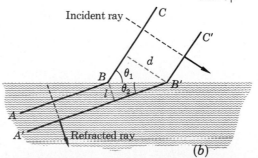

(a)

(b)

$$d = v_1 t \quad \text{and} \quad l = v_2 t$$

which yields, after division of one equation by the other,

$$\frac{d}{l} = \frac{v_1}{v_2}$$

Moreover, from Fig. 22.20b we see that

$$\frac{d}{BB'} = \sin \theta_1 \quad \text{and} \quad \frac{l}{BB'} = \sin \theta_2$$

which, after division of one by the other, gives the relation

$$\frac{d}{l} = \frac{\sin \theta_1}{\sin \theta_2}$$

But since $d/l = v_1/v_2$, the relation becomes

$$\frac{\sin \theta_1}{\sin \theta_2} = \frac{v_1}{v_2} \tag{22.3}$$

Although this equation describing the refraction phenomenon is helpful, a more useful form can be obtained if we define a quantity called the index of refraction.

Definition   *The* **index of refraction** *n of a material is*

$$n = \frac{\text{speed of light in vacuum}}{\text{speed of light in material}} = \frac{c}{v} \tag{22.4}$$

Because light has its largest speed in vacuum, the index of refraction is unity or larger. Typical values for n are listed in Table 22.1. As you see, the refractive index for air is very close to unity, while the index for diamond is large, 2.42. Of course, the index of refraction for vacuum is exactly 1. The index of refraction changes slightly with wavelength as we shall see later. The index is larger for blue light than for red light.

Because $v = c/n$ from the definition of refractive index, we can rewrite Eq. (22.3) as

$$\frac{\sin \theta_1}{\sin \theta_2} = \frac{c/n_1}{c/n_2} = \frac{n_2}{n_1}$$

which can be rewritten as

Snell's Law

$$n_1 \sin \theta_1 = n_2 \sin \theta_2 \tag{22.5}$$

which we refer to as **Snell's law.**

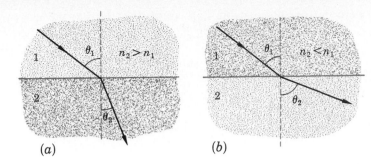

(a)　　　　　　　　　(b)

We see that if n_2 is greater than n_1, then $\sin\theta_1$ is larger than $\sin\theta_2$. In this case θ_1 would be larger than θ_2. This is the instance illustrated in Fig. 22.21a and is the most common one. Sometimes, however, we are interested in the reverse case, where n_2 is smaller than n_1. This would be applicable to a beam of light going from glass to air, for example. Under these circumstances Eq. (22.5) predicts that θ_2 is larger than θ_1, as shown in Fig. 22.21b.

FIGURE 22.21
If $n_2 > n_1$, the beam bends toward the normal, while if $n_2 < n_1$, the reverse is true.

Illustration 22.4 A diver beneath the surface of the ocean shines a bright searchlight up at an angle of 37° to the vertical. At what angle does the light emerge into the air?

Reasoning The situation is shown in Fig. 22.22. Notice that material 1 is water and material 2 is air. Applying Snell's law and using $n_1 = 1.33$ and $n_2 = 1.00$ gives

$$1.33 \sin 37° = 1.00 \sin\theta$$

$$\sin\theta = 0.80 \quad \text{or} \quad \theta = 53°$$

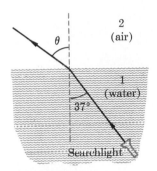

FIGURE 22.22
The underwater searchlight sends out a beam which bends away from the normal as it passes into the air.

Illustration 22.5 Light is incident upon the surface of the water in a level, flat-bottomed glass dish, as shown in Fig. 22.23. At what angle will the light emerge from the bottom of the dish?

Reasoning At the air-water surface we have

$$1.00 \sin\theta_1 = n_w \sin\theta_2$$

At the water-glass interface we have

$$n_w \sin\theta_2 = n_g \sin\theta_3$$

Since quantities equal to the same thing are equal to each other, we find

$$1.00 \sin\theta_1 = n_g \sin\theta_3$$

Notice that this equation is exactly the relation we would find if the water

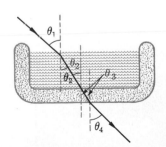

Note: $\theta_2 > \theta_3$

FIGURE 22.23
The direction of travel of a light beam is not altered by a parallel plate of transparent material.

were not present and the light went directly into the glass from the air. Proceeding, at the lower glass-air surface we have

$$n_g \sin \theta_3 = 1.00 \sin \theta_4$$

Combining this with the previous equation gives

$$\sin \theta_1 = \sin \theta_4$$

from which $\theta_1 = \theta_4$.

This important result shows that *a uniform layer of transparent material does not change the direction of a beam of light.* The beam is usually slightly displaced sideways, however. (Why?)

22.10 Total Internal Reflection

Aside from their romantic value, diamonds owe a great deal of their beauty to the phenomenon of total internal reflection. It is this property which causes diamonds to sparkle in all directions. What happens is that a beam of light becomes trapped within the diamond. When it finally does emerge, the ray of light can be emitted in any one of many directions. Hence the crystal gives off light (or sparkles) in random directions. There are many other cases where total internal reflection is of importance. We now investigate this type of behavior of light in some detail.

Consider a light source O below the surface of a lake, as shown in Fig. 22.24a. Snell's law tells us that

$$\sin \theta_2 = \frac{n_1}{n_2} \sin \theta_1$$

and since we are assuming n_1 to be larger than n_2, we know that θ_2 is larger than θ_1, as shown. Notice that ray OC is bent nearly parallel to the water surface as it emerges. Clearly, the critical case, shown in Fig. 22.24b, occurs

FIGURE 22.24

When θ_1 is greater than the critical angle θ_c, the beam is totally internally reflected.

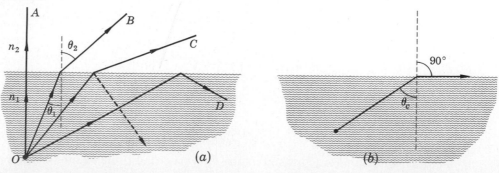

(a)   (b)

when $\theta_2 = 90°$ and the emerging beam just skims the surface. Thus, when $\theta_2 = 90°$,

$$\sin 90° = \frac{n_1}{n_2} \sin \theta_c$$

or

$$1.00 = \frac{n_1}{n_2} \sin \theta_c \tag{22.6}$$

giving

$$\sin \theta_c = \frac{n_2}{n_1}$$

If θ_1 is larger than θ_c, the critical angle, Snell's law tells us that $\sin \theta_2$ is greater than 1.00. However, the sine of 90° is 1.00, and this is the largest that the sine can become. Hence Snell's law tells us that it is impossible to find the refracted beam if θ_1 is greater than θ_c. This is easily understood by reference to Fig. 22.24b. When $\theta_1 = \theta_c$, the refracted ray is just barely getting out of the water. If θ_1 is larger than θ_c, the beam will not leave the water: it will all be reflected as shown by the beam *OD* in the figure.

We see, therefore, that *a beam of light trying to travel from an optically dense material* (a material of high index of refraction) *to an optically less*

FIGURE 22.25
(*a*) Light is caused to follow a glass fiber by total internal reflection. (*b*) A glass fiber gastroscope attached to a camera. A light source outside the picture at the left supplies light to the fiber bundle at the bottom. This light pipe is inserted through the throat to the stomach. Light reflected from the stomach wall is reflected up through the central fibers of the bundle and forms an image on the film of the camera. Visual observation is also possible. (*American Optical Corp., Fiber Optics Division.*)

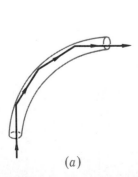

(*a*)

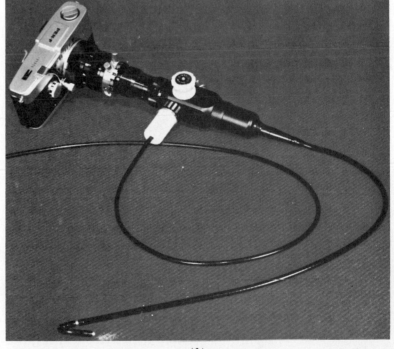

(*b*)

dense material will be totally internally reflected if the angle of incidence exceeds the critical angle given by Eq. (22.6). Notice that total internal reflection can occur only if the beam is going from the water to air and not if it is going from air to water. In most cases we are concerned with air as the second material, and so $n_2 = 1.00$. For those cases, $\sin \theta_c = 1/n_1$. Typical critical angles are 49° for water, 42° for crown glass, and 24° for diamond. Since diamond has such a high index of refraction, the critical angle is quite small. Hence the beam of light within a diamond must hit the surface nearly straight on if it is to emerge into the air. The jeweler cuts the crystal in such a way that once a beam of light gets inside it, the chance that it will strike a surface at an angle of 24° or less is quite small. As a result, the trapped beam reflects many times within the crystal before it is able to escape.

Total internal reflection makes it possible to "pipe" light around corners. By using a gently curved rod of glass, light which enters one end is totally internally reflected around the curve, as shown in Fig. 22.25. By using very narrow glass fibers in a bundle, the composite picture of an object can be piped from place to place; such a device is called a **light pipe.** See Fig. 22.25*b* for an example of this use.

22.11 Focal Point of a Lens

The phenomenon of refraction finds its most useful application in lenses. *A properly construced lens is capable of focusing a beam of parallel light into a small region at a focal point.* The mechanism by which this is done is illustrated in Fig. 22.26*a* and *b*. We recall that a light wave travels more slowly in glass than in air. Hence, the central portion of the incident plane wave shown in Fig. 22.26*a* is found to have fallen behind the outer portions of the wave because it traveled a greater distance in the glass. The emergent wave is therefore curved as shown. Since the rays, i.e., the directions of travel of the light, are perpendicular to the wavefronts, the light is converged toward point F on the axis.

FIGURE 22.26
Parallel rays are converged to the focal point by the converging lens. They are diverged and appear to come from the focal point in a diverging lens.

Although we do not prove it here, the various rays will converge to a single point, as shown, if the surfaces of the lens are portions of spheres. However, this is only an approximation, and it becomes a poor one if the

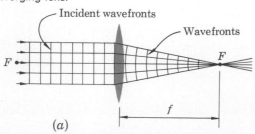

(a)

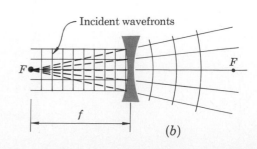

(b)

Converging   Diverging

FIGURE 22.27
The converging lenses are thickest at the center; the reverse is true for the diverging lenses.

surfaces are highly curved, in other words, if they constitute more than a very small portion of a sphere. Moreover, we assume the lens to be relatively thin. In these circumstances, the rays parallel to the axis are converged nearly to a point. *The point to which parallel rays are converged by a converging lens is called the* **focal point** *of the lens.*

    There is a simple way to determine the position of the focal point of a lens such as the one shown in Fig. 22.26a. If the light from the sun is passed through it, an image of the sun is formed. Since the sun is very far away, the waves from it are essentially flat and the rays are parallel. Hence, the place where the image of the sun is formed is the focal point of the lens.

    A different type of lens is shown in Fig. 22.26b. Notice that since this lens is thinner in the middle than on the edges, the outer portion of the wave will fall behind. Now the emergent wave is spherical in form but diverging from the lens. Under the same general restrictions as stated for the other type of lens, this diverging wave appears to come from the point F in the figure. *This point, from which the original parallel beam appears to diverge, is called the* **focal point of the diverging lens.**

    We see, then, that two general types of lens are possible. Converging lenses, which are thickest in the middle, cause a parallel beam of light to converge to the focal point. Diverging lenses are thickest near the edges and cause a parallel beam of light to spread out (or diverge) as though from a point, the focal point. Even if the lenses were turned around, they would still behave in this way. Hence each lens has two focal points, one on each side of it, and both are at the same distance from its center. This distance *from the center of the lens to the focal point is called the* **focal length** *f* **of the lens.** Several different shapes of common lenses are shown in Fig. 22.27. Notice that the converging lenses are all thickest at the center.

    Suppose that a glass lens is immersed in a liquid having an index of refraction which is the same as that of the glass. In this case, the wave travels with the same speed in the glass and in the liquid. For this reason, the wave is not disturbed by passing through the lens, and the lens does not cause it to focus. Thus, the focal length of a lens depends on the medium in which it is used. Most commonly this medium is air. However, this is not always the case. The focal length of a lens immersed in a medium other than air is given by

$$f_{\text{medium}} = \frac{f_{\text{air}} n_m (n_g - 1)}{n_g - n_m}$$

Definition

Definition

Lens Focal Length

609

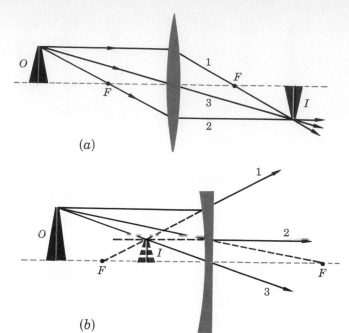

(a)

(b)

FIGURE 22.28
Graphical location of the
images formed by thin
lenses consists of drawing
the three rays shown.

where n_m is the refractive index of the medium and n_g is the index of refraction of the glass. Can you show that this relation is reasonable for the limiting cases of $n_m \rightarrow n_g$ and $n_g \rightarrow 1.00$?

22.12 Ray Diagrams for Thin Lenses

We saw in the last section that a ray traveling parallel to the lens axis is bent by the lens. It is converged to the focus by a converging lens and diverged from the focus by a diverging lens. These rays are shown as ray 1 in Fig. 22.28a and b.

The second ray is just the reverse of ray 1. For the convex lens, it comes from the source, through the other focal point, and is converged parallel to the axis by the lens. Notice that it is exactly the same as ray 1 traveling in the reverse direction. In the diverging case, ray 2 heads toward the other focal point but never reaches it. The lens diverges it parallel to the axis as shown. It, too, is ray 1 traced in the reverse direction.

The third ray comes from the source and goes straight through the center of the lens without deflection. It is easy to see why it behaves this way by referring to Fig. 22.29. Notice that the rays of light which go through the center of the lens enter and leave the lens at surfaces that are parallel to each other. Hence, the ray behaves as though it had gone through a flat plate of glass. It will be recalled that a ray of light is not deviated in direction by a flat plate which has parallel faces. Therefore, the rays of light passing through the lens center proceed undeviated.

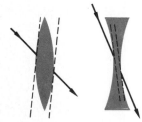

FIGURE 22.29
The ray passing through the
center of the lens
essentially passes through a
flat plate and is therefore
not deviated. A small
displacement of the ray
occurs, but this is not
shown in the figure. Why is
it negligible for a thin lens?

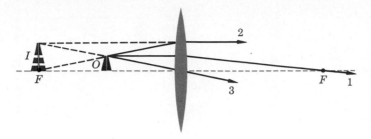

FIGURE 22.30
Virtual images are formed
by convex lenses when the
object is inside the focal
point.

We are now able to draw three of the many rays of light which pass through a lens. Any two of these allow us to locate the image of an object. Two examples of this construction have already been given in Fig. 22.28. Notice that the image in part *a* is real, since the three rays actually converge at the image. A screen placed there would catch the light and show an image of the object. In part *b*, however, the image is virtual, or imaginary, since the three rays merely appear to come from the image position. A screen placed at that position does not show an image, because the three rays actually do not meet at that place.

Illustration 22.6 A convex lens of focal length 10.0 cm is used to form an image of an object placed 5.0 cm in front of it. Draw a ray diagram to locate the image.

Reasoning The appropriate ray diagram is shown in Fig. 22.30. We notice that the eye will assume that the three rays come from the image position indicated. As we see, the image is virtual, erect, and enlarged.

Illustration 22.7 A concave lens of focal length 10.0 cm is used to form an image of an object placed 5.0 cm in front of the lens. Find the image position by means of a ray diagram.

Reasoning The appropriate ray diagram is shown in Fig. 22.31. Here, too, the image is virtual. It is erect and diminished in size.

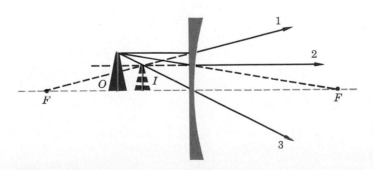

FIGURE 22.31
Is the image formed by a
diverging lens always virtual
if the object is real?

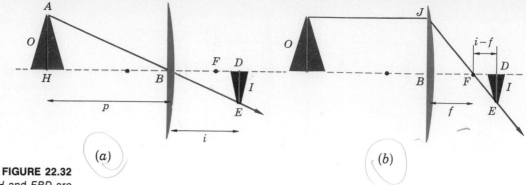

FIGURE 22.32
Triangles *ABH* and *EBD* are
similar, as are *JFB* and *EDF*.

22.13 Thin-Lens Formula

Consider the image formed by the converging lens shown in Fig. 22.32*a* and
b. In part *a,* triangles *ABH* and *EBD* are similar, and so we can write

$$\frac{I}{O} = \frac{i}{p}$$

From the two similar triangles *JFB* and *EDF* shown in part *b*, we also have

$$\frac{I}{O} = \frac{i - f}{f}$$

Equating these two expressions and simplifying yields

$$\frac{1}{p} + \frac{1}{i} = \frac{1}{f} \qquad (22.1)$$

This relation is exactly the same as the mirror equation (22.1). We have
obtained it by taking *p* and *i* positive when the object and image are in their
normal positions. That is, the object is on the side from which the light
comes, and the image is on the side to which the light is going. Also, Eq.
(22.2) for the magnification applies to lenses as well, since one of the above
relations is identical to it.

We can derive the relation applicable to concave lenses by referring to
the sets of similar triangles illustrated in Fig. 22.33*a* and *b*. We have

$$\frac{I}{O} = \frac{i}{p} \qquad \text{and} \qquad \frac{I}{O} = \frac{f - i}{f}$$

Equating and simplifying, we find that

$$\frac{1}{p} - \frac{1}{i} = -\frac{1}{f}$$

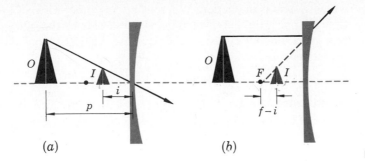

(a) (b)

FIGURE 22.33
A consideration of the
similar triangles leads to the
lens equation.

In order to make this relation appear the same as Eq. (22.1), we must
take both i and f negative. If we then agree to say that concave (diverging)
lenses have negative focal lengths, the negative sign preceding the $1/f$ term
may be omitted. Moreover, if we agree to call image distances negative
when the image lies on the side of the lens from which the light came, then
the negative sign on the $1/i$ term may be omitted. In that case, Eq. (22.1)
will apply to all lenses and mirrors. Notice also that Eq. (22.2) for the
magnification applies to all mirrors and lenses.

Let us briefly review our sign conventions in using Eq. (22.1),

$$\frac{1}{p} + \frac{1}{i} = \frac{1}{f} \qquad (22.1)$$

*Lens Equation and Its Sign
Conventions*

1 The object distance is always positive if the object is on the side of
   the lens or mirror from which the light is coming.
2 The image distance is positive:
   a In the case of mirrors if the image is on the side of the mirror
      where the light is, i.e., in front of the mirror.
   b In the case of lenses if the image is on the side of the lens to
      which the transmitted light goes.
3 The focal length of converging mirrors and lenses is positive, while
   the focal length of diverging mirrors and lenses is negative

Notice that the natural or normal positions for the object and image are
positive. It is normal for the object to be on the side from which the light
comes. We usually accept an image formed by a lens as normal when it is
on the side of the lens to which the light is going. On the other hand, for
mirrors we usually think of the image as normal when it is on the same side
as the object, since the light is reflected, not transmitted. Of course, converging lenses and mirrors usually seem more normal to us than diverging lenses
and mirrors do. Hence, the normal things are positive, while the others must
carry a negative sign.

Before leaving this section, we should point out that several different
sign conventions are sometimes used. Negative signs are often employed in
the lens and mirror equations. There seems to be no general agreement on

which system is the better. When using any other form of the lens equation than the one given here, you must always abide by the sign rules pertinent to that equation. Of course, we always use Eq. (22.1) and the rules we have just stated.

Illustration 22.8 A diverging lens with a 20-cm focal length forms an image of a 3.0-cm-high object placed 40 cm from it. Find the image position and size.

Reasoning From Eq. (22.1) we have, using all lengths in centimeters,

$$\frac{1}{40} + \frac{1}{i} = -\frac{1}{20} \qquad \text{or} \qquad i = -\tfrac{40}{3}\text{ cm}$$

The image is on the "wrong" side of the lens, i.e., on the same side as the light is coming from. It is imaginary, of course, as a ray diagram will easily show. To find the image size, we use Eq. (22.2):

$$\frac{I}{O} = \frac{i}{p} \qquad \text{or} \qquad I = \frac{(3)(\frac{40}{3})}{40} = 1\text{ cm}$$

Notice that we do not carry the signs along when finding the image size.

22.14 Combinations of Lenses

Most optical instruments contain more than one lens. These systems are easily dealt with if we proceed in systematic fashion. Let us consider the final image formed by the two lenses shown in Fig. 22.34*a*. The object is 20 cm from the first lens, which in turn is 30 cm from the second lens. Both lenses are converging. To solve this by a ray diagram as well as by formula, we proceed as in Fig. 22.34*b* and *c*.

First, *ignore the second lens completely, and find the image of the original object cast by the first lens.* The ray diagram locates the image I_1 in part *b* of the figure. By formula we have, using centimeters,

$$\frac{1}{20} + \frac{1}{i_1} = \frac{1}{10} \qquad \text{or} \qquad i_1 = 20\text{ cm}$$

where i_1 is shown in the figure.

We now use the image formed by the first lens as an object for the second. This is possible since an eye or any other instrument placed to the right of the image will see the image as though there were really an object at that place. Ignoring the first lens completely and using I_1 as the object for

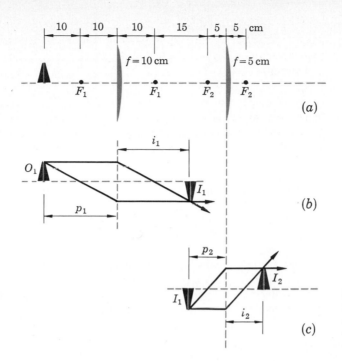

FIGURE 22.34
In order to find the image
cast by a lens combination,
we consider each lens in
turn, as in (b) and (c).

the second lens, we draw the ray diagram in Fig. 22.34c. The final image is
located at I_2, or, in equation form,

$$\frac{1}{p_2} + \frac{1}{i_2} = \frac{1}{f_2} \quad \text{and} \quad \frac{1}{30 - 20} + \frac{1}{i_2} = \frac{1}{5}$$

or

$$i_2 = 10 \text{ cm}$$

where i_2 is shown in the figure. Clearly, the final image formed by the two
lenses in combination, I_2, is real and erect.

To find the size of the final image in terms of O_1, the height of the
object, we apply Eq. (22.2) twice. For the situation of Fig. 22.34b we have

$$\frac{I_1}{O_1} = \frac{20}{20} \quad \text{so} \quad I_1 = O_1$$

Now, using Fig. 22.34c and remembering that the height of I_1 is the same as
O_1 and that I_1 is really the object for the second lens, we have

$$\frac{I_2}{I_1} = \frac{i_2}{p_2} \quad \text{or} \quad I_2 = (O_1)(\tfrac{10}{10}) = O_1$$

Hence, we find the unusual situation where the image is the same height as
the original object.

Summary

Light consists of EM waves in the visible wavelength range, 4×10^{-7} to 7×10^{-7} m. The speed of light in vacuum is $c = 2.998 \times 10^8$ m/s. It travels less fast in all other media. We characterize a material optically by its index of refraction n. It is the ratio of the speed in vacuum to the speed in the material. For air, $n = 1.0003$; for water, $n = 1.33$.

When light reflects from a smooth surface, the angle of incidence equals the angle of reflection. When light passes from one material (index of refraction n_1) to another (index n_2), it refracts and obeys Snell's law: $n_1 \sin \theta_1 = n_2 \sin \theta_2$. If $n_2 > n_1$, the ray bends toward the normal. If $n_2 < n_1$, then θ_1 is smaller than θ_2. This gives rise to total internal reflection for $\theta_2 = 90°$. The critical angle for total internal reflection is given by $\sin \theta_c = n_2/n_1$.

Real images are places where light rays focus to give optical duplicates of an object. Such images can be displayed on a screen placed at the image position. Vir-tual (or imaginary) images are places from which the rays only appear to come. A screen placed there will not display the image.

Plane mirrors give rise to virtual images of objects placed in front of them. The image is the same distance behind the mirror as the object is in front. The image is virtual, upright, and the same size as the object.

Parallel rays are converged to the focal point by a converging mirror or lens. Parallel rays appear to di-verge from the focal point when reflected by a convex mirror or after passing through a diverging lens. The object distance p is related to the image distance i and the focal length f by $1/p + 1/i = 1/f$. Sign conventions for use of this relation are given in Sec. 22.13. Also, the ratio of image length to object length (the magnifica-tion) is given by i/p.

Images for mirrors and lenses can be found graphi-cally by tracing three rays. These rays are described in Secs. 22.6 and 22.8 for mirrors and Sec. 22.12 for lenses.

Minimum Learning Goals

Upon completion of this chapter, you should be able to do the following:

1 Give the approximate wavelength limits of the visi-ble spectrum and arrange a list of colors in order of their wavelengths.
2 State the speed of light in vacuum. Compute n for a material when the speed of light in it is given, or vice versa.
3 Explain why a distant source gives rise to parallel rays. Distinguish between wavefront and ray; give the relation of one to the other.
4 Draw the reflected ray when the incident ray on a smooth surface is given.
5 Explain the meaning of refraction and be able to use Snell's law for a refracted beam.
6 Using a diagram, show why total internal reflection

occurs only if $n_2 < n_1$. Use Snell's law to find the critical angle for total internal reflection. List a few uses of this phenomenon.
7 State whether a given image is real or virtual.
8 Use ray diagrams to locate images for both single mirrors and single lenses.
9 Use the lens and mirror equation to obtain p, i, or f if two of the three are given or described to you. Relate f to R for a mirror. State the sign for f in any given case.
10 Find the size of an image when the object size is given, or vice versa.
11 Tell whether a lens is diverging or converging in air when its shape is given to you.
12 Explain how the focal point and focal length of a concave mirror and converging lens can be ob-tained by experiment.

Important Terms and Phrases

You should be able to define or explain each of the following:

Visible spectrum

$c = 3 \times 10^8$ m/s
Index of refraction
Wavefront; ray; parallel light
Incidence angle = reflection angle

Refraction; Snell's law
Specular and diffuse reflection
A parallel plate does not deviate a beam
Total internal reflection; critical angle
Virtual versus real image

Focal point; focal length; $f = R/2$
Three rays for image location
Lens and mirror equation; sign conventions
$I/O = i/p$
Convex, concave, diverging, converging

Questions and Guesstimates

1 Consider a concave mirror and an object at infinity. Where is the image formed? Is it upright or inverted? Is it larger or smaller than the object? Answer these questions as the object is slowly moved in toward the mirror. In particular, note the positions of the object where any of the answers change.

2 Repeat Question 1 for a convex mirror.

3 Repeat Question 1 for a converging lens.

4 Repeat Question 1 for a diverging lens.

5 Suppose the surface of a piece of tissue paper is greatly magnified so the roughness of the paper is easily seen. Sketch what the surface would look like, and use your sketch to explain why diffuse reflection does not negate the law of reflection.

6 Explain why an object under water does not appear to be as far below the surface as it really is when one looks at it from above the water.

7 Using a wavefront diagram, explain why a lens can be either converging or diverging depending on the material in which it is embedded.

8 Can an empty water glass focus a beam of light? A full water glass? Is it possible to start a fire by accident if a bowl of water is set in a sunlit window?

9 How does the mirror equation fit in with the results we learned for a plane mirror? Does it apply to a plane mirror?

10 Why does the drinking straw shown in Fig. P22.1 appear bent at the water surface?

11 A spherical air bubble in a piece of glass acts as a small lens. Explain. Is it converging or diverging?

12 How can you determine the focal length of a converging lens? Of a diverging lens?

13 Repeat Question 12 for mirrors.

14 Two plane mirrors are placed together so that they form a right angle. An object is then placed between them. How many images are formed? Repeat for an angle between the mirrors of 30°.

15 About how much longer does it take for a pulse of light from the moon to reach the earth because of the presence of air than if there were a vacuum above the earth? (E)

16 A "solar furnace" can be constructed by using a concave mirror to focus the sun's rays on a small region, the furnace region. How would you expect the temperature of the furnace to vary with area and focal length of the mirror? (E)

17 Newton believed light consisted of a stream of particles and that the "light corpuscles" were strongly attracted by the water surface as light went from air to water. How would this lead to the observed refraction effect? Why did the observed speed of light negate this idea?

18 In various science museums (as well as in some unexpected places), a room is designed so that a person can whisper at one particular point in the room and be heard clearly at a certain distant point. How must the room be constructed to achieve this effect? (Sound waves can be focused by reflection just as light waves are.)

FIGURE P22.1

Problems

1 A radar beam is used to locate an airplane. The time taken for a radar pulse to travel from the transmitter to plane and back to transmitter is 5.0×10^{-4} s. How far away is the plane?

2 A radar pulse reflected by the moon requires 2.6 s for the round trip from earth to moon. How far away is the moon?

3 How far will a beam of light travel in water in the time it takes to travel exactly 1 cm through air?

4* Suppose length D in Fig. 22.3 is 10.0 m. Give the two lowest possible speeds of rotation of the mirrored cube shown for which the beam will be properly reflected.

5 Many cameras have a focusing mark that must be set to read the distance from the camera to the object being photographed. Suppose you wish to photograph yourself in a plane mirror. If you and the camera are 2.0 m from the mirror, at what value should you set the distance indicator on the camera?

6 An interior decorator wishes to mount a plane wall mirror in such a way that a person 165 cm (5 ft 5 in) tall will be able to see his or her full length in it. What is the shortest length of mirror possible for this use and how high should it be mounted on the wall?

7 A girl has an earring in her left ear. She looks at herself in a plane mirror and sees an earring in the ear of her image. Is it in the right or the left ear of the image?

8* If an object is placed between two parallel mirrors, an infinite number of images result. Suppose the mirrors are a distance $2a$ apart and the object is midway between the mirrors. Find the distances of the images from the object.

9* Two plane mirrors make an angle of 90° with each other. Show that a light ray reflected first by one mirror and then by the other will end up going in a direction opposite to that from which it came.

10** Show that the result stated in Prob. 9 is a special case of the following: When a ray is reflected successively from two plane mirrors that make an angle θ with each other, the angle between the incoming and outgoing rays is 2θ.

11 A 0.50-cm-high object is placed 30 cm in front of a concave mirror that has a 15-cm radius of curvature. (a) Find the position and size of the image and state whether it is real or virtual and upright or inverted. Repeat for object distances of (b) 15, (c) 12, and (d) 6 cm. Check your calculations with ray diagrams.

12 A concave mirror having a 10-cm radius of curvature forms an image of an object 2.0 cm high that is placed 20 cm in front of the mirror. (a) Find the position and the size of the image. Is it real or virtual? Is it erect or inverted? Repeat for object distances of (b) 10, (c) 8, and (d) 4 cm. (Check with a ray diagram.)

13 (a) Find the position, size, and nature of the image formed of a 3-cm-high object placed 50 cm in front of a convex mirror having a 20-cm radius of curvature. Repeat for object distances of (b) 20 and (c) 5 cm. (Check with a ray diagram.)

14 (a) If an object is placed 40 cm in front of a convex mirror having a focal length of 10 cm, where is the image formed? What is the magnification? Is the image real or virtual? Repeat for object distances of (b) 20 and (c) 8 cm. (Check with a ray diagram.)

15* A concave mirror having a 200-cm radius of curvature is used to form a real image of an object. (a) Where must the object be placed if the image distance is to equal the object distance? (b) Are the object and image superimposed? (c) Compare the object and image sizes. (Check with a ray diagram.)

16* (a) Given a concave mirror having a 20-cm focal length, where must an object be placed if the image is to be real and twice the size of the object? (b) Repeat for the case where the image is to be virtual.

17* A virtual image is formed by a convex mirror having a 10-cm focal length. (a) Where must the object be placed if the image is to be half the size of the object? (b) Is it possible to obtain a virtual image that is larger than the object for this type of object?

18* A virtual image one-third the size of the object is formed by a convex mirror that has a 20-cm focal length. (a) Where is the object? (b) Where is the image?

19** (a) Where must the object be placed if the image formed by a convex mirror is to be half as far from the mirror as the object? (b) How large is the magnification in this case?

20** (a) Where must the object be placed if the image formed by a concave mirror is to be half as far from the mirror as the object? (b) Is the image real or virtual?

21 (a) What is the speed of light in an oil that has an index of refraction of 1.43? (b) What is the wavelength of sodium yellow light in the oil? (For this light in vacuum, $\lambda = 589$ nm.)

22 The index of refraction of a certain piece of transparent plastic is 1.48 for mercury blue light that has a vacuum wavelength of 436 nm. (*a*) What is the speed of this light in the plastic? (*b*) What is its wavelength in the plastic? (*c*) What is its frequency in the plastic?

23 Light at an incidence angle of 37° enters a flat glass plate (*n* = 1.50). (*a*) What is the angle of refraction inside the glass? (*b*) After the beam leaves the plate, what is the angle between it and the beam incident on the plate?

24 The index of refraction of glass is different for different wavelengths. Flint glass has an index of refraction of 1.650 for blue light (λ = 430 nm) and 1.615 for red light (λ = 680 nm). A beam of light consisting of these two colors is shone into flint glass at an incidence angle of 50°. Find the angles between the two color beams in the glass.

25 A beam of light is incident from air at an angle of 53° on a layer of water floating on a layer of carbon disulfide. Find the angle the beam makes in each liquid.

26 At what angle must a fish look to see an insect sitting on the shore at the water's edge? Assume the fish to be below the surface of a still lake.

27 If a beam of light is traveling inside a solid cube in a plane parallel to the base, it will keep traveling around inside the cube forever provided that it strikes each surface at an incident angle of 45°. (*a*) What must the refractive index of the cube be if the beam is to be totally internally reflected at each surface? (*b*) What practical difficulties appear in such a scheme?

28 A light beam from a source 2.0 m below the surface of a calm pool of water is shone upward toward the surface. What is the maximum angle that the beam in the water can make with the vertical if part of it is to be able to escape into the air above?

29* A layer of water is placed on top of a glass plate that has *n* = 1.560. If a beam of light is to enter the water from the glass, what maximum angle can the beam in the glass make with the perpendicular to the glass?

30 A crown-glass lens has a focal length of 20 cm in air. What is its focal length when submerged in water?

31 A 30-cm-focal-length converging lens forms an image of a 2.0-cm-high object. Find the position, size, and nature of the image for the following object distances: (*a*) 90, (*b*) 40, and (*c*) 10 cm. Check your answers with ray diagrams.

32 A converging lens of focal length 20 cm is to form an image of a 3.0-cm-high object. Find the position, size, and nature of the image for the following object distances: (*a*) 100, (*b*) 40, and (*c*) 10 cm. Check your answers with ray diagrams.

33 Find the position, size, and nature of the image formed by a diverging lens having a focal length of −30 cm if the 5.0-cm-high object is at distances of (*a*) 90, (*b*) 60, and (*c*) 20 cm. Check your answers with ray diagrams.

34 A 2.0-cm-high object is imaged by a diverging lens with *f* = −20 cm. Find the position, size, and nature of the image for the following object distances: (*a*) 60, (*b*) 20, and (*c*) 5.0 cm. Check your answers by sketches.

35* A virtual image that is half as far from the lens as the object is to be formed by a diverging lens for which *f* = −30 cm. (*a*) Where should the object be placed? (*b*) What will be the magnification?

36* (*a*) Where must an object be placed with respect to a converging lens if the image is to be the same size as the object? (*b*) Is the image real or virtual? Express your answer in terms of *f*.

37* If a diverging lens is to be used to form an image which is half the size of the object, where must the object be placed?

38** Two identical converging lenses with *f* = 30 cm are 90 cm apart. (*a*) Find the final image position for an object placed 120 cm in front of the first lens. (*b*) How large is the magnification of the system?

39** An object is placed 12 cm in front of a converging lens (*f* = 8 cm). At a distance of 36 cm beyond the first lens is a diverging lens having *f* = −6 cm. (*a*) Find the position and magnification of the final image. (*b*) Is it real or virtual? (*c*) Is it erect or inverted?

40** A converging lens of *f* = 20 cm is followed by a diverging lens of *f* = −30 cm, and their separation is 30 cm. If an object is placed 10 cm in front of the first lens, find (*a*) the final image position and (*b*) its magnification.

41** An object is placed 40 cm in front of a converging lens of *f* = 20 cm, which in turn is 50 cm in front of a plane mirror. Find all the images formed by this system.

42** An 8-cm-focal-length diverging lens is placed 16 cm to the left of a concave spherical mirror having a radius of 20 cm. If an object is placed 8 cm to the left of the lens, find all the images formed by the system.

43** An alternate form of the lens and mirror equations is

$$s_I s_O = f^2$$

where s_o is the distance of the object from the focal point and s_I is the distance of the image from the focal point. Derive this relation. Show s_o and s_I in a sketch for mirrors and for lenses.

44** In certain cases when an object and screen are separated by a distance D, two positions x of the converging lens relative to the object will give an image on the screen. Show that these two values are

$$x = \tfrac{1}{2}D\left(1 \pm \sqrt{1 - \frac{4f}{D}}\right)$$

Under what conditions will no image be found?

23
Optical Devices

Now that we understand lenses and mirrors, we are in a position to discuss some common optical devices. We discuss the operation of such widely different devices as the human eye and the telescope, as well as the microscope and other instruments which are of importance to us. In so doing not only do we obtain practice in the use of lenses and mirrors, but also we become more intelligent operators of the optical devices we are frequently called on to use.

23.1 The Eye

The most familiar of all optical devices, the eye, is also one of the most complicated. Although the actual lens system used in it is not very complex, the associated interpretive equipment is as complex as humans themselves. The exact way in which the image formed on the retina of the eye is transformed to our sensation of sight is a problem still challenging biophysicists. It is necessary for us to restrict our discussion to the lens system and the formation of the image on the retina of the eye.

A simplified diagram of the eye is shown in Fig. 23.1. As you probably already know, the cornea is a protective cover for the eye, the iris diaphragm controls the amount of light entering, and the retina is a sensitive surface that changes the image formed on it to electrical energy which is then transmitted to the brain. When a ray of light enters the eye, it is

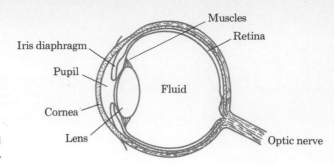

FIGURE 23.1
Diagram of the human eye.

refracted at the cornea. Lesser refraction effects occur in the pupil and lens, since the refractive indices of the cornea, pupil, lens, and fluid portions of the eye are quite similar.

For the normal relaxed eye, these combined refraction effects form an image of distant objects on the retina. Hence, the focal length of the eye is about the distance from the retina to the lens. If you draw ray diagrams for a converging lens, you will find that the image distance increases as the object is brought closer to the lens. In the eye, however, the image distance must always be such that the image is formed at the retina. This will be true only if the eye lens is made more converging as the object viewed is brought closer to it. The muscles of the eye alter the shape of the rather deformable eye lens so as to make it thicker (more converging) when viewing close objects.

Myopia (Nearsightedness)

Normally, one is able to relax the eye lens to the state where a distant object is focused on the retina. A person unable to do this is said to be **nearsighted,** or **myopic.** *The myopic eye is able to focus only objects which are less than a certain distance from the eye. This distance is called the* **far point** *of the eye.* Since the eye remains too converging to allow proper focus for very distant objects, a myopic person must be fitted with spectacles having a diverging lens in order to see objects far away clearly. We illustrate this effect in Fig. 23.2a, where the dashed lines indicate the position to which distant objects would be focused without the spectacle lens.

Hyperopia (Farsightedness)

People who have **hyperopia,** *i.e.,* **farsighted** *people, are able to relax the eye to see distant objects, but are unable to make the eye lens fat enough to focus nearby objects onto the retina. The normal person cannot make the lens*

FIGURE 23.2
A diverging lens is used to correct myopia, whereas a converging lens is needed to correct hyperopia. The diagrams are exaggerated.

converging enough to see objects closer than about 25 cm, the normal **near point** *of the eye.* A farsighted person must wear a converging spectacle lens

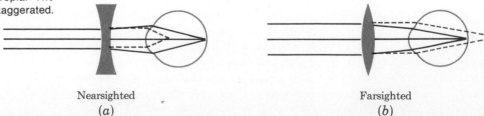

Nearsighted
(a)

Farsighted
(b)

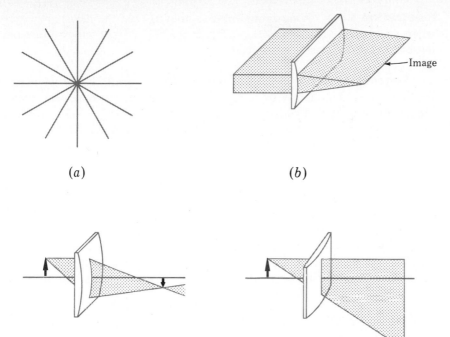

(a)   (b)

(c)   (d)

FIGURE 23.3
Cylindrical lenses form line
images rather than point
images. Rays in various
planes are focused
differently, as shown in (c)
and (d).

in order to aid the eye to bring nearby objects to focus on the retina. This situation is shown in Fig. 23.2b.

If the eye has very little ability to alter the shape of the lens, we say that the eye has lost its **accommodation.** Such an eye may be able to focus neither very distant nor very close objects. The use of bifocal spectacles allows one to look through diverging lenses when looking straight ahead or converging lenses when looking down. Some people actually have three types of lenses built into a single spectacle lens.

Another common type of eye defect is **astigmatism.** When a person with this type of eye defect views the test pattern shown in Fig. 23.3a, some of the lines appear darker than others. This is caused by a nonspherical shape of the eye lens. As you know, a spherical lens focuses parallel rays at a single point, the focal point. But a slightly misshapen eye lens often acts as a spherical lens with a cylindrical lens superimposed on it. We can see the effect of the cylindrical lens by examining Fig. 23.3b, c, and d.

Notice in part *b* that the lens focuses parallel rays into a line image rather than a point image. Rays in a plane perpendicular to the cylinder axis, as in c, are focused, but rays in a plane parallel to the axis, as in *d,* are not focused at all. The result is to cause the eye lens to focus the lines in part *a* at different positions. When a vertical line is in proper focus, a horizontal line, for example, may not be in focus. Consequently, lines oriented at different angles appear different to a person who has this type of eye defect. To compensate for this, an eyeglass which exactly cancels the eye's cylindrical lens part must be used.

Astigmatism

Illustration 23.1 A farsighted man is able to read the newspaper only when it is held at least 75 cm from his eyes. What focal length must the lenses of reading glasses have for him?

Reasoning We want him to be able to see print clearly when the reading material is 25 cm from his eyes, i.e., at the normal near point. Hence, a lens is needed which will give a *virtual* image at $i = -75$ cm when the actual object distance is $p = 25$ cm. Using the lens equation, we get

$$\frac{1}{25} + \frac{1}{-75} = \frac{1}{f}$$

from which

$$f = 37.5 \text{ cm}$$

Notice that a converging lens is needed.

Illustration 23.2 What must be the focal length of a corrective lens for a woman whose far point is 50 cm?

Reasoning The corrective lens must form an image of a distant object in such a way that the image will be virtual and 50 cm from the eye. Therefore, $i = -50$ cm, $p \to \infty$, and so the lens equation is

$$\frac{1}{\infty} + \frac{1}{-50} = \frac{1}{f}$$

from which

$$f = -50 \text{ cm}$$

Notice that the required lens is diverging.

23.2 Simple Camera

A camera operates very much as the human eye. It uses a lens to produce an image of an object on a film. The film serves the purpose of the retina in the eye. A schematic diagram of a typical simple camera is shown in Fig. 23.4. The image is inverted on the film, and its size I is related to the object size by the usual relation

$$\frac{I}{O} = \frac{i}{p}$$

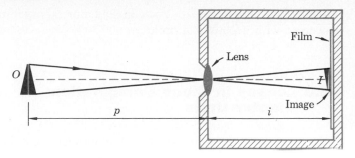

FIGURE 23.4
The simple box camera forms an inverted image on the film. In practice, the lens hole would be smaller than shown, and a shutter would cover the opening except when the picture was being taken.

Unlike the eye, the lens of a camera cannot be made with variable focal length. Hence, to achieve good focus on the film, the lens must be moved back and forth as the distance to the object changes. Cameras which do not have movable lenses usually have only a very small hole open in front of the lens; they operate much like a pinhole camera, which has no lens at all but merely uses a small pinhole to admit the light to the film. The operation of such a camera is treated as a question at the end of this chapter.

Expensive cameras possess very complex systems instead of a single lens. The complexity is necessary if a camera is to give very sharp images and fast shutter speeds. It is clear why the first condition is advantageous. The second, fast shutter speeds, allows one to take pictures of swiftly moving objects. Any moving object will blur the image somewhat. But the shorter the time the camera shutter is open, the less blurred the image will be. Since the shutter must be open long enough to allow sufficient light to hit the film, fast shutter speeds mean that the lens must be large so that a large amount of light enters the camera.

As we saw when discussing the lens equation, only the central portion of a spherical lens can be used if a clear image is desired. This becomes even more important if a camera is to be used to take close-up pictures, since then the lens must be very convex. It is only by making a complicated combination lens that the focusing errors inherent to a single lens can be eliminated. We say that such a lens has been corrected for **spherical aberration.**

Another lens defect causes images to have colored edges. This is called **chromatic aberration.** It results from the fact that the speed of light in glass varies with wavelength. As a result, the index of refraction of the glass is not the same for all colors. Blue light is refracted more strongly by the lens than red light is. This causes the colors in a beam of ordinary light to separate, and the image is therefore colored.

Lens Defects

To correct for this defect, two or more types of glass must be layered together to form the lens. Expensive lenses consist of several individual lenses cemented together. A lens which has been partly corrected for chromatic aberration is called an **achromatic lens.** However, it is impossible to free a lens of this defect completely. Indeed, lenses used in expensive optical systems, such as a very good microscope, are very complex. Not only do they correct the system for spherical and chromatic aberration, but also corrections are made for other lens defects. The design of precise optical

625

instruments is very involved. You can read more about it in advanced texts concerned with geometrical optics.

23.3 Lenses in Close Combination: Diopter Units

Perhaps you have had your eyes tested and noticed that the examiner often places several spectacle lenses, one in front of the other, in front of your eye at once. In order to make use of the observed best combination, he or she needs to know how to add the effects of thin lenses in close combination. The necessary simple formula can be derived readily, as we now show. We consider only the case where the focal lengths of the lenses are much longer than the lens separations.

Referring to Fig. 23.5, we find first the image formed by the first lens. Thus,

$$\frac{1}{p_1} + \frac{1}{i_1} = \frac{1}{f_1}$$

This image is then used as the object for the second lens. The object distance is $p_2 = -i_1$ if we ignore the small separation of the lenses. We use a negative sign because the object is on the wrong side of the lens and our sign conventions require a negative object distance in such cases. Now, writing the lens equation for the second lens yields

$$\frac{1}{-i_1} + \frac{1}{i_2} = \frac{1}{f_2}$$

If we add these last two equations, we have

$$\frac{1}{p_1} + \frac{1}{i_2} = \frac{1}{f_1} + \frac{1}{f_2} \qquad \text{or} \qquad \frac{1}{p_1} + \frac{1}{i_2} = \frac{1}{f}$$

where the combined focal length f of the two lenses is given by

f for Lens Combinations

$$\frac{1}{f} = \frac{1}{f_1} + \frac{1}{f_2} \qquad \text{close combination} \tag{23.1}$$

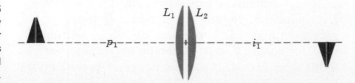

FIGURE 23.5
If the two lenses are very close together, their combined effect is to act as a single lens having a focal length $1/f = 1/f_1 + 1/f_2$.

As we see, *two thin lenses in close combination behave as a single lens having a focal length given by Eq. (23.1).*

In order to save the eye examiner the trouble of adding reciprocals, a new unit using the reciprocal of the focal length is defined. *The* **power** *of a lens in* **diopters** *is the reciprocal of the focal length in meters.* For example, the power of a 20-cm focal-length diverging lens is $1/(-0.20)$, or -5 diopters.

Illustration 23.3 Three lenses of focal lengths 20, -30, and 60 cm are placed in contact. Find the focal length of the combination.

Reasoning The powers of the lenses are 5, $-\frac{10}{3}$, and $\frac{10}{6}$ diopters. Their combined power is

$$\tfrac{30}{6} - \tfrac{20}{6} + \tfrac{10}{6} = \tfrac{20}{6} \text{ diopters}$$

Hence the combined focal length is $\frac{6}{20}$, or 0.30, m. It is interesting to notice that if the $\frac{10}{6}$-diopter lens had been negative, the three lenses would have substantially the same effect as a flat plate of glass.

23.4 Magnifying Glass (Simple Magnifier)

The normal person can see an object clearly only if it is a distance γ from the eye or farther.[1] We show in Fig. 23.6 the image formed on the retina of the eye when the object is placed at that distance. If we could properly focus the image when the object was closer, the image on the retina would be much larger, as indicated in part *b* of Fig. 23.6. It would then be possible to see fine detail on the object better, since it would appear much larger on the retina.

In order to see an object clearly, i.e., focus it on the retina, when it is closer than γ, we need to use a converging lens to aid the eye lens. The effect of such a magnifying lens is shown in Fig. 23.7. We see that if we place the object inside the focal point of the lens, a virtual image of the object is formed much farther away from the lens. If the eye is placed just behind the

---

[1]As stated previously, this distance is about 25 cm for the normal eye.

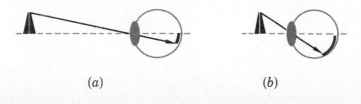

(a)　　　　　(b)

FIGURE 23.6
When an object is brought closer to the eye, the image on the retina is increased in size.

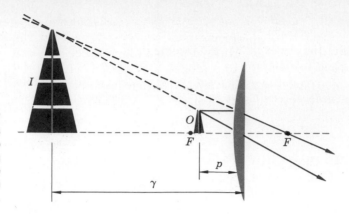

FIGURE 23.7
A magnifying glass
effectively allows one to
place the object being
examined closer to the eye
than would be possible with
the eye alone.

lens, the image can be seen clearly if it is γ or more away from the lens as shown.

Let us now see what effect this lens used in this way has on the size of the image on the retina. In Fig. 23.8a we show the image formed on the retina when the object is γ from the eye without the use of the magnifying lens. Part b of the same figure is redrawn from Fig. 23.7, with the eye shown in addition. The eye is actually looking at the virtual image formed by the magnifying glass. However, the size of the image on the retina is the same as if the eye were focused on the actual object, which is now much closer than γ. Hence, with the lens, the image on the retina is larger than without. This is purely the result of the fact that the lens has allowed us to view the object closer than at the normal closest distance, γ.

Definition   *The* **magnifying power** *M of an optical device is defined to be the ratio of the angle subtended by the image on the retina when the device is used to the angle subtended by the image on the retina when the object is viewed directly at a distance γ, or (refer to Fig. 23.8),*

Magnifying Power

$$M = \frac{\phi'}{\phi}$$

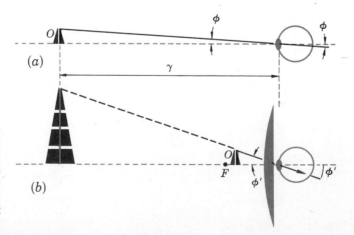

FIGURE 23.8
In both cases, the eye is
focused on the triangle a
distance γ from it. The
actual object is much closer
to the eye than γ in (b), and
so the angle subtended by it
is much larger.

The exact value of M will differ somewhat depending on how the magnifying glass is held and how the eye is focused. A fair approximation in this case may be had by saying that the eye is just behind the lens and that the object is located close to the focal point of the lens. It turns out that for small angles such as we are usually concerned with, the angle in radians is nearly equal to its tangent or its sine.[2] Then from Fig. 23.8 we have

$$\phi = \frac{O}{\gamma} \quad \text{and} \quad \frac{O}{f} \approx \phi'$$

which gives

$$M = \frac{\phi'}{\phi} = \frac{\gamma}{f}$$

A more exact treatment shows that $M = (\gamma/f) + 1$, but for most purposes the approximate form given is adequate. (Why?)

M for a Magnifier

We therefore see that a lens of high magnifying power should have a short focal length—it should be highly converging. If f is 4 cm and $\gamma = 25$ cm, the magnifying power would be 6.25 and the object would appear about 6 times larger with than without the magnifying glass.

23.5 Microscope

As seen in the previous section, the magnifying power of the simple magnifying glass is about γ/f. Since the object must be placed within the focal length of the lens, it is not possible to make f extremely short. Hence, the power is limited by this as well as other practical considerations. The compound microscope allows larger magnification by the use of two converging lenses. A diagram of the microscope is shown in Fig. 23.9.

---

[2] You can check this by referring to Appendix 5 and recalling that $1° = 2\pi/360 = 0.0175$ rad.

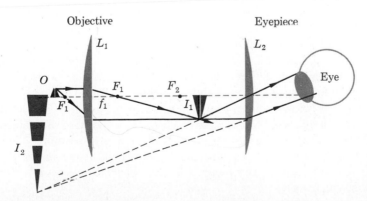

FIGURE 23.9
The eye sees the final image I_2. The microscope has greatly increased the angle subtended at the eye.

In this device the first lens, the **objective lens,** forms a real image I_1 of the object. Notice that the object is placed just outside the focal point of the objective. Since $I/O = i/p$, I_1 is a factor i_1/p_1 larger than the object. Since the object is usually placed fairly close to the focal point, $p_1 \approx f_1$. In addition, the image I_1 will be close to the second lens, as we shall soon see, and therefore i_1 is essentially equal to the length of the barrel of the microscope, L, about 18 cm in most cases. Therefore, I_1 is about L/f_1 times larger than the original object.

The second lens, the **eyepiece lens,** is used as a magnifying glass to look at the image formed by the objective. As such, I_1 must fall inside its rather small focal length. Moreover, for good viewing, the virtual image I_2 formed by the eyepiece must be about γ from the eyepiece. As shown in the last section, the magnifying power of the eyepiece used as a magnifying glass is γ/f_2. However, the objective lens had already magnified the object by a factor L/f_1, and so the angle subtended to the naked eye by I_1 was already L/f_1 times as large as the angle subtended by the original object. Therefore, the total magnifying power of the microscope is

M for a Simple Microscope

$$M = \left(\frac{L}{f_1}\right)\left(\frac{\gamma}{f_2}\right)$$

As we see, both f_1 and f_2 should be made small for highest magnification. In order to achieve small f_1 and f_2, a good microscope uses quite complex combinations of lenses for the objective lens and for the eyepiece. Great care must be taken in the lens design, or various lens aberrations will so seriously distort and color the image as to make the instrument nearly worthless.

23.6 Astronomical Telescope

Two basic problems confront astronomers when they design a telescope to look at the moon or some distant planet. They would like the object to appear larger than when viewed with the naked eye. Also, they would like to increase the faint amount of light which reaches the naked eye directly from the planet. To overcome this latter difficulty they use an objective, or first, lens which is very large so that it collects a great amount of light from the star. One telescope, at the Yerkes Observatory, has an objective lens which is 100 cm in diameter and has a 19-m focal length. Clearly, such a large lens will gather much more light than the very small opening in the eye. Such a telescope is shown in Fig. 23.10.

As shown, the objective lens forms an image I_1 of the distant object. This image will be very close to F_1, the focal point of the objective lens, since the light from the distant object will be nearly parallel. The eyepiece lens acts, as usual, as a magnifying glass to look at the image cast by the first lens. Of course, the image formed by the eyepiece is virtual and about 25 cm from the eyepiece lens.

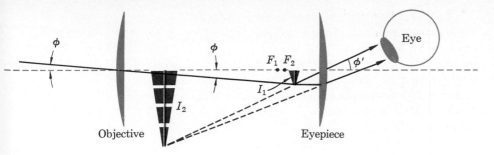

Objective   Eyepiece

To find the magnifying power of this device, we notice that if the naked eye looked at the distant planet, the angle ϕ subtended would be the same as the angle subtended at the eyepiece. (Why?) Hence, if ϕ is small, so that it may be replaced by its tangent, we have

$$\phi \approx \frac{I_1}{f_1}$$

since the image is very close to the focal point. Also, since the focal point of the magnifying eyepiece is placed close to I_1, we have

$$\phi' \approx \frac{I_1}{f_2}$$

If we divide this latter equation by the former, we find the magnifying power of the telescope to be

$$M = \frac{\phi'}{\phi} = \frac{f_1}{f_2}$$

M for a Telescope

We see that the focal length of the objective should be long and that of the eyepiece should be small.

Since it is very difficult to make large, perfect lenses, many large astronomical telescopes use a concave mirror in place of an objective lens. One arrangement for such a telescope is shown in Fig. 23.11. The size of the flat deflecting mirror is exaggerated. A reflecting telescope is not affected by

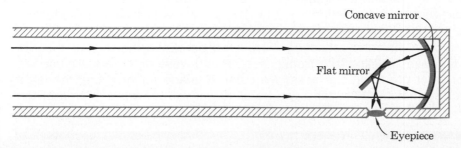

FIGURE 23.10
The two-lens telescope shown has considerably increased the angle subtended at the eye, but the final image is inverted.

FIGURE 23.11
A schematic diagram of a reflecting telescope. The flat mirror is much smaller than shown.

(*Australian News & Information Bureau; Photograph by D. Moore.*)

Radio Astronomy

Optical telescopes, until recently, were our only sensitive means for looking into the depths of space. They are limited in their sight to wavelengths of EM radiation which are visible to the eye or which can be photographed. This is, as we know, only a small portion of the total electromagnetic spectrum. As a consequence, optical telescopes are able to observe only those bodies in outer space which strongly emit light. These bodies must, of course, be white-hot in order to appear luminous.

We are led to wonder whether we are not missing much that exists in outer space since we with optical telescopes can see only the hot suns, the visibly glowing bodies. If we were able to "see" much cooler bodies in space, perhaps we would find new objects, such as suns which had cooled or suns which had not yet been born. To do this, we must have devices which can see longer wavelengths than visible light. In particular, telescopes which can see radio waves would be of tremendous value—hence the incentive for radio astronomy.

As we shall see in Chap. 24, in order to reflect EM waves, a surface should be smooth to within a wavelength of the radiation. The size of the reflector must be many wavelengths large. These restrictions require that a reflecting-type radio telescope use metal mirrors which are much larger than the wavelengths that it is to see. However, these mirrors need not be as smooth as those used for shorter wavelengths. In fact, wire-mesh surfaces can be and have been used. The larger the mirror, of course, the more energy it can reflect and bring to focus. Therefore, a larger reflector is preferred.

A picture of a large radio telescope is shown in the figure. It is located in an isolated spot in Australia far from sources of extraneous, earth-generated radio noise. This, as well as others elsewhere in the world, has found many previously unknown objects in the far reaches of space. The device illustrated makes use of wavelengths in the 10- to 20-cm range and is capable of detecting radio-wave sources which are 5 billion light-years away—10 times farther than is possible with present optical telescopes.

The 64-m-diameter reflecting mirror, the parabolic metal grid, focuses the radio waves coming to it from space. Radio-wave detectors are used to monitor what the telescope sees. Provision is made for orienting the large reflector so that the telescope can be aimed at will toward any area of the heavens which is 30° above the horizon.

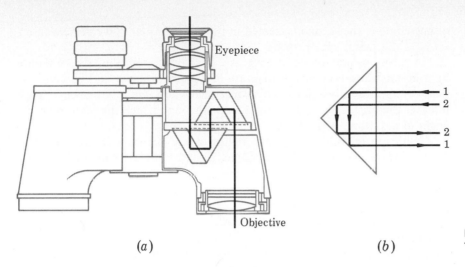

Eyepiece

Objective

(a)

(b)

FIGURE 23.12
The prism binocular.

chromatic aberration except for the eyepiece. Moreover, spherical aberration in the mirror is eliminated by using parabolic mirrors. Since it is considerably easier to produce a good large mirror than a large lens, it is clear why reflecting telescopes are sometimes preferred.

23.7 Binocular

As you can see in Fig. 23.10, the telescope gives an inverted image of a distant object. Moreover, right and left are reversed in the image, as you can confirm with a ray diagram. Because of this, a simple telescope is not very suitable for viewing objects on the earth. Usually we use binoculars for this purpose.

A diagram of a binocular is shown in Fig. 23.12a. Each half of the binocular consists of an objective lens, an eyepiece, and two prisms mounted at right angles to each other. The prisms are used to reflect the light by total internal reflection, as shown in part b of the figure. Notice how the reflection process inverts rays 1 and 2 in part b. As a result, the prisms invert the image and reverse right and left in it, thereby counteracting the reversals caused by the objective lens. The final image formed by the binocular for viewing by the eye is therefore erect and preserves the normal right and left.

23.8 Prism Spectroscope

In later chapters we speak frequently about the spectrum (or colors of light) given off by various atoms. These spectra are observed by means of instruments called **spectroscopes.** These are of two types, only one of which is

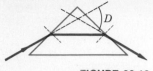

FIGURE 23.13
The prism shown deviates
the light beam through the
angle *D*.

discussed now; the second is treated in the next chapter. Our present interest is in the prism spectroscope.

A prism, usually made of glass, is a device frequently used to separate light into its various colors. A diagram of a prism is shown in Fig. 23.13. A beam of light usually will be bent twice, once when it enters and once when it leaves the prism. We call the total angle through which the ray is bent the **angle of deviation.** It is shown as angle *D* in the figure.

With the incident angle, the angles of the prism, and the refractive index of the glass known, it is possible, though difficult, to compute *D* by using Snell's law. However, it can be seen that the higher the index of refraction of the glass, the larger the deviation of the beam. This has important consequences, as we now see.

We mentioned earlier that the speed of light in most materials varies depending on the wavelength of the light. This is equivalent to saying that the index of refraction of the material depends on the color of the light. For most materials, the index of refraction for violet light is larger than for red light. Hence, violet light is bent more by a glass prism than red light is. Consequently, if a beam of white light enters a prism, as in Fig. 23.14, the light will be dispersed into its colors, as shown in the figure. The ability to disperse light varies from material to material. For high dispersion, the index of refraction must change markedly with wavelength.

In a prism spectroscope, a prism is used to disperse the wavelengths of light coming from the light source which is to be examined. In a typical device, such as the one shown in Fig. 23.15, the light from the source is used to illuminate a narrow slit. (Let us assume that a yellow sodium arc is shining on the slit. This gives off wavelengths of light very close to 5.89×10^{-5} cm, a characteristic yellow light.) Since the slit is at the focal point of the collimating lens, the light is parallel after leaving the lens. It is then refracted, as shown, by the prism. (Since we are assuming monochromatic light, i.e., single-color light, all the light is bent the same.) The objective lens forms an image of the slit at its focal point. If a photographic plate is placed as shown, a single line, or an image of the slit, is photographed on the plate. Alternatively, the objective lens and photographic plate can be replaced by a telescope (since the light is parallel), and the yellow line, or image of the slit, can be seen visually.

If light from a mercury arc is used instead of sodium light, several lines are photographed on the plate. They look as shown in Fig. 23.16. (This spectrum is also shown in color in the color plate insert.) Since each of the several discrete colors emitted by a mercury arc is deviated differently by the

FIGURE 23.14
The angle of deviation by a
prism is not the same for all
wavelengths of light. Hence
the prism disperses white
light into its constituent
colors.

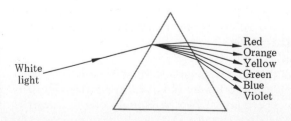

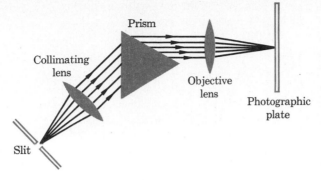

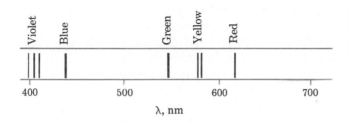

prism, several images of the slit, or lines, appear on the plate. These slit images are called **spectral lines.** Each line is the result of a certain color emitted by the light source. Every type of atom gives spectral lines which are characteristic of that type of atom. Not only does the emitted light, when observed through a spectroscope, allow us to tell what atoms are present, but also the lines or colors emitted tell us a good deal about the structure of the atom itself. We pursue this point further in a later chapter.

Spectral Lines

23.9 Polarized Light

Many optical devices use the fact that light is a transverse vibration. As we shall see, this fact is important when light is transmitted through certain materials. It is also a factor when light is reflected. Although our previous discussion has not been concerned with this feature of light waves, it is fundamental to the behavior of light which we discuss now.

We saw in previous chapters that light is EM radiation. It consists of waves such as that shown in Fig. 23.17. The electric field vector is sinusoidal and perpendicular to the direction of propagation of the wave, as shown. If the wave is traveling along the x axis of the figure, the electric field vibrates up and down at a given point in space as the wave passes by. There is a magnetic field wave perpendicular to the page and in step with the electric field. We call a wave such as this a **plane-polarized wave.** It derives its name from the fact that the electric vector vibrates in only one plane, the plane of the page in this case.

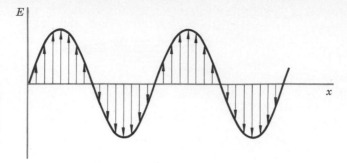

E

x

FIGURE 23.17
The electric vector vibrates
in a single plane when a
beam of light is plane-
polarized.

Most light consists of many, many waves such as that shown in Fig. 23.17. If the direction of propagation is to the right, the electric vectors must all vibrate perpendicular to the x axis shown in Fig. 23.17. However, they need not all vibrate in the plane of the page, and actually most of them do not. Let us stand at the end of the x axis in Fig. 23.17 and look back along it toward the point O, in other words, with the wave traveling straight toward us. The great multitude of waves coming toward us give rise to many individual electric vectors randomly oriented, as shown in Fig. 23.18a, where many more vectors than shown should actually be drawn. If the waves were plane-polarized vertically, i.e., in the plane of the page in Fig. 23.17, the approaching electric vectors would appear as shown in Fig. 23.18b. For a horizontally plane-polarized wave, the vectors would appear as in Fig. 23.18c.

Unpolarized light can be conveniently plane-polarized by use of a Polaroid sheet. This is a sheet of transparent plastic in which special needlelike crystals of iodoquinine sulfate have been embedded and oriented. The resulting sheet will allow light to pass through it only if the electric vector is vibrating in a specific direction. Hence, if unpolarized light is incident upon the sheet, the transmitted light will be plane-polarized and will consist of the sum of the electric vector components parallel to the permitted direction. (Before the invention of Polaroid other methods were used, but because of its convenience and low cost, Polaroid has displaced them except in certain very exacting situations.)

Any vector can be thought of as consisting of two perpendicular components. Hence, if the electric field is oriented as shown in Fig. 23.19a, it can be thought of as consisting of a vertical and horizontal component, as shown

FIGURE 23.18
If a pencillike beam of light
is coming straight out of the
page, the electric field
vibration will be as shown
for three types of beams.

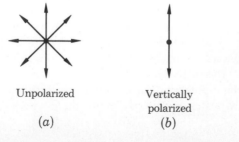

Unpolarized

Vertically
polarized

Horizontally
polarized

(a)

(b)

(c)

in part *b* of that figure. If we pass light vibrating at the angle shown through a Polaroid whose transmission direction is vertical, the vertical component of the vibration will pass through and the horizontal component will be stopped.

Consider what happens as unpolarized light is passed through two Polaroids as shown in Fig. 23.20. In part *a*, the polarizer (the first Polaroid) allows only the vertical vibrations to pass. These are also transmitted by the analyzer (the second Polaroid), since it, too, is vertical. However, in part *b*, the polarizer has been rotated through 90° and allows only horizontal vibrations to pass. These are completely stopped by the vertically oriented Polaroid. Therefore (almost) no light comes through the combination. We say that the polarizer and analyzer are **crossed** in this latter case.

Polarization of light is used in many technical and scientific applications. One use is in the determination of the concentration of optically active substances. **Optically active substances** *rotate the plane of polarization when polarized light passes through them.* For example, if a sugar solution is placed between the crossed polarizer and analyzer of Fig. 23.20*b*, it is found that the light is no longer stopped by the analyzer. However, by rotating the analyzer so that it is no longer oriented vertically, the light can be stopped. The sugar solution has rotated the original horizontal orientation of the direction of polarization somewhat toward the vertical. The amount of rotation of the plane of polarization has been found to be a direct measure of the concentration of the sugar in the solution. Hence we can measure the sugar concentration by this method. Many other substances are optically active. This activity can be used as a clue to the structure of the molecules compos-

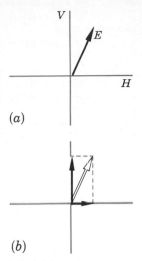

(*a*)

(*b*)

FIGURE 23.19
The electric field vector can be split into *x* and *y* components.

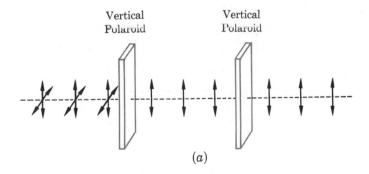

(*a*)

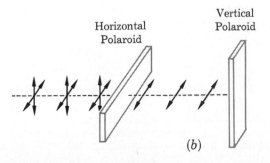

(*b*)

FIGURE 23.20
(*a*) The unpolarized light is polarized by the first Polaroid, the polarizer.
(*b*) The second Polaroid, the analyzer, and the polarizer are crossed, and the beam is completely stopped by the analyzer.

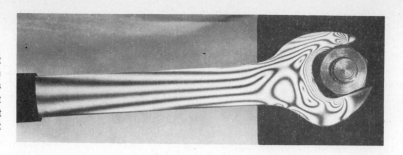

FIGURE 23.21
A strained, transparent object viewed through crossed Polaroids shows alternate dark and bright bands. The stress is greatest where the bands are closest together.

ing the substance. Whether a molecule rotates the polarization plane depends on the exact way in which the atoms are located relative to one another in the molecule. Some kinds of sugar exist in two forms; one rotates the polarization plane to the left, and the other rotates it to the right. Both types of molecule have the same chemical composition, but one is the isomer of the other; i.e., the structures differ only in spatial relationships on two or more chemically equivalent bonding sites within the molecule. There are many fascinating examples of this left- and right-handedness in nature. Some enzymes consume only one isomer of a molecule, on which they thrive. Certain types of protein in particular are notable for their left- and right-handed structures in assuming a helical configuration which influences their physical behavior.

Under a microscope the details of objects often are seen more clearly by examining them between crossed Polaroids. Portions of the object which appear the same in ordinary light may differ considerably in their ability to change the polarization of the transmitted light. Hence, these otherwise unobservable details can easily be seen. When a transparent object is under high stress, it often rotates the plane of polarization of the transmitted light. As a result, a nonuniformly stressed object observed between crossed Polaroids will show alternate dark and bright bands, as in Fig. 23.21. Where the bands are closest together, the stress is most uneven. By examining plastic models of strained objects like that in Fig. 23.21, it is possible to tell exactly how the stress is distributed. This is of great importance in the design of various parts for machines.

We seldom realize that the light coming from the blue sky is largely plane-polarized. Nor do we often remember that the glare of light from the surface of a lake or white concrete is considerably polarized. (This latter fact forms the basis for Polaroid sunglasses.) Since polarized light can be recognized only by use of an analyzer, our eyes do not tell us of its presence. The whole subject of polarized light is an interesting one, and you may wish to pursue it further.[3]

---

[3] A readable booklet on the subject in the Momentum Book Series is W. A. Shurcliff and S. S. Ballard, "Polarized Light," D. Van Nostrand Company, Inc., Princeton, N.J., 1965.

Summary

In the human eye, the eye lens focuses an image on the retina. By changing the shape of the lens, the muscles of the eye can bring objects at various distances into focus. Nearsighted people are unable to focus well on distant objects; farsighted ones cannot focus nearby objects well. Astigmatism occurs when the eye lens system has a cylindrical distortion.

Rays passing through the outer portions of highly curved spherical lenses are not focused properly. This effect is called spherical aberration. Different colors are not focused exactly the same by a lens, and this effect gives rise to chromatic aberration.

The power of a lens in diopters is the reciprocal of the focal length in meters. For lenses in close combination, the power of the combination is equal to the sum of the powers of the individual lenses. As a result, $1/f = 1/f_1 + 1/f_2$.

The magnifying power of an optical device is the ratio of the angle subtended at the naked eye by the object to the angle subtended when viewed through the device. A magnifying glass allows one to view an object when it is closer than the near point of the eye. Its approximate magnifying power is γ/f, where γ is the near-point distance, about 25 cm.

In a simple microscope, the objective lens forms a real image of the object. This image is then viewed by an eyepiece lens which acts as a magnifier. The magnifying power of such a microscope is $(L/f_0)(\gamma/f_e)$. In a good microscope, each of the two lenses is actually a complex combination of lenses.

The prism spectroscope separates a beam of light into its constituent wavelengths (or colors). An image of the entrance slit is formed for each wavelength. Each image is called a spectral line and represents a distinct wavelength in the original light beam.

Ordinarily, light in a beam is unpolarized. Its electric vector vibrates in a complex way at all angles possible perpendicular to the direction of the beam. When unpolarized light passes through a Polaroid, vibrations in only one direction are passed through. This transmitted beam, in which the electric vector always vibrates in the same direction, is called a polarized beam. Optically active substances have the ability to change the direction of polarization of a beam as the beam passes through them.

Minimum Learning Goals

Upon completion of this chapter, you should be able to do the following:

1 Sketch the important features of the eye and explain the function of each.
2 Explain what is meant by myopia, hyperopia, astigmatism, and lack of accommodation. Tell how each of these defects can be eased by use of lenses. Work problems such as Illustrations 23.1 and 23.2.
3 Sketch the construction of a simple camera. Explain how focusing is accomplished in a movable-lens camera.
4 Give the meaning of spherical aberration and of chromatic aberration.
5 Define the diopter unit. Find the focal length of several given thin lenses in close combination.
6 Explain the operation of a magnifying glass. In relation to it, give the meaning of magnifying power. State the approximate magnifying power of a given converging lens.
7 Show how the two-lens microscope operates by sketching its optical system and drawing a ray diagram for it. Distinguish between the object lens and the eyepiece lens.
8 Sketch the optical system for the astronomical telescope and locate the images it produces.
9 Explain what is meant by a prism spectroscope and show why it gives rise to line images. Describe how it separates colors and can be used to analyze a beam of light.
10 Distinguish between unpolarized and polarized light. Explain how polarized light can be produced. Define the following terms: Polaroid, crossed Polaroids, optically active substances.

Important Terms and Phrases

You should be able to define or explain each of the following:

Parts of the eye: iris, pupil, cornea, lens, retina
Myopia, hyperopia, astigmatism, lost accommodation

Near point; far point
Spherical aberration; chromatic aberration
Power of a lens in diopters
Magnifying power
Objective; eyepiece

Prism spectroscope; spectral line
Polarized versus unpolarized light; plane of polarization
Crossed Polaroids; optically active substances

Questions and Guesstimates

1 Show that a real image formed of a woman by a converging lens is inverted, but she and her image still have the same right hand. Show that exactly the reverse is true for an image formed by a plane mirror.

2 Clearer images are obtained in optical instruments when only a small portion of the lens is used. In the case of the pinhole camera, no lens is needed. To see how this is possible, draw a small, bright object about 1 mm high at 10 cm from a 1-cm opening in a large, opaque screen. Show how the bright spot cast by the object on a screen 5 cm behind the opening decreases in size as the opening is made smaller. Show that in the limit of a pinhole opening, two objects 1 cm apart and both 10 cm from the opening will give rise to well-defined images on the screen.

3 Show why a pinhole placed in front of a lens leads to a good image even when the image is not quite in focus. (See Question 2.)

4 A glass prism deviates a beam of blue light somewhat more than a beam of red light. Show by the means of wavefronts how this leads us to conclude that red light travels faster in glass.

5 Which of the following, as normally used, form real images: (a) eye, (b) camera, (c) microscope, (d) telescope, (e) binocular, (f) projection lantern, (g) plane mirror, (h) concave shaving mirror. (i) searchlight mirror?

6 Who, if either, will be able to see better without glasses under water, a nearsighted swimmer or one who is farsighted? Assume the swimmer (a) is not wearing and (b) is wearing goggles.

7 The wavelengths of light emitted by hot mercury vapor, for example, are called spectral lines. Explain clearly why they are called lines.

8 One can buy a cheap microscope for use by children. Invariably, the images seen in such a microscope have colored edges. Why is this so?

9 Suppose that the inside of a box camera is filled with water and that the lens is made stronger so that a focus will still occur at the location of the film. Will the pictures that the camera takes be changed in any way? Repeat for a box with only a pinhole and no lens.

10 Why is the speed of a camera important? What design factors influence the speed? (Consider both lens and shutter speeds.)

11 What happens to the light energy which is not transmitted by a Polaroid when unpolarized light is incident on it? Can you think of any drawback this might pose in using a Polaroid?

12 How can one determine whether a beam of light is polarized? Whether it is composed of two beams, one polarized and the other not?

13 With a commercial camera with a 5-mm-diameter lens opening, i.e., aperture diameter, the proper exposure time for a scene is $\frac{1}{60}$ s. About what would be the exposure time for a pinhole camera with a 0.50-mm-diameter pinhole and the same type of film? (E)

14 You have available a long, cylindrical, cardboard mailing tube and two lenses with focal lengths 60 and 10 cm that can be fitted into the tube. Use these to design a toy telescope.

Problems

1 (a) When the eye-lens system is adjusted to view a distant object, what is its focal length? Assume the distance from eye lens to retina is 2.0 cm. (b) Repeat for the case where the object being viewed is at the near point, 25 cm from the eye.

2 An object which is 0.010 cm long is viewed by the naked eye when the object is at the near point, 25 cm. How long is the image on the retina? (Take the eye-lens-to-retina distance as 1.0 cm.)

3 A certain woman is able to see printing in a book

clearly when the book is 60 cm away but not when it is closer. (*a*) Is she near- or farsighted? (*b*) What type and focal-length lens should she use to correct her sight?

4 A certain man is not able to see objects clearly unless they are closer than about 1.50 m. (*a*) Is he myopic or hyperopic? (*b*) What type and focal-length lens should he use to correct his sight?

5* A student is able to see clearly the title on the outside of this book when the book is no more than 90 cm away. What focal-length eyeglasses should the student use to see distant objects?

6* The eyeglass prescription for a certain person is listed as −1.50 diopters. Describe the eye defect that the lens is designed to correct.

7* A little girl wears thick, magnifying-glass-type eyeglasses. Her older brother holds the eyeglasses in sunlight and obtains images of the sun. He finds that each lens gives an image 30 cm from the lens. What are the girl's probable far point and near point without glasses?

8* A teacher notices that a child in his class holds pages very close to her eyes when reading. The usual position for this child is 11.0 cm. (*a*) Is the child near- or farsighted? (*b*) What kind and what focal-length lens should the child probably use in eyeglasses?

9 If a certain single-lens camera has a lens-to-film distance of 10.0 cm, and if it takes pictures which are 8×6 cm, how far from a painting which is 100×100 cm must the camera be placed if the image of the painting is just to fit on the photograph?

10 When the camera from Prob. 9 is used to photograph a tree from a distance of 40 m, the image on the film turns out to be 2.0 cm high. How tall is the tree?

11* In a simple box camera, the fixed distance from the lens to the film is 10.0 cm. If the focal length of the lens is 9.50 cm, where must an object be placed to give the best image on the film?

12* Prove that, for a simple, fixed-lens-position camera, the length on the film of an image of a given object varies inversely with the distance of the object from the camera lens.

13 Find (*a*) the power and (*b*) the focal length of the following lenses when placed in close combination: $f_1 = 20$, $f_2 = -60$, and $f_3 = 40$ cm.

14 Find (*a*) the power and (*b*) the focal length of the following lenses when placed in close combination: $f_1 = 80$, $f_2 = -25$, and $f_3 = -60$ cm.

15 What focal-length lens is needed in close contact with a converging 50-cm lens to form a diverging lens of −100-cm focal length?

16 An optometrist finds that the person being examined can see best when lenses with the following three focal lengths are stacked one in front of the other before the person's eye: 20, 50, and −100 cm. What focal length should be prescribed?

17 (*a*) About what magnification could a person with a 25-cm near-point distance achieve by use of a 12-cm-focal-length converging lens? (*b*) Repeat for a person whose near-point distance is 15 cm.

18 A certain magnifying glass gives an image of the sun 3.0 cm from the center of the lens. What is the approximate magnifying power of the lens?

19* A millimeter scale is viewed at 25 cm by a person's eye. The scale is now viewed through a magnifying glass that has a 20-cm focal length. What is the approximate ratio of the image size on the retina in the two cases?

20* Show that if a magnifying glass is used in such a way that the image of a tiny object is at infinity, then the magnifying power is given almost exactly by $M = \gamma/f$.

21** Show that if a magnifying glass is used in such a way that the image of a tiny object is at a distance γ from the lens, then the magnifying power is given almost exactly by $M = (\gamma/f) + 1$.

22* What is the approximate magnifying power of two lenses with $f = 30$ cm when they are used in close combination?

23 A certain microscope has an eyepiece whose magnifying power is 5.0. It is used in conjunction with a 3.0-cm-focal-length objective lens. What is the magnifying power of the microscope?

24 In a simple microscope, the objective has a 3.0-cm focal length, and the eyepiece has a 5.0-cm focal length. What is the magnifying power of the microscope?

25* A boy makes a simple microscope by cementing a 5.0-cm-focal-length lens to one end of a 10-cm-long tube and a 3.0-cm-focal-length lens to the other. (*a*) If he uses the 3.0-cm lens as the eyepiece, about how far in front of the objective must he place the specimen he is observing? (*b*) What will the approximate magnifying power of his microscope be?

26** Two marks 0.0100 mm apart are viewed in a microscope that has a magnifying power of 400. What angle do the marks subtend at the eye placed at the eyepiece? Give your answer in degrees.

27 A certain reflecting telescope uses a 70-cm-focal-length mirror and an eyepiece whose focal length is

4.0 cm. What is the magnifying power of the telescope?

28 What is the magnifying power of a reflecting telescope that has a mirror with a 160-cm radius of curvature and a 12-cm-focal-length eyepiece?

29* What is the magnifying power of a telescope that uses a 1.25-diopter objective lens and an eyepiece that has a magnifying power of 4?

30* What is the magnifying power of an astronomical telescope having a 1.50-diopter objective and an eyepiece lens having a magnifying power of 5?

31 A telescope at the Yerkes Observatory has an objective lens with a focal length of about 19 m. When it is used to observe the moon, how many kilometers on the moon corresponds to a 1.0-cm length on the image cast by the objective lens? Distance to the moon is 3.8×10^8 m.

32** (a) By what factor is the light intensity increased in a telescope if the diameter of the objective lens is changed from 0.50 to 4.0 cm? (Assume that the other dimensions remain constant.) (b) Suppose now, instead, that the focal length of the objective was tripled in going from the small to the large lens. By what factor is the light intensity at a given point in the image changed?

33** You will notice in Fig. 23.10 that the telescope inverts the object. This is an objectionable point if one wishes to view the opera from a distant seat in an opera house. Instead, one can use an opera glass (i.e., galilean telescope) such as the one pictured in Fig. P23.1. For the one shown there, locate the position of the final image of a distant object. Is it real or virtual? Erect or inverted? What is the magnifying power of this telescope?

34** Show that if the apex angle (i.e., the top angle) of a prism A is very small (i.e., the prism is very thin) and a beam of light strikes the prism perpendicular to one of the faces, the deviation D of the beam is given by $D = (n - 1)A$.

35** In Fig. 23.13, take the apex angle of the prism, i.e., the top angle, to be 60°. If $n = 1.50$ and the angle of incidence is 53°, find (a) the angle at which the beam leaves the prism and (b) the angle of deviation D.

36** A certain piece of equipment contains a 48-cm-focal-length lens made of glass having $n = 1.50$. For purposes of cooling the lens, it is found expedient to immerse the portion of the apparatus containing the lens in a rectangular vessel of water. What focal-length lens made of the same type of glass must be placed in the water in contact with the original lens if the apparatus is to operate properly?

FIGURE P23.1

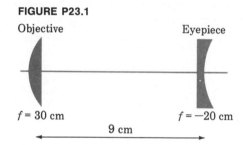

24
Interference and Diffraction

In the last two chapters we discussed the behavior of lenses and mirrors, using the concept of light rays. We did not need to know whether the light consisted of particles or waves for these discussions. This is not true of the topics treated in this chapter. We shall see that the wave nature of light gives rise to interference phenomena much like the interference effects we encountered in our study of wave motion and sound. The mere existence of these, as well as other effects discussed in this chapter, led to the final acceptance of the wave nature of light, as we shall see.

24.1 Huygens' Principle and Diffraction

Have you ever watched carefully as gentle waves lap against a post or some other obstacle in their path? If you have, you have noticed that the waves seem to bend around the post instead of casting a clear shadow of the post. A related situation is shown in Fig. 24.1, where we see plane water waves generated in a ripple tank. They strike a barrier that has a small hole in it. Notice how the waves pass through the hole and spread out so as to fill the whole region behind the barrier.

This general type of behavior is noticed not only for water waves, but also for sound waves, electromagnetic waves, and light waves. It is a characteristic behavior of waves, and we give it a special name, diffraction:

Waves are capable of bending around obstacles, a phenomenon called **diffraction.**

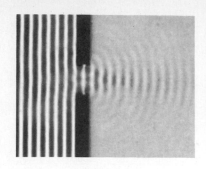  

FIGURE 24.1

The water waves at the left are incident on a hole in a barrier. Note that the waves spread out from the hole in the region at the right and that the waves spread to fill the whole region when the wavelength is comparable to the size of the hole. (*From U. Haber-Schaim et al., PSSC Physics, D. C. Heath and Co., Lexington, Mass., 1965*).

As we shall see, diffraction causes waves of all types to behave in seemingly peculiar ways.

To explain the phenomenon of diffraction, Christian Huygens, a contemporary of Newton, postulated what is now known as **Huygens' principle:**

Each point on a wave crest acts as a new source of waves.

For example, in Fig. 24.1, the portion of the wave crest striking the tiny hole in the barrier acts as a new wave source. As a result, waves spread outward in all directions from the hole, and so the waves fill the whole region behind the barrier.

We see from this that obstacles in the path of waves cannot be expected to cast clear shadows. This seems to disagree with what we know about light waves because objects in the path of light cast shadows that are easily observed. A clue to the resolution of this discrepancy can be found in Fig. 24.1. We notice there that the shorter the distance between wave crests, i.e., the smaller the wavelength, the less tendency the waves have to spread out behind the barrier. Later in this chapter we investigate this point in greater detail.

During the centuries since Huygens proposed his principle, both experimental work and theoretical work have established the basis for the principle and confirmed its validity. We therefore assume, when convenient, that each point on a wave crest acts as a new wave source. This simple artifice is useful in explaining the most complex wave behavior, and you will see examples of this use as our discussion proceeds.

24.2 Interference

An interesting experiment involving water waves is shown in Fig. 24.2. We see there two vibrators that send two sets of identical water waves out along the water surface. Notice what happens as the waves from the two sources interact with one another. Along certain lines (labeled B) they create very large wave crests, while along other lines (labeled D) that radiate from the sources no wave crests are seen at all. Apparently the water waves from the

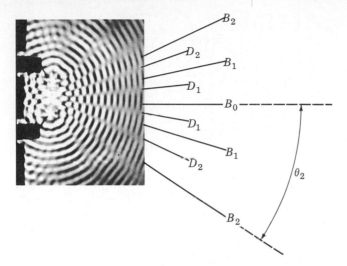

FIGURE 24.2
The two sources send out coherent water waves much as the waves sent out from the slits in Fig. 24.1. (*From U. Haber-Schaim et al., PSSC Physics, D. C. Heath and Co., Lexington, Mass., 1965.*)

two sources reinforce one another at certain points, while at other points they seem to cancel one another. Let us now investigate such phenomena.

You will recall from our work with sound waves and waves on a string that identical waves can reinforce or cancel each other. To review this fact, consider the two waves *A* and *B* shown in Fig. 24.3. In part *a*, the two waves are in phase, crest on crest and trough on trough. When they are added, the resultant wave is large as shown in the lower part of the figure.

FIGURE 24.3
Identical waves can reinforce or cancel each other, depending on their relative phases.

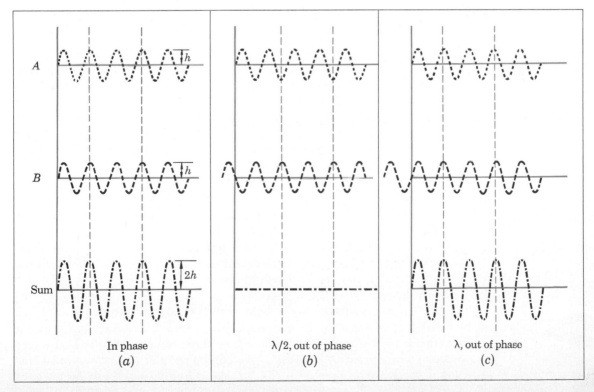

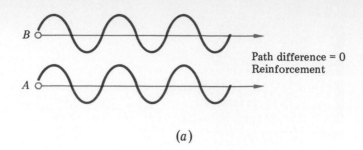

Path difference = 0
Reinforcement

(a)

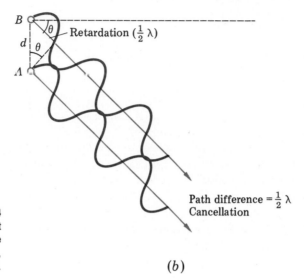

Retardation ($\frac{1}{2}\lambda$)

Path difference = $\frac{1}{2}\lambda$
Cancellation

FIGURE 24.4
Reinforcement occurs at
those angles for which the
path difference is λ, 2λ,
3λ, etc.

(b)

But the situation in *b* is quite different. There wave *B* has been held back through $\frac{1}{2}\lambda$, so now crest falls on trough for the two waves. The waves are $\frac{1}{2}\lambda$, or 180°, out of phase. Now when they are added, they exactly cancel each other, and so their sum wave is zero. The waves interfere destructively.

In part *c* we see what happens when wave *B* is λ behind wave *A*. The two waves are now λ, or 360°, out of phase. Now crest falls on crest, and so the two waves add to give a large resultant wave. The waves interfere constructively.

In general, we conclude (as we did in the case of sound waves) that *two identical waves will interfere constructively if the two waves are in phase.* If one of the waves is retarded by a distance λ, 2λ, 3λ, etc., relative to the other wave, the waves will still reinforce when combined because crest will still fall on crest. But if the relative retardation is $\frac{1}{2}\lambda$, $3(\frac{1}{2}\lambda)$, $5(\frac{1}{2}\lambda)$, etc., then crest will fall on trough and the two waves will interfere destructively; they will cancel each other.

Let us now return to the water wave situation shown in Fig. 24.2. We wish to find out why the waves reinforce in certain regions and cancel in others. This question is easily settled if we refer to Fig. 24.4. The two sources are sketched at points *A* and *B*. They send out identical waves as indicated. Notice in part *a* that the waves are in step, crest on crest and trough on trough. Hence reinforcement occurs in the direction shown in *a*. This is the direction of line B_0 in Fig. 24.2.

But in Fig. 24.4*b* the situation is quite different. In this direction the wave from source *B* is retarded by $\frac{1}{2}\lambda$ relative to the wave from *A*. Now the wave crests of one fall on the troughs of the other. As a result, the waves leaving the sources in this direction cancel. That is why the water waves in Fig. 24.2 cancel along the two lines labeled D_1. (You should be able to explain why the other D_1 line shown in Fig. 24.2 occurs.)

If you consider what would happen if the angle θ in Fig. 24.4 were increased still further, you see that wave *B* would be held back still more relative to wave *A*. But if θ and the relative retardation were increased until the relative retardation was λ, then the two waves would reinforce each other again. That is what is happening along lines B_1 in Fig. 24.2. Along these two lines the relative retardation is λ.

Reasoning in this way, you can convince yourself that the relative retardation along cancellation lines D_2 is $3(\frac{1}{2}\lambda)$. Moreover, along reinforcement lines B_2 the relative retardation is 2λ. As we see, lines B_0, B_1, B_2, and similar lines represent lines along which the waves reinforce. Along them, the relative retardation between the waves from the two sources is 0, λ, 2λ, and so on. The waves from the two sources therefore reinforce each other.

Let us now obtain a mathematical relation for the angles at which these reinforcing lines occur. To do this, examine the small triangle in Fig. 24.4. Notice that the angle labeled θ in it is equal to the angle θ that the beams make with the horizontal. We see at once that in the small triangle

$$\text{Relative retardation} = d \sin \theta$$

where d is the distance between the sources.

To find the angles at which reinforcement occurs, the angles to the points labeled *B* in Fig. 24.2, we recall that for reinforcement the relative retardation must be 0, λ, 2λ, 3λ, or, in general, $n\lambda$, where n is an integer. Therefore, if θ_n is the angle for which the relative retardation is $n\lambda$, we have

$$n\lambda = d \sin \theta_n \qquad (24.1)$$

This equation tells us at what angles θ_n brightness occurs.

For example, along line B_0 in Fig. 24.2 we have $n = 0$ (because the waves are not retarded relative to each other), and so θ_0 is given by

$$0 = d \sin \theta_0 \qquad \text{or} \qquad \theta_0 = 0$$

Similarly, along line B_2 we have $n = 2$, and so

$$2\lambda = d \sin \theta_2$$

Let us now illustrate this with a numerical example.

Illustration 24.1 Suppose in Fig. 24.2 that the sources are 2.0 cm apart and the waves have a crest-to-crest distance (a wavelength) of 0.70 cm. What is the angle at which the reinforcement line B_2 occurs?

Reasoning We are told that $d = 2.0$ cm and $\lambda = 0.70$ cm, and we are interested in the situation for which $n = 2$. Substituting in Eq. (24.1) gives

$$\sin \theta_2 = \frac{(2)(0.70 \text{ cm})}{2.0 \text{ cm}} = 0.70$$

from which $\theta_2 = 44°$. The lines B_2 in Fig. 24.2 would make $44°$ angles with the horizontal.

24.3 Young's Double-Slit Experiment

The experiment described in Sec. 24.2, dealing with the interference of water waves from two sources, is not peculiar to water waves. You will recall that the prongs of a tuning fork can give rise to interference of sound waves. The explanation for this phenomenon is similar to the description of the interfering water waves, except that the waves are compressional sound waves rather than transverse water waves. It should be clear that any identical waves, transverse or longitudinal, are capable of exhibiting interference phenomena.

As we mentioned previously, Newton believed light to be corpuscular. He pictured light to be a stream of particles shot out from light sources. Naturally, these particles must travel in straight lines, much as baseballs would do. Although Grimaldi had shown as early as 1660 that light can be diffracted, i.e., bent around objects, Newton was able to devise an explanation of this fact in terms of his light corpuscles. His explanation was not very satisfying, but most people accepted his pronouncements concerning the nature of light. It was not until after 1803 that the wavelike nature of light became widely accepted.

In 1803 and 1807 Thomas Young published the results of his experiments demonstrating the phenomenon of interference of light. Using a narrow beam of sunlight passing through a hole in a window shutter, he allowed the light to fall on two narrow, parallel slits in a piece of cardboard, as illustrated schematically in Fig. 24.5. To the right of the slits he placed a screen. He observed a pattern of alternating bright and dark regions, called *fringes,* on the screen as shown. His observations of these fringes, and his inference that light is a wave phenomenon, allowed him to compute the wavelength of light for the first time. Let us see how he did it.

If you look back at Fig. 24.2, you will notice that a vertical wall placed at the right hand edge of the figure would show a water pattern. At the points labeled B, the water waves would be high. But where the lines labeled D intersect the wall, the water would be calm. This interference pattern on the wall corresponds to the interference pattern shown in Fig. 24.5.

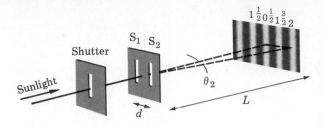

The bright fringes in the pattern correspond to the positions labeled B in the water wave interference pattern. As you might suspect, the D positions correspond to the dark positions in Young's double-slit pattern.

Using this correspondence with the water interference experiment, we can interpret Young's pattern as follows. The two slits act as two light sources that send out identical waves. Brightness occurs on the screen at the fringe labeled 0 because the waves traveling to this position reinforce each other; their relative retardation is zero.

At fringe 2, for example, the two waves again reinforce. These fringes correspond to regions B_2 in Fig. 24.2, and so the relative retardation between the waves reaching there is 2λ, where λ is the wavelength of the light waves. As we see, the situations in Figs. 24.2 and 24.5 are completely analogous. We can therefore apply Eq. (24.1) to Young's double-slit fringes and write

$$n\lambda = d \sin \theta_n \qquad (24.1)$$

In this expression, n is the fringe number as labeled in Fig. 24.5. The slit separation is d, and the wavelength of the light is λ. Of course, θ_n is the angle to the nth fringe as typified by θ_2 in the figure. Frequently we call n the *order number* of the fringe. Using this terminology, we call the θ_2 fringe the *second-order fringe*.

Young used Eq. (24.1) to compute the wavelength of light. Because he used sunlight, his light beam consisted of all the colors of the rainbow, and so his beam contained an infinite number of wavelengths. Each wavelength had its bright fringe at the particular angle that satisfies Eq. (24.1). Consequently, his fringes consisted of a multitude of interlocking fringes, one for each wavelength in his light. Because of this, his fringes were colored; the portion of the fringe closest to the center was blue, while the outer part was red. (Why?)

In a typical experiment, consider the distance L in Fig. 24.5 to be 120 cm, the slit separation $d = 0.025$ cm, and the distance from the center of the pattern to the approximate center of the fringe labeled 2 is 0.50 cm. To find θ_2, we see from Fig. 24.5 that

$$\tan \theta_2 = \frac{\text{distance } 0 \to 2}{D} = \frac{0.50 \text{ cm}}{120 \text{ cm}} = 0.00417$$

from which $\theta_2 = 0.24°$.

Thomas Young
1773–1829

An English physicist and physician, Young was a child prodigy whose marvelous intellect served science throughout his life. By the age of 4 he had twice read the Bible from cover to cover. During his youth he learned several languages including not only Latin and Greek but also Hebrew, Arabic, Persian, and Turkish. He studied medicine, but his poor bedside manner hindered his progress in his chosen profession. Luckily, he inherited considerable wealth from an uncle and became financially independent. He was therefore free to follow his interest in science.

Young's earliest research involved the human eye. While still a medical student, he discovered the mechanism of the accommodation ability of the eye. The cause of astigmatism was another of his discoveries. His interest in the human eye led him to investigate the nature of light, as explained in this chapter. Other investigations culminated in the discovery of how the eye perceives color; he was also one of the first to realize that light waves are transverse.

In addition to his studies involving light, Young made notable discoveries in the area of heat, elasticity, and surface tension. He wrote many articles for the *Encyclopedia Britannica* and contributed much to the scientific discussions of his day. Impelled by his wide-ranging interests, he became an expert on ancient Egyptian hieroglyphics and was among the first to decipher them.

4 3 2 1 0 1 2 3 4

I

FIGURE 24.6
Interference fringes
produced by a double-slit
system. (*After Jenkins and
White.*)

Young made use of data such as these to compute the wavelength of the light near the center of the fringe. Substituting in Eq. (24.1), he would obtain

$$\lambda = \frac{d}{n} \sin \theta_n = \frac{0.025 \times 10^{-2}\,\text{m}}{2} \sin 0.24 = 5.2 \times 10^{-7}\,\text{m}$$

He was therefore able to conclude that light had a wavelength of about 500 nm, with the wavelength of blue light being somewhat smaller than this and that of red light somewhat longer.

Now, of course, we are not restricted to sunlight in performing a double-slit experiment. Laser light, for example, is monochromatic (i.e., of a single wavelength). Even though the laser beam consists of only a single wavelength, the interference fringes obtained in a double-slit experiment are still quite wide. A photograph of a typical fringe pattern using monochromatic light is shown in Fig. 24.6. The numbers above each fringe are analogous to those in Fig. 24.5. Below the photograph is shown a graph of the light intensity in the fringe pattern.

As you see, the double-slit fringes are quite diffuse. Thus it is difficult to locate their centers exactly, and so wavelengths cannot be measured very precisely from the fringe pattern. Fortunately, the pattern can be sharpened tremendously by using thousands of parallel slits rather than two. Such a device is called a *diffraction grating,* and we discuss it later on.

24.4 Coherent Waves and Sources

You might suppose that a pattern similar to that of the double slit could be obtained using two tiny light bulbs in place of the two slits. The bulbs should act as two sources, and the waves from them should interfere. But this would be a false assumption for reasons that we point out now.

The difficulty arises because steady, *easily observed interference patterns can be obtained only if the waves combine in such a way that they reinforce or cancel each other steadily.* For example, if you refer to Fig. 24.3, you see that the waves we were concerned with are identical in shape and wavelength. As a result, if the crests coincide at one point on the waves, the waves coincide everywhere. Consequently, the combined wave striking a screen or entering your eye does not fluctuate between reinforcement and cancellation as time goes on. If the waves reinforce at a certain point in the pattern at one instant, they will continue to reinforce at that point. The pattern does not fluctuate with time.

This would not be the case if the two waves were of different wavelength and shape. Then sometimes they would reinforce, but an instant later they might cancel at the same point. Therefore a stable, easily observed interference pattern would not be visible.

We see from this that *interference effects are easily visible between two waves only if the waves are identical in wavelength and shape.* Moreover, *the phases of the waves relative to each other must be preserved.*[1] We call waves of this type **coherent waves:**

> Coherent waves have the same form, wavelength, and phase relation to one another.

The sources of such waves are called *coherent sources.* More is said about coherency in Sec. 26.13.

Two light bulbs are not coherent sources. Each bulb emits its own waves depending on what the atoms in the filament of the bulb happen to be doing at that instant. Consequently, the emitted waves are not of the same shape, and they do not preserve a definite phase relation to each other even though both may consist of the same wavelengths. The two light bulbs do not send out coherent waves, and so the interference pattern typical of the double slit will not be observed.

Because two sources of light are almost always noncoherent, usually it is necessary to divide a single light beam into two parts to obtain an interference pattern. For example, in the double-slit experiment, the two slits are illuminated by the same beam, the same light wave. This wave is divided into two distinct parts by the two slits. Because the two waves thus generated are part of the same wave, they are coherent waves and give rise to the interference effects discussed.

24.5 Michelson's Interferometer

Interference between beams of coherent light forms the basis for one of the most precise methods available for measuring lengths. The Michelson interferometer is shown schematically and simplified in Fig. 24.7. A beam

---

[1] This would not be the case, for example, if a small length of one wave were snipped out of the wave and the wave were rejoined so as to slightly decrease the length of that section.

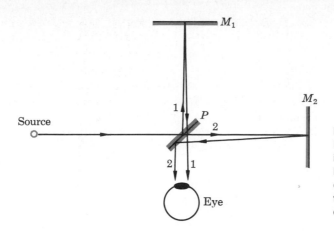

FIGURE 24.7
In the Michelson
interferometer, the two
halves of the beam travel
different paths and may
therefore interfere with
each other.

of light from the monochromatic source is split into two parts by the semi-transparent mirror P. About half the beam is reflected up to mirror M_1 and back, as shown. The other half is transmitted to mirror M_2 and reflected to P, where it is reflected down, as shown. (Of course, other reflected beams occur too, but we are interested in only the two shown.)

If beam 1 travels exactly as far as beam 2, they will be in phase when they enter the eye of the observer and brightness will be seen. Suppose, now, that mirror M_2 is moved to the right through a distance $\lambda/4$. Since beam 2 must travel this added distance twice (going down and coming back), beam 2 will now be traveling a distance $\lambda/2$ farther than beam 1. When they are joined at the eye, they will completely cancel each other, since they are $\lambda/2$ out of phase. The eye will see only darkness.

Moreover, if mirror M_2 is moved another $\lambda/4$ to the right, beam 2 will then be traveling a distance λ longer than that traveled by beam 1. As a result, the waves will reinforce each other, and brightness will be seen once again. Clearly, as M_2 is moved slowly to the right, alternate brightness and darkness will be seen by the observer. Each time that the mirror is moved through $\lambda/2$, the bright fringe will give way to darkness and then the brightness will return. If the observer counts 1000 bright fringes appearing to the eye as mirror M_2 is slowly moved to the right, the observer will know that the mirror has been moved $1000(\lambda/2)$.

We see, then, that we can easily measure the distance moved by mirror M_2 to within an accuracy of $\lambda/2$. Since λ might be blue light of wavelength 4×10^{-5} cm, it is possible to measure movements as small as 0.00002 cm. Actually, by special techniques, this device can be used to measure lengths to an accuracy of nearly one-hundredth this value. Although we have considerably oversimplified the pattern seen by the eye of the observer, this instrument is basically quite simple. Not only is it of value in the precise measurement of lengths, but also it can be used for the measurement of refractive indices of gases.

24.6 Interference from Thin Films

The beautiful colors reflected from a thin film of oil on a water puddle and the colorful appearance of a soap bubble in a bright light are both the result of interference of light caused by a thin film. We now investigate how these interference fringes arise and see how they can be used to measure small distances.

Suppose that two very flat glass plates have a wedge of air between them, as shown in Fig. 24.8. Actually the angle between the plates would be smaller than shown. If the plates are illuminated from above by monochromatic light, an observer looking at the plates from above will see alternate dark and bright fringes, as shown by the letters D and B in the figure.

These fringes are the result of the interference of beams reflected from the upper and lower sides of the air wedge, as shown by the beams a and b of Fig. 24.8. When brightness is seen at B, rays a and b are either in phase or a whole number of wavelengths out of phase. In going from fringe B to B', we have held back ray b through one whole wavelength. We did this by making ray b at B' travel a distance $2(t_2 - t_1)$ farther than did ray b at B. (We neglect the small angle which the rays make with the vertical.) Hence we see that from one bright fringe to the next in an air wedge, the thickness has changed by $\lambda/2$. (The distance is $\lambda/2$, and not λ, because ray b travels the extra distance twice, once down and once up.) Similarly, the dark fringes occur at $\lambda/2$ differences in thickness.

If multicolored or white light is used in this type of experiment, the different colors reinforce at different places. As a result, the fringes are highly colored, an effect one often observes for soap and oil films in sunlight. (Can you show that the blue fringes will be more closely spaced than the red?)

If the wedge of Fig. 24.8 had been filled with liquid instead of vacuum (or air to a good approximation), ray b would have been delayed even longer relative to ray a. Since the speed of light in a liquid of index of refraction n is c/n, it takes longer for a beam to travel a given distance in liquid compared with vacuum. In the time taken for a beam to travel d in the liquid, the beam would have moved nd in vacuum. For this reason,

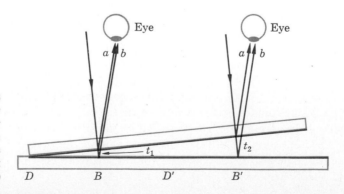

FIGURE 24.8
The two rays reflected by the two sides of the air wedge can interfere with each other. The diagram is only schematic. (In what respect is it not correct?)

when light travels a distance d in a material of refractive index n, we say that the **equivalent optical path length** *in the material is nd*. As we see, beam *b* in Fig. 24.8 will appear to have traveled a distance (*n*)(twice the thickness) farther than beam *a*.

Optical Path Length

Previously we discussed the case of the air-filled wedge. For that case *n* = 1.00, and so the optical path length difference between beams *b* and *a* was simply (2)(thickness). However, for a liquid with refractive index *n* filling the wedge, the optical path difference will be (*n*)(2)(thickness). As a result, if t_1 and t_2 are at adjacent bright spots, we have

$$2(t_2 - t_1)n = \lambda$$

In other words, the change in thickness between bright fringes is $\frac{1}{2}\lambda/n$ instead of $\frac{1}{2}\lambda$.

Notice also that the point of contact of the glass plates in Fig. 24.8 is dark. Since ray *b* travels essentially the same distance as ray *a* at this point, we would ordinarily expect it to be bright. Early investigators tried in vain to polish the surfaces in such a way that brightness would exist at this point. However, the better the contact between the plates, the more perfectly dark this region becomes. We now know that *a beam reflected by an optically dense material, i.e., high value of n, undergoes a 180° phase change in the process*. However, this does not concern us, since we are primarily interested in differences of thickness such as are shown in the illustration below.

Illustration 24.2   When a plane convex lens is placed on a flat glass plate, as shown in Fig. 24.9*a*, a phenomenon known as **Newton's rings** is demonstrated. If the system is illuminated and viewed from above, a series of interference rings are observed, as shown in part *b*. If light of 589 nm is used, what is the thickness of the air gap at the position of the tenth dark ring? Repeat if the gap is filled with water with *n* = 1.33.

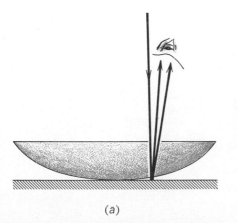

(a)

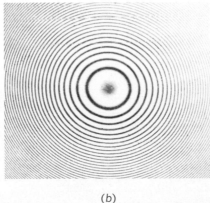

(b)

FIGURE 24.9
The interference fringes in (*b*), Newton's rings, are formed by the interference of the light rays reflected by the air wedge shown in (*a*). (Not to scale.) [(*b*) *Bausch & Lomb Optical Co.*]

Reasoning The center spot is dark, and the gap width there is zero. Going out to the first dark ring, the ray reflected from the lower surface must have been held back through a distance λ. Hence the thickness of the gap there is $\lambda/2$. Similarly, going from the first to the second ring, the wedge must increase once again by a thickness $\lambda/2$, and the wedge thickness is $2(\lambda/2)$. Clearly, at the tenth dark ring, the wedge thickness will be $10(\lambda/2)$. We therefore find the gap to have a thickness

$$10\frac{5.890 \times 10^{-7}\text{ m}}{2} = 2.945 \times 10^{-6}\text{ m}$$

at the site of the tenth dark ring. In practice, the smoothness of the circles or other interference fringes for an air wedge can be used to determine how perfectly flat the surfaces are.

For the water-filled wedge, the equivalent optical path length is $(1.33)(2)(\text{thickness})$. As a result, the gap thickness will be $1/1.33$ times the value found above. It is therefore 2.21×10^{-6} m.

24.7 Diffraction Grating

In order to measure the wavelength of light accurately with a practical instrument, a diffraction grating is often used. A device constructed for this purpose is called a **grating spectrometer.** *The grating is basically a large number of parallel, evenly spaced slits in an opaque screen.* A common grating might have 10,000 slits in 1 cm. Hence the separation d between the centers of adjacent slits would be 0.0001 cm. We shall see that the operation of this device is not too different from Young's double slits.

A schematic diagram of a common grating spectrometer (or spectrograph) is shown in Fig. 24.10. The light source Sc illuminates the slit S at the end of a tube C, called a **collimator,** which is used to make the rays of light parallel. Lens L_1 is placed with S at its focus. Hence the light coming from the slit is made parallel, i.e., is collimated, by lens L_1. This parallel beam of light then passes through the grating and enters a telescope T. The telescope consists of lens L_2 that forms an image of the slit at I. By means of

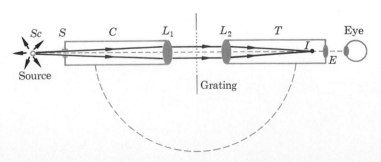

FIGURE 24.10
A schematic diagram of a grating spectrometer.

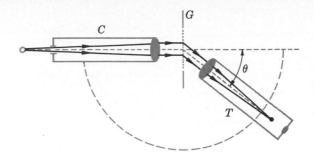

FIGURE 24.11
When the telescope is rotated on the arc of a circle, as shown, an image of the slit is formed by interference at an angle θ to the straight-through beam.

the eyepiece E, the observer views the image I. This image appears as a sharp, bright line, an image of the brightly lit slit at S. Provision is made for the telescope to be rotated through accurately known angles, as shown in Fig. 24.11.

Of course, when θ is zero, one sees a clear image of the collimator slit. This direct-through image is called by various names: the **central maximum,** the **zeroth-order maximum,** and the **central image.** It has the same color as the light source. If white light is being used, the central image is also white. When the cross hairs of the telescope are coincident with this image, θ in Fig. 24.11 is zero.

We would expect in a multiple-slit device such as this to observe interference effects somewhat like those seen with a double slit. This is actually the case. However, with a grating having thousands of slits, the interference pattern observed is much sharper than that found for double slits. If monochromatic light is used to illuminate the slit of the collimator, one observes the light-intensity pattern shown in Fig. 24.12 as a function of the angle θ.

For a good instrument, the images at 0, 1, and 2 would be much sharper than shown. This figure should be compared with Fig. 24.6 showing a similar diagram for the double slit. *The interference pattern of a grating is much wider in angle and considerably sharper than the double-slit pattern.* We call the images seen through the telescope at equal angles on both sides of the central maximum the **high-order maxima.** For example, the images indicated by 1 in Fig. 24.12 are called the **first-order maxima,** or spectrum. Similarly, images 2 are called the **second-order maxima.** Depending on the grating, only one or perhaps several orders may be seen before the 90°

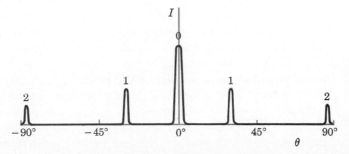

FIGURE 24.12
The diffraction grating gives a central image and symmetrical side images. For a good grating, the lines would be much sharper than shown.

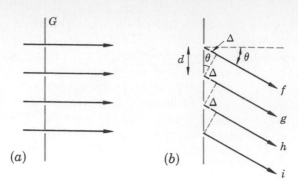

position is reached. We now derive the relation between the positions of the orders and the wavelength of the light.

Referring to Fig. 24.13, we consider four of the thousands of slits in the grating. Clearly, the direct, or straight-through, rays shown in part *a* are in phase and will reinforce each other when brought together within the telescope. However, the rays shown coming from the slits in part *b* will travel different distances to the telescope. As shown, ray *f* travels Δ farther than *g*. Ray *g* travels a distance Δ greater than *h*, and so on. If Δ is exactly a whole number of wavelengths long, the light from all the slits will reinforce. The first-order maximum occurs for $\Delta = \lambda$; the second-order occurs for $\Delta = 2\lambda$; and, in general, the *n*th-order maximum (if possible) occurs for $\Delta = n\lambda$.

We see at once from the little triangle in part *b* of Fig. 24.13 that

$$\sin \theta = \frac{\Delta}{d}$$

Since $\Delta = n\lambda$ at the *n*th-order maximum, *we have the following equation for the angular position θ_n of the nth maximum:*

Grating Equation
$$\sin \theta_n = \frac{n\lambda}{d} \tag{24.2}$$

This is called the **grating equation.** It is identical to Eq. (24.1) for the double slit.

Since we can measure with high precision the angle θ_n at which the *n*th-order maximum occurs, it is necessary to know only the grating spacing *d* in order to determine λ accurately. For example, if the yellow light from a sodium-arc lamp is used in even a simple spectrometer, it is not difficult to see that the sodium light gives *two* slit images (or lines) at each order position. These lines are very close together and have wavelengths of 589.0 and 589.6 nm. The mere fact that one is able to see these two lines as distinct images provides some measure of the potential accuracy of such a device.

If mercury light is used in a grating spectrometer, several different colored lines are seen in each order. As pointed out previously, the mercury light consists of a yellow line (579.0 nm), a greenish-yellow line (546.1 nm), a blue line (435.8 nm), and a violet line (404.7 nm). Other fainter lines are also visible. (*As pointed out in the last chapter, these lines in the spectrum of* *a light source, called* **spectral lines,** *are actually images of the spectrograph slit.* These images appear as bright lines in a photograph, and hence the name *line* is used.) In later chapters we see that other types of light sources also exist. While *the light from a mercury arc consists of several easily observed discrete lines (called a* **bright-line spectrum**), *the light from an incandescent bulb contains all the colors, and no sharp lines are observed. The incandescent source gives off a* **continuous spectrum,** *since a continuous band of color is seen when it is used in a spectrometer.* These facts are shown clearly in the color plate.

Illustration 24.3 A particular grating has 10,000 lines per centimeter. At what angles will the 5890-Å line appear?

Reasoning The grating space d is $\frac{1}{10,000}$ cm $= 10^{-6}$ m. Using the grating equation, we have

$$\sin \theta_1 = \frac{5.89 \times 10^{-7}}{10^{-6}} = 0.589$$

From the tables of sines, $\theta_1 = 36°$, and so the first-order images will be found at this angle on each side of the central maximum. For the second order we have

$$\sin \theta_2 = 2 \frac{5.89 \times 10^{-7}}{10^{-6}} = 1.178$$

Because it is impossible for the sine of an angle to be greater than unity, this and higher-order images will not exist.

24.8 Sharpness of the Maxima

In the last section we stated that the maxima in the diffraction-grating interference pattern are much sharper than those for the double slit. It is quite easy to show why this is true. For simplicity, let us base our explanation on the grating shown in Fig. 24.14 that contains only 10 slits, although a practical grating would have thousands of slits. Take the distance between adjacent slits to be d, and let us call the relative retardation between waves from adjacent slits Δ. Then the relative retardation between rays 1 and 6 is 5Δ.

You will recall that the first-order maximum occurs when $\Delta = \lambda$. We pointed this out in connection with our discussion of Fig. 24.13. The second-order maximum occurs for $\Delta = 2\lambda$. These are the same results we obtain for a double slit, and so, in this regard, the patterns for the double slit and the diffraction grating are identical. The angles at which the maxima occur are identical as long as the slits have the same separation. But the similarity ends there.

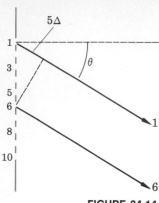

FIGURE 24.14
Maxima occur when $\Delta = n\lambda$, while darkness occurs if $\Delta = n\lambda + \lambda/10$ for this grating.

In the case of the double slit, the light intensity in the interference pattern first becomes zero midway between maxima. This is not the case for the diffraction grating. Let us now see how wide the diffraction-grating fringes are.

Suppose the angle θ shown in Fig. 24.14 is that of the first maximum. Then $\Delta = \lambda$, and the waves from all the slits reinforce one another. Now let us increase θ slightly so that $\Delta = \lambda + \lambda/10$. (Note in passing that the second-order maximum occurs at a much larger θ, namely, when $\Delta = \lambda + \lambda$.) In this case, the relative retardation between rays 1 and 6 in Fig. 24.14 is 5Δ, or $5\lambda + \frac{1}{2}\lambda$. But this means that the crests of the wave from slit 1 will fall on the troughs of the waves from slit 6. As a result, the waves from these two slits cancel.

Similar reasoning will show you that for this same value of Δ, namely, $\lambda + \lambda/10$, the wave from slit 2 will cancel that from slit 7. In the same way, waves from slits 3 and 8 will cancel, as will those from 4 and 9 and those from 5 and 10. As we see, no light will be seen at this angle because the waves from all the slits cancel pairwise. We can therefore say that the first-order maximum is only wide enough for Δ to change from λ to $\lambda + \lambda/10$. In terms of angles, this means that the angle between the position of the maximum and the edge of the maximum is only about one-tenth the angular distance between the central maximum and the first-order maximum. If the first-order maximum occurs at 30°, the maximum itself would be only about 3° wide.

Notice that the factor 10 occurred in our calculation because it was the number of slits in the grating. A practical grating would contain perhaps 10,000 slits, and so its maxima would be only $\frac{1}{1000}$ as wide as we calculated for a 10-line grating. As you see, *the widths of the maxima vary inversely with the number of lines on the grating*. For this reason, the diffraction grating gives such well-defined maxima.

24.9 Diffraction by a Single Slit

Until now we assumed that the width of the opening for our slits was negligible in comparison to the wavelength of light being used. If you look at Fig. 24.1, you will see that the smaller the slit in comparison to the wavelength, the better the waves diffract into the region of normal shadow. Now we wish to investigate the reasons for this effect and to describe more fully how a single slit diffracts light. In so doing, we will learn that interference fringes exist in the single-slit diffraction pattern. Indeed, if you look at Fig. 24.1, you can notice faint evidence for angular variation in the intensity of the diffracted wave.

To see this effect in the case of light waves, light can be sent through a single slit, and the transmitted light can be recorded on a photographic film, as shown in Fig. 24.15. The central bright spot is considerably wider than the slit. Moreover, bright bands occur on each side of the central image and

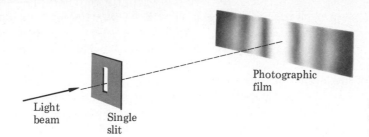

Photographic
film

Light
beam

Single
slit

FIGURE 24.15
Single-slit diffraction
pattern. (Not to scale.)

must result from some sort of interference effect. Let us now see what is involved in this situation.

As shown in Fig. 24.16a, the light rays which go straight through the single slit will all be in phase with one another. For this reason, the straight-through position is bright and gives rise to the central bright spot shown in Fig. 24.15. However, at an angle θ to the straight-through beam, rays from various parts of the slit will travel different distances to the film or photographic plate. The most important situations are shown in Fig. 24.16b, c, and d.

In part b, ray B from the middle of the slit is half a wavelength behind ray A. As a result, these two rays cancel each other. But that is not all, because we see that rays leaving the slit from positions just above A and B will also cancel since they, too, will have a path difference of $\lambda/2$. In fact, each ray leaving the lower half of the slit has a corresponding ray leaving the upper half which will cancel it. Hence, at this angle θ, no light will reach the film from the slit, and therefore you will observe darkness. As seen from the figure, this situation occurs when $\sin\theta = \lambda/b$, where b is the slit width. Notice that if slit width b is equal to the wavelength of the light, the dark spot occurs at $\theta = 90°$. In other words, *if the slit width is decreased until it is as small as λ, the image of the slit spreads to become infinitely wide.*

If b is considerably larger than λ, as pictured in Fig. 24.16, a side bright fringe will occur for the angle θ shown in part c. In this case, the rays from the bottom third of the slit cancel those from the center third, while the top third is left uncanceled. Darkness is again achieved at the larger angle

FIGURE 24.16
In analyzing the single-slit pattern qualitatively, we section the slit into portions whose rays differ by $\lambda/2$ in the path length. Why?

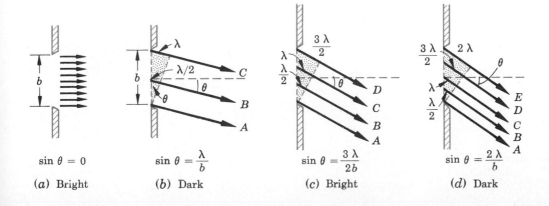

$\sin\theta = 0$

(a) Bright

$\sin\theta = \dfrac{\lambda}{b}$

(b) Dark

$\sin\theta = \dfrac{3\lambda}{2b}$

(c) Bright

$\sin\theta = \dfrac{2\lambda}{b}$

(d) Dark

shown in part *d*. Here the slit can be thought of as being divided into fourths. The bottom one-fourth slit is canceled by the portion just above it. Similarly, the two upper sections also cancel. Hence darkness is observed at this angle. These, then, are the interpretations of the various features of the single-slit diffraction pattern seen in Fig. 24.15.

The most important feature of the single-slit pattern, for our purposes, is the position of the first minimum next to the central maximum. Calling the angle between the central maximum and the first minimum θ_c, we have found that

$$\sin \theta_c = \frac{\lambda}{b} \qquad (24.3)$$

This relation is used in the next section.

24.10 Diffraction and the Limits of Resolution

One of the most important consequences of diffraction is that it limits our ability to observe very fine details of objects. We can appreciate this difficulty by referring to Fig. 24.17. Shown there are two light sources that send light through a slit to a screen. If the slit is quite small, the images cast on the screen are accompanied by nonnegligible diffraction fringes as shown. These fringes are the result of the light having passed through the slit whose width we specify to be *b*.

You can begin to understand the difficulty this presents if you consider the following rough analogy, which we pursue further later. The pupil of the eye corresponds roughly to the slit, and two lines on an object viewed by the eye are the two details to be seen. The retina of the eye acts as a screen. But because the images on the retina are made fuzzy by the diffraction effects of the slit (the pupil in this case), the eye is hindered in seeing fine detail in the object being viewed.

Returning to the situation shown in Fig. 24.17, we can see the images of the two objects on the screen as separate entities as long as the angle θ is not

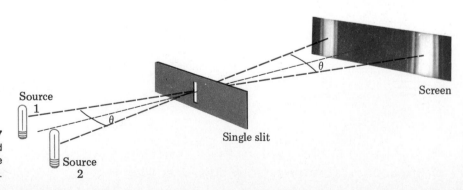

FIGURE 24.17
In the situation diagramed here, the two sources are well resolved on the screen.

too small. Difficulty arises when θ is so small that the diffraction patterns overlap appreciably. They can no longer be seen as separate (i.e., they can no longer be *resolved*) when they are close enough together that the central maximum of one pattern falls on the first minimum of the other pattern. In that situation, the case of *minimum resolution,* angle $\theta = \theta_c$, where θ_c is as defined in Sec. 24.9.

Using the result of the previous section, we can resolve the sources only if their angular separation θ is larger than θ_c, where

$$\sin \theta_c = \frac{\lambda}{b} \qquad (24.3)$$

As we expect, the smaller the slit width b, the farther apart the objects must be if they are to be resolved, for the interference pattern broadens with decreasing slit width.

Although our discussion has been in terms of objects viewed through a slit, a similar expression applies if the slit is replaced by a circular hole, such as the pupil of the eye or the objective lens of a microscope. For those situations it is found that

The limiting angle θ_c for resolution of two objects by use of a circular aperture of diameter D is

$$\sin \theta_c = 1.22 \frac{\lambda}{D} \qquad (24.4)$$

Now let us see what sort of limit this diffraction effect places on our ability to view objects with a microscope.

We show in Fig. 24.18 the objective lens of a microscope and two details S_1 and S_2 of an object being viewed by it. The details are a distance s apart, where s is much smaller than shown, and the objective lens has a diameter D. As indicated, the details are a distance d from the lens. How close together can the details be and still be resolved?

According to Eq. (24.4), the details can just be resolved if the angle θ_c that they subtend is given by

$$\sin \theta_c = 1.22 \frac{\lambda}{D}$$

But from the figure we see that

$$\sin \tfrac{1}{2}\theta_c = \frac{\tfrac{1}{2}s}{\sqrt{d^2 + (\tfrac{1}{2}s)^2}} \cong \frac{s}{2d}$$

because s is actually very much smaller than d.

However, for small angles such as θ_c will be, the angle in radians is equal to its sine. Therefore, for θ_c in radians,

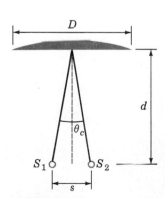

FIGURE 24.18
The two sources can just be resolved if $\theta = \theta_c$.

$$\theta_c = \frac{s}{d}$$

Making this same approximation in Eq. (24.4), we have

$$\theta_c = 1.22\frac{\lambda}{D}$$

Equating these two expressions for θ_c yields, after a little arithmetic,

$$s = 1.22\left(\frac{d}{D}\right)\lambda$$

If we look at Fig. 24.18, we see that d/D is the ratio of the object distance to the lens diameter. In all normal uses of microscopes, this ratio is approximately unity. As a result, to a rough approximation,

$$s \approx \lambda$$

In other words, *the smallest detail that can be seen in a microscope is of the same size as the wavelength of light being used.* This is a fundamental restriction imposed by diffraction and cannot be circumvented by use of perfect lenses or ingenious microscope design.

As we see, diffraction effects cause images to be fuzzy. Another example of this fact is shown in Fig. 24.19. The shadow of the washer shown there is surrounded by diffraction fringes, and the situation becomes even worse for a smaller object. In the case of objects of size comparable to the wavelength of light being used, details of the object are completely obscured by diffraction.

We must therefore conclude that it is impossible to obtain images of objects with detail comparable in size to the wavelength of radiation being used. For this reason, even the best microscopes cannot discern details comparable to the wavelength of light or smaller. Although a precise statement of the ability of optical devices to resolve the details of very small objects is beyond the scope of this book, *a rough rule of thumb is that detail smaller than a few wavelengths of light cannot be seen.*

24.11 Diffraction of X-Rays by Crystals

One of the most important applications of interference and diffraction is in the study of the structure of crystals. Because the atom layers in crystals are only about a nanometer apart, it is necessary to use very-short-wavelength EM radiation to study these atomic layers. As we learned in Chap. 21, these very-short-wavelength waves are called x-rays. Let us now see how x-rays can give an interference pattern when they are incident on a crystal.

For the purposes of our discussion, suppose we are using a crystal of sodium chloride, table salt. Its lattice is shown in Fig. 8.2, and we show it in cross section in Fig. 24.20a. As we see, the atoms of the crystal are uniformly spaced in planes. They are a distance d apart. If a beam of x-rays is incident on the crystal as shown in part b of Fig. 24.20, beam a, reflected from the top layer of atoms, will not travel as far as beam b. If the excess distance traveled by beam b, $2d \sin \theta$, is a whole number of wavelengths, the beams will reinforce. When that happens, all the layers will reflect beams which reinforce one another. Hence, strong reflection will occur when

$$n\lambda = 2d \sin \theta \qquad (24.5)$$

Bragg Equation

where n is an integer. This is called the **Bragg equation** after W. H. Bragg and his son W. L. Bragg, who first made extensive use of it in 1913.

Notice that Eq. (24.5) is similar to the grating equation (24.2). However, it differs by a factor of 2. In addition, the angle θ is defined differently in the two cases.

Equation (24.5) has been basic to many fundamental measurements.

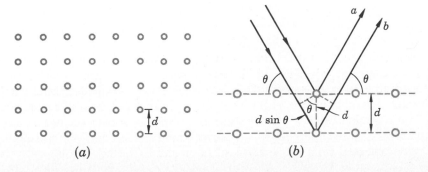

(a)   (b)

FIGURE 24.20
The atoms in crystals lie in evenly spaced planes. A simple example is shown in (a). When the atomic planes reflect x-rays, as in (b), the reflected rays give rise to interference effects.

FIGURE 24.21
Many possible parallel-layer systems of atoms can exist in a crystal. Three such systems are shown in the figure.

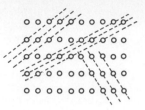

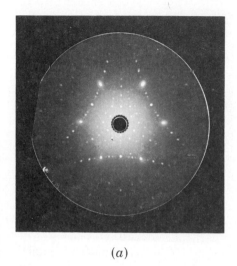

FIGURE 24.22
(a) When a beam of x-rays is shone on a single crystal, a Laue diffraction picture is obtained. (b) If a crystalline powder is used instead of a single crystal, a Debye-Scherrer diffraction pattern is obtained.

(a)

(b)

For example, consider what happens in Bragg reflection from a salt crystal. Since the spacing of the atom layers in salt can be found from the density of salt and the mass of its atoms, the distance d is known in that case. Since both n and θ can be measured, the wavelength of the x-rays can be found by using Eq. (24.5). This is one way in which we have learned that the wavelengths of x-rays are in the range near 0.1 nm. Of course, if λ is known, the distance d can be measured. This is the basis for the field of x-ray crystallography, in which the structure of crystals is measured by using x-rays.

Actual x-ray techniques for crystal-structure determination are quite involved. Notice that many possible layers of planes exist in the crystal. A few are shown in Fig. 24.21. When one considers the crystal in three dimensions, the situation is even more complex. Depending on how the crystal is used, the photograph of the diffraction pattern may be a series of spots, a set of circular rings, parallel lines, and so on. Two types of photographs obtained are shown in Fig. 24.22. The photograph in part *a* was taken by sending a beam of x-rays through a single crystal, while the other photograph was obtained in a similar way by using a polycrystalline material. Interpretation of such photographs is a rather involved science in itself. However, most of our knowledge concerning the structure of crystals has been obtained from the analysis of photographs such as these.

Summary

Waves can bend around obstacles into regions shadowed by the obstacle. This phenomenon is called diffraction. It is a result of the interference property of waves.

Two or more identical waves will give rise to interference effects when brought together. Waves which are in phase or $n\lambda$ out of phase (where n is an integer) will interfere constructively. Two waves which are 180°, or $\frac{1}{2}\lambda$, out of phase interfere destructively. Destructive interference also occurs if the phase difference is $\frac{1}{2}\lambda + n\lambda$, where n is an integer.

Identical waves, coherent waves, can be obtained by splitting a single wave disturbance into two parts. In Young's double-slit experiment, a single wave incident on two slits gives rise to two coherent waves. These then interfere with each other to produce a series of bright and dark fringes. The positions of the fringes depend on wavelength and can therefore be used to measure λ.

Two coherent waves can be obtained by partial reflections of a single beam from two different surfaces. In the Michelson interferometer, two portions of a beam produced in this way follow different paths and are then brought back together. By changing the length of one path, extremely accurate distance measurements can be made. Reflections from the two surfaces of a thin film or gap can also produce two interfering beams. Since the interference effects depend on wavelength, the fringes produced by white light in such phenomena are highly colored.

The equivalent optical path length for a distance d in a material of index of refraction n is nd. A beam will travel a distance nd in vacuum during the time it takes to travel a distance d in the material.

A diffraction grating consists of thousands of parallel, closely spaced slits. The interference pattern formed by it is much sharper than the double-slit pattern. Use is made of diffraction gratings in spectrometers for analyzing light. In such a spectrometer, images of the slit (called spectral lines) are formed at angles θ_n to the original beam given by $n\lambda = d \sin \theta_n$. The wavelength λ incident on a grating with slit separation d gives images at angles θ_n, where n is the order number, an integer.

Light from hot gases contains waves of only certain wavelengths. Each wavelength gives rise to a line, or an image, in the spectrometer. This series of lines is called a bright-line spectrum. Hot, incandescent solids give off a continuous range of wavelengths which results in a continuous band of color in the spectrometer. This is called a continuous spectrum.

Diffraction places a limit on our ability to discern detail in optical devices. As a rule of thumb, it is impossible to see detail smaller than the wavelength of radiation being used.

Interference of x-rays can be carried out by using crystals. The atomic planes within the crystal act as reflecting surfaces. Use is made of the interference patterns so obtained in studying the structure of crystals.

Minimum Learning Goals

Upon completion of this chapter, you should be able to do the following:

1 Describe a water-wave experiment which illustrates the phenomenon of diffraction.

2 Show the phase relation of two identical waves if they are to interfere constructively or destructively.

3 Explain how two coherent beams are obtained in Young's experiment. Using a diagram, show why these two beams can interfere destructively and constructively at various points. Consider the diagram and justify the relation $n\lambda = d \sin \theta_n$ for the bright fringe positions.

4 Use a double-slit interference pattern to determine λ if sufficient data are given.

5 Outline the construction and operation of the Michelson interferometer.

6 Explain how interference can be produced by a thin film or wedge. Tell why the fringes formed in white light are colored. Compute the thickness difference between two adjacent bright or dark fringes in an air wedge.

7 Tell what a diffraction grating is and show how it is used in a grating spectrometer. Explain why brightness is observed at angles for which $n\lambda = d \sin \theta_n$. Relate n to the order of an image.

8 Explain what is meant by a spectral line. Distinguish between a bright-line spectrum and a continuous spectrum.

9 Describe what happens to a beam of light transmitted through a slit as the slit is made very narrow. Pay particular attention to what happens when the slit width approaches λ. Explain the importance of this effect in our ability to observe details of objects.

10 Given the Bragg relation, explain the parameters in it. From a consideration of reflection from crystal planes, show how the relation arises.

Important Terms and Phrases

You should be able to define or explain each of the following:

Diffraction
Constructive interference; destructive interference
In phase; out of phase; $180°$ or $\frac{1}{2}\lambda$ out of phase
Young's double-slit experiment
First-, second-, zeroth-order fringes

Michelson interferometer
Equivalent optical path length nd
Newton's rings
Diffraction grating; grating spectrometer; $n\lambda = d\sin\theta_n$
Spectral line; line spectrum; continuous spectrum
Detail smaller than λ cannot be seen
Bragg equation

Questions and Guesstimates

1 The two loudspeakers shown in Fig. P24.1 are connected to the same oscillator. They therefore send out identical sound waves. Under what conditions would you be able to notice an interference effect as you walked along line AB? What if the loudspeaker were replaced by light bulbs?

2 Two cars sit side by side in a large, vacant parking lot with their horns blowing. Would you expect to be able to notice interference effects from the two sound sources? What if the horns were replaced by two clarinets playing the same note?

3 A telephone pole casts a clear shadow in the light from a distant source. Why is no such effect noticed for the sound from a distant car horn?

4 Why is it impossible to obtain interference fringes in a double-slit experiment if the slit separation is less than the wavelength of light being used?

5 Devise a Young's double-slit experiment for sound, using a single loudspeaker as a wave source.

6 Mercury light consists of several distinct wavelengths. Suppose that in a double-slit experiment, filters are placed over the slits so that $\lambda = 436$ nm (blue) light goes through one slit and $\lambda = 546$ nm (green) light goes through the other. Will it be possible to see an interference pattern on the screen?

7 What change will occur in a Young's double-slit experiment if the whole apparatus is immersed in water rather than air? What change would be observed in an arrangement of Newton's rings if the space between plate and lens were filled with water?

8 Very thin films are placed on the surface of coated lenses of expensive cameras to reduce reflection. Suppose a film of transparent ($n = 1.3$) material

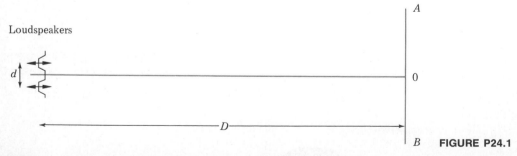

Loudspeakers

d

D

A

0

B **FIGURE P24.1**

(a)

(b)

FIGURE P24.2

having an equivalent optical path length of 1250 Å is coated onto a lens. Why will this reduce reflection from the lens? What other feature of the coating is important if it is to be most effective? Why does reflected light from coated lenses appear colored?

9 Very thin films are sometimes deposited on glass plates. The thickness of the film can be controlled by observing the change in color of white light reflected from the surface as the film's thickness is increased. Explain.

10 Why does a glass or metal surface which has a thin oil film on it often reflect a rainbow of color when white light is reflected from it?

11 Figure P24.2 shows interference fringes observed when glass plates are placed on top of optically flat surfaces (called an optical flat). Tell as much as you can about the surface of the glass plate in parts *a* and *b* of the figure.

12 Suppose that two additional slits are added to the two slits in a Young's double-slit experiment, one on each side of the two, so that there are four equally spaced slits. For a certain slit-to-screen distance, it is noticed that the center point of the fringe pattern is dark. Explain how this could occur.

13 Explain the following statement: The difference in thickness between two bright fringes in a thin-film

interference pattern is zero or $\frac{1}{2}\lambda/n$, where λ is the wavelength of light used and n is the index of refraction of the film.

14 Should a microscope have any better resolving power when blue rather than red light is used? Explain.

15 Suppose that you are given a diffraction grating whose characteristics are unknown. How can it be used to determine the wavelength of an unknown spectral line?

16 Using two pieces of flat glass (microscope slides are ideal), press them together in various ways and estimate how close together the surfaces are from observation of the interfering reflected light. (You can see the interference pattern easily in any lighted room *provided* you get the plates close enough together.)

17 How could one use a Michelson interferometer to measure the index of refraction of air?

18 When viewed in reflected light, the coated lens of a camera appears violet with a reddish hue (magenta). The coating is magnesium fluoride, which has an index of refraction of 1.25. Estimate the thickness of the coating. (E)

19 Assuming that diffraction caused by the pupil of the eye is the limiting factor, about how far away could a car be for its headlights to be resolved? (E)

Problems

1 Two identical sound sources are at the coordinate origin and send 75-cm-wavelength waves out along the *x* axis. One source is now moved slowly to negative *x*-axis values. For an observer at a point

$x = 20$ m on the *x* axis, what positions of the movable source give rise to (*a*) loudness and (*b*) least sound? Assume the sources vibrate in phase with each other.

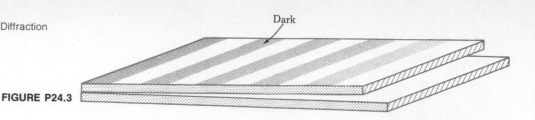

Dark

FIGURE P24.3

2 Two identical sources on the x axis send sound waves to an observer still farther out on the axis. One of the sources is now moved back along the axis, and the observer hears alternate loud and weak sounds as the motion proceeds. If the source moves 30 cm between observed loud sounds, what is the wavelength of the sound?

3* Two identical sound sources vibrate in phase and send 60-cm-wavelength waves toward each other along the x axis. One source is at $x = 0$ while the other is at $x = 3.0$ m. At what points on the x axis between them will the combined sound be (a) maximum and (b) minimum?

4 Suppose the two sound sources in Fig. P24.1 send out identical in-phase waves with $\lambda = 50$ cm. Maxima and minima of sound are heard as one walks along line AB. What is the path difference from the two sources at (a) point 0, (b) the first maximum away from 0, (c) the third maximum, (d) the third minimum?

5 In a Young's double-slit experiment, light with $\lambda = 546$ nm is used, and the first-order maximum is found to be at 2.5°. (a) What is the slit separation? (b) At what angle will the second-order maximum occur?

6 The slits in a Young's double-slit experiment are 0.015 cm apart, and light with wavelength 633 nm is used. (a) At what angle does the fourth-order maximum occur? (b) The fifth-order?

7* In Fig. P24.1 the sound sources send out identical, in-phase waves with $\lambda = 50$ cm. If $d = 5.0$ m and $D = 25$ m, how far along AB from 0 is (a) the first-order maximum and (b) the first-order minimum?

8* In a double-slit experiment, the slit separation is 0.20 cm and the slit-to-screen distance is 100 cm. Calling the position of the central bright fringe zero, locate the positions of the first three maxima on both sides of the central maximum. The wavelength is 500 nm.

9* For the double-slit experiment described in Prob. 8, find the positions of the first three minima.

10* What slit separation in a double-slit experiment would give a second-order maximum 1.00 cm from the central bright spot? The screen-to-slit distance is 2.0 m, and $\lambda = 500$ nm.

11* When sodium yellow light ($\lambda = 589$ nm) is used in a double-slit experiment, the first-order maximum is found at an angle of 3.0×10^{-4} rad. When the light is replaced by a source of an unknown wavelength, the second maximum occurs at 4.0×10^{-4} rad. (a) What is the wavelength of this latter light? (b) In what region of the spectrum is it found?

12 Two parallel glass plates are originally in contact and viewed from directly above with 500-nm light (green) reflected nearly perpendicularly by the surfaces. As the plates are slowly separated, darkness is observed at certain separations. What are the first four values? *Hint:* Darkness is observed when the plate separation is zero.

13 Referring to Fig. P24.3, we see that the left-hand edge is in contact. If the light used is blue light with $\lambda = 440$ nm, about how far apart are the plates at the last dark band on the right? Assume the space between the plates is air-filled.

14* When blue light of 440-nm wavelength is reflected from an air wedge formed by two flat plates of glass, the bright fringes are found to be 0.50 cm apart. (a) How thick is the air wedge 4.0 cm from the line of contact of the plates? Assume that the wedge is viewed at normal incidence. (b) Repeat if the wedge is filled with water rather than air.

15* Suppose in Fig. 24.9b that the light used to produce Newton's rings is the mercury green line, 546 nm, and that the radius of the seventh dark ring is 1.50 cm. (a) How large is the air gap at this position? (b) If the gap is now filled with water, how big is the gap at the new position of this ring? The center point of the pattern is dark.

16* A very thin wedge of plastic shows interference fringes when illuminated perpendicularly with white light. Two adjacent blue fringes ($\lambda = 450$ nm) are separated by 0.40 cm. (a) If the index of refraction of the plastic is 1.48, what is the difference in thickness of the wedge between the positions of these two fringes? Two answers are possible. (b) How could you determine which is correct?

17* Expensive cameras have nonreflecting coated lenses. Typically, the lens surface is covered with a thin layer of magnesium fluoride ($n = 1.25$). How thick must this layer be for destructive interference to occur between the light reflected from the two surfaces of the layer when 5000-Å light is used? Why, qualitatively, would a layer of Lucite ($n = 1.48$) be unsuitable for this purpose? *Hint:* Compare the amounts of light reflected by the Lucite and by the glass.

18 To determine the wavelength of a certain light, the light is used in a Michelson interferometer. It is found that when the movable mirror is moved 2.00 mm, 8500 fringes pass the field of view. What is the wavelength of the light?

19 To determine the pitch of a high-precision screw, the screw is used to move one mirror in a Michelson interferometer. It is found that 1 rev of the screw results in 2023 fringes ($\lambda = 546$ nm, green) passing the field of view. How far does one turn of the screw move the mirror?

20** One leg of a Michelson interferometer has in it an evacuated glass tube 2.0 cm long. A gas is slowly let into the tube, and the number of times that the field of view changes from bright to dark back to bright is counted to be 210 (that is, 210 fringes pass). If yellow light with $\lambda = 579$ nm is being used, what is the refractive index of the gas?

21 To calibrate a diffraction grating, a student sends red light from a helium-neon laser (6328 Å) through the grating. The first-order maximum occurs at an angle of 38°. What is the grating spacing? At what angle does the second-order maximum occur?

22 The sodium-arc yellow light is actually a doublet composed of two wavelengths, 5889.95 and 5895.92 Å. Compute the angular separation between these two lines in first order for a grating with 5000 slits per centimeter. Repeat for the second order.

23 A certain grating has 4000 lines per centimeter. What is the angular separation between the blue (4358-Å) and green (5461-Å) mercury lines in (a) the first-order spectrum and (b) the second-order spectrum?

24 A certain diffraction grating has 5000 lines per centimeter. At what angle does the second-order

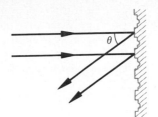

FIGURE P24.4

spectrum of the sodium yellow line occur ($\lambda = 5890$ Å)?

25 For a particular grating it is found that the second-order mercury blue line (4358 Å) lies exactly at 30°. At what angle will the first-order yellow line (5790 Å) be found?

26* Modern grating spectrometers often use reflection gratings. These are mirrors on the surfaces of which there are a series of reflecting lines (equivalent to the slits of the transmission grating). This situation is shown in Fig. P24.4, where the distance between the centers of the reflecting lines is d. Find the grating equation for this device; i.e., the angles θ at which interference maxima occur.

27** (a) Is it possible to design a grating so that the first-order 600-nm red line will lie on top of the second-order 400-nm violet line? (b) If so, how? (c) If not, could it be done for any other combination of orders? (d) If so, how?

28** A diffraction grating (10,000 lines per centimeter) is used in a large tank of water. At what angles (in the water) will the blue mercury line (435.8 nm) appear?

29** Steel sheds often have a corrugated metal surface with corrugation repeating every 10 cm or so. Under appropriate conditions, this type of wall can act as a reflecting diffraction grating for sound waves (see Prob. 26). What wavelength waves at normal incidence will give rise to a first-order maximum at an angle of 30° to the normal?

30 Find the angular width of the central maximum (i.e., the angle between the two side minima) for a 0.50-mm-wide single slit illuminated with 600-nm light.

31 A single slit is illuminated with 500-nm light, and it is found that its first diffraction minimum occurs at an angle of 3.0° from the center of the diffraction pattern. What is the width of the slit?

32* To determine the width of a very narrow slit, a girl sends red light from a helium-neon laser ($\lambda = 633$ nm) through the slit in a single-slit diffraction experiment. Since she (rightly) does not wish to view the laser light directly, the diffracted beam is allowed to strike a screen 5.0 m from the slit. She measures the distance from the central maximum to the first minimum to be 3.0 cm. How wide is the slit?

33* The headlights of a distant truck are viewed by a man's eye. If the eye opening has a diameter of 0.30 cm, how far away will the truck be if the two headlights are just resolved? Assume that the limiting factor is diffraction caused by the eye opening, that the effective $\lambda = 500$ nm, and that the truck light separation is 150 cm. What can you conclude from your result?

34* An image of a photographic slide is projected onto a screen 3.0 m away by a lens with 2.0-cm diameter. The lens is 12 cm from the slide. Assume the lens is perfect, so that diffraction limits its imaging ability. How close together can two tiny spots on the film be if they are to be resolved on the screen? Assume $\lambda = 500$ nm.

35* A known wavelength of light (600 nm) falls on two slits (separation unknown), together with an unknown wavelength. It is found that the fourth-order maximum of the known wavelength falls at the same position on a screen as the fifth-order maximum of the unknown. What is the wavelength of the unknown?

36* A double-slit system immersed in water is illuminated by 600-nm light. An interference pattern is formed on a screen at 2.0-m distance in the same tank of water. What is the distance from the central maximum to the first-order maximum on the screen if the separation of the slits is 0.040 cm?

37 A beam of x-rays is reflected from a crystal of NaCl by using the crystal planes which are separated by a distance of 2.820 Å. The angle of incidence for strong reflection is 40°. (a) What are the possible wavelengths of the x-rays? (b) How could you determine which of these alternatives was correct?

38 In a Bragg x-ray reflection type of experiment, x-rays of wavelength 0.148 nm are found to reflect strongly from a crystal at a Bragg angle of 23.0°. What are the possible values of the crystal spacing that give rise to this reflection?

25
Birth of
Modern Physics

By 1900 many scientists felt that most of the great discoveries in physics had already been made. To be sure, a few vexing problems still remained to be solved, but it appeared that nearly all the fundamental physical laws had been found. As we shall see in this chapter, such a view was completely incorrect. Vast areas of nature's physical behavior were still unknown at that time.

25.1 Two Men, Three Discoveries

As we look through the history of science, we see that each truly great scientific advance is associated with the name of a single person. Galileo is recognized as the leader in our understanding of how objects undergo translational motion. Newton's name is enshrined in his three laws of motion and in the law of gravitation. Faraday pioneered the way to an understanding of magnetism, while Maxwell unified all electricity in his four fundamental equations. These, as well as many other similar examples, attest to the fact that the intellect of a single individual has the power to illuminate large areas of science for us all.

That is not to say that these individuals made their discoveries in isolation. Quite the contrary. Historians of science show clearly that each of these discoveries is the culmination of years of work by many others. Indeed, Newton once wrote, "If I have seen further than other men, it is because I stood on the shoulders of giants." Even so, other people stood on the shoulders of these same giants and saw nothing. While we must pay due respect to their predecessors, the insight and genius of these great scientists should not be underestimated. But we should not stand in such awe of our scientific ancestors that we underestimate our own capabilities. The discoveries we discuss in this chapter and those that follow often came from unexpected sources.

We begin our discussions of what is called "modern physics," as opposed to the "classical physics" known prior to 1900, by examining the discoveries of two men, Albert Einstein and Max Planck. Planck's discovery was made earlier, in 1900, but we postpone its discussion until we outline one of the two extraordinary discoveries made by Einstein in 1905. You will see that these three discoveries opened the way to vast areas of nature then unknown.

25.2 Postulates of Relativity

Albert Einstein is best known for his theory of relativity. This famous theory, presented in 1905 when he was only 26, is a magnificent example of momentous deductions from a clear analysis of experimental fact. *Einstein recognized the following two statements as being compatible with all known experimental facts:*

Two Postulates

1 The speed of light in vacuum is always measured to be the same ($c = 2.998 \times 10^8$ m/s) no matter how fast the light source or observer may be moving. (Of course, accurate measurements are assumed.)

2 Absolute speeds cannot be measured. Only speeds relative to some other object can be determined.

These are the two basic postulates of Einstein's theory of relativity.

It is not possible to prove these postulates directly. They are the consensus of all the experimental facts known. We consider it possible, though unlikely, that some experiment will sometime be found to disprove one of them. But they are supported by many unsuccessful attempts to disprove them. Moreover, as we shall see, they lead to astounding conclusions which have been well verified by experiment.

The second postulate needs some explanation, perhaps. It is easy to measure the relative speeds of objects. A car's speedometer tells us at once how fast the car is moving relative to the roadway. But this is not an absolute speed. The earth is moving because of both its rotation on its axis and its motion around the sun. Since we know both these speeds, we could, if required, find the car's motion relative to the sun.

But the sun itself is moving in our galaxy, the Milky Way. And the center of the galaxy is in motion relative to the more distant stars. There seems to be no way to define a definite, absolute speed of an object since everything appears to be moving. We can state only how fast one object is moving relative to another.

There is another way to state the second postulate which gives us an inkling of its fundamental importance. This alternate statement is usually made in terms of reference frames. *A* **reference frame** *is any coordinate system relative to which measurements are taken.* For example, the position of a sofa, table, and chairs can be described relative to the walls of a room. The room is then the reference system or frame used. Or, perhaps a fly is sitting on a window in a moving car. We can describe the fly's position in the car, using the walls of the car as a reference frame. Alternatively, we can describe the position of a spaceship relative to the positions of the distant stars. A coordinate system based on these stars is then the reference frame.

The second postulate can be stated in terms of reference frames in the following way:

2′ The basic laws of nature are the same in all reference frames moving with constant velocity relative to each other.[1]

Often this statement is shortened by using the term *inertial reference frame.* *An* **inertial reference frame** *is a coordinate system in which the law of inertia applies:* a body at rest remains at rest unless an unbalanced force on it causes it to be accelerated. The other laws of nature also apply in such a system. *To a very good approximation, all reference systems moving with constant velocity relative to the distant stars are inertial frames.*

Inertial Reference Frame

2″ The basic laws of nature are the same in all inertial reference frames.

You can understand the relation between these two alternate ways of stating the second postulate by considering the following. When we say that only relative speeds can be measured, a lack of bias in reference frames is being assumed. For example, a spaceship may be heading for the moon at a speed of 10^5 km/day relative to the moon. But it is also true that the moon is heading toward the ship at 10^5 km/day relative to the ship. The fact that one is moving relative to the other is easily ascertained. But both statements are equivalent, and neither object can be said to be at rest.

Suppose, though, that some law of nature depended on the speed of the reference frame. The people in the spaceship could use the law to obtain an indication of their speed. People on the moon could do likewise. The two measured speeds would be different. As a result, they would be capable of measuring more than just their relative speeds. In fact, the law could be used to set up an absolute ranking of speeds. But this would contradict the second postulate. We therefore conclude that all nature's laws must be the same in all inertial reference frames.

---

[1] For example, Newton's second law is expressed as $F = ma$ in any of these frames. However, as we shall see, m (or F or a) may not have the same value in each.

25.3 *c* Is a Limiting Speed

By use of Einstein's two postulates, we can prove by logic alone that

No material object can be accelerated to speeds in excess of the speed of light in vacuum.

The validity of this statement is easily demonstrated in the following simple way. We prove it by the technique called **reductio ad absurdum,** in which we disprove a proposition (in this case, that an object can travel faster than *c*) by showing that the proposition leads to a known false result (in this case, that an observer will measure a value different from *c* for the speed of light).

Suppose we have two stations in space, shown as *A* and *B* in Fig. 25.1. The inertial observers at *A* and *B* have instructed the spaceship operator to follow a straight-line path between *A* and *B*. The ship is to travel at its top constant speed and is to send a light pulse from the front of the ship toward *B* as it passes *A*. Of course, *A* and *B*, working in partnership, can determine the speed of the spaceship by timing its flight from *A* to *B*. Let us now make the false assumption that they find the speed of the ship to be 2*c*.

The spaceship sent out a pulse of light as it passed *A*, and since the laws of nature must apply to all three inertial observers (*A, B,* and the person in the ship), the light pulse must behave in a normal way to each of them. In particular, the light pulse must precede the ship and must reach *B* before the ship does. Therefore *A* and *B,* working together, would find that the light pulse is moving faster than the ship. But they measure the ship as moving with speed 2*c*, and so they find that the speed of the light pulse is greater than 2*c*. But this is an impossible result, since it contradicts the known fact that all observers will obtain *c* for the speed of light. We therefore conclude that our original assumption was false; the spaceship could not have been moving between *A* and *B* with a speed of 2*c*.

This experiment will always lead to this contradiction as long as we insist that the speed of the ship exceed *c*. We therefore conclude that the spaceship cannot exceed the measured speed of light *c*. Indeed, we can enlarge this line of reasoning to include all material objects and signals that carry energy. As a result we can state:

A Limiting Speed Exists   Nothing that carries energy can be accelerated to the speed of light *c*.

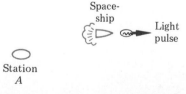

FIGURE 25.1
What is the maximum speed with which the ship can pass between the two space stations?

Space-
ship

Light
pulse

Station
A

Station
B

As we proceed, we see that this result of Einstein's theory repeatedly has been tested carefully and has been found correct in every test.

25.4 Simultaneity

We see in this section that the basic postulates of relativity force us to conclude that events which are simultaneous in one inertial reference frame may not be simultaneous in another. To show this simply, we again resort to a thought experiment. The progress of a light pulse as noted by two inertial observers forms the basis for our experiment.

As shown in Fig. 25.2a, suppose a boxcar is traveling to the right at a very high constant velocity. At the exact center of the car is a high-speed flashbulb which has reflectors so it will send out light pulses to the right and left when it flashes. The boxcar is fitted with photocells at each end, so a man in the boxcar can detect when the light pulses strike the ends of the car. By some ingenious device, a woman at rest on the earth is also able to measure the progress of the two pulses. Notice that both observers are in inertial reference frames (one is the moving boxcar, the other is the earth),

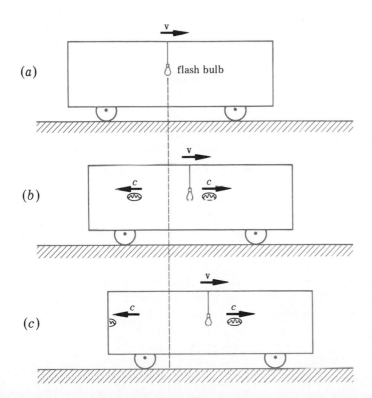

FIGURE 25.2
Unlike the inertial observer in the moving frame, the observer stationary on earth does not see the light pulses strike the ends of the car simultaneously.

Albert Einstein
(1879–1955)

As a student, Einstein was unhappy in the rigid, militaristic school system of his native Germany. His exceptional abilities and interest in physics flourished after he moved to Zurich to complete his undergraduate studies. Despite excellent references, he could not get a teaching job and found work in the patent office instead, continuing his studies in theoretical physics in his spare time. At the age of 26, he published papers on Brownian motion, the photoelectric effect, and relativity. Any one of these would have merited a Nobel Prize (he won it for his paper on the photoelectric effect). These famous studies were but the beginning of a long and brilliant career. He fled Hitler's Germany, and in 1940 he became a U.S. citizen. Partly through his efforts, the United States assembled many of the world's best scientists to develop atomic bombs before Nazi Germany could do so. In addition to being the most famous modern scientist, Einstein was noted for his warm and compassionate personality and his deep interest in peace and social justice.

and so they must both see the light pulses behave "normally" in their reference frames. Of course, "normal" for the woman on earth is that the light pulses travel with speed c to the right and left from the flashbulb. "Normal" for the man in the car is that the two light pulses strike the detectors at opposite ends of his car simultaneously.

Consider first the man in the car. To him, the experiment is very simple. The flashbulb is at rest relative to him in the center of his car. When the bulb flashes, two pulses travel the equal distances to the two ends of the car in equal times. (Remember, for him the experiment must be the same whether or not the car is moving, since he cannot tell.) Hence *the light pulses hit the two ends of the car simultaneously.*

Consider now how the woman stationary on the earth sees the experiment. Her measurements show the experiment to proceed "normally" (for her), and so the situation progresses as shown in parts *b* and *c* of Fig. 25.2.

Notice that the pulses travel equal distances in equal time to the right and left. But since the boxcar is moving to the right, the distance to the left end is shortened. As a result, the observer stationary on the earth measures the pulse on the left to strike the end of the boxcar before the other pulse strikes the opposite end. According to her, *the light pulses do not hit the two ends of the car simultaneously.*

We must therefore conclude that time is not a simple quantity because

Events which are simultaneous in one inertial system may not be simultaneous in another.

Further considerations show that this situation exists only if the two events occur at different locations. In this case, one event took place at one end of the car and the other was at the opposite end.

25.5 Moving Clocks Run Too Slowly

As you might suspect from the results of the previous section, time is not a simple quantity. Einstein pointed this out when he showed that a clock ticks out time differently for an observer who holds the clock and for one moving past it. We demonstrate this effect in a thought experiment using a very special clock. But it was proved to be true in general by Einstein.

Consider the clock held by the woman in Fig. 25.3. It consists of a pulse of light reflecting between two mirrors in a cylindrical vacuum tube. Each time the light pulse strikes the lower mirror, it clicks out a unit of time which we shall call a "click." If the tube is 1.5 m long, the woman can compute easily that

$$1 \text{ click} = \frac{2d}{c} = \frac{3.0 \text{ m}}{3.0 \times 10^8 \text{ m/s}} = 10^{-8} \text{ s}$$

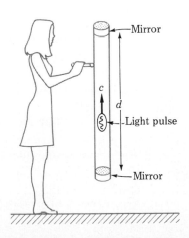

FIGURE 25.3
The light clock registers one click each time the light pulse is reflected from the lower mirror.

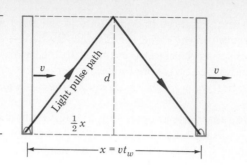

FIGURE 25.4
The light pulse in the
moving clock must travel a
distance larger than *2d*
during one click interval.
The light-pulse path length
is $2\sqrt{d^2 + (\frac{1}{2}vt_w)^2}$.

Suppose several copies of this clock are made and one is being used by a man in a spaceship. The woman with her identical clock looks out the window of her laboratory (which is in another spaceship) and sees the man shoot past her with speed v. She is pleased to see that he is using a clock similar to hers and contacts the man by radio. He tells her the clock is functioning well and is ticking out time as usual, one click each $2d/c$ seconds.

After thinking about it a bit, the woman discovers there is something very peculiar about this. She concludes that the man's clock must be ticking out time more slowly than hers. We can understand her reasoning as follows.

Since the man's clock is operating properly for him, she knows it must be operating as shown in Fig. 25.4. We see there the clock in its position at two consecutive clicks. Notice that the woman knows that the light pulse moves along the path indicated. Although the man sees the pulse move straight up and down in the clock, the woman knows that the pulse moves to the right as well, because of the movement of the clock to the right.[2] The woman computes the time between clicks on the man's clock as follows.

According to the woman, the pulse moves a distance given by the colored line in the figure. From the pythagorean theorem and the dimensions given in the figure, we see that

$$\text{Pulse path length} = 2\sqrt{d^2 + (\tfrac{1}{2}x)^2}$$

But the woman knows that the man's clock is traveling with speed v past her. Further, according to her clock, a time t_w will be taken to move from one position to the other. Therefore, she knows that $x = vt_w$. As a result, according to the woman,

$$\text{Pulse path length} = 2\sqrt{d^2 + (\tfrac{1}{2}vt_w)^2}$$

Further, she knows that a light pulse always travels through vacuum

---

[2] You might ask, Who is right? They both are, as we shall soon see. Both are describing the behavior correctly as measured in their own reference frames.

with speed c. According to her, then, the time taken for the change in position shown in the figure should be

$$t_w = \frac{\text{pulse path length}}{c} = \frac{2\sqrt{d^2 + (\frac{1}{2}vt_w)^2}}{c}$$

We can solve for t_w in this equation and find (after squaring both sides, rearranging, and taking square roots)

$$t_w = \frac{2d/c}{\sqrt{1 - (v/c)^2}}$$

But we recognize $2d/c$ to be the time that the man insists it takes for his clock to make one click. We therefore have the following result:

$$\frac{\text{Time interval on}}{\text{stationary clock}} = \left[\frac{1}{\sqrt{1 - (v/c)^2}}\right]\left(\frac{\text{time interval on}}{\text{moving clock}}\right)$$

For example, suppose the man is moving at a speed of $0.75c$ past the woman. Then $\sqrt{1 - (v/c)^2}$ has a value 0.66, and the inverse of this is 1.51. Under these conditions the woman's clock will tick out 1.51 clicks during the time she knows the man's clock takes to tick out 1 click. As we see, the moving clock ticks out time more slowly than the stationary clock.

A clock moving with speed v ticks out a time of $\sqrt{1 - (v/c)^2}$ second during 1 s on a stationary clock.

After arriving at this unexpected result, the woman contacts the man by radio and informs him that she has discovered that moving clocks tick out time too slowly. Before she can give him the details, he states that he has been thinking along the same lines. He has discovered that her clock, which was moving past him with speed v, was ticking out time too slowly. Then they both recall that only relative motion has meaning. Neither clock is special.

Any clock moving relative to an observer will appear to tick out time too slowly compared to a clock stationary with respect to the observer.

Time Dilation

We call this effect **time dilation** since time is stretched out, so to speak, for moving clocks.
   This astonishing result applies to all timing mechanisms, no matter how complex. If the man had been using the growth rate of a fungus as a clock, the woman would have found the fungus growth rate to be slowed by its motion. Even aging of the human body will be slowed by motion at high speed, as we see in one of the following illustrations.
   But there is one point we should always remember. A good clock always behaves normally to a person at rest relative to the clock. Observers

moving past the clock may claim it ticks out time too slowly. In spite of this, the clock still ticks out time properly as viewed by an observer stationary relative to it.

Illustration 25.1 One striking example of time dilation is obtained by measuring how long unstable particles "live." For example, a particle called the *pion* lives on the average only about 1.8×10^{-8} s when at rest in the laboratory. It then changes to another form. How long would such a particle live when shooting through the laboratory at $0.95c$?

Reasoning In the second case, the pion is moving with speed $0.95c$ relative to the observers in the laboratory. Experiments should show that the internal clock of the pion, which controls how long it lives, should be slowed because of its motion. A time of 1.8×10^{-8} s read by the moving clock should be as follows when timed by the laboratory clock:

$$\text{Life according to lab clock} = \frac{1.8 \times 10^{-8} \text{ s}}{\sqrt{1 - (0.95)^2}}$$

which turns out to be 5.8×10^{-8} s. As we see, the moving pion should live about 3 times as long as a stationary one. This experiment and variations of it have been carried out. The results found by experiment agree with the computed results.

Illustration 25.2 The star closest to our solar system is Alpha Centauri, which is 4.3×10^{16} m away. Since light moves with a speed of 3×10^8 m/s, it would take a pulse of light 1.43×10^8 s, or 4.5 yr, to reach there from the earth. (We say that the distance to the star is 4.5 light-years.) How long would it take, according to earth clocks, for a spaceship to make the round trip if its speed is $0.9990c$? According to clocks on the spaceship, how long would it take?

Reasoning To a good approximation, we can take the spaceship speed to be c for this computation, and so the round trip would require 9.0 yr according to earth clocks.

The spaceship clocks will appear to run too slow by the relativistic factor

$$\sqrt{1 - (0.999)^2} \approx 0.045$$

Therefore the spaceship clocks will read $(0.045)(9.0) = 0.40$ yr instead of the 9.0 yr read from the earth clocks. As a result, the journey would seem to take only about 5 months according to the crew of the spaceship—far more tolerable than the 9.0 yr which people on earth would record.

Incidentally, the twin of one of the crew who was left behind on the

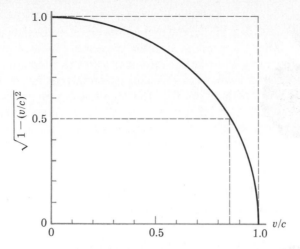

FIGURE 25.5
The relativistic factor differs
appreciably from unity only
at speeds which approach
the speed of light.

earth would age 9.0 yr during the voyage. The twin in the spaceship, however, would age only 5 months. This phenomenon, the **twin paradox,** has been discussed at length by scientists. They generally agree that this result is valid and that the two twins actually will age differently.[3]

Twin Paradox

Illustration 25.3 Graph the relativistic factor $\sqrt{1-(v/c)^2}$ as a function of v/c. Explain why we do not observe relativistic time dilation in everyday phenomena.

Reasoning The appropriate graph is shown in Fig. 25.5. Notice that the relativistic factor departs from unity only at extremely high speeds. For most purposes, speeds below $0.10c$ are too small to show appreciable relativistic effects. In everyday life, our clocks never come anywhere close to such high speeds. However, as we learned in our study of electricity, the electrons in a beam such as that in a TV tube are easily accelerated to relativistic speeds.

25.6 Relativistic Length Contraction

The time-dilation effect implies a peculiar effect involving measured lengths. To see what this effect is, consider once again the man and woman of the previous section. Let us say that the woman is on the earth while the man is traveling with speed v along a straight line from earth to the nearest star, Alpha Centauri. Astronomers based on earth tell us the star is

---

[3] To test this effect with actual clocks, an extremely precise clock transported around the earth by plane has been compared with a stationary "twin." The expected result was found. For a discussion of the experiment, see J. Hafele, *Physics Teacher* **9**:416 (1971).

$d = 4.3 \times 10^{16}$ m away from the earth. Because relative speeds can be measured easily, the man and woman agree that their speed relative to each other is v as the man in the spaceship shoots from the earth to the star. The woman is at rest in a reference frame in which the earth and the star are also at rest.[4] She sees the man shooting past her at speed v.

The man, on the other hand, is at rest relative to his spaceship, and he takes the ship itself to be his reference frame. Relative to the ship, both the earth and the star are moving with speed v. Let us now examine the man's flight from the earth to the star from the woman's vantage point.

The woman knows that the distance from the earth to the star, both at rest relative to her reference frame, is $d_e = 4.3 \times 10^{16}$ m, where the subscript e stands for "earth." Using $x = vt$, she computes that the time registered by her earth clock for the man's trip to the star will be

$$t_e = \text{earth time} = \frac{d_e}{v}$$

Indeed, when the ship turns around at the star and returns to earth, the total time the ship has been in flight is $2t_e = 2d_e/v$.

However, the man's computation will be somewhat different. Using spaceship clocks, he times his flight from the earth to the star and finds it takes a time t_s. He can then compute the distance to the star by use of $x = vt$, to obtain

$$d_s = vt_s$$

where the subscript s refers to measurements in a reference frame at rest relative to the spaceship. A similar computation for the return trip tells him that his total flight covered a distance $2d_s$ in $2t_s$.

We therefore have the following equations that are undeniably correct for the two observers who formulated them:

$$2d_e = v(2t_e)$$
and
$$2d_s = v(2t_s)$$

But we know the time-dilation effect influences the spaceship clock in such a way that when it is compared to the earth clock upon return of the spaceship to earth, we have

$$t_s = \sqrt{1 - (v/c)^2} \; t_e$$

The spaceship clock ticked out time too slowly.

Substituting this value for t_s in the expression for d_s yields

---

[4] We neglect the comparatively small relative motion of the earth and star.

$$d_s = v \sqrt{1 - (v/c)^2}\, t_e$$

However, $d_e = vt_e$, and so $t_e = d_e/v$. Using this value for t_e gives

$$d_s = \sqrt{1 - (v/c)^2}\, d_e$$

In other words, the distance from earth to the star measured by the man in the spaceship is smaller than the distance measured by astronomers on earth. Apparently, if you are in motion along a line between two points, the distance between the two points appears shorter than if you were at rest relative to them. The ratio of the two distances is the relativistic factor, $\sqrt{1 - (v/c)^2}$.

Einstein found this to be a general result. We can summarize it as follows:

> If an object and an observer are in relative motion with speed v, then the observer will measure the object to have contracted along the line of motion. The contraction factor is $\sqrt{1 - (v/c)^2}$.

Notice that the contraction occurs only along the line of motion. No such contraction is observed perpendicular to the direction of motion.

Illustration 25.4 A woman traveling at high speed in a spaceship holds a meterstick in her hand. What does she notice about the length of the stick as she rotates it from a position parallel to the line of motion to a position perpendicular?

Reasoning She notices no change in the stick's length. The length-contraction effect concerns objects moving at high speed relative to the observer. The meterstick is at rest relative to the woman.

25.7 Relativistic Mass-Energy Relation

The postulates of relativity tell us that no object can be accelerated to speeds in excess of the speed of light. It can be seen at once that this conflicts with prerelativity ideas. For example, consider an object of mass m being accelerated by a constant force F. We would usually state that the acceleration is $a = F/m$. As a result, the object's final velocity after time t is simply

$$v = v_0 + at = v_0 + \frac{F}{m}t$$

This relation must be wrong. It predicts that the velocity can increase without limit. As long as the force keeps acting, the velocity will keep on increasing. This contradicts the conclusion that the object's speed cannot exceed c. Something is obviously wrong with our nonrelativistic ideas for objects moving at very high speeds.

A consideration of the postulates of relativity leads to the source of the difficulty. Using reasoning too lengthy for us to give here, Einstein was able to show that his postulates together with the momentum-conservation law yield the fact that the mass of an object varies with the speed of the object. *At rest, the object has a mass m_0, its* **rest mass.** *But, at high speeds, the object's mass is larger.*

Relativistic Mass

An object of rest mass m_0 has an apparent mass $m = m_0/\sqrt{1 - (v/c)^2}$ when moving with speed v past the observer.

This variation of mass with speed is shown in Fig. 25.6.

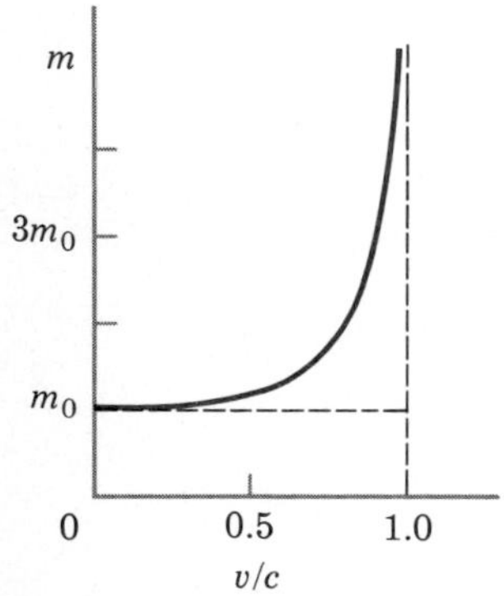

As we see from the graph, the apparent mass m is close to the rest mass m_0 as long as v/c is less than a few tenths. But when v becomes close to the speed of light, so that v/c approaches 1, the apparent mass becomes much larger than m_0. In terms of the equation for the apparent mass

$$m = \frac{m_0}{\sqrt{1 - (v/c)^2}} \tag{25.1}$$

FIGURE 25.6
$m \to \infty$ as $v/c \to 1$.

we notice that as v approaches c, the quantity $v/c \to 1$.[5] Therefore, for $v \to c$,

$$m = \frac{m_0}{\sqrt{1 - (v/c)^2}} \to \frac{m_0}{\sqrt{1 - 1}} \to \frac{m_0}{0} \to \infty$$

The apparent mass becomes infinite when the object's speed reaches the speed of light.

The variation of mass with speed can be used to justify the fact that no object can be accelerated to a speed in excess of the speed of light. An infinite mass would require an infinite force to accelerate it. Because infinite forces are not available, it is apparent that an object with speed $v \to c$ cannot be accelerated to the speed of light, a speed at which the object's mass would be infinite.

The force acting to accelerate an object gives energy to the object. At low speeds we know that the work done by the applied force equals the increase in KE of the object, provided changes in PE and friction work are negligible. This is still true at speeds close to c. But now the KE of an object is no longer given by $\frac{1}{2}m_0v^2$. Nor is it, as one might guess, $\frac{1}{2}mv^2$. Instead it is found that the kinetic energy of an object is given by

Relativistic KE

$$KE = (m - m_0)c^2 \tag{25.2}$$

---

[5] The symbolism $a \to b$ is to be read "a approaches b in magnitude."

When $v \ll c$, Eq. (25.2) reduces[6] to KE $= \frac{1}{2}m_0 v^2$. We see, then, that the correct expression for KE is given by Eq. (25.2). The expression KE $= \frac{1}{2}m_0 v^2$ is very accurate for speeds low enough to ensure that $m \approx m_0$. But when $v \rightarrow c$, the value $(m - m_0)c^2$ must be used for KE.

It is possible to show that a relation similar to Eq. (25.2) applies to all types of energy. Einstein showed that *for any change in energy of an object, there is a corresponding change in mass*. The result is that *for a change in energy ΔE, the object's mass changes by an amount Δm, given by*

$$\Delta E = \Delta m \, c^2 \qquad\qquad (25.3)$$

Relativistic Mass-Energy Relation

(This is often written as $E = mc^2$.) As an example, if you increase the PE of an object, its mass increases in accordance with Eq. (25.3). In fact, this equation predicts that mass can be created by providing energy (we see examples of this later). More spectacularly, mass can be destroyed to provide energy. In either case, the change in mass Δm is equivalent to a change of energy in the amount $\Delta m \, c^2$. Perhaps you already know that the nuclear energy of a reactor or nuclear bomb results from the fact that an amount of mass Δm is destroyed to produce an energy $\Delta m \, c^2$.

Illustration 25.5 The available chemical energy in a 100-g apple is about 100 kcal (the nutritionists leave off the prefix *kilo* and call them calories). We learned in our study of heat that 1 cal is 4.184 J of energy, and so an apple contains about 420 kJ of available energy. Compare this with the energy one could obtain by changing all the mass to energy.

Reasoning According to the mass-energy relation,

$$\text{Energy} = \Delta m \, c^2$$

In this case $\Delta m = 0.10$ kg, and $c = 3 \times 10^8$ m/s, giving

$$\text{Energy} = 9 \times 10^{15} \text{ J}$$

We see from this that when we eat an apple, we obtain only a very small fraction (5×10^{-11}) of its total energy.

---

[6]To show this, make use of the mathematical fact that, for $x \ll 1$, the quantity $1/\sqrt{1-x} \cong 1 + \frac{1}{2}x$. Then, if we call $(v/c)^2$ the quantity x, in the case when $(v/c)^2 \ll 1$,

$$m = \frac{m_0}{\sqrt{1 - (v/c)^2}} = m_0\left(1 + \frac{1}{2}\frac{v^2}{c^2}\right)$$

Therefore Eq. (25.2) becomes

$$\text{KE} = mc^2 - m_0 c^2 = m_0 c^2 + \frac{m_0}{2}\frac{v^2}{c^2}c^2 - m_0 c^2 = \frac{1}{2}m_0 v^2.$$

Illustration 25.6 The light given off by a TV tube comes from electrons shooting down the tube and hitting a fluorescent screen at its end. Their speeds are of the order of one-third the speed of light. What is the apparent mass of such a high-speed electron ($m_0 = 9.1 \times 10^{-31}$ kg)?

Reasoning The electrons are moving with speed $v = c/3$ relative to the person watching the TV set. As a result,

$$m = \frac{m_0}{\sqrt{1 - (\frac{1}{3})^2}} = \frac{m_0}{\sqrt{0.89}} = 1.06 m_0$$

and so $m = 9.6 \times 10^{-31}$ kg. Even at this very high speed the electron mass has increased by only 6 percent.

25.8 Planck's Discovery

In 1900, five years before Einstein proposed his theory of relativity, Max Planck (1858–1947) made a discovery that seemed less than earth-shaking at the time, but which now we recognize as the first of a Pandora's box of surprises. Planck, along with others, had been trying to interpret the radiation given off by hot, nonreflecting objects, so-called **blackbodies.** Careful measurements of the intensity of light (as well as infrared and ultraviolet radiation) given off by red-hot objects indicated that the intensity varies with wavelength as shown in Fig. 25.7. As we see, only a small fraction of the emitted radiation has wavelengths in the visible range. Most is in the infrared (or heat) wavelength range. Furthermore, the curves show that as the temperature is increased, the radiation maximum shifts from the infra-

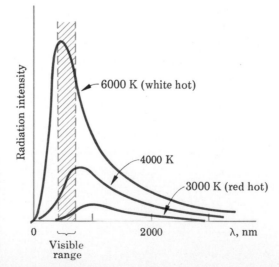

FIGURE 25.7
Blackbody radiation. For comparative purposes, the temperatures correspond as follows: 6000 (sun's surface), 4000 (carbon arc), 3000 K (very hot tungsten lamp).

red to the visible.[7] This agrees with our experience that a white-hot body is hotter than a red-hot one.

In order to interpret these curves, we are led to ask what sort of transmitting antenna could be sending out EM radiation from the hot object. Since the wavelengths involved are very short, the frequency of the vibrating charges must be very large. For example, at a wavelength of 1000 nm we have

$$\text{Frequency} = \nu = \frac{c}{\lambda}$$

or

$$\nu = \frac{3 \times 10^8 \text{ m/s}}{10^{-6} \text{ m}} = 3 \times 10^{14} \text{ Hz}$$

where, following custom in atomic physics, we have used the symbol ν (Greek nu) for frequency in place of f. Notice how very high this frequency is. Only in atomic-size antennas can charges be oscillated this fast. As a result, we expect that the EM radiation is being emitted by the vibrating charges within atoms and molecules composing the hot object.

There are many models we could postulate for these atomic or molecular vibrators. For example, if the object were composed of diatomic polar molecules, the vibrating molecule would be represented as shown in Fig. 25.8. The two atoms are held together by a springlike force, and since the molecule is polar, its two atoms carry equal and opposite charges. As the atoms vibrate back and forth, they act as vibrating charges on an antenna and therefore emit EM radiation of frequency ν_0, where ν_0 is the natural frequency of vibration of the molecular spring system. At least, this is the way Planck and his contemporaries reasoned.

FIGURE 25.8
As the dipolar molecule vibrated back and forth, it was thought to send out EM waves.

It turns out, however, that all theories of radiation based on this model failed to describe the radiation from hot objects accurately. The theories were capable of duplicating the curves of Fig. 25.7 at long wavelengths but gave completely incorrect predictions at small wavelengths. It was Max Planck who discovered how the theory could be modified to agree with experiment. His modification is easily understood but difficult to justify. In fact, his only justification for it was that it gives the correct answer. Let us now see what he had to assume to get agreement between theory and experiment.

As we know, the amplitude of vibration of a system such as that shown in Fig. 25.8 depends on the energy of the system. This is true for a mass on a spring, a pendulum, and all other oscillators. In all cases, the more energy the system has, the bigger the amplitude of the vibration. Although the frequency of vibration is always ν_0, the natural vibration frequency, the amplitude increases as the energy increases. *Planck found that he had to assume that a system could not vibrate with just any energy:*

---

[7] The position of the maximum in the curve is given by *Wien's law:* $\lambda_m T = 2.90 \times 10^{-3}$ m $\cdot$ K, where T is the kelvin temperature and λ_m is the wavelength at which the maximum occurs.

A vibrator of natural frequency ν_0 can vibrate only with energies $h\nu_0$, $2h\nu_0$, $3h\nu_0$, ..., and with no others.[8]

The quantity h is simply a constant (it is not the amplitude of vibration!) that has to have the value

$$h = 6.626 \times 10^{-34} \text{ J} \cdot \text{s}$$

if agreement with experiment is to be achieved. It is now called **Planck's constant.**

Planck's assumption is truly astonishing. It means that a vibrator can oscillate only with certain amplitudes and with no others. For example, since the total energy of a pendulum is mgH, where H is the height to which the bob swings, Planck says that mgH can be only $h\nu_0$, $2h\nu_0$, etc., and nothing in between. To see what this means, let us consider a pendulum which has a natural frequency $\nu_0 = 1$ Hz and a bob of mass 100 g. Then the heights to which this pendulum could swing would be

$$H_1 = \frac{h\nu_0}{mg} = \frac{(6.6 \times 10^{-34} \text{ J} \cdot \text{s})(1 \text{ s}^{-1})}{(0.10 \text{ kg})(9.8 \text{ m/s}^2)} = 6.7 \times 10^{-34} \text{ m}$$

or $\quad H_2 = 2H_1 = 13 \times 10^{-34} \text{ m}$

or $\quad H_3 = 3H_1 = 20 \times 10^{-34} \text{ m}$

etc. No intermediate maximum vibration heights are possible. It could not be induced to vibrate to a height of 16×10^{-34} m, for example.

Notice that the difference between successive heights of vibration as predicted by Planck is only about 10^{-33} m. This is far too small a difference to measure. As a result, we can never tell whether Planck is correct by observing the vibration of a pendulum. The gaps between the allowed energies are too small to be measured. This turns out to be true for all common vibrating systems. Hence, *we can neither prove nor disprove Planck's assumption using laboratory-type vibrating systems.*

Planck was therefore faced with a rather disturbing situation. He could obtain a suitable theory for the radiation from a hot object provided he was willing to make the assumption outlined above. The experimental test of it for other vibrating systems appeared to be impossible. Therefore, at the time, it was viewed by both Planck and his contemporaries as a rather curious result but one of doubtful validity. We shall see, however, that it appears to be correct and of extreme importance.

25.9 Einstein's Use of Planck's Concept

Five years after Planck's discovery, another natural phenomenon was shown to involve Planck's constant h. This discovery was made by Albert Einstein

---

[8] Zero energy is not allowed, since this leads to conflict with the uncertainty principle mentioned later.

in the same year (1905) that he proposed his theory of relativity. As we shall see, he was able to explain in a detailed way the results of an experiment first performed by Heinrich Hertz. In so doing, he postulated that light had corpuscular as well as wave properties. Einstein's postulate, later verified, has become an integral part of modern physics.

It was discovered in 1887 by Hertz (who also produced and detected the first radio waves) *that light could dislodge electrons from a metal plate.* We now know this to be a general phenomenon: *short-wavelength EM energy incident on a solid can cause the solid to emit electrons. This is called the* **photoelectric effect,** *and the emitted electrons are called* **photoelectrons.**

An experiment for observing the photoelectric effect is shown in Fig. 25.9. A metal plate *P* is sealed in a glass vacuum tube, together with a small wire *C*, the collector. These elements are connected in a battery and galvanometer circuit, as shown. When the tube is covered so that no light enters it, no current flows through the galvanometer, since the portion of the circuit from *P* to *C* inside the bulb lacks a connection. The vacuum space between *P* and *C* has essentially infinite resistance.

If short-wavelength light is incident on the plate *P*, as shown, the galvanometer needle deflects. The direction of flow of the current shows that electrons are leaving the plate *P* and traveling through the vacuum tube to the collector *C*. It may appear that the light heats up the plate and that when it becomes hot, thermionic emission occurs as in an ordinary diode. This is not the case, however. Careful experiments have shown that *no matter how feeble the light and no matter how massive the metal plate, a stream of electrons is emitted from the plate the instant the light reaches it. No heating is required.*

It is further observed that, for a given light source, *the number of electrons emitted from the plate is proportional to the intensity of the light.* If the battery voltage is large enough to attract all the emitted electrons to the collector, then the current in the galvanometer is directly proportional to the light intensity. (It is for this reason that a photoelectric cell, such as this, is used to measure light intensity.)

A more startling feature is shown in Fig. 25.10. Suppose the wavelength of a light beam can be varied while its intensity (i.e., energy per unit

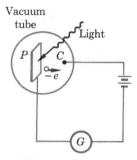

FIGURE 25.9
When light strikes the plate
P of the photocell, electrons
are emitted from it.

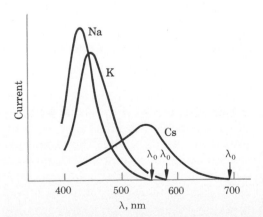

FIGURE 25.10
The current in the circuit of
Fig. 25.9 varies with
wavelength as shown. Data
for three different metals
(sodium, potassium, and
cesium) are shown. What is
the meaning of the λ_0 value
indicated in each case?

AIP Niels Bohr Library

Max Planck
(1858–1947)

As a young German student, Planck came close to choosing music rather than physics as his major interest; his long scientific career was devoted almost entirely to the fields of heat and thermodynamics. He discovered the quantized nature of energy while a professor in Berlin, publishing his findings in a series of papers between 1897 and 1901. He developed a formula that agreed with his experimental data on heat radiation, and then he had to introduce the quantum concept to explain his results. At first, Planck thought his quantum concept might be nothing more than a happy mathematical accident. As others made further important progress based on the concept, his doubts diminished, but he continued to look in vain for ways to reconcile quantum physics with the classical newtonian view. By 1918, when he received the Nobel Prize, his preeminence as a scientist and teacher were beyond question. In 1930, he became head of the Kaiser Wilhelm Society, the most prestigious scientific organization in Germany, but was forced to resign in 1937 because of his opposition to Hitler's persecutions. His son was executed in 1944 after an attempt on Hitler's life. After the war, when he was nearly 90, he was reappointed head of the society, which was renamed the Max Planck Society in his honor.

---

area per second) is kept constant. Then the current in the circuit of Fig. 25.9 can be monitored as the variable-wavelength beam is incident on the plate of the photocell. It is found that the current varies with wavelength in the way shown in Fig. 25.10. The three curves shown are for plates made of sodium metal, potassium metal, and cesium metal. Other plate materials yield similar curves but with different values for λ_0, the wavelength at which the current in the circuit becomes zero.

The most startling feature of these curves is that *no electrons are emitted if the wavelength of the light is larger than* λ_0. *This wavelength is called the* **photoelectric-threshold wavelength.** No matter how intense the light, if its wavelength is even just slightly longer than λ_0, no electrons are emitted. No matter how weak the light, if the light has wavelength shorter than λ_0, electrons are emitted essentially as soon as the light is turned on. *The particular value of* λ_0, *the critical wavelength for electron emission, depends on the material from which the plate is made.*

Another experiment involving the circuit of Fig. 25.9 yields further important data. In this experiment, a beam of known wavelength and intensity is directed at the plate. The energy of the fastest electron emitted by the plate is then measured. To carry out that measurement, the battery of Fig. 25.9 is replaced by a variable-voltage source *of reversed polarity.* Because the collector is now negative instead of positive, the photoelectrons are repelled by the collector. The current in the circuit drops to zero when this reverse voltage is made large enough. At the voltage V_0 (the stopping potential) which results in zero current, the work done by the fastest photoelectron as it travels from the plate to the collector is $V_0 e$ because it moves through a voltage difference V_0. But this work must equal the kinetic energy of the most energetic photoelectron. We can therefore determine $(\mathrm{KE})_{\mathrm{max}}$, the maximum KE of the photoelectrons, by measuring the stopping potential V_0. They are related through

$$(\mathrm{KE})_{\mathrm{max}} = V_0 e$$

If now V_0, and therefore $(\mathrm{KE})_{\mathrm{max}}$, is measured for various-wavelength beams of light incident on the plate, an interesting result is found. When $(\mathrm{KE})_{\mathrm{max}}$ is plotted against $1/\lambda$, a straight-line relation is found, as shown in Fig. 25.11. Moreover, the value of λ at which $(\mathrm{KE})_{\mathrm{max}}$ becomes zero is the threshold wavelength, λ_0. The equation of the straight-line relation is

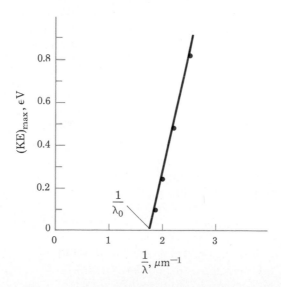

FIGURE 25.11
The photoelectron energy varies inversely as the wavelength. This particular graph is for sodium.

$$(KE)_{max} = \frac{A}{\lambda} - B \qquad (25.4)$$

The constant B varies from substance to substance, but A, the slope of the line, is the same for all materials and has a value of 2.0×10^{-25} J · m.

Many attempts have been made to explain all these observations in terms of the wave nature of light. None has been successful. Two basic difficulties are encountered by any wave interpretation:

1   How can one conceive of waves giving rise to a threshold wavelength? Light with λ just slightly less than λ_0 does not differ appreciably from light with λ just slightly greater than λ_0. Yet wavelengths slightly shorter than λ_0 cause electrons to be emitted, while those just slightly longer do not.

2   How can even the weakest possible beam of light cause electrons to be emitted as soon as the light is turned on? The light energy seems to localize on one electron instantaneously and causes the electron to break free from the solid.

Thus it appeared that a new approach was needed to explain the photoelectric effect. This bold, imaginative step was taken by Einstein.

In order to resolve this dilemma, Einstein seized on Planck's ideas of quantized oscillator energies. Planck, it will be recalled, postulated that an oscillator with natural frequency ν_0 could take on only certain discrete energies, namely, $h\nu_0, 2h\nu_0, \ldots$, where $h = 6.626 \times 10^{-34}$ J · s. We say that the oscillator's energies are **quantized** and that the possible energies differ by an **energy quantum** $h\nu_0$. As we have seen, the EM radiation (including light) emitted by a hot object was considered to be emitted by atomic and molecular oscillators composing the object.

Einstein reasoned that if these atomic oscillators were to emit radiation in the way Planck visualized, the energy must be emitted in little bursts or packets. For example, since EM radiation carries energy, an oscillator emitting light must be sending out energy. However, since an oscillator can have only certain discrete energies, it cannot throw out energy continuously. It must throw out the energy in bursts of magnitude $h\nu_0$ because this is the spacing between the allowed energies of the oscillator.

Photon Concept   To be specific, suppose an oscillator has energy $37h\nu_0$. If it loses energy by sending out radiation, its energy can change to $36h\nu_0$ but not to anything in between since the oscillator's energies are quantized. But in so doing, the oscillator must have thrown out a pulse of light or other radiation, the energy of the pulse being $h\nu_0$. *We call such a pulse of EM energy a* **light quantum** *or* **photon.** Hence we see that there is some justification for thinking that a beam of light consists of a series of energy packets or photons; each photon has energy $h\nu_0$.

Einstein therefore postulated the following character for light:

A beam of light with wavelength λ (and frequency $\nu = c/\lambda$) consists of a stream of photons. Each photon carries energy $h\nu$.

We sec later how the photon energy is related to the structure of atoms and molecules. Let us now apply Einstein's model for a light beam to the photoelectric effect.

If light does consist of little particles of energy, these quanta, or photons, will collide with individual electrons as the light beam strikes a substance. When the energy of the photon is larger than the energy needed to tear an electron loose from the substance, electrons are emitted the instant the light is turned on. When the energy of a light quantum, or photon, is less than that value, no electrons are emitted no matter how intense the light. (The chance of *two* photons hitting the same electron simultaneously is practically zero.) We see at once that the energy needed to tear an electron out of the plate is exactly equal to the energy of a light quantum having the threshold wavelength. Hence, the work function ϕ, which is the minimum work needed to tear an electron loose from a solid, as discussed in Chap. 20, is given by

$$\phi = \frac{hc}{\lambda_0} = h\nu_0$$

In the event that the light quantum has more energy than this, i.e., if λ is smaller than λ_0, not only can an electron be knocked out of the plate, but also it can have KE to spare. That is, the energy hc/λ of the photon (or light quantum) is lost partly in doing work ϕ, in tearing the electron loose, and the remainder of the energy appears as the KE of the electron. We may therefore write

$$\tfrac{1}{2}mv^2 = \frac{hc}{\lambda} - \phi \qquad (25.5) \qquad \textit{Photoelectric Equation}$$

This is called the **photoelectric equation.**

If we compare this equation with Eq. (25.4), we notice that $\tfrac{1}{2}mv^2$ in this equation is the same as $(KE)_{max}$ in Eq. (25.4).[9] Moreover, the constant A corresponds to hc, where h is Planck's constant and c is the speed of light. When we evaluate hc, its numerical value is equal to the value of A found by experiment. As a final confirmation of Eq. (25.5), the work function ϕ, determined by equating it to the experimental value of B in Eq. (25.4), is the same as the work function determined by entirely different experiments.

Thus we can conclude that photoelectrons are ejected from a material if an incident photon has enough energy to eject one. The photon energy is $h\nu$, which is the same as hc/λ. A photon with the threshold wavelength λ_0 has an energy hc/λ_0, and this energy is equal to ϕ, the work function. Such a photon is just barely capable of ejecting photoelectrons. Photons of wavelength shorter than λ_0 have more than enough energy to eject photoelectrons, and the excess energy appears as the KE of the photoelectron.

---

[9] Most of the photoelectrons will have less than this energy because of collisions as they leave the surface.

Illustration 25.7 When light of wavelength 5×10^{-5} cm is incident on a particular surface, the stopping potential is 0.60 V. What is the value of the work function for this material?

Reasoning We make use of the photoelectric equation (25.5):

$$\phi = \frac{hc}{\lambda} - \tfrac{1}{2}mv^2$$

But we had $(KE)_{max} = V_0 e$, where V_0 is the stopping potential. And so

$$\phi = \frac{hc}{\lambda} - V_0 e$$

$$= \frac{(6.63 \times 10^{-34}\ \text{J} \cdot \text{s})(3 \times 10^8\ \text{m/s})}{5 \times 10^{-7}\ \text{m}} - (0.60\ \text{V})(1.60 \times 10^{-19}\ \text{C})$$

from which $\phi = 3.0 \times 10^{-19}$ J $= 1.9$ eV. Most metals have a work function several times larger than this value. However, various oxides and more complex compounds have work functions in this range.

25.10 Compton Effect

Since light and x-rays are both EM waves, the photon concept should apply to x-rays as well. Direct evidence for the x-ray photon was first provided by A. H. Compton in 1923. He noticed that when a monochromatic, i.e., single-wavelength, beam of x-rays was shone on a graphite block, two kinds of x-rays were scattered from the block. One kind had the same wavelength as the incoming radiation and constituted most of the scattered radiation. But some of the scattered x-rays had a longer wavelength than that of the incident rays. The unchanged wavelength portion of the beam can be pictured as arising in the following way: The oscillating electric field in the incident beam causes the charges in the atoms to oscillate with the same frequency as the wave itself. These oscillating charges act as antennas, radiating waves of the same frequency and wavelength. Hence the scattered x-rays are reradiated waves from the oscillating atomic charges.

As we stated, in addition to this relatively intense beam of scattered x-rays, there is another type of scattered x-ray which has a slightly longer wavelength. The exact wavelength of these anomalous x-rays varies in a precise and relatively simple way, depending on the angle at which they are scattered. No explanation for their existence appeared possible by using simply the wave picture of x-rays.

A simple explanation for this phenomenon was simultaneously and independently presented by Compton and Peter Debye. They considered that the x-ray beam consists of photons, each of energy $h\nu$, and that a photon collides with an electron much as two balls would collide, as shown in

Fig. 25.12. The photon gives up some of its energy to the electron and bounces off as shown in part *b* of the figure. Since its energy is less after the collision, its wavelength must be longer.

They then treated the problem mathematically exactly as we would treat the problem of the elastic collision of two balls; i.e., both energy and momentum must be conserved. Thus,

Energy before collision = energy after collision

and

Momentum before collision = momentum after collision

These equations, then, could be solved directly for the loss in energy of the photon as a function of the angle at which it was scattered. The result could be used to compute the wavelength of the scattered photon, by using the fact that a photon's energy is hc/λ. When this was done, the computed wavelengths of the scattered photons were found to coincide exactly with the measured values. Here again was a striking confirmation of the particle properties of EM waves.

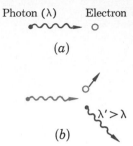

FIGURE 25.12
In the Compton effect, a photon collides with an electron. Both energy and momentum are conserved in the collision.

25.11 Momentum of the Photon

Since a photon has energy, we would expect it to have momentum and mass as well. We can see, however, that the rest mass of a photon must be zero. Since it travels with speed c in vacuum, we have

$$m = \frac{m_0}{\sqrt{1 - (v/c)^2}} = \frac{m_0}{\sqrt{1 - 1}} = \frac{m_0}{0}$$

If m_0 were anything but zero, the photon would have infinite mass. But since $E = mc^2$, infinite mass implies infinite photon energy, and we know this to be untrue. Therefore, we must conclude that *the rest mass of the photon is zero.*

To find the momentum of the photon, we use the two relations for photon energy:

Photon energy $= (m - m_0)c^2 = mc^2$

and <center>Photon energy $= h\nu$</center>

Photon's Rest Mass

If we equate these two expressions, we can solve for the momentum of the photon, which is $mv = mc$ in this instance. Doing so, we find

$$\text{Photon momentum} = p = \frac{h\nu}{c} = \frac{h}{\lambda} \qquad (25.6)$$

Momentum of Photon

This is the value used by Compton and Debye in their theories of the Compton effect.

We therefore have a method for assigning both energy and momentum to the photons of electromagnetic waves. Although these waves appear to possess a particlelike nature under certain circumstances, the *photons owe all their mass to their kinetic energy. They have no rest mass* and always travel with speed c in vacuum.

A striking example of what happens to a photon when it undergoes change is found in the phenomenon of **pair production.** By means of experiments described in Chap. 27, it is sometimes observed that a high-energy photon disappears with the creation of two electrons. This is pictured in Fig. 25.13.

A photon traveling through matter is found to disintegrate at point x into a negative electron and a positron (a particle having the mass of an electron and a charge $+e$). Notice that no net charge is created in this process, so that charge for the whole system is conserved. Energy and momentum must be conserved as well, of course. These conservation conditions can be satisfied only if the disintegration occurs near a nucleus.[10] Although charge, momentum, and energy are conserved in pair production, rest mass is actually created in the process, since the original photon had zero rest mass, while the final products have a rest mass equal to twice the mass of an electron. The created rest mass came from the conversion of energy to mass through $\Delta E = \Delta m \, c^2$. Energy is conserved in this process only because we agree to attribute an energy mc^2 to a mass m.

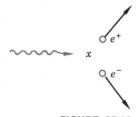

FIGURE 25.13
The photon transforms into a positron and electron when it reaches point x. This is called pair production.

25.12 Particle Waves

As we have seen, under certain conditions EM waves act as though they were composed of particles. This wave-particle duality may seem rather strange. However, there is no known case where contradictions arise because of it. In any given experiment it is always possible to say which type of behavior will exist. The general behavior can be summarized by the statement that wavelike behavior predominates during propagation of radiation, while particlelike behavior predominates during interaction with matter.

A somewhat analogous situation can exist among people. No matter how well we think we know a man, we are unable to state a priori what his actions will be when he is confronted by a new situation. If we are surprised by his actions, we rationalize our surprise by saying that we did not know him as well as we thought we did. Similarly with light waves: In 1900 we believed that we knew them quite well. We were therefore surprised to find

---

[10] Another object, a nucleus or some other particle, is needed to take on some of the momentum of the photon. The electron-positron pair cannot alone satisfy both the energy- and momentum-conservation conditions.

that they behaved in unexpected ways in new situations, as in the Compton and photoelectric experiments.

Once the particle nature of light waves had been discovered, it was perhaps reasonable that someone should ask whether particles might sometimes act as waves. This question was first treated by Louis de Broglie in 1923. (He was at that time 31 years old, and this work constituted his doctoral thesis.) He reasoned that since the momentum of a photon was given by h/λ, it might be possible that the momentum of a material particle, an electron, for example, would be equal to a similar quantity. In that case, λ would be the wavelength of an electron—or at least of some wavelike property associated with the electron. He therefore wrote $mv = h/\lambda$, from which

$$\text{Wavelength associated with a particle} = \frac{h}{mv} \qquad (25.7)$$

De Broglie Wavelength

This is called the **de Broglie wavelength** *of a particle.*

We see in the next chapter that he had available a method for checking this assumption. In particular, Niels Bohr had postulated a model for the hydrogen atom in which the electron was assumed to behave in a very definite fashion. This model had proved exceptionally successful in explaining the properties of the atom, but Bohr was unable to offer any explanation of why the electron behaved in the way he had to assume it did. De Broglie was able to show that if the electron had wave properties with wavelength given by Eq. (25.7), Bohr's postulate followed at once from this wavelike property. We return to this problem after a discussion of the Bohr atom in the next chapter.

A direct proof of the wavelike character of material particles was obtained by Davisson and Germer in 1927. Investigating the scattering of electrons from metal crystals, they fired a beam of electrons at a crystal of nickel and observed the number of electrons scattered at various angles after the beam collided with the metal. They found that under certain circumstances the electron beam was scattered very selectively, many electrons coming off at certain angles and very few at others. At first they had no interpretation for these results and reported them as unexplained.

When it was suggested to Davisson and Germer that these results might be wave-interference effects resulting from the electron-wave character postulated by de Broglie, additional measurements were begun to check this point. It was soon confirmed by various investigators that electrons are reflected from crystals in the same fashion as x-rays are reflected. We saw in Chap. 24 how the angles at which x-rays will be strongly reflected from a crystal can be computed from the crystal-lattice spacing and the wavelength of the x-rays. By using de Broglie's expression for the wavelength to be associated with an electron, the angles of strong reflection of electrons were predicted, and excellent agreement was found with the experimental results. Hence we have a reasonably direct proof of the fact that electrons sometimes behave as waves and that their wavelike character can be predicted from Eq. (25.7).

Illustration 25.8   An electron is accelerated through a potential difference of 182 V. How large is its associated wavelength?

   Reasoning   We first find the speed of the electron from the usual relation

$$Vq = \tfrac{1}{2}mv^2$$

(It is allowable to use nonrelativistic methods here, since v will be much smaller than c.) Substituting, we find $v = 8 \times 10^6$ m/s. The associated wavelength is

$$\lambda = \frac{h}{mv} = \frac{6.6 \times 10^{-34}\,\text{J} \cdot \text{s}}{(9.1 \times 10^{-31}\,\text{kg})(8 \times 10^6\,\text{m/s})}$$
$$= 0.91 \times 10^{-10}\,\text{m} = 0.91\,\text{Å}$$

This is of the same order of magnitude as x-ray wavelengths. It is therefore clear why electrons will show diffraction effects similar to x-rays.

Illustration 25.9   A 50-g ball rolls along a table with a speed of 20 cm/s. How large is its associated wavelength?

   Reasoning   Applying Eq. (25.7) and using SI units, we find $\lambda = 6.6 \times 10^{-32}$ m. In order to perform an interference experiment with such a ball, we would have to pass it through slits about this far apart. Since this length is many orders of magnitude smaller than atomic dimensions, it is obviously impossible to carry out such an interference experiment with the ball.
   Similar calculations for any laboratory-sized object will convince you that the wavelengths associated with the object are far too small to allow observation of the wave nature of the object. It is only when m is very small for electrons and atoms, that the wavelength becomes large enough for these effects to be measurable.

25.13 Uncertainty Principle

Since the time of the discovery of the wavelike nature of the electron, many experiments have been carried out to see whether other particles also exhibit this behavior. As we saw in Illustration 25.9, it is impossible to devise an experiment to observe the wave nature of large objects. However, particles of atomic size can be investigated relatively easily for wavelike effects. No exception to de Broglie's wavelength equation has ever been found. In fact, it is now commonplace to use electrons and neutrons as well as x-rays in diffraction experiments designed to investigate crystal structure.
   The wave nature of all particles leads to a great philosophical principle.
Prior to this discovery, philosophers had often argued about whether the

fate of the universe was completely determined. Could we, in principle at least, determine the position, speed, and energy of all the particles in the universe and then predict the course of all future events? It appears that the wave nature of all particles requires us to give a negative answer to this question. This fact is embodied in the *Heisenberg uncertainty principle,* which we now examine.

Let us consider how we would locate the position, speed, and KE of an object or a particle. In order to locate the particle, we must either touch it with another particle or look at it in a beam of light. Let us make the light beam as weak as possible so that its momentum will not disturb the object at which we are looking. To that end, we look at the object by using a single photon. Or if we choose, we touch the object with a single, extremely small particle. We call the photon, or particle, which we use to investigate the object the **probe particle.**

To minimize the disturbance which the probe particle will cause, we use as low an energy as we possibly can. There is, however, a lower limit on this energy, because the wavelength of the probe particle must be smaller than the object we are looking at. Otherwise, as we saw in Chap. 24, interference and diffraction effects cause the waves associated with the probe to cast extremely blurred images of the object. In particular, we saw in Chap. 24 that the finest detail we can see using waves (either light or particle waves) is detail of the same size as the wavelength. Hence, the position of the object at which we are looking may be in error by an amount

$$\Delta x \approx \lambda$$

Moreover, the momentum of the probe particle (whether photon or material) is given by $p = h/\lambda$. When it touches the object we are looking at, some of this momentum is transferred to the object, and the momentum of the object may be altered because of this disturbance. The uncertainty in the momentum of the object therefore is given by

$$\Delta p \approx \frac{h}{\lambda}$$

If we multiply the expressions for Δp and Δx, we find

$$\Delta p \, \Delta x \approx h$$

In other words, when we use the most precise experiment imaginable to locate the position of an object and measure its momentum simultaneously, the product of the errors in these two measurements will be approximately as large as Planck's constant h. This appears to be a perfectly general relation, and it is one form of Heisenberg's uncertainty principle.

A second form of the uncertainty principle can be obtained through similar reasoning. As the probe particle passes by the object we are looking at, the position of the object is uncertain to within a distance of about λ, as we have just seen. Of course, λ is the wavelength associated with the probe particle. If the speed of the probe particle is v, the time taken for the particle to pass through this distance of uncertainty is λ/v. Therefore, the exact time when the object is at a particular position is uncertain by an amount

$$\Delta t \approx \frac{\lambda}{v}$$

In addition, the energy of the probe particle is partly lost to the object under observation when the two come into contact. As a result, the uncertainty in the energy of the object is of the order of the probe particle's energy. Therefore,

$$\Delta E \approx \frac{hv}{\lambda}$$

Multiplying these two expressions yields[11]

$$\Delta E \, \Delta t \approx h$$

We have arrived therefore at two uncertainty relations, one involving momentum and the other involving energy. These relations were first proposed by Werner Heisenberg in 1927. Let us now restate them in a more exact form.

Heisenberg Uncertainty Principle *In a simultaneous measurement of coordinate x and momentum p of a particle,*

Uncertainty Principle

$$\Delta x \, \Delta p \geq \frac{h}{2\pi} \tag{25.8}$$

where Δx and Δp are the errors (or uncertainties) in x and p. Similarly, if the energy E of a particle at time t is measured, then the errors ΔE and Δt are such that

$$\Delta E \, \Delta t \geq \frac{h}{2\pi} \tag{25.9}$$

It is impossible, then, even in principle, to know everything about an object. There will always be uncertainty about its exact energy at a given time and its exact momentum at a given place. This is one of the fundamental results inherent in the concepts of light quanta and particle waves. Clearly, a new formalism is needed to describe atomic particles and light quanta in situations where these effects are important. The methods of **quantum** or **wave mechanics** were devised to handle these phenomena.

---

[11] More exact calculation shows that h should be replaced by $h/(2\pi)$ in each case.

Let us examine a simple experiment to see exactly where the classical methods of mechanics do and do not apply. For this purpose, consider a modified double-slit experiment, as shown in Fig. 25.14. A uniform beam of light is incident on the two slits, as indicated, and the light passing through the slits strikes the screen behind them. As we learned in Chap. 24, if the slit widths and slit separation d are much larger than the wavelength of the light, clear shadows are cast. Two bright spots would be noticed on the screen, and these would be rather clear-cut images of the slits. This is shown in Fig. 25.14a.

Similarly, if a parallel *beam of electrons* is incident on two slits, it will behave as shown in Fig. 25.14a provided that the wavelength associated with the electrons is much smaller than the slit separation. Hence, the situation shown in Fig. 25.14a is exactly what we would predict from classical particle mechanics. A beam of baseballs or a beam of electrons would equally well pass through the holes and hit the screen within a well-defined region. Thus, classical newtonian mechanics is valid when the particle wavelength is much smaller than the geometrical dimensions involved in the experiment.

If we consider, however, the behavior of a light beam when the slit separation is comparable to the wavelength of the light, we observe a wide interference pattern on the screen. As shown in Fig. 25.14b, images of the slits are no longer observed. Similarly with the electron beam: If the wavelength associated with the particles is comparable to the slit separation, the electron beam spreads and hits the screen in an interference pattern, as illustrated in part b of Fig. 25.14. The intensity of the interference pattern in the case of the light waves is analogous to the number of electrons hitting

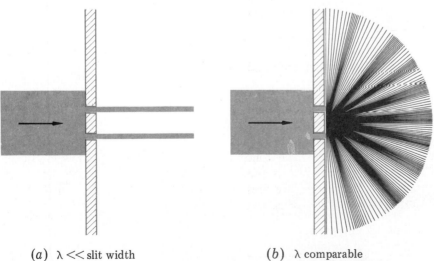

(a) $\lambda \ll$ slit width

(b) λ comparable to slit dimension

FIGURE 25.14
(a) When the wavelength associated with a particle is much smaller than the slit width, clear images of the slit are formed.
(b) However, when λ is comparable to the slit width, typical wave interference phenomena are observed.

the screen in the case of the particles. No particles strike where the intensity of light with identical λ is zero, and a maximum number of particles strike the screen where the light intensity would be maximum. This behavior is completely different from what newtonian mechanics would predict. Hence, classical mechanics is not applicable in this situation. *Classical mechanics becomes invalid when the particle wavelength becomes comparable to the geometrical dimensions involved in the experiment.*

It would appear, however, that the behavior of the light beam is always adequately described in terms of wave phenomena, at least in this experiment. Since the particle behavior can be described in terms of the associated wave in both cases, while newtonian mechanics can describe only the case shown in part *a*, it is apparent that the wave viewpoint is more generally applicable.

Soon after de Broglie suggested the wave nature of particles, Erwin Schrödinger developed an equation to describe the behavior of particles in terms of their wave nature. **Schrödinger's equation,** which is analogous to the equation used to describe the behavior of electromagnetic waves, forms the basis of quantum mechanics. It is now believed that, with slight modification, Schrödinger's equation will predict the observed behavior of particles under all nonrelativistic conditions. Therefore we should perhaps discard all the mechanics we learned in this text and start once again from this new basic equation. However, that would be foolhardy. Schrödinger's equation is exceptionally difficult to use to obtain practical answers except in the simplest problems. *Physicists therefore retain newtonian and classical concepts and use them to solve most problems. They worry about relativistic effects only when particle speeds approach the speed of light or when very accurate results are required. They replace newtonian mechanics by quantum mechanics only when dealing with dimensions comparable to particle wavelengths.* In the latter case, despite the difficulties involved, quantum mechanics must be used in order to obtain reliable answers. We see in the next chapter that quantum mechanics must be used in the discussion of the internal workings of atoms.

25.15 Electron Microscope

As we have seen, electrons and other atomic-size particles possess a wavelike character. For ordinary beams of electrons, the associated wavelength is in the very short x-ray range, as calculated in Illustration 25.8. If we could use a beam of electrons rather than light in a microscope, the microscope should be able to distinguish particles of very small size since the limiting size is comparable to the wavelength of the waves used. Whereas light microscopes are capable of showing detail larger than about 1000 nm, a microscope using a beam of electrons should be able to show detail as small as a few tens of nanometers.

The basic principle of an electron microscope is nearly identical to that

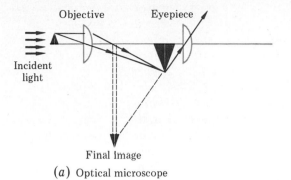

Incident
light

Final Image

(*a*) Optical microscope

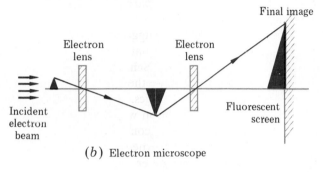

(*b*) Electron microscope

FIGURE 25.15
In the electron microscope
the final image is real, while
in the optical microscope it
is usually virtual.

of a light microscope. You will recall that in a light microscope an objective lens forms an image of the object and then this image is examined by using an eyepiece, which is basically a magnifying glass. This situation is reviewed in part *a* of Fig. 25.15. A similar situation exists in the electron microscope, as shown in Fig. 25.15*b*. However, the second lens focuses the electron beam onto a fluorescent screen so that an image (much like that on a TV screen) is formed. The operator then photographs or observes this image on the screen. Of course, the electron-beam path must be in vacuum so that the electrons will not undergo collisions with atoms of the air, which would deflect and stop the electrons. A typical microscope and a photograph of an image produced by it are shown in Fig. 25.16. A light microscope would be incapable of showing any of the detail given there since the wavelength of light is larger than the objects being viewed.

The electron lenses in Fig. 25.15 can be of two types, electrostatic or magnetic. Basically a lens is a device to focus the beam, electrons (charged particles) in this case. These two types of lenses are shown schematically in Fig. 25.17. You should examine each diagram to see exactly how the focusing action is achieved. Unfortunately, it is difficult to correct these devices for lens defects, and so at present the theoretical limits of resolution for the electron microscope have not been achieved. Detail smaller than a few tenths of a nanometer (the size of atoms) is still not readily observed, although in principle the electron microscope should be capable of doing so.

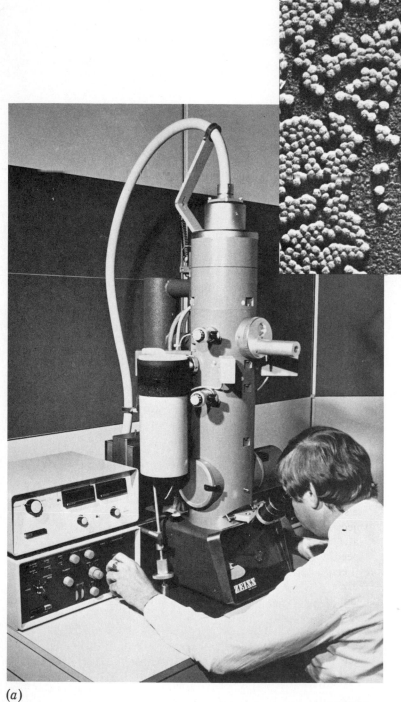

(a)

(b)

FIGURE 25.16
(a) This electron microscope uses electrons accelerated through about 70,000 V and is capable of a resolution of 0.35 nm. The source of electrons is at the top, and the final, magnified image is cast on a fluorescent screen at the bottom. Photographic plates can be inserted in this plane to obtain a photographic record. (*Courtesy of Carl Zeiss, Inc.*) (b) Each of the molecules of poliomyelitis virus photographed here has a diameter of about 20 nm. (*R. C. Williams, Virus Laboratory, University of California, Berkeley.*)

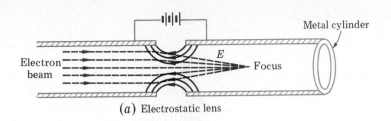

(a) Electrostatic lens

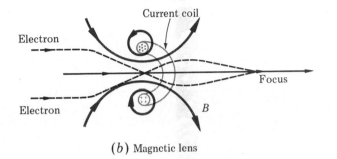

(b) Magnetic lens

FIGURE 25.17
In the magnetic lens, the B field of the coil causes the electrons to follow a path that loops around the coil axis in a spiral of decreasing radius.

Summary

Relativity is based on two apparently inescapable experimental results: the speed of light in vacuum is always measured to be c, and only relative speeds can be measured. The second of these is equivalent to stating that the laws of nature are the same in all inertial reference frames.

Among the important conclusions arrived at from these facts are the following: No material object can be accelerated to speeds in excess of c; events which are simultaneous in one inertial frame may not be simultaneous in another; a moving clock appears to be slowed by a factor of $\sqrt{1 - (v/c)^2}$; dimensions of objects are shortened along the line of motion by a factor $\sqrt{1 - (v/c)^2}$; an object's mass varies with speed according to the relation $m/m_0 = 1/\sqrt{1 - (v/c)^2}$; energy and mass are related through $\Delta E = \Delta m\, c^2$; the kinetic energy of an object is $(m - m_0)c^2$, and this reduces to $\frac{1}{2}m_0 v^2$ at low speeds.

Planck found that an oscillator with natural frequency v_0 can have only energies nhv_0, where n is an integer. For oscillators that are much larger than atomic size, these energies are so close together that they appear continuous.

Einstein showed that a beam of EM radiation is equivalent to a stream of photons. Each photon has an energy hc/λ. In the photoelectric effect, a photon collides with an electron in a solid. If the photon has energy in excess of the solid's work-function energy, it may eject the electron from the solid. In the Compton effect, a photon scattered by collision with a nearly free electron loses part of its energy. The scattered photons have larger λ than the incident photons.

A photon has zero rest mass, but its momentum is h/λ. De Broglie applied this relation to material objects and found that an object moving with speed v has a wavelength $\lambda = h/mv$ associated with it. For atomic particles, the wavelength is large enough to create easily measured interference and diffraction effects.

The wave nature of particles gives rise to the law of nature summarized by the Heisenberg uncertainty principle. The principle is represented by the relations $\Delta x\, \Delta p \geq h/(2\pi)$ and $\Delta E\, \Delta t \geq h/(2\pi)$.

Quantum mechanics takes account of the wave nature of matter. It is therefore applicable even when classical, newtonian mechanics is not. Classical mechanics is no longer valid when the particle wavelength becomes comparable to the geometrical dimensions involved in an experiment.

Minimum Learning Goals

Upon completion of this chapter, you should be able to do the following:

1 State the two basic postulates of relativity. Explain the meaning of inertial reference frame and express the second postulate in terms of laws of nature as applied to such a frame.

2 State the conclusions relativity gives rise to in respect to the following: maximum speed of objects; simultaneous events; time dilation; length contraction; mass variation with speed; KE; mass-energy conversion. Compute the answers to simple problems involving these conclusions.

3 Compute the allowed energies (according to Planck) for an oscillator with known natural frequency provided Planck's constant is given. Explain why the energy of a pendulum appears to be continuous.

4 Sketch a graph of radiation intensity versus λ for a hot object. Show how the graph changes with temperature.

5 Describe the photoelectric effect and point out what is meant by the photoelectric threshold. State the energy of a photon in terms of its wavelength. Explain how the photon concept applies to the photoelectric effect. Compute the threshold wavelength from a knowledge of the work function.

6 Describe the Compton effect and explain how it can be interpreted in terms of photon scattering.

7 State the rest mass and momentum of the photon.

8 Give the de Broglie wavelength of a particle of known mass moving with a known speed. Explain why the wave properties of electrons are easily noticed while those of a baseball are not noticeable.

9 State both forms of the Heisenberg uncertainty principle in both words and symbols.

10 Explain under what conditions newtonian classical mechanics must be replaced by quantum mechanics. Reasoning from the interference effects observed for light, show why newtonian mechanics breaks down under these conditions.

Important Terms and Phrases

You should be able to define or explain each of the following:

Postulates of relativity
Reference frame; inertial frame
Time dilation
Relativistic factor
Length contraction (relativistic)
Rest mass: $m = m_0 / \sqrt{1 - (v/c)^2}$
$\Delta E = \Delta m\, c^2$

Light quantum; photon
Photoelectric effect; threshold wavelength
Work function
Compton effect
Photon momentum $= h/\lambda$
de Broglie wavelength $= h/(mv)$
Davisson-Germer experiment
Uncertainty principle
Electron microscope

Questions and Guesstimates

1 Suppose you are in a spaceship traveling away from earth at a speed $0.90c$. A laser beam is directed at the ship from the earth. If you measure the speed of the laser light relative to your ship, what will be the light's speed?

2 An egg carton sits on the seat of a car as the car speeds down the road. Explain what the law of inertia states about the carton according to (a) an observer in the car and (b) an observer at rest on the earth. (c) Under what conditions will the law not appear to be obeyed by the carton?

3 Suppose an astronaut has perfect pitch so that she can recognize at once that a particular tuning fork gives off a sound of middle C when struck. What would she hear if she listened to the tuning fork inside her spaceship while traveling through space at a speed of $0.9c$?

4 Before leaving the earth, a vibrating tuning fork in

a spaceship is used to mark time intervals by printing out a number on a chart each time it passes its centerpoint. A similar tuning-fork system with an identical period is kept on earth as the rocketship goes out into space, travels around for a few years at nearly the speed of light, and returns to earth. How will the two charts for the two tuning forks compare when the ship returns to earth?

5 Most human beings live less than 100 yr. Since the maximum velocity one can acquire relative to the earth is c, the speed of light, a person from earth can travel no farther than 100 light-years into space in 100 yr. Does this necessarily mean that no person from earth will ever be able to travel farther from earth than 100 light-years? (One light-year is the distance light travels in one year, or 9.46×10^{15} m.)

6 Suppose the speed of light were only 20 m/s and all the relativistic results applied if you use this speed for c. Discuss how our lives would be changed.

7 A spaceship moving past the earth at speed v_{earth} shoots a projectile straight ahead with a speed u_{ship} relative to the ship as viewed by a person within the ship. At ordinary speeds, the speed of the projectile relative to the earth is $v_{earth} + u_{ship}$. Show that this cannot be correct at relativistic speeds. (The correct relation at high speeds is

$$\frac{v_{earth} + u_{ship}}{1 + v_{earth}u_{ship}/c^2}$$

You can find how it is derived by referring to a more advanced text.)

8 It should be clear from this chapter that the statement "matter can neither be created nor destroyed" is false. What can one say instead?

9 Discuss how our world would be affected if nature suddenly changed in such a way that Planck's constant became 10^{32} times larger than it is. Consider the situation from two different aspects: (a) quantization of energy of oscillators; (b) the uncertainty principle.

10 How does the photon picture of light explain the following features of the photoelectric effect: (a) critical wavelength; (b) stopping potential is inversely proportional to wavelength?

11 How can the work function of a metal be measured? Planck's constant?

12 Make a list of experiments in which light behaves as a wave and a list of experiments in which its quantum character is important. Is there any experiment in your list which can be explained from both standpoints?

13 When light is shone on a reflecting surface in vacuum, a pressure is exerted on the surface by the light. Explain. Would the pressure be different if the surface were black so that it absorbed the light?

14 If all the mass energy of a fuel could be utilized, about how many kilograms of fuel would be needed to furnish the energy required by a city of 300,000 people for 1 yr? (E)

15 If we believe the uncertainty relations, why must we refuse to believe that all molecular motion ceases at absolute zero of temperature?

16 Estimate the power change for a local radio-station antenna system as it changes from one quantized oscillation energy state to an adjacent state. What energy photons does the station radiate? What wavelength photons? What frequency? (E)

17 Ultraviolet light causes sunburn while visible light does not. Explain why. Some people insist they sunburn easiest when their skin is wet. Do you see any reason for this?

18 In view of the material you learned in this chapter, what new insight can you give to your answer of Question 15 of Chap. 6?

Problems

1 Suppose you are in a railroad boxcar that moves along a perfectly straight, horizontal track with speed 0.20c. If you drop a ball from a height of 1.50 m above the floor, how far along the floor does it strike relative to the point on the floor directly below the point from which it was dropped?

2 (a) Suppose you are in an elevator that is rising at a constant speed of 2.0 m/s. You drop a penny from your hand, which is 1.50 m above the floor. How long does it take for the penny to reach the floor? (b) Repeat if the elevator is standing still.

3 An unstable particle at rest has an average life of only 2.6×10^{-8} s before it changes to a new form. We say "its lifetime is 2.6×10^{-8} s." How long a lifetime will it appear to have as it moves through the laboratory with a speed of 2.70×10^8 m/s?

4 How fast must the particles described in Prob. 3 be moving if they are to appear to have a lifetime of 7.0×10^{-8} s?

5 Although neutrons within a nucleus have an infinite lifetime, an isolated neutron's lifetime is only about 650 s. Will a neutron shot out from the sun

at a speed of $0.995c$ be able to reach Saturn before it changes form? Saturn is about 1.50×10^{12} m from the sun.

6 When at rest in the laboratory, a certain particle designated K_L^0 has a lifetime of 5.7×10^{-8} s. If such a particle is moving with a speed of $0.9980c$ just after it is created, how far is it likely to go in the laboratory before it changes form?

7* A certain particle has a lifetime of 1.00×10^{-8} s when at rest. How fast must a beam of these particles be moving if they are to travel a distance of 15 m in the laboratory before changing form?

8* It is found that a newly created beam of particles travels 2.5 m in the laboratory before they change form. Their speed in the laboratory is $0.9970c$. What is the lifetime of these particles when they are at rest relative to the laboratory?

9 The length of a spaceship is measured by observers on earth, and it is found to be 26.38 m when the ship is at rest on earth. What will observers on earth measure the ship's length to be as the ship flies past the earth with a speed of (a) $0.20c$ and (b) $0.9920c$?

10 A spaceship has an insignia on its side that consists of a square with one side along the direction in which the ship moves. How fast must the ship be moving past the earth if the insignia appears to earthlings to be a rectangle that has one side 3 times longer than the other?

11* The roads in rural Iowa are directed mostly north-south (N-S) or east-west (E-W) and are 1.00 mi apart. (a) How fast must a westward-directed plane be moving if passengers on it are to see the N-S roads half as far apart as the E-W roads? (b) An Iowan looking up at the plane as it flies overhead measures its length to be 18 m. What is the plane's length when at rest at the airport? (c) A passenger uses her elaborate watch to time the flight from one road to the next. What value does she find from her watch? (d) The Iowan with the aid of a friend 1 mi to the west measures how long it takes for the plane to move from one road to the next. What value do they obtain?

12* According to astronomers on earth, the essentially constant distance to the star Arcturus is about 40 light-years. How long, in years, would it take for a spaceship to reach there if its speed were $0.990c$ relative to the earth according to (a) earth clocks and (b) spaceship clocks?

13* The essentially constant, straight-line distance between the earth and the star Alpha Centauri is about 4.3×10^{16} m. Suppose a spaceship could be

sent to the star with a speed of 2.0×10^8 m/s. (a) How long will the trip take according to earth clocks? (b) How large a time will the spaceship clocks record this journey to take? (c) How large will the spaceship occupants measure the earth-to-star distance to be? (d) How fast will the spaceship occupants compute their speed to be from the results of parts b and c? (e) How does this compare to the earth-measured speed?

14 A 90-g mass hangs at the end of a spring whose spring constant is 0.020 N/m. (a) What is the natural vibration frequency for this system? (b) What is the energy gap between allowed energies for this oscillator? Give your answer in both joules and electronvolts.

15 The hydrogen chloride molecule acts in many ways as two balls joined by a spring, as shown in Fig. P25.1. It has a back-and-forth vibration (stretching and compressing the spring) with a natural frequency of 8.5×10^{13} Hz. What is the energy gap between allowed energies for this oscillator? Express your answer in both joules and electronvolts.

16 The helium-neon laser emits red light with $\lambda = 6.328 \times 10^{-7}$ m. What is the energy of a photon in such a beam? Express your answer in both joules and electronvolts.

17 It is convenient to remember that photons with 1-eV energy have a wavelength of 1240 nm. (a) Prove this fact. (b) What region of the spectrum is this? (c) What wavelength corresponds to a 3.0-eV photon?

18 Find the work function for a material which has a threshold wavelength of 3.3×10^{-7} m. Express your answer in electronvolts.

19 (a) Find the threshold wavelength for a material (gold) which has a work function of 4.8 eV. (b) What region of the spectrum is this in?

20 (a) Find the energy of the EM wave quanta sent out by a radio station operating at 10^6 Hz. Express the answer in electronvolts. (b) What would be the speed of an electron which has this amount of kinetic energy?

21 The average thermal translational KE of a particle is $\frac{3}{2}kT$. (a) What photon wavelength is equivalent to this average thermal energy at 27°C? (b) What type of radiation is this?

22 Light with wavelength 400 nm is shone on a sur-

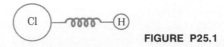

FIGURE P25.1

face having a work function of 2.0 eV. Find the speed of the fastest photoelectron emitted from the surface.

23 The stopping potential measured when 400-nm light is shone on a certain substance is 0.30 V. What is the work function of the substance?

24* The energy of the carbon-carbon bond in organic molecules is about 80 kcal/mol. If all the energy of a photon could be utilized in breaking this bond, photons of what wavelength could just accomplish it? *Hint:* 80 kcal/mol = $80,000/(6 \times 10^{23})$ cal per bond.

25 An energy of 13.6 eV is needed to tear loose the electron from a hydrogen atom, i.e., to ionize the atom. If this is to be done by striking the atom with a photon, what is the longest-wavelength photon which can accomplish it? (Assume all the photon energy to be effective.)

26 (a) What is the momentum of a 20-eV photon? (b) How does this compare to the momentum of a 20-eV electron?

27 How large an impulse does a 500-nm photon exert on a surface if the photon is (a) absorbed by the surface and (b) reflected straight back by the surface?

28* In modern nuclear accelerators, particles are sometimes accelerated to energies of billions of electronvolts. (a) What is the mass of a proton of 2×10^9 eV? (b) How fast is it moving ($m_0 = 1.67 \times 10^{-27}$ kg)?

29 To melt 1.00 g of ice at 0°C requires an energy of 80 cal. By what percentage does the mass of the ice increase because of the energy added to melt it?

30 Chemists sometimes say "the mass of the reactants equals the mass of the products" in a chemical reaction. When 2 g of hydrogen is burned with 16 g of oxygen to form 18 g of water, the reaction gives off about 60,000 cal of heat energy. How much mass is lost in the process?

31 At low speeds, if a man moving with speed v relative to the earth shoots a projectile out along his line of motion with speed u relative to himself, then the speed of the projectile relative to earth will be simply $v + u$. This cannot be correct at speeds near c since speeds in excess of c would be predicted. (For example, if $v = 0.7c$ and $u = 0.7c$, the speed relative to the earth would be $1.4c$, an impossibility.) Einstein showed the relative speed to be given by the formula

$$\frac{v + u}{1 + vu/c^2}$$

If a spaceship has a speed $0.7c$ past the earth and shoots a projectile out from itself in its line of motion with speed $0.9c$, what is the speed of the projectile relative to the earth?

32 Repeat Prob. 31, replacing the projectile by a light pulse. Before working the problem mathematically, can you give the answer from a consideration of what you already know?

33 What is the wavelength associated with an electron which has fallen through a potential difference of 6 V?

34 Atoms are of the order of 0.10 nm in radius. How fast must an electron be moving if its associated wavelength is to be smaller than 0.10 nm?

35** A photon with wavelength λ strikes an electron (originally at rest) head on, and both rebound elastically. (a) Assuming that all the motion occurs along the same straight line, show that the velocity v of the electron after impact and the final wavelength of the photon are given by

$$v = \frac{1.46 \times 10^{-3}}{\lambda} \text{ m/s} \qquad \text{provided that } v \ll c$$

and

$$\frac{1}{\lambda'} = \frac{1}{\lambda} - \frac{mv^2}{2hc}$$

(b) Evaluate both v and λ' if $\lambda = 1.00$ nm.

36** A police officer sitting alongside a roadway aims a radar beam (frequency $= \nu$) at approaching cars. Suppose the beam is reflected by a car of speed v and mass M. Considering the waves to consist of photons, show that the laws of conservation of momentum and energy lead to the following equations:

$$h(\nu + \nu') = cM(v - v')$$

and

$$h(\nu - \nu') = \tfrac{1}{2}M(v' - v)(v' + v)$$

where the primed quantities represent conditions after the photons are reflected.

Use these two relations to show that the fractional change in frequency of the radar wave reflected by the car is

$$\frac{\Delta \nu}{\nu} = \frac{1}{2c}\left(1 + \frac{\nu'}{\nu}\right)(v' + v) \simeq \frac{2v}{c}$$

Find $\Delta \nu$ for a radar beam of 1.5×10^{10} Hz reflected from a car moving at 20 m/s.

26
Atomic Structure and the Emission of Light

We saw in the last chapter that the years following 1900 were years of rapid advances in physics. In this chapter we see how these developments were extended to explain the structure of atoms and their interaction with EM waves. Before doing so, however, we first discuss an important experiment which contributed greatly to our understanding of atomic structure.

26.1 Nuclear Atom

At the beginning of this century, the internal structure of atoms was still a mystery. It was recognized that each atom has associated with it a number of electrons Z equal to the atomic number of the element in question. Since each atom is electrically neutral, the charge $-Ze$ carried by the electrons in an atom must be balanced by a positive charge $+Ze$. Moreover, the Z electrons provide only a small fraction of the total mass of an atom. Hence, in addition to its electrons, the atom must have within it a mass nearly equal to its total mass, with a positive charge Ze.

The arrangement of the electrons, positive charge, and mass within atoms was first learned from the definitive experiments of Ernest Rutherford and his associates in 1911. We illustrate the basic idea behind these experiments in Fig. 26.1. As we see there, a beam of α particles (alpha particles) is incident on a thin film of gold atoms. The α particles are shot out from the radioactive element radium. These particles were known to carry a charge

+2e and to have a mass about equal to that of the next to smallest mass atom, helium. Since a gold atom has a mass about 50 times larger than an α particle, one might first picture the situation to be that shown in Fig. 26.1a.

Shown there is the so-called **Thomson model** of the atom (after its proposer, J. J. Thomson). It assumes the atom to consist of a more or less uniform spherical mass having the atom's positive and negative charges distributed throughout it. Since the target film used was of the order of 100 atoms thick, the Thomson atom model would predict the α particle to be slowed and perhaps even stopped by the film. The experimental results, however, turned out to be quite different.

First, it was found that only a small percentage of the α particles were affected in any way by the presence of the film. Despite the fact that the films were hundreds of atom layers thick and were known to be free of holes, most of the α particles behaved as though the film were not there. The second unexpected result had to do with what happened to the few particles which apparently did hit something. Some of these were deflected through very large angles as though they had struck a very massive object. For example, a very few of the particles were deflected nearly straight backward.

In order to explain these results, Rutherford postulated the **nuclear atom.** His measurements showed that the atom must contain a very tiny massive core which carries all the atom's positive charge and nearly all its mass. The diameter of this positively charged ball (or **nucleus**) at the center of the atom had to be of the order of 10^{-14} m or less. However, the diameter of atoms as computed from the density of crystals was known to be of order 10^{-10} m.[1] Therefore, if the nucleus were pictured to be a positively charged ball about 1 mm in diameter, the atom drawn to this same scale would have a radius of about 10 m. It is impossible, therefore, for us to draw a scaled diagram of the nuclear atom since, if the atom is to fit on the page, the nucleus will be too small to be seen.

The atoms of the periodic table were known from their chemical behavior to contain a number of electrons equal to the atomic number for that element. For example, hydrogen is the first element in the table, and since its atomic number is 1, it must possess one electron. Similarly, gold, being the seventy-ninth element in the table, must possess 79 electrons. Rutherford postulated that since the atom must be electrically neutral, the nucleus must carry a charge of $+79e$. The immense region outside the nucleus contains the 79 electrons of the atom, as shown schematically in Fig. 26.2. Since the electrons are also very small, the major portion of the atom behaves as empty space and does not deflect the bombarding particles. Since Rutherford's time, his nuclear picture of the atom has been fully confirmed. We use this model to explain other facets of atomic behavior.

A schematic picture of the carbon atom which is consistent with Rutherford's result is shown in Fig. 26.3. The nucleus and the electrons would be smaller than pinpricks if the radius of the atom were as large as shown in the

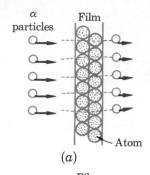
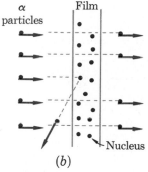

(a)

(b)

FIGURE 26.1
The films used by Rutherford and his associates were hundreds of atom layers thick. (a) Prediction based on Thomson's model. (b) Typical experimental results.

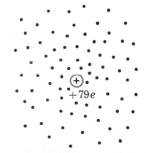

FIGURE 26.2
The Rutherford atom model pictured the positive nucleus at the center of a spherical region containing the electrons. Both the electrons and nucleus should be much, much smaller than shown.

---

[1] The number of atoms per unit volume can be obtained by dividing the mass per unit volume, i.e., the density, by the mass of an atom. The volume taken up by each atom can then be found by taking the reciprocal of the number of atoms per unit volume.

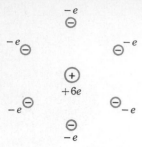

FIGURE 26.3
A possible but incorrect picture of the carbon atom.

figure. This picture is not acceptable as it stands, however, because it is not stable mechanically. The positive nucleus will attract the six negative electrons, and they will fall into the nucleus and be neutralized. In fact, it can be proved that electrostatic forces by themselves are not capable of holding a stationary group of charges in static equilibrium. Hence, there is no possible way in which stationary electrons and a nucleus could be arranged in space so that they would remain fixed in place.

One might speculate that perhaps the electrons are circling the nucleus much as the planets circle the sun. The inverse-square gravitational force supplies the centripetal force needed to hold a planet in its orbit. A similar situation could apply to the atom, with the inverse-square Coulomb attraction replacing the gravitational force. Even though this idea had much to recommend it, there was a nearly overwhelming reason for not believing it.

When an electron rotates about a positive nucleus, it creates, in effect, an oscillating pair of charges. Maxwell's equations predict that an oscillating-charge system will radiate EM waves in the manner we discussed for an antenna. For an atom, the waves would be of very high frequency, probably light waves. But if the atom oscillator radiated energy, it would be bound to run down. The electrons would spiral into the nucleus, and the atom would collapse. Since this does not happen, the whole idea appears to be wrong.

On the other hand, if the atom did radiate energy, this would be a convenient explanation for the fact that atoms sometimes give off light. Unfortunately, though, the frequency would change as the electrons fell in toward the nucleus, contradicting the fact that a given atom radiates only very well-defined wavelengths. As you can see, the situation was far from satisfactory in 1913, when people were still groping for an answer to this problem.

26.2 Spectrum of Hydrogen

In spite of the fact that the mechanics of the atom were not well understood in 1913, detailed and precise data were available concerning the light given off by atoms. Grating and prism spectrometers had been widely used for years, and the field of optical spectra (light radiation by atoms and molecules) had been the subject of diligent experimental research. In principle, researchers found any gaseous mass of atoms could be made to emit light by sending a spark, or discharge, through it by means of two high-voltage probes. For materials which are normally solid, a gaseous mass could be created by first vaporizing the material in a hot arc. The wavelengths of light given off by these hot gases, i.e., their spectrum, could be investigated by use of a spectrometer, as discussed in Sec. 24.7.

Since hydrogen is the simplest of all atoms, having only one electron, it was natural that great interest should be manifested in its spectrum. As it turns out, this atom does have a simple spectrum. When measured, it was found to consist of the series of spectral lines shown in Fig. 26.4. (Recall from Sec. 24.7 that a spectral line is actually an image of the slit of the

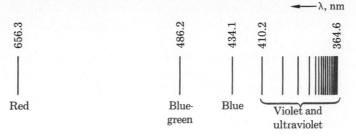

FIGURE 26.4
The Balmer series of lines
for hydrogen.

spectrometer. Each wavelength gives a separate image.) The lines in the near ultraviolet were visible only in photographs, of course.

Notice that the lines at shorter wavelengths are closer and closer together. However, no lines of wavelength shorter than $\lambda = 364.6$ nm occur (in this region), and this shortest wavelength of the series is called the **series limit.** According to the theory we shall present shortly, there should be an infinite number of lines in this series. About 40 have actually been resolved. The remainder are too close together to be seen distinctly.

Series Limit

Since these spectral lines seem to have a definite sort of order, it is perhaps natural to try to fit their wavelengths to an empirical formula. This was first done by Balmer in about 1885, and this series is now known as the Balmer series. He found that the wavelengths of the lines could be expressed by the following remarkably simple formula:

$$\frac{1}{\lambda} = R\left(\frac{1}{2^2} - \frac{1}{n^2}\right) \tag{26.1}$$

Balmer Series Formula

where R is a constant of value 1.0974×10^7 m^{-1} and $n = 3, 4, 5$, and so on. Of course, 2^2 is simply 4. If n is set equal to 3, λ turns out to be 656 nm according to the formula. This is also the first line in the Balmer series, shown in Fig. 26.4. For $n = 4$, λ is given as 486.2 nm, and so on. The integers from 3 to infinity, when placed in Eq. (26.1), yield the lines of the Balmer series. When n is set equal to infinity, the formula yields the series limit, 364.6 nm. The empirical constant R is called the **Rydberg constant** in honor of the man who accurately determined its value. It is customary to denote the lines in this series as the H$_\alpha$ line, H$_\beta$ line, H$_\gamma$ line, etc.

Later, it was found that hydrogen atoms emit wavelengths other than those found in the Balmer series. The Lyman series occurs in the far ultraviolet, while the Paschen series is in the infrared. These series are illustrated in Fig. 26.5. Other series still farther in the infrared have also been found. Amazingly enough, these series also follow formulas very much like Balmer's. It is found that

Lyman: $\quad\quad \frac{1}{\lambda} = R\left(\frac{1}{1^2} - \frac{1}{n^2}\right) \quad\quad n = 2, 3, \ldots$

$$\tag{26.2}$$

Balmer: $\quad\quad \frac{1}{\lambda} = R\left(\frac{1}{2^2} - \frac{1}{n^2}\right) \quad\quad n = 3, 4, \ldots$

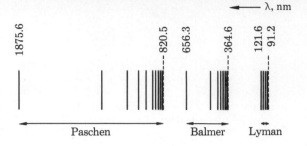

FIGURE 26.5
The three shortest spectral
series of lines given off by
hydrogen atoms.

Paschen: $$\frac{1}{\lambda} = R\left(\frac{1}{3^2} - \frac{1}{n^2}\right) \qquad n = 4, 5, \ldots \qquad (26.2)$$

and so on, with $R = 1.0974 \times 10^7$ m^{-1}.

It is apparently more than mere coincidence that such simple formulas should apply to a phenomenon as complicated as light emission. Clearly some great simplicity in atomic behavior must be responsible for this remarkable set of relations. They should therefore furnish a simple test of atomic theories. As we shall see, they furnished Niels Bohr with the clues he needed to provide us with the first workable picture of the atom.

26.3 Bohr Atom

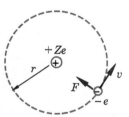

In 1913, when he was a 28-year-old research student at Cambridge, Niels Bohr presented a novel picture of the hydrogen atom. He built it on the concept of an electron revolving in a circular orbit about the nucleus as center. The centripetal force needed to hold the electron in orbit was furnished by the electrical attraction between the nucleus and the electron. The concept is shown in Fig. 26.6, where the charge on the nucleus is taken to be Ze. In the case of hydrogen the atomic number Z is unity.

FIGURE 26.6
In the Bohr atom, the
electron is assumed to
travel in a circle about the
nucleus. The centripetal
force is furnished by the
attraction to the nucleus.

Coulomb's law tells us that the electrical force on the electron is $k(Ze^2/r^2)$ toward the center, where $k = 9 \times 10^9$ N $\cdot$ m^2/C^2. This force furnishes the centripetal force mv^2/r, and hence we can write

$$\frac{kZe^2}{r^2} = \frac{mv^2}{r}$$

Solving for mv^2 yields

$$mv^2 = \frac{kZe^2}{r} \qquad (26.3)$$

These ideas were nothing new, of course. As stated earlier in this chapter, an oscillating electron should radiate energy, and hence the atom would run down and collapse. For this reason, this general picture had not been

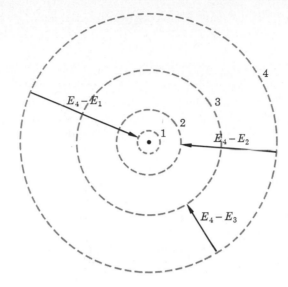

widely accepted. To circumvent this difficulty, Bohr took the attitude that perhaps the atom did not obey ordinary electric laws. He tentatively assumed that the oscillating electron did not radiate energy, even though this was contradictory to the behavior observed for the oscillating electrons in antennas and in other nonatomic systems.

Let us suppose, with Bohr, that there are certain special orbits in which the electron is stable, i.e., in which it does not radiate energy.[2] These orbits would be somewhat as shown in Fig. 26.7. If the electron is in orbit 2, it might, under certain circumstances, fall to orbit 1. Actually, since the electron is attracted by the nucleus, it will fall "downhill" and will lose PE as it falls from orbit 2 to orbit 1. *If the total energy of the electron in orbit n is E_n and in orbit p is E_p, the electron will lose energy $E_n - E_p$ when it falls from orbit n to orbit p.*

Stable Orbits Assumed

Since energy can be neither created nor destroyed, the energy lost $E_n - E_p$ must go somewhere. Bohr postulated it to be radiated as a single quantum of light energy, hv. He therefore wrote for the energy of the emitted photon,

Photon Emission Assumed

$$hv = \frac{hc}{\lambda} = E_n - E_p \qquad (26.4)$$

Let us now see what conclusions we can draw from Eqs. (26.3) and (26.4).

To find E_n, we notice that the electron has both KE and PE. Its KE is $\frac{1}{2}mv^2$, or, from Eq. (26.3), we have

---

[2] We do not preclude the possibility, however, that the electron might move from one orbit to another under appropriate conditions.

$$KE = \frac{Ze^2k}{2r}$$

where $k = 9 \times 10^9 \, \text{N} \cdot \text{m}^2/\text{C}^2$.

The electrostatic PE of the electron is negative, however. This is so because we define the potential energy of two charges to be zero when the charges are infinitely far apart. But as the electron comes closer to the nucleus, it is going "downhill" in potential energy because the nucleus attracts it. It therefore moves to potential energies less than zero, to negative potential energy. (The situation is analogous to the situation in gravitational PE where we define the zero of PE to be at the ceiling of a room. Then everything in the room has negative PE.) In the present electrical case, the PE of an electron a distance r from a positive charge Ze is

$$PE = \frac{-Ze^2k}{r}$$

Adding these two contributions gives us the total energy, namely,

$$E = \frac{-Ze^2k}{2r} \tag{26.5}$$

Notice that the energy becomes less, i.e., more negative, as r becomes smaller. This is the result of the fact that PE is lost by the electron as it falls toward the nucleus.

Substitution into Eq. (26.4) yields

$$\frac{hc}{\lambda} = \frac{Ze^2k}{2}\left(\frac{1}{r_p} - \frac{1}{r_n}\right)$$

or

$$\frac{1}{\lambda} = \frac{Ze^2k}{2hc}\left(\frac{1}{r_p} - \frac{1}{r_n}\right) \tag{26.6}$$

Equation (26.6) is very similar to the experimental relation, Eq. (26.2). Both equations equate the reciprocal of λ to a difference of two terms. Since the radii of the orbits r_p and r_n are not known, it should be possible to choose theoretical expressions for them so that Eq. (26.6) would be exactly the same as Eq. (26.2). In view of Planck's and Einstein's success in picturing energy to be quantized (i.e., it comes in small packets or quanta), Bohr looked for some other quantity which, when quantized, would provide the proper values for the orbital radii. He found that the proper choice was to say that the angular momentum of the electron mvr must be a whole-number multiple of $h/(2\pi)$, where h is Planck's constant. He then wrote

Angular Momentum Quantized

$$mvr_n = \frac{nh}{2\pi} \tag{26.7}$$

where n is any integer and r_n is the radius of the nth orbit.

Solving for v in Eq. (26.7) and substituting for v in Eq. (26.3), we find r_n to be given as

$$\frac{1}{r_n} = \frac{4\pi^2 Z e^2 m k}{h^2} \frac{1}{n^2} \qquad (26.8)$$

Substituting this (and a similarly derived expression for r_p) in Eq. (26.6) yields

$$\frac{1}{\lambda} = \frac{2\pi^2 Z^2 e^4 m k^2}{h^3 c} \left(\frac{1}{p^2} - \frac{1}{n^2} \right) \qquad (26.9)$$

According to this, when an electron falls from orbit n to orbit p, light is emitted and its wavelength is given by Eq. (26.9). We notice at once that Eq. (26.9) is exactly the same in form as Eq. (26.2), the experimental relation. Furthermore, when the constant in Eq. (26.9) is evaluated, it proves to be almost exactly equal to the Rydberg constant R of Eq. (26.2).

Bohr's derivation of Eq. (26.9), an equation which predicts almost exactly the known wavelengths emitted by hydrogen atoms, was a remarkable success. For the first time physicists were able to derive a quantitative picture of light emission by atoms. In deriving it, *Bohr had made two unorthodox assumptions, in addition to using Planck's and Einstein's ideas concerning the energy of a quantum of light. He assumed that certain stable electron orbits exist in which the electron circles the nucleus without emitting energy. Moreover, he postulated that these orbits are such that the angular momentum of the electron is equal to an integer multiple of $h/(2\pi)$.*

If pressed for a justification of these assumptions, Bohr could support them only with the argument that they seemed to work. The reason for their validity was not known. However, while the assumptions seem to contradict our experience with ordinary objects, an atom is not an ordinary object. We have no reason not to believe that electrons in atoms behave as Bohr postulated. The fact that his postulates lead to valid results forces us to conclude that they may perhaps be correct. As we see later in this chapter, the modern trend is to accept Bohr's postulates in a general sort of way but to replace his simple picture of electron orbits with more complicated spatial patterns.

26.4 Emission of Light by Bohr's Atom

As we saw in the last section, according to Bohr the hydrogen electron can revolve in any one of a large number of stable orbits about the nucleus. The radii of these orbits are given by Eq. (26.8). If we evaluate the constant n in that relation, we find that the stable (or allowed) electronic orbits in hydrogen are as follows:

$n = 1$: $\qquad\qquad r_1 = 0.53 \times 10^{-10}$ m

$n = 2$: $r_2 = (4)(0.53 \times 10^{-10})$ m
$n = 3$: $r_3 = (9)(0.53 \times 10^{-10})$ m

and so on. These orbits were pictured in Fig. 26.7. Notice that the radii of the orbits are in the ratio $1:2^2:3^2:4^2$, etc.

We learned in mechanics and thermodynamics that a system will continue to lose energy until further energy loss is impossible. This is why a pendulum eventually stops swinging and why a ball falls to the earth. Similarly, the electron in a hydrogen atom will eventually lose as much energy as it can. As we have seen, the electron loses energy as it falls closer to the nucleus. However, it can fall no lower than orbit 1, since there is no orbit smaller than this to which it can fall. An atom in which the electron has fallen to its lowest energy level, orbit 1, is said to be **unexcited.** The atom is stable in this energy state and does not emit light. Its total energy is E_1.

Work is required to raise (or excite) the electron from this lowest energy state (orbit) to one of the higher orbits or energy levels. The electron cannot by itself move to orbit 2 or any of the higher orbits. However, if the atom is bombarded by other particles or atoms, a collision with the electron may give it enough energy that it is lifted to an orbit of higher energy.

Suppose, for example, that the electron has been knocked out to orbit 4, as shown in Fig. 26.7. Although the electron is relatively stable in this state, it will fall to a lower energy state (or orbit) within a very small fraction of a second. When it does so, it loses energy, and a quantum of light, i.e., a photon, is emitted. For example, the electron could fall directly from orbit 4 to orbit 2. According to Eq. (26.9), the wavelength of the light emitted as a result of this transition is

$$\frac{1}{\lambda} = R\left(\frac{1}{2^2} - \frac{1}{4^2}\right)$$

where the constant has been replaced by its measured value, the Rydberg constant. Notice that this wavelength, according to Eq. (26.2), is the second line of the Balmer series.

Similarly, if the electron had fallen directly to orbit 1, the wavelength would have been

$$\frac{1}{\lambda} = R\left(\frac{1}{1^4} - \frac{1}{4^2}\right)$$

which is the third line in the Lyman series. Notice that the electron has fallen through a greater energy difference in this case, and so the light emitted will be of higher energy and shorter wavelength. This fits in very well, since we know that the Lyman series is in the far-ultraviolet portion of the spectrum, while the Balmer series is in the visible portion of the spectrum.

Of course, the electron might also fall from state 4 to state 3. This would be a line in the Paschen series. We see that if an electron falls from an outer orbit to the first orbit, a line in the Lyman series is emitted. If it falls to

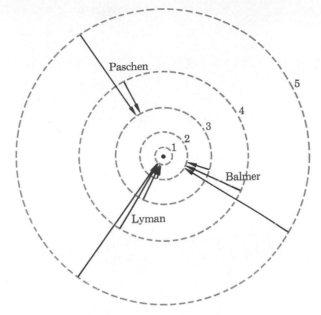

FIGURE 26.8
In a single spectral series, the electron always falls to the same inner orbit.

energy state 2, a wavelength in the Balmer series is given off. For a transition down to level 3, a line in the Paschen series results. This is shown in Fig. 26.8.

The series-limit line is emitted when the electron falls from outside the atom to the lowest energy state for that series. When the electron falls from orbit 3 to 2, the first line in the Balmer series is emitted. From orbit 4 to 2, the second line of the Balmer series results. Actually, the energy difference between the levels decreases rapidly as we go to higher and higher orbits. Hence, nearly as much energy is emitted when the electron falls from orbit 10 to 2 as when it falls from orbit 100 to 2. This means that the lines in the Balmer series become very closely spaced as we go to the wavelengths emitted by transitions from the outermost orbits to orbit 2. Of course, the most energy is emitted if the electron falls from outside the atom ($n \rightarrow \infty$) to orbit 2. This results in the emission of the series-limit wavelength.

Origin of Series Limit

26.5 Energy-Level Diagrams

It is often convenient to focus attention on the energies of the electron (and the atom) without bothering about the geometry of orbits and so on. In fact, usually we are interested in only the energies, since they alone influence the wavelengths of light emitted by the atom. We therefore use an **energy-level diagram** for an atom (and for many other systems as well). Let us now see what such a diagram is for the hydrogen atom.

As we have seen in Eq. (26.5), the total energy of the hydrogen atom when the electron is in the *n*th orbit is

$$-E_n = \frac{Ze^2k}{2r_n}$$

But $1/r_n$ can be found from Eq. (26.8). When it is substituted in Eq. (26.5), we find

$$E_n = -\frac{2\pi^2 k^2 e^4 Z^2 m}{h^2}\frac{1}{n^2} \tag{26.10}$$

Evaluating the constant, you find the energy, in electronvolts, to be

Bohr-Atom Energies

$$E_n = -\frac{13.6Z^2}{n^2} \quad \text{eV} \tag{26.11}$$

In the case of the hydrogen atom, $Z = 1$.

From this we see that the energy of the atom is zero when $n \to \infty$. This, you will recall, is the result of the way we defined our zero for potential energy: the PE is zero when the nucleus and electron are infinitely far apart. When $n \to \infty$, $r_n \to \infty$. Further, when $n = 1$ and the electron is in the first orbit, $E_1 = -13.6$ eV for hydrogen. We call this the **ground state** of the hydrogen atom. Similarly, when $n = 2$, the electron is in the second orbit and $E_2 = -3.4$ eV, and so on. The energy-level diagram used to illustrate these energies for the hydrogen atom is shown in Fig. 26.9a. It is simply a series of horizontal lines drawn to show the energies which the atom can assume.

Part *b* of the figure shows how the various spectral lines emitted by hydrogen can be represented on such a diagram. Each arrow shown in the diagram indicates a transition between two energy levels. For example, the

FIGURE 26.9
(*a*) The energy-level diagram for hydrogen. (*b*) Origin of the spectral series.

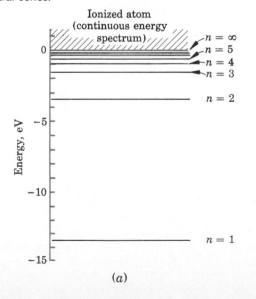
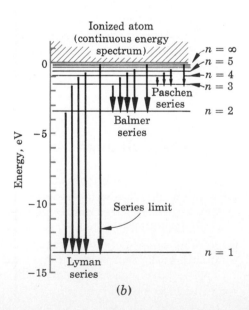

(*a*)   (*b*)

transition arrow on the left in Fig. 26.9b represents an atom that undergoes a transition from the $n = 2$ to $n = 1$ energy level. This transition gives rise to the emission of the first line in the Lyman series.

On this type of diagram, the length of the transition arrow is a direct measure of the energy involved. Hence the Lyman series arrows (not all are shown) are longer than those of the Balmer series, telling us at once that the Lyman series wavelengths will be the shorter. It is also easily seen from this diagram that the spectral lines in a series corresponding to transitions from the higher n values will lie very close together since these energy levels have nearly the same values.

Illustration 26.1 Singly ionized helium is a helium atom which has lost one of its two electrons. Draw the energy-level diagram for this ion.

Reasoning The singly ionized helium atom will be much like a hydrogen atom except that the charge on the nucleus is $+2e$, and so $Z = 2$. From Eqs. (26.10) and (26.11) we have, after setting $Z = 2$,

$$E_n = -\frac{54.4}{n^2} \quad \text{eV}$$

This then yields $E_1 = -54.4$ eV, $E_2 = -13.6$ eV, $E_3 = -6.04$ eV, etc. These values are shown in the energy diagram of Fig. 26.10.

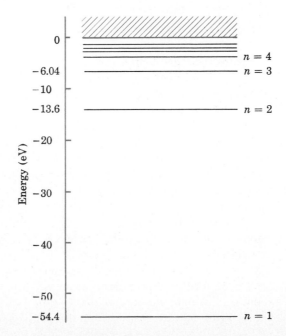

FIGURE 26.10
The energy-level diagram for singly ionized helium.

Illustration 26.2 Find the wavelength of the first line of the series equivalent to the Lyman series in singly ionized helium.

Reasoning The Lyman series corresponds to the electron falling from the higher levels to the $n = 1$ level. In this case, we are interested in the first line of the series, and so we need the $n = 2$ to $n = 1$ transition. From Fig. 26.10 we have $E_1 = -54.4$ eV and $E_2 = -13.6$ eV. Therefore the energy of the emitted photon will be $54.4 - 13.6 = 40.8$ eV. This is equivalent to 65×10^{-19} J. Since the photon energy is $h\nu$ or $h(c/\lambda)$, we have

$$\lambda \approx \frac{hc}{65 \times 10^{-19} \text{ J}} = 30.0 \text{ nm}$$

We could have done the computation much more easily if we had known that 1 eV corresponds to $\lambda = 1240$ nm. Then, since λ varies inversely as energy, the wavelength can be found by proportion:

$$\frac{\lambda}{1240 \text{ nm}} = \frac{1 \text{ eV}}{40.8 \text{ eV}}$$

from which

$$\lambda \approx 30.0 \text{ nm}$$

This conversion, that 1 eV corresponds to 1240 nm, often proves convenient in computations such as this.

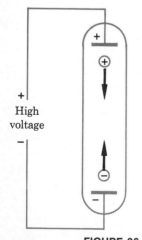

+
**High
voltage**
−

FIGURE 26.11
The high voltage across the tube causes the electrons and ions to accelerate. If the voltage is high enough, these moving charges will ionize other atoms by collision.

26.6 Absorption of Light by Bohr's Atom

Unexcited hydrogen atoms do not give off light, because the electron is in the lowest energy level at orbit 1; that is, the atom is in the ground state. As noted previously, *the atom will emit light only if it has somehow been excited into one of its higher energy levels. Then the electron can fall back toward the nucleus, emitting light in the process.*

There are several ways in which the electrons can be excited to higher energy levels. At very high temperatures the atoms will have so much KE that when they collide with one another, the electrons will become excited to a higher level. Very hot gases give off light because the atoms are excited by this means.

Alternatively, if the atoms of a gas are placed in a discharge tube, the electrons can be excited to higher levels by collisions with fast-moving ions or electrons. As illustrated in Fig. 26.11, the high voltage between the two electrodes accelerates the small number of ions and free electrons which have been formed from neutral atoms by stray cosmic or nuclear radiation shooting through the gas. If the voltage is high enough, the accelerated charges will collide with other atoms and ionize them, i.e., tear electrons loose from them. A large number of high-speed charged particles are soon created, and an electrical discharge occurs. When the electrons recombine with the ions, light is given off. Other levels of excitation occur, as well as ionization. These, too, give off light when the electron falls to lower energy levels. The emission of light by atoms in a gas-discharge tube is the source of light in neon signs and similar devices.

Atoms can also be excited by shining light on them. We can think of a beam of light as consisting of photons with energy $h\nu$. If one of the photons collides with an atom, it might give its energy to the electron and excite it to a higher orbit. Except in rare instances such as the Compton effect, a photon must lose all its energy in a collision, or it loses none whatever. Hence, a photon will give energy to the electron in an atom only if the photon energy $h\nu$ is just equal to the difference in energy between the original level and some higher level.

Because of this fact, atoms of hydrogen and other gases are excited only by particular wavelengths of light. Since most of the electrons will be in the lowest energy level or state, the ground state of the atom, the electron will require at least an energy $E_2 - E_1$ to be excited. Exactly this amount of energy is required to lift the electron from level 1 to level 2 (see Fig. 26.9). Photon energies less than this will not be able to raise the electron to level 2, and hence they cannot be absorbed.

If the photon energy is slightly greater than $E_2 - E_1$, it cannot be absorbed, for the electron would have to be raised to a nonexistent energy level, i.e., a level different from those which Bohr found to be allowed. The next-higher photon energy capable of exciting the atom is $E_3 - E_1$. This photon would raise the electron to level 3. Similarly photon energies of $E_4 - E_1$, $E_5 - E_1$, and so on are capable of exciting the electron. These photons (or the equivalent wavelengths of light) will be strongly absorbed by hydrogen atoms. No other photons will be absorbed.

A photon of energy $E_\infty - E_1$ will tear the electron completely loose from the atom. In other words, the atom will be ionized by this wavelength of radiation. We call this difference in energy the **ionization energy** of the atom. Bombarding photons, electrons, or any other particles must have an energy $E_\infty - E_1$ or larger to ionize the atom. From Fig. 26.9 we see that this energy is 13.6 eV for hydrogen. If the atom is given more energy than this, the electron will be thrown loose and will have energy to spare. This energy appears as KE of the freed electron. Such energy is not quantized and is represented by the continuous region of available energies (the *continuum*) for $E > 0$ in Fig. 26.9.

There is a very precise relation between the wavelengths of light which an atom will absorb and those which it emits. For example, unexcited hydrogen atoms cannot absorb photons of energy less than $E_2 - E_1$. However, a photon of exactly this energy will be strongly absorbed, since it is just correct for raising the electron to level 2. In fact, this transition gives rise to the longest-wavelength line in the Lyman series, as we have seen. Clearly, the wavelength of the first line of the Lyman series not only is emitted by an excited atom but also may be absorbed by an unexcited atom. Similarly, all the other wavelengths we have shown to be strongly absorbed are also members of the Lyman series. As we would expect from this discussion, excited atoms emit the wavelengths of light that are highly absorbed by the unexcited atoms.

Illustration 26.3 What is the longest wavelength of light capable of ionizing a hydrogen atom?

Reasoning The desired wavelength has enough energy to tear the electron loose from the atom, i.e., to raise it to the $n \to \infty$ energy level. This same energy is emitted when the electron falls from orbit ∞ to orbit 1. Clearly, the wavelength is given by Eq. (26.9) with $p = 1$, $n = \infty$, and the constant $R = 1.097 \times 10^5$ cm^{-1},

$$\frac{1}{\lambda} = 1.097 \times 10^7 \left(\frac{1}{1} - \frac{1}{\infty}\right) \text{m}^{-1}$$

or $\qquad \lambda = 0.912 \times 10^{-7}$ m $= 91.2$ nm

This, of course, is also the series limit for the Lyman series.

This answer could have been obtained more quickly by noting from Fig. 26.9 or Eq. (26.11) that $E_\infty - E_1 = 13.6$ eV. Then using the proportion discussed in Illustration 26.2, we have

$$\frac{\lambda}{1240 \text{ nm}} = \frac{1 \text{ eV}}{13.6 \text{ eV}}$$

from which $\lambda = 91.2$ nm.

26.7 De Broglie's Interpretation of Bohr's Orbits

As we have seen, Bohr's theory for the hydrogen atom was very successful in its ability to describe the behavior of the atom. Moreover, since singly ionized helium and doubly ionized lithium (for which $Z = 3$) are also compatible with Bohr's model, it is no surprise to find that they, too, correspond very well to the predictions of the model. In addition, we shall see that Bohr's model serves to correlate observations concerning more complex atoms. Despite these successes, the Bohr model is seriously deficient in several ways. One of its most serious weaknesses has to do with Bohr's assumption concerning orbits. Let us now see how de Broglie's concept of the wave nature of particles leads to a more satisfactory picture of atoms.

You will recall that a particle of mass m and speed v is characterized by its de Broglie wavelength, $\lambda = h/(mv)$. It seemed to de Broglie that the puzzling quantization phenomena found by Planck and later by Bohr had an analogy in our common experience. In the vibration of strings, tubes, etc., the vibrations are quantized in the sense that resonance occurs only at certain special frequencies. Invariably these frequencies are those for which the wave involved "fits" properly on the string, in the tube, or whatever. With this fact in mind, *de Broglie thought of Bohr's orbits as being the resonance paths for the electron wave.*

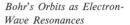
Bohr's Orbits as Electron-Wave Resonances

In order for a wave to resonate on a circular path, the path must be exactly one wavelength long, or two wavelengths long, or three, etc. Only then will the wave crests appear always at the same points on the path so that they will always resonate crest on crest and trough on trough. This situation is shown in Fig. 26.12. For that case the path length is four wavelengths long. In general, if the path is n wavelengths long, we have

$$2\pi r_n = n\lambda$$

or since $\lambda = h/(mv)$,

$$2\pi r_n = n\frac{h}{mv}$$

But this can be written as

$$mvr_n = n\frac{h}{2\pi}$$

and *this is exactly Bohr's assumption,* Eq. (26.7)!

We find, therefore, that *the Bohr orbits can be considered the resonances of the electron's wave circling the nucleus.* As such, the orbits are no longer the result of an ad hoc assumption, but are traceable to the more fundamental fact that particles have wave properties. Although the Bohr theory is made more plausible by this explanation of the orbits, we see in the next section that Bohr's theory is grossly oversimplified.

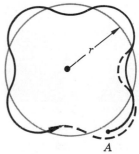

FIGURE 26.12
If the orbit length $2\pi r$ is an integral number of wavelengths, the wave will reinforce itself when it returns to the starting point A. In the case shown, $2\pi r = 4\lambda$.

26.8 Resonances of Waves in Atoms

Before continuing with our discussion of the role of de Broglie waves in an atom, let us examine the meaning of the de Broglie wave itself. For many years there was a great deal of discussion as to what the wave properties of particles really told us. It is now generally accepted that the *de Broglie waves are related to material particles in basically the same way that electromagnetic waves are related to photons.*

For example, when a beam of light is sent through two slits, an interference pattern is observed on a screen beyond the slits. In fact, this type of phenomenon is noticed for all types of waves, as we have seen. In past chapters we were content to compute the positions of the maxima and minima in the interference patterns. However, photons strike all positions where any light is observed; the number of photons striking a given point in the pattern is proportional to the intensity of the light at that point. Hence, *the light-intensity pattern can be thought of as a diagram showing the propor-*

Niels Bohr
(1885–1962)

Bohr received his Ph.D. from the University of Copenhagen in 1911 and went to Manchester the following spring to work in Rutherford's laboratory. He brought with him a good understanding of the quantum ideas of Planck and Einstein, and he proceeded to apply them to Rutherford's nuclear atom. His daring assumption that electron orbits are quantized revolutionized atomic theory, although most physicists accepted it slowly. In 1913, he returned to Copenhagen to become the first director of the Institute for Theoretical Physics. This "Copenhagen school" soon became a meeting place for theoretical physicists from all over the world; it was particularly influential in forming our present interpretation of quantum mechanics. During World War II, he escaped from German-occupied Denmark and took part in the U.S. effort to develop the fission reactor and the nuclear bomb. He was greatly concerned about the need to control the enormous military and political power of the atomic bomb. Following the war, he devoted himself wholeheartedly to the development of peaceful uses of atomic power and to attempts to control the threat posed by atomic and hydrogen weapons.

tions of the photons in the incoming beam which hit the screen at various positions.

To compute the light intensity for the complete pattern, we must solve a rather complex mathematical problem by use of the **wave equation.** The solution of this equation results in finding a quantity called the **wave function,** which is represented by the symbol ψ (Greek letter psi, pronounced "psigh"). It is a quantity which, when squared, gives the intensity of light we should find at any point in space. But since the light intensity is proportional to the number of photons, the wave function also tells us how many photons will be found at each point in space. In summary, solution of the wave equation for light waves passing through a double slit or some similar device yields the wave function ψ for the light waves. This wave function gives the light intensity everywhere in the interference pattern. In turn, this tells us how many photons exist in each portion of the pattern.

A similar situation applies to de Broglie waves. The wave equation which governs their behavior is the Schrödinger equation, already mentioned in Sec. 25.14. Like the wave equation for light, Schrödinger's equation can be solved to yield a wave function ψ. When squared, this wave function gives the intensity of the de Broglie waves. As before, *the intensity tells us the distribution of particles in space.* Whereas the light intensity and its wave function describe the motion of photons, the Schrödinger-equation wave function describes the motion of electrons (or whatever other particle is considered).

To see what this means in an actual situation, consider the hydrogen atom. At best, Bohr's theory can be telling us only where the maxima in the interference pattern will be. As in the case of the double-slit or any other interference experiment, this is only a small part of the story since photons (for light) or electrons (for de Broglie waves) also exist elsewhere. *When Schrödinger's equation is solved for the hydrogen atom* (not a very simple chore), *the intensity pattern is found to depend on just how the electron wave is resonating within the atom. To describe these resonances, we use sets of integers called* **quantum numbers.**

The *principal quantum number n* is identical to the *n* used in Bohr's theory. It tells us the energy of the atom. In both the Bohr theory and the wave theory of the atom, the energy is

Principal Quantum Number

$$E_n = -\frac{13.6Z^2}{n^2} \qquad \text{eV}$$

When $n = 1$, Bohr assumed the electron to be in its first orbit, a circle. However, the wave solution gives us the whole intensity pattern and tells us that the electron is likely to be found in many different positions but is most likely to be found in those regions where the wave function ψ is large. This intensity pattern corresponding to $n = 1$ for the hydrogen atom is shown in Fig. 26.13a. We show there the chance (or relative probability) that the electron will be found at given radii from the nucleus. Notice that the horizontal axis is r/r_1, the radius divided by the radius of the first Bohr orbit, $r_1 = 0.053$ nm. Therefore $r/r_1 = 1$ is the position of the first Bohr orbit. Bohr's theory, of course, says the electron will be found only at the radius of the first Bohr orbit.

As we expected, *the wave theory shows that the Bohr's orbit radius is the radius at which the electron is most likely to be found.* It is, so to speak, the location of the maximum in the intensity pattern. *However, there is a chance that the electron will be found at many other places in the atom as well.* (Recall that in a light-interference pattern, the photons can be found elsewhere than at the maxima.) So we see that Bohr's picture of an electron

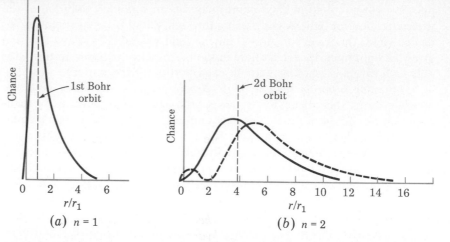

FIGURE 26.13
(a) The wave theory
predicts the relative
probabilities shown that the
electron will be found at the
various radii (r is measured
in ratio to r_1, the radius of
the first Bohr orbit). (b)
Two different resonance
forms for $n = 2$.

circling the nucleus in orbit is not realistic. In fact, the intensity pattern
found from the wave theory is not even a ring about the nucleus. Instead, it
constitutes a sphere centered on the nucleus. An attempt is made to show
this in Fig. 26.14a. To get the three-dimensional picture of the physical
situation, the pattern should be rotated about a vertical axis through the
center; the intensity of the pattern in space gives the chance of finding the
electron there. Although Bohr's theory correctly predicts the energy of the
$n = 1$ configuration, it fails badly in predicting where the electron is likely to
be found in the atom.

Let us now go on to consider the resonance pattern within the atom
when $n = 2$. You will recall that Bohr's theory predicts the electron to be in
an orbit of radius $4\,r_1$ in this case. The wave-theory prediction is shown in
Fig. 26.13b by the solid curve. Notice that, as before, the electron is most
likely to be found at the Bohr orbit but can also be found elsewhere. This is
not unexpected. However, as in the $n = 1$ case, the actual intensity pattern
in space is not shaped like a circular orbit. Instead, it consists of an hour-
glass shape as shown in Fig. 26.14b.[3] Recall that the space diagram is ob-

---

[3] A more accurate drawing would show a faint connection between the two halves.

FIGURE 26.14
To obtain the electron
distribution in space, the
figures shown must be
rotated about a central
vertical axis. All the
distributions are constant on
any circle with the axis as
center.

(a) $n = 1$ (b) $n = 2$ (c) $n = 2$ (d) $n = 2$ (e) $n = 5$
$l = 0$ $l = 1$ $l = 1$ $l = 0$ $l = 3$
$m_l = 0$ $m_l = 0$ $m_l = 1$ $m_l = 0$ $m_l = 1$

tained by rotating the diagram about a vertical axis. As you can see, the electron is restricted to two lobes on the axis.

However, this is not the only way in which the electron's de Broglie wave can resonate for the energy with $n = 2$. Two other resonance forms are also found; all three belong to the same group since their principal quantum number n and hence their energies are the same. The intensity patterns for these other two resonances are shown in Fig. 26.14c and d. Together they give rise to the electron distribution shown by the broken curve in Fig. 26.13b.

When we go to $n = 3$, again there turn out to be several resonances possible with this energy. As we saw, for $n = 1$ there was one resonance form. For $n = 2$ there are three resonance forms. If $n = 3$, there are six different resonance patterns. At larger n's some of them become quite complicated; one is shown in Fig. 26.14e.

We see from all this that *the Bohr theory is indeed a gross oversimplification for the electron behavior in the hydrogen atom. In particular, Bohr's concept of fixed orbits is untenable. However, the energy levels of the atom are predicted correctly by the Bohr theory,* and the principal quantum number n introduced by Bohr has great importance. Although we should always keep the limitations of the Bohr model in mind, it offers a framework for a systematic description of atoms, and we make frequent reference to it.

Illustration 26.4 Estimate the position uncertainty for the electron in the hydrogen atom when it is in the $n = 1$ energy state.

Reasoning To do this, let us use the uncertainty principle in the form

$$\Delta p \, \Delta x \geq \frac{h}{2\pi}$$

We know the particle energy in the $n = 1$ state is -13.6 eV, or about -2.2×10^{-18} J. Moreover,

$$KE = \tfrac{1}{2}mv^2 = \frac{p^2}{2m}$$

But we found in the development of Eq. (26.5) that the KE of the Bohr electron was numerically equal to the total energy of the atom. Therefore $KE = 2.2 \times 10^{-18}$ J. Equating this to $p^2/(2m)$ and solving for p, we find

$$p = \sqrt{(2)(9 \times 10^{-31})(2.2 \times 10^{-18})} = 2.0 \times 10^{-24} \text{ kg} \cdot \text{m}$$

If we assign the maximum value to the momentum uncertainty Δp, we obtain the least possible value for Δx. Let us therefore equate Δp to p itself. Then from the uncertainty relation we find

$$\Delta x \geq 0.53 \times 10^{-10} \text{ m} = 0.53 \text{ Å}$$

Since the radius of the atom is about 0.5 Å, we see that there is no possibility at all that we can pinpoint the position of the electron in the atom. The wave theory of the atom reflects this fact by showing us that the electron can be found throughout the atom.

26.9 Quantum Numbers and the Pauli Exclusion Principle

As we have seen in the previous sections, the hydrogen atom and its electron can exist in certain discrete energy levels characterized by an integer n. In particular, we have seen that the energy levels are given by the relation

$$E_n = \frac{-13.6Z^2}{n^2} \quad \text{eV}$$

The integer n ranges from 1 to infinity as the atom assumes its various allowed energies. Although we arrived at this result by use of Bohr's model, the wave picture, based on the solution of the Schrödinger equation, leads to the same result. Hence it is seen that n is a fundamental parameter needed to describe the state of a hydrogen atom. As mentioned earlier, it is called the *principal quantum number*. Notice that it characterizes the energy level in which the electron is to be found. Bohr pictured each value of n to be associated with a particular orbit for the electron, but this proves untenable, as pointed out in the last section. Nevertheless, it is common to say that each value of n corresponds to a particular **shell** (rather than orbit) about the nucleus. For example, when the atom is in the $n = 3$ energy level, it is customary to say that the electron is in the $n = 3$ shell.

We also saw in the last section that more than one wave resonance form is possible for the same energy, i.e., for the same value of the principal quantum number n. The wave theory shows that two other quantum numbers must be specified in order to designate a particular wave resonance
Orbital Quantum Number within the atom. One of these, the **orbital quantum number,** is related to the angular momentum of the Bohr electron in its resonance orbit. It is represented by the letter l and can assume integer values from 0 to $n - 1$. For example, when $n = 1$, the possible values for l are limited to a single value, namely $l = 0$. When $n = 2$, it is apparent that l can take on the values 0 and 1 since $n - 1 = 1$ in this case. Notice that l is always less than n.
Magnetic Quantum Number The third quantum number, the **magnetic quantum number,** can assume the values $0, \pm 1, \pm 2, \ldots, \pm l$. It is represented by m_l. When $n = 4$, for example, the largest possible value for l is 3, and hence m_l can take on the values $-3, -2, -1, 0, +1, 2, 3$. This means that when the atom is in the $n = 4$ energy level, seven different resonance forms are possible which have $l = 3$. In addition, there are five resonance forms with $l = 2$, three resonance forms with $l = 1$, and one resonance form with $l = 0$. Hence the atom can exist in $7 + 5 + 3 + 1 = 16$ different resonances, each having the same energy, namely, the energy of the $n = 4$ level.

Finally, a quantum condition exists for the electron itself. As we mentioned before, the electron acts as a small magnet because of its spin about an axis through its center. This magnet can take up only two orientations relative to an external magnetic field in which the atom may find itself. It can align either parallel or antiparallel to the field line direction. We characterize this by assigning a **spin quantum number** designated $m_s = \pm\frac{1}{2}$; the two signs represent the aligned and antialigned positions.

Spin Quantum Number

We see therefore that four quantum numbers are needed to describe the state of an electron in an atom:

Principal: $n = 1, 2, \ldots$
Orbital: $l = 0, 1, \ldots, n - 1$
Magnetic: $m_l = 0, \pm1, \ldots, \pm l$
Spin: $m_s = \pm\frac{1}{2}$

Or, in summary,

$$0 \leqslant |m_l| \leqslant l \leqslant n - 1 < n < \infty$$

and

$$m_s = \pm\frac{1}{2}$$

For a given set of n, l, and m_l values, there is a very definite wave resonance pattern for the electron wave within the atom. To specify the electron completely, however, we must know whether the electron spin is aligned with $(m_s = \frac{1}{2})$ or opposite to $(m_s = -\frac{1}{2})$ the magnetic field. We call each combination of the quantum numbers n, l, m_l, and m_s an electronic **state** of the atom. We shall now see that an extremely important law of nature applies to the behavior of electrons in the available states.

The importance of designating these states as we have done was first appreciated fully by Wolfgang Pauli in 1925. He wished to extend these concepts to atoms other than hydrogen. In order to properly assign states to the various electrons in multielectron atoms, he arrived at the following conclusion, which is known as the *Pauli exclusion principle:*

No two electrons in an atom can have the same four quantum numbers; i.e., no two electrons can exist in the same state.

Pauli Exclusion Principle

This principle is basic to an understanding of the electronic structure of atoms, as we see in the next section.

26.10 Periodic Table

Until now we have been primarily concerned with an atom which has only one electron. This might be hydrogen, singly ionized helium, doubly ionized lithium, and so on. We are now in a position to discuss how the additional electrons are arranged in the multielectron atoms found in nature and listed in the periodic table To do this, we once again use the concept of electron shells about the nucleus; each value of n has associated with it a shell. Moreover, we assume that the same resonances found for the single-electron atom can be carried over qualitatively to more complex atoms. That is, we use electronic states specified by the n, l, m_l, and m_s quantum numbers described in the previous section.

The question we must now answer is: How do the electrons arrange themselves in the various atomic states when more than one electron exists in an atom? For example, there are six electrons in each carbon atom. In which energy levels and electronic states are they to be found? This question can be answered by using the following three rules, which we already discussed:

1   A neutral atom has a number of electrons equal to its atomic number Z.
2   In an unexcited atom, the electrons are in the lowest possible energy states.
3   No two electrons in an atom can have the same four quantum numbers (the exclusion principle).

Let us now use these rules to determine the electronic structure of the unexcited atoms in the periodic table.

Hydrogen ($Z = 1$)   Its single electron will be in the $n = 1$ level. This is the lowest possible energy level, and no violation of the exclusion principle occurs.

Helium ($Z = 2$)   Its two electrons can both exist in the $n = 1$ level since they can have the following nonidentical quantum numbers:

Electron	n	l	m_l	m_s
1	1	0	0	$\frac{1}{2}$
2	1	0	0	$-\frac{1}{2}$

However, since these are the only combinations of quantum numbers possible for $n = 1$, a third electron cannot enter this shell. The shell is filled.

Lithium ($Z = 3$)   This atom has three electrons, and so the third must go into the $n = 2$ shell. We have

Electron	n	l	m_l	m_s
1	1	0	0	$\frac{1}{2}$
2	1	0	0	$-\frac{1}{2}$
3	2	0	0	$\frac{1}{2}$

Since this third electron is in the second energy level, it is much more easily removed from the atom than the first two are. Hence lithium loses one electron in chemical reactions and is univalent.

Larger Z Atoms   Obviously there are quite a few possible combinations for the quantum numbers when $n = 2$. If you count them, you will find there are eight:

n	l	m_l	m_s
2	0	0	$\pm\frac{1}{2}$
2	1	0	$\pm\frac{1}{2}$
2	1	+1	$\pm\frac{1}{2}$
2	1	-1	$\pm\frac{1}{2}$

Therefore eight electrons can exist in the $n = 2$ shell. This means that the shell will not become closed until element $Z = 10$, neon, is reached. You probably know that this is an unreactive gas, unreactive because it has a closed shell. The next element, $Z = 11$, is sodium. This is univalent since its extra electron is alone out in the $n = 3$ shell and is rather easily removed.

As one proceeds to the very-high-Z elements in the table, the concept of shells becomes less useful. The trouble arises primarily because the separation between energy levels is relatively small at high n values. In these cases the repulsions between the various electrons in the atom contribute energies large enough to sometimes cancel the influence of energy differences between shells. Despite this complication, the shell approach still proves useful for qualitative considerations.

26.11 Production of X-Rays

We know that atoms emit light or other EM radiation when an electron falls from a high to a lower energy level. For hydrogen and one-electron ions, the spectra are quite simple, since their energy levels are given by the relation

$$E_n = -\frac{13.6Z^2}{n^2} \quad \text{eV}$$

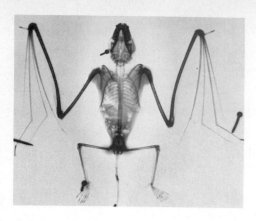

An x-ray of a bat.
(*Hewlett-Packard.*)

Discovery of X-Rays

Many of the laws of physics have been discovered by interpreting the results of experiments carefully designed to determine these laws. This was the case in Galileo's discovery of the law of falling bodies. Newton's laws are also of this same general type. However, sometimes in physics a curious, unexpected laboratory circumstance leads to the discovery of an important phenomenon. This was how Wilhelm Konrad Röntgen (1845–1923) discovered x-rays.

Carrying out experiments with high-voltage discharges or sparks, he applied a potential difference of several thousand volts to electrodes in the two ends of a partly evacuated tube. Under such conditions, a discharge or glow much like that observed in neon signs occurs. However, if the pressure of the gas in the tube is reduced enough, the glow nearly ceases. While carrying out experiments with a highly evacuated discharge tube in his darkened laboratory in 1895, Röntgen observed that a nearby fluorescent screen (much like that on the end of present-day TV tubes) was also glowing in the darkness. By moving the screen about the room, he was able to show that the light given off by the fluorescent screen resulted from something taking place in the discharge tube. Since a lighttight cover could be placed over the tube without greatly affecting the glow on the screen, it was clear that the fluorescent glow was caused by something other than the light given off by the tube. Röntgen named this unknown radiation striking the screen and coming from the tube **x-rays.**

He carried out many experiments with the rays from the tube and found that they were highly penetrating, even being able to pass through a book. Heavy metals and bone (among other materials) were not nearly so transparent to the rays as such materials as wood and paper. He was even able to cast a shadow with these rays. Shadows could be produced showing the bones of his wife's hand and the ring on her finger. Because of this ability to produce shadows, Röntgen held open the possibility that these rays were short-wavelength light. However, he was reluctant to state this as fact because, unlike light, the rays were not deflected, i.e., refracted, appreciably in passing from air to water.

We know today that the x-rays discovered by Röntgen are indeed similar to light waves but have much shorter wavelengths. Unfortunately, x-rays damage human tissue and cause severe burns through overexposure. Even when visible burns do not occur, other damage may be present. Many early workers in the field of x-rays suffered deteriorating health and even death because of their ignorance of the harmful properties of these little-understood rays.

In other atoms, the innermost electrons are largely influenced by the nearby nucleus, and so this relation also applies rather well for them. Notice, in particular, that the charge of the nucleus is very influential in determining these inner levels. For example, in the case of $Z = 100$, the energies involved are 10,000 times larger than for hydrogen. Consequently, *photons emitted when an electron falls from the $n = 2$ shell to the $n = 1$ shell in a high-Z element will have energies on the order of 100,000 eV. This corresponds to a wavelength of 1240 nm/100,000, or about 0.012 nm, a wavelength in the x-ray region.*

Although the atoms of heavy elements are capable of emitting x-rays, they do not do so unless properly excited. In order for an x-ray photon to be emitted, an electron must fall from an outer shell of the atom to a vacancy in an inner shell. Since unexcited atoms have no vacancies in these shells, x-rays are not emitted by them. The atom must first be excited in such a way that an electron in the $n = 1$ or $n = 2$ shell is thrown out of the shell, thereby providing a vacancy into which an outer electron can fall. This is usually accomplished with an x-ray tube like the one shown in Fig. 26.15 or the more elaborate tube shown previously in Fig. 20.23.

As shown, electrons emitted from the hot filament are accelerated through potential differences of the order of 10^5 V. When these high energy electrons strike the high-Z atoms in the target, electrons are knocked out of the inner shells of the atoms. As other electrons fall into the vacancies, x-ray photons are emitted. *The x-rays so generated have wavelengths characteristic of the energy differences between the various shells within the atom.* That is, the emitted photons carry an energy equal to the difference in energies between the two shells which act as starting point and end point for the electron that falls into the vacancy. *X-rays emitted by this process are referred to as* **characteristic x-rays.**

Characteristic X-Rays

Another type of x-ray emitted from a target when it is bombarded by electrons is referred to as **bremsstrahlung,** *from the German "braking radiation." As the name implies, these x-rays are emitted by the bombarding electrons as they are suddenly slowed on impact with the target.* We know that any accelerating charge emits electromagnetic radiation (a charge oscillating on an antenna, for example). Hence these impacting electrons also emit radiation as they are strongly decelerated by the target. Since the rate of deceleration is so large, the emitted radiation is correspondingly of short

Bremsstrahlung

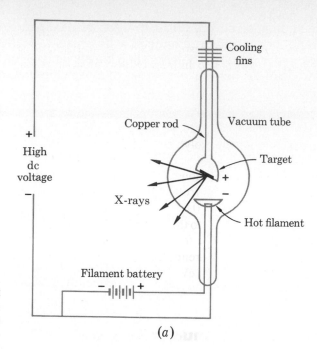

FIGURE 26.15

Electrons emitted by the hot
filament bombard the target.
The target then emits
x-rays.

(a)

wavelength, and so the bremsstrahlung is in the x-ray region. However, unlike the characteristic x-rays, the bremsstrahlung has a continuous range of wavelengths. This reflects the fact that the deceleration process can occur in a nearly infinite number of different ways, so that the energy released varies widely from one impact to another.

In Fig. 26.16 is shown a graph of the radiation emitted from a molybdenum target bombarded by 35,000-eV electrons. The two sharp peaks are the characteristic x-rays emitted as electrons fall to the $n = 1$ shell of this atom from its $n = 2$ and $n = 3$ shells. The shorter wavelength, of course, corresponds to the higher-energy transition, i.e., the $n = 3$ to $n = 1$ transition. Bremsstrahlung is the cause of the lower-intensity radiation spread over all wavelengths longer than λ_m. Since the energy of the electrons in the impacting beam was 35,000 eV, the emitted photons cannot have energies

FIGURE 26.16

X-rays emitted from a
molybdenum target when
bombarded by 35,000-eV
electrons.

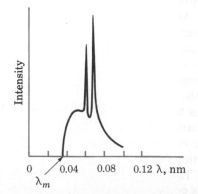

larger than this value. Using our conversion based on 1240 nm being equivalent to 1 eV, we find that 35,000 eV corresponds to (1240 nm)/35,000 ≈ 0.035 nm. As we see from Fig. 26.16, the highest-energy bremsstrahlung does indeed have this wavelength.

Illustration 26.5 From the data in Fig. 26.16, find the energy difference between the $n = 1$ and $n = 2$ levels in molybdenum.

Reasoning As we saw in our discussion of Fig. 26.16, the long-wavelength peak in that figure, 0.70 Å, results from the $n = 2$ to $n = 1$ transition. Therefore the photon of wavelength 0.07 nm carries away the energy lost by an electron as it falls from the $n = 2$ to the $n = 1$ shell. Since 1240 nm corresponds to 1 eV, 0.070 nm corresponds to an energy of 1240/0.070, or about 18,000 eV. Therefore the energy difference between these two shells in molybdenum atoms must be about 18,000 eV.

26.12 Line, Band, and Continuous Spectra

The energy levels in multielectron atoms are not as simple as the Bohr energy levels. This is the result of the fact that the various electrons in the same atom exert forces on one another and these forces cause changes in the energies of the electrons. As a result, the electrons have energies that are somewhat different from those predicted by Bohr's theory. Moreover, because each electron wave has its own characteristic resonance pattern in the atom, two electrons in the same atom seldom have the same energy. Consequently, each Bohr energy level must be replaced by a number of levels in a multielectron atom.

This fact is of little concern for the lowest energy levels in a high-Z atom. For these atoms, the innermost shells are separated by energies of tens of thousands of electronvolts. In comparison to such large energy differences, the tenths-of-electronvolt energies of interaction between the electrons are negligible. But for the outer electrons of the atom where the energy between shells is only a few electronvolts, the interaction energies between electrons are no longer negligible. As a result, the energy levels of the outer electrons are quite unlike those given by Bohr's theory. This effect causes the outer electron energy levels to become almost hopelessly complex for atoms such as iron and gold. Since the energy-level diagrams of such atoms lack the basic simplicity of the hydrogen diagram, the spectrum of emitted wavelengths from these atoms is also far less simple.

A tremendous number of spectral lines are observed in the radiation emitted by high-Z atoms. No obvious regularities such as the Balmer series are discernible in the spectra. An example of this complexity is shown in Fig. 26.17, where we give a very small portion of the spectrum emitted by iron atoms vaporized in a hot electric arc. This is typical of the spectra of high-atomic-number atoms. In spite of this complexity, the line spectra of

FIGURE 26.17
A small portion of the iron spectrum.

300 nm   310 nm

the elements are of great practical utility. Since each element gives off its own distinctive spectral lines, its presence is easily detected by spectroscopic means. For example, a material can be analyzed for its atomic composition by vaporizing it in an arc and measuring the spectral lines given off by the excited atoms in its vapor. This technique, emission spectroscopy, is used routinely to determine the quantitative as well as qualitative composition of inorganic materials. Notice that in this technique the atoms are vaporized, and their spectrum consists of discrete spectral lines, called a **line spectrum.** See the color plate insert for examples of the line spectra.

Line Spectrum

Before proceeding to the spectrum emitted (and absorbed) by atoms in molecules and solids, let us briefly summarize the requirements for a line spectrum. It originates in the case of widely separated atoms, a situation in which it is possible to speak of distinct electronic shells about the nucleus. When an electron falls from one shell to another, a very definite amount of energy is lost. Hence, a distinct wavelength of light is emitted, the same for all like atoms, and the light appears as a line in the spectrometer. *The spectrum of isolated atoms is therefore a line spectrum.* Once we begin to pack atoms togther in a liquid or solid, however, this picture is no longer valid.

Consider the two atoms and their shells illustrated in Fig. 26.18. The electrons in the third shell are seriously affected by the neighbor atom, and we can no longer say that the electron energy in this shell is the same as it

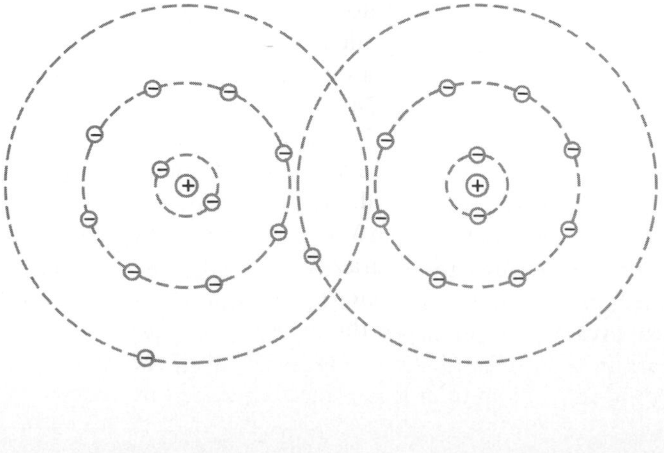

FIGURE 26.18
In a solid, the atoms tend to overlap. Hence the energies of the outer orbits are poorly defined, and the spectral lines become diffused.

740

is in a single, isolated atom. For simplicity, however, we assume that the inner shells are not too seriously overlapped by the neighbor atom and that their energies are not changed much.

Transitions to lower shells give rise to short-wavelength emission, usually ultraviolet or x-ray. Visible and infrared radiation is the result of transitions in the outer shells. It follows, therefore, that the emitted visible light is seriously dependent on the closeness of the atoms, while the ultraviolet and x-ray emission is not much affected by this factor. Exact calculations confirm this reasoning. As the separated atoms are brought closer, the possible energies of the electrons in outer shells become smeared out. Hence, *electron transitions between these outer shells in a solid or liquid no longer give rise to the emission of sharp wavelength lines.*

When a solid or liquid is heated enough to give off light, the light is emitted by the outer atomic electrons as they fall from one outer shell to another. Since these shells and their energies are diffuse, a whole range of wavelengths are emitted. The spectrum no longer contains definite, visible spectral lines but consists instead of a broad, continuous blur of color, with no individual lines discernible. *This type of spectrum, called a* **continuous spectrum,** *is emitted by incandescent objects such as molten metals or the hot filament of a light bulb.* An example of a continuous spectrum can be seen on the color plate insert. Since no individual spectral lines can be seen in a continuous spectrum, it is not useful for analysis of the molecular composition of the solid.

Continuous Spectrum

A third type of spectrum, a **band spectrum,** *is emitted by molecules.* This case is intermediate between those of a solid and free atoms. To a first approximation, the atomic emission spectrum is not much changed by inclusion of the atom in a molecule. Only the energy levels characteristic of the valence electrons are modified seriously. However, inclusion of the atoms within a molecule gives rise to a whole new group of energy levels. These are levels characteristic of vibrations involving bonds holding the atoms together in the molecule.

Band Spectrum

You will recall that Planck found that any system which vibrates with natural frequency ν_0 possesses a series of energy levels spaced an energy $h\nu_0$ apart. To a good approximation, two adjacent atoms within a molecule act as two masses held together by a spring. They have a natural frequency of vibration ν_0 relative to each other and therefore give rise to a series of energy levels $h\nu_0$ apart. In fact, each different chemical bond has its own characteristic vibration frequency and gives rise to a distinctive series of energy levels. Typical examples are frequencies of 3.24×10^{12} Hz for the C—N bond, 3.4×10^{12} Hz for the C—C bond, and 6.5×10^{12} Hz for the C=C bond.

As a vibrating molecule falls from one vibrational energy level to another, a photon with energy $h\nu_0$ is emitted. This energy is 0.014 eV for a C—C bond and gives rise to a photon with $\lambda = 88,000$ nm, a wavelength in the infrared. The emitted photons are characteristic of the molecular bond and can be used to identify bonds within molecules. This technique, called **infrared spectroscopy,** is used routinely for the analysis and identification of organic compounds.

26.13 Coherency

As we have seen, light and other EM radiations are emitted by atoms and molecules as they fall from an excited state to a lower-energy state. In each transition, a photon is emitted. The frequency of the wave in the energy pulse is given by energy $= h\nu$, but the wave in the pulse extends only a short distance in space. We represent the situation schematically in Fig. 26.19a.

As we show there, a light beam consists of a stream of light quanta. Each photon is a limited number of wavelengths long (more than shown, however). These pulses all have nearly the same wavelengths since they result from a single atomic transition. However, each is emitted by a different atom. As a result, there is no definite relation in either time or position between the pulses. They were emitted at random by the atoms, and so their waves have random phase relations with respect to one another. *We call a beam such as this, where the photons have random phases, an* **incoherent beam** *or* **pulse.**

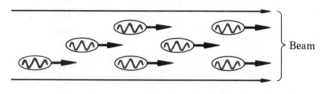

(*a*) Incoherent photons

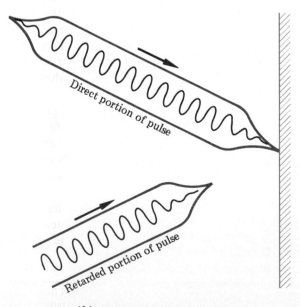

FIGURE 26.19
(*a*) Schematic representation of an ordinary light beam (incoherent photons). (*b*) A portion of a single pulse can be retarded relative to the remainder of the pulse so as to cause interference. (The diagram is highly schematic.)

(*b*) Coherent photons

Consider what happens as such a beam strikes a screen or the retina of your eye. Two or more photons may strike the same spot at once. If their waves are in phase, their amplitudes will be added and will be reinforced. But if they are 180° out of phase, they will cancel each other. Since the pulses are at a random phase relative to each other, both reinforcement and cancellation (and everything in between) will occur as time goes on and the photons strike the screen. If we were able to record the light intensity and show the details of its variation to a time as short as 10^{-10} s, we would record a highly variable light intensity on the screen. This would be a record of the cancellations and reinforcements of the photons as they strike the screen.

But most light-measuring devices, including the eye, record only an average intensity over a period much longer than 10^{-10} s. As a result, an incoherent beam of photons appears bright, but not nearly as bright as it could be if all the photons reinforced one another. And even though cancellation within the beam is taking place, it occurs for such a short time that darkness is not perceived. Even so, it is possible to produce destructive interference with such a beam, as we shall now see.

Suppose the beam of Fig. 26.19a is split into two parts by means of an interferometer or by reflection from the two surfaces of a thin film. Let us further suppose that each photon consists of a wave pulse which is 30 cm long[4] (i.e., the pulses in the figure are 30 cm long and contain about 6×10^5 cycles). If the beams are now rejoined after one has traveled only a short distance farther than the other, the two halves of each photon will be rejoined, slightly out of phase. (See Fig. 26.19b.) We can think crudely of each photon being sliced in half lengthwise. The amplitude is smaller in each but otherwise the photons are unchanged. One half has been held back relative to the other; then the two halves are rejoined. As long as the path-length difference is small in comparison with the length of a wave pulse, the half beams will show obvious interference effects when they are rejoined. If the path difference is $n\lambda$ (where n is an integer), constructive interference and brightness will exist at the screen. If the path difference is $n\lambda + \frac{1}{2}\lambda$, destructive interference will occur and darkness will be observed on the screen. This is, of course, exactly what happens in the interference experiments described in Chap. 24.

The two beams shown in Fig. 26.19b give rise to visible interference effects at the screen. This is a result of the fact that the wave pulses hitting the screen from the two beams are identical and have a fixed phase relation to each other. We say that the two beams are coherent. **Coherent beams** *Coherent Beams* *consist of wave pulses that correspond identically in the two beams and maintain a fixed phase relation to each other.*

It is interesting to note that the beams lose their coherency if one beam is retarded too far. Each wave pulse is limited in length. If the path-length difference for the two beams is too large, the pulses will no longer be joined

---

[4] For example, suppose it takes an atom 10^{-9} s to fall from one state to another. During this time, a pulse is being emitted. The leading end of the pulse will travel a distance $= ct = 0.30$ m in this time, and so the pulse will be 30 cm long.

on top of each other when the beams are rejoined. Instead, unrelated photons will be brought together, and only random phase differences will exist. Long-term cancellation and reinforcement will no longer be observed.

As we see, *the individual wave pulses in an ordinary beam of light are not coherent; the wave in one pulse has phase relations that differ widely from those in the waves in the other pulses.* The light we observe from such a beam is the average result of random, very fast interference effects. But if the beam is split into two parts, the two parts can be given a definite phase relation to each other. If the separated beams are rejoined in such a way that the two beams are coherent, pronounced interference effects are visible. Although reinforcement can occur between the two parts of the beam, the beam's intensity is still limited by the fact that its individual wave pulses are not coherent. In the next section we learn about a light source which gives out coherent wave pulses.

26.14 Laser

Let us now turn our attention to a remarkable type of device which makes use of the fact that, under very special circumstances, atoms can be made to emit light waves which are all in phase with one another. In almost all light sources, the atoms act independently; the emission of a photon by one atom is not coordinated with the emission by other atoms. As a result, the light beam consists of a complex mixture of electromagnetic waves from the various atoms. Of course, these waves are not all in phase with one another, and so they sometimes cancel and sometimes add. We saw in the last section that this causes the light beam to be much less intense than it would be if all the atoms emitted their waves in phase. A very intense beam would result if all the atoms could be persuaded to emit their waves together in phase. One light source comes close to achieving this: it is called a **laser.**

Many types of lasers are available, but they all operate on the principle from which they get their name, *l*ight *a*mplification by *s*timulated *e*mission of *r*adiation. The concept of stimulated emission, which we soon discuss, was first presented clearly by Einstein in 1917. It was not until 1953, however, that the concept was applied to laserlike devices. In that year a microwave laser, called a *maser,* was developed. Lasers, which use light rather than microwaves, were first conceived and constructed in 1958.

Let us begin our discussion by explaining what is meant by *stimulated emission.* Suppose a monochromatic beam of light is incident on a hydrogen atom as shown schematically in Fig. 26.20. If the atoms are unexcited as in part *a*, the incident photons can excite the atom as shown, provided the photon energy exactly equals the energy difference between the two levels. This is the light absorption process we discussed previously. We conclude that incident photons can be absorbed by an atom if they have proper energy.

Now let us suppose that similar photons are incident on an excited atom, as shown in Fig. 26.20*b*. The energy of the incident photon equals

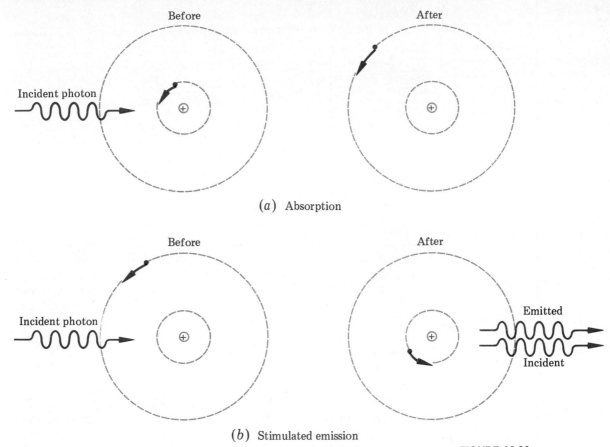

Before　　　　　　　　　　　After

Incident photon

(a) Absorption

Before　　　　　　　　　　　After

Incident photon

Emitted

Incident

(b) Stimulated emission

FIGURE 26.20
The photon resulting from
stimulated emission is in
phase with the incident
photon.

exactly the energy that the atom must eject if it falls to its lower energy state. Einstein showed that these incident photons *stimulate* the excited atom to fall to the lower state. Moreover, as shown in the figure, *the photon ejected by the atom as it falls to the lower state is in phase with the incident photon that stimulated it to make the transition.* This process is fundamental to the operation of the laser.

There is a subtle point involved here, however, that will become important for us later. Absorption and stimulated emission are competing processes. If the proportion of excited atoms is small, then most of the photons incident on atoms will be absorbed. Only if there are more excited than unexcited atoms will the incident beam increase in intensity as it passes by the atoms. Only then will stimulated emission exceed absorption. We therefore require a *population inversion* (more excited than unexcited atoms) if stimulated emission is to predominate.

The laser we describe here is the helium-neon type you will probably encounter in your laboratory work. Its basic element is a glass tube filled with a low-pressure mixture of two gases, helium and neon, in the ratio of about 15 percent helium to 85 percent neon. The ends of the tube are sealed

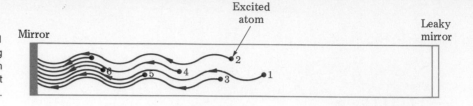

FIGURE 26.21
Schematic diagram showing how stimulated emission builds up a strong coherent wave in a laser tube.

with two accurately parallel mirrors, as shown in Fig. 26.21. One of these mirrors is only lightly silvered, and so it allows about 1 percent or less of the light incident on it to leak through. The gas atoms in the tube are excited by a gas discharge caused by high-voltage electrodes in the tube.

The gas discharge causes many of the helium and neon atoms to be excited. As they fall back to their lower energy states, they give off the light we normally associate with neon-type signs. The atoms are emitting their waves in an uncoordinated way, and so the light waves are incoherent. To obtain predominantly coherent waves, we must have a population inversion so that stimulated emission will predominate. The gases helium and neon are used in the laser tube because they are capable of achieving a population inversion in the following simple way.

Because of the gas discharge in the laser tube, many helium (and neon) atoms are excited to various energy states. As shown in the energy-level diagrams of Fig. 26.22, helium has an energy state A that is 20.6 eV higher than the ground state of the atom. This state is what is called a *metastable state*. In such a state, the atom resists falling to lower states, and so the atom exists in state A for an abnormally long time. As a result, excited helium atoms make transitions to state A and remain there. Over a certain time, a large number of atoms become semilocked in state A, and so a population inversion is possible.

The second type of atom, neon, chosen for the laser has energy states B and C (in Fig. 26.22) whose energies are close to the energy of state A in the helium atom. When a helium atom in state A collides with an unexcited neon atom, it can excite the neon atom to state B by giving the neon atom its excitation energy. (Actually, the excited neon atom in state B has slightly

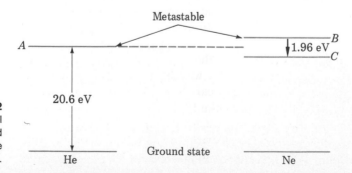

FIGURE 26.22
Portions of the energy-level diagrams for helium and neon. States A and B are metastable (long-lived).

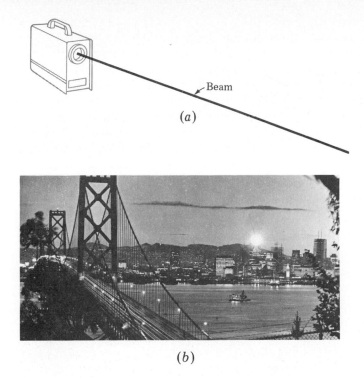

Beam

(a)

(b)

FIGURE 26.23
Even though the Electro
Optics Associates laser (a)
has an output of only
0.0005 W, its narrow,
pencillike beam shines
brightly in the distance in
(b). For comparison, notice
the less bright light coming
from the high-intensity
lamps on the San
Francisco-Oakland Bay
Bridge. (Electro Optics
Associates, Palo Alto, Calif.)

more energy than the helium atom in state A. This additional energy is
acquired by the neon atom from the KE of the collision participants.) But
state B in the neon atom is also a metastable state. Therefore, as time
passes, many neon atoms become semilocked in state B, and so a population
inversion occurs for the neon atoms.

Let us now return to Fig. 26.21, where we now picture the tube to have
enough neon atoms in excited state B that a population inversion exists.
Eventually one atom (labeled 1) falls to state C (Fig. 26.22) and emits a
photon with energy 1.96 eV ($\lambda = 632.8$ nm, red). This photon stimulates
atom 2 to emit a like photon. Then, in succession, a whole multitude of
excited neon atoms emit identical photons, and these photons are all in
phase, as indicated in Fig. 26.21. In a short time the tube is filled with
coherent waves moving back and forth between the two mirrors at the ends
of the tube. A very intense, monochromatic, coherent beam is set up in the
tube. A small fraction of the beam exits from the tube through the leaky
mirror at one end.

Because all the waves issuing from the end of the laser tube are in
phase, the beam is of high intensity. Its wavelength is sharply defined,
632.8 nm, because all the waves are identical. Not only is the beam intense
and coherent, but also it forms a pencillike beam that diverges very little.
Any rays within the tube that diverged much from the axis of the tube are
lost out the sides during the many trips back and forth through the tube.
The fact that the beam does not diverge appreciably is of great practical

importance. Unlike light from a bulb, the laser beam's energy does not fan out into space. Instead, it flows out into space through a thin cylinder and maintains its intensity over very long distances. For example, the laser shown in Fig. 26.23 has a power output of only 0.0005 W, but the light energy is confined to such a narrow pencil that the intercepted beam far outshines the high-power bulbs in the foreground. Laser beams sent to the moon and reflected to the earth are currently used to take measurements on the moon. On earth, the pencillike quality of laser beams makes such beams of value in surveying and in monitoring slight strains in the earth's crust near possible earthquake sites. These beams are also used for the precise alignment of industrial and scientific equipment.

Laser beams from some types of laser are capable of extremely high intensities. Beams of tiny cross section are powerful enough to cut through metals and are widely used as precision cutting and boring devices. They find use in medicine for eye surgery, for tumor destruction, and for an increasing number of other applications. More powerful lasers are used for welding metals by means of the intense heat generated as the beam is absorbed in matter. They also appear to have military application as beams that can destroy aircraft and tanks. Perhaps the most important application of lasers will be in the nuclear fusion-energy source, a topic we discuss in the next chapter. Many other uses could be cited, and the list grows almost daily. It is interesting to realize that it would have been impossible to conceive this very useful light source without the vast amount of basic knowledge of atoms, their energy levels, and behavior amassed by many scientists over nearly a half century. This is a typical example of how increasing knowledge of nature leads us to better ways of utilizing its laws.

Summary

The atom is of the order of 10^{-10} m in diameter. It consists of a positively charged nucleus of diameter about 10^{-15} m. If the nuclear charge is $+Ze$, there are Z electrons in that portion of the atom outside the nucleus.

Excited hydrogen atoms emit light. The wavelengths of the light constitute several series of spectral lines. These series can be represented by the relation $1/\lambda = R(1/p^2 - 1/n^2)$, where $n > p$ and both are integers. When $p = 1$, the Lyman series in the ultraviolet is found. For $p = 2$, the Balmer series results; this series is partly in the visible region. When $p = 3$, the Paschen series in the infrared is found.

Bohr was able to explain the hydrogen atom spectrum by making the following assumptions: The electron revolves about the nucleus in certain stable orbits in which no energy is radiated. The radii of these orbits are given by a quantum condition on the angular momentum, $mvr_n = nh/(2\pi)$. The atom loses energy as the electron falls from an outer orbit to an inner one, and this energy is emitted as a photon with energy $h\nu$. Bohr was able to show that the energy levels of the hydrogen atom are given by $E_n = -13.6/n^2$ eV.

Atoms can be excited to higher energy states by collisions at high temperature or in an electric discharge. Incident photons can also excite an atom. But the photon must have an energy exactly equal to that needed to lift the atom to the higher level. Hydrogen atoms absorb the Lyman series of wavelengths.

De Broglie postulated that an electron has an associated wavelength $\lambda = h/(mv)$. The Bohr orbits can be interpreted as the resonance paths for the electron waves. This interpretation justifies Bohr's quantum condition for angular momentum.

Schrödinger's equation formulates the wave nature of particles in a quantitative way. The solution of it tells us how electrons are distributed within atoms. It shows that Bohr's circular orbits are only qualitatively correct. The Schrödinger solution shows that four quantum numbers are of importance: n, l, and m_l can assume integer values restricted by $0 \leq |m_l| \leq l \leq n - 1 < n \lesssim \infty$, while $m_s = \pm\frac{1}{2}$.

The structure of multielectron atoms is strongly influenced by the Pauli exclusion principle. It states that no two electrons in an atom can have the same set of quantum numbers. Because of this, only two electrons can exist in the $n = 1$ shell, only eight in the $n = 2$ shell, and so on. The shell structure of an atom greatly influences its chemical properties.

X-rays are produced when high-energy electrons bombard solids. Characteristic x-rays are emitted by high-Z elements when electronic transitions occur in the innermost shells. Gaseous atoms emit line spectra whereas hot, incandescent solids and liquids emit a continuous spectrum.

Two waves are coherent if they are of identical form and maintain the same relative phase. Coherent waves give rise to interference effects which persist for extended lengths of time. Ordinary light consists of incoherent wave pulses. Laser light-wave pulses are not only coherent but also in phase with one another. As a result, laser beams can be made very intense. They can also be made to have nearly zero divergence of the rays.

Minimum Learning Goals

Upon completion of this chapter, you should be able to do the following:

1   Describe Rutherford's experiment and explain how it leads to the concept of the nuclear atom.
2   Give the approximate diameter of an atom.
3   Sketch the lines of the Balmer series and write the Balmer formula. Compute the wavelength of a given line of the Balmer series provided the Rydberg constant is given. Repeat for both the Lyman series and the Paschen series. State which series are partly in the visible region.
4   Describe Bohr's model for the hydrogen atom. In your description, give his postulates in regard to stable orbits and emission of light. Explain on the basis of the model how the various spectral series arise.
5   Draw the energy-level diagram for a simple one-electron atom when the formulas for its energies are given. Show on the energy-level diagram for hydrogen how the various spectral series arise.
6   Explain why hydrogen atoms normally absorb the wavelengths of the Lyman series but not those of the Balmer series.
7   Describe how the de Broglie wave concept can be used to explain Bohr's choice of stable orbits.
8   Explain the meaning of an electron-distribution diagram such as those shown in Fig. 26.14.
9   State the Pauli exclusion principle and show why it predicts lithium to be univalent.
10 Describe how x-rays are produced in an x-ray tube. Distinguish between characteristic x-rays and bremsstrahlung. Compute the shortest-wavelength x-rays emitted by a target impacted by electrons of a given high energy.
11 Distinguish between a line and a continuous spectrum. Point out which kinds of light sources emit each.
12 Explain the difference between coherent and incoherent waves.
13 State the important features of a laser beam in regard to coherency, phase, and pencillike quality. Point out how these features lead to specialized uses for lasers.

Important Terms and Phrases

You should be able to define or explain each of the following:

Nuclear atom
Spectral series; series limit
Lyman, Balmer, and Paschen series
Bohr atom and its postulates
$$\frac{1}{\lambda} = R\left(\frac{1}{p^2} - \frac{1}{n^2}\right)$$

Energy-level diagram
Ionization energy; ground state
Pauli exclusion principle
Atomic shells
Characteristic x-rays; bremsstrahlung
Line spectrum; continuous spectrum
Coherent waves and beams
Laser

Questions and Guesstimates

1 Why doesn't the hydrogen gas prepared by students in the laboratory glow and give off light?

2 Given a glass tube containing two electrodes sealed through its two ends. The gas inside is either hydrogen or helium. How can you tell which it is without breaking the tube? If the gas is at high pressure, what difficulty might you have?

3 Hydrogen gas at room temperature absorbs light of wavelength equal to the lines in the Lyman series, but does not absorb the wavelengths of the Balmer series. Why not?

4 When white light is passed through a vessel containing hydrogen gas, it is found that wavelengths of the Balmer series as well as the Lyman series are absorbed. We conclude from this that the gas is very hot. Why can we draw this conclusion? (This is actually the basis for one method of measuring the temperature of a hot gas.)

5 The spectral lines emitted from an arc formed in hydrogen gas at very low pressure are much sharper than those generated in a high-pressure discharge tube. Explain.

6 Explain clearly why x-ray emission lines in the range of 1 Å are not observed from an x-ray tube using a low-atomic-number metal as the target in the tube.

7 Why do hot solids give off a continuous spectrum, while hot gases give off a line spectrum?

8 A steel company suspects that one of its competitors is adding a fraction of a percent of a rare-earth element to its (the competitor's) product. How can the element quickly be identified and its concentration determined?

9 It is suspected that a certain lot of benzene contains a trace of acetone as impurity. What practical methods exist for testing this suspicion?

10 In the helium atom, the two electrons are in the same shell but avoid each other well enough that their interaction is of only secondary importance. Estimate the ionization energy (in electronvolts) for helium, i.e., the energy required to tear one electron loose. Also, estimate the energy needed to tear the second electron loose. Which of these two values is most reliable? (E)

11 The ionization energies for lithium, sodium, and potassium are 5.4, 5.1, and 4.3 eV, respectively, while those for helium, neon, and argon are 24.6, 21.6, and 15.8 eV, respectively. Explain in terms of atomic structure why these values are to be expected.

12 Estimate how much energy a photon must have if it is to be capable of expelling an electron from the innermost shell of a gold atom. (E)

13 The diameter of a nucleus is about 10^{-15} m. Estimate the least momentum a proton must have if it is to be a part of the nucleus. (E)

Problems

1 Ten thousand tiny projectiles are shot at random places into a 4000-cm^2 window whose pane is partly broken out. Only 500 of the projectiles go through the window. (a) How large is the area of the hole in the pane? (b) The window pane is now completely removed. Suspended by tiny threads, 200 tiny spheres are placed in the window opening. Now 9400 of the 10,000 projectiles go straight through the window. About what is the cross-sectional area of each sphere? (c) To what in Rutherford's experiment do the spheres of part b correspond?

2 Some of the α particles used in the Rutherford experiment had a KE of 4.8 MeV. For these particles $m_0 = 4 \times 1.66 \times 10^{-27}$ kg. Find the speed and m for these particles.

3* Rutherford and his coworkers shot α particles ($q = +2e$) at gold atoms ($Z = 79$). Some of the particles had a KE of 4.8 MeV. (a) What is the PE (in terms of r) of an α particle a distance r from the gold nucleus? (b) How close can Rutherford's α particles come to the center of the gold nucleus? Assume that the gold nucleus remains essentially stationary, and neglect the effect of the (relatively distant) atomic electrons.

4* The density of gold is 19.3 g/cm^3, and its atomic mass is 197 kg/kmol. (a) What is the mass of a gold atom? (b) How many gold atoms are there in a 1-cm^2 area of gold film which is 0.010 cm thick? (c) The diameter of a gold nucleus is about 10^{-14} m. Assuming no overlap, how much area of the 1.0-cm^2 total area do the gold nuclei cover? (d) If he used a film of this thickness, about what fraction of the α particles would Rutherford have observed to be strongly deflected?

5 Compute the wavelength of (a) the fourth line in

the Lyman series and (b) the fifth line of the Balmer series. (Use $R = 1.097 \times 10^7$ m^{-1}.)

6 Find the wavelength of the fifteenth line in the Balmer series and compare it to the wavelength of the sixteenth line. (Use $R = 1.097 \times 10^7$ m^{-1}.)

7 Call $\Delta\lambda$ the wavelength difference between the seventh line of the Balmer series and the series limit. Find the ratio of $\Delta\lambda$ to the series-limit wavelength. Use $R = 1.097 \times 10^7$ m^{-1}.

8 Confirm the series limit given for the Paschen series in Fig. 26.5 by computing it from the Paschen series formula. Use $R = 1.097 \times 10^7$ m^{-1}.

9* Bohr assumed the electron in the hydrogen atom to be orbiting with nonrelativistic speed. What speed does the electron have in the $n = 1$ orbit according to his theory? Compare it to c.

10* In the singly ionized helium atom, a single electron orbits the nucleus which has a charge $+2e$. What is the radius of the first Bohr orbit for this ion?

11* Suppose that the Bohr theory can be applied to the innermost electron in a gold atom ($Z = 79$) by neglecting the presence of all the other electrons. (This is really not too bad an approximation.) (a) Show that the energy needed to remove this electron from the atom is $(13.6)(79)^2$ eV. (b) What is the radius of the first Bohr orbit for this atom?

12* Suppose an electron revolves in a circular path of radius 0.10 nm about the hydrogen nucleus. (a) What speed must the electron have if the Coulomb force is to furnish the centripetal force? (b) What is the frequency of the electron in the orbit? (c) On the basis of classical theory, to what wavelength of radiation should this give rise?

13* (a) What is the formula for the energy levels of doubly ionized lithium, written in the form $E_n = -(\text{const})/n^2$ electronvolts? (b) How much energy (in electronvolts) is needed to remove the last electron from doubly ionized lithium? (c) What wavelength photon would just be capable of knocking this electron free?

14* Repeat Prob. 13 for the case of triply ionized beryllium. The beryllium nucleus has a charge of $+4e$.

15* At extremely high temperatures, collisions strip nearly all the electrons from the atoms. In a high-temperature gas of sodium atoms, many of the atoms are stripped of all but one electron. (a) What is the wavelength of the first line of the Lyman series for such an ion? (b) What is the Lyman series limit? (c) In what portion of the EM spectrum are these wavelengths found?

16 What are the energies (in electronvolts) of the three lowest-energy photons that unexcited hydrogen gas atoms will absorb?

17 Repeat Prob. 16 for absorption by singly ionized helium atoms. You may wish to refer to Illustration 26.2 in answering this.

18* A beam of ultraviolet light with $\lambda = 80$ nm is incident on unexcited hydrogen atoms. When a photon strikes one of the atoms and ejects its electron, what is the kinetic energy of the electron once it is free of the atom? This is called the *atomic photoelectric effect*.

19* A beam of 5.0-nm wavelength x-rays shines on a gas of unexcited hydrogen atoms. It expels electrons (photoelectrons) from the hydrogen atoms. (a) What is the energy of the expelled photoelectrons? (b) What is their speed?

20* The ionization energy of unexcited helium atoms is 24.6 eV. Suppose 30-nm ultraviolet light is incident on such atoms. (a) What is the energy of the fastest electron ejected from the atoms by the U.V. light? (b) What is the speed of the electron?

21 How many electrons can exist in the $n = 3$ shell of a Bohr-type atom?

22 How many electrons can exist in the $n = 4$ shell of a Bohr-type atom?

23 An atomic subshell consists of those electrons in an atom which have the same n and l but different m_l and m_s. How many electrons exist in the $n = 3$, $l = 2$ subshell of gold?

24 How many electrons are needed to fill (a) the $n = 4$ shell and (b) the $n = 3$, $l = 1$ subshell? (See Prob. 23 for definition of subshell.)

25 Modern color TV sets often have electron beams accelerated through more than 20,000 V. What are the shortest-wavelength x-rays generated by a 20,000-V beam as it hits the end of the TV tube? Some early TV sets were not properly shielded and leaked appreciable amounts of x-rays outside the set.

26 To reach tumors deep within a person's body, so-called "hard" x-rays are used. These are generated by use of very high voltages. What is the shortest-wavelength x-ray generated by an x-ray tube operating at 200 kV?

27* An ordinary x-ray tube might easily operate at 30,000 V with a current of 10 mA bombarding the plate. (a) How much heat is produced in the plate per second by this bombardment? (b) If the specific-heat capacity of the metal of the 200-g plate is 0.10 cal/(g)(°C), what temperature rise will occur in 1 min if no heat is lost?

28* Some very-high-energy x-ray tubes have water circulated through the inside of the metal plate in order to cool it. Suppose that a certain tube is operating at 30 mA and 200,000 V. How much would the cooling water be heated if it circulated through the plate at 1 liter/min?

29* A room-temperature gas of hydrogen atoms is bombarded by a beam of electrons which has been accelerated through a potential difference of 12.9 V. What wavelengths of light will the gas emit as a result of the bombardment?

30 The series limit for a certain x-ray series in copper is about 1.3 Å. What is the energy required to expel an electron from the atom when the electron is in the lowest energy level of this series?

31* (a) From the data of Fig. 26.16, determine the energy difference between the $n = 2$ and $n = 3$ levels in molybdenum. (b) If you wished to construct the energy-level diagram for this atom, what further data would be needed?

32* An atom may take 10^{-9} s to emit a photon. Suppose the atom emits a wave pulse during this time with the front end of the pulse emitted 10^{-9} s earlier than the rear end. If the wavelength of the wave is 500 nm, how many wavelengths long is the photon?

33** Two different laser beams of the same type will be coherent if the lasers emit exactly the same wavelength. Even if the wavelengths are slightly different, the two beams will show an interference effect. When joined, they will give a resultant beam which fluctuates in time from brightness to darkness. This is similar to the phenomenon of beats discussed in Chap. 14 for sound waves. If one beam has a wavelength of exactly 600 nm, what must the wavelength of the other beam be to produce maximum brightness once each second? [You may want to use the fact that, for $x \ll 1$, $1/(1 \pm x) \cong 1 \mp x$.]

27
The Atomic Nucleus

As we saw in the last chapter, Rutherford's experiments showed that the atomic nucleus is a very small entity at the center of the atom. Within it is concentrated all the positive charge of the atom as well as most of its mass. In this chapter we examine the details of the nucleus. Furthermore, we become acquainted with some of the properties of nuclei such as radioactivity, fission, and fusion. Various applications of nuclear physics are also given.

27.1 Structure of the Nucleus

In the last chapter we saw how Rutherford was able to infer the structure of the atom from the results of scattering experiments. Since that time, the nuclear model of the atom has been confirmed in numerous ways. *We therefore picture the atom as follows. At its center is a tiny nucleus; it is here that all the positive charge of the atom is located, together with more than 99.9 percent of its mass. The atom itself has a radius of the order of 10^5 times larger than that of the nucleus. In this outer region the electrons of the atom are found.* Their behavior was discussed in the last chapter. We now proceed to a discussion of the atomic nucleus.

The nucleus is composed of neutrons (n) and protons (p). As we learned previously, *a proton is simply the nucleus of the hydrogen atom. It has a charge of $+e = 1.602 \times 10^{-19}$ C and a mass of 1.673×10^{-27} kg. The neutron has no charge, and its mass is 1.675×10^{-27} kg.* We make frequent use

of a more tractable mass unit called the **atomic mass unit** (u), where
1 u = 1.6606 × 10⁻²⁷ kg. The exact definition of the atomic mass unit is
given later. In this unit, the proton's mass is 1.007 u, the mass of the neutron
is 1.009 u, and the electron's mass is 0.00055 u.

Since the charge on a nucleus is Ze for an element of atomic number Z,
we see that *there are Z protons in the nucleus.* These Z protons furnish Z
units of mass (in atomic mass units) to the nucleus. However, the mass M of
a nucleus exceeds Z for all elements except hydrogen, and we attribute this
extra mass to the neutrons in the nucleus. *We refer to the neutrons and
protons in a nucleus as* **nucleons.** *The total number of nucleons within the
nucleus is called the* **atomic mass number** *of the element in question and is
represented by the symbol A.* Since the masses of the proton and neutron are
close to 1 u, the value of A will be close to the mass of the nucleus measured
in atomic mass units. For example, the oxygen nucleus has eight neutrons
and eight protons within it; it has a mass of 15.995 u, and its mass number
is 16.

Mass Number A

The size of the nucleus can be measured by shooting high-energy particles at it. Such measurements show that *the nucleus can be approximated
roughly as a sphere with radius R given by*

Nuclear Radius

$$R = (1.2 \times 10^{-15})(A^{1/3}) \quad \text{m}$$

The fact that $R \sim A^{1/3}$ allows us to infer that the nuclei all have about the
same density ρ. For if we assume the nuclei of the various elements to have
the same density, then

$$A \sim \text{mass} = (\tfrac{4}{3}\pi R^3)\rho$$

and so it follows that $R \sim A^{1/3}$.

Returning now to the simplest of all atoms, hydrogen, we note that its
mass number is unity. Its nucleus is simply a proton. For convenience, we
usually express atomic masses in atomic units and charges in multiples of
the charge quantum, $+e$. Hence the hydrogen nucleus has approximately
unit mass and unit charge.

Following hydrogen in the periodic table,[1] the next heavier element is
helium. Helium has two electrons outside its nucleus and is a chemically
inert gas. Since the nuclear charge must exactly balance the electronic
charge, helium nuclei must carry a charge of $+2e$, and so there must be two
protons within each nucleus. But the mass of the helium nucleus is found to
be 4.0 u. A mass of 2.0 u is contributed by the two protons. The remaining
2.0 u of mass is due to the presence of two neutrons.

As we proceed through the periodic table, we see that all nuclei can be
thought of as composed of protons and neutrons. A shorthand method for
describing *the nucleus of an element X is given by the symbolism*

[1] See Appendixes 3 and 4 for the periodic table and a table of isotopes.

That is, one would symbolize hydrogen as ${}^1_1H$ and helium as ${}^4_2He$. For fluorine, the nucleus of which contains nine protons ($Z = 9$) and has a mass number of 19, the symbol is ${}^{19}_9F$.

For the higher elements in the periodic table, we select uranium as an example. All uranium atoms have a nuclear charge of $92e$, and uranium is therefore listed as element 92 in the periodic table. (Note that the nature of the element is determined by its chemical reactivity. This in turn is determined by the electron structure of its atom. Hence the chemist is concerned with only the charge on the nucleus, because this determines how many electrons the atom will have available for chemical reactions.) As we see in the next section, not all nuclei of the same element have the same mass. However, let us consider uranium 238, the uranium nucleus having a mass of about 238 u. We designate this nucleus by the symbol ${}^{238}_{92}U$. The symbol tells us that this nucleus contains 92 protons. They furnish a charge of $+92e$ and 92 of the total 238 mass units. The remaining $238 - 92 = 146$ units of mass must be furnished by neutrons. We therefore see that the nucleus ${}^{238}_{92}U$ contains 92 protons and 146 neutrons.

27.2 Isotopes

Early investigators thought that nuclear masses were essentially whole-number multiples of the hydrogen mass, but as time passed, more accurate measurements of the atomic masses indicated that this was not true. We should understand that the atomic mass (nearly the same as the nuclear mass, since the electrons have such small mass) was determined first by chemists. For example, if 1 g of hydrogen combined with 35 g of chlorine to form 36 g of hydrogen chloride, HCl, the chemist reasoned that the chlorine atom was 35 times as massive as the hydrogen atom. The atomic masses in the periodic table have been determined in much this way.

However, if we examine the atomic masses of the elements, we find values such as 10.81 u for boron, 24.31 u for magnesium, and 35.45 u for chlorine. Many elements have atomic masses that are far from being integers. This is a very strange fact if we believe the atoms to be composed of neutrons, protons, and electrons. The masses of these particles are 1.009, 1.007, and 0.00055 u, respectively. There is no way that the observed mass of the atoms quoted above can be obtained by adding multiples of the near integer masses of the proton and neutron to the nearly negligible mass of the electron. This state of affairs caused a considerable amount of difficulty in early attempts to interpret nuclear structure. It was not until the invention of the mass spectrograph that the difficulty was resolved.

The mass spectrograph (or spectrometer) is used to measure the masses of atomic-size particles by deflecting their ions in a magnetic field. A typical arrangement is shown in Fig. 27.1, where it is assumed that the entire apparatus is in vacuum. Atoms are ionized in the ion source. When a positive

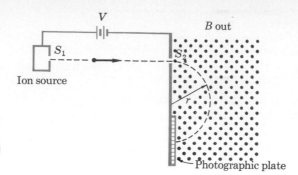

FIGURE 27.1
In the mass spectrograph, positive ions are deflected by a magnetic field.

ion strays out through slit S_1, it is attracted to the negative plate and slit S_2. If it is a singly charged ion, its charge will be $+e$, and upon reaching S_2, its kinetic energy will be Ve, where V is the potential through which the ion falls. We therefore have

$$\tfrac{1}{2}mv^2 = Ve \tag{27.1}$$

When the ion passes through slit 2, it enters a magnetic field perpendicular to the paper. The ion will be deflected in a circular path, as shown. As discussed in Chap. 18, the radius of the path can be found from the fact that the centripetal force is furnished by the magnetic force on the ion,

$$Bev = \frac{mv^2}{r} \tag{27.2}$$

We can now find an expression for m, the ion mass, by eliminating v from these two equations. The result is

$$m = \frac{B^2 e}{2V} r^2 \tag{27.3}$$

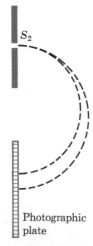

S_2

Photographic plate

FIGURE 27.2
The two isotopes of chlorine describe different paths in the mass spectrograph as shown here (with some exaggeration).

Because B, e, and V are constants, m can be determined from a measurement of r, the radius of the path that the ion follows in the magnetic field. In the apparatus illustrated, r is measured by noticing where the ion beam impacts the photographic plate. More often, an electronic detector scans the plate region to determine the location of the beam.

When the mass spectrograph was put to use, an interesting discovery was made. Many high-purity elements were found to act like several species appearing under the same chemical identity. For example, pure chlorine was found to consist of two types of ions. Two separate beams of ions were found in the mass spectrometer, as illustrated schematically in Fig. 27.2. The masses of the ions in the two beams could be found from the path radii. In addition, the relative abundance of the two types of ions could be found from the intensities of the two beams. The results are summarized in the following table:

	Type 1	Type 2
Mass, u	35.0	37.0
Relative abundance	0.754	0.246

If we multiply the masses of these two species by the relative abundance of each and add, we obtain an average mass of 35.5 u. This is exactly the mass determined by the chemists. We conclude that *the atomic masses listed in the periodic table are the average masses of the individual species as found mixed in nature.*

We see at once that the numerical values in atomic mass units of the individual species determined by the mass spectrometer are essentially integers. This proves to be the case for all the elements examined in this way. Since we know that all types of chlorine have a nuclear charge of $17e$ (because the atomic number of chlorine is 17), the chlorine nucleus must contain 17 protons. It is therefore concluded that the nucleus of species 1 given above contains $35 - 17 = 18$ neutrons, while species 2 contains 20 neutrons. *We call nuclei of the same element which contain different numbers of neutrons* **isotopes** *of the element.* In this terminology, chlorine found in nature contains two isotopes designated $^{35}_{17}\text{Cl}$ and $^{37}_{17}\text{Cl}$. We often refer to isotopes such as these as, for example, chlorine 35 and chlorine 37 or as ^{35}Cl and ^{37}Cl.

Definition

All isotopes of an element behave similarly in chemical reactions. This follows from the fact that all the isotopes have the same nuclear charge and thus the same number of electrons in their atoms. The atomic mass of each isotope is very close to an integer, since its nucleus is made up of neutrons and protons, each of which has essentially unit mass. Most of the elements occurring in nature are mixtures of isotopes, and these mixtures are used in ordinary chemical reactions. The chemically determined atomic masses in the periodic table are nonintegers because they are the average of the isotopic masses found in nature. In Appendix 4 you will find an abbreviated list of the isotopes.

Before leaving this section, let us give the precise definition of the atomic mass unit:

One **atomic mass unit** *(1 u) is exactly one-twelfth the mass of the carbon 12 atom.*

Definition

As we stated previously, $1\text{ u} = 1.6606 \times 10^{-27}$ kg. Carbon 12 was taken as the reference atom for a very practical reason. This isotope of carbon constitutes 98.9 percent of the carbon found in nature. It occurs widely in molecules analyzed by the mass spectrometer and acts as a readily available material for calibrating the spectrometer.

27.3 Mass Defect and Binding Energy

With the advent of extremely sensitive mass spectrographs, the masses of nuclei were determined to high precision. The very exact results thus obtained provided a striking confirmation of Einstein's concept of mass-energy interchange. Indeed, these measurements showed the possibility for the utilization of nuclear energy, an energy source previously not even suspected. We can appreciate the principle involved by examining a specific situation.

To take a simple case, let us analyze data for the mass of the helium nucleus. Its total mass should consist of two neutron masses plus two proton masses:

$$
\begin{aligned}
2n &= 2.01734 \text{ u} \\
2p &= \underline{2.01456} \\
\text{Computed He mass} &= 4.03190 \\
\text{Measured He mass} &= \underline{4.00150} \\
\text{Mass difference} &= 0.03040 \text{ u}
\end{aligned}
$$

Clearly, the computed and measured masses do not agree. The discrepancy, 0.030 u, is far too large to be attributed to experimental error. Where did this extra mass go when the neutrons and protons joined to form the nucleus? This is the question we now investigate.

The helium discrepancy is not the only one. Similar discrepancies occur for other nuclei as well, as illustrated in Fig. 27.3, where the mass change per nucleon $\Delta m/A = (m - m_{np})/A$ is plotted against atomic number. By mass change per nucleon we mean the true nuclear mass m minus the combined mass of the protons and neutrons m_{np}, all divided by the total number of neutrons and protons A. Of course Δm is zero for hydrogen, which lies off the top of the graph in Fig. 27.3. The quantity Δm is called the *mass defect*.

We know of only one way in which mass can be lost in a situation like this. In Chap. 25 we saw that mass can be converted to energy and that the relation between them is

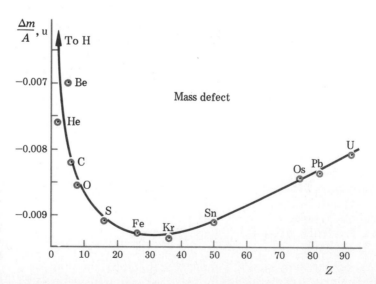

FIGURE 27.3
The mass discrepancy per nucleon for nuclei near the center of the periodic table is larger than that for elements at either end of the table.

$$\text{Energy} = \Delta m\, c^2 \qquad\qquad (27.4)$$

where Δm is the lost mass and c is the speed of light, 3×10^8 m/s. Hence it would appear that when two protons and two neutrons are joined to form a helium nucleus, energy must be given off. The amount of energy given off is related to the mass loss by Eq. (27.4).

As we see from Fig. 27.3, all the nuclei have less mass than the neutrons and protons which were joined to form the nuclei. Mass was lost and energy was given off as the nuclei were formed. This also means that energy is required to tear the nucleons apart from one another since mass would need to be created in the process. *The energy needed to tear a nucleus apart into its constituent isolated nucleons is called the* **binding energy** *of the nucleus.*

Definition

We can compute the binding energy per nucleon directly from the data of Fig. 27.3. For example, the mass loss per nucleon in forming a beryllium (Be) nucleus is 0.007 u. This is equivalent (from energy $= \Delta m\, c^2$) to an energy loss of 6.5 MeV. Hence the binding energy per nucleon of the Be nucleus is 6.5 MeV. That much energy multiplied by the number of nucleons in the Be nucleus must be furnished to tear this nucleus completely apart. In Fig. 27.4 we show a graph of the binding energy per nucleon for typical elements. Notice that *a large binding energy per nucleon implies a very stable nucleus.* We return to this topic when we discuss nuclear power.

Binding Energy Related to Stability

Illustration 27.1 How much energy is given off when one helium nucleus is formed from neutrons and protons?

Reasoning We saw in the last section that the mass lost in this process is 0.030 u. Making use of the conversion factor 1 u = 1.66×10^{-27} kg, we find that the mass lost in forming one helium nucleus is

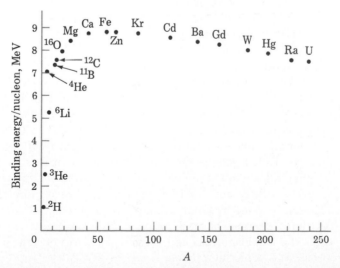

FIGURE 27.4
Binding energy per nucleon for representative elements. Note the very high stability of ^{4}He.

$$\Delta m = (0.030 \text{ u})(1.66 \times 10^{-27} \text{ kg/u}) = 5.0 \times 10^{-29} \text{ kg}$$

Because energy $= \Delta m \, c^2$, we have

$$\text{Energy} = (5.0 \times 10^{-29} \text{ kg})(9 \times 10^{16} \text{ m}^2/\text{s}^2)$$
$$= 4.47 \times 10^{-12} \text{ J} = 2.79 \times 10^7 \text{ eV} = 27.9 \text{ MeV}$$

It is convenient to notice that since a mass loss of 0.030 u gives rise to 27.9 MeV of energy,

$$1 \text{ u} = (1 \text{ u})\left(\frac{27.9 \text{ MeV}}{0.030 \text{ u}}\right) = 930 \text{ MeV}$$

We also notice that since 4 kg of helium contains 6×10^{26} nuclei, the energy liberated in the formation of 4 kg of helium is

$$(6 \times 10^{26})(4.5 \times 10^{-12} \text{ J}) = 2.7 \times 10^{15} \text{ J}$$

This is a tremendous amount of energy. For comparison, the work done in lifting a 100-kg person a distance of 100 m is only about 10^4 J. The amount of energy liberated in forming 4 kg of helium nuclei from protons and neutrons is large enough to lift 200 million 100-kg people through a distance of over 100 km.

27.4 Radioactivity

Certain naturally occurring elements are not stable, but slowly decompose by throwing away a portion of each nucleus. We say that they are **radioactive.** The first discovery of a radioactive element was made in 1896 by Henri Becquerel when he found that uranium atoms ($Z = 92$) give off radiation which fogs photographic film and plates. Two years later, Marie and Pierre Curie succeeded in chemically isolating two more radioactive and previously unknown elements, polonium ($Z = 84$) and radium ($Z = 88$). It has since been found that all the elements having Z greater than 83 are radioactive to some extent.

When a radioactive element decomposes (we say it *decays*), the original nucleus emits a particle of some sort. We will find later that the high-Z radioactive elements that occur naturally on earth emit either a beta particle (which is an ordinary electron) or an alpha particle (a helium 4 nucleus). In addition, the nucleus that remains after the decay often emits high-energy photons (called *gamma rays*) and may decay further.

In addition to the radioactive elements that we find occurring naturally on earth, nearly any stable element can be made into a radioactive one. This can be done by bombarding the nonradioactive element with particles that have energies of many million electronvolts. The bombardment disrupts the stable nucleus and transforms it to one that is radioactive. We discuss this topic more later.

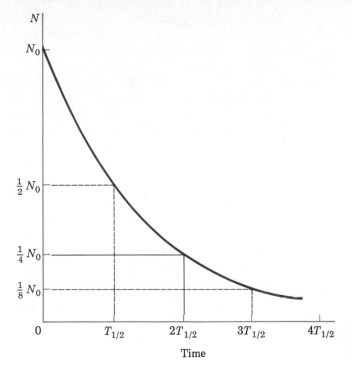

FIGURE 27.5
A radioactive element decays exponentially.

No matter what the source of a radioactive element may be, it is found to decay as a function of time in the way shown in Fig. 27.5. It is assumed for the figure that you start with N_0 nuclei of a radioactive element and monitor how many nuclei remain undecayed as time t elapses. We designate the number of undecayed nuclei by N. As you see from the figure, N starts at N_0 when $t = 0$ and decreases steadily afterward. This type of curve is called an *exponential decay* curve, and we give its equation in the next section.

Each radioactive element takes its own characteristic time to decay even though the general appearance of the decay curve is the same for all elements. We describe the rate at which a substance decays by stating the time that it takes for half the nuclei to decay. In Fig. 27.5 we see that N has decreased to $\frac{1}{2}N_0$ after a time that is designated as the half-life:

*The **half-life** $(T_{1/2})$ of a substance is the time taken for half of the material   **Definition**
to decay.*

Half-lives of radioactive substances differ widely. The half-life of uranium 238 is 4.5 billion years, while the half-life of radium is 1620 yr. Radon gas, the element to which radium decays, has a half-life of only 3.8 days. Many artificially produced radioactive substances have half-lives of only a fraction of a second. Even so, all these elements decay in conformity with the exponential decay law.

An interesting consequence of this mode of decay is the following.

Suppose you start with N_0 nuclei that have a half-life $T_{1/2}$. After a time $T_{1/2}$, only $\frac{1}{2}N_0$ undecayed nuclei remain. If you wait another time $T_{1/2}$, then half of these nuclei will have decayed, thereby leaving $\frac{1}{2}\cdot\frac{1}{2}N_0$ undecayed. After still another time $T_{1/2}$, half of these remaining nuclei will have decayed, leaving $\frac{1}{2}\cdot\frac{1}{2}\cdot\frac{1}{2}N_0$ remaining. We conclude that after n half-lives have transpired, the number of undecayed nuclei remaining will be $(\frac{1}{2})^n N_0$. What fraction of N_0 will be left after a time of $4T_{1/2}$?[2] By asking yourself a similar question for various values of n, you should be able to construct the decay curve of Fig. 27.5.

We can obtain a mathematical relation for the number ΔN of nuclei that decay in time Δt if we recognize the following experimental fact: the number of nuclei that decay in time Δt is proportional to the number N of nuclei available for decay and the time Δt allowed for decay. We therefore have

$$\Delta N \sim N \Delta t$$

This can be written as an equation if we introduce a proportionality constant which is given the symbol λ and is called the **decay constant.** The proportion becomes

$$\Delta N = \lambda N \Delta t \qquad (27.5)$$

We call the quantity $\Delta N/\Delta t$ the **activity** of the sample. It is the number of decays that occur in unit time, and we discuss it further in Sec. 27.13.

The decay constant λ has a simple meaning that we can understand if we rearrange the equation to give

$$\lambda = \frac{\Delta N/N}{\Delta t}$$

Because $\Delta N/N$ is the fraction of the nuclei that decay in time Δt, we see that the decay constant λ is the fraction of nuclei that decay in unit time. The equation also shows that the unit of λ is one divided by a time unit.

We now have two ways to characterize the decay of a substance. Its rate of decay can be described by either λ or $T_{1/2}$. Of course, these two quantities must be related. By using calculus, the relation can be shown to be

$$\text{Half-life} = \frac{0.693}{\text{decay constant}} \qquad T_{1/2} = \frac{0.693}{\lambda} \qquad (27.6)$$

Illustration 27.2 Iodine 131 is a radioactive isotope made in nuclear reactors for use in medicine. It becomes concentrated in the thyroid gland when taken into the body. There, it acts as a radiation source in the treat-

---

[2] Answer: $N_0/16$.

ment of hyperthyroidism. Its half-life is 8 days. Suppose a hospital orders 20 mg of ^{131}I and stores it for 48 days. How much of the original ^{131}I will still be present?

Reasoning Each 8 days, the iodine decays by one-half. We can therefore make the following table:

Time, days	Iodine present, mg	Time, days	Iodine present, mg
0	20	32	1.25
8	10	40	0.625
16	5	48	0.312
24	2.5		

After 48 days only 0.312 mg of the original 20 mg will remain.

Illustration 27.3 We see later when we discuss the units used in radioactivity that the amount of material in which 3.7×10^{10} disintegrations occur in 1 s is said to be a **curie** (Ci) of the material. What fraction of a curie is present in 1 g of radium? The half-life of radium is 1620 yr, or 5.1×10^{10} s.

Reasoning The number of atoms decaying in time Δt is

$$\Delta N = \lambda N \Delta t$$

Since there are Avogadro's number of atoms in one atomic mass of radium, 226 kg, the number of atoms N in 1 kg is

$$N = \frac{6 \times 10^{26} \text{ atoms}}{226 \text{ kg}} = 2.66 \times 10^{24} \text{ atoms/kg} = 2.66 \times 10^{21} \text{ atoms/g}$$

We still need to find λ. This can be done using Eq. (27.6). We have

$$\lambda(5.1 \times 10^{10} \text{ s}) = 0.693 \quad \text{or} \quad \lambda = 1.36 \times 10^{-11} \text{ s}^{-1}$$

Placing this in the first expression for ΔN yields

$$\Delta N = (1.36 \times 10^{-11} \text{ s}^{-1})(2.66 \times 10^{21} \text{ atoms/g}) \Delta t$$
$$= 3.6 \times 10^{10} \Delta t \quad \text{atoms/g}$$

when Δt is expressed in seconds. The number of disintegrations per second in 1 g of radium is therefore 3.6×10^{10}. Since 1 Ci of material is defined to have 3.7×10^{10} disintegrations per second, we see that 1 g of radium is equivalent to 3.6/3.7, or 0.97, Ci. Actually, the curie was originally thought to be exactly the number of disintegrations per second from 1 g of radium. However, more precise measurements showed that the curie as defined was slightly larger than this experimental value.

27.5 Exponential Decay

The decay curve shown in Fig. 27.5 is well known in science; it is called an *exponential curve.* As we saw in the last section, it has the property that the curve height decreases by one-half for each half-life along the horizontal axis. The curve can be stated in mathematical terms as

Decay Equation

$$N = N_0 e^{-\lambda t} \tag{27.7}$$

where λ is the decay constant. It is related to the half-life through $\lambda T_{1/2} = 0.693$, as we stated in Eq. (27.6). The function $e^{-\lambda t}$ is called an *exponential function,* and e is the base for natural logarithms, 2.7183.

Use of Eq. (27.7) is facilitated by the fact that many hand calculators have a key for this function. If your calculator does, you may wish to test your use of it by checking the following typical values:

x	e^x	e^{-x}
0	1.000	1.0000
2	7.389	0.1353
3	20.085	0.0498
4	54.598	0.0183
10	22,026	4.5×10^{-5}

In the event you do not have a calculator with this capability, most handbooks have a table of exponential functions.

Often it is necessary to solve Eq. (27.7) for λ or t when the ratio N/N_0 is known. This can be done in the following way. If we take the logarithm (to base 10) of Eq. (27.7) and recall the following facts

$$\log ab = \log a + \log b \qquad \text{and} \qquad \log a^n = n \log a$$

we find that

$$\log N = \log N_0 - \lambda t \log e$$

But e is 2.7183, and so this becomes

$$\log N = \log N_0 - 0.434\lambda t$$

which gives

$$\lambda t = 2.303(\log N_0 - \log N) = 2.303 \log(N_0/N)$$

Those who know about natural logarithms and possess an appropriate calculator or tables can save work by taking natural logarithms at the outset. Their result would be

$$\lambda t = \ln \frac{N_0}{N}$$

Let us now see how we can make use of the equation for the exponential decay law.

Illustration 27.4 Uranium 238 has a half-life of 4.5×10^9 yr. It is believed that the earth solidified about 4.0×10^9 yr ago. What fraction of the uranium 238 then found on the earth remains undecayed today?

Reasoning We use the decay law, Eq. (27.7), with

$$\lambda = \frac{0.693}{T_{1/2}} = \frac{0.693}{4.5 \times 10^9 \text{ yr}} = 1.54 \times 10^{-10} \text{ yr}^{-1}$$

This gives

$$\frac{N}{N_0} = e^{-\lambda t} = e^{-(1.54 \times 10^{-10} \text{ yr}^{-1})(4 \times 10^9 \text{ yr})}$$
$$= e^{-0.616} = 0.54$$

A fraction 0.54 of the original uranium 238 is still in existence.

Illustration 27.5 Ten percent of a certain radioactive substance decays in 12.0 hr. What are the decay constant and the half-life for the substance?

Reasoning We are told that in Eq. (27.7) the value of N/N_0 is 0.90 at a time $t = 12$ hr. Substitution gives

$$0.90 = e^{-\lambda(12 \text{ hr})}$$

To eliminate the necessity of taking logarithms of numbers less than unity, let us invert both sides of this equation. Then, because $1/e^{-x} = e^x$, we have

$$1.111 = e^{\lambda(12 \text{ hr})}$$

Taking logs of both sides of this equation gives

$$\log 1.111 = \lambda(12 \text{ hr}) \log e$$

or

$$0.0458 = \lambda(12 \text{ hr})(0.434)$$

Solving for λ gives

$$\lambda = 0.0088 \text{ hr}^{-1}$$

The half-life is found in the following way:

$$T_{1/2} = \frac{0.693}{\lambda} = \frac{0.693}{0.0088 \text{ hr}^{-1}} = 79 \text{ hr}$$

27.6 Decay Products and Radioactive Series

When a nucleus decays, it does so by emitting a particle or radiation. We discuss what happens to the emitted entity in Sec. 27.12. In this section we are concerned with the nucleus that remains after the emission process.

Radioactive nuclei found in nature emit α particles (alpha particles, helium 4 nuclei), β particles (beta particles,[3] ordinary electrons), and γ rays (gamma rays, very-short-wavelength photons). Let us begin our discussion with γ ray emission.

Because a γ ray is simply a photon having no rest mass and no charge, the nucleus that emits it has merely adjusted its internal structure in such a way as to lose some energy. This is similar to the emission of light by an atom. In an atom, the energy is lost when the electron falls to a lower energy state. When a nucleus throws out a γ ray, the nucleus also falls to a lower energy state. However, we do not yet know enough about the structure of the nucleus to describe this type of transition nearly as precisely as in the atomic case. In any event, if nucleus $_Z^A X$ loses a γ ray, the reaction is usually written as

$$_Z^A X \rightarrow \gamma + {}_Z^A X$$

The final charge Z and mass number A of the nucleus are the same as before the γ ray was emitted.

When a nucleus emits an α particle, on the other hand, the mass number of the nucleus decreases by 4 and its charge decreases by $2e$. This is the result of the fact that the emitted α particle is actually a helium-atom nucleus, $_2^4 \text{He}$. For this type of emission we would write

$$_Z^A X \rightarrow \alpha + {}_{Z-2}^{A-4} Y$$

To take an actual example, radium emits an α particle, and so

$$_{88}^{226} \text{Ra} \rightarrow \alpha + {}_{86}^{222} \text{Rn}$$

where the product nucleus is the inert gas radon.

Sometimes it is convenient to write mass and charge numbers for the particles:

---

[3] We see later that another type of beta particle exists. It is identical in most ways to an electron, but its charge is $+e$, not $-e$. It is called a *positron*. We distinguish the two by β^- and β^+.

$$^{226}_{88}\text{Ra} \rightarrow {}^{4}_{2}\alpha + {}^{222}_{86}\text{Rn}$$

Notice, in such an equation, that the sum of the mass numbers must be the same on each side. The same is true for the charges.

Negative beta particles are simply electrons emitted by the nucleus. This poses a conceptual difficulty since no electrons are in the nucleus. We can imagine, however, that the neutron is a proton plus an electron. (This is purely an imaginary device which has utility but no basic validity.) When the β particle is emitted, the original neutron in the nucleus is replaced by a proton. Therefore the emission of a β particle increases the net charge of the nucleus by 1, but changes the mass hardly at all (since the electron mass is so small). In a typical case

$$^{A}_{Z}\text{X} \rightarrow {}^{0}_{-1}\beta + {}^{A}_{Z+1}\text{Y}$$

To illustrate this type of reaction, we discuss one of the less abundant isotopes of lead, $^{214}_{82}\text{Pb}$. It has a half-life of 26.8 min and decays by β emission. The decay reaction is given on the next page.

Atomic mass number

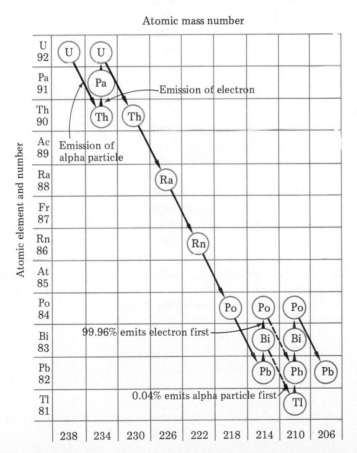

FIGURE 27.6
A typical radioactive series.
It is called the uranium
series since the parent
nucleus is uranium.

$$^{214}_{82}\text{Pb} \rightarrow {}^{0}_{-1}\beta + {}^{214}_{83}\text{Bi}$$

The bismuth isotope formed in this reaction also decays by β emission. This reaction, which is typical of numerous others, is only a part of a whole series of decay reactions.

The uranium series is a dramatic example of the type of decay in which one nucleus decays to another, which decays to still another, and so on. Uranium 238 is slightly radioactive, having a half-life of 4.5×10^9 yr, and decays according to the scheme shown in Fig. 27.6. You should follow this scheme step by step so that you understand its meaning. (Notice that emission of a beta particle causes a vertically upward change of one unit; alpha-particle emission causes a mass decrease of four units and a downward change of two units.) In any event, the final nucleus of this series is lead. Two other series such as this exist in nature. They all start with a long-half-life nucleus. Perhaps others once existed and have long since decayed to imperceptible traces. In any event, the last member of the series is a stable nucleus. For each series the terminal nucleus is lead, but it is a different isotope in each case.

27.7 Nuclear Reactions and Transmutations

When we wrote the radioactive decay equation

$$^{226}_{88}\text{Ra} \rightarrow \alpha + {}^{222}_{86}\text{Rn}$$

in the previous section, we were writing a **nuclear reaction.** Several factors determine whether a postulated nuclear reaction will actually occur. This particular reaction does occur, of course. We know that the total energy, rest energy included, of the original reactants (in this case, just $^{226}_{88}\text{Ra}$) is equal to the final energy of the products. For this to be true, the rest mass of the original radium nucleus must be greater than or equal to the rest masses of the two products, the radon nucleus and the α particle. If this were not true, we would be creating mass in the reaction. Since we are assuming that no external source of energy is acting on the system, we would be creating mass from nothing and this is obviously impossible.

The fact that the total energy before reaction (including the equivalent energy of the rest masses) must equal the total energy after reaction is a useful tool in the study of nuclear reactions. For example, when Rutherford performed one of the very first induced nuclear reactions in 1918, he shot α particles at nitrogen nuclei and observed the reaction

$$^{14}_{7}\text{N} + {}^{4}_{2}\alpha \rightarrow {}^{17}_{8}\text{O} + {}^{1}_{1}\text{H}$$

In other words, the α particle entered the ^{14}N nucleus, which then disintegrated by ejecting a proton. The original nitrogen nucleus was *transmuted* into oxygen. We now ask whether even very slow α particles could cause this reaction.

To learn more about this reaction, we notice that the masses of the reactants are

$$\text{Mass of } {}^{14}\text{N} = 14.0031 \text{ u}$$
$$\text{Mass of } \alpha = \underline{4.0026}$$
$$\text{Total mass before reaction} = 18.0057 \text{ u}$$

In the same way, we examine the masses after reaction,

$$\text{Mass of } {}^{17}\text{O} = 16.9991 \text{ u}$$
$$\text{Mass of } {}^{1}\text{H} = \underline{1.0078}$$
$$\text{Total mass after reaction} = 18.0069 \text{ u}$$

We find that the products have more mass than the original reactants, the difference being 0.0012 u. This mass could be created only if additional energy had been added to the reaction. Since 1.0 u is equivalent to 931 MeV, as shown in Illustration 27.1, we see that the increase in mass in this reaction required an external energy of $\left(\frac{931}{1}\right)(0.0012) = 1.1$ MeV. The incident α particle must have had at least this amount of KE in order to make the reaction occur. Actually, since momentum must also be conserved in such a reaction, the end products will not be standing still. As a result, the particle must have more than 1.1 MeV of KE if the reaction is to be feasible.

Computations like this tell us a great deal about the feasibility of proposed nuclear reactions. Of course, a large variety of reactions are possible. In fact, the field of nuclear reactions in physics is as involved as the subject of organic reactions in chemistry.

27.8 Nuclear Force and Nuclear Stability

Thus far in our study of physics we learned about two fundamental forces. One of these, *the gravitational force,* is of great importance when massive objects are concerned. This force *is always attractive, but it is quite weak.* Two objects on a table attract each other, but the force is so small that it can be measured only with very delicate equipment.

The second fundamental force we encountered is the electric force. It can be either attractive or repulsive depending on the charges involved. *This force is quite large* and is easily demonstrated with electrostatic charges. For example, two charged balls exert easily measured forces on each other. At one time it was thought that magnetic forces were different from electric forces. However, by use of the theory of relativity, these two forces can be shown to be basically the same. It is the electric force which holds atoms, as well as liquids and solids, together. Most of the phenomena we encounter in everyday life are intimately related to this force.

There is yet another force which is probably not familiar to you. *The third force is the one which holds together the nucleons in the nucleus. Unlike the other two forces, it is not an inverse-square-law force.* You will recall that

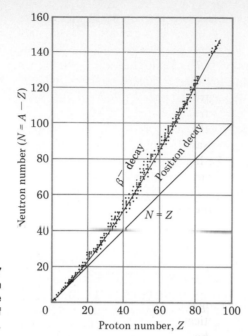

FIGURE 27.7
The nuclei shown as dots in
this chart are stable or have
half-lives in excess of
1000 yr.

both the gravitational and electrical forces vary as $1/r^2$. They are long-range forces in the sense that they are finite and nonzero even at very large values of r.

The nuclear force, on the other hand, *is a very-short-range force. Two nucleons exert negligible force on each other until they are brought closer than about a nuclear diameter (about 5×10^{-15} m) from each other. They attract each other strongly as they are brought closer than this. The nuclear attraction force is so large at small separation that it completely overpowers electrical repulsive forces.* Two protons, although repelling each other with their like charges, are held together strongly by the nuclear force. To a first approximation at least, the nuclear attractive force is the same between proton and proton, neutron and neutron, and proton and neutron.

In spite of its strength, the nuclear force is unable to hold together many protons without a sufficient number of neutrons to "dilute" the positive charge. The electrostatic repulsion force is long-range, and each proton "feels" the repulsion of every other proton in the nucleus. But the short-range nuclear attractive force is felt by only nearest-neighbor particles in the nucleus. For this reason, among others, only certain combinations of protons and neutrons form stable nuclei. Figure 27.7 charts the known stable nuclei.

Notice the axes in Fig. 27.7. If the nuclei had equal numbers of protons and neutrons, they would lie on the $N = Z$ line. As we see, however, the larger nuclei require more neutrons to dilute their charge in order to remain stable. Also shown in the figure are the modes of decay of unstable nuclei which deviate from those indicated. For example, unstable nuclei that lie above the region of stable nuclei (i.e., in the region labeled "β^- decay") will decay in such a way as to reach the stable region. They can do this by ejecting a beta particle, because this will increase Z for the nucleus and thereby shift it toward the region of stability. Transmuted unstable nuclei that reside in the region labeled "positron (β^+) decay" can reach the stable region by ejecting positive particles. They thereby decrease Z for the nucleus. The positive particle ejected is a positron, a particle that is identical to an electron except for its charge; the positron's charge is $+e$.

Another parameter that provides us with a criterion for nuclear stability is the binding energy per nucleon. As we saw in Fig. 27.4, the nuclei near the center of the list of elements have the highest binding energy per nucleon. These nuclei are intrinsically harder to tear apart than those near the two ends of the list of elements. As we see in the following sections, this has extreme practical implications and consequences.

27.9 Nuclear Fission

Nuclear power sources are one of the leading options available in our search for energy. Let us now investigate this practical application of nuclear physics. If we refer to Fig. 27.4, we notice that the binding energy per nucleon is greatest for the intermediate-size nuclei. Nucleons in the smallest nuclei and in the very largest nuclei are much less tightly bound than those in nuclei of moderate size. This is a reflection of the fact, shown in Fig. 27.3, that nucleons lose the most mass when they are packed into nuclei of intermediate size. As we shall see, this forms the basis for the production of nuclear energy.

Suppose it were possible to split a high-Z nucleus such as uranium in half. Two equal-mass particles, each with charge $\frac{1}{2}Z$, would be formed. But because each nucleon in the parent nucleus has more mass than a nucleon would normally have in a nucleus of atomic number $\frac{1}{2}Z$, each half must emit the energy equivalent of this excess mass to become stable. As a result, when the uranium nucleus is split in half, a large amount of energy is given off as the two halves settle into a stable form of nucleus. The energy emitted per nucleon is equal to the energy equivalent of the difference in mass defect between the original large and final intermediate-size nuclei.

It is therefore apparent that for a very-high-Z nucleus to divide in half, a large amount of energy must be liberated as the system returns to stability. This liberated energy could then be used as an energy source. Unfortunately, however, *this splitting of high-Z nuclei into intermediate-sized nuclei, a process we call* **fission,** *seldom occurs spontaneously.* Even though the products of the fission reaction have less energy than the original material, the reaction requires external influences if it is to proceed.

Fission

We have a similar situation in chemistry. Heat is given off when wood is burned in oxygen. Obviously, then, the reactants have more energy than the products, and hence the reaction should be possible. However, wood does not combine spontaneously with oxygen at room temperature—at least not to any great extent. The wood-oxygen chemical reaction must first be started by some external means, a hot flame from a match, for example. Once it has started, the reaction produces enough heat to keep itself going provided that the geometrical arrangement of the piece of wood is such that the heat generated does not escape too easily.

As it turns out, the nuclei of most of the heavy elements cannot be easily split. Striking them with very-high-energy particles sometimes causes splitting. Even though more energy is given off when the fission reaction occurs than was needed to start it, this energy is not easily utilized to keep the reaction going. *Hence, even though all heavy nuclei are potential sources of energy, it is impractical to liberate the energy given off by the fission reaction in nearly all cases.*

There are, however, a few instances in which the fission reaction proves to be of practical importance. The first was discovered by Hahn and Strassmann in 1939. With the aid of others, *they found that an isotope of uranium, ^{235}U, would capture (or attach itself to) a slowly moving neutron and would then undergo spontaneous fission.*[4] *Here was a fission reaction which did not require high-energy bombardment. Moreover, when each ^{235}U nucleus splits, the reaction products include an average of about three neutrons,* and of course a large amount of energy is given off. Each ^{235}U nucleus gives off about 200 MeV of energy when it splits.

We see, then, that *we can visualize a* **chain reaction** *in* ^{235}U *initiated by a single slow neutron.* This is shown diagrammatically in Fig. 27.8. Only the first two steps in the reaction are shown, but if all the neutrons are successfully captured by ^{235}U nuclei, the original neutron has already generated 3^2 neutrons. In an ideal situation, after n steps in the chain reaction have occurred, 3^n neutrons will be available. If each step of the reaction takes 0.01 s, at the end of 1 s the total number of neutrons would be 3^{100}, or about 10^{48}. Since 235 g of ^{235}U contains only 6×10^{23} atoms, you can see that such a reaction could occur with explosive violence.

If the reaction is to proceed smoothly and be self-sustaining, a **critical size** and **mass** of ^{235}U are needed, as we can see by reference to Fig. 27.9. When fission occurs in the thin film shown in part *a*, the neutrons produced

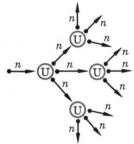

FIGURE 27.8
A chain reaction can be initiated by a single neutron.

---

[4] This isotope of uranium constitutes only 0.7 percent of the mixture of uranium isotopes occurring in nature.

FIGURE 27.9
The efficiency of neutron utilization depends on the size and shape of the piece of ^{235}U.

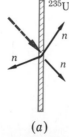

(a)

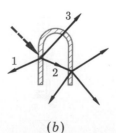

(b)

(c)

easily escape into the air and so the reaction does not build up. In part *b*, the situation is somewhat more favorable, but here, too, the reaction stops after two steps because of the loss of neutrons to the surroundings. In part *c* we have a sphere of ^{235}U. If the sphere is large enough, most of the neutrons produced will be captured before they can get away. *The critical mass is that amount of ^{235}U in which, on the average, exactly one neutron from each reaction causes a further reaction. In this case the reaction will proceed smoothly at its initial slow rate.*

If a much larger than critical mass is involved, the reaction will build up at a fast rate and an explosion will occur. This is desirable, of course, in making a nuclear weapon. However, in the nuclear reactor, we wish the reaction to proceed smoothly so that a steady but nonexplosive source of energy results. In practice, the number of reacting neutrons in a reactor is controlled by the use of neutron-absorbing rods. For example, cadmium rods readily absorb neutrons, thereby removing them from the reaction. Hence, if such rods are put into the reactor, the nuclear reaction slows down. The reaction rate is readily adjusted by positioning rods of this sort in the reactor.

27.10 Nuclear Reactors

The reactor in a nuclear power station serves the same purpose as the furnace in a steam generator. It acts as an intense source of heat, and that heat is used to generate steam. The steam in turn is used to drive the turbines of the electric-generator system. A schematic diagram of a typical reactor is shown in Fig. 27.10.

The heart of the reactor consists of the fissionable material, the fuel, sealed in cylindrical tubes. Originally, uranium 235 was the principal reactor fuel. Now, however, other fissionable materials are also in use as fuel rods. These rods are immersed in a material such as water, carbon, a hydrocarbon, or some similar low-atomic-mass material. This material, the *moderator,* slows the fission-produced neutrons and reflects them into the fissionable material. (Heavy water, water made from the isotope ^{2}H rather than

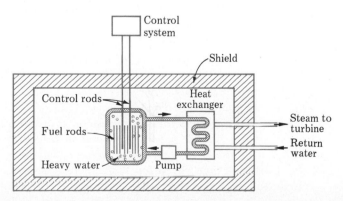

FIGURE 27.10
Schematic diagram of a reactor.

¹H, is sometimes used because it is less apt to remove neutrons from the reaction.) In the design shown, the moderator also acts as the heat-exchange fluid to carry heat away from the fuel rods.

When a nucleus undergoes fission within the fuel rod, highly unstable intermediate Z nuclei are formed. These undergo extensive radioactive decay and eject high-energy particles in the process. As these particles are slowed, their energy is changed to heat, thereby heating the reactor system. This heat is then carried away to a heat exchanger by the heavy water.

In the heat exchanger, the heat is transferred to ordinary water in a steam-boiler system. Steam is generated and then used to power electric turbines. As we see, the steam does not come in direct contact with the reactor core. For this reason, its level of radioactivity is low. But the fluid which circulates through the core is bombarded by radiation from the fission products. Like all other portions of the core material, it is often highly radioactive.

When the material in the fuel rods has been used for many months, its original fissionable material is much depleted. The rods are then removed and replaced by new ones. Unfortunately, we still have no universally accepted disposal method for the waste material in the old rods. This material consists of highly radioactive, fairly long-lived fission products. It takes centuries for the radioactivity to decay to harmless levels. Finding an accepted method for disposal of this waste is one of the major problems associated with nuclear reactors.

However, reactors also can provide us with radioactive materials for medical, industrial, and other uses. Many of the radiation sources presently used by hospitals, industry, and research laboratories are made by placing suitable materials within the core of the reactor. In addition, research reactors exist in many parts of the world. The intense radiation produced in their cores can be "piped" outside the reactor to act as powerful beams of radiation. As we see, the fission process has vast potential for good as well as hazards for human beings.

27.11 Fusion Reaction

The fission reaction makes use of the fact that high-Z nuclei possess more mass per nucleon than intermediate-Z nuclei. Let us now consider another energy-releasing process, one that involves nuclei with small atomic numbers. As we saw in our discussion of the mass defect, *energy is released as protons and neutrons are joined (fused) to form helium nuclei. A similar process is possible when any light nuclei are fused to form nuclei with intermediate Z.* As we pointed out in our discussions of Figs. 27.3 and 27.4, intermediate-size nuclei have less mass per nucleon than do nuclei with small Z. Therefore, *if two small-Z nuclei are fused in a reaction called the* **fusion reaction,** *mass must be destroyed and its equivalent energy must be released.* We see from this that *the fusion reaction can serve as an energy source.*

Unfortunately, nuclear fusion is more difficult to induce than fission.

The electrostatic charge on the nucleus aids the fission process because like charges repel. But when we try to fuse two protons, for example, *the Coulomb repulsion of the two nuclei impedes the fusion process. Only with great difficulty can we bring two light nuclei close enough together for the nuclear attraction force to hold them together.* Even though the overall fusion reaction gives off more energy than was put in, tremendous energies are needed to get the reaction started.

As we see in the next chapter, the sun is heated by a fusion reaction. The extremely high temperatures in its core cause protons to collide with such high energy that they get close enough to fuse. Even though the protons and other particles have enough thermal energy to escape any ordinary container, the huge gravitational force exerted on them by the sun's mass confines them to the sun. In effect, the sun's gravitational force furnishes a container in which the fusion reaction proceeds.

At present, the only practical use of the fusion reaction on earth is in the fusion-type nuclear bomb, the hydrogen bomb, for example. In such a bomb, the fusion reaction is ignited by a fission-type bomb. The temperatures generated by the fission bomb are high enough to give the low-Z nuclei enough energy to fuse. Of course, once the bomb ignites, it explodes; its energy cannot be utilized easily.

To build a useful fusion energy source, we must learn how to produce extremely high reactant temperatures and we must find some way to hold the reactants together long enough to utilize the energy they give off. Various schemes involving huge electric discharges and superpowerful lasers are under development as heat sources. Because no material is capable of withstanding the necessary reaction temperatures, confinement of the reactants presents a formidable problem. The reactants compose an extremely hot gas of charged particles, a gas that is called a **plasma.** At present, a major effort is being made to confine the plasma by magnetic fields. Just as a charged beam can be confined to a circular path by a magnetic field, more complex systems of moving charges can also be held in place by magnetic fields. But the high energies involved and the erratic motions of the particles make this a task of enormous difficulty, and we are still uncertain whether success will be achieved.

If our quest proves successful, the fusion reaction will replace the fission reaction in producing nuclear power. In all likelihood, fusion power would then become an energy source that would be available to us for at least millions of years to come. However, present projections are that this power source will not be available before the year 2000 at the earliest. In fact, we have no certainty that it will ever be commercially feasible. Even so, much of the energy we use on the earth comes from a fusion power source, the sun.

27.12 Radiation: Effects and Detection

As we use nuclear power and other sources of radiation, the effects of radiation on the human body and on materials we use become important. A good deal can be learned about high-energy radiations by studying the more or less typical properties of α particles, β particles, neutrons, and γ rays. For example, the interaction of protons with atoms is closely akin to the behavior of α particles, and similarities exist within the other groups mentioned.

An α particle is relatively massive (4 u) and carries a double charge ($+2e$). When such a particle is shot at atoms, we would expect it to have a fair chance of collision. In fact, it is found that α particles ionize air very rapidly. As they travel through the air, they occasionally strike an atom and tear an electron loose from it. (The word *strike* is used rather inexactly here. Even if the positive α particle passes close to an atom, the electrical attraction between it and the negative electron can cause ionization.)

By using methods discussed later in this section, the path of charged particles through air can actually be seen. In fact, even the individual ions they produce can be counted. It is found that even quite energetic α particles cannot travel far through air before being stopped. For example, the range of the 7.7-MeV α particles emitted from bismuth 214 is only about 7 cm in air. The range is much smaller in denser materials. In aluminum the same α particle would penetrate only about 0.004 cm. As we see, α particles are quite easily stopped.

We should point out that an α particle does not stop after one collision with an atom. Measurements show that about 35 eV of energy is lost for each atom ionized in air. Hence a 7.7-MeV particle would create about 200,000 ions before coming to rest. The number of ions created by the particle, or alternatively its range, can be used as a measure of the energy of the particle.

A proton has properties similar to the α particle. Since it is only one-fourth as large in mass and one-half as large in charge, the proton would be expected not to ionize air as much as an α particle would. This turns out to be true. A proton will travel about 5 to 10 times as far through matter as an α particle of the same energy. From this it is apparent that protons are less than one-fifth as effective as α particles in ionizing atoms.

The β particle is considerably different from the two particles thus far discussed. It has only about $\frac{1}{1830}$ as much mass as the proton. A bombarding β particle is capable of tearing an atomic electron loose, but in this case it is the result of a repulsive force. In spite of the difference in mechanism, there is really not too much difference between the overall ionization effect of a positive and a negative bombarding particle provided that the bombarding particles have the same mass.

A great difference between the action of β and α particles is apparent, however, because of their mass differences. An α particle striking an electron is much like a 10,000-kg truck striking a 2-kg toy truck. The α particle continues on almost as though it had hit nothing. On the other hand, an electron or a β particle colliding with another electron in an atom is like two equal-sized objects colliding. The β particle is deflected considerably when it undergoes a near-head-on collision. Depending on how the particles collide, a large share of the β-particle energy may be lost in just one collision. However, head-on collisions are rather rare events, and so particle deflections are usually not very large. The β particle will undergo relatively few ionizing collisions, because its mass and charge are small.

The range of a β particle in air is considerably larger than that of an α particle with the same energy. As a rough order-of-magnitude estimate, a β particle will penetrate matter hundreds of times farther than an α particle of similar energy. Although a piece of paper will stop many α particles, it is not uncommon for a β particle to pass through absorbers as thick as a book. Compared with neutrons and γ rays, though, β particles are relatively easily stopped.

Neutrons have no charge and a mass very close to that of a proton. As a result, there is no ordinary electrostatic repulsion or attraction between these particles and the various portions of the atom. Consequently, a neutron will undergo collision only rarely. A direct hit on an electron or nucleus must occur before any disturbance to its travel is noticed. A β or an α particle can ionize an atom by a near miss, since the electrostatic forces between the charges act strongly on the atom even though a true collision does not occur. This is impossible for a neutron, however. Hence *the neutron is a highly penetrating, very slightly ionizing particle.*

Until now we discussed the behavior of particles only. Gamma rays are quite different in character from any of these, since they are electromagnetic radiation, having neither charge nor rest mass. Nevertheless, we already discussed instances where light and x-rays interact with matter, namely, in the photoelectric and Compton effects. *Since γ rays are merely short-wavelength x-rays, we would expect them to show similar characteristics.* This is indeed true.

A γ ray, being a photon, will ordinarily lose all its energy in one event, except in the case of Compton scattering. When a beam of γ rays, or photons, passes through a gas, many of them are stopped when they strike atomic electrons and eject them from the atom. This is simply the photoelectric effect, of course. As we know from our experience with x-rays, however, short-wavelength or high-energy electromagnetic radiation is extremely penetrating. The x-rays which penetrate and pass through our body when an x-ray photograph is taken are, in comparison with γ rays, relatively low-energy radiation, usually less than 0.10 MeV.

The γ rays emitted by nuclei are often several million electronvolts in energy. Since the penetrating ability of γ rays is generally greater for shorter wavelengths, it is clear that γ *rays are highly penetrating.* Very-short-wavelength, i.e., high-energy, γ rays are capable of penetrating several inches of concrete. Since the photoelectric effect accounts for a large fraction of the loss of γ rays from a beam at energies less than a few million electronvolts, materials containing a large number of electrons per unit volume should be the best γ-ray absorbers. For this reason, the very heavy elements, lead in particular, are used as shielding against γ rays and x-rays.

We see from this discussion that most of these types of radiation cause ionization of atoms in their path. This fact serves as the basis for most methods of detecting them.

Photographic Emulsions   When an ionizing particle or a photon passes through the emulsion layer on a photographic plate or film, the ions formed

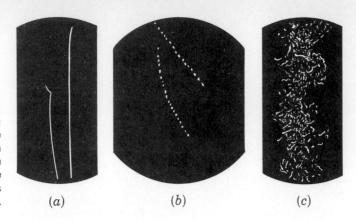

(a) (b) (c)

in the emulsion act as loci for the formation of silver specks when the emulsion is developed. As a result, the ions formed along the particle's path leave a photographic record of the particle's passage.

Cloud and Bubble Chambers When a fog is about to form in a supersaturated vapor, or when a bubble is about to form in a superheated liquid, the droplets and bubbles form preferentially on ions. Hence the ions along the paths of a particle can be discerned as the droplets or bubbles form. Typical patterns observed in a cloud chamber are shown in Fig. 27.11. The patterns seen in a bubble chamber are much the same except that the path lengths are much shortened because of difference in density between liquid and gas. Often a magnetic field is superimposed on the chamber, causing the paths of charged particles to be curved and thereby yielding additional information about the particles.

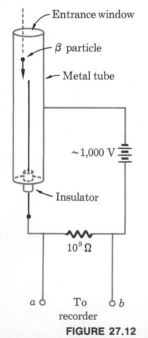

Geiger Counter This portable device is widely used for detection of radiation. Its basic construction is shown in Fig. 27.12. A high potential difference exists between a metal tube and a wire on its axis. The tube is filled with a special gas mixture at reduced pressure. When a particle enters the counter through the thin entrance window, it forms ions in the tube. The current which flows because of these ions alone is extremely small, but the high voltage causes the ions to initiate a discharge within the tube, and so, momentarily, the tube allows a current to pass. For reasons we cannot pursue here, the current stops in a small fraction of a second. Hence, each time a particle passes through the counting tube, a pulse of current passes through the circuit. Any recording device (a loudspeaker, meter, etc.) connected across terminals ab will register each time such a pulse occurs, and so the particles passing through the counter tube can be counted.

FIGURE 27.12
The Geiger counter.

Scintillation Counter When a fast-moving particle strikes a fluorescent material, a pulse of light is given off. This fact is basic to picture formation in a TV tube and is the principle on which the scintillation counter is based.

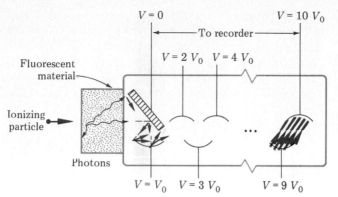

$V = 0$ $V = 10\,V_0$

To recorder

$V = 2\,V_0$ $V = 4\,V_0$

Fluorescent material

Ionizing particle

Photons

$V = V_0$ $V = 3\,V_0$ $V = 9\,V_0$

FIGURE 27.13
In the scintillation counter, light from the scintillator causes photoemission of electrons in the phototube. In the first stage, electrons accelerate through the potential V_0, and each liberates several secondary electrons. This multiplicative process continues and results in a comparatively large output current pulse, much larger than indicated in the figure.

As shown in Fig. 27.13, a phototube is used to detect the light pulse, but since the light pulse from a single particle is so small, a special phototube called a **photomultiplier** tube is used. Its principle of operation is shown in the figure. This device, like the Geiger counter, is used widely for radiation detection.

Solid-State Detectors These devices are basically solid-state semiconductor diodes. As you recall, no current will pass through such a device in the reverse direction. However, if a fast-moving particle passes through the junction region between the *n*- and *p*-type semiconductor, a current pulse passes through the diode. This pulse can be used to count the particles in the same way as for the previous two devices.

27.13 Radiation Units

As we saw in the last section, high-energy radiation is capable of ionizing atoms, and it can tear apart molecules. The resultant damage is of importance, and we discuss it in due course. But first let us learn some of the units used to describe the effects of radiation.

Four different classes of units are designed to measure four different facets of radiation. They measure source activity, exposure, absorbed dose, and biological effectiveness. Let us now take up each of these in turn.

Source Activity The activity of a radioactive source is equal to the number of disintegrations that take place in the source in unit time. As an equation,

$$\text{Activity} = \frac{\Delta N}{\Delta t} \qquad (27.8) \qquad \textbf{Definition}$$

where ΔN is the number of nuclei that decay in time Δt.

The SI unit for activity is the **becquerel** *(Bq); it is the number of disintegrations that take place in one second.* If in a sample 80 nuclei decay in 1 s, then the activity of the sample is 80 Bq. Or, as we learned in Illustration 27.3, the number of decays that occur in 1 g of radium each second is 3.6×10^{10}. Hence the activity of a 1-g sample of radium is 3.6×10^{10} Bq. Another unit in common use for activity is the curie (Ci), where by definition

$$1 \text{ Ci} = 3.7 \times 10^{10} \text{ Bq}$$

There is a simple relation between the activity of a sample, its decay constant λ, and the number N of atoms in it. To find it, we use Eq. (27.5), which was

$$\Delta N = \lambda N \Delta t$$

Substituting this value for ΔN into Eq. (27.8) gives

$$\text{Activity} = \lambda N = \frac{0.693N}{T_{1/2}} \tag{27.9}$$

where the latter form is obtained by use of the fact that $\lambda = 0.693/T_{1/2}$.

As an example of the use of this relation, let us find the activity of a 1-g sample of strontium 90. This substance has a half-life of 28 yr and is a dangerous product of nuclear explosions.

To use Eq. (27.9), we must first find the number N of atoms in 1 g of ^{90}Sr. We can do this easily if we recall that 1 kmol of ^{90}Sr, which is 90 kg, contains 6.02×10^{26} atoms. Then

$$N = \text{number in 1 g} = \frac{0.001 \text{ kg}}{90 \text{ kg}} (\text{number in 90 kg})$$

$$= \frac{0.001}{90} (6.02 \times 10^{26}) = 6.7 \times 10^{21}$$

But we already knew that

$$T_{1/2} = 28 \text{ yr} = 8.8 \times 10^8 \text{ s}$$

and so Eq. (27.9) yields

$$\text{Activity} = \frac{0.693N}{T_{1/2}} = \frac{(0.693)(6.7 \times 10^{21})}{8.8 \times 10^8 \text{ s}} = 5.3 \times 10^{12} \text{ Bq}$$

Exposure *The exposure is a measure of the ionization caused by x-ray and γ-ray beams.* It is defined only for these beams and only if their photon energies are less than 3 MeV. Its unit is the roentgen (R), which is defined in the following way.

The beam is sent through air at standard temperature and pressure (STP) for a certain time. Positive ions are formed in the air by the beam. If Q coulombs of positive ions are formed in m kilograms of the air through which the beam passes, then the exposure during that length of time is

$$\text{Exposure (in R)} = \frac{Q}{2.58 \times 10^{-4}\, m} \qquad \text{C/kg}$$

Therefore *an exposure of 1 R will produce 2.58×10^{-4} C of positive ions per kilogram of standard air.*

Absorbed Dose *The absorbed dose is a measure of the energy absorbed from the radiation beam in unit mass of biological material through which the beam is passing. Its SI unit is the* **gray** (*Gy*). Suppose a beam passes through a mass of m kilograms of tissue and loses an energy E joules in that mass during the duration of the beam. Then the absorbed dose (in grays) given to the material constituting the mass is defined by

$$\text{Absorbed dose (Gy)} = \frac{E}{m} \qquad \text{J/kg}$$

In other words, *1 Gy is equivalent to an absorbed energy of 1 J/kg.*

Frequently used is another unit that measures absorbed dose. It is the **rad** (rd) and is defined as follows:

$$1\ \text{rd} = 0.01\ \text{Gy}$$

The gray and rad units are applicable to absorbed dose of all types of ionizing radiation. However, the definition was originally chosen in such a way that an exposure of flesh to 1 R of x-rays gives rise to an absorbed dose of 1 rd. As a result, the rad and roentgen are often used interchangeably when we deal with the very special case of x-rays absorbed in flesh.

Biological Effectiveness As we have seen, the gray (and rad) measures the beam energy absorbed in a unit mass of biological material. This type of measure is satisfactory from a physical standpoint, but it is inappropriate for many biological applications. The difficulty arises because the biological changes caused by 1 Gy of absorbed radiation vary depending on the type of radiation involved and the material being irradiated. To circumvent this difficulty, a quantity called the *relative biological effectiveness* (RBE) of a radiation is defined.[5]

A beam of 200-keV x-rays is taken as a standard for defining RBE. We define the RBE of a beam interacting with a biological system as follows:

---

[5] The RBE is sometimes called the *quality factor* (QF).

TABLE 27.1
TYPICAL RBE VALUES

Radiation	Typical RBE
200-keV x-rays	1.00000
γ-rays	
1-MeV	0.7
4-MeV	0.6
Electrons	1.0
Protons (1 to 10 MeV)	2
Neutrons	
Slow	~3
Fast	~10
α particles	10–20

$$\text{RBE} = \frac{\text{biological effect of 1 Gy of radiation}}{\text{biological effect of 1 Gy of 200-keV x-rays}}$$

By its very nature, the RBE is a rather imprecise quantity. Despite that fact, typical values for various types of radiation can be stated, and a few such values are given in Table 27.1. The appropriate RBE values for a given type of radiation also depend on the object being irradiated. For example, neutrons are particularly effective in causing eye cataracts, and an RBE of about 30 is found in that instance.

To quantify the biological effect of a given absorbed dose, a unit called the **rem** is defined as follows:

$$\text{Biologically equivalent dose (in rem)} = \text{RBE} \times \text{dose (in rd)}$$

As an example, we see from the table that a 1 rd absorbed dose of 4-MeV γ rays is equivalent to 0.6 rem. Similarly, 3 rd of 5-MeV protons is equivalent to 6 rem. Moreover, 3 rd of 5-MeV protons is equivalent to 6 rd of 200-keV x-rays.

Illustration 27.6 Cobalt 60 γ rays have RBE = 0.7. A tumor that is ordinarily given a dose of 1000 rd from a cobalt source is to be treated with neutrons having RBE = 3. How many rads are needed from the neutron beam?

Reasoning By definition, the dose (in rem) equals RBE times the dose (in rad). Therefore the cobalt-60 dose is equivalent to

$$(0.7 \text{ rem/rd})(1000 \text{ rd}) = 700 \text{ rem}$$

Applying the same procedure to the neutron beam gives

and so the biologically equivalent neutron dose would be $700/3 = 233$ rd.

27.14 Radiation Damage

Since radiation can tear apart molecules, it is capable of damaging materials. One of the most common types of radiation damage is due to the ultraviolet rays in sunlight. These lead to sunburn and tanning of the skin. The high-energy photons disrupt skin molecules on impact and cause these easily observed effects. In this case, the damage is usually of little importance. Most of the sun's ultraviolet rays are absorbed by the ozone in the upper atmosphere, so normal exposure to the sun's rays need not be avoided. However, in recent years we have become aware that a serious hazard could arise if we deplete the ozone layer with manufactured chemicals. There is danger then that the increased ultraviolet radiation reaching us could increase the incidence of skin cancer.

We are continuously exposed to other radiation in addition to sunlight. Nearly all materials contain a slight amount of radioactive substances. As a result, your body is unavoidably exposed to a low level of background radiation. Typically, each person experiences a background radiation dose of about 0.1 rem each year. Let us now examine the effects of different levels of radiation dose on the body.

Background Radiation

High levels of radiation covering the whole body disrupt the blood cells so seriously that life cannot be maintained. For whole-body doses in excess of 500 rem, death is likely to occur. Even a whole-body dose of 100 rem can cause radiation sickness of a very serious, although nonfatal, nature. Blood abnormalities occur for doses in the range of 30 rem and above. At still lower whole-body doses, the overall effects on the body are less apparent, but nevertheless can cause serious consequences.

Radiation Damage to Human Beings

Even very low radiation doses reaching the reproductive regions of the body are potentially dangerous. The giant molecules in our bodies which carry reproductive information can be disrupted by a single radiation impact. If enough of these molecules are damaged, defective reproduction information will be furnished to a fetus as it develops. As a result, birth abnormalities occur. Even though there is some evidence that a low level of reproduction abnormalities may be beneficial to people, most birth defects are not desirable. For this reason, *no one of child-bearing age should be exposed to unnecessary radiation of the reproductive organs.* Of course, a properly given arm x-ray, for example, presents no such danger.

In addition to causing birth abnormalities, low levels of radiation present two other hazards. First, there appears to be a delayed cancer effect. Although cancer may not appear at once, low levels of radiation may cause cancer to develop many years later. Second, a child is particularly vulnerable to radiation. Because the child is growing rapidly, any cell mutations caused by radiation could have serious consequences. For this reason, most doctors are reluctant to prescribe x-ray scans for children unless absolutely necessary.

There is no "safe" limit of body exposure to radiation. It can only be said that radiation should be kept to the least value possible within reason. For example, since we are all subjected to a background radiation of about 0.1 rem/yr, there is no reason to disrupt our lives to avoid radiation doses lower than this. Even though a person who lives in the mountains may experience an annual background dose 0.05 rem higher than at sea level, the difference is not large enough to warrant moving. In the last analysis, people must often make a compromise between radiation safety and other considerations. To guide us in making that compromise, maximum occupational radiation doses have been specified. As a rough rule, the maximum yearly dose, except for the eyes and reproductive organs, is about 5 rem.

27.15 The Particle Zoo

In ancient times, the 1950s, physics appeared to be much simpler than it now seems to be—at least in regard to the physics of fundamental particles. At that time, there seemed to be six fundamental particles: the photon, electron, proton, neutron, neutrino, and meson.[6] You will recall from Sec. 5.6 that the neutrino (ν) is emitted in beta decay. A typical reaction is

$$^{12}_{5}\text{B} \rightarrow {}^{12}_{6}\text{C} + {}^{0}_{-1}\beta + {}^{0}_{0}\bar{\nu}$$

where the reason for the bar above ν is explained later.

The last particle mentioned, the meson,[7] was predicted to exist by Hideki Yukawa (1935) in his theory of the nuclear force. After World War II, physicists began a search for it and in 1947 discovered a particle they thought to be the one they sought. However, it soon became apparent that this was a new type of particle, and Yukawa's meson was not found until a few years later. Indeed, as very-high-energy nuclear accelerators became available in the 1950s, nuclear bombardment reactions produced a multitude of new particles. This process of discovery has still not ended, and as we shall see, more massive new particles are expected to be found as the energy capabilities of accelerators continue to increase. A partial list of the known particles is given in Table 27.2. We discuss this table as we proceed.

Already in 1928 Dirac had presented a theory that predicted that there should be two types of electrons. Both have the same mass, but one is the ordinary negative electron while the other carries an equal *positive* charge. This latter particle is called the *positron* and is represented by e^+, β^+, and $^{0}_{+1}\beta$. The positron has the ability to annihilate a negative electron; the end

---

[6] Various antiparticles were also known at the time. We discuss them subsequently.
[7] The name *meson* comes from the Greek *mesos,* "middle," because the particle's mass is intermediate to the electron and proton masses.

TABLE 27.2
LIST OF PARTICLES

Type	Name	Symbol	Antiparticle	Rest mass, MeV	Lifetime, s
Photon	Photon	γ	Self	0	∞
Leptons	Electron	e^- (β^-)	e^+ (β^+)	0.51	∞
	Muon	μ^-	μ^+	106	2.2×10^{-6}
	Tau	τ^-	τ^+	1.8	—
	Neutrino (electron)	ν_e	$\bar{\nu}_e$	~ 0	$\sim \infty$
	Neutrino (muon)	ν_μ	$\bar{\nu}_\mu$	~ 0	$\sim \infty$
	Neutrino (tau)	ν_τ	$\bar{\nu}_\tau$	~ 0	$\sim \infty$
Hadrons					
Baryons	Proton	p	$\bar{p}$	938	$\sim \infty$
	Neutron	n	$\bar{n}$	940	700
	Lambda	Λ^0	$\bar{\Lambda}^0$	1116	3×10^{-10}
	Sigma	Σ^+	$\bar{\Sigma}^-$	1189	0.8×10^{-10}
		Σ^0	$\bar{\Sigma}^0$	1192	$<10^{-14}$
		Σ^-	$\bar{\Sigma}^+$	1197	1.5×10^{-10}
	Plus many more-massive particles				
Mesons	Pion	π^+	$\bar{\pi}^-$	140	2.6×10^{-8}
		π^0	Self	135	0.8×10^{-16}
	Kaon	K^+	K^-	494	1×10^{-8}
		K_s^0	$\bar{K}_s^0$	498	1×10^{-10}
		K_L^0	$\bar{K}_L^0$	498	6×10^{-8}
	Eta	η^0	Self	549	$<10^{-16}$
	Plus many more-massive particles				

result is a photon. We say that the positron is an antielectron, the antiparticle for the electron. Positrons are now known to be the products of many nuclear decay reactions.

As time went on, it became apparent that every particle has a corresponding antiparticle. The antiproton has a charge $-e$ and is capable of annihilating an ordinary proton. Even the chargeless neutron has an antiparticle which is easily recognized by the fact that it annihilates an ordinary neutron. We use the symbols shown in Table 27.2 for the antiparticles. The earth and regions of the universe close to it are made of regular matter, not antimatter. You might speculate what would happen if a spaceship from an antimatter universe were to visit the earth.

It now appears that many massive particles that have been discovered recently are simply excited states of less massive particles. Tremendous energy differences exist among the excited states, energy differences that give rise (through the equivalence of mass and energy, $\Delta E = \Delta m\, c^2$) to widely different masses for different excitation states. We therefore expect to find even more massive particles as accelerators used to produce them achieve higher energies.

Most of the newly discovered particles are unstable. They decay with the lifetimes given in the last column of the table. Notice that a free neutron is unstable. This instability is masked, however, when the neutron is combined within a nucleus. At present there is some question as to the stability of the free proton. If it is unstable, its lifetime is probably greater than

10^{30} yr. Because of its very long lifetime, free proton decay will be very difficult to detect. Even so, experiments are currently underway to measure the decay of the proton, if it is in fact unstable.

Leptons These particles, listed in Table 27.2, are thought to be structure-less, point particles. To the best of our current knowledge, they are primary particles in the sense that they are not composed of still smaller particles. The leptons are not subject to the strong nuclear force, the force that holds the nucleons in the nucleus. They apparently do not "feel" the strong nuclear force.

Hadrons This group contains the particles that respond to the strong nuclear force. We see in the next section that there is another distinctive feature of this group of particles: each hadron appears to be composed of still tinier particles called *quarks*. As you see in Table 27.2, the hadrons are subdivided into two classifications, baryons and mesons. This distinction also has to do with the assumed fact that hadrons are composed of quarks.

27.16 Quarks

As more and more supposedly fundamental particles were discovered in the 1960s and 1970s, it became apparent that not all could be simple particles. Moreover, by scattering leptons from protons and neutrons, it became increasingly clear that the proton and neutron have diameters of order 10^{-15} m and internal structure. The internal structure was compatible with the idea of still smaller, pointlike particles within the hadron. These constituent particles are called *quarks*.

Initially, in 1963, the quark theory[8] for hadron structure assumed there were only three quarks (plus three antiquarks). The quarks were given the names *up, down,* and *strange*. Their charges are given in the following table:

Quark	Symbol	Charge
Up	u	$+\frac{2}{3}e$
Down	d	$-\frac{1}{3}e$
Strange	s	$-\frac{1}{3}e$

The antiparticles are represented by $\bar{u}$, $\bar{d}$, and $\bar{s}$, and their charges are opposite in sign to those of the corresponding particles. Notice that these

---

[8] This theory was proposed independently by Murray Gell-Mann and George Zweig. Gell-Mann provided the name for these particles in a whimsical allusion to the phrase "Three quarks for Muster Mark" found in Joyce's *Finnegan's Wake*.

particles carry charges less than the charge quantum, and so they should be recognizable by their fractional charge. In spite of an extensive search for them, free quarks have never been found. More is said about the observability of quarks later.

If we accept the concept of quarks, then all the low-mass hadrons can be explained as being quark combinations. For example, a proton is a combination of three quarks, $u + u + d$, while a neutron is a combination of $d + d + u$. Not only do such combinations give the correct charges for the hadrons, but also they give their other properties. For example, the hadron spin[9] (angular momentum and magnetic moment) is the sum of the spins of its constituent quarks. The quark hypothesis was a stunning success since all the hadrons known at the time could be explained in terms of quark combinations. Indeed, one hadron not yet discovered but predicted as a possible quark combination was soon found to exist.

The quark concept also explained the difference between the two classifications of hadrons, namely, baryons and mesons. Baryons such as the proton are combinations of three quarks; mesons, however, consist of only two quarks. As you see, except for the fact that no one had succeeded in isolating a quark, the quark theory was a striking success.

If you have the time, you can easily enumerate the various possible combinations of these quarks and antiquarks. Moreover, it is possible to compute their excited states in a way analogous to the computation of the excited states of an atom. At the high energies involved, the mass equivalent of the excitation energy is appreciable and gives rise to what appear to be new particles at the various excitation states. It was therefore possible to compute the nature of the various hadrons one should expect from the quark model by using the three quarks and three antiquarks listed above. These computations furnished the incentive for the discovery of new hadrons.

Experiment after experiment confirmed the existence of these predicted hadrons. But in 1974 a new particle was discovered that did not fit into this scheme. Its existence could be explained only if a new quark (and antiquark) existed. This new quark was named *charm* and is designated by *c*. Eventually, for reasons we do not delve into here, it was realized that fifth and sixth quarks must exist; their names are *top* (*t*) and *bottom* (*b*), or, as some people prefer, *truth* and *beauty*. In any event, the quarks that we presently believe to exist are as follows:

Quark	Up	Down	Strange	Charm	Top	Bottom
Symbol	u	d	s	c	t	b

Each has a corresponding antiquark. By using these six quarks and their antiquarks, it is possible to account for the existence of all presently known hadrons.

---

[9] Quarks, like electrons, have spins of $\pm\frac{1}{2}$.

27.17 Fundamental Forces

The particles we have been discussing exert forces on one another. You are already familiar with three of these forces: the gravitational force, the electrostatic force, and the nuclear attractive force. This latter force holds together the nucleons in the nucleus, and it is also the attractive force among all hadrons. The strengths of these forces are widely different. We show this in Fig. 27.14, where the force between two protons is shown as a function of the distance between their centers. (Since the diameter of a proton is about 10^{-15} m, proton separations less than this are meaningless on the graph.) Notice that the gravitational force between two protons is a factor of about 10^{-36} smaller than the electrostatic force between them. Because the nuclear attraction force between the two protons has a range less than 10^{-14} m, the nuclear force is negligible for separations greater than this.

The force between quarks is attractive and is believed to be much like the nuclear force between hadrons provided the separation distance is small. However, as shown in the figure, the quark force does not drop at large separations. Instead, the attractive force actually appears to increase with separation distance at large separations.[10] This has a very important consequence. As you try to separate two quarks, the force holding them together increases. The situation is analogous to the stretching of a rubber

---

[10] Evidence for this is sketchy, and so we cannot be too certain of the material in this paragraph.

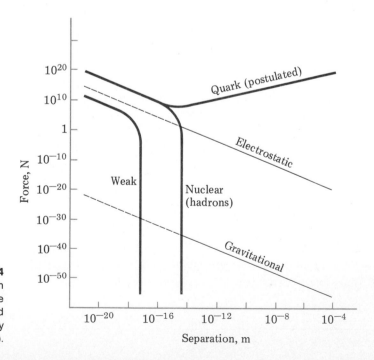

FIGURE 27.14
Force of one proton on another compared to the force between quarks and the weak force (partly schematic).

band; the more you stretch it, the greater is the required stretching force. We therefore surmise that perhaps nearly infinite forces are required to separate quarks. In addition, tremendous energy will be stored within the stretched bond between the quarks. Both these postulated properties of the quark bond help us to rationalize the fact that quarks have not yet been found in isolation.

The remaining force shown in Fig. 27.14 is the *weak force*. This force was first recognized to exist in 1934. In that year Enrico Fermi[11] presented the currently accepted theory of beta decay, i.e., the theory of how a radioactive nucleus emits a beta particle. He assumed that this previously unknown force was responsible for the nucleus throwing out an electron. As you can see in Fig. 27.14, the weak force has a range less than about 10^{-17} m and is considerably smaller than the nuclear force.

In 1958 R. Feynman greatly enlarged our understanding of the weak force (which he called the "universal Fermi interaction") by showing how the force could cause particles to transform to electrons and neutrinos. For example, in the radioactive decay of a free neutron by beta-particle emission, the weak force causes the following reaction:

$$n \rightarrow p + e^- + \bar{\nu}_e$$

Reactions such as this are characteristic of the weak interaction. This fact is often summarized in the whimsical statement that "the weak interaction is a pernicious disease that transforms elementary particles into electrons and neutrinos."

One of the loftiest ideals of theoretical physicists is to devise a theory that unifies the forces shown in Fig. 27.14. In part this has already been done. Long ago Maxwell showed that the electrostatic and electromagnetic forces are related. Today we treat them as a single basic force because given one, we can find the other. Similarly, the weak force and the electrostatic (or electromagnetic) force were shown to be related in 1967 in nearly identical theories presented independently by Weinberg and Salam. Their theory, called the Weinberg-Salam theory, has been tested in many ways and has survived each test.

A moderately successful theory has been devised to unify the nuclear force and the quark force. It, however, has not been nearly as well established as the Weinberg-Salam theory. Finally, steps have been taken to unify the nuclear-quark force theory with the Weinberg-Salam theory. This final unifying theory is referred to as the *grand unified theory*. One of its predictions is that the proton is unstable, a prediction mentioned earlier. But no one yet claims that we achieved the ultimate goal, a theory that satisfactorily unifies the electromagnetic, weak, nuclear, and quark forces. Moreover, relating these forces to the gravitational force is a completely unsolved problem for the future.

---

[11] Fermi is famous for many things. It was he who led the group that first successfully carried out a controlled nuclear fission reaction.

Summary

The nucleus of an atom with atomic number Z and atomic mass number A contains Z protons and $A - Z$ neutrons. Its radius is given approximately by $R = 1.2 \times 10^{-15} A^{1/3}$ m.

On the atomic-mass scale, the carbon 12 atom has a mass of exactly 12 u. The proton and neutron masses are close to 1.0 u.

A nucleus of element X with mass number A and atomic number Z is represented by $^A_Z X$. Two nuclei having the same Z but differing in A are isotopes of the same element. They act the same in chemical reactions but differ in the number of neutrons in their nuclei. The masses of isotopes (in atomic mass units) are close to integers. The atomic masses given in the periodic table are averages of the isotopes existing in nature.

All nuclei have less mass than the combined masses of their individual protons and neutrons. The mass Δm lost in assembling the nucleus is related to the binding energy (BE) of the nucleus through $BE = \Delta m \, c^2$. In effect, this lost mass must be created when the nucleus is separated into its parts.

Nuclei of intermediate-Z elements have the highest binding energy per nucleon. As a result, energy is given off in the fission and fusion reactions. Nuclear reactors use the fission reaction to generate energy.

All elements with $Z > 83$ are radioactive. Radioactive elements follow an exponential decay law. The time taken for half of the nuclei to decay is called the half-life. The half-life $= 0.693/\lambda$, where λ is the decay constant. During a short time interval Δt the number of nuclei which decay, ΔN, is related to the number of nuclei available for decay N by $\Delta N = \lambda N \Delta t$.

Radioactive materials formed in nature emit α and β particles and γ rays; α particles are helium nuclei, β particles are electrons, γ rays are the same as x-rays. In order of decreasing ionizing ability and increasing penetration ability, these are α, β, and γ.

The nuclear force holds together the nucleons. Unlike the electric and gravitational forces, it is a short-range force. It acts only over distances of less than about 10^{-14} m. At such small distances, however, it is much larger than the Coulomb force.

Absorbed radiation dose is measured in terms of a unit called the gray. One gray is equivalent to an absorbed dose of one joule in each kilogram of material. The dose in rems = (RBE) (dose in rads), where RBE is the relative biological effectiveness of the radiation in question.

Minimum Learning Goals

Upon completion of this chapter, you should be able to do the following:

1 Give an order-of-magnitude value for the radius of a nucleus.

2 Distinguish between atomic number Z and atomic mass number A. Relate them to the number of protons and neutrons in a nucleus.

3 Interpret symbols such as $^6_3 Li$.

4 Explain what is meant by an isotope.

5 Define the atomic mass unit and give the approximate masses of the proton and neutron in this unit.

6 Sketch a graph of the mass defect per nucleon as a function of Z.

7 Sketch a graph of the binding energy per nucleon versus Z and explain why it is related to the graph in item 6.

8 Sketch a graph of N versus t for a radioactive substance. Show the half-life on the graph. Compute the fraction of material not decayed after a length of time equal to a given integer number of half-lives.

9 Define each quantity in the equation $\Delta N = \lambda N \Delta t$ and be able to use the equation in simple situations. Give the relation between λ and the half-life.

10 Define α particle, β particle, γ ray, and positron.

11 Write the nuclear reaction equation for a given nucleus which emits one of the following: α particle, β particle, γ ray.

12 Prepare a diagram such as Fig. 27.6 for a series in which the starting nucleus and the emitted particles or rays (α, β, and γ) are given.

13 Compare the nuclear force to the gravitational and Coulomb forces in regard to strength, range, and dependence on r.

14 Explain why the nuclear fission reaction should release energy by reference to the mass-defect and binding-energy graphs. State what is meant by a fission chain reaction and relate this to why ^{235}U, but not ^{238}U, is usable in a nuclear bomb.

15 Sketch a schematic diagram of a nuclear power reactor showing fuel rods, moderator, control rods,

heat exchanger, and output to turbine; explain the function of each.

16 Explain why the nuclear fusion reaction should release energy by reference to the mass-defect and binding-energy graphs. State why fusion is much more difficult to achieve in a laboratory reactor than fission.

17 Compare the range and ionization effect of α, β, and γ radiation when passing through matter.

18 Define each of the following: becquerel, curie, gray, rad, rem, and relative biological effectiveness.

19 Describe the following: Geiger counter, scintillation counter, cloud chamber, bubble chamber.

20 Explain why radiation can be harmful to people. In your explanation, point out which regions of the body and which type of person should be particularly well shielded from radiation.

Important Terms and Phrases

You should be able to define or explain each of the following:

Atomic number Z; atomic mass number A
Isotope
Atomic mass unit u
Mass defect; binding energy
Radioactive; half-life
Decay constant; $\Delta N = \lambda N \Delta t$
Activity; becquerel unit, curie unit

Decay series
α particle; β particle; γ ray; positron
Nuclear force
Fission; chain reaction
Nuclear reactor
Fusion reaction
Gray, rad, rem, and RBE
Quarks
Lepton, hadron

Questions and Guesstimates

1 How many neutrons are there in the nucleus of $^{39}_{19}K$? How many protons? How many electrons does this atom have?

2 Why do chemists consider different isotopes to be the same element even though their nuclei are not the same?

3 Would the optical spectra of ^{35}Cl and ^{37}Cl atoms differ in any major way? Explain.

4 Even though an α particle is attracted rather than repelled by an electron, it is still capable of knocking an electron loose from the atom without a direct collision. Explain.

5 Why is a given thickness of lead a better absorber of 1-MeV α, β, and γ rays than the same thickness of water? Why is water a better shielding barrier against neutrons than lead?

6 The artificially produced isotope ^{102}Ag has a half-life of 73 min. It decays in two alternative ways. Part of the nuclei emit a positive electron, a positron. The rest capture one of the electrons in the first Bohr orbit and take it into the nucleus. In what way are the two equivalent? What would one notice in the laboratory for each process?

7 Tritium is the ^3H isotope of hydrogen. Its atomic mass is 3.016; the atomic mass of ^1H is 1.0078.

Using the fact that the neutron mass is 1.00867, what do you predict about the stability of tritium? Repeat for ^2H, deuterium, which has an atomic mass of 2.0141.

8 Before an x-ray survey is made of the gastrointestinal tract, the patient must drink a solution of a barium compound. Why?

9 When an x-ray photograph is taken of a person's arm, the bones are clearly shown on the photograph. Why do they show up in this way? After all, the arm is no thicker where the bone is than elsewhere.

10 A small amount of radium is sealed in an evacuated glass tube. When it is later broken open, the tube is found to contain some gas. The mass spectrometer shows the gas to consist of a very-low-molecular-weight species and a very-high-atomic-weight species. What are they? Estimate their relative abundance.

11 In atomic mass units, the mass of an electron is 0.000549, of a proton 1.00728, and of a neutron 1.00867. Is it feasible for a proton and electron both at rest to combine and form a neutron? Explain. Would it matter if they were close together or far apart when they were at rest?

12 Why is the fusion reaction so much more difficult to initiate than the fission reaction?

13 Is there any possibility of causing the fusion of two small nuclei without making them collide with extremely high energy? Defend your answer.

14 A nuclear reactor produces energy chiefly in the form of heat. Explain how this heat is generated as a result of nuclear fission.

15 The fission and fusion reactions appear to create energy. How can we reconcile this fact with the statement that energy can be neither created nor destroyed?

16 It is possible for a man working with x-rays to burn his hand so seriously that he must have it amputated, and yet the man may suffer no other conse-

quences. However, an x-ray overexposure so slight as to cause no observable damage to his body could cause one of his subsequent offspring to be seriously deformed. Explain why.

17 Most radiologists feel that women beyond child-bearing age can safely be exposed to much more x-radiation than young women. How can they justify such an opinion?

18 Low-energy (soft) x-rays are used to treat skin cancer, while high-energy (hard) x-rays are used to treat cancer deep within the body. Why are soft x-rays not used for this latter purpose, even though they can penetrate deeply enough to kill the cancerous region?

Problems

1 Estimate the radius of a ^{27}Al nucleus and determine its average density in kilograms per cubic meter.

2 Compute the approximate radius of the ^{235}U nucleus and determine its density in kilograms per cubic meter.

3 Assuming the density of nuclear matter to be 2×10^{17} kg/m^3, about what would be the radius of the earth if it were compacted to this density? See the inside cover of the text for data about the earth.

4 If the sun were compacted so its density was about that of nuclei, namely 2×10^{17} kg/m^3, what would be the radius of the sun? See the inside cover of the text for data about the sun.

5* A certain mass spectrograph accelerates ions through 2000 V and deflects them in a magnetic field of 0.070 T. When a certain univalent ion is examined in the spectrograph, it follows a path with $r = 13.0$ cm. What is the mass of the ion in both kilograms and atomic mass units?

6* In a mass spectrograph like that shown in Fig. 27.1, it is found that r for ^{12}C is 10.00 cm. How large would r be for ^{16}O? Assume identical charges and accelerating potentials.

7 Potassium found in nature contains essentially only two isotopes. One has an atomic mass of 38.964 u and constitutes 93.3 percent of the whole. The other 6.7 percent has a mass of 40.975 u. Compute from these data the atomic mass which the chemists list in the periodic table.

8 The two isotopes of carbon found in quantity on earth have the following masses and percentage abundance: 12.0000 u, 98.892 percent; 13.0034 u,

1.108 percent. Calculate the mass of carbon as measured by chemists.

9* Two isotopes exist in commercial copper. One constitutes 70 percent of the whole and has an atomic mass of 62.96 u. The chemists assign a mass of 63.58 u to copper. What is the mass of the second isotope?

10* By using a mass spectrograph like the one in Fig. 27.1, the mass of what are thought to be singly ionized particles is measured. The result is 22.0 u. In fact, however, the particles are doubly charged. What is their true mass?

11 Use Fig. 27.3 to state how much mass is lost as a krypton 84 nucleus is assembled from free protons and neutrons. What is the percentage mass loss?

12 Use Fig. 27.4 to state how much energy is required to tear a barium 138 nucleus apart into free neutrons and protons. To how much mass (in atomic mass units) is this energy equivalent?

13* Compute the mass defect per nucleon and the binding energy per nucleon for the ^{12}C nucleus. *Hint:* Recall that the mass of the *atom* is exactly 12 u.

14* Compute the mass defect per nucleon and the binding energy per nucleon for the ^{20}Ne nucleus. Its atomic mass is 19.99244 u.

15* Use Fig. 27.3, together with the proton and neutron masses, to find the mass of an *atom* of sulfur 32.

16* In Appendix 4 we find that the atomic mass of ^{16}O is 15.994915 u while it is 16.999133 u for ^{17}O. Use these data to find the binding energy of the extra neutron in the ^{17}O nucleus.

17 The following data are taken for the time dependence of the number of counts per minute on a Geiger counter placed above a radioactive sample. What is the half-life of the material?

Time (h)	0	1.0	2.0	3.0	4.0	5.0
Counts per minute	860	637	472	350	258	190

18 A Geiger counter placed above a radioactive sample registers 2580 counts per minute. How many counts per minute will it register after four half-lives have passed?

19 A sample that contains 6.2×10^{11} atoms has a half-life of 18 yr. (a) What is its decay constant? (b) How many nuclei in it undergo decay in 30 s? (c) What is its activity in curies?

20 A tiny ampoule of radon gas contains 3.0×10^{12} atoms of radon. The half-life of radon is 3.8 days. How many disintegrations occur in the ampoule each second?

21 Watches with numerals visible in the dark often have radioactive material in the paint used for the numerals. A student estimates from measurements using a Geiger counter that 1000 disintegrations occur each second on the watch face. How many curies of radioactivity exist on the watch if the student's figures are correct?

22 A tiny piece of rock is radioactive, and a Geiger counter placed above it registers 87 counts in 1 min. Assuming that the counter intercepts radiation from half of the decaying nuclei, what is the activity of the rock?

23** Uranium 238 has a half-life of 4.5×10^9 yr. What is the activity of a 1-kg sample of pure ^{238}U?

24** The half-life of ^{60}Co, a radioactive element produced in reactors for use as a medical and commercial radiation source (1.33- and 1.17-MeV γ rays), is 5.3 yr. How many atoms of ^{60}Co are there in 1 g of the material? What is the decay constant for the material? How many disintegrations occur each second in 1 g of the material?

25* What fraction of a radioactive sample decays in 200 yr if the half-life of the material is 140 yr?

26* Strontium 90 is a radioactive fission product from nuclear fission reactors and bombs. Since its half-life is quite long (about 28 yr, or 8.8×10^8 s), it is a persistent contaminant and presents serious disposal problems. What fraction of the original strontium still remains 100 yr after a nuclear bomb explodes?

27** Measurements show that only 15 percent of a radioactive material remains after 6.0 hr. What is the half-life of this material?

28** The half-life of ^{90}Sr, a radioactive contaminant left in the wake of fission-type bombs, is about 28 yr. How long does it take before 99 percent of the original strontium 90 has decayed?

29** The end decay product of ^{238}U is ^{206}Pb. The half-life of ^{238}U is about 4.5×10^9 yr. On earth we find ancient rocks which have ^{238}U and ^{206}Pb intermixed in such a way that it appears that about 45 percent of the original ^{238}U has decayed. How old are the rocks? (This is one way in which we can estimate the time since the earth solidified.)

30** The age of an ancient piece of wood can be found by comparing the amount of ^{14}C in it to the carbon-14 content of new wood. This isotope of carbon is radioactive ($T_{1/2} = 5730$ yr) and is taken on by the wood as it grows. If a piece of wood found in an ancient cave contains only 10 percent as much ^{14}C as an equal quantity of new wood, about how old is the ancient wood? This subject, radiocarbon dating, is discussed in greater detail in the next chapter.

31 The radioactive element $^{90}_{39}$Y (yttrium 90) emits a beta particle. What is the resultant isotope?

32 The radioactive element $^{66}_{29}$Cu emits a beta particle. What is the resultant isotope?

33 Plutonium 246 ($^{246}_{94}$Pu) emits in succession two beta particles and two alpha particles and several γ rays. What is the resultant isotope?

34 Actinium 231 ($^{231}_{89}$Ac) emits in succession two beta particles, four alphas, one beta, and one alpha plus several γ rays. What is the resultant isotope?

35* Cobalt 60 emits a 1.33-MeV gamma ray. By what fraction does its nuclear mass decrease in the process?

36* A radioactive isotope of mercury, $^{203}_{80}$Hg, emits a 0.279-MeV gamma ray. By what fraction does the nuclear mass decrease in the process?

37* A radioactive series in addition to that shown in Fig. 27.6, the *thorium series,* starts with $^{232}_{90}$Th and emits in succession one α, two β, four α, one β, one α, and one β particle(s). What is the final product of the series?

38* A radioactive series in addition to that shown in Fig. 27.6, the *actinium series,* starts with $^{235}_{92}$U and emits in succession one α, one β, two α, one β, three

α, two β, and one α particle(s). What is the final product of the series?

39* One milligram of radium (^{226}Ra) is sealed in a 2.0-cm^3 bottle. Assuming that no helium gas can escape from the bottle, how many helium atoms will there be in the bottle after 1 yr? How many grams is this? (The half-life of radium is 1620 yr, while the half-lives of the remaining members of the series do not exceed a few days, until ^{210}Pb is reached. Since ^{210}Pb has a half-life of 21 yr, one can approximate the situation by assuming the series to terminate with this isotope.)

40* One milligram of radium is sealed in a bottle. It decays to radon, and the radon in turn decays with a half-life of 3.8 days. After about a month, how much radon (in grams) will there be in the bottle? (*Hint:* The total amount of radium remains essentially constant. At equilibrium, the decay rate of the radon must equal the decay rate of the radium.)

41* Consider the following reaction:

$$^{1}_{1}H \quad + \quad ^{13}_{6}C \quad \rightarrow \quad ^{13}_{7}N \quad + \quad ^{1}_{0}n$$
$$1.007825 \quad 13.00336 \quad 13.00574 \quad 1.008665$$

where n represents a neutron and atomic masses of the particles (including electrons) are given below each reactant. Can this reaction be initiated by a proton ($^{1}_{1}H$) that has kinetic energy of 1.80 MeV?

42* From an energy standpoint, is the following reaction possible if the incident proton has a KE of 1.48 MeV?

$$^{1}_{1}H \quad + \quad ^{7}_{3}Li \quad \rightarrow \quad ^{7}_{4}Be \quad + \quad ^{1}_{0}n$$
$$1.007825 \quad 7.01600 \quad 7.01693 \quad 1.008665$$

The masses indicated include the atomic electrons.

43* Suppose 1 kg of deuterium (heavy hydrogen, ^{2}H) is combined to form 1 kg of helium according to the reaction

$$^{2}_{1}H \quad + \quad ^{2}_{1}H \quad \rightarrow \quad ^{4}_{2}He$$
$$2.0141 \quad 2.0141 \quad 4.0026$$

where the atomic masses are given. (*a*) How much energy (in joules) is liberated? (*b*) If the confined helium has a specific heat capacity of 0.75 cal/(g)($^\circ$C), by how much does its temperature increase as this energy is added to it?

44* Neutrons are frequently detected by allowing them to be captured by a boron nucleus. This reaction is

$$^{1}_{0}n \quad + \quad ^{10}_{5}B \quad \rightarrow \quad ^{7}_{3}Li \quad + \quad ^{4}_{2}He$$
$$1.0087 \quad 10.0129 \quad 7.0160 \quad 4.0026$$

where the mass of the atom is given without subtracting the electron masses. (Since there are the same number of electrons on each side of the equation, this will be of no consequence. Why?) Energy given off in the reaction appears as KE of the products. The resulting swiftly moving helium ions (α particles) are then counted, by using ordinary techniques. Compute the speed of the α particle, assuming the original reactants to be essentially at rest. *Hint:* Compute the energy liberated in the reaction, and then write the laws of conservation of energy and momentum.

45* A neutron with speed 10^6 m/s hits a stationary deuterium atom (heavy hydrogen, $^{2}_{1}H$) head on in a perfectly elastic collision. (*a*) Find its speed after collision. (*b*) Repeat if the deuterium atom is replaced by an oxygen atom, $^{16}_{8}O$. Notice that light nuclei are most effective in slowing neutrons.

46 Iodine 131 (^{131}I) is used to treat thyroid disorders because when ingested, it localizes in the thyroid gland. Its half-life is 8.1 days. What is the activity of 1 μg of ^{131}I?

47 Phosphorous 32 (^{32}P) has a half-life of 14.3 days and is used in medicine because it tends to localize in bone. What is the activity of 1 g of ^{32}P?

48 How many grams of iron 59 (^{59}Fe) are there in a 1-mCi sample of it? Its half-life is 46.3 days.

49 The isotope tritium, $^{3}_{1}H$, has a half-life of 4600 days. How many grams of tritium are there in a sample that has an activity of 1×10^9 Bq?

50* How much is the temperature of water raised when given a radiation dose of 3 rd?

51* How large a radiation dose must be deposited in lead to raise its temperature 3°C? For lead, $c = 0.031$ cal/(g)($^\circ$C).

52 The yearly average exposure to medical x-rays in the United States is 70 mrem per person. (*a*) To how many rads of 200-keV x-rays is this equivalent? (*b*) To how many rads of 5-MeV protons?

53 To give a lethal dose of 500 rem, how long would one need to be exposed to 4-MeV gamma rays that give a dose of 20 rd/s?

54 The background radiation at Dallas, Texas, can be divided into two types. It furnishes a yearly dose of about 30 mrd of γ radiation and about 1.5 mrd of particles having an RBE of about 20. How many rems will a resident of Dallas experience each year from this cause?

55 Some authorities believe that the maximum allowed dose to a person's hands should be 70 rem/yr. Express this as dose in rads if the radiation being used is (*a*) x-rays; (*b*) γ rays; (*c*) fast electrons; (*d*) fast neutrons.

28
Physics of
the Universe

The past several chapters have been concerned with the smallest bits of matter found in nature —atoms, nuclei, and the basic particles. Now we turn our attention to the other extreme, the universe itself. In this chapter we examine its major features and some theories of its origin.

28.1 The Primordial Fireball

Scientists discover how nature behaves; they correlate these discoveries to bring order to a multitude of seemingly disconnected facts; they construct theories to explain the order found; they test the theories by predicting the results of experiments still not performed. Experimental results are at the heart of all science. Without experiment, theories cannot be tested, and so they can be neither verified nor rejected.

Astrophysics is greatly hampered by the lack of our ability to perform the really crucial experiments. The most important act of all, the formation of the universe, was performed several billions of years ago and is still going on. We cannot duplicate this process. All our knowledge of the origin of the universe must be gleaned from this one "experiment" over which we have no control. Many of the pertinent data should have been noted billions of years before people existed. Other data will not become available until billions of years in the future. We have no alternative but to work with the few data we can acquire by the limited means available. For that reason, most of what is said in this chapter must be viewed with caution; many of our interpretations may later be proved incorrect.

Until recently there were several competing theories for the earliest history of the universe. At present, only one is widely accepted, that formulated in its current form in 1948 by G. Gamow. It is called the **big bang theory.** According to this theory, *about 15 billion years ago the universe existed in a highly compressed state.* The mass within it had a density greater than 5×10^{12} kg/m^3, more than 5 billion times the density of water. This highly compressed universe was at an incredibly high temperature, probably higher than 10^{12} K.

At such extreme temperature and pressure, atoms and molecules could not exist. They would have been collapsed to near nothingness by the high compression if they had existed. Moreover, at temperatures that exceed those in the center of the sun by a factor of 10^5, all particles except the most primitive were torn apart. It is believed that only photons, electrons, positrons, neutrinos, and a comparatively few protons and neutrons then existed. The universe was then a fiery cauldron just beginning a violent expansion. No force could successfully resist the expansion driven by the high-energy thermal motions and the shattering pressure of the particles and radiation within it.

As this incredibly hot fireball began to expand, work was performed against gravitational forces. The huge pressure within the fireball did work during the expansion against the tremendous gravitational forces holding the fireball together. As a result, the material of the fireball lost thermal KE as the gravitational PE increased. *Consequently, the temperature of the fireball dropped rapidly as the expansion continued.* Nevertheless, the expansion process accelerated the outgoing matter to speeds near the speed of light. The universe appears to be expanding even now with a speed close to *c*.

It is interesting to speculate on the eventual fate of our expanding universe. Although the expansion is still being slowed by gravitational forces, we do not know enough about the universe to say whether it is expanding with energies large enough to provide the escape velocity for the matter within it. If it has enough energy, the universe will continue to expand forever. But if not, the expansion will eventually stop. Gravitational forces will begin to draw the universe back again into the primordial fireball from which it came. If this second alternative applies, eventually the fireball will be recreated and the whole process will be repeated. In fact, the fireball we have visualized may simply be the latest of a series of pulsations of a pulsating universe. It would appear that we can never learn the history of the universe during previous pulsations even if they existed. Can you say why?

28.2 Galaxies and Stars

As the original fireball expanded into space, its temperature was soon reduced enough to permit particles to be formed. A typical reaction in which this occurs is the process called **pair production,** in which a photon is changed to an electron and a positron. (This is the reverse of the antiparticle annihilation reaction.) There are other reactions of this general type where, in effect, energy is changed into rest mass and equal but opposite charges are created. The end result, as the fireball cooled, was to form a gas composed primarily of protons, neutrons, electrons, and other high-energy particles. The temperature was still too high to permit formation of hydrogen atoms, and, of course, nuclei for larger atoms were still unformed.

We can estimate the temperature of the fireball when hydrogen atoms began to form. The ionization energy of hydrogen is 13.6 eV. Clearly, thermal energy kT could not be much larger than this if hydrogen atoms were to survive. Since 300 K is equivalent to $\frac{1}{40}$ eV, 13.6 eV would correspond to a temperature of about 160,000 K. As its temperature fell below this value, the fireball became a cloud of hot hydrogen gas together with neutrons and other basic particles. As we shall see on page 802, the cloud of gas filling the universe has become so cold its temperature is now of the order of 3 K.

So far we have discussed the expanding cloud as a smooth, homogeneous entity. In all probability, the gas was not completely uniformly distributed throughout space. Certain regions must have had higher densities than others as a result of random factors, such as thermal motion. Although we have no direct proof of the hypothesis, it seems reasonable that regions of unusually high density acted as focal points for what might best be described as a gravitational condensation. Over huge regions of space, unbalanced gravitational forces were produced and made the matter start moving toward regions of higher density. Of course, superimposed on this was the continued radial motion of the material of the expanding fireball. *These huge, noncompact regions into which matter began to collect are presently huge* **galaxies,** *aggregate systems containing many stars.*

Formation of Galaxies

As the cloud began to form regions of higher density, the masses had to obey the usual laws of motion. One, the law of conservation of angular momentum, had a profound effect on the behavior of the condensing cloud. Any small net angular motion ω_0 of the mass within the huge region originally occupied by the galaxy must be multiplied by the moment of inertia of that region I_0 to obtain the original angular momentum $I_0\omega_0$. However, as aggregation into a galaxy took place, the radius r of gyration for the galaxy's mass became much smaller, thereby decreasing I since $I \propto r^2$. Since the angular momentum must be conserved, $I\omega = I_0\omega_0$; and as I decreases, ω must increase. Therefore, *as condensation took place, the material began to spin about the center of the condensation.*

Spinning of the condensing cloud mass would be fastest for regions with the largest rotation at the outset. We can therefore expect some galaxies to be spinning very little while others have relatively large angular speeds ("large" in this context still means millions of years per rotation). As with any spinning system (a stone on a string or a twirling pizza, for example), the spinning causes the system to flatten into a disk with the axis of rotation perpendicular to its plane. For the fastest-spinning galaxies, the spiral galaxies, this effect is very pronounced (Fig. 28.1). *Our own solar system is part of a somewhat less spiraling galaxy, the Milky Way, and is located about two-thirds of the way out from the galactic center.* (There are about 100 billion stars in our galaxy, the diameter of which is of the order of

FIGURE 28.1
The spiral galaxy M81.
(*Lick Observatory photograph.*)

100,000 light-years. It takes our sun about 200 million years to rotate once around the galactic center with its approximately 250-km/s orbital speed.)

While a galaxy as a whole is in the process of formation, localized regions in it are condensing much more rapidly into centers of mass, which eventually become the stars in the galaxy. It is easy to see how they become intensely hot. As matter is pulled toward the center by gravitational forces, PE is converted to KE. This means that a tremendous amount of energy is carried to the condensation center by the aggregating mass. Hence the temperature of the aggregate can become extremely high and in fact exceeds the temperatures needed for complete ionization of all the hydrogen atoms in the aggregate. This huge mass, perhaps many times larger than our sun, is now a white-hot, very dense "soup," called a **plasma,** composed of protons, neutrons, electrons, and other basic particles. This is the birth of a star.

28.3 Stellar Evolution

Pulled radially inward by gravitational forces, the star will continue to contract until its internal pressure balances the gravitational pressure. As in any gas, the internal pressure increases with rising temperature. For small aggregates of matter, the gravitational forces are small, and so equilibrium is reached at rather low temperatures. For the large stars of interest, however, contraction continues until the temperature in the star reaches a few million degrees. At this high temperature, the average thermal energy of a proton is about 1000 eV. In spite of this rather low average energy, enough high-energy protons exist for a fusion reaction to become possible.

At the earliest stages, the fusion reaction of importance is

$$_{1}^{1}\text{H} + _{1}^{1}\text{H} \rightarrow _{1}^{2}\text{H} + \text{positron} + \text{neutrino}$$

Fusion Reactions in Stars

where the neutrino is a zero-charge particle which has no rest mass. The deuteron, $_{1}^{2}\text{H}$, reacts again with a proton:

$$_{1}^{2}\text{H} + _{1}^{1}\text{H} \rightarrow _{2}^{3}\text{He} + \gamma$$

and this isotope of helium reacts as follows:

$$_{2}^{3}\text{He} + _{2}^{3}\text{He} \rightarrow _{2}^{4}\text{He} + 2\,_{1}^{1}\text{H}$$

In other words, protons are fused in this reaction to form helium; six protons react to form a helium nucleus and two protons. As with any fusion reaction of this type, large amounts of energy are released. The star is now capable of supplying energy to itself without further gravitational collapse.

As soon as the fusion reaction has become powerful enough to increase the thermal pressure within the star to a point where it balances the gravitational pressure, the star is stabilized. The temperature at the very center of the sun, the hottest point, is estimated to be close to 15 million kelvin. Although the proton reaction in the sun has been proceeding for about 4.5 billion years, apparently enough protons remain for the reaction to continue steadily for about that long in the future. In stars more massive than our own sun, the interior temperature is higher. (Why?) At these higher temperatures, other nuclear fusion reactions are possible.

Eventually the proton reaction uses most of the available protons, and the reaction slows down. As a result, the thermal pressure decreases and the unbalanced portion of the gravitational pressure causes the star to begin to contract again. When this happens, the outer (formerly cooler) regions of the star are heated enough for the protons there to begin reacting through the fusion process previously outlined. The resulting thermal pressure in this portion of the star causes the outer layers to expand. Hence the outer portion of the star enlarges while simultaneously cooling. At this stage the star changes in appearance from a white-hot star to a considerably larger, redder star, called a **red giant.**

Red-Giant Stars

Although the exterior surface of the star is cooled during this transition, the interior is heated as a result of the inner contraction. The core is largely ^4He, a resultant product of the burned-out proton-fusion reaction. Only after the star has reached a temperature of about 100 million kelvin does the helium begin to fuse. At that stage begins the reaction

$$\ce{^4_2He} + \ce{^4_2He} \rightarrow \ce{^8_4Be}$$

The beryllium then combines with helium:

$$\ce{^8_4Be} + \ce{^4_2He} \rightarrow \ce{^{12}_6C}$$

This reaction is called the **triple alpha** reaction because the stripped helium atom is an α particle.

Later stages of development are uncertain, but laboratory experiments using large accelerators show that reactions between ^{12}C and the other high-energy particles in the star can lead to the formation of larger nuclei. In fact, it seems likely that the stable elements of the periodic table are formed at this stage of star life. Since our own sun and the planets contain reasonably comparable amounts of the heavy elements, a clue exists as to the evolutionary stage of our own solar system.

Eventually the fusion reactions in the interior of a red giant must die out as the available fuel becomes exhausted. At that stage the gravitational forces are no longer balanced by sufficient thermal pressure, and the star will contract, becoming further heated because of the conversion of potential to thermal energy. After a few tens of millions of years, the star will have shrunk to a very dense, white-hot body called a **white dwarf.**

We have good theoretical reasons to believe that a white dwarf cannot be stable if its mass is greater than about 1.2 times the mass of our sun. Although definitive evidence is lacking for this hypothesis because the masses of only a few white dwarfs have been measured, the existing data do not contradict it. However, since many red giants have masses much in excess of this limiting mass, they must somehow lose mass as they contract to the white-dwarf stage. Exactly how this is accomplished is not known.

One possibility consistent with observation is that, through processes observed as novas and supernovas, the collapsing red giant undergoes explosions as it approaches the white-dwarf stage. These explosions send out into space great masses of gas composed of nuclei. As a result, the galaxy acquires a cloud composed mainly of hydrogen and helium but also containing the nuclei of the heavier elements. This cloud could then undergo gravitational condensation, and new stars would be formed. Consequently the whole process of star evolution outlined above could be repeated. According to this hypothesis, our own sun appears to be a star of this type since it contains the nuclei of heavy elements. Eventually (perhaps in 4 or 5 billion years), our sun should become a red giant; still later, it should contract to become a white dwarf. Of course, during the red-giant stage, the earth would become too hot for human habitation.

28.4 The Expanding Universe

According to the big bang theory, the universe is expanding. If so, we should notice a Doppler effect in the light of receding stars. Let us compute the relation between the wavelength of, say, a line in the hydrogen spectrum from a source on earth λ_0 as compared to the wavelength as observed for a source on a star. Suppose the star to be receding from the earth with speed v.

Consider a crest of the wave emitted by the star source. It will travel toward earth a distance ct in a time t, where c is the velocity of light. However, the stellar source will emit a wave crest once every τ_0 seconds as measured by a timer on the star. According to relativity theory, this time τ_0 will actually be read as $\tau_0/\sqrt{1 - (v/c)^2}$ by an earth clock. As a result, the distance the first crest will move during the time between its emission and the emission of the next crest will be (according to an earth observer)

$$\text{Distance} = ct = \frac{c\tau_0}{\sqrt{1 - (v/c)^2}}$$

During this time, the stellar source will have moved a distance vt, and so the second crest will actually be a distance $ct + vt$ behind the first. This, then, is the wavelength that would be observed on earth, namely,

$$\lambda = ct + vt \qquad \text{or} \qquad \lambda = \frac{(c + v)\tau_0}{\sqrt{1 - (v/c)^2}}$$

But since the stellar source is itself in an inertial system, the usual laws must apply there. Hence, $\tau_0 = \lambda_0/c$, and so *the wavelength λ of the star's line as observed on earth is related to the wavelength λ_0 of the line from a stationary source by the equation*[1]

$$\lambda = \lambda_0 \frac{1 + v/c}{\sqrt{1 - (v/c)^2}}$$

We see that λ *is larger than* λ_0. *In other words, the wavelengths of the spectral lines from a receding star will appear lengthened, i.e., shifted from blue toward the red. This is referred to as the* **red shift.**

Red Shift

If we examine the light reaching us from the stars, we do find the wavelengths emittted by the atoms to be shifted to the red. We interpret this to mean that all stars are moving away from us. By use of the red-shift equation, it is possible to compute their recession speeds. It is found that the more distant a star is from the earth, the faster it is receding from us. For the most distant stars, v is quite close to c, the speed of light. For them, lines in the blue are shifted into the infrared.

A rather simple experimental relation is found between the recession speed v of a star and its distance s from earth. It is

$$s = 3.6 \times 10^{17}v \qquad \text{m}$$

Let us now interpret this relation in terms of the big bang theory. We shall see that it allows us to compute the age of the universe.

---

[1] If you compare this with the Doppler effect equation for sound obtained in Chap. 14, you will see that they are different. At low v/c values, however, where relativistic effects can be ignored, they do coincide.

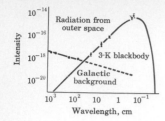

Radiation from outer space

3-K blackbody

Galactic background

Intensity

Wavelength, cm

Temperature of Space

We can consider the whole universe to be like the interior of a huge oven; the oven's temperature is the average temperature of the universe. As we saw in Chap. 25, the radiation emitted from a blackbody has a wavelength dependence characteristic of the temperature of the blackbody. Knowing the wavelength distribution of the electromagnetic radiation striking us from outer space, we can determine the temperature of the oven about us, the universe. The rather sketchy experimental data presently available indicate that the average temperature of the universe is close to 3 K. This is seen in the figure, where the points represent the data and the solid curve is predicted for a 3-K blackbody by Planck's radiation law. The maximum in the intensity curve occurs at a wavelength near 0.1 cm; this is in the short radar or microwave range. Measurements in this range are extremely difficult for two reasons: (1) the instrumentation needed is near the limit of that available for electronic detection; (2) the earth's atmosphere absorbs radiation in this short-wavelength range so strongly that it is almost impossible to obtain reliable data on earth. When an observatory is established on the moon or in space, this difficulty will be eliminated. We see, however, that the presently available data strongly suggest that the average temperature of the universe is close to 3 K.

According to the big bang theory, the earth is part of this now quite cold expanding cloud. It is a property of such an expanding system that everything within it is separating from everything else. The situation is much like that of the blueberries in a blueberry muffin. As the muffin bakes, the dough expands and each blueberry recedes from its neighbors. We can further illustrate this idea by the diagram of Fig. 28.2. If two objects are moving along the same line but at different speeds v_1 and v_2, they will be separating at a speed $v_2 - v_1$. If we assume they started from the same point at time $t = 0$, that is, when the fireball first exploded, their present separations should be given by

$$s = (v_2 - v_1)t$$

But this is identical in form to the experimental relation found from the red-shift data, namely,

$$s = (v)(3.6 \times 10^{17})$$

From our definition of $v_2 - v_1$, it is the same as v. Hence we have

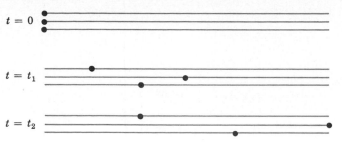

$$t = 0$$

$$t = t_1$$

$$t = t_2$$

$$t = 3.6 \times 10^{17}\ \text{s} = 1.1 \times 10^{10}\ \text{yr}$$

as the time from the first explosion of the fireball until now. This gives an age of 11 billion years for the universe.[2] Although this age has been arrived at from rather tentative assumptions, other data tend to support it at least approximately. One such supporting evidence is given in the next section.

Age of the Universe

28.5 Age Determination from Radioactivity

In principle, it is a simple matter to determine the age of a rock or other substance if it contains radioactive material together with its decay products. For example, at various places on the earth one can find rocks which still contain measurable amounts of ^{238}U. This element has a half-life of 4.5 billion years and decays to the end product ^{206}Pb, one of the less abundant isotopes of lead. Intermixed with the original ^{238}U is the final decay product ^{206}Pb.

It so happens that on earth we find about as much ^{238}U in such rocks as we find ^{206}Pb. We must conclude, therefore, that half the uranium has decayed to lead in the time since the rock was formed. Noting that half the uranium has decayed, we conclude that the rock is about one ^{238}U half-life old, namely 4.5 billion years.[3] This then gives us the time since the earth cooled sufficiently to allow rocks to solidify near its surface. How much older the earth is than this can be estimated only if it is known how the earth was formed. This we do not know for certain. The other natural radioactive elements which can be used to estimate the age of the earth's rocks all agree within reason with the result just given.

Radioactive dating methods can also be used to date much younger substances. Carbon-containing substances can be placed within a certain time scale provided the substance came originally from plant or animal life. This method, called **radiocarbon dating,** is based on the use of the rare isotope ^{14}C.

Radiocarbon Dating

Cosmic radiation striking the upper portion of the earth's atmosphere

[2] Analysis of these data is quite error-prone. Estimates for the age range from about 10 to 20 billion years.

[3] More accurately, about 3.9×10^9 yr.

leads to the formation of ^{14}C through the reaction

$$^{14}_{7}N + ^{1}_{0}n \rightarrow ^{14}_{6}C + ^{1}_{1}H$$

This unstable isotope of carbon has a half-life of 5730 yr. After formation, it circulates through the atmosphere, sometimes as carbon dioxide, CO_2. Plants take up CO_2 and incorporate it in their leaves and other structures. As a result, all plant life has some ^{14}C in it. Hence all plant life is slightly radioactive. Before the use of atomic weapons, the ratio of ^{14}C to ^{12}C in plant life was 1.5×10^{-12}.

If you have an old piece of wood and with very delicate measuring techniques find it to be only half as radioactive from ^{14}C as presently grown objects, you could assume that the wood had been dead for a long enough time for half its ^{14}C to have decayed. Since this takes one half-life, 5730 yr, you can conclude that to be the age of the wood. Of course, we have used easy numbers for computation purposes, but the method is not limited to them. A very old artifact cannot be dated this way since its radioactivity has decreased to too low a level. At the other end of the time scale, the object must be at least old enough for its radioactivity to have decreased measurably.

This gives only a glimpse of dating methods based on radioactivity. Many other systems exist, and often checks can be made of one method against the other. As far as the earth and moon are concerned, we can now say with some certainty that the ages of the oldest rocks on both are about the same, 3.9 billion years.

28.6 Origin of the Earth

Although the age of the oldest rocks on the earth is now fairly well agreed to be about 3.9 billion years, the circumstances which led to the creation of the solar system and earth are largely the subject of speculation. As we have seen, there is reason to believe that our sun is the product of a second-stage condensation; it appears to have condensed from the remnants of the explosion associated with the formation of one or more white dwarfs. If we accept that hypothesis, several alternate theories for the formation of the planets can be suggested. The most likely one appears to be the following.

As the cloud began to contract to form the sun, conservation of angular momentum required the cloud to rotate more and more swiftly. In order to preserve the angular momentum of the solar system as it is now, if the planets had not formed, the angular speed of the sun would need to be much larger than it is at present. Apparently, during the contraction process, it proved more advantageous from a kinetic and energetic standpoint for the rotating cloud to separate into rather distinct parts while only the center-most part contracted to form the sun. The outer portions underwent local condensation to form the planets. This picture predicts that all the planets would be rotating in the same way about the sun since they must preserve the original angular momentum of the cloud. Such is indeed the case.

Gravitational forces caused various portions of the cloud to coalesce into what were to become the sun and planets. Intense heat was generated as the cloud fragments impacted one another. The more massive the final object formed, the greater the gravitational forces and therefore the higher the temperature of the final object. In the case of the sun, the temperature rose high enough to ignite the fusion reaction. But since the planets are much less massive objects, their final temperatures were much lower than that of the sun, far too low to ignite the fusion reaction. We are not certain whether the initial temperature of the earth was high enough to cause it to be a molten mass.

There is much evidence to show that the earth was molten at some stage in its development, but this could easily have occurred later. In fact, much of the core of the earth is liquid even now. The very high temperatures generated there are thought to be the result of energy given off in the radioactive decay of the small amount of unstable nuclei still found on the earth. When the earth was formed, its radioactivity was considerably higher than now. It is believed that the earth was maintained in a molten state for a time by the energy released in radioactive decay.

Molten Earth

At this early time, the earth consisted of all the elements, but lighter ones predominated since they were formed in larger quantities in the fireball. Since atoms of many light elements had speeds larger than the escape speed at this high temperature, they escaped from the earth unless they were bound chemically into large molecules. The major portion of the earth left behind is iron, silicon, oxygen, magnesium, and aluminum. In view of the fact that oxygen is a rather light gas, it may seem surprising that it remains in large quantities. The reason for this is not far to seek; oxygen combines with silicon to form rather heavy silicate ions and molecules, which were retained.

As time went on, the earth cooled by radiation of heat out into space. The crust of the earth solidified. In this process of cooling, material rose to the surface. This process was instrumental in apportioning the materials inside the earth. The molten iron and silicon moved toward the center while the lighter silicates and similar compounds moved to the surface.

The situation within the earth today is known to us (albeit sketchily) from several sources. The reflection and penetration into the earth of waves sent out from sites of earthquakes and diagnostic explosions show that the surface of the earth has a thin crust of the order of 30 km thick, as indicated in Fig. 28.3. Below that is a thick solid region called the **mantle.** Still lower is the molten core, with perhaps a rather small solid inner core. Let us talk about each in turn.

The composition of the crust is varied, but is characterized by the fact that about 45 percent is oxygen in combination. Most of the rocks on earth are primarily compounds of oxygen and silicon as in silicates, although compounds of oxygen with aluminum, iron, calcium, etc. are also found.

Extending about 3000 km inward below the crust is the mantle, which is now also solid. Like the crust, it is still warmed from the heat that radioactive materials have produced in the earth. (Only the surface of the earth is greatly heated by the sun.) We believe the mantle to be composed mainly of oxygen, magnesium, and silicon in the compound magnesium silicate. Al-

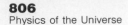

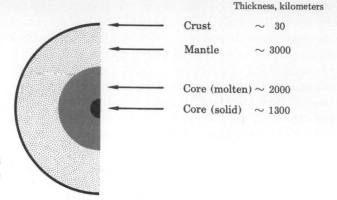

Thickness, kilometers

Crust ~ 30

Mantle ~ 3000

Core (molten) ~ 2000

Core (solid) ~ 1300

FIGURE 28.3
Tentative diagram of the
structure of the earth.

though the temperature in the lower mantle is red-hot, the material is under great pressure and is solid.

Beneath the mantle and extending to nearly the center of the earth is the molten core. It is primarily iron (we believe), but probably contains appreciable quantities of nickel and silicon as well. Contemporary theory supposes that electrical currents flow in the molten metallic core, and these currents are thought to be the source of the earth's magnetic field, but this process is still conjectural.

Our knowledge of the earth is still woefully incomplete. The moon and the planets are even less well known. At this writing, and despite the recent landings, we still have only tentative theories for the origin of the moon. And the structure and history of the planets are almost a total mystery. In view of our lack of understanding of the earth itself, it is not likely that the full character of the moon will be known in the near future. Clearly, a great deal remains to be learned about our own solar system.

28.7 The End of the Beginning

In the previous sections we obtained a glimpse of the wide areas of astrophysics and geophysics. These, together with many others, such as biophysics and particle physics, are only a few of the topics we would like to be able to discuss in detail. But a book must have an end. Even though the field of physics and its importance to you are much too large to be contained within the covers of a single book, you have at least begun to understand their scope.

Looking forward to the years ahead, we know that the physics of today is only the beginning of the physics which will be known a century from now. We expect that, in years to come, physics, biology, medicine, chemistry, and psychology will more and more merge into one vast science. As we learn more about molecules and how they combine to form cells, we shall be in a better position to learn the basis of life itself.

Yet this is but one avenue along which the physics of the future will advance. Our knowledge of the universe is still in its infancy. We are only beginning to understand the mechanism by which it was created. The behavior of distant galaxies, the history and future of the stars are but a few of the questions which intrigue the astrophysicists of our day. We cannot even begin to guess what course their investigations will follow in the years ahead.

And while some are investigating the immense reaches of the universe, many others will be answering questions about the smallest entities of which we know. Our understanding of the nucleus is still incomplete. We have only begun to learn of structure *within* the "primary" particles. Many people think we shall never know what (if anything) goes on inside a neutron, for example. The scientists of 1900 could hardly have been expected to predict the discoveries so important to present-day physics—the photon, relativity, the wave character of matter. We, too, can predict only that the days of exciting discoveries have not yet ended.

You, of course, will be intimately concerned with these developments, since you will be living in a world greatly influenced by them. It is to be hoped that you will use the knowledge gained from this book as a footing on which to build in the future and that you will relate the discoveries of science to the needs of society. Many people speak of science as a monster which will eventually destroy all society. This is a definite possibility which we must all face, scientist and nonscientist alike.

As human beings, we cannot hide our heads in the sand and hope that scientific discoveries will cease or peacefully melt away. Science will continue to advance. Whether these advances will be for the improvement or destruction of humans is dependent on how we, as members of society, react to them. To make proper use of our scientific discoveries will require the active cooperation of scientist, politician, philosopher, theologian—all citizens. We must understand one another if we are to help one another in this monumental task. Even those of you who do not proceed further in physics will now be better prepared to meet this challenge. We hope that, with the additional knowledge and understanding you have obtained from this book, you will be better able to take your rightful place in the universe of your future.

Summary

According to the big bang theory of the universe, our universe existed as a highly dense, extremely hot fireball about 11 billion years ago. The ball expanded rapidly and was cooled. It is still expanding. Whether it will continue to expand forever or will eventually contract to its original form is not known.

As the universe expanded and cooled, local fluctuations in mass gave rise to regions of condensing matter which formed galaxies. Within the galaxies, hot suns (or stars) were formed as the material pulled into a condensation point by huge gravitational forces generated heat upon collision. The temperatures so generated were often high enough to ignite the nuclear fusion reaction, and this then furnished additional energy to the star. As the star contracted and went through several stages of development, the higher-atomic-number elements found in the universe were formed and ejected.

The age of ancient rocks on the earth can be determined from their radioactivity. Such measurements tell us that the earth's crust solidified about 3.9 billion years ago. At one stage, the earth was molten. Part of the necessary heat was the result of the gravitational forces which formed it, and part was due to energy furnished by radioactive decay. Even now the center of the earth is molten, and the heat is furnished by radioactive substances.

Minimum Learning Goals

Upon completion of this chapter, you should be able to do the following:

1 Give an outline of the big bang theory including the following: original situation; expansion and cooling; galaxy formation; rotation; star formation; energy source of stars; production of He and higher-Z elements; time span since the big bang.
2 Describe the process by which the material of stars is heated to the point where the fusion reaction can ignite. State at least one major way that our sun and the stars maintain their high temperatures.
3 Explain how the red shift implies that the universe is expanding.
4 Explain how the age of a uranium-bearing rock can be determined from an analysis of its constitution. Give the basic idea of radiocarbon dating.
5 Sketch a cross section of the earth and show its major layers. Give a rough estimate of the thickness of the crust. State which parts are molten.

Important Terms and Phrases

You should be able to define or explain each of the following:

Big-bang theory
Age of the universe
Galaxy

Sun and stars
Red giant; white dwarf
Temperature of space at present
Red shift
Crust, mantle, core of earth

Questions and Guesstimates

1 What factors must be known in order to compute whether our universe will pulsate or expand forever?
2 Why can't one determine whether the universe has pulsated several times before?
3 Our present knowledge indicates that the primordial fireball took about an hour to cool to the range 10^9 K. At that time, how large was the average thermal energy kT, in electronvolts? Compare this to the binding energy of a ^{4}He nucleus. What can you say about nuclear stability at that time?
4 The temperature of the surface of the sun can be estimated by measuring the wavelength dependence of the light given off by it. Explain how such an estimate can be made.
5 We know that charge is conserved. Does this mean that there is the same number of positive and negative charges in the universe now as in the original fireball?
6 When we look into space, we find that all the galaxies are flying away from us. Does this mean that we are at the center of the universe?
7 Since there is no sign of life in our solar system other than that on earth, the nearest neighbors we could have would live on a planet of the nearest star, Alpha Centauri, which is 4×10^{16} m away. About how long would it take one of our present spaceships to reach there? A spaceship with speed $0.99c$? (E)
8 When it was formed about 5 billion years ago, why was the earth much more radioactive than it is now? Would you expect the radioactivity types today to be the same as then?
9 Ordinarily we assume that the heating of a radioactive substance will not alter its half-life. Estimate how hot the substance must be in order for this assumption to be wrong. (You may wish to use the fact that at 300 K thermal energy is about $\frac{1}{40}$ eV.) (E)
10 If the temperature of outer space is actually about $-270°$C, why don't astronauts freeze to death during space journeys? On the moon, they actually require refrigeration. Why?
11 Why doesn't the moon have any atmosphere?
12 The temperature inside deep mines is uniform throughout the year. What does this tell us about the source of the heat in the mine?
13 About how old is a piece of wood if its ^{14}C content is about one-tenth that in trees today? What factors could make your estimate be in error?
14 The earth receives energy from the sun at an average rate of about 2 cal/(m^2)(min). Estimate how much hydrogen is being converted to helium in the sun each day.

Appendix 1
Conversion of Units

It often happens that we wish to convert a quantity expressed in one set of units to another set of units. Typically, we might want to know how many kilometers are equivalent to 20 mi. For conversions like this, we make use of conversion factors. Let us see what a conversion factor is.

We know, for example, that

$$100 \text{ cm} = 1 \text{ m}$$

Division by 100 cm gives

$$1 = \frac{1 \text{ m}}{100 \text{ cm}} = 0.010 \text{ m/cm}$$

This is a conversion factor between meters and centimeters. Notice that the conversion factor is equal to unity. When we multiply any quantity by a conversion factor, it is the same as multiplying by unity. It does not change the value of the quantity in question. We now give a few examples of the use of conversion factors.

1 How many meters are there in 30 mi? The appropriate conversion factor is obtained from 1.60 km = 1 mi to be 1.60 km/mi. Since this is simply unity, we can multiply or divide by it without changing the value of a quantity. Therefore

$$30 \text{ mi} = (30 \text{ mi})\left(1.60 \frac{\text{km}}{\text{mi}}\right) = 48 \text{ km}$$

2 How many miles are there in 6000 km? Because 1 mi/1.60 km = 1, we have

$$6000 \text{ km} = (6000 \text{ km})\left(\frac{1 \text{ mi}}{1.60 \text{ km}}\right) = 3750 \text{ mi}$$

3 How many hours are there in 200,000 s? We know that 1 h = 60 min and 60 s = 1 min. Therefore

$$200{,}000 \text{ s} = (200{,}000 \text{ s})\left(\frac{1 \text{ min}}{60 \text{ s}}\right) = \frac{20{,}000}{6} \text{ min}$$

$$= \left(\frac{20{,}000}{6} \text{ min}\right)\left(\frac{1 \text{ h}}{60 \text{ min}}\right) = 55.6 \text{ h}$$

Typical conversion factors are given on the inside front cover of this book.

Appendix 2
Powers of Ten

It is more convenient to write numbers such as 1,420,000,000 and 0.00031 in a form called *scientific notation*. Their representations in this notation are 1.42×10^9 and 3.1×10^{-4}, respectively. Any number can be written as a number smaller than 10 times a power of 10, as we now show. Consider the identities

$$621 = (62.1)(10) = (6.21)(10)(10) = 6.21 \times 10^2$$

$$86,435 = (8.6435)(10)(10)(10)(10) = 8.6435 \times 10^4$$

As we see, if we move the decimal point of a number n places to the left, we must multiply the number by 10^n if it is to remain unchanged. Check your understanding of this concept by showing that

$$96,400 = 9.6 \times 10^4$$

$$1,420,000,000 = 1.42 \times 10^9$$

The procedure for writing numbers less than unity is similar. But here we divide rather than multiply by powers of 10. It is illustrated by the following examples:

$$0.163 = (1.63)(\tfrac{1}{10}) = (1.63)(10^{-1}) = 1.63 \times 10^{-1}$$

$$0.0604302 = (6.04302)(\tfrac{1}{10})(\tfrac{1}{10}) = 6.04302 \times 10^{-2}$$

Clearly, when we move the decimal point n places to the right, we must multiply the number by 10^{-n} if its value is to remain unchanged. Test yourself by showing that

$$0.00031 = 3.1 \times 10^{-4}$$

$$0.00000625 = 6.25 \times 10^{-6}$$

People sometimes summarize this notation by the following statements:

1 To evaluate a number such as 6.53×10^n, move the decimal point in 6.53 a total of n places to the right. Thus

$$6.53 \times 10^7 = 65,300,000$$

2 To evaluate a number such as 3.70×10^{-n}, move the decimal point in 3.70 a total of n places to the left. Thus

$$3.70 \times 10^{-4} = 0.000370$$

Periodic Table of the Elements

The values listed are based on $^{12}_{6}C = 12$ u exactly. For radioactive elements, the approximate atomic weight of the most stable isotope is given in brackets.

Period	I$_A$	II$_A$	III$_B$	IV$_B$	V$_B$	VI$_B$	VII$_B$	VIII	VIII	VIII	I$_B$	II$_B$	III$_A$	IV$_A$	V$_A$	VI$_A$	VII$_A$	0
1	1 H 1.00797																	2 He 4.003
2	3 Li 6.939	4 Be 9.012											5 B 10.81	6 C 12.011	7 N 14.007	8 O 15.9994	9 F 19.00	10 Ne 20.183
3	11 Na 22.990	12 Mg 24.31											13 Al 26.98	14 Si 28.09	15 P 30.974	16 S 32.064	17 Cl 35.453	18 Ar 39.948
4	19 K 39.102	20 Ca 40.08	21 Sc 44.96	22 Ti 47.90	23 V 50.94	24 Cr 52.00	25 Mn 54.94	26 Fe 55.85	27 Co 58.93	28 Ni 58.71	29 Cu 63.54	30 Zn 65.37	31 Ga 69.72	32 Ge 72.59	33 As 74.92	34 Se 78.96	35 Br 79.909	36 Kr 83.80
5	37 Rb 85.47	38 Sr 87.62	39 Y 88.905	40 Zr 91.22	41 Nb 92.91	42 Mo 95.94	43 Tc [99]	44 Ru 101.1	45 Rh 102.905	46 Pd 106.4	47 Ag 107.870	48 Cd 112.40	49 In 114.82	50 Sn 118.69	51 Sb 121.75	52 Te 127.60	53 I 126.90	54 Xe 131.30
6	55 Cs 132.905	56 Ba 137.34	†	72 Hf 178.49	73 Ta 180.95	74 W 183.85	75 Re 186.2	76 Os 190.2	77 Ir 192.2	78 Pt 195.09	79 Au 196.97	80 Hg 200.59	81 Tl 204.37	82 Pb 207.19	83 Bi 208.98	84 Po [210]	85 At [210]	86 Rn [222]
7	87 Fr [223]	88 Ra [226]	‡															

† *Lanthanide series*

57 La 138.91	58 Ce 140.12	59 Pr 140.91	60 Nd 144.24	61 Pm [147]	62 Sm 150.35	63 Eu 152.0	64 Gd 157.25	65 Tb 158.92	66 Dy 162.50	67 Ho 164.93	68 Er 167.26	69 Tm 168.93	70 Yb 173.04	71 Lu 174.97

‡ *Actinide series*

89 Ac [227]	90 Th 232.04	91 Pa [231]	92 U 238.03	93 Np [237]	94 Pu [242]	95 Am [243]	96 Cm [247]	97 Bk [247]	98 Cf [251]	99 Es [254]	100 Fm [253]	101 Md [256]	102 No [254]	103 Lw [257]

Appendix **4**
An Abbreviated Table of Isotopes

The values listed are based on $^{12}_{6}C = 12$ u exactly. Electron masses are included.

Atomic number Z	Symbol	Average atomic mass	Element	Mass number A	Relative abundance, %	Mass of isotope
0	*n*	1.008665	Neutron	1		
1	H	1.00797	Hydrogen	1	99.985	1.007825
				2	0.015	2.014102
2	He	4.0026	Helium	3	0.00015	3.016049
				4	100 —	4.002603
3	Li	6.939	Lithium	6	7.52	6.015123
				7	92.48	7.016005
4	Be	9.0122	Beryllium	9	100 —	9.012183
5	B	10.811	Boron	10	19.78	10.012939
				11	80.22	11.009305
6	C	12.01115	Carbon	12	98.892	12.0000000
				13	1.108	13.003355
7	N	14.0067	Nitrogen	14	99.635	14.003074
				15	0.365	15.000109
8	O	15.9994	Oxygen	16	99.759	15.994915
				17	0.037	16.999133
				18	0.204	17.999160
9	F	18.9984	Fluorine	19	100	18.998403
10	Ne	20.183	Neon	20	90.92	19.992440
				22	8.82	21.991384
11	Na	22.9898	Sodium	23	100 —	22.989770
12	Mg	24.312	Magnesium	24	78.60	23.985045
13	Al	26.9815	Aluminum	27	100	26.981541
14	Si	28.086	Silicon	28	92.27	27.976927
				30	3.05	29.973761
15	P	30.9738	Phosphorus	31	100	30.973763
16	S	32.064	Sulfur	32	95.018	31.972072
17	Cl	35.453	Chlorine	35	75.4	34.968854
				37	24.6	36.965903
18	Ar	39.948	Argon	40	99.6	39.962384
19	K	39.102	Potassium	39	93.08	38.963708
20	Ca	40.08	Calcium	40	96.97	39.962589
21	Sc	44.956	Scandium	45	100	44.955914
22	Ti	47.90	Titanium	48	73.45	47.947948
23	V	50.942	Vanadium	51	99.76	50.943963
24	Cr	51.996	Chromium	52	83.76	51.940510
25	Mn	54.9380	Manganese	55	100	54.938046
26	Fe	55.847	Iron	56	91.68	55.934939

Atomic number Z	Symbol	Average atomic mass	Element	Mass number A	Relative abundance, %	Mass of Isotope
27	Co	58.9332	Cobalt	59	100	58.93319
28	Ni	58.71	Nickel	58	67.7	57.93534
				60	26.23	59.93079
29	Cu	63.54	Copper	63	69.1	62.92959
30	Zn	65.37	Zinc	64	48.89	63.92914
31	Ga	69.72	Gallium	69	60.2	68.92558
32	Ge	72.59	Germanium	74	36.74	73.92118
33	As	74.9216	Arsenic	75	100	74.92160
34	Se	78.96	Selenium	80	49.82	79.91652
35	Br	79.909	Bromine	79	50.52	78.91834
36	Kr	83.30	Krypton	84	56.90	83.91150
37	Rb	85.47	Rubidium	85	72.15	84.91180
38	Sr	87.62	Strontium	88	82.56	87.90562
39	Y	88.905	Yttrium	89	100	88.90586
40	Zr	91.22	Zirconium	90	51.46	89.90471
41	Nb	92.906	Niobium	93	100	92.90638
42	Mo	95.94	Molybdenum	98	23.75	97.90541
43	Tc	*	Technetium	98		97.90721
44	Ru	101.07	Ruthenium	102	31.3	101.90435
45	Rh	102.905	Rhodium	103	100	102.90550
46	Pd	106.4	Palladium	106	27.2	105.90348
47	Ag	107.870	Silver	107	51.35	106.90509
48	Cd	112.40	Cadmium	114	28.8	113.90336
49	In	114.82	Indium	115	95.7	114.90388
50	Sn	118.69	Tin	120	32.97	119.90220
51	Sb	121.75	Antimony	121	57.25	120.90382
52	Te	127.60	Tellurium	130	34.49	129.90623
53	I	126.9044	Iodine	127	100	126.90448
54	Xe	131.30	Xenon	132	26.89	131.90416
55	Cs	132.905	Cesium	133	100	132.90543
56	Ba	137.34	Barium	138	71.66	137.90524
57	La	138.91	Lanthanum	139	99.911	138.90636
58	Ce	140.12	Cerium	140	88.48	139.90544
59	Pr	140.907	Praseodymium	141	100	140.90766
60	Nd	144.24	Neodymium	144	23.85	143.90998
61	Pm	*	Promethium	145		144.91275
62	Sm	150.35	Samarium	152	26.63	151.91974
63	Eu	151.96	Europium	153	52.23	152.92124
64	Gd	157.25	Gadolinium	158	24.87	157.92410
65	Tb	158.924	Terbium	159	100	158.92535
66	Dy	162.50	Dysprosium	164	28.18	163.92918
67	Ho	164.930	Holmium	165	100	164.93030
68	Er	167.26	Erbium	166	33.41	165.93031
69	Tm	168.934	Thulium	169	100	168.93423
70	Yb	173.04	Ytterbium	174	31.84	173.93887
71	Lu	174.97	Lutetium	175	97.40	174.94079
72	Hf	178.49	Hafnium	180	35.44	179.94656
73	Ta	180.948	Tantalum	181	100	180.94801
74	W	183.85	Tungsten	184	30.6	183.95099
75	Re	186.2	Rhenium	187	62.93	186.95577
76	Os	190.2	Osmium	192	41.0	191.96149
77	Ir	192.2	Iridium	193	61.5	192.96294
78	Pt	195.09	Platinum	195	33.7	194.96482

Atomic number Z	Symbol	Average atomic mass	Element	Mass number A	Relative abundance, %	Mass of Isotope
79	Au	196.967	Gold	197	100	196.96655
80	Hg	200.59	Mercury	202	29.80	201.97063
81	Tl	204.37	Thallium	205	70.50	204.97446
82	Pb	207.19	Lead	208	52.3	207.97664
83	Bi	208.980	Bismuth	209	100	208.98042
84	Po	210	Polonium	210		209.98287
85	At	*	Astatine	211		210.98750
86	Rn	*	Radon	211		210.99060
87	Fr	*	Francium	221		221.01418
88	Ra	226.054	Radium	226		226.02541
89	Ac	*	Actinium	225		225.02314
90	Th	232.038	Thorium	232	100	232.03805
91	Pa	231.036	Protactinium	231		231.03594
92	U	238.03	Uranium	233		233.03963
				235	0.715	235.04393
				238	99.28	238.05079
93	Np	*	Neptunium	239		239.05294
94	Pu	*	Plutonium	239		239.05216
95	Am	*	Americium	243		243.06138
96	Cm	*	Curium	245		245.06549
97	Bk	*	Berkelium	248		248.070305
98	Cf	*	Californium	249		249.07485
99	Es	*	Einsteinium	254		254.08801
100	Fm	*	Fermium	252		252.08265
101	Md	*	Mendelevium	255		255.0911
102	No	*	Nobelium	254		254
103	Lw	*	Lawrencium	257		257

*The atomic masses of unstable elements are not listed unless the isotope given constitutes the major isotope.

Appendix 5
Trigonometric Functions

Angle, deg	Sine	Co-sine	Tan-gent	Angle, deg	Sine	Co-sine	Tan-gent	Angle, deg	Sine	Co-sine	Tan-gent
0°	0.000	1.000	0.000	31°	.515	.857	.601	61°	.875	.485	1.804
1°	.018	1.000	.018	32°	.530	.848	.625	62°	.883	.470	1.881
2°	.035	0.999	.035	33°	.545	.839	.649	63°	.891	.454	1.963
3°	.052	.999	.052	34°	.559	.829	.675	64°	.899	.438	2.050
4°	.070	.998	.070	35°	.574	.819	.700	65°	.906	.423	2.145
5°	.087	.996	.088	36°	.588	.809	.727	66°	.914	.407	2.246
6°	.105	.995	.105	37°	.602	.799	.754	67°	.921	.391	2.356
7°	.122	.993	.123	38°	.616	.788	.781	68°	.927	.375	2.475
8°	.139	.990	.141	39°	.629	.777	.810	69°	.934	.358	2.605
9°	.156	.988	.158	40°	.643	.766	.839	70°	.940	.342	2.757
10°	.174	.985	.176	41°	.658	.755	.869	71°	.946	.326	2.904
11°	.191	.982	.194	42°	.669	.743	.900	72°	.951	.309	3.078
12°	.208	.978	.213	43°	.682	.731	.933	73°	.956	.292	3.271
13°	.225	.974	.231	44°	.695	.719	.966	74°	.961	.276	3.487
14°	.242	.970	.249	45°	.707	.707	1.000	75°	.966	.259	3.732
15°	.259	.966	.268	46°	.719	.695	1.036	76°	.970	.242	4.011
16°	.276	.961	.287	47°	.731	.682	1.072	77°	.974	.225	4.331
17°	.292	.956	.306	48°	.743	.669	1.111	78°	.978	.208	4.705
18°	.309	.951	.325	49°	.755	.656	1.150	79°	.982	.191	5.145
19°	.326	.946	.344	50°	.766	.643	1.192	80°	.985	.174	5.671
20°	.342	.940	.364	51°	.777	.629	1.235	81°	.988	.156	6.314
21°	.358	.934	.384	52°	.788	.616	1.280	82°	.990	.139	7.115
22°	.375	.927	.404	53°	.799	.602	1.327	83°	.993	.122	8.144
23°	.391	.921	.425	54°	.809	.588	1.376	84°	.995	.105	9.514
24°	.407	.914	.445	55°	.819	.574	1.428	85°	.996	.087	11.43
25°	.423	.906	.466	56°	.829	.559	1.483	86°	.998	.070	14.30
26°	.438	.899	.488	57°	.839	.545	1.540	87°	.999	.052	19.08
27°	.454	.891	.510	58°	.848	.530	1.600	88°	.999	.035	28.64
28°	.470	.883	.532	59°	.857	.515	1.664	89°	1.000	.018	57.29
29°	.485	.875	.554	60°	.866	.500	1.732	90°	1.000	.000	∞
30°	.500	.866	.577								

Appendix 6
Answers

Chapter 1

1. (a) 6.04×10^5; (b) 1.63×10^{-4};
   (c) 4.7×10^{-8}; (d) 2.95×10^{-9};
   (e) 7.3×10^7
3. 4.30×10^{-5}
5. 3.2×10^6
7. $\lambda = v/f$
9. $a = \frac{1}{2}$; $b = -\frac{1}{2}$
11. 10 blocks; 53° north of east
13. 71 km, 71 km
15. 14.2 cm at 102°
17. 276 km at 56°
19. 1540 km, 890 km
21. 183 lb, 114°; −183 lb, 246°
23. 24.4 cm, 35°
25. 3.4 N, −29°
27. 105 paces at 50°
29. 1400 km at 36° north of east
31. 55 lb at −69°
33. 103 m, 34°
35. 24.4 cm at 255°
37. 53.9 cm; 0.1 cm
39. (a) 441 cm²; (b) 1280 cm³;
    (c) 3.7×10^{-3} cm
41. 13.1 lb; 10.5 lb

Chapter 2

1. 91.4 mi/h; 147 km/h
3. 73.5 ms
5. 0.49 cm/yr; 49 cm per century upward
7. (a) 34.9 cm/s; (b) zero
9. (a) +1.00 cm/s; (b) +0.86 cm/s;
   (c) −0.40 cm/s; (d) −1.00 cm/s; (e) zero
11. 26.7 s
13. (a) 15.2 cm; (b) 91.4 cm
15. (a) 1.39 m/s; (b) 2.61×10^{-4} m/s;
    (c) 11.2 m/s
17. (a) zero; (b) 2.4 cm/s; (c) −0.91 cm/s
19. 0.56 m/s²

21. 1.79 m/s²; 175 m
23. 3.2×10^5 m/s²; 4.7×10^{-4} s
25. 4.3 s; 64 m
27. 2.4×10^{17} m/s²; 3.33×10^{-11} s
29. (a) 418 s; (b) 0.58 m/s
31. 42 m
33. 1.50 min; 0.30 min
35. (a) 0.20 m/s²; (b) 830 m
37. 1.54 m/s²
39. (a) 1.01 s; (b) 9.9 m/s
41. 1.25 s; 60 ft/s
43. 11.0 m
45. 10.1 m
47. (a) 3.0 m; (b) 2.0 s
49. 17.7 m/s

Chapter 3

1. Proof
3. 280 N
5. 140 N
7. 1980 N
9. 229 N
11. 1.86×10^{-40} N; $F/W = 1.14 \times 10^{-14}$
13. $F_e/F_0 = 1.34$
15. 25 h
17. (a) 40 N; (b) 28 N; (c) 34.6 N
19. (a) 8.0, 17.6, 6.4 N; (b) 10.8, 0.20, 5.7 m/s²
21. 1000 N
23. (a) 0.272; (b) 1.42 m/s²
25. 5.4 m
27. 1×10^5 N
29. 29.4 m/s²
31. (a) 768 N; (b) 408 N
33. 5.4 m/s²; 5.9 m
35. (a) 3.6 m/s², 10.9 N; (b) 0.70 m/s², 10.9 N
37. (a) 3.82 m/s², 9.6 N; (b) 0.90 m/s², 14.2 N
39. (a) 47 N; (b) 1.43 s
41. 3.9 m/s²
43. 1.73 m/s²; no

45. $F/(2m)$; $(F - 2f)/(2m)$
47. 0.735 N; 2.39 N; 1.84 m/s²; 3.68 m/s²
49. 127 m; 55 m
51. 2.71 m/s²; 50 N
53. 0.98 m/s²; 17.6 N; 105 N
55. 470 lb
57. (*a*) 1.8, 3.94, 1.44 lb; (*b*) 35, 0.65, 18.2 ft/s²
59. 45 ft

Chapter 4

1. 4330 J; 2600 ft · lb
3. 88 J
5. −343 J
7. 147 kJ
9. 6.7 hp
11. 0.47 hp
13. 16.9 km/gal
15. 3.6×10^{-14} N
17. 5000 N; 8.0 s; 0.255
19. 0.455 W
21. 470 kJ; no; 0.117 hp
23. 14.2 m/s
25. 15.5 N
27. (*a*) 180 kJ; (*b*) 0.022
29. 2.3×10^{-4} hp
31. 1940 N; 0.33 ms
33. 13.5 m/s
35. 4.4 m/s; 2.42 m/s; 3.1 m/s
37. 8.6×10^{-4} N
39. Proof
41. 7.2 m/s
43. (*a*) $2\sqrt{Lg/3}$; (*b*) $2\sqrt{Lg/3}$
45. 6.2 N; 0.98 m/s
47. (*a*) 20; (*b*) 15; (*c*) 75 percent
49. (*a*) 8.3; (*b*) 11.1; (*c*) 75 percent
51. 0.59 hp
53. (*a*) 265 N; (*b*) 0.0265 N

Chapter 5

1. (*a*) 20,000 kg · m/s; (*b*) 2.25×10^{-4} kg · m/s
3. $m\sqrt{2gh}$
5. 8570 N; 60,000 N · s
7. 7000 N; 62.5 m
9. 7200 N
11. (*a*) 400 N · s; (*b*) 1330 N
13. (*a*) 8.5 N · s; (*b*) 21,200 N

15. (*a*) 1.2 ms; (*b*) 0.50 N · s; (*c*) 417 N
17. $vM_1/(M_1 + M_2)$
19. 5.0 m/s to right
21. 0.0200 m/s
23. 0.10 m
25. 162 m/s
27. 225 s
29. (*a*) $\sqrt{2gh}$; (*b*) $\frac{1}{4}h$
31. Zero
33. 3.34×10^{-27} kg
35. 21 m/s
37. $\frac{1}{2}v_0$ and $3v_0/2$ both to right
39. 2.16×10^4 m/s
41. $v_0\sqrt{2}$ at 225°
43. $2.01v_0$ at 207°; $KE_{before} < Ke_{after}$
45. $\frac{1}{2}$
47. 479 m/s

Chapter 6

1. (*a*) 0.0694 rev, 0.436 rad; (*b*) 464°, 1.29 rev; (*c*) 263°, 4.58 rad
3. (*a*) 9.16×10^{-3} rad; (*b*) 0.0337 rad
5. (*a*) 200 deg/s; (*b*) 3.49 rad/s
7. 1.74×10^{-3} rad/s
9. 0.47 rev/s²; 2.94 rad/s²
11. (*a*) 0.80 rev/s²; (*b*) 160 rev
13. 1.13 rev/s
15. (*a*) 3.49 rad/s; (*b*) 17.4 cm/s
17. 17.4 rev
19. (*a*) 463 m/s; (*b*) zero
21. 3.77 m
23. 7.96 rev/s, 50 rad/s, 2870 deg/s
25. (*a*) 50 m; (*b*) 26.5 rev
27. (*a*) 2.0 rev/s², 225 rev; (*b*) 70.7 m
29. 1.67 rad/s²
31. 21,600 N
33. 6040 N
35. 0.394 rev/s
37. 14 m/s
39. 9.5×10^{-8} N
41. 6.0×10^{24} kg
43. 3.55×10^7 m/s²; $F_c/(mg) = 3.6 \times 10^6$
45. 10.1 m/s
47. 15.7 N; 27.4 N
49. Proof
51. $\sqrt{rg/(\tan \theta)}$
53. 0.99m; 4.85 m/s, 2.00 m/s

55. 7.84×10^{-13} m
57. (a) 28.7 m; (b) 1.66 s
59. 13.9 m/s

Chapter 7
1. 450 N at 249°
3. 433 N
5. 676 N
7. (a) 730 N; (b) 2130 N
9. 28 N; 80 N
11. 522 N; 336 N
13. 882, 383, 939, 321 N
15 (a) I, 0.766L, $L/3$, 0.289L; (b) $-FL$, 0.766FL, $FL/3$, 0.289FL
17. $(700b)/(a + b)$
19. 196 N; 309 N
21. 351 N
23. (a) 1175 N; (b) 755 N, 100 N
25. (a) 2.76F; (b) $-0.44F$, 3.46F
27. 975, 294, 604 N
29. 1.35×10^{47} kg·m^2
31. (a) 3.37 kg·m^2; (b) 4.4 kg·m^2
33. 1.00×10^{-3} kg·m^2
35. 0.80 N·m
37. 0.0080 N·m
39. (a) 10 s; (b) 3.18 rev
41. 0.060 kg·m^2; 0.489 N
43. 4.8 rev/s
45. 0.65 rev/s
47. 1.71 m/s; 5.5 rev/s
49. 0.286 cm
51. (a) 1.50 kg·m^2; 28 rev/s
53. (a) 0.75 rev/s; (b) yes
55. Proof
57. 0.388 rev/s

Chapter 8
1. 2.70 g/cm^3
3. 0.0697 m^3
5. 1.222 g/cm^3
7. (a) 7.35×10^6 N/m^2; (b) 7.35×10^6 Pa; (c) $P_h/P_a = 73$
9. 0.91 mm
11. 1.43×10^{-4}
13. Proof
15. 3750 Pa

17. (a) 6×10^{-7}; (b) 1.0×10^7 Pa
19. 6080 N, 1370 lb
21. 1.47×10^5 Pa; 2.47×10^5 Pa
23. 163 cm^2
25. 68.6 cmHg; 9.1×10^4 N/m^2
27. 6.3×10^8 N
29. 10 N
31. 0.0169 N
33. 4.67
35. 2.00 cm^3; 5.00 g/cm^3
37. 950 kg/m^3
39. 0.895
41. 21.60 g
43. 12.00 cm^3; 0.833 g/cm^3
45. 1.34 mPl
47. 5.7×10^{-4} m
49. 4.2×10^{-4} N
51. (a) rise; (b) 12.1 m/s
53. (a) 17.1 m/s; (b) 102 cm^3
55. 1.96×10^5 Pa
57. 22 kPa

Chapter 9
1. (a) 25°C, 298 K; (b) -35°C, 238 K
3. (a) -38°F. 675°F; (b) 234 K, 630 K
5. -40°C
7. 5.89×10^{-26} kg
9. 2.83×10^{-26} kg
11. 3.3×10^{23}
13. (a) 4.15×10^{-23} kg; (b) 2.41×10^{19}; (c) 2.29×10^{19}
15. 2.12×10^5 N/m^2, 2.12×10^5 Pa, 2.10 atm, 160 cmHg
17. (a) 132°C; (b) 10.5 atm
19. 177°C
21. 2.63V_0
23. 580°C
25. $V_0(P_a + \rho gh)/P_a$
27. 251 kg
29. 13.6×10^5 Pa
31. 1.15×10^{11} atm
33. 1910 m/s
35. (a) 10,000 K; (b) 141,000 K
37. 2.1×10^{-16} J; 5×10^5 m/s
39. (a) 5.5×10^{-11} kmol; (b) 3.3×10^{10}
41. 11,200 cm^3
43. Proof

45. 0.019

47. 42,300 kg/kmol

Chapter 10

1. 35,000 cal

3. (*a*) 93 cal; (*b*) 0.37 Btu

5. 34

7. 7.75 h

9. 43°C

11. 2400 cal

13. 3075 cal

15. 78 g

17. 8.0 g

19. 10,200 cal

21. 14.3 g

23. 68°C

25. 0.172 cal/(g)(°C)

27. 24.3°C

29. 7.85 g

31. 0.55 cal/(g)(°C)

33. 0°C

35. (*a*) 25 J; (*b*) 6.0 cal; (*c*) 9.6°C

37. (*a*) 0.070 J; (*b*) 84°C

39. 0.0094°C

41. (*a*) 450 cal; (*b*) 790 m/s

43. 0.25 cm

45. 0.63 cm

47. 1.0016

49. 100.009 cm^3

51. 3.4×10^5 N

53. Proof

55. 3.4×10^{-5} cal/(g)(°C)

57. 800 cal/s

59. 29 kcal

61. 2.0×10^5 cal

63. 98.78°C, 1.22°C

65. 0.043

67. Proof

69. (*a*) 13.50 g/m^3; (*b*) 79 percent

71. 21 percent

73. 29 percent

Chapter 11

1. 2960 J

3. (*a*) zero; (*b*) −10.8 cal

5. (*a*) 785 J, zero; (*b*) 785 J, 520 J

7. (*a*) 120 J, zero, −40 J, zero; (*b*) 80 J

9. (*a*) 879 K; (*b*) 628 K

11. (*a*) 22.5 J, −12.5 J; (*b*) 6.0 cal

13. (*a*) 2.46×10^{-5} kmol; (*b*) 879 K, 1465 K;
  (*c*) 100 cal, 71.6 cal; (*d*) 0.111

15. 1350 cal

17. −150°C

19. 0.088

21. 0.58

23. 41 hp

25. (*a*) 23.3×10^8 J/s; (*b*) 16.3×10^8 J/s;
  (*c*) same values

27. 734 cal/s

29. Proof

31. 8; 0.125

33. (*a*) 8; (*b*) 32; (*c*) 1.13×10^{15}

35. 36.8 J/K

37. 0.43 J/K

39. (*a*) zero; (*b*) 2.22×10^{-23} J/K

Chapter 12

1. (*a*) $\frac{1}{3}$ Hz; (*b*) 3.0 s; (*c*) 7.0 cm

3. 10.9 N/m

5. (*a*) 1.96 N/m; (*b*) 1.00 s

7. (*a*) 2.69×10^4 N/m; (*b*) 2.55 cm

9. (*a*) 0.150 J; (*b*) 0.024 J

11. 0.0294 J

13. (*a*) 0.41 m/s; (*b*) 6.67 m/s^2

15. (*a*) 14.0 m/s^2; (*b*) 0.75 m/s

17. (*a*) 3.0 Hz; (*b*) 15 cm; (*c*) 2.83 m/s; (*d*) zero

19. (*a*) 53 m/s^2; (*b*) zero; (*c*) $53 \cos\theta$ m/s^2

21. (*a*) zero; (*b*) 4.2 m/s; (*c*) 3.7 m/s

23. (*a*) 0.094 m/s; (*b*) 0.30 m/s^2; (*c*) 0.070 m/s

25. (*a*) 1.68 s; (*b*) 0.60 Hz

27. (*a*) 1.00 Hz; (*b*) 0.99 m/s

29. (*a*) 1.38 Hz; (*b*) 0.87 m/s

31. 39 cm

33. Proof

35. 1.58 cm

Chapter 13

1. 1.80 s

3. (*a*) 25; (*b*) 4.0 cm; (*c*) 25 Hz; (*d*) equal

5. 400 m

7. 70 m/s

9. 156 N

11.

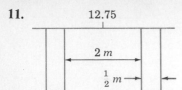

12.75

2 m

$\frac{1}{2}$ m

13. (a) 0.86 m; (b) 103 m/s
15. (a) 30 m; (b) 600 m/s
17. (a) 0.53 m; (b) 267 m/s
19. 200, 300, 400 Hz
21. 70 Hz
23. 0.75
25. Proof
27. (a) 2500 Hz; (b) 1250 Hz
29. 3.2 m/s
31. 2400 m/s
33. 0.50, 1.50, 2.50, 3.50 Hz, and so on.

Chapter 14
1. 2.04 km
3. (a) 1.55 m; (b) 0.34 m
5. 4.4×10^{-10} m²/N
7. 315 m/s
9. Proof
11. 3 percent
13. 2.5 W/m²; 0.20 percent
15. 73 dB
17. (a) 1×10^{-6} W/m²; (b) 1.2×10^{-8} J/s;
(c) 1.2×10^{-8} W
19. About 8
21. 120 dB
23. 28, 57, 85 Hz
25. $380n$ hertz with $n = 2, 3, 4, \ldots$; yes
27. (a) $850n$ hertz; (b) $850(n - \frac{1}{2})$ hertz with
$n = 1, 2, 3, \ldots$
29. $x = 0.1, 0.5, 0.9, 1.3, 1.7, 2.1, 2.5, 2.9$ m
31. $f_0 \pm 0.70$ Hz
33. 2202 and 2198 Hz
35. 113, 340, 567 Hz
37. 1700 Hz
39. $0.065n$ hertz with $n = 1, 2, 3, \ldots$
41. 34 m/s
43. 2990 Hz
45. 3.45 Hz
47. Proof

Chapter 15
1. (a) 2.11 N, attractive; (b) 2.11 N, attractive

3. (a) 3.27×10^{-6} C; (b) same sign
5. -1.04×10^{-3} C
7. (a) -0.40 N; (b) $+1.80$ N
9. -5.17 m
11. 4.8 N at $-45°$
13. 3×10^{-6} and 1.5×10^{-6} C
15. 1.67×10^5 N/C in $-x$ direction
17. mg/q upward
19. (a) zero; (b) 3.05×10^5 N/C straight down
21. 5.27×10^6 m/s; -2.64 cm
23. (a) 1.03×10^{13} m/s²; (b) 454 m/s;
(c) 4.8×10^{-21} J
25. (a) 4.5×10^{22}; (b) 5.8×10^{23};
(c) 3.2×10^{-11}

Chapter 16
1. (a) 3.6×10^{-5} J; (b) -3.6×10^{-5} J
3. (a) 3000 V/m; (b) 4.8×10^{-16} N
5. (a) 2.0 cm; (b) 4.8×10^{-17} N
7. (a) -800 V; (b) 1200 V; (c) zero
9. (a) 1.225×10^5 V/m; (b) $+3.2 \times 10^{-19}$ C
11. (a) 2.05×10^6 m/s; (b) 12.0 eV
13. 5.4×10^5 m/s
15. (a) 30 eV; (b) 130 eV
17. 5.66×10^6 m/s
19. 90,000 V; 0.27 J
21. -360 kV
23. 450 kV
25. 1.80×10^{-7} C
27. 4.7×10^{-8} F
29. (a) 3.6×10^{-8} C; (b) 13.3 V;
(c) 2.4×10^{-8} J; (d) 2.4×10^{-8} J

Chapter 17
1. (a) 2400 C; (b) 1.50×10^{22}
3. 1.88×10^{20}
5. 360 Ω
7. 0.079 Ω
9. 115 m
11. (a) 480 Ω; (b) 1.06×10^{-4} Ω · m
13. 1600°C
15. 7.1 Ω, 92.9 Ω
17. 5.5 Ω, 2.18 A
19. (a) 10.9 Ω; (b) 1.10 A; (c) zero
21. 4.0 μF
23. -4.1 A, 0.86 A. -6.4 V, 12.4 V
25. (a) 2.0 A; (b) 0.67 A; (c) 32 W
27. 12.0 V

29. 15 A

31. 2.0 A, 5.0 A, 54 V

33. 1.0 A, 3.0 A, 26 V, 7.6×10^{-5} C

35. 8.4 A, 6.4 A, 8.4 A, 74 V, 5.3×10^{-4} C

37. (a) 4.0 V; (b) 8.0 and 16.0 μC

39. Proof

41. 0.103 Ω

43. 11.9 V, 0.85 Ω

Chapter 18

1. (a) 0.024 N; (b) toward earth

3. 0.80 T

5. (a) 8.3×10^{-3} N; (b) westward

7. 1.34×10^4 m/s

9. 0.167 m

11. Circle perpendicular to B, $r = 1.14$ m

13. 8.0×10^{-6} T westward

15. (a) zero; (b) 1.33×10^{-5} T

17. 2.36×10^{-3} T

19. 1250 A

21. 1.20×10^{-3} Wb; zero

23. Zero

25. 11.2 G

27. Along line $y = -2x$

29. $r = 0.165$ cm, pitch = 5.9 cm

31. (a) 6.3×10^{14}; (b) 1.00×10^{-9} T; (c) clockwise

33. 3.5×10^6 m/s

35. Proof

37. 0.100-Ω shunt

39. 19960 Ω in series

41.

$$1150\ \Omega \qquad 10{,}800\ \Omega$$
$$+ \qquad 12\text{ V} \qquad 120\text{V}$$

43. (a) $0.094B$ A · m²; (b) decrease

Chapter 19

1. (a) 1.85 mV; (b) zero; (c) counterclockwise

3. (a) zero; (b) zero

5. (a) 0.024 V; (b) clockwise

7. 0.70 V

9. (a) $0.063a^2N$ volts; (b) $18.8a^2N$ volts

11. (a) 0.16 V; (b) 2.7 mH

13. (a) 0.21 mH; (b) 0.21 mH; (c) 29 mH

15. (a) 0.034 J; (b) 0.021 T

17. Proof

19. (a) 0.40 V; (b) south; (c) not easily

21. (a) $0.8Bav$; (b) $0.8Bav/R$; (c) clockwise; (d) $0.64B^2a^2v/R$; (e) stop

23. $0.938mgR/(B^2a^2)$

25. (a) $3V_0$; (b) $0.40\ V_0$

27. $0.23T$

29. (a) 110 V; (b) 24 A

31. (a) 13.3; (b) 1600 V

33. (a) 0.97; (b) 0.030

Chapter 20

1. (a) 1.00×10^{-4} s; (b) 1×10^{-4} s

3. (a) 8.0 s; (b) 8 s; (c) 1.6×10^{-5} C; (d) 2×10^{-6} A

5. 2.83 A

7. 17.7 V

9. (a) 67.5 W; (b) 16.1 cal/s

11. (a) 8000 Ω; (b) 9.9×10^{-8} F

13. 250 W

15. (a) 0.45 A; (b) 0.64 A

17. (a) 3.0 Ω; (b) $3.0 \times 10^6\ \Omega$

19. (a) 13.3 Hz; (b) 1.0 A

21. (a) 5.0 A; (b) 5.0 A; (c) 500 W

23. (a) 2.8 A; (b) 0.028 A; (c) zero

25. 33 Ω

27. 30 mH

29. (a) 0.663 or 0.743 H; (b) 0.47 or 0.33 mH

31. (a) 0.100 A; (b) 0.10 W; (c) 0.010

33. (a) 180 W; (b) 92 mH

35. (a) 5.2×10^{-20} J; (b) 0.072 times as large

37. (a) 3.5×10^8 Pa; (b) 3500 times larger

39. 0.20

Chapter 21

1. 3.0 cm

3. (a) 1.25 Hz; (b) 2.4×10^8 m

5. 14.7 s

7. 3.2×10^{-5} H

9. (a) 3.75 V/m; (b) yes

11. (a) 1.67×10^{-14} T; (b) no; (c) out of page

13. (a) 3×10^8 m/s; (b) 30,000 V/m; (c) yes

15. 0.80 mV; 3.3×10^{-12} T

17. 1.26×10^{-6} V

19. 137 V/m; 4.6×10^{-7} T

21. 8.3×10^{-8} m; ultraviolet

Chapter 22

1. 75 km
3. 0.75 cm
5. 4.0 m
7. Right ear
9. Proof
11. (a) 10 cm, 0.17 cm, real, inverted
    (b) 15 cm, 0.50 cm, real inverted
    (c) 20 cm, 0.83 cm, real, inverted
    (d) -30 cm, 2.5 cm, virtual, errect
13. (a) -8.3 cm, 0.50 cm, virtual, erect
    (b) -6.7 cm, 1.0 cm, virtual, erect
    (c) -3.3 cm, 2.0 cm, virtual, erect
15. (a) 200 cm; (b) no; (c) same
17. (a) 10 cm; (b) no
19. (a) f; (b) $\frac{1}{2}$
21. (a) 2.098×10^8 m/s; (b) 412 nm
23. (a) $23.6°$; (b) zero
25. $37°$ and $29.3°$
27. At least 1.414
29. $58.5°$
31. (a) 45 cm, 1.0 cm, real inverted
    (b) 120 cm, 6.0 cm, real, inverted
    (c) -15 cm, 3.0 cm, virtual, erect
33. (a) -22.5 cm, 1.25 cm, virtual, erect
    (b) -20.0 cm, 1.67 cm, virtual, erect
    (c) -12.0 cm, 3.0 cm, virtual, erect
35. (a) 30 cm; (b) 0.50
37. f
39. (a) 4.0 cm before second lens, $\frac{2}{3}$; (b) virtual; (c) inverted
41. Relative to mirror: 80, 10, -10 cm
43. Proof

Chapter 23

1. (a) 2.0 cm; (b) 1.85 cm
3. (a) Farsighted; (b) converging, 43 cm
5. -90 cm
7. Far point $= \infty$, near point $= 150$ cm
9. 167 cm
11. 190 cm
13. (a) 5.83 diopters; (b) 17.1 cm
15. -33 cm
17. (a) 2.1; (b) 1.3
19. 1.25
21. Proof
23. 30

25. (a) 17.5 cm; (b) 3.3
27. 17
29. 13
31. 200 km
33. 411 cm in front; virtual; erect; 1.43
35. (a) $45°$; (b) $38°$

Chapter 24

1. (a) $-75n$ centimeters; (b) $-37.5 - 75n$ centimeters
3. (a) $x = 30n$ centimeters; (b) $x = 15 + 30n$ centimeters
5. (a) 1.25×10^{-3} cm; (b) $5.0°$
7. (a) 2.5 m; (b) 1.25 m
9. ± 0.0125, ± 0.0375, ± 0.0625 cm
11. (a) 393 nm; (b) ultraviolet
13. 880 nm
15. (a) 1911 nm; (b) 1437 nm
17. 100 nm
19. 0.0552 cm
21. 1.028×10^{-4} cm; does not occur
23. (a) $2.58°$; (b) $5.50°$
25. $19.4°$
27. (a) no; (c) true for $n_{400}/n_{600} = 1.50$
29. 5.0 cm
31. 9.6×10^{-4} cm
33. 7.4 km
35. 480 nm
37. $4.32/n$ angstroms

Chapter 25

1. Zero
3. 6.0×10^{-8} s
5. Yes
7. 2.94×10^8 m/s
9. (a) 25.85 m; (b) 3.33 m
11. (a) 2.60×10^8 m/s; (b) 36 m;
    (c) 3.1×10^{-6} s; (d) 6.2×10^{-6} s
13. (a) 2.15×10^8 s; (b) 1.60×10^8 s;
    (c) 3.21×10^{16} m; (d) 2.0×10^8 m/s;
    (e) same
15. 5.6×10^{-20} J, 0.35 eV
17. (b) infrared; (c) 413 nm
19. (a) 258 nm; (b) ultraviolet
21. (a) 32,000 nm; (b) infrared
23. 2.8 eV

25. 91 nm
27. (a) 1.33×10^{-27} N · s; (b) 2.65×10^{-27} N · s
29. 3.7×10^{-10} percent
31. $0.982c$
33. 0.50 nm
35. (b) 1.46×10^6 m/s, 1.005 nm

Chapter 26

1. (a) 200 cm^2; (b) 1.20 cm^2
3. (a) $(3.64 \times 10^{-26})/r$ joules;
   (b) 4.7×10^{-14} m
5. (a) 95 nm; (b) 397 nm
7. 0.052
9. 2.18×10^6 m/s; $v/c = 0.0073$
11. (b) 6.7×10^{-13} m
    (c) 10.2 nm
13. (a) $E_n = -122/n^2$ electron volts;
    (b) 122 eV; (c) 10.2 nm
15. (a) 1.00 nm; (b) 0.75 nm; (c) x-ray
17. 40.8, 48.4, 51.0 eV
19. (a) 234 eV; (b) 9.1×10^6 m/s
21. 18
23. 10
25. 0.062 nm
27. (a) 300 J, 71.7 cal; (b) 215°C
29. 97, 103, 122, 486, 656, 1880 nm
31. 3000 eV
33. $600 \pm 1.2 \times 10^{-12}$ nm

Chapter 27

1. 2.3×10^{17} kg/m^3
3. 193 m
5. 3.3×10^{-27} kg; 2.0 u
7. 39.10 u
9. 65.03 u
11. 0.78 u; 0.93 percent
13. 8.24×10^{-3} u; 7.7 MeV
15. 31.97 u
17. 2.35 min
19. (a) 0.0385 yr^{-1}; (b) 22,700; (c) 20 mCi
21. 2.7×10^{-8} Ci
23. 1.23×10^7 Bq $= 3.3 \times 10^{-4}$ Ci
25. 0.37
27. 2.19 h
29. 3.9×10^9 yr
31. $^{90}_{40}$Zr

33. $^{238}_{92}$U
35. 2.4×10^{-5}
37. $^{208}_{82}$Pb
39. 4.6×10^{15} atoms; 3.1×10^{-8} g
41. No
43. (a) 5.7×10^{14} J; (b) 1.82×10^{11} °C
45. (a) 3.3×10^5 m/s; (b) 8.8×10^5 m/s
47. 1.06×10^{16} Bq
49. 2.8×10^{-6} g
51. 38,900 rd
53. 42 s
55. (a) 70 rd/yr; (b) 110 rd/yr; (c) 70 rd/yr;
    (d) 7 rd/yr

Index